# STUDENT'S SOLUTIONS MANUAL

## JUDITH A. PENNA
*Indiana University Purdue University Indianapolis*

# BASIC COLLEGE MATHEMATICS WITH EARLY INTEGERS
## SECOND EDITION

## Marvin Bittinger
*Indiana University Purdue University Indianapolis*

## Judith A. Penna
*Indiana University Purdue University Indianapolis*

**Addison-Wesley**
is an imprint of

PEARSON

ISBN-13: 978-0-321-60544-3
ISBN-10: 0-321-60544-6

2 3 4 5 6 BRR 14 13 12 11

**Addison-Wesley**
is an imprint of

www.pearsonhighered.com

# Contents

# Chapter 1
# Whole Numbers

**1.** 2 3 $\boxed{5}$ , 8 8 8

The digit 5 means 5 thousands.

**3.** 1, 4 8 8, $\boxed{5}$ 2 6

The digit 5 means 5 hundreds.

**5.** 1 2 1, 6 2 $\boxed{9}$ , 2 7 0

The digit 9 names the number of thousands.

**7.** 1 2 1, 6 2 9, 2 $\boxed{7}$ 0

The digit 7 names the number of tens.

**9.** 5702 = 5 thousands + 7 hundreds + 0 tens + 2 ones, or 5 thousands + 7 hundreds + 2 ones

**11.** 93,986 = 9 ten thousands + 3 thousands + 9 hundreds + 8 tens + 6 ones

**13.** 2058 = 2 thousands + 0 hundreds + 5 tens + 8 ones, or 2 thousands + 5 tens + 8 ones

**15.** 1576 = 1 thousand + 5 hundreds + 7 tens + 6 ones

**17.** 1,424,161,948 = 1 billion + 4 hundred millions + 2 ten millions + 4 millions + 1 hundred thousand + 6 ten thousands + 1 thousand + 9 hundreds + 4 tens + 8 ones

**19.** 99,886,568 = 9 ten millions + 9 millions + 8 hundred thousands + 8 ten thousands + 6 thousands + 5 hundreds + 6 tens + 8 ones

**21.** 617,249 = 6 hundred thousands + 1 ten thousand + 7 thousands + 2 hundreds + 4 tens + 9 ones

**23.** A word name for 85 is eighty-five.

**25.**

Eighty-eight thousand ⟶ 88,000

**27.**

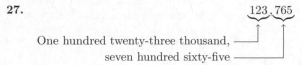

One hundred twenty-three thousand, ⟶
seven hundred sixty-five ⟶ 123,765

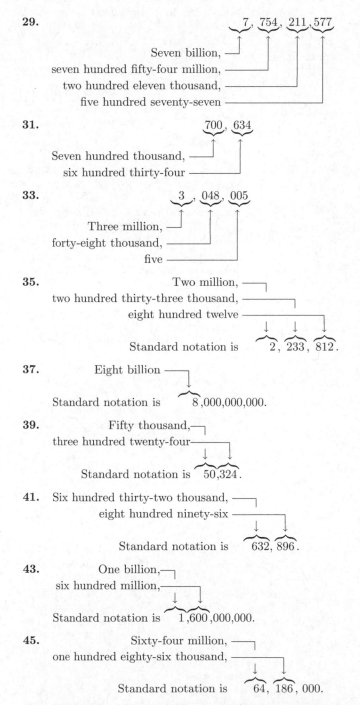

**29.** 7, 754, 211, 577

Seven billion, ⟶
seven hundred fifty-four million, ⟶
two hundred eleven thousand, ⟶
five hundred seventy-seven ⟶

**31.** 700, 634

Seven hundred thousand, ⟶
six hundred thirty-four ⟶

**33.** 3, 048, 005

Three million, ⟶
forty-eight thousand, ⟶
five ⟶

**35.** Two million, ⟶
two hundred thirty-three thousand, ⟶
eight hundred twelve ⟶

Standard notation is 2, 233, 812.

**37.** Eight billion ⟶

Standard notation is 8,000,000,000.

**39.** Fifty thousand, ⟶
three hundred twenty-four ⟶

Standard notation is 50,324.

**41.** Six hundred thirty-two thousand, ⟶
eight hundred ninety-six ⟶

Standard notation is 632, 896.

**43.** One billion, ⟶
six hundred million, ⟶

Standard notation is 1,600,000,000.

**45.** Sixty-four million, ⟶
one hundred eighty-six thousand, ⟶

Standard notation is 64, 186, 000.

**47.** First consider the whole numbers from 100 through 199. The 10 numbers 102, 112, 122, ... , 192 contain the digit 2. In addition, the 10 numbers 120, 121, 122, ... , 129 contain the digit 2. However, we do not count the number 122 in this group because it was counted in the first group of ten numbers. Thus, 19 numbers from 100 through 199 contain the digit 2. Using the same type of reasoning for the whole numbers from 300 to 400, we see that there are also 19 numbers in this group that contain the digit 2.

Finally, consider the 100 whole numbers 200 through 299. Each contains the digit 2.

Thus, there are $19 + 19 + 100$, or 138 whole numbers between 100 and 400 that contain the digit 2 in their standard notation.

## Exercise Set 1.2

**1.**
$$\begin{array}{r} 3\,6\,4 \\ +\ \ 2\,3 \\ \hline 3\,8\,7 \end{array}$$
Add ones, add tens, then add hundreds.

**3.**
$$\begin{array}{r} \overset{1}{\phantom{0}}\ \ \\ 8\,6 \\ +\,7\,8 \\ \hline 1\,6\,4 \end{array}$$
Add ones: We get 14 ones, or 1 ten + 4 ones. Write 4 in the ones column and 1 above the tens. Add tens: We get 16 tens.

**5.**
$$\begin{array}{r} \overset{1}{\phantom{0}}\ \ \ \\ 1\,7\,1\,6 \\ +\,3\,4\,8\,2 \\ \hline 5\,1\,9\,8 \end{array}$$
Add ones: We get 8. Add tens: We get 9 tens. Add hundreds: We get 11 hundreds, or 1 thousand + 1 hundred. Write 1 in the hundreds column and 1 above the thousands. Add thousands: We get 5 thousands.

**7.**
$$\begin{array}{r} \overset{1}{\phantom{0}}\ \\ 9\,9 \\ +\ \ 1 \\ \hline 1\,0\,0 \end{array}$$
Add ones: We get 10 ones, or 1 ten + 0 ones. Write 0 in the ones column and 1 above the tens. Add tens: We get 10 tens.

**9.**
$$\begin{array}{r} \overset{1}{\phantom{0}}\ \ \ \\ 8\,1\,1\,3 \\ +\ \ 3\,9\,0 \\ \hline 8\,5\,0\,3 \end{array}$$
Add ones: We get 3. Add tens: We get 10 tens, or 1 hundred + 0 tens. Write 0 in the tens column and 1 above the hundreds. Add hundreds: We get 5. Add thousands: We get 8.

**11.**
$$\begin{array}{r} \overset{1}{\phantom{0}}\ \ \ \\ 3\,5\,6 \\ +\,4\,9\,1\,0 \\ \hline 5\,2\,6\,6 \end{array}$$
Add ones: We get 6. Add tens: We get 6. Add hundreds: We get 12 hundreds, or 1 thousand + 2 hundreds. Write 2 in the hundreds column and 1 above the thousands. Add thousands: We get 5.

**13.**
$$\begin{array}{r} \overset{1}{}\overset{2}{}\overset{1}{}\ \\ 3\,8\,7\,0 \\ 9\,2 \\ 7 \\ +\ \ 4\,9\,7 \\ \hline 4\,4\,6\,6 \end{array}$$
Add ones: We get 16 ones, or 1 ten + 6 ones. Write 6 in the ones column and 1 above the tens. Add tens: We get 26 tens, or 2 hundreds + 6 tens. Write 6 in the tens column and 2 above the hundreds. Add hundreds: We get 14 hundreds, or 1 thousand + 4 hundreds. Write 4 in the hundreds column and 1 above the thousands. Add thousands: We get 4.

**15.**
$$\begin{array}{r} \overset{1}{}\overset{1}{}\ \ \\ 4\,8\,2\,5 \\ +\,1\,7\,8\,3 \\ \hline 6\,6\,0\,8 \end{array}$$
Add ones: We get 8. Add tens: We get 10 tens. Write 0 in the tens column and 1 above the hundreds. Add hundreds: We get 16 hundreds. Write 6 in the hundreds column and 1 above the thousands. Add thousands: We get 6 thousands.

**17.**
$$\begin{array}{r} \overset{1}{}\overset{1}{}\overset{1}{}\ \ \\ 2\,3,4\,4\,3 \\ +\,1\,0,9\,8\,9 \\ \hline 3\,4,4\,3\,2 \end{array}$$
Add ones: We get 12 ones, or 1 ten + 2 ones. Write 2 in the ones column and 1 above the tens. Add tens: We get 13 tens. Write 3 in the tens column and 1 above the hundreds. Add hundreds: We get 14 hundreds. Write 4 in the hundreds column and 1 above the thousands. Add thousands: We get 4 thousands. Add ten thousands: We get 3 ten thousands.

**19.**
$$\begin{array}{r} \overset{1}{}\overset{1}{}\overset{1}{}\overset{1}{}\ \ \\ 7\,7,5\,4\,3 \\ +\,2\,3,7\,6\,7 \\ \hline 1\,0\,1,3\,1\,0 \end{array}$$
Add ones: We get 10 ones, or 1 ten + 0 ones. Write 0 in the ones column and 1 above the tens. Add tens: We get 11 tens. Write 1 in the tens column and 1 above the hundreds. Add hundreds: We get 13 hundreds. Write 3 in the hundreds column and 1 above the thousands. Add thousands: We get 11 thousands. Write 1 in the thousands column and 1 above the ten thousands. Add ten thousands: We get 10 ten thousands.

**21.** We look for pairs of numbers whose sums are 10, 20, 30, and so on.
$$\begin{array}{r} \overset{2}{}\ \ \\ 4\,5 \\ 2\,5 \\ 3\,6 \\ 4\,4 \\ +\,8\,0 \\ \hline 2\,3\,0 \end{array}$$

**23.**
$$\begin{array}{r} \overset{1}{}\ \overset{1}{}\ \\ 1\,2,0\,7\,0 \\ 2\,9\,5\,4 \\ +\ \ 3\,4\,0\,0 \\ \hline 1\,8,4\,2\,4 \end{array}$$
Add ones: We get 4. Add tens: We get 12 tens, or 1 hundred + 2 tens. Write 2 in the tens column and 1 above the hundreds. Add hundreds: We get 14 hundreds, or 1 thousand + 4 hundreds. Write 4 in the hundreds column and 1 above the hundreds. Add thousands: We get 8 thousands. Add ten thousands: We get 1 ten thousand.

**25.**
$$\begin{array}{r} \overset{3}{}\overset{1}{}\overset{2}{}\ \\ 4\,8\,3\,5 \\ 7\,2\,9 \\ 9\,2\,0\,4 \\ 8\,9\,8\,6 \\ +\,7\,9\,3\,1 \\ \hline 3\,1,6\,8\,5 \end{array}$$
Add ones: We get 25. Write 5 in the ones column and 2 above the tens. Add tens: We get 18 tens. Write 8 in the tens column and 1 above the hundreds. Add hundreds: We get 36 hundreds. Write 6 in the hundreds column and 3 above the thousands. Add thousands: We get 31 thousands.

**27.** Perimeter = 50 yd + 23 yd + 40 yd + 19 yd
We carry out the addition.
$$\begin{array}{r} \overset{1}{}\ \ \\ 5\,0 \\ 2\,3 \\ 4\,0 \\ +\,1\,9 \\ \hline 1\,3\,2 \end{array}$$
The perimeter of the figure is 132 yd.

---

**29.** Perimeter = 402 ft + 298 ft + 196 ft + 100 ft + 453 ft + 212 ft

We carry out the addition.

```
    2 2
    4 0 2
    2 9 8
    1 9 6
    1 0 0
    4 5 3
  + 2 1 2
  -------
  1 6 6 1
```

The perimeter of the figure is 1661 ft.

**31.** Perimeter = 200 ft + 85 ft + 200 ft + 85 ft

We carry out the addition.

```
    1 1
    2 0 0
      8 5
    2 0 0
  +   8 5
  -------
    5 7 0
```

The perimeter of the hockey rink is 570 ft.

**33.** 4 [8] 6, 2 0 5

The digit 8 tells the number of ten thousands.

**35.** One method is described in the answer section in the text. Another method is: $1 + 100 = 101$, $2 + 99 = 101$, ..., $50 + 51 = 101$. Then the sum of 50 101's is 5050.

# Exercise Set 1.3

**1.**
```
    6 5
  - 2 1
  -----
    4 4
```
Subtract ones, then subtract tens.

**3.**
```
    8 6 6
  - 3 3 3
  -------
    5 3 3
```
Subtract ones, subtract tens, then subtract hundreds.

**5.**
```
    7 16
    8 6̸
  - 4 7
  -----
    3 9
```
We cannot subtract 7 ones from 6 ones. Borrow 1 ten to get 16 ones. Subtract ones, then subtract tens.

**7.**
```
    4 11
    5̸ 1̸
  - 3 7
  -----
    1 4
```
We cannot subtract 7 ones from 1 one. Borrow 1 ten to get 11 ones. Subtract ones, then subtract tens.

**9.**
```
       15
    4 5̸ 13
    5̸ 6̸ 3̸
  - 1 9 4
  -------
    3 6 9
```
We cannot subtract 4 ones from 3 ones. Borrow 1 ten to get 13 ones. Subtract ones. We cannot subtract 9 tens from 5 tens. Borrow 1 hundred to get 15 tens. Subtract tens, then subtract hundreds.

**11.**
```
       8 11
    3 9̸ 1̸
  - 3 6 5
  -------
      2 6
```
We cannot subtract 5 ones from 1 one. Borrow 1 ten to get 11 ones. Subtract ones, then tens, then hundreds. Note that 3 hundreds − 3 hundreds = 0 hundreds, but we do not write the 0 when it is the first digit of a difference.

**13.**
```
       7 11
    9 8̸ 1̸
  - 7 4 7
  -------
    2 3 4
```
We cannot subtract 7 ones from 1 one. Borrow 1 ten to get 11 ones. Subtract ones, subtract tens, then subtract hundreds.

**15.**
```
       7 13
    6 8̸ 3̸
  - 2 6 6
  -------
    4 1 7
```
We cannot subtract 6 ones from 3 ones. Borrow 1 ten to get 13 ones. Subtract ones, then tens, then hundreds.

**17.**
```
         6 16
    7 7 6̸ 9
  - 2 3 8 7
  ---------
    5 3 8 2
```
Subtract ones. We cannot subtract 8 tens from 6 tens. Borrow 1 hundred to get 16 tens. Subtract tens, subtract hundreds, then subtract thousands.

**19.**
```
      14 10
    3 4̸ 0̸ 12
    4̸ 5̸ 1̸ 2̸
  - 1 7 3 4
  ---------
    2 7 7 8
```
We cannot subtract 4 ones from 2 ones. Borrow 1 ten to get 12 ones. Subtract ones. We cannot subtract 3 tens from 0 tens. Borrow 1 hundred to get 10 tens. Subtract tens. We cannot subtract 7 hundreds from 4 hundreds. Borrow 1 thousand to get 14 hundreds. Subtract hundreds, then thousands.

**21.**
```
         10
      2 0̸ 18
    5 3̸ 1̸ 8̸
  - 2 2 4 9
  ---------
    3 0 6 9
```
We cannot subtract 9 ones from 8 ones. Borrow 1 ten to get 18 ones. Subtract ones. We cannot subtract 4 tens from 0 tens. Borrow 1 hundred to get 10 tens. Subtract tens, then hundreds, then thousands.

**23.**
```
         13
      8 3̸ 17
    3 9̸ 4̸ 7̸
  - 2 8 5 8
  ---------
    1 0 8 9
```
We cannot subtract 8 ones from 7 ones. Borrow 1 ten to get 17 ones. Subtract ones. We cannot subtract 5 tens from 3 tens. Borrow 1 hundred to get 13 tens. Subtract tens, then hundreds, then thousands.

**25.**
```
    11 15 13
     1̸  5̸  3̸ 17
    1̸ 2̸, 6̸  4̸ 7̸
  -      4 8 9 9
  --------------
         7 7 4 8
```

**27.**
$$
\begin{array}{r}
{\scriptstyle 4\ 11\ 2\ \overset{13}{3}\ 12} \\
\cancel{5}\ \cancel{1},\ \cancel{3}\ \cancel{4}\ \cancel{2} \\
-\ 4\ 7,\ 1\ 9\ 8 \\
\hline
4\ 1\ 4\ 4
\end{array}
$$

**29.**
$$
\begin{array}{r}
{\scriptstyle 7\ 10} \\
\cancel{8}\ \cancel{0} \\
-\ 2\ 4 \\
\hline
5\ 6
\end{array}
$$

**31.**
$$
\begin{array}{r}
{\scriptstyle 8\ 10} \\
6\ \cancel{9}\ \cancel{0} \\
-\ 2\ 3\ 6 \\
\hline
4\ 5\ 4
\end{array}
$$

**33.**
$$
\begin{array}{r}
{\scriptstyle 6\ 16\ 3\ 10} \\
7\ \cancel{6}\ \cancel{4}\ 0 \\
-\ 3\ 8\ 0\ 9 \\
\hline
3\ 8\ 3\ 1
\end{array}
$$

**35.**
$$
\begin{array}{r}
{\scriptstyle 7\ 9\ 18} \\
6\ \cancel{8}\ \cancel{0}\ \cancel{8} \\
-\ 3\ 0\ 5\ 9 \\
\hline
3\ 7\ 4\ 9
\end{array}
$$
We have 8 hundreds or 80 tens. We borrow 1 ten to get 18 ones. We then have 79 tens. Subtract ones, then tens, then hundreds, then thousands.

**37.**
$$
\begin{array}{r}
{\scriptstyle 2\ 9\ 10} \\
2\ \cancel{3}\ \cancel{0}\ \cancel{0} \\
-\ \ \ 1\ 0\ 9 \\
\hline
2\ 1\ 9\ 1
\end{array}
$$
We have 3 hundreds or 30 tens. We borrow 1 ten to get 10 ones. We then have 29 tens. Subtract ones, then tens, then hundreds, then thousands.

**39.**
$$
\begin{array}{r}
{\scriptstyle 5\ 9\ 9\ 17} \\
\cancel{6}\ \cancel{0}\ \cancel{0}\ \cancel{7} \\
-\ 1\ 5\ 8\ 9 \\
\hline
4\ 4\ 1\ 8
\end{array}
$$
We have 6 thousands, or 600 tens. We borrow 1 ten to get 17 ones. We then have 599 tens. Subtract ones, then tens, then hundreds, then thousands.

**41.**
$$
\begin{array}{r}
{\scriptstyle 8\ 10\ \ \ 2\ 17} \\
\cancel{9}\ \cancel{0},\ 2\ \cancel{3}\ \cancel{7} \\
-\ 4\ 7,\ 2\ 0\ 9 \\
\hline
4\ 3,\ 0\ 2\ 8
\end{array}
$$

**43.**
$$
\begin{array}{r}
{\scriptstyle 10\ 16} \\
{\scriptstyle 9\ 0\ 6\ 13} \\
\cancel{1}\cancel{0}\ \cancel{1},\ 7\ \cancel{3}\ 4 \\
-\ \ \ \ \ 5\ 7\ 6\ 0 \\
\hline
9\ 5,\ 9\ 7\ 4
\end{array}
$$

**45.**
$$
\begin{array}{r}
{\scriptstyle 6\ 9\ 9\ 10} \\
\cancel{7}\ \cancel{0}\ \cancel{0}\ \cancel{0} \\
-\ 2\ 7\ 9\ 4 \\
\hline
4\ 2\ 0\ 6
\end{array}
$$
We have 7 thousands or 700 tens. We borrow 1 ten to get 10 ones. We then have 699 tens. Subtract ones, then tens, then hundreds, then thousands.

**47.**
$$
\begin{array}{r}
{\scriptstyle 8\ 9\ 9\ 10} \\
3\ \cancel{9},\ \cancel{0}\ \cancel{0}\ \cancel{0} \\
-\ 3\ 7,\ 6\ 9\ 5 \\
\hline
1\ 3\ 0\ 5
\end{array}
$$
We have 9 thousands or 900 tens. We borrow 1 ten to get 10 ones. We then have 899 tens. Subtract ones, then tens, then hundreds, then thousands. Note that 3 thousands − 3 thousands = 0 thousands, but we do not write the 0 when it is the first digit of a difference.

**49.**
$$
\begin{array}{r}
{\scriptstyle 9\ 9\ 9\ 18} \\
1\ \cancel{0},\ \cancel{0}\ \cancel{0}\ \cancel{8} \\
-\ \ \ \ \ \ \ 1\ 9 \\
\hline
9\ 9\ 8\ 9
\end{array}
$$
We have 1 ten thousand, or 1000 tens. We borrow 1 ten to get 10 ones. We then have 999 tens. Subtract ones, then tens, then hundreds, then thousands.

**51.**
$$
\begin{array}{r}
{\scriptstyle 4\ 9\ 9\ 9\ 11} \\
\cancel{5}\ \cancel{0},\ \cancel{0}\ \cancel{0}\ \cancel{1} \\
-\ \ \ \ \ 1\ 9\ 8\ 4 \\
\hline
4\ 8,\ 0\ 1\ 7
\end{array}
$$
We have 5 ten thousands, or 5000 tens. We borrow 1 ten to get 11 ones. We then have 4999 tens. Subtract ones, then tens, then hundreds, then thousands.

**53.**
$$
\begin{array}{r}
{\scriptstyle 1\ 1} \\
9\ 4\ 6 \\
+\ \ 7\ 8 \\
\hline
1\ 0\ 2\ 4
\end{array}
$$
Add ones: We get 14. Write 4 in the ones column and 1 above the tens. Add tens: We get 12. Write 2 in the tens column and 1 above the hundreds. Add hundreds: We get 10 hundreds.

**55.**
$$
\begin{array}{r}
{\scriptstyle 1\ 1\ \ \ 1} \\
5\ 7,\ 8\ 7\ 7 \\
+\ 3\ 2,\ 4\ 0\ 6 \\
\hline
9\ 0,\ 2\ 8\ 3
\end{array}
$$
Add ones: We get 13. Write 3 in the ones column and 1 above the tens. Add tens: We get 8. Add hundreds: We get 12. Write 2 in the hundreds column and 1 above the thousands. Add thousands: We get 10. Write 0 in the thousands column and 1 above the ten thousands. Add ten thousands: We get 9 ten thousands.

**57.**
$$
\begin{array}{r}
{\scriptstyle 1\ 1} \\
5\ 6\ 7 \\
+\ 7\ 7\ 8 \\
\hline
1\ 3\ 4\ 5
\end{array}
$$
Add ones: We get 15. Write 5 in the ones column and 1 above the tens. Add tens: We get 14. Write 4 in the tens column and 1 above the hundreds. Add hundreds: We get 13 hundreds.

**59.**
$$
\begin{array}{r}
{\scriptstyle 1\ 1\ \ \ 1} \\
1\ 2,\ 8\ 8\ 5 \\
+\ \ 9\ 8\ 0\ 7 \\
\hline
2\ 2,\ 6\ 9\ 2
\end{array}
$$
Add ones: We get 12. Write 2 in the ones column and 1 above the tens. Add tens: We get 9. Add hundreds: We get 16. Write 6 in the hundreds column and 1 above the thousands. Add thousands: We get 12. Write 2 in the thousands column and 1 above the ten thousands. Add ten thousands. We get 2 ten thousands.

**61.**

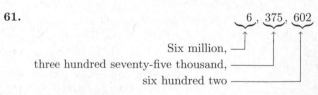

Six million, ——
three hundred seventy-five thousand, ——
six hundred two ——

**63.**
$$
\begin{array}{r}
9,\ \_\ 4\ 8,\ 6\ 2\ 1 \\
-\ 2,\ 0\ 9\ 7,\ \_\ 8\ 1 \\
\hline
7,\ 2\ 5\ 1,\ 1\ 4\ 0
\end{array}
$$

To subtract tens, we borrow 1 hundred to get 12 tens.

$$
\begin{array}{r}
{\scriptstyle 5\ 12} \\
9,\ \_\ 4\ 8,\ \cancel{6}\ \cancel{2}\ 1 \\
-\ 2,\ 0\ 9\ 7,\ \_\ 8\ 1 \\
\hline
7,\ 2\ 5\ 1,\ 1\ 4\ 0
\end{array}
$$

In order to have 1 hundred in the difference, the missing digit in the subtrahend must be 4 (5 − 4 = 1).

$$\begin{array}{r} 9,\_4\,8,\overset{5}{\cancel{6}}\overset{12}{\cancel{2}}\,1 \\ -\,2,0\,9\,7,4\,8\,1 \\ \hline 7,2\,5\,1,1\,4\,0 \end{array}$$

In order to subtract ten thousands, we must borrow 1 hundred thousand to get 14 ten thousands. The number of hundred thousands left must be 2 since the hundred thousands place in the difference is 2 (2 − 0 = 2). Thus, the missing digit in the minuend must be 2 + 1, or 3.

$$\begin{array}{r} 9,\overset{2}{\cancel{3}}\overset{14}{\cancel{4}}\,8,\overset{5}{\cancel{6}}\overset{12}{\cancel{2}}\,1 \\ -\,2,0\,9\,7,4\,8\,1 \\ \hline 7,2\,5\,1,1\,4\,0 \end{array}$$

## Exercise Set 1.4

**1.**
$$\begin{array}{r} \overset{4}{\phantom{0}}6\,5 \\ \times\quad 8 \\ \hline 5\,2\,0 \end{array}$$  Multiplying by 8

**3.**
$$\begin{array}{r} \overset{2}{\phantom{0}}9\,4 \\ \times\quad 6 \\ \hline 5\,6\,4 \end{array}$$  Multiplying by 6

**5.**
$$\begin{array}{r} \overset{2}{\phantom{0}}5\,0\,9 \\ \times\quad 3 \\ \hline 1\,5\,2\,7 \end{array}$$  Multiplying by 3

**7.**
$$\begin{array}{r} \overset{1}{\phantom{0}}\overset{2}{\phantom{0}}\overset{6}{\phantom{0}}9\,2\,2\,9 \\ \times\quad 7 \\ \hline 6\,4,6\,0\,3 \end{array}$$  Multiplying by 7

**9.**
$$\begin{array}{r} \overset{2}{\phantom{0}}5\,3 \\ \times\quad 9\,0 \\ \hline 4\,7\,7\,0 \end{array}$$  Multiplying by 9 tens (We write 0 and then multiply 53 by 9.)

**11.**
$$\begin{array}{r} \overset{2}{\phantom{0}}\overset{3}{\phantom{0}}8\,5 \\ \times\quad 4\,7 \\ \hline 5\,9\,5 \\ 3\,4\,0\,0 \\ \hline 3\,9\,9\,5 \end{array}$$  Multiplying by 7 / Multiplying by 40 / Adding

**13.**
$$\begin{array}{r} 8\,7 \\ \times\,1\,0 \\ \hline 8\,7\,0 \end{array}$$  Multiplying by 1 ten (We write 0 and then multiply 87 by 1.)

**15.**
$$\begin{array}{r} 9\,6 \\ \times\quad 2\,0 \\ \hline 1\,9\,2\,0 \end{array}$$  Multiplying by 2 tens (We write 0 and then multiply 96 by 2.)

**17.**
$$\begin{array}{r} \overset{3}{\phantom{0}}\overset{2}{\phantom{0}}6\,4\,3 \\ \times\quad 7\,2 \\ \hline 1\,2\,8\,6 \\ 4\,5\,0\,1\,0 \\ \hline 4\,6,2\,9\,6 \end{array}$$  Multiplying by 2 / Multiplying by 70 / Adding

**19.**
$$\begin{array}{r} \overset{1}{\phantom{0}}\overset{1}{\phantom{0}}\overset{1}{\phantom{0}}\overset{1}{\phantom{0}}4\,4\,4 \\ \times\quad 3\,3 \\ \hline 1\,3\,3\,2 \\ 1\,3\,3\,2\,0 \\ \hline 1\,4,6\,5\,2 \end{array}$$  Multiplying by 3 / Multiplying by 30 / Adding

**21.**
$$\begin{array}{r} \overset{2}{\phantom{0}}\overset{1}{\phantom{0}}\overset{3}{\phantom{0}}\overset{2}{\phantom{0}}\overset{5}{\phantom{0}}\overset{3}{\phantom{0}}5\,6\,4 \\ \times\,4\,5\,8 \\ \hline 4\,5\,1\,2 \\ 2\,8\,2\,0\,0 \\ 2\,2\,5\,6\,0\,0 \\ \hline 2\,5\,8,3\,1\,2 \end{array}$$  Multiplying by 8 / Multiplying by 50 / Multiplying by 400 / Adding

**23.**
$$\begin{array}{r} 8\,5\,3 \\ \times\,9\,3\,6 \\ \hline 5\,1\,1\,8 \\ 2\,5\,5\,9\,0 \\ 7\,6\,7\,7\,0\,0 \\ \hline 7\,9\,8,4\,0\,8 \end{array}$$  Multiplying by 6 / Multiplying by 30 / Multiplying by 900 / Adding

**25.**
$$\begin{array}{r} 6\,4\,2\,8 \\ \times\,3\,2\,2\,4 \\ \hline 2\,5\,7\,1\,2 \\ 1\,2\,8\,5\,6\,0 \\ 1\,2\,8\,5\,6\,0\,0 \\ 1\,9\,2\,8\,4\,0\,0\,0 \\ \hline 2\,0,7\,2\,3,8\,7\,2 \end{array}$$  Multiplying by 4 / Multiplying by 20 / Multiplying by 200 / Multiplying by 3000 / Adding

**27.**
$$\begin{array}{r} 3\,4\,8\,2 \\ \times\quad 1\,0\,4 \\ \hline 1\,3\,9\,2\,8 \\ 3\,4\,8\,2\,0\,0 \\ \hline 3\,6\,2,1\,2\,8 \end{array}$$  Multiplying by 4 / Multiplying by 1 hundred (We write 00 and then multiply 3482 by 1.)

**29.**
$$\begin{array}{r} 8\,7\,6 \\ \times\,3\,4\,5 \\ \hline 4\,3\,8\,0 \\ 3\,5\,0\,4\,0 \\ 2\,6\,2\,8\,0\,0 \\ \hline 3\,0\,2,2\,2\,0 \end{array}$$  Multiplying by 5 / Multiplying by 40 / Multiplying by 300 / Adding

**31.**
$$
\begin{array}{r}
\overset{\overset{\overset{\overset{5\ 5\ 5}{1\ 1\ 1}}{1\ 1\ 1}}{3\ 3\ 3}}{}\\
7\ 8\ 8\ 9\\
\times\ 6\ 2\ 2\ 4\\
\hline
3\ 1\ 5\ 5\ 6\\
1\ 5\ 7\ 7\ 8\ 0\\
1\ 5\ 7\ 7\ 8\ 0\ 0\\
4\ 7\ 3\ 3\ 4\ 0\ 0\ 0\\
\hline
4\ 9,1\ 0\ 1,1\ 3\ 6\\
\end{array}
$$
    Multiplying by 4
    Multiplying by 20
    Multiplying by 200
    Multiplying by 6000
    Adding

**33.**
$$
\begin{array}{r}
\overset{\overset{2\quad 3}{3\quad 4}}{}\\
5\ 6\ 0\ 8\\
\times\ 4\ 5\ 0\ 0\\
\hline
2\ 8\ 0\ 4\ 0\ 0\ 0\\
2\ 2\ 4\ 3\ 2\ 0\ 0\ 0\\
\hline
2\ 5,2\ 3\ 6,0\ 0\ 0\\
\end{array}
$$
Multiplying by 5 hundreds (We write 00 and then multiply 5608 by 5.)
Multiplying by 4000
Adding

**35.**
$$
\begin{array}{r}
\overset{\overset{2}{4}}{}\\
5\ 0\ 0\ 6\\
\times\ 4\ 0\ 0\ 8\\
\hline
4\ 0\ 0\ 4\ 8\\
2\ 0\ 0\ 2\ 4\ 0\ 0\ 0\\
\hline
2\ 0,0\ 6\ 4,0\ 4\ 8\\
\end{array}
$$
Multiplying by 8
Multiplying by 4 thousands (We write 000 and then multiply 5006 by 4.)

**37.** $A = 728 \text{ mi} \times 728 \text{ mi} = 529,984$ square miles

**39.** $A = l \times w = 90 \text{ ft} \times 90 \text{ ft} = 8100$ square feet

**41.**
$$
\begin{array}{r}
\overset{1\quad\ 1}{}\\
4\ 9\ 0\ 8\\
5\ 6\ 6\ 7\\
+\ 2\ 1\ 1\ 0\\
\hline
1\ 2,6\ 8\ 5\\
\end{array}
$$
Add ones: We get 15. Write 5 in the ones column and 1 above the tens. Add tens: We get 8. Add hundreds: We get 16. Write 6 in the hundreds column and 1 above the thousands. Add thousands: We get 12 thousands.

**43.**
$$
\begin{array}{r}
\overset{1\ \ \ 1\ 1\ 1}{}\\
3\ 4\ 0,7\ 9\ 8\\
+\ \ \ 8\ 6,6\ 7\ 9\\
\hline
4\ 2\ 7,4\ 7\ 7\\
\end{array}
$$
Add ones: We get 17. Write 7 in the ones column and 1 above the tens. Add tens: We get 17. Write 7 in the tens column and 1 above the hundreds. Add hundreds: We get 14. Write 4 in the hundreds column and 1 above the thousands. Add thousands: We get 7. Add ten thousands: We get 12. Write 2 in the ten thousands column and 1 above the hundred thousands. Add hundred thousands: We get 4 hundred thousands.

**45.**
$$
\begin{array}{r}
\overset{8\ 10}{}\\
4\ 9\ 0\ 8\\
-\ 3\ 6\ 6\ 7\\
\hline
1\ 2\ 4\ 1\\
\end{array}
$$
Subtract ones. We cannot subtract 6 tens from 0 tens. We have 9 hundreds or 90 tens. We borrow 1 hundred to get 10 tens. We have 8 hundreds. Subtract tens, hundreds, and thousands.

**47.**
$$
\begin{array}{r}
\overset{\overset{13}{2\ \ 8\ \ 10\ \ \ 8\ \ 18}}{}\\
3\ 4\ 0,7\ 9\ 8\\
-\ \ \ 8\ 6,6\ 7\ 9\\
\hline
2\ 5\ 4,1\ 1\ 9\\
\end{array}
$$
We cannot subtract 9 ones from 8 ones. Borrow 1 ten to get 18 ones. Subtract ones. Then subtract tens and hundreds. We cannot subtract 6 thousands from 0 thousands. We have 4 ten thousands or 40 thousands. We borrow 1 ten thousand to get 10 thousands. Subtract thousands. We cannot subtract 8 ten thousands from 3 ten thousands. We borrow 1 hundred thousand to get 13 ten thousands. Subtract ten thousands and then hundred thousands.

**49.** Use a calculator to perform the computations in this exercise.

First find the total area of each floor:

$A = l \times w = 172 \times 84 = 14,448$ square feet

Find the area lost to the elevator and the stairwell:

$A = l \times w = 35 \times 20 = 700$ square feet

Subtract to find the area available as office space on each floor:

$14,448 - 700 = 13,748$ square feet

Finally, multiply by the number of floors, 18, to find the total area available as office space:

$18 \times 13,748 = 247,464$ square feet

## Exercise Set 1.5

**1.**
$$
\begin{array}{r}
1\ 2\\
6\ \overline{)\ 7\ 2}\\
6\\
\hline
1\ 2\\
1\ 2\\
\hline
0\\
\end{array}
$$
Think: 7 tens ÷ 6. Estimate 1 ten.
Think: 12 ones ÷ 6. Estimate 2 ones.

The answer is 12.

**3.** $\dfrac{23}{23} = 1$     Any nonzero number divided by itself is 1.

**5.** $22 \div 1 = 22$     Any number divided by 1 is that same number.

**7.** $\dfrac{0}{7} = 0$     Zero divided by any nonzero number is 0.

**9.** $\dfrac{16}{0}$ is not defined, because division by 0 is not defined.

**11.** $\dfrac{48}{8} = 6$ because $48 = 8 \cdot 6$.

**13.**
$$
\begin{array}{r}
5\ 5\\
5\ \overline{)\ 2\ 7\ 7}\\
2\ 5\\
\hline
2\ 7\\
2\ 5\\
\hline
2\\
\end{array}
$$
Think: 27 ÷ 5. Try 5.
Think: 27 ÷ 5 again. Try 5.

The answer is 55 R 2.

**15.**

```
    1 0 8
8 ) 8 6 4     8 ÷ 8 = 1
    8
    ‾‾
    6 4      There are no groups of 8 in 6.
    6 4      Write 0 above 6. Bring down 4.
    ‾‾‾
      0      64 ÷ 8 = 8
```
The answer is 108.

**17.**

```
    3 0 7
4 ) 1 2 2 8     12 ÷ 4 = 3
    1 2
    ‾‾‾
      2 8      There are no groups of 4 in 2.
      2 8      Write 0 above the second 2.
      ‾‾‾
        0      Bring down 8. 28 ÷ 4 = 7
```
The answer is 307.

**19.**

```
    7 5 3
6 ) 4 5 2 1     Think: 45 ÷ 6. Try 7.
    4 2
    ‾‾‾
      3 2       Think: 32 ÷ 6. Try 5.
      3 0
      ‾‾‾
        2 1     Think: 21 ÷ 6. Try 3.
        1 8
        ‾‾‾
          3
```
The answer is 753 R 3.

**21.**

```
    7 4
4 ) 2 9 7     Think: 29 ÷ 4. Try 7.
    2 8
    ‾‾‾
    1 7       Think 17 ÷ 4. Try 4.
    1 6
    ‾‾‾
      1
```
The answer is 74 R 1.

**23.**

```
    9 2
8 ) 7 3 8     Think: 73 ÷ 8. Try 9.
    7 2
    ‾‾‾
    1 8       Think: 18 ÷ 8. Try 2.
    1 6
    ‾‾‾
      2
```
The answer is 92 R 2.

**25.**

```
    1 7 0 3
5 ) 8 5 1 5     Think: 8 ÷ 5. Try 1.
    5
    ‾
    3 5         35 ÷ 5 = 7
    3 5
    ‾‾‾
      1 5       Bring down 1. There are no
      1 5       groups of 5 in 1. Write 0 above
      ‾‾‾
        0       1. Bring down 5. 15 ÷ 5 = 3
```
The answer is 1703.

**27.**

```
    9 8 7
9 ) 8 8 8 8     Think: 88 ÷ 9. Try 9.
    8 1
    ‾‾‾
    7 8         Think: 78 ÷ 9. Try 8.
    7 2
    ‾‾‾
      6 8       Think: 68 ÷ 9. Try 7.
      6 3
      ‾‾‾
        5
```
The answer is 987 R 5.

**29.**

```
       1 2,7 0 0
1 0 ) 1 2 7,0 0 0    Think: 12 ÷ 10. Try 1.
      1 0
      ‾‾
        2 7          Think: 27 ÷ 10. Try 2.
        2 0
        ‾‾
          7 0        70 ÷ 10 = 7. Since the last two
          7 0        numbers in the dividend are 0
          ‾‾         and 0 ÷ 7 = 0, the last two
            0        digits of the quotient are 0.
```
The answer is 12,700.

**31.**

```
           1 2 7
1 0 0 0 ) 1 2 7,0 0 0    Think: 1270 ÷ 1000. Try 1.
          1 0 0 0
          ‾‾‾‾‾
            2 7 0 0       Think: 2700 ÷ 1000. Try 2.
            2 0 0 0
            ‾‾‾‾‾
              7 0 0 0     7000 ÷ 1000 = 7
              7 0 0 0
              ‾‾‾‾‾
                    0
```
The answer is 127.

**33.**

```
      5 2
7 0 ) 3 6 9 2    Think: 369 ÷ 70. Try 5.
      3 5 0
      ‾‾‾‾‾
        1 9 2    Think: 192 ÷ 70. Try 2.
        1 4 0
        ‾‾‾‾‾
          5 2
```
The answer is 52 R 52.

**35.**

```
      2 9
3 0 ) 8 7 5     Think: 87 ÷ 30. Try 2.
      6 0
      ‾‾
      2 7 5     Think: 275 ÷ 30. Try 9.
      2 7 0
      ‾‾‾‾‾
          5
```
The answer is 29 R 5.

**37.**

```
      4 0
2 1 ) 8 5 2    21 is close to 20. Think: 85 ÷ 20. Try 4.
      8 4
      ‾‾
      1 2      There are no groups of 21 in 12.
               Write a 0 above 2.
```
The answer is 40 R 12.

**39.**

```
          8
8 5 ) 7 6 7 2    85 is close to 90. Think: 767 ÷ 90.
      6 8 0      Try 8.
      ‾‾‾‾‾
      | 8 7 |
```
Since 87 is larger than the divisor 85, the estimate is too low. We try 9.

```
          9 0
8 5 ) 7 6 7 2
      7 6 5
      ‾‾‾‾‾
          2 2    There are no groups of 85 in 22.
                 Write 0 above 2.
```
The answer is 90 R 22.

**41.**
```
          3
 1 1 1 ) 3 2 1 9     111 is close to 100. Think: 321 ÷ 100.
         3 3 3       Try 3.
```
Since we cannot subtract 333 from 321, 3 is too large. Try 2.
```
              2 9
 1 1 1 ) 3 2 1 9
         2 2 2
         9 9 9       Think: 999 ÷ 100. Try 9.
         9 9 9
             0
```
The answer is 29.

**43.**
```
          1 0 5
 8 ) 8 4 3           8 ÷ 8 = 1
     8
     4 3             There are no groups of 8 in 4.
     4 0             Write 0 above 4. Bring down 3.
       3             Think: 43 ÷ 8. Try 5.
```
The answer is 105 R 3.

**45.**
```
          1 6 0 9
 5 ) 8 0 4 7         Think: 8 ÷ 5. Try 1.
     5
     3 0             30 ÷ 5 = 6
     3 0
       4 7           There are no groups of 5 in 4. Write
       4 5           0 above 4. Bring down 7.
         2           Think: 47 ÷ 5. Try 9.
```
The answer is 1609 R 2.

**47.**
```
          1 0 0 7
 5 ) 5 0 3 6         5 ÷ 5 = 1. The next number in the
     5               dividend is 0, and 0 ÷ 5 = 0.
       3 6           There are no groups of 5 in 3. Write
       3 5           0 above 3. Bring down 6.
         1           Think: 36 ÷ 5. Try 7.
```
The answer is 1007 R 1.

**49.**
```
            2 2
 4 6 ) 1 0 5 8       46 is close to 50. Think: 105 ÷ 50.
       9 2           Try 2.
       1 3 8         Think: 138 ÷ 50. Try 2.
         9 2
         4 6
```
Since the difference, 46, is not smaller than the divisor, 46, 2 is too small. Try 3.
```
            2 3
 4 6 ) 1 0 5 8
       9 2
       1 3 8
       1 3 8
           0
```
The answer is 23.

**51.**
```
          1 0 7
 3 2 ) 3 4 2 5       32 is close to 30. Think: 34 ÷ 30.
       3 2           Try 1.
       2 2 5         There are no groups of 32 in 22.
       2 2 4         Write 0 above 2. Bring down 5.
           1         Think: 225 ÷ 30. Try 7.
```
The answer is 107 R 1.

**53.**
```
          4
 2 4 ) 8 8 8 0       24 is close to 20. Think: 88 ÷ 20.
       9 6           Try 4.
```
Since we cannot subtract 96 from 88, 4 is too large. Try 3.
```
          3 8
 2 4 ) 8 8 8 0
       7 2
       1 6 8         Think: 168 ÷ 20. Try 8.
       1 9 2
```
Since we cannot subtract 192 from 168, 8 is too large. Try 7.
```
          3 7 0
 2 4 ) 8 8 8 0
       7 2
       1 6 8
       1 6 8
           0         The last digit in the dividend is 0,
                     and 0 ÷ 24 = 0.
```
The answer is 370.

**55.**
```
            5
 2 8 ) 1 7, 0 6 7    28 is close to 30. Think: 170 ÷ 30.
       1 4 0         Try 5.
         3 0
```
Since 30 is larger than the divisor, 28, 5 is too small. Try 6.
```
            6 0 8
 2 8 ) 1 7, 0 6 7
       1 6 8
         2 6 7       There are no groups of 28 in 26.
         2 2 4       Write 0 above 6. Bring down 7.
           4 3       Think 267 ÷ 30. Try 8.
```
Since 43 is larger than the divisor, 28, 8 is too small. Try 9.
```
            6 0 9
 2 8 ) 1 7, 0 6 7
       1 6 8
         2 6 7
         2 5 2
           1 5
```
The answer is 609 R 15.

**57.**
```
            3 0 4
 8 0 ) 2 4, 3 2 0
       2 4 0
          3 2 0
          3 2 0
              0
```
The answer is 304.

**59.**

```
              3 5 0 8
    2 8 5 ) 9 9 9, 9 9 9
              8 5 5
            ─────────
            1 4 4 9
            1 4 2 5
            ─────────
              2 4 9 9
              2 2 8 0
            ─────────
                2 1 9
```

The answer is 3508 R 219.

**61.**

```
              8 0 7 0
    4 5 6 ) 3, 6 7 9, 9 2 0
            3 6 4 8
            ─────────
              3 1 9 2
              3 1 9 2
            ─────────
                    0
```

The answer is 8070.

**63.** The distance around an object is its <u>perimeter</u>.

**65.** For large numbers, <u>digits</u> are separated by commas into groups of three, called <u>periods</u>.

**67.** In the sentence $10 \times 1000 = 10,000$, 10 and 1000 are called <u>factors</u> and 10,000 is called the <u>product</u>.

**69.** The sentence $3 \times (6 \times 2) = (3 \times 6) \times 2$ illustrates the <u>associative</u> law of multiplication.

**71.**

| $a$ | $b$ | $a \cdot b$ | $a + b$ |
|-----|-----|-------------|---------|
|     | 68  | 3672        |         |
| 84  |     |             | 117     |
|     |     | 32          | 12      |

To find $a$ in the first row we divide $a \cdot b$ by $b$:

$$3672 \div 68 = 54$$

Then we add to find $a + b$:

$$54 + 68 = 122$$

To find $b$ in the second row we subtract $a$ from $a + b$:

$$117 - 84 = 33$$

Then we multiply to find $a \cdot b$:

$$84 \cdot 33 = 2772$$

To find $a$ and $b$ in the last row we find a pair of numbers whose product is 32 and whose sum is 12. Pairs of numbers whose product is 32 are 1 and 32, 2 and 16, 4 and 8. Since $4 + 8 = 12$, the numbers we want are 4 and 8. We will let $a = 4$ and $b = 8$. (We could also let $a = 8$ and $b = 4$).

The completed table is shown below.

| $a$ | $b$ | $a \cdot b$ | $a + b$ |
|-----|-----|-------------|---------|
| 54  | 68  | 3672        | 122     |
| 84  | 33  | 2772        | 117     |
| 4   | 8   | 32          | 12      |

**73.** We divide 1231 by 42:

```
              2 9
    4 2 ) 1 2 3 1
            8 4
          ───────
            3 9 1
            3 7 8
          ───────
              1 3
```

The answer is 29 R 13. Since 13 students will be left after 29 buses are filled, then 30 buses are needed.

---

## Chapter 1 Mid-Chapter Review

**1.** The statement is false. For example, $8 - 5 = 3$, but 5 is not equal to $8 + 3$.

**2.** The statement is true. See page 19 in the text.

**3.** The statement is true. See page 19 in the text.

**4.** The statement is false. For example, $3 \cdot 0 = 0$ and 0 is not greater than 3. Also, $1 \cdot 1 = 1$ and 1 is not greater than 1.

**5.** It is true that zero divided by any nonzero number is 0.

**6.** The statement is false. Any number divided by 1 is the number itself. For example, $\frac{27}{1} = 27$.

**7.**

$$\overbrace{95}, \overbrace{406}, \overbrace{237}$$

Ninety-five million, ——
four hundred six thousand, ——
two hundred thirty-seven ——

**8.**

```
      5  9 14
      6̶ 0̶ 4̶
    −  4 9 7
    ─────────
      1 0 7
```

**9.** $2 \boxed{6} 9 8$

The digit 6 names the number of hundreds.

**10.** $\boxed{6} 1, 2 0 4$

The digit 6 names the number of ten thousands.

**11.** $1 4 \boxed{6}, 2 3 7$

The digit 6 names the number of thousands.

**12.** $5 8 \boxed{6}$

The digit 6 names the number of ones.

**13.** $3 0 6, 4 5 8, 1 \boxed{2} 9$

The digit 2 names the number of tens.

**14.** $3 0 \boxed{6}, 4 5 8, 1 2 9$

The digit 6 names the number of millions.

**15.** $3 0 6, 4 \boxed{5} 8, 1 2 9$

The digit 5 names the number of ten thousands.

**16.** $3\,0\,6,4\,5\,8,\boxed{1}\,2\,9$

The digit 1 names the number of hundreds.

**17.** $5602 = 5$ thousands $+ 6$ hundreds $+ 0$ tens $+ 2$ ones, or 5 thousands $+ 6$ hundreds $+ 2$ ones

**18.** $69{,}345 = 6$ ten thousands $+ 9$ thousands $+ 3$ hundreds $+ 4$ tens $+ 5$ ones

**19.** A word name for 136 is one hundred thirty-six.

**20.** A word name for 64,325 is sixty-four thousand, three hundred twenty-five.

**21.** Standard notation for three hundred eight thousand, seven hundred sixteen is 308,716.

**22.** Standard notation for four million, five hundred sixty-seven thousand, two hundred ninety-one is 4,567,291.

**23.**
```
   3 1 6
 + 4 8 2
 ───────
   7 9 8
```

**24.**
```
   1 1
   5 9 3
 + 4 3 7
 ───────
 1 0 3 0
```

**25.**
```
   1 1
   2 6 3 8
 + 5 2 8 4
 ─────────
   7 9 2 2
```

**26.**
```
   1 1 1
   4 6 1 7
   2 4 3 6
 +   4 8 1
 ─────────
   7 5 3 4
```

**27.**
```
   7 8 6
 - 3 2 1
 ───────
   4 6 5
```

**28.**
```
      11
   5  /  14
   6̶  2̶  4̶
 - 2  8  5
 ─────────
   3  3  9
```

**29.**
```
      15
   2  5̶  9  12
   3  6̶  0̶  2̶
 - 1  7  4  8
 ────────────
   1  8  5  4
```

**30.**
```
   4 9 9 14
   5̶ 0̶ 0̶ 4̶
 -     6 7 6
 ───────────
   4 3 2 8
```

**31.**
```
   3
   3 6
 × 6
 ─────
 2 1 6
```

**32.**
```
   1 1
   5 5
     5 6 7
 ×    2 8
 ─────────
   4 5 3 6
 1 1 3 4 0
 ─────────
 1 5,8 7 6
```

**33.**
```
       2
       1
       3
       4 0 7
 ×   3 2 5
 ─────────
     2 0 3 5
   8 1 4 0
 1 2 2 1 0 0
 ───────────
 1 3 2,2 7 5
```

**34.**
```
       2 2 3
           1
       9 4 3 5
 ×       6 0 2
 ─────────────
   1 8 8 7 0
 5 6 6 1 0 0 0
 ─────────────
 5,6 7 9,8 7 0
```

**35.**
```
          2 5 3
      ┌─────────
    4 │ 1 0 1 2
          8
        ─────
          2 1
          2 0
          ───
            1 2
            1 2
            ───
              0
```
The answer is 253.

**36.**
```
             1 1 2
        ┌───────────
   3 8  │ 4 2 6 1
          3 8
          ───
            4 6
            3 8
            ───
              8 1
              7 6
              ───
                5
```
The answer is 112 R 5.

**37.**
```
              2 3
        ┌───────────
   6 0  │ 1 3 9 9
          1 2 0
          ─────
            1 9 9
            1 8 0
            ─────
              1 9
```
The answer is 23 R 19.

**38.**
```
              1 4 4
        ┌───────────
   5 6  │ 8 0 9 5
          5 6
          ───
          2 4 9
          2 2 4
          ─────
            2 5 5
            2 2 4
            ─────
              3 1
```
The answer is 144 R 31.

**39.** Perimeter $= 10$ m $+ 4$ m $+ 8$ m $+ 3$ m $= 25$ m

**40.** $A = 4$ in. $\times 2$ in. $= 8$ sq in.

**41.** When numbers are being added, it does not matter how they are grouped.

**42.** Subtraction is not commutative. For example, $5 - 2 = 3$, but $2 - 5 \neq 3$.

**43.** Answers will vary. Suppose one coat costs \$150. Then the multiplication $4 \cdot \$150$ gives the cost of four coats.

Suppose one ream of copy paper costs \$4. Then the multiplication $\$4 \cdot 150$ gives the cost of 150 reams.

**44.** Using the definition of division, $0 \div 0 = a$ such that $a \cdot 0 = 0$. We see that $a$ could be *any* number since $a \cdot 0 = 0$ for any number $a$. Thus, we cannot say that $0 \div 0 = 0$. This is why we agree not to allow division by 0.

## Exercise Set 1.6

**1.** Round 48 to the nearest ten.

4 $\boxed{8}$
↑

The digit 4 is in the tens place. Consider the next digit to the right. Since the digit, 8, is 5 or higher, round 4 tens up to 5 tens. Then change the digit to the right of the tens digit to zero.

The answer is 50.

**3.** Round 463 to the nearest ten.

4 6 $\boxed{3}$
↑

The digit 6 is in the tens place. Consider the next digit to the right. Since the digit, 3, is 4 or lower, round down, meaning that 6 tens stays as 6 tens. Then change the digit to the right of the tens digit to zero.

The answer is 460.

**5.** Round 731 to the nearest ten.

7 3 $\boxed{1}$
↑

The digit 3 is in the tens place. Consider the next digit to the right. Since the digit, 1, is 4 or lower, round down, meaning that 3 tens stays as 3 tens. Then change the digit to the right of the tens digit to zero.

The answer is 730.

**7.** Round 895 to the nearest ten.

8 9 $\boxed{5}$
↑

The digit 9 is in the tens place. Consider the next digit to the right. Since the digit, 5, is 5 or higher, we round up. The 89 tens become 90 tens. Then change the digit to the right of the tens digit to zero.

The answer is 900.

**9.** Round 146 to the nearest hundred.

1 $\boxed{4}$ 6
↑

The digit 1 is in the hundreds place. Consider the next digit to the right. Since the digit, 4, is 4 or lower, round down, meaning that 1 hundred stays as 1 hundred. Then change all digits to the right of the hundreds digit to zeros.

The answer is 100.

**11.** Round 957 to the nearest hundred.

9 $\boxed{5}$ 7
↑

The digit 9 is in the hundreds place. Consider the next digit to the right. Since the digit, 5, is 5 or higher, round up. The 9 hundreds become 10 hundreds. Then change all digits to the right of the hundreds digit to zeros.

The answer is 1000.

**13.** Round 9079 to the nearest hundred.

9 0 $\boxed{7}$ 9
↑

The digit 0 is in the hundreds place. Consider the next digit to the right. Since the digit, 7, is 5 or higher, round 0 hundreds up to 1 hundred. Then change all digits to the right of the hundreds digit to zeros.

The answer is 9100.

**15.** Round 32,839 to the nearest hundred.

3 2, 8 $\boxed{3}$ 9
↑

The digit 8 is in the hundreds place. Consider the next digit to the right. Since the digit, 3, is 4 or lower, round down, meaning 8 hundreds stays as 8 hundreds. Then change all digits to the right of the hundreds digit to zero.

The answer is 32,800.

**17.** Round 5876 to the nearest thousand.

5 $\boxed{8}$ 7 6
↑

The digit 5 is in the thousands place. Consider the next digit to the right. Since the digit, 8, is 5 or higher, round 5 thousands up to 6 thousands. Then change all digits to the right of the thousands digit to zeros.

The answer is 6000.

**19.** Round 7500 to the nearest thousand.

7 $\boxed{5}$ 0 0
↑

The digit 7 is in the thousands place. Consider the next digit to the right. Since the digit, 5, is 5 or higher, round 7 thousands up to 8 thousands. Then change all the digits to the right of the thousands digit to zeros.

The answer is 8000.

**21.** Round 45,340 to the nearest thousand.

$$4\,5,\boxed{3}\,4\,0$$
$$\uparrow$$

The digit 5 is in the thousands place. Consider the next digit to the right. Since the digit, 3, is 4 or lower, round down, meaning that 5 thousands stays as 5 thousands. Then change all the digits to the right of the thousands digit to zeros.

The answer is 45,000.

**23.** Round 373,405 to the nearest thousand.

$$3\,7\,3,\boxed{4}\,0\,5$$
$$\uparrow$$

The digit 3 is in the thousands place. Consider the next digit to the right. Since the digit, 4, is 4 or lower, round down, meaning that 3 thousands stays as 3 thousands. Then change all the digits to the right of the thousands digit to zeros.

The answer is 373,000.

**25.**
|          | Rounded to the nearest ten |
|---------|---------|
| 7 8     | 8 0     |
| + 9 2   | + 9 0   |
|         | 1 7 0 ← Estimated answer |

**27.**
|          | Rounded to the nearest ten |
|---------|---------|
| 8 0 7 4   | 8 0 7 0   |
| − 2 3 4 7 | − 2 3 5 0 |
|           | 5 7 2 0 ← Estimated answer |

**29.**
|          | Rounded to the nearest ten |
|---------|---------|
| 4 5     | 5 0     |
| 7 7     | 8 0     |
| 2 5     | 3 0     |
| + 5 6   | + 6 0   |
| 3 4 3   | 2 2 0 ← Estimated answer |

The sum 343 seems to be incorrect since 220 is not close to 343.

**31.**
|          | Rounded to the nearest ten |
|---------|---------|
| 6 2 2   | 6 2 0   |
| 7 8     | 8 0     |
| 8 1     | 8 0     |
| + 1 1 1 | + 1 1 0 |
| 9 3 2   | 8 9 0 ← Estimated answer |

The sum 932 seems to be incorrect since 890 is not close to 932.

**33.**
|          | Rounded to the nearest hundred |
|---------|---------|
| 7 3 4 8   | 7 3 0 0   |
| + 9 2 4 7 | + 9 2 0 0 |
|           | 1 6, 5 0 0 ← Estimated answer |

**35.**
|          | Rounded to the nearest hundred |
|---------|---------|
| 6 8 5 2   | 6 9 0 0   |
| − 1 7 4 8 | − 1 7 0 0 |
|           | 5 2 0 0 ← Estimated answer |

**37.**
|          | Rounded to the nearest hundred |
|---------|---------|
| 2 1 6   | 2 0 0   |
| 8 4     | 1 0 0   |
| 7 4 5   | 7 0 0   |
| + 5 9 5 | + 6 0 0 |
| 1 6 4 0 | 1 6 0 0 ← Estimated answer |

The sum 1640 seems to be correct since 1600 is close to 1640.

**39.**
|          | Rounded to the nearest hundred |
|---------|---------|
| 7 5 0   | 8 0 0   |
| 4 2 8   | 4 0 0   |
| 6 3     | 1 0 0   |
| + 2 0 5 | + 2 0 0 |
| 1 4 4 6 | 1 5 0 0 ← Estimated answer |

The sum 1446 seems to be correct since 1500 is close to 1446.

**41.**
|          | Rounded to the nearest thousand |
|---------|---------|
| 9 6 4 3   | 1 0, 0 0 0   |
| 4 8 2 1   | 5 0 0 0   |
| 8 9 4 3   | 9 0 0 0   |
| + 7 0 0 4 | + 7 0 0 0 |
|           | 3 1, 0 0 0 ← Estimated answer |

**43.**
|          | Rounded to the nearest thousand |
|---------|---------|
| 9 2, 1 4 9   | 9 2, 0 0 0   |
| − 2 2, 5 5 5 | − 2 3, 0 0 0 |
|              | 6 9, 0 0 0 ← Estimated answer |

**45.**
|          | Rounded to the nearest ten |
|---------|---------|
| 4 5     | 5 0     |
| × 6 7   | × 7 0   |
|         | 3 5 0 0 ← Estimated answer |

**47.**
|          | Rounded to the nearest ten |
|---------|---------|
| 3 4     | 3 0     |
| × 2 9   | × 3 0   |
|         | 9 0 0 ← Estimated answer |

**49.**
|          | Rounded to the nearest hundred |
|---------|---------|
| 8 7 6   | 9 0 0   |
| × 3 4 5 | × 3 0 0 |
|         | 2 7 0, 0 0 0 ← Estimated answer |

**51.**

|              | Rounded to the nearest hundred |
|--------------|--------------------------------|
| 4 3 2        | 4 0 0                          |
| × 1 9 9      | × 2 0 0                        |
|              | 8 0, 0 0 0 ← Estimated answer  |

**53.** Rounding to the nearest ten, we have $347 \div 73 \approx 350 \div 70$.

```
        5 0
7 0 ) 3 5 0
      3 5
        0
        0
        0
```

**55.** $8452 \div 46 \approx 8450 \div 50$

```
          1 6 9
5 0 ) 8 4 5 0
      5 0
      3 4 5
      3 0 0
        4 5 0
        4 5 0
            0
```

**57.** $1165 \div 236 \approx 1200 \div 200$

```
              6
2 0 0 ) 1 2 0 0
        1 2 0 0
              0
```

**59.** $8358 \div 295 \approx 8400 \div 300$

```
            2 8
3 0 0 ) 8 4 0 0
        6 0 0
        2 4 0 0
        2 4 0 0
              0
```

**61.** We round the cost of each option to the nearest hundred and add.

| 7 4 5 0 | 7 5 0 0 |
|---------|---------|
| 1 5 9 5 | 1 6 0 0 |
| 1 5 4 0 | 1 5 0 0 |
| + 6 2 5 | + 6 0 0 |
|         | 1 1, 2 0 0 |

The estimated cost is $11,200.

**63.** We round the cost of each option to the nearest hundred and add.

| 8 8 2 0 | 8 8 0 0 |
|---------|---------|
| 2 8 7 0 | 2 9 0 0 |
| 6 2 4 5 | 6 2 0 0 |
| + 9 8 5 | + 1 0 0 0 |
|         | 1 8, 9 0 0 |

The estimated cost is $18,900. Since this is more than Sara and Ben's budget of $17,700, they cannot afford their choices.

**65.** Answers will vary depending on the options chosen.

**67.** a) First we round the cost of the car and the destination charges to the nearest hundred and add.

| 2 1, 1 6 0 | 2 1, 2 0 0 |
|------------|------------|
| + 6 7 0    | + 7 0 0    |
|            | 2 1, 9 0 0 |

The number of sales representatives, 112, rounded to the nearest hundred is 100. Now we multiply the rounded total cost of a car and the rounded number of representatives.

```
    2 1, 9 0 0
×        1 0 0
2, 1 9 0, 0 0 0
```

The cost of the purchase is approximately $2,190,000.

b) First we round the cost of the car to the nearest thousand and the destination charges to the nearest hundred and add.

| 2 1, 1 6 0 | 2 1, 0 0 0 |
|------------|------------|
| + 6 7 0    | + 7 0 0    |
|            | 2 1, 7 0 0 |

From part (a) we know that the number of sales representatives, rounded to the nearest hundred, is 100. We multiply the rounded total cost of a car and the rounded number of representatives.

```
    2 1, 7 0 0
×        1 0 0
2, 1 7 0, 0 0 0
```

The cost of the purchase is approximately $2,170,000.

**69.** $2716 \div 28 \approx 2700 \div 30$

```
          9 0
3 0 ) 2 7 0 0
      2 7 0
          0
          0
          0
```

We estimate that 90 people attended the banquet.

**71.**

Since 0 is to the left of 17, $0 < 17$.

**73.**

Since 34 is to the right of 12, $34 > 12$.

**75.**

Since 1000 is to the left of 1001, $1000 < 1001$.

**77.**

Since 133 is to the right of 132, $133 > 132$.

**79.**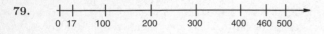

Since 460 is to the right of 17, $460 > 17$.

**81.**

Since 37 is to the right of 11, $37 > 11$.

**83.** Since 190,078 lies to the right of 172,000 on the number line, we can write $190,078 > 172,000$.

Conversely, since 172,000 lies to the left of 190,078 on the number line, so we could also write $172,000 < 190,078$.

**85.** Since 1694 lies to the left of 5249 on the number line, we can write $1694 < 5249$.

Conversely, since 5249 lies to the right of 1694 on the number line, we could also write $5249 > 1694$.

**87.**
$$
\begin{array}{r}
{\scriptstyle 1\ \ 1\ \ 1\ 1} \\
6\,7,7\,8\,9 \\
+\,1\,8,9\,6\,5 \\
\hline
8\,6,7\,5\,4
\end{array}
$$
Add ones. We get 14. Write 4 in the ones column and 1 above the tens. Add tens: We get 15 tens. Write 5 in the tens column and 1 above the hundreds. Add hundreds: We get 17 hundreds. Write 7 in the hundreds column and 1 above the thousands. Add thousands: We get 16 thousands. Write 6 in the thousands column and 1 above the ten thousands. Add ten thousands: We get 8 ten thousands.

**89.**
$$
\begin{array}{r}
{\scriptstyle 16} \\
{\scriptstyle 5\ \ \not{6}\ \ 17} \\
\not{6}\,7,\not{7}\,8\,9 \\
-\,1\,8,9\,6\,5 \\
\hline
4\,8,8\,2\,4
\end{array}
$$
Subtract ones: We get 4. Subtract tens: We get 2. We cannot subtract 9 hundreds from 7 hundreds. We borrow 1 thousand to get 17 hundreds. Subtract hundreds. We cannot subtract 8 thousands from 6 thousands. We borrow 1 ten thousand to get 16 thousands. Subtract thousands, then ten thousands.

**91.**
$$
\begin{array}{r}
{\scriptstyle 1} \\
{\scriptstyle 4} \\
4\,6 \\
\times\,3\,7 \\
\hline
3\,2\,2 \\
1\,3\,8\,0 \\
\hline
1\,7\,0\,2
\end{array}
$$

**93.**
$$
\begin{array}{r}
5\,4 \\
6\,\overline{)\,3\,2\,8} \\
3\,0 \\
\hline
2\,8 \\
2\,4 \\
\hline
4
\end{array}
$$

The answer is 54 R 4.

**95.** Using a calculator, we find that the sum is 30,411. This is close to the estimated sum found in Exercise 41.

**97.** Using a calculator, we find that the difference is 69,594. This is close to the estimated difference found in Exercise 43.

## Exercise Set 1.7

**1.** $x + 0 = 14$

We replace $x$ by different numbers until we get a true equation. If we replace $x$ by 14, we get a true equation: $14 + 0 = 14$. No other replacement makes the equation true, so the solution is 14.

**3.** $y \cdot 17 = 0$

We replace $y$ by different numbers until we get a true equation. If we replace $y$ by 0, we get a true equation: $0 \cdot 17 = 0$. No other replacement makes the equation true, so the solution is 0.

**5.** $x = 12,345 + 78,555$

To solve the equation we carry out the calculation.
$$
\begin{array}{r}
1\,2,3\,4\,5 \\
+\,7\,8,5\,5\,5 \\
\hline
9\,0,9\,0\,0
\end{array}
$$

We can check by repeating the calculation. The solution is 90,900.

**7.** $908 - 458 = p$

To solve the equation we carry out the calculation.
$$
\begin{array}{r}
9\,0\,8 \\
-\,4\,5\,8 \\
\hline
4\,5\,0
\end{array}
$$

We can check by repeating the calculation. The solution is 450.

**9.** $16 \cdot 22 = y$

To solve the equation we carry out the calculation.
$$
\begin{array}{r}
2\,2 \\
\times\,1\,6 \\
\hline
1\,3\,2 \\
2\,2\,0 \\
\hline
3\,5\,2
\end{array}
$$

We can check by repeating the calculation. The solution is 352.

**11.** $t = 125 \div 5$

To solve the equation we carry out the calculation.
$$
\begin{array}{r}
2\,5 \\
5\,\overline{)\,1\,2\,5} \\
1\,0 \\
\hline
2\,5 \\
2\,5 \\
\hline
0
\end{array}
$$

We can check by repeating the calculation. The solution is 25.

**13.**
$$
\begin{aligned}
13 + x &= 42 \\
13 + x - 13 &= 42 - 13 \quad &&\text{Subtracting 13 on both sides} \\
0 + x &= 29 \quad &&\text{13 plus } x \text{ minus 13 is } 0 + x. \\
x &= 29
\end{aligned}
$$

Check: $\dfrac{13 + x = 42}{13 + 29 \ ? \ 42}$

$\phantom{13 + 29 \ ?}42 \ \Big| \quad$ TRUE

The solution is 29.

**15.** $\quad 12 = 12 + m$

$12 - 12 = 12 + m - 12 \qquad$ Subtracting 12 on both sides

$\phantom{12 - 12 =} 0 = 0 + m \qquad\qquad$ 12 plus $m$ minus 12 is $0 + m$.

$\phantom{12 - 12 =} 0 = m$

Check: $\dfrac{12 = 12 + m}{12 \ ? \ 12 + 0}$

$\phantom{12 \ ? \ 12 +}12 \ \Big| \quad$ TRUE

The solution is 0.

**17.** $\quad 10 + x = 89$

$10 + x - 10 = 89 - 10$

$\phantom{10 + }x = 79$

Check: $\dfrac{10 + x = 89}{10 + 79 \ ? \ 89}$

$\phantom{10 + 79 \ ?}89 \ \Big| \quad$ TRUE

The solution is 79.

**19.** $\quad 61 = 16 + y$

$61 - 16 = 16 + y - 16$

$\phantom{61 - }45 = y$

Check: $\dfrac{61 = 16 + y}{61 \ ? \ 16 + 45}$

$\phantom{61 \ ? \ 16 +}61 \ \Big| \quad$ TRUE

The solution is 45.

**21.** $\quad 3 \cdot x = 24$

$\dfrac{3 \cdot x}{3} = \dfrac{24}{3} \qquad$ Dividing by 3 on both sides

$\phantom{\dfrac{3 \cdot x}{3} =}x = 8 \qquad$ 3 times $x$ divided by 3 is $x$.

Check: $\dfrac{3 \cdot x = 24}{3 \cdot 8 \ ? \ 24}$

$\phantom{3 \cdot 8 \ ?}24 \ \Big| \quad$ TRUE

The solution is 8.

**23.** $\quad 112 = n \cdot 8$

$\dfrac{112}{8} = \dfrac{n \cdot 8}{8} \qquad$ Dividing by 8 on both sides

$\phantom{\dfrac{112}{8} =}14 = n$

Check: $\dfrac{112 = n \cdot 8}{112 \ ? \ 14 \cdot 8}$

$\phantom{112 \ ?}112 \quad$ TRUE

The solution is 14.

**25.** $\quad 3 \cdot m = 96$

$\dfrac{3 \cdot m}{3} = \dfrac{96}{3} \qquad$ Dividing by 3 on both sides

$\phantom{\dfrac{3 \cdot m}{3} =}m = 32$

Check: $\dfrac{3 \cdot m = 96}{3 \cdot 32 \ ? \ 96}$

$\phantom{3 \cdot 32 \ ?}96 \ \Big| \quad$ TRUE

The solution is 32.

**27.** $\quad 715 = 5 \cdot z$

$\dfrac{715}{5} = \dfrac{5 \cdot z}{5} \qquad$ Dividing by 5 on both sides

$\phantom{\dfrac{715}{5} =}143 = z$

Check: $\dfrac{715 = 5 \cdot x}{715 \ ? \ 5 \cdot 143}$

$\phantom{715 \ ?}715 \quad$ TRUE

The solution is 143.

**29.** $\quad 8322 + 9281 = x$

$\phantom{8322 + }17,603 = x \qquad$ Doing the addition

The number 17,603 checks. It is the solution.

**31.** $\quad 47 + n = 84$

$47 + n - 47 = 84 - 47$

$\phantom{47 + n - 47 =}n = 37$

The number 37 checks. It is the solution.

**33.** $45 \times 23 = x$

To solve the equation we carry out the calculation.

$$\begin{array}{r} 4\,5 \\ \times\ 2\,3 \\ \hline 1\,3\,5 \\ 9\,0\,0 \\ \hline 1\,0\,3\,5 \end{array}$$

The number 1035 checks. It is the solution.

**35.** $\quad x + 78 = 144$

$x + 78 - 78 = 144 - 78$

$\phantom{x + 78 - 78 =}x = 66$

The number 66 checks. It is the solution.

**37.** $\quad 6 \cdot p = 1944$

$\dfrac{6 \cdot p}{6} = \dfrac{1944}{6}$

$\phantom{\dfrac{6 \cdot p}{6} =}p = 324$

The number 324 checks. It is the solution.

**39.** $\quad 5 \cdot x = 3715$

$\dfrac{5 \cdot x}{5} = \dfrac{3715}{5}$

$\phantom{\dfrac{5 \cdot x}{5} =}x = 743$

The number 743 checks. It is the solution.

**41.** $\quad x + 214 = 389$

$x + 214 - 214 = 389 - 214$

$\phantom{x + 214 - 214 =}x = 175$

The number 175 checks. It is the solution.

**43.** $\quad 567 + x = 902$

$567 + x - 567 = 902 - 567$

$\phantom{567 + x - 567 =}x = 335$

The number 335 checks. It is the solution.

**45.** $234 \cdot 78 = y$

$18,252 = y$   Doing the multiplication

The number 18,252 checks. It is the solution.

**47.** $18 \cdot x = 1872$

$$\frac{18 \cdot x}{18} = \frac{1872}{18}$$

$$x = 104$$

The number 104 checks. It is the solution.

**49.** $40 \cdot x = 1800$

$$\frac{40 \cdot x}{40} = \frac{1800}{40}$$

$$x = 45$$

The number 45 checks. It is the solution.

**51.** $\qquad 2344 + y = 6400$

$2344 + y - 2344 = 6400 - 2344$

$$y = 4056$$

The number 4056 checks. It is the solution.

**53.** $m = 7006 - 4159$

To solve the equation we carry out the calculation.

$$\begin{array}{r} {}^{6}\;{}^{9}\;{}^{9}\;{}^{16} \\ \cancel{7\,0\,0\,6} \\ -\;4\,1\,5\,9 \\ \hline 2\,8\,4\,7 \end{array}$$

The number 2847 checks. It is the solution.

**55.** $165 = 11 \cdot n$

$$\frac{165}{11} = \frac{11 \cdot n}{11}$$

$$15 = n$$

The number 15 checks. It is the solution.

**57.** $58 \cdot m = 11,890$

$$\frac{58 \cdot m}{58} = \frac{11,890}{58}$$

$$m = 205$$

The number 205 checks. It is the solution.

**59.** $491 - 34 = y$

To solve the equation we carry out the calculation.

$$\begin{array}{r} {}^{8}\;{}^{11} \\ 4\,\cancel{9}\,\cancel{1} \\ -\;\;3\,4 \\ \hline 4\,5\,7 \end{array}$$

The number 457 checks. It is the solution.

**61.**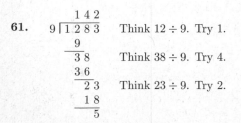

Think $12 \div 9$. Try 1.

Think $38 \div 9$. Try 4.

Think $23 \div 9$. Try 2.

The answer is 142 R 5.

**63.**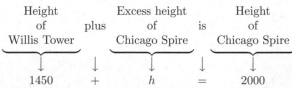

17 is close to 20. Think: $56 \div 20$. If we try 2, we find that it is too small. Try 3.

Think $57 \div 20$. Again, 2 is too small. Try 3.

Think $68 \div 20$. If we try 3, we find that it is too small. Try 4.

The answer is 334.

**65.** Since 123 is to the left of 789 on the number line, $123 < 789$.

**67.** Since 688 is to the right of 0 on the number line, $688 > 0$.

**69.** Round $6,\,3\,7\,5,\,\boxed{6}\,0\,2$ to the nearest thousand.

The digit 5 is in the thousands place. Consider the next digit to the right. Since the digit 6 is 5 or higher, round 5 thousands to 6 thousands. Then change all digits to the right of the thousands digit to zero.

The answer is 6,376,000.

**71.** $23,465 \cdot x = 8,142,355$

$$\frac{23,465 \cdot x}{23,465} = \frac{8,142,355}{23,465}$$

$$x = 347 \quad \text{Using a calculator to divide}$$

The number 347 checks. It is the solution.

## Exercise Set 1.8

**1.** *Familiarize.* We visualize the situation. Let $h =$ the number of feet by which the height of the Chicago Spire exceeds the height of the Willis Tower.

| Height of Willis Tower | Excess height of Chicago Spire |
|---|---|
| 1450 ft | $h$ |
| Height of Chicago Spire 2000 ft | |

*Translate.* We translate to an equation.

| Height of Willis Tower | plus | Excess height of Chicago Spire | is | Height of Chicago Spire |
|---|---|---|---|---|
| ↓ | ↓ | ↓ | ↓ | ↓ |
| 1450 | + | $h$ | = | 2000 |

*Solve.* We subtract 1450 on both sides of the equation.

$$1450 + h = 2000$$

$$1450 + h - 1450 = 2000 - 1450$$

$$h = 550$$

*Check.* We can add the difference, 550, to the height of the Willis Tower, 1450: $1450 + 550 = 2000$. We can also estimate:

$$2000 - 1450 \approx 2000 - 1500 = 500 \approx 550.$$

The answer checks.

*State.* The Chicago Spire is 550 ft taller than the Willis Tower.

3. *Familiarize*. We visualize the situation. Let $h =$ the height of the Empire State Building, in feet.

| Height of Aon Center 1136 ft | Excess height of Empire State Building 114 ft |
|---|---|
| Height of Empire State Building $h$ | |

*Translate*. We translate to an equation.

| Height of Aon Center | plus | Excess height of Empire State Building | is | Height of Empire State Building |
|---|---|---|---|---|
| ↓ | ↓ | ↓ | ↓ | ↓ |
| 1136 | + | 114 | = | $h$ |

*Solve*. We carry out the calculation.

$$\begin{array}{r} \overset{1}{1}\,1\,3\,6 \\ +\,1\,1\,4 \\ \hline 1\,2\,5\,0 \end{array}$$

*Check*. We can repeat the calculation. We can also estimate:

$$1136 + 114 \approx 1140 + 110 = 1250$$

The answer seems reasonable.

*State*. The height of the Empire State Building is 1250 ft.

5. *Familiarize*. We visualize the situation. Let $c =$ the amount of caffeine in an 8-oz serving of coffee, in milligrams.

| Caffeine in Red Bull 76 milligrams | Excess caffeine in coffee 19 milligrams |
|---|---|
| Caffeine in coffee $c$ | |

*Translate*. We translate to an equation.

| Caffeine in Red Bull | plus | Excess caffeine in coffee | is | Caffeine in coffee |
|---|---|---|---|---|
| ↓ | ↓ | ↓ | ↓ | ↓ |
| 76 | + | 19 | = | $c$ |

*Solve*. We carry out the calculation.

$$\begin{array}{r} \overset{1}{7}\,6 \\ +\,1\,9 \\ \hline 9\,5 \end{array}$$

Thus, $95 = c$.

*Check*. We can repeat the calculation. We can also estimate: $76 + 19 \approx 80 + 20 = 100 \approx 95$. The answer seems reasonable.

*State*. An 8-oz serving of coffee contains 95 milligrams of caffeine.

7. *Familiarize*. We first make a drawing. Let $r =$ the number of rows.

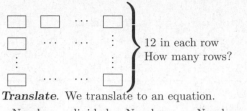

*Translate*. We translate to an equation.

| Number of holes | divided by | Number per row | is | Number of rows |
|---|---|---|---|---|
| ↓ | ↓ | ↓ | ↓ | ↓ |
| 216 | ÷ | 12 | = | $r$ |

*Solve*. We carry out the division.

$$\begin{array}{r} 1\,8 \\ 1\,2\,\overline{)2\,1\,6} \\ \underline{1\,2\phantom{\,6}} \\ 9\,6 \\ \underline{9\,6} \\ 0 \end{array}$$

Thus, $18 = r$, or $r = 18$.

*Check*. We can check by multiplying: $12 \cdot 18 = 216$. Our answer checks.

*State*. There are 18 rows.

9. *Familiarize*. We visualize the situation. Let $h =$ the number of hours that people age 12 and older watched TV, on average, in 2000.

| TV-watching hours in 2000 $h$ | Excess hours in 2008 202 |
|---|---|
| TV-watching hours in 2008 1704 | |

*Translate*. We translate to an equation.

| TV-watching hours in 2000 | plus | Excess hours in 2008 | is | TV-watching hours in 2008 |
|---|---|---|---|---|
| ↓ | ↓ | ↓ | ↓ | ↓ |
| $h$ | + | 202 | = | 1740 |

*Solve*. We subtract 202 on both sides of the equation.

$$h + 202 = 1704$$
$$h + 202 - 202 = 1704 - 202$$
$$h = 1502$$

*Check*. We can add the difference, 1502, to 202: $202 + 1502 = 1704$. We can also estimate:

$$1704 - 202 \approx 1700 - 200 = 1500 \approx 1502.$$

The answer checks.

*State*. On average, people age 12 and older spent 1502 hours watching TV in 2000.

11. *Familiarize*. We visualize the situation. Let $m =$ the number of miles by which the Canadian border exceeds the Mexican border.

| Mexican border 1933 mi | Excess miles in Canadian border $m$ |
|---|---|
| Canadian border 3987 mi | |

*Translate.* We translate to an equation.

| Length of Mexican border | plus | Excess length of Canadian border | is | Length of Canadian border |
|---|---|---|---|---|
| $\downarrow$ | $\downarrow$ | $\downarrow$ | $\downarrow$ | $\downarrow$ |
| 1933 | + | $m$ | = | 3987 |

*Solve.* We subtract 1933 on both sides of the equation.

$$1933 + m = 3987$$
$$1933 + m - 1933 = 3987 - 1933$$
$$m = 2054$$

*Check.* We can add the difference, 2054, to the subtrahend, 1933: $1933 + 2054 = 3987$. We can also estimate:

$$3987 - 1933 \approx 4000 - 2000$$
$$\approx 2000 \approx 2054$$

The answer checks.

*State.* The Canadian border is 2054 mi longer than the Mexican border.

**13.** *Familiarize.* We visualize the situation. Let $p =$ the number of pixels on the screen. Repeated addition works well here.

$$\underbrace{\boxed{1920 \text{ pixels}} + \boxed{1920 \text{ pixels}} + \cdots + \boxed{1920 \text{ pixels}}}_{1080 \text{ addends}}$$

*Translate.* We translate to an equation.

| Number of pixels in a row | times | Number of rows | is | Total number of pixels |
|---|---|---|---|---|
| $\downarrow$ | $\downarrow$ | $\downarrow$ | $\downarrow$ | $\downarrow$ |
| 1920 | $\times$ | 1080 | = | $p$ |

*Solve.* We carry out the multiplication.

$$\begin{array}{r} 1\,9\,2\,0 \\ \times\; 1\,0\,8\,0 \\ \hline 1\,5\,3\,6\,0\,0 \\ 1\,9\,2\,0\,0\,0\,0 \\ \hline 2,0\,7\,3,6\,0\,0 \end{array}$$

Thus, $2,073,600 = p$.

*Check.* We can repeat the calculation. The answer checks.

*State.* There are 2,073,600 pixels on the screen.

**15.** *Familiarize.* We visualize the situation. Let $m =$ the number of associate's degrees earned by men in 2005.

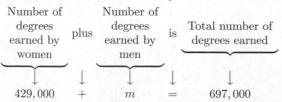

| Number of degrees earned by women 429,000 | Number of degrees earned by men $m$ |
|---|---|
| Total number of degrees earned 697,000 | |

*Translate.* We translate to an equation.

| Number of degrees earned by women | plus | Number of degrees earned by men | is | Total number of degrees earned |
|---|---|---|---|---|
| $\downarrow$ | $\downarrow$ | $\downarrow$ | $\downarrow$ | $\downarrow$ |
| 429,000 | + | $m$ | = | 697,000 |

*Solve.* We subtract 429,000 on both sides of the equation.

$$429,000 + m = 697,000$$
$$429,000 + m - 429,000 = 697,000 - 429,000$$
$$m = 268,000$$

*Check.* We can add the number of degrees earned by women and the number of degrees earned by men.

$$429,000 + 268,000 = 697,000$$

We get the total number of degrees earned, so the answer checks.

*State.* 268,000 associate's degrees were earned by men in 2005.

**17.** *Familiarize.* We first draw a picture. Let $h =$ the number of hours in a week. Repeated addition works well here.

$$\underbrace{\boxed{24 \text{ hours}} + \boxed{24 \text{ hours}} + \cdots + \boxed{24 \text{ hours}}}_{7 \text{ addends}}$$

*Translate.* We translate to an equation.

| Number of hours in a day | times | Number of days in a week | is | Number of hours in a week |
|---|---|---|---|---|
| $\downarrow$ | $\downarrow$ | $\downarrow$ | $\downarrow$ | $\downarrow$ |
| 24 | $\times$ | 7 | = | $h$ |

*Solve.* We carry out the multiplication.

$$\begin{array}{r} 2\,4 \\ \times\; 7 \\ \hline 1\,6\,8 \end{array}$$

Thus, $168 = h$, or $h = 168$.

*Check.* We can repeat the calculation. We an also estimate:

$$24 \times 7 \approx 20 \times 10 = 200 \approx 168$$

Our answer checks.

*State.* There are 168 hours in a week.

**19.** *Familiarize.* We visualize the situation. Let $r =$ the number of dollars by which the average monthly rent in Houston exceeds the average monthly rent in Dallas/Fort Worth.

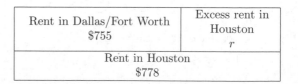

| Rent in Dallas/Fort Worth $755 | Excess rent in Houston $r$ |
|---|---|
| Rent in Houston $778 | |

*Translate.* We translate to an equation.

| Rent in Dallas/Forth Worth | plus | Excess rent in Houston | is | Rent in Houston |
|---|---|---|---|---|
| $\downarrow$ | $\downarrow$ | $\downarrow$ | $\downarrow$ | $\downarrow$ |
| 755 | + | $r$ | = | 778 |

*Solve.* We subtract 755 on both sides of the equation.

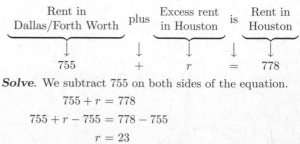

$$755 + r = 778$$
$$755 + r - 755 = 778 - 755$$
$$r = 23$$

**Check**. We can add: $755 + 23 = 778$. We get the rent in Houston, so the answer checks.

**State**. The average monthly rent in Houston is $23 higher than in Dallas/Fort Worth.

**21. Familiarize**. We first draw a picture. We let $r =$ the average monthly rent each person will pay.

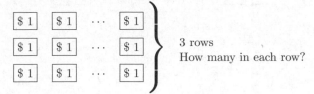

**Translate**. We translate to an equation.

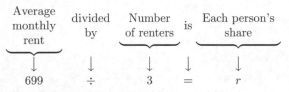

**Solve**. We carry out the division.

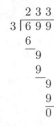

Thus, $233 = r$.

**Check**. We can check by multiplying: $233 \cdot 3 = 699$. The answer checks.

**State**. Each person can expect to pay an average monthly rent of $233.

**23. Familiarize**. We draw a picture. Let $t =$ the total amount of rent a person would pay in Atlanta during a 12-month period. Repeated addition works well here.

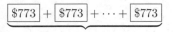

$$\underbrace{\$773 + \$773 + \cdots + \$773}$$

12 addends

**Translate**. We translate to an equation.

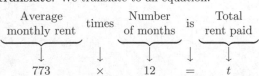

**Solve**. We carry out the multiplication.

$$
\begin{array}{r}
7\,7\,3 \\
\times\quad 1\,2 \\
\hline
1\,5\,4\,6 \\
7\,7\,3\,0 \\
\hline
9\,2\,7\,6
\end{array}
$$

Thus, $9276 = t$.

**Check**. We repeat the calculation. The answer checks.

**State**. On average, a tenant would pay $9276 in rent during a 12-month period in Atlanta.

**25. Familiarize**. We visualize the situation. Let $p =$ the Colonial population in 1680.

| Population in 1680 | Increase in population |
|---|---|
| $p$ | 2,628,900 |
| Population in 1780 | |
| 2,780,400 | |

**Translate**. We translate to an equation

$$
\begin{array}{ccccc}
\underbrace{\text{Population}}_{\text{in 1680}} & \text{plus} & \underbrace{\text{Increase in}}_{\text{population}} & \text{is} & \underbrace{\text{Population}}_{\text{in 1780}} \\
\downarrow & \downarrow & \downarrow & \downarrow & \downarrow \\
p & + & 2,628,900 & = & 2,780,400
\end{array}
$$

**Solve**. We subtract 2,628,900 on both sides of the equation.

$$p + 2,628,900 = 2,780,400$$
$$p + 2,628,900 - 2,628,900 = 2,780,400 - 2,628,900$$
$$p = 151,500$$

**Check**. Since $2,628,900 + 151,500 = 2,780,400$, the answer checks.

**State**. In 1680 the Colonial population was 151,500.

**27. Familiarize**. We visualize the situation. Let $m =$ the number of motorcycles that were sold in 2006.

| Motorcycles sold in 2000 | Excess sales in 2006 |
|---|---|
| 710,000 | 480,000 |
| Motorcycles sold in 2006 | |
| $m$ | |

**Translate**. We translate to an equation.

$$
\begin{array}{ccccc}
\underbrace{\text{Motorcycles}}_{\text{sold in 2000}} & \text{plus} & \underbrace{\text{Excess sales}}_{\text{in 2006}} & \text{is} & \underbrace{\text{Motorcycles}}_{\text{sold in 2006}} \\
\downarrow & \downarrow & \downarrow & \downarrow & \downarrow \\
710,000 & + & 480,000 & = & m
\end{array}
$$

**Solve**. We carry out the addition.

$$
\begin{array}{r}
7\,1\,0,0\,0\,0 \\
+\,4\,8\,0,0\,0\,0 \\
\hline
1,1\,9\,0,0\,0\,0
\end{array}
$$

Thus, $1,190,000 = m$.

**Check**. We can repeat the addition. We can also estimate:

$710,000 + 480,000 \approx 700,000 + 500,000 = 1,200,000 \approx 1,190,000$

The answer seems reasonable.

**State**. 1,190,000 motorcycles were sold in 2006.

**29. _Familiarize_.** We draw a picture. Let $g =$ the amount each grandchild received.

4 rows
How many in each row?

**_Translate_.** We translate to an equation.

| Total amount earned | divided by | Number of grandchildren | is | Amount each grandchild received |
|---|---|---|---|---|
| ↓ | ↓ | ↓ | ↓ | ↓ |
| 312 | ÷ | 4 | = | $g$ |

**_Solve_.** We carry out the division.

```
      7 8
  4 ⟌ 3 1 2
      2 8
      ───
        3 2
        3 2
        ───
          0
```

Thus, $78 = g$.

**_Check_.** We can check by multiplying the amount each grandchild received by the number of grandchildren:

$\$78 \cdot 4 = \$312$. The answer checks.

**_State_.** Each grandchild received $78.

**31. _Familiarize_.** We visualize the situation. Let $b =$ the average monthly parking rate in Bakersfield.

| Rate in Bakersfield $b$ | Additional cost in New York City $545 |
|---|---|
| Rate in New York City $585 | |

**_Translate_.** We translate to an equation.

| Rate in Bakersfield | plus | Additional cost in New York City | is | Rate in New York City |
|---|---|---|---|---|
| ↓ | ↓ | ↓ | ↓ | ↓ |
| $b$ | + | 545 | = | 585 |

**_Solve_.** We subtract 545 on both sides of the equation.

$$b + 545 = 585$$
$$b + 545 - 545 = 585 - 545$$
$$b = 40$$

**_Check_.** We can add the rate in Bakersfield to the additional cost in New York City: $\$40 + \$545 = \$585$. The answer checks.

**_State_.** The average monthly parking rate in Bakersfield is $40.

**33. _Familiarize_.** We draw a picture of the situation. Let $c =$ the total cost of the purchase. Repeated addition works well here.

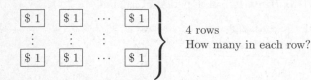

24 addends

**_Translate_.** We translate to an equation.

| Number purchased | times | Cost of each refrigerator | is | Total cost |
|---|---|---|---|---|
| ↓ | ↓ | ↓ | ↓ | ↓ |
| 24 | × | 1019 | = | $c$ |

**_Solve_.** We carry out the multiplication.

```
        1
        3
    1 0 1 9
  ×     2 4
  ─────────
    4 0 7 6
  2 0 3 8 0
  ─────────
  2 4, 4 5 6
```

Thus, $24,456 = c$.

**_Check_.** We can repeat the calculation. We can also estimate: $24 \times 1019 \approx 24 \times 1000 \approx 24,000 \approx 24,456$. The answer checks.

**_State_.** The total cost of the purchase is $24,456.

**35. _Familiarize_.** We first draw a picture. Let $w =$ the number of full weeks the episodes can run.

5 in each row

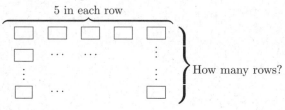

How many rows?

**_Translate_.** We translate to an equation.

| Number of episodes | divided by | Number shown per week | is | Number of weeks |
|---|---|---|---|---|
| ↓ | ↓ | ↓ | ↓ | ↓ |
| 177 | ÷ | 5 | = | $w$ |

**_Solve_.** We carry out the division.

```
      3 5
  5 ⟌ 1 7 7
      1 5
      ───
        2 7
        2 5
        ───
          2
```

**_Check_.** We can check by multiplying the number of weeks by 5 and adding the remainder, 2:

$$5 \cdot 35 = 175, \qquad 175 + 2 = 177$$

**_State_.** 35 full weeks will pass before the station must start over. There will be 2 episodes left over.

**37. _Familiarize_.** We first draw a picture of the situation. Let $g =$ the number of gallons that will be used in 5900 mi of city driving.

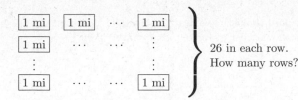

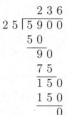 26 in each row.
How many rows?

**Translate.** We translate to an equation.

$$\underbrace{\text{Number}\atop\text{of miles}} \quad \underbrace{\text{divided}\atop\text{by}} \quad \underbrace{\text{Number of}\atop\text{mpg}} \quad \text{is} \quad \underbrace{\text{Number of}\atop\text{gallons}}$$

$$5900 \quad \div \quad 25 \quad = \quad g$$

**Solve.** We carry out the division.

$$\begin{array}{r} 236 \\ 25\overline{)5900} \\ \underline{50} \\ 90 \\ \underline{75} \\ 150 \\ \underline{150} \\ 0 \end{array}$$

Thus, $236 = g$.

**Check.** We can check by multiplying the number of gallons by the number of miles per gallon: $25 \cdot 236 = 5900$. The answer checks.

**State.** The Hyundai Tucson GLS will use 236 gal of gasoline in 5900 mi of city driving.

**39. Familiarize.** First we draw a picture. Let $c =$ the number of columns. The number of columns is the same as the number of squares in each row.

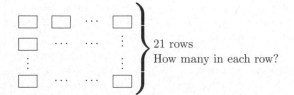
21 rows
How many in each row?

**Translate.** We translate to an equation.

$$\underbrace{\text{Number}\atop\text{of squares}} \quad \underbrace{\text{divided}\atop\text{by}} \quad \underbrace{\text{Number}\atop\text{of rows}} \quad \text{is} \quad \underbrace{\text{Number}\atop\text{of columns.}}$$

$$441 \quad \div \quad 21 \quad = \quad c$$

**Solve.** We carry out the division.

$$\begin{array}{r} 21 \\ 21\overline{)441} \\ \underline{42} \\ 21 \\ \underline{21} \\ 0 \end{array}$$

Thus, $21 = c$.

**Check.** We can check by multiplying the number of rows by the number of columns: $24 \cdot 21 = 441$. The answer checks.

**State.** The puzzle has 21 columns.

**41. Familiarize.** We first draw a picture. Let $A =$ the area and $P =$ the perimeter of the court, in feet.

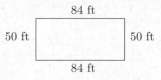

**Translate.** We write one equation to find the area and another to find the perimeter.

a) Using the formula for the area of a rectangle, we have

$$A = l \cdot w = 84 \cdot 50$$

b) Recall that the perimeter is the distance around the court.

$$P = 84 + 50 + 84 + 50$$

**Solve.** We carry out the calculations.

a)
$$\begin{array}{r} 50 \\ \times\,84 \\ \hline 200 \\ 4000 \\ \hline 4200 \end{array}$$

Thus, $A = 4200$.

b) $P = 84 + 50 + 84 + 50 = 268$

**Check.** We can repeat the calculation. The answers check.

**State.** a) The area of the court is 4200 square feet.

b) The perimeter of the court is 268 ft.

**43. Familiarize.** We first draw a picture. We let $x =$ the amount of each payment.

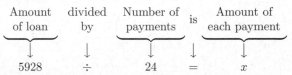
24 rows
How many in each row?

**Translate.** We translate to an equation.

$$\underbrace{\text{Amount}\atop\text{of loan}} \quad \underbrace{\text{divided}\atop\text{by}} \quad \underbrace{\text{Number of}\atop\text{payments}} \quad \text{is} \quad \underbrace{\text{Amount of}\atop\text{each payment}}$$

$$5928 \quad \div \quad 24 \quad = \quad x$$

**Solve.** We carry out the division.

$$\begin{array}{r} 247 \\ 24\overline{)5928} \\ \underline{48} \\ 112 \\ \underline{96} \\ 168 \\ \underline{168} \\ 0 \end{array}$$

Thus, $247 = x$, or $x = 247$.

**Check.** We can check by multiplying 247 by 24: $24 \cdot 247 = 5928$. The answer checks.

**State.** Each payment is $247.

**45. Familiarize.** First we find the distance in reality between two cities that are 3 in. apart on the map. We make a drawing. Let $d =$ the distance between the cities, in miles. Repeated addition works well here.

$$\underbrace{\boxed{215 \text{ miles}} + \boxed{215 \text{ miles}} + \boxed{215 \text{ miles}}}_{3 \text{ addends}}$$

**Translate.**

$$\underbrace{\text{Number of}}_{\text{miles per inch}} \quad \underset{\text{times}}{\downarrow} \quad \underbrace{\text{Number}}_{\text{of inches}} \quad \underset{\text{is}}{\downarrow} \quad \underbrace{\text{Distance,}}_{\text{in miles}}$$

$$215 \quad \times \quad 3 \quad = \quad d$$

**Solve.** We carry out the multiplication.

$$\begin{array}{r} 2\,1\,5 \\ \times \quad 3 \\ \hline 6\,4\,5 \end{array}$$

Thus, $645 = d$.

**Check.** We can repeat the calculation or estimate the product. Our answer checks.

**State.** Two cities that are 3 in. apart on the map are 645 miles apart in reality.

Next we find the distance on the map between two cities that, in reality, are 1075 mi apart.

**Familiarize.** We visualize the situation. Let $m =$ the distance between the cities on the map.

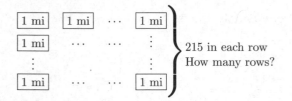

**Translate.**

$$\underbrace{\text{Number}}_{\text{of miles}} \quad \underset{\text{by}}{\underbrace{\text{divided}}} \quad \underbrace{\text{Number of}}_{\text{miles per inch}} \quad \underset{\text{is}}{\downarrow} \quad \underbrace{\text{Distance,}}_{\text{in inches}}$$

$$1075 \quad \div \quad 215 \quad = \quad m$$

**Solve.** We carry out the division.

$$\begin{array}{r} 5 \\ 215\overline{\smash{\big)}\,1\,0\,7\,5} \\ \underline{1\,0\,7\,5} \\ 0 \end{array}$$

Thus, $5 = m$.

**Check.** We can check by multiplying: $215 \cdot 5 = 1075$. Our answer checks.

**State.** The cities are 5 in. apart on the map.

**47. Familiarize.** We draw a picture of the situation. Let $c =$ the number of cartons that can be filled.

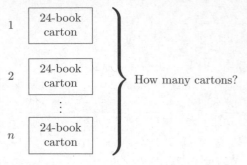

**Translate.**

$$\underbrace{\text{Number}}_{\text{of books}} \quad \underset{\text{by}}{\underbrace{\text{divided}}} \quad \underbrace{\text{Number}}_{\text{per carton}} \quad \underset{\text{is}}{\downarrow} \quad \underbrace{\text{Number of}}_{\text{full cartons.}}$$

$$1344 \quad \div \quad 24 \quad = \quad c$$

**Solve.** We carry out the division.

$$\begin{array}{r} 5\,6 \\ 24\overline{\smash{\big)}\,1\,3\,4\,4} \\ \underline{1\,2\,0} \\ 1\,4\,4 \\ \underline{1\,4\,4} \\ 0 \end{array}$$

**Check.** We can check by multiplying the number of cartons by 24: $24 \cdot 56 = 1344$. The answer checks.

**State.** 56 cartons can be filled.

**49. Familiarize.** This is a multistep problem.

We must find the total price of the 5 video games. Then we must find how many 10's there are in the total price. Let $p =$ the total price of the games.

To find the total price of the 5 video games we can use repeated addition.

$$\underbrace{\boxed{\$64} + \boxed{\$64} + \boxed{\$64} + \boxed{\$64} + \boxed{\$64}}_{5 \text{ addends}}$$

**Translate.**

$$\underbrace{\text{Price}}_{\text{per game}} \quad \underset{\text{times}}{\downarrow} \quad \underbrace{\text{Number}}_{\text{of games}} \quad \underset{\text{is}}{\downarrow} \quad \underbrace{\text{Total price}}_{\text{of games}}$$

$$64 \quad \cdot \quad 5 \quad = \quad p$$

**Solve.** First we carry out the multiplication.

$$64 \cdot 5 = p$$
$$320 = p$$

The total price of the 5 video games is $320. Repeated addition can be used again to find how many 10's there are in $320. We let $x =$ the number of $10 bills required.

$$\begin{array}{|c|} \hline \$320 \\ \hline \end{array}$$
$$\begin{array}{|c|c|c|c|} \hline \$10 & \$10 & \cdots & \$10 \\ \hline \end{array}$$

Translate to an equation and solve.

$$10 \cdot x = 320$$
$$\frac{10 \cdot x}{10} = \frac{320}{10}$$
$$x = 32$$

**Check**. We repeat the calculations. The answer checks.

**State**. It takes 32 ten dollar bills.

**51. Familiarize**. This is a multistep problem. We must find the total amount of the debits. Then we subtract this amount from the original balance and add the amount of the deposit. Let $a =$ the total amount of the debits. To find this we can add.

**Translate**.

| First debit | plus | Second debit | plus | Third debit | is | Total amount |
|---|---|---|---|---|---|---|
| ↓ | ↓ | ↓ | ↓ | ↓ | ↓ | ↓ |
| 46 | + | 87 | + | 129 | = | $a$ |

**Solve**. First we carry out the addition.

```
  1 2
    4 6
    8 7
+ 1 2 9
-------
  2 6 2
```

Thus, $262 = a$.

Now let $b =$ the amount left in the account after the debits.

| Amount left | is | Original amount | minus | Amount of debits |
|---|---|---|---|---|
| ↓ | ↓ | ↓ | ↓ | ↓ |
| $b$ | = | 568 | − | 262 |

We solve this equation by carrying out the subtraction.

```
  5 6 8
− 2 6 2
-------
  3 0 6
```

Thus, $b = 306$.

Finally, let $f =$ the final amount in the account after the deposit is made.

| Final amount | is | Amount after debits | plus | Amount of deposit |
|---|---|---|---|---|
| ↓ | ↓ | ↓ | ↓ | ↓ |
| $f$ | = | 306 | + | 94 |

We solve this equation by carrying out the addition.

```
  1 1
  3 0 6
+   9 4
-------
  4 0 0
```

Thus, $f = 400$.

**Check**. We repeat the calculations. The answer checks.

**State**. There is $400 left in the account.

**53. Familiarize**. This is a multistep problem. We begin by visualizing the situation.

| One pound 3500 calories | | | |
|---|---|---|---|
| 100 cal | 100 cal | . . . | 100 cal |
| 15 min | 15 min | | 15 min |

Let $x =$ the number of hundreds in 3500. Repeated addition applies here.

**Translate**. We translate to an equation.

| 100 calories | times | How many 100's | is 3500? | |
|---|---|---|---|---|
| ↓ | ↓ | ↓ | ↓ | ↓ |
| 100 | · | $x$ | = | 3500 |

**Solve**. We divide by 100 on both sides of the equation.

$$100 \cdot x = 3500$$
$$\frac{100 \cdot x}{100} = \frac{3500}{100}$$
$$x = 35$$

From the chart we know that doing aerobic exercise for 15 min burns 100 calories. Thus we must do 15 min of exercise 35 times in order to lose one pound. Let $t =$ the number of minutes of aerobic exercise required to lose one pound. We translate to an equation.

| Number of times | times | Number of minutes | is | Total time |
|---|---|---|---|---|
| ↓ | ↓ | ↓ | ↓ | ↓ |
| 35 | × | 15 | = | $t$ |

```
    1 5
  × 3 5
  -----
    7 5
  4 5 0
  -----
  5 2 5
```

Thus, $525 = t$.

**Check**. $525 \div 15 = 35$, so there are 35 15's in 525 min, and $35 \cdot 100 = 3500$, the number of calories that must be burned in order to lose one pound. The answer checks.

**State**. You must do aerobic exercise for 525 min, or 8 hr, 45 min, in order to lose one pound.

**55. Familiarize**. This is a multistep problem. We begin by visualizing the situation.

| One pound 3500 calories | | | |
|---|---|---|---|
| 100 cal | 100 cal | . . . | 100 cal |
| 20 min | 20 min | | 20 min |

From Exercise 53 we know that there are 35 100's in 3500. From the chart we know that golfing for 20 min will burn 100 calories. This must be done 35 times in order to lose one pound. Let $t =$ the time it takes to lose one pound. We have:

$$t = 35 \times 20$$
$$t = 700$$

**Check**. $700 \div 20 = 35$, so there are 35 20's in 700 min, and $35 \cdot 100 = 3500$, the number of calories that must be burned in order to lose one pound. The answer checks.

**State**. You must golf for 700 min, or 11 hr, 40 min, walking, in order to lose one pound.

**57. Familiarize**. This is a multistep problem. We will find the number of seats in each class and then add to find the total seating capacity. Let $F$ = the number of first-class seats, $E$ = the number of economy-class seats, and $T$ = the total number of seats.

**Translate**. We translate to three equations.

$$
\underbrace{\text{Rows in first class}}_{\downarrow \atop 2} \underset{\downarrow \atop \times}{\text{times}} \underbrace{\text{Seats in each row}}_{\downarrow \atop 4} \underset{\downarrow \atop =}{\text{is}} \underbrace{\text{Total first-class seats}}_{\downarrow \atop F}
$$

$$
\underbrace{\text{Rows in economy class}}_{\downarrow \atop 16} \underset{\downarrow \atop \times}{\text{times}} \underbrace{\text{Seats in each row}}_{\downarrow \atop 6} \underset{\downarrow \atop =}{\text{is}} \underbrace{\text{Total economy-class seats}}_{\downarrow \atop E}
$$

$$
\underbrace{\text{Number of first-class seats}}_{\downarrow \atop F} \underset{\downarrow \atop +}{\text{plus}} \underbrace{\text{Number of economy-class seats}}_{\downarrow \atop E} \underset{\downarrow \atop =}{\text{is}} \underbrace{\text{Total number of seats}}_{\downarrow \atop T}
$$

**Solve**. We solve th first two equations.

$$2 \times 4 = F \qquad 16 \times 6 = E$$
$$8 = F \qquad\quad 96 = E$$

Now we substitute 8 for F and 96 for E in the third equation and add to find T.

$$F + E = T$$
$$8 + 96 = T$$
$$104 = T$$

**Check**. We repeat the calculations. The answer checks.

**State**. The total seating capacity of the plane is 104.

**59. Familiarize**. This is a multistep problem. We will find the number of bones in both hands and the number in both feet and then the total of these two numbers. Let $h$ = the number of bones in two human hands, $f$ = the number of bones in two human feet, and $t$ = the total number of bones in two hands and two feet.

**Translate**. We translate to three equations.

$$
\underbrace{\text{Number of bones in one hand}}_{\downarrow \atop 27} \underset{\downarrow \atop \cdot}{\text{times}} \underset{\downarrow \atop 2}{2} \underset{\downarrow \atop =}{\text{is}} \underbrace{\text{Number of bones in both hands}}_{\downarrow \atop h}
$$

$$
\underbrace{\text{Number of bones in one foot}}_{\downarrow \atop 26} \underset{\downarrow \atop \cdot}{\text{times}} \underset{\downarrow \atop 2}{2} \underset{\downarrow \atop =}{\text{is}} \underbrace{\text{Number of bones in both feet}}_{\downarrow \atop f}
$$

$$
\underbrace{\text{Number of bones in both hands}}_{\downarrow \atop h} \underset{\downarrow \atop +}{\text{plus}} \underbrace{\text{Number of bones in both feet}}_{\downarrow \atop f} \underset{\downarrow \atop =}{\text{is}} \underbrace{\text{Total number of bones}}_{\downarrow \atop t}
$$

**Solve**. We solve each equation.

$$27 \cdot 2 = h \qquad 26 \cdot 2 = f$$
$$54 = h \qquad\quad 52 = f$$

$$h + f = t$$
$$54 + 52 = t$$
$$106 = t$$

**Check**. We repeat the calculations. The answer checks.

**State**. In all, a human has 106 bones in both hands and both feet.

**61.** Round 234,562 to the nearest hundred.

$$2\,3\,4,\,5\,\boxed{6}\,2$$
$$\uparrow$$

The digit 5 is in the hundreds place. Consider the next digit to the right. Since the digit, 6, is 5 or higher, round 5 hundreds up to 6 hundreds. Then change all digits to the right of the hundreds place to zeros.

The answer is 234,600.

**63.** Round 234,562 to the nearest thousand.

$$2\,3\,4,\,\boxed{5}\,6\,2$$
$$\uparrow$$

The digit 4 is in the thousands place. Consider the next digit to the right. Since the digit, 5, is 5 or higher, round 4 thousands up to 5 thousands. Then change all digits to the right of the thousands place to zeros.

The answer is 235,000.

**65.**

|  | Rounded to the nearest thousand |
|---|---|
| $\begin{aligned} 28,&430 \\ -11,&977 \end{aligned}$ | $\begin{aligned} 28,&000 \\ -12,&000 \\ \hline 16,&000 \end{aligned}$ ← Estimated answer |

**67.**

|  | Rounded to the nearest thousand |
|---|---|
| $\begin{aligned} 2100 \\ +5800 \end{aligned}$ | $\begin{aligned} 2000 \\ +6000 \\ \hline 8000 \end{aligned}$ ← Estimated answer |

**69.**

|  | Rounded to the nearest hundred |
|---|---|
| $\begin{aligned} 799 \\ \times 887 \end{aligned}$ | $\begin{aligned} 800 \\ \times\;\;900 \\ \hline 720,000 \end{aligned}$ ← Estimated answer |

**71. Familiarize**. This is a multistep problem. First we will find the differences in the distances traveled in 1 second. Then we will find the differences for 18 seconds. Let $d$ = the difference in the number of miles light would travel per second in a vacuum and in ice. Let $g$ = the difference in the number of miles light would travel per second in a vacuum and in glass.

*Translate*. We translate to two equations.

| Distance in ice | plus | Additional distance | is | Distance in a vacuum. |
|---|---|---|---|---|
| ↓ | ↓ | ↓ | ↓ | ↓ |
| $142,000$ | $+$ | $d$ | $=$ | $186,000$ |

| Distance in glass | plus | Additional distance | is | Distance in a vacuum. |
|---|---|---|---|---|
| ↓ | ↓ | ↓ | ↓ | ↓ |
| $109,000$ | $+$ | $g$ | $=$ | $186,000$ |

*Solve*. We begin by solving each equation.

$$142,000 + d = 186,000$$
$$142,000 + d - 142,000 = 186,000 - 142,000$$
$$d = 44,000$$

$$109,000 + g = 186,000$$
$$109,000 + g - 109,000 = 186,000 - 109,000$$
$$g = 77,000$$

Now to find the differences in the distances in 18 seconds, we multiply each solution by 18.

For ice: $18 \cdot 44,000 = 792,000$

For glass: $18 \cdot 77,000 = 1,386,000$

*Check*. We repeat the calculations. Our answers check.

*State*. In 18 seconds light travels 792,000 miles farther in ice and 1,386,000 miles farther in glass than in a vacuum.

## Exercise Set 1.9

**1.** Exponential notation for $3 \cdot 3 \cdot 3 \cdot 3$ is $3^4$.

**3.** Exponential notation for $5 \cdot 5$ is $5^2$.

**5.** Exponential notation for $7 \cdot 7 \cdot 7 \cdot 7 \cdot 7$ is $7^5$.

**7.** Exponential notation for $10 \cdot 10 \cdot 10$ is $10^3$.

**9.** $7^2 = 7 \cdot 7 = 49$

**11.** $9^3 = 9 \cdot 9 \cdot 9 = 729$

**13.** $12^4 = 12 \cdot 12 \cdot 12 \cdot 12 = 20,736$

**15.** $3^5 = 3 \cdot 3 \cdot 3 \cdot 3 \cdot 3 = 243$

**17.** $12 + (6 + 4) = 12 + 10$    Doing the calculation inside the parentheses
        $= 22$    Adding

**19.** $52 - (40 - 8) = 52 - 32$    Doing the calculation inside the parentheses
        $= 20$    Subtracting

**21.** $1000 \div (100 \div 10)$
     $= 1000 \div 10$    Doing the calculation inside the parentheses
     $= 100$    Dividing

**23.** $(256 \div 64) \div 4 = 4 \div 4$    Doing the calculation inside the parentheses
        $= 1$    Dividing

**25.** $(2 + 5)^2 = 7^2$    Doing the calculation inside the parentheses
        $= 49$    Evaluating the exponential expression

**27.** $(11 - 8)^2 - (18 - 16)^2$
   $= 3^2 - 2^2$    Doing the calculations inside the parentheses
   $= 9 - 4$    Evaluating the exponential expressions
   $= 5$    Subtracting

**29.** $16 \cdot 24 + 50 = 384 + 50$    Doing all multiplications and divisions in order from left to right
        $= 434$    Doing all additions and subtractions in order from left to right

**31.** $83 - 7 \cdot 6 = 83 - 42$    Doing all multiplications and divisions in order from left to right
        $= 41$    Doing all additions and subtractions in order from left to right

**33.** $10 \cdot 10 - 3 \times 4$
   $= 100 - 12$    Doing all multiplications and divisions in order from left to right
   $= 88$    Doing all additions and subtractions in order from left to right

**35.** $4^3 \div 8 - 4$
   $= 64 \div 8 - 4$    Evaluating the exponential expression
   $= 8 - 4$    Doing all multiplications and divisions in order from left to right
   $= 4$    Doing all additions and subtractions in order from left to right

**37.** $17 \cdot 20 - (17 + 20)$
   $= 17 \cdot 20 - 37$    Carrying out the operation inside parentheses
   $= 340 - 37$    Doing all multiplications and divisions in order from left to right
   $= 303$    Doing all additions and subtractions in order from left to right

**39.** $6 \cdot 10 - 4 \cdot 10$
   $= 60 - 40$    Doing all multiplications and divisions in order from left to right
   $= 20$    Doing all additions and subtractions in order from left to right

**41.** $300 \div 5 + 10$
   $= 60 + 10$    Doing all multiplications and divisions in order from left to right
   $= 70$    Doing all additions and subtractions in order from left to right

**43.**  $3 \cdot (2 + 8)^2 - 5 \cdot (4 - 3)^2$

$= 3 \cdot 10^2 - 5 \cdot 1^2$  Carrying out operations inside parentheses

$= 3 \cdot 100 - 5 \cdot 1$  Evaluating the exponential expressions

$= 300 - 5$  Doing all multiplications and divisions in order from left to right

$= 295$  Doing all additions and subtractions in order from left to right

**45.**  $4^2 + 8^2 \div 2^2 = 16 + 64 \div 4$

$= 16 + 16$

$= 32$

**47.**  $10^3 - 10 \cdot 6 - (4 + 5 \cdot 6) = 10^3 - 10 \cdot 6 - (4 + 30)$

$= 10^3 - 10 \cdot 6 - 34$

$= 1000 - 10 \cdot 6 - 34$

$= 1000 - 60 - 34$

$= 940 - 34$

$= 906$

**49.**  $6 \times 11 - (7 + 3) \div 5 - (6 - 4) = 6 \times 11 - 10 \div 5 - 2$

$= 66 - 2 - 2$

$= 64 - 2$

$= 62$

**51.**  $120 - 3^3 \cdot 4 \div (5 \cdot 6 - 6 \cdot 4)$

$= 120 - 3^3 \cdot 4 \div (30 - 24)$

$= 120 - 3^3 \cdot 4 \div 6$

$= 120 - 27 \cdot 4 \div 6$

$= 120 - 108 \div 6$

$= 120 - 18$

$= 102$

**53.**  $2^3 \cdot 2^8 \div 2^6 = 8 \cdot 256 \div 64$

$= 2048 \div 64$

$= 32$

**55.**  We add the numbers and then divide by the number of addends.

$$\frac{\$64 + \$97 + \$121}{3} = \frac{\$282}{3} = \$94$$

**57.**  We add the numbers and then divide by the number of addends.

$$\frac{320 + 128 + 276 + 880}{4} = \frac{1604}{4} = 401$$

**59.**  $8 \times 13 + \{42 \div [18 - (6 + 5)]\}$

$= 8 \times 13 + \{42 \div [18 - 11]\}$

$= 8 \times 13 + \{42 \div 7\}$

$= 8 \times 13 + 6$

$= 104 + 6$

$= 110$

**61.**  $[14 - (3 + 5) \div 2] - [18 \div (8 - 2)]$

$= [14 - 8 \div 2] - [18 \div 6]$

$= [14 - 4] - 3$

$= 10 - 3$

$= 7$

**63.**  $(82 - 14) \times [(10 + 45 \div 5) - (6 \cdot 6 - 5 \cdot 5)]$

$= (82 - 14) \times [(10 + 9) - (36 - 25)]$

$= (82 - 14) \times [19 - 11]$

$= 68 \times 8$

$= 544$

**65.**  $4 \times \{(200 - 50 \div 5) - [(35 \div 7) \cdot (35 \div 7) - 4 \times 3]\}$

$= 4 \times \{(200 - 10) - [5 \cdot 5 - 4 \times 3]\}$

$= 4 \times \{190 - [25 - 12]\}$

$= 4 \times \{190 - 13\}$

$= 4 \times 177$

$= 708$

**67.**  $\{[18 - 2 \cdot 6] - [40 \div (17 - 9)]\}+$
$\{48 - 13 \times 3 + [(50 - 7 \cdot 5) + 2]\}$

$= \{[18 - 12] - [40 \div 8]\}+$
$\{48 - 13 \times 3 + [(50 - 35) + 2]\}$

$= \{6 - 5\} + \{48 - 13 \times 3 + [15 + 2]\}$

$= 1 + \{48 - 13 \times 3 + 17\}$

$= 1 + \{48 - 39 + 17\}$

$= 1 + 26$

$= 27$

**69.**  $x + 341 = 793$

$x + 341 - 341 = 793 - 341$

$x = 452$

The solution is 452.

**71.**  $7 \cdot x = 91$

$\dfrac{7 \cdot x}{7} = \dfrac{91}{7}$

$x = 13$

The solution is 13.

**73.**  $3240 = y + 898$

$3240 - 898 = y + 898 - 898$

$2342 = y$

The solution is 2342.

**75.**  $25 \cdot t = 625$

$\dfrac{25 \cdot t}{25} = \dfrac{625}{25}$

$t = 25$

The solution is 25.

**77.**  *Familiarize*. We first make a drawing.

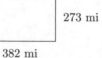
273 mi

382 mi

*Translate*. We use the formula for the area of a rectangle.

$$A = l \cdot w = 382 \cdot 273$$

*Solve*. We carry out the multiplication.

$$A = 382 \cdot 273 = 104,286$$

*Check*. We repeat the calculation. The answer checks.

*State*. The area is 104,286 square miles.

**79.** $1 + 5 \cdot 4 + 3 = 1 + 20 + 3$
$\qquad = 24 \qquad$ Correct answer

To make the incorrect answer correct we add parentheses:
$\qquad 1 + 5 \cdot (4 + 3) = 36$

**81.** $12 \div 4 + 2 \cdot 3 - 2 = 3 + 6 - 2$
$\qquad\qquad\qquad = 7 \qquad$ Correct answer

To make the incorrect answer correct we add parentheses:
$\qquad 12 \div (4 + 2) \cdot 3 - 2 = 4$

## Chapter 1 Concept Reinforcement

**1.** The statement is true. See page 42 in the text.

**2.** $a \div a = \dfrac{a}{a} = 1$, $a \neq 0$; the statement is true.

**3.** $a \div 0$ is not defined, so the statement is false.

## Chapter 1 Important Concepts

**1.** $4\ 3\ \boxed{2}, 0\ 7\ 9$

The digit 2 names the number of thousands.

**2.**
```
    1   1 1
   3 6, 0 4 7
 + 2 9, 2 5 5
 ─────────────
   6 5, 3 0 2
```

**3.**
```
        7 9 15
     4 8̶ 0̶ 5̶
   − 1 5 6 8
   ─────────
     3 2 3 7
```

**4.**
```
       2 1
       1
       7 3
       6 8 4
     × 3 2 9
   ─────────
     6 1 5 6    Multiplying by 9
   1 3 6 8 0    Multiplying by 20
 2 0 5 2 0 0    Multiplying by 300
 ───────────
 2 2 5, 0 3 6
```

**5.**
```
        3 1 5
   2 7 ) 8 5 1 9
        8 1
        ───
          4 1
          2 7
          ───
          1 4 9
          1 3 5
          ─────
            1 4
```

The answer is 315 R 14.

**6.** Round 36,468 to the nearest hundred.

$3\ 6, 4\ \boxed{6}\ 8$
$\qquad\quad \uparrow$

The digit 4 is in the hundreds place. Consider the next digit to the right. Since the digit, 6, is 5 or higher, round 4 hundreds up to 5 hundreds. Then change the digits to the right of the hundreds digit to zeros.

The answer is 36,500.

**7.** Round 36,468 to the nearest thousand.

$3\ 6, \boxed{4}\ 6\ 8$
$\qquad \uparrow$

The digit 6 is in the thousands place. Consider the next digit to the right. Since the digit, 4, is 4 or lower, round down, meaning that 6 thousands stays as 6 thousands. Then change the digits to the right of the thousands digit to zeros.

The answer is 36,000.

**8.** Since 78 is to the left of 81 on the number line, $78 < 81$.

**9.** $24 \cdot x = 864$
$\qquad \dfrac{24 \cdot x}{24} = \dfrac{864}{24} \qquad$ Dividing by 24
$\qquad\qquad x = 36$

Check: $\quad\underline{\phantom{24 \cdot x = 864}}$
$\qquad\quad 24 \cdot x = 864$
$\qquad\quad 24 \cdot 36\ ?\ 864$
$\qquad\qquad\quad 864\ \big|\qquad$ TRUE

The solution is 36.

**10.** $6^3 = 6 \cdot 6 \cdot 6 = 216$

## Chapter 1 Review Exercises

**1.** $4, 6\ 7\ \boxed{8}, 9\ 5\ 2$

The digit 8 means 8 thousands.

**2.** $1\ \boxed{3}, 7\ 6\ 8, 9\ 4\ 0$

The digit 3 names the number of millions.

**3.** $2793 = 2$ thousands + 7 hundreds + 9 tens + 3 ones

**4.** $56,078 = 5$ ten thousands + 6 thousands + 0 hundreds + 7 tens + 8 ones, or 5 ten thousands + 6 thousands + 7 tens + 8 ones

**5.** $4,007,101 = 4$ millions + 0 hundred thousands + 0 ten thousands + 7 thousands + 1 hundred + 0 tens + 1 one, or 4 millions + 7 thousands + 1 hundred + 1 one

**6.**

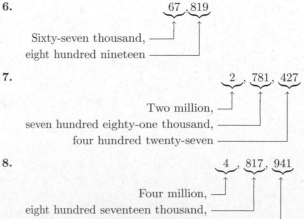

**7.**

**8.**

9.  Four hundred seventy-six thousand, ⌐
        five hundred eighty-eight ⌐

            Standard notation is   $\overbrace{476}$ , $\overbrace{588}$ .

10.                          One billion, ⌐
        five hundred sixty-three million, ⌐

            Standard notation is   1 ,563,000,000.

11.
```
      1   1
      7 3 0 4
    + 6 9 6 8
    ---------
    1 4, 2 7 2
```

12.
```
      1 1   1
      2 7, 6 0 9
    + 3 8, 4 1 5
    -----------
      6 6, 0 2 4
```

13.
```
      1   1
      2 7 0 3
      4 1 2 5
      6 0 0 4
    + 8 9 5 6
    ---------
    2 1, 7 8 8
```

14.
```
          1 1
      9 1, 4 2 6
    +   7, 4 9 5
    -----------
      9 8, 9 2 1
```

15.
```
          13
      7 9 3̸ 15
      8̸ 0̸ 4̸ 5̸
    - 2 8 9 7
    ---------
      5 1 4 8
```

16.
```
      8 9 9 11
      9̸ 0̸ 0̸ 1̸
    - 7 3 1 2
    ---------
      1 6 8 9
```

17.
```
      5 9 9 13
      6̸ 0̸ 0̸ 3̸
    - 3 7 2 9
    ---------
      2 2 7 4
```

18.
```
        16 13
      2 6̸ 3̸ 9 15
      3̸ 7, 4̸ 0̸ 5̸
    - 1 9, 6 4 8
    -----------
      1 7, 7 5 7
```

19.
```
            2
      1 7, 0 0 0
    ×       3 0 0
    -----------
    5, 1 0 0, 0 0 0
```
Multiplying by 300
(Write 00 and then
multiply 17,000 by 3.)

20.
```
      6 3 4
      7 8 4 6
    ×     8 0 0
    -----------
    6, 2 7 6, 8 0 0
```
Multiplying by 800
(Write 00 and then
multiply 7846 by 8.)

21.
```
      1 3
      2 5
      2 4
        7 2 6
    ×     6 9 8
    -----------
        5 8 0 8    Multiplying by 8
      6 5 3 4 0    Multiplying by 9
    4 3 5 6 0 0    Multiplying by 6
    -----------
    5 0 6, 7 4 8
```

22.
```
      3 2
      6 4
        5 8 7
    ×     4 7
    -----------
      4 1 0 9    Multiplying by 7
    2 3 4 8 0    Multiplying by 4
    -----------
    2 7, 5 8 9
```

23.
```
          8 3 0 5
    ×       6 4 2
    -------------
      1 6 6 1 0
      3 3 2 2 0 0
    4 9 8 3 0 0 0
    -------------
    5, 3 3 1, 8 1 0
```

24.
```
         1 2
    5 )6 3
       5
      ---
       1 3
       1 0
      ---
         3
```
The answer is 12 R 3.

25.
```
          5
    1 6 )8 0
         8 0
        ----
          0
```
The answer is 5.

26.
```
        9 1 3
    7 )6 3 9 4
       6 3
       ---
          9
          7
         ---
          2 4
          2 1
         ----
            3
```
The answer is 913 R 3.

27.
```
        3 8 4
    8 )3 0 7 3
       2 4
       ---
        6 7
        6 4
        ---
          3 3
          3 2
          ---
            1
```
The answer is 384 R 1.

28.
```
          4
    6 0 )2 8 6
         2 4 0
         -----
           4 6
```
The answer is 4 R 46.

**29.**
$$\begin{array}{r} 54 \\ 79\,\overline{\smash{)}4\,2\,6\,6} \\ 3\,9\,5 \\ \hline 3\,1\,6 \\ 3\,1\,6 \\ \hline 0 \end{array}$$

The answer is 54.

**30.**
$$\begin{array}{r} 452 \\ 38\,\overline{\smash{)}1\,7{,}1\,7\,6} \\ 1\,5\,2 \\ \hline 1\,9\,7 \\ 1\,9\,0 \\ \hline 7\,6 \\ 7\,6 \\ \hline 0 \end{array}$$

The answer is 452.

**31.**
$$\begin{array}{r} 5008 \\ 14\,\overline{\smash{)}7\,0{,}1\,1\,2} \\ 7\,0 \\ \hline 1\,1\,2 \\ 1\,1\,2 \\ \hline 0 \end{array}$$

The answer is 5008.

**32.**
$$\begin{array}{r} 4389 \\ 12\,\overline{\smash{)}5\,2{,}6\,6\,8} \\ 4\,8 \\ \hline 4\,6 \\ 3\,6 \\ \hline 1\,0\,6 \\ 9\,6 \\ \hline 1\,0\,8 \\ 1\,0\,8 \\ \hline 0 \end{array}$$

The answer is 4389.

**33.** Round 345,759 to the nearest hundred.

$$3\,4\,5{,}7\,\boxed{5}\,9$$
$$\uparrow$$

The digit 7 is in the hundreds place. Consider the next digit to the right. Since the digit, 5, is 5 or higher, round 7 hundreds up to 8 hundreds. Then change the digits to the right of the hundreds digit to zero.

The answer is 345,800.

**34.** Round 345,759 to the nearest ten.

$$3\,4\,5{,}7\,5\,\boxed{9}$$
$$\uparrow$$

The digit 5 is in the tens place. Consider the next digit to the right. Since the digit, 9, is 5 or higher, round 5 tens up to 6 tens. Then change the digit to the right of the tens digit to zero.

The answer is 345,760.

**35.** Round 345,759 to the nearest thousand.

$$3\,4\,5{,}\,\boxed{7}\,5\,9$$
$$\uparrow$$

The digit 5 is in the thousands place. Consider the next digit to the right. Since the digit, 7, is 5 or higher, round 5 thousands up to 6 thousands. Then change the digits to the right of the thousands digit to zero.

The answer is 346,000.

**36.** Round 345,759 to the nearest hundred thousand.

$$3\,\boxed{4}\,5{,}\,7\,5\,9$$
$$\uparrow$$

The digit 3 is in the hundred thousands place. Consider the next digit to the right. Since the digit, 4, is 4 or lower, round down, meaning that 3 hundred thousands stays as 3 hundred thousands. Then change the digits to the right of the hundred thousands digit to zero.

The answer is 300,000.

**37.** Since 67 is to the right of 56 on the number line, $67 > 56$.

**38.** Since 1 is to the left of 23 on the number line, $1 < 23$.

**39.**

$$\begin{array}{c} \text{Rounded to} \\ \text{the nearest hundred} \end{array}$$

$$\begin{array}{r} 4\,1{,}3\,4\,8 \\ +\,1\,9{,}7\,4\,9 \end{array} \qquad \begin{array}{r} 4\,1{,}3\,0\,0 \\ +\,1\,9{,}7\,0\,0 \\ \hline 6\,1{,}0\,0\,0 \leftarrow \text{Estimated answer} \end{array}$$

**40.**

$$\begin{array}{c} \text{Rounded to} \\ \text{the nearest hundred} \end{array}$$

$$\begin{array}{r} 3\,8{,}6\,5\,2 \\ -\,2\,4{,}5\,4\,9 \end{array} \qquad \begin{array}{r} 3\,8{,}7\,0\,0 \\ -\,2\,4{,}5\,0\,0 \\ \hline 1\,4{,}2\,0\,0 \leftarrow \text{Estimated answer} \end{array}$$

**41.**

$$\begin{array}{c} \text{Rounded to} \\ \text{the nearest hundred} \end{array}$$

$$\begin{array}{r} 3\,9\,6 \\ \times\,7\,4\,8 \end{array} \qquad \begin{array}{r} 4\,0\,0 \\ \times\,7\,0\,0 \\ \hline 2\,8\,0{,}0\,0\,0 \leftarrow \text{Estimated answer} \end{array}$$

**42.**
$$46 \cdot n = 368$$
$$\frac{46 \cdot n}{46} = \frac{368}{46}$$
$$n = 8$$

Check: $46 \cdot n = 368$
$$46 \cdot 8 \;?\; 368$$
$$368 \quad | \quad \text{TRUE}$$

The solution is 8.

**43.**
$$47 + x = 92$$
$$47 + x - 47 = 92 - 47$$
$$x = 45$$

Check: $47 + x = 92$
$$47 + 45 \;?\; 92$$
$$92 \quad | \quad \text{TRUE}$$

The solution is 45.

**44.**  $1 \cdot y = 58$
$$y = 58 \qquad (1 \cdot y = y)$$
The number 58 checks. It is the solution.

**45.**  $24 = x + 24$
$$24 - 24 = x + 24 - 24$$
$$0 = x$$
The number 0 checks. It is the solution.

**46.** Exponential notation for $4 \cdot 4 \cdot 4$ is $4^3$.

**47.** $10^4 = 10 \cdot 10 \cdot 10 \cdot 10 = 10,000$

**48.** $6^2 = 6 \cdot 6 = 36$

**49.**  $8 \cdot 6 + 17 = 48 + 17 \quad$ Multiplying
$$= 65 \qquad\qquad \text{Adding}$$

**50.**  $10 \cdot 24 - (18 + 2) \div 4 - (9 - 7)$
$$= 10 \cdot 24 - 20 \div 4 - 2 \qquad \text{Doing the calculations}$$
$$\text{inside the parentheses}$$
$$= 240 - 5 - 2 \qquad\qquad \text{Multiplying and dividing}$$
$$= 235 - 2 \qquad\qquad\quad \text{Subtracting from}$$
$$= 233 \qquad\qquad\qquad \text{left to right}$$

**51.**  $(80 \div 16) \times [(20 - 56 \div 8) + (8 \cdot 8 - 5 \cdot 5)]$
$$= 5 \times [(20 - 7) + (64 - 25)]$$
$$= 5 \times [13 + 39]$$
$$= 5 \times 52$$
$$= 260$$

**52.** We add the numbers and divide by the number of addends.
$$\frac{157 + 170 + 168}{3} = \frac{495}{3} = 165$$

**53. *Familiarize*.** Let $x =$ the additional amount of money, in dollars, Natasha needs to buy the desk.

***Translate*.**

| Money available | plus | Additional amount | is | Price of desk |
|:---:|:---:|:---:|:---:|:---:|
| ↓ | ↓ | ↓ | ↓ | ↓ |
| 196 | + | $x$ | = | 698 |

***Solve*.** We subtract 196 on both sides of the equation.
$$196 + x = 698$$
$$196 + x - 196 = 698 - 196$$
$$x = 502$$

***Check*.** We can estimate.
$$196 + 502 \approx 200 + 500 \approx 700 \approx 698$$
The answer checks.

***State*.** Natasha needs $502 dollars.

**54. *Familiarize*.** Let $b =$ the balance in Tony's account after the deposit.

***Translate*.**

| Original balance | plus | Deposit | is | New balance |
|:---:|:---:|:---:|:---:|:---:|
| ↓ | ↓ | ↓ | ↓ | ↓ |
| 406 | + | 78 | = | $b$ |

***Solve*.** We add on the left side.
$$406 + 78 = b$$
$$484 = b$$

***Check*.** We can repeat the calculation. The answer checks.

***State*.** The new balance is $484.

**55. *Familiarize*.** Let $y =$ the year in which the copper content of pennies was reduced.

| Original year | plus 73 yr is | Year of copper reduction |
|:---:|:---:|:---:|
| ↓ | ↓    ↓    ↓ | ↓ |
| 1909 | +    73    = | $y$ |

***Solve*.** We add on the left side.
$$1909 + 73 = y$$
$$1982 = y$$

***Check*.** We can estimate.
$$1909 + 73 \approx 1910 + 70 \approx 1980 \approx 1982$$
The answer checks.

***State*.** The copper content of pennies was reduced in 1982.

**56. *Familiarize*.** We first make a drawing. Let $c =$ the number of cartons filled.

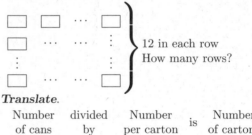

***Translate*.**

| Number of cans | divided by | Number per carton | is | Number of cartons |
|:---:|:---:|:---:|:---:|:---:|
| ↓ | ↓ | ↓ | ↓ | ↓ |
| 228 | ÷ | 12 | = | $c$ |

***Solve*.** We carry out the division.
$$
\begin{array}{r}
19 \\
12\overline{\smash{)}228} \\
\underline{12\phantom{8}} \\
108 \\
\underline{108} \\
0
\end{array}
$$

Thus, $19 = c$, or $c = 19$.

***Check*.** We can check by multiplying: $12 \cdot 19 = 228$. Our answer checks.

***State*.** 19 cartons were filled.

**57. *Familiarize*.** This is a multistep problem. Let $s =$ the cost of 13 stoves, $r =$ the cost of 13 refrigerators, and $t =$ the total cost of the stoves and refrigerators.

***Translate*.**

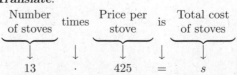

| Number of stoves | times | Price per stove | is | Total cost of stoves |
|:---:|:---:|:---:|:---:|:---:|
| ↓ | ↓ | ↓ | ↓ | ↓ |
| 13 | · | 425 | = | $s$ |

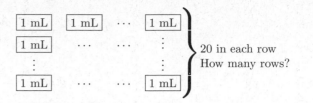

| Number of refrigerators | times | Price per refrigerator | is | Total cost of refrigerators |
|---|---|---|---|---|
| ↓ | ↓ | ↓ | ↓ | ↓ |
| 13 | · | 620 | = | $r$ |

| Cost of stoves | plus | Cost of refrigerators | is | Total cost |
|---|---|---|---|---|
| ↓ | ↓ | ↓ | ↓ | ↓ |
| $s$ | + | $r$ | = | $t$ |

*Solve.* We first carry out the multiplications in the first two equations.

$$13 \cdot 425 = s \qquad 13 \cdot 620 = r$$
$$5525 = s \qquad\qquad 8060 = r$$

Now we substitute 5525 for $s$ and 8060 for $r$ in the third equation and then add on the left side.

$$s + r = t$$
$$5525 + 8060 = t$$
$$13,585 = t$$

*Check.* We repeat the calculations. The answer checks.

*State.* The total cost was $13,585.

**58. Familiarize.** Let $b$ = the number of beehives the farmer needs.

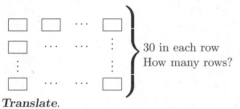

$$\left.\begin{array}{c}\phantom{x}\end{array}\right\} \text{30 in each row How many rows?}$$

*Translate.*

| Number of trees | divided by | Number of trees pollinated by each hive | is | Number of hives needed |
|---|---|---|---|---|
| ↓ | ↓ | ↓ | ↓ | ↓ |
| 420 | ÷ | 30 | = | $b$ |

*Solve.* We carry out the division.

```
        1 4
   30 ⟌ 4 2 0
        3 0
        1 2 0
        1 2 0
            0
```

Thus, $14 = b$, or $b = 14$.

**Check.** We can check by multiplying: $30 \cdot 14 = 420$. The answer checks.

**State.** The farmer needs 14 beehives.

**59.** $A = l \cdot w = 14 \text{ ft} \cdot 7 \text{ ft} = 98 \text{ square ft}$

Perimeter $= 14 \text{ ft} + 7 \text{ ft} + 14 \text{ ft} + 7 \text{ ft} = 42 \text{ ft}$

**60. Familiarize.** We make a drawing. Let $b$ = the number of beakers that will be filled.

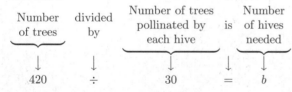

$$\left.\begin{array}{c}\phantom{x}\end{array}\right\} \text{20 in each row How many rows?}$$

*Translate.*

| Amount of alcohol | divided by | Amount per beaker | is | Number of beakers filled |
|---|---|---|---|---|
| ↓ | ↓ | ↓ | ↓ | ↓ |
| 2753 | ÷ | 20 | = | $b$ |

*Solve.* We carry out the division.

```
        1 3 7
   20 ⟌ 2 7 5 3
        2 0
          7 5
          6 0
          1 5 3
          1 4 0
            1 3
```

Thus, 137 R 13 = $b$.

*Check.* We can check by multiplying the number of beakers by 137 and then adding the remainder, 13.

$$137 \cdot 20 = 2740 \text{ and } 2740 + 13 = 2753$$

The answer checks.

*State.* 137 beakers can be filled; 13 mL will be left over.

**61. Familiarize.** This is a multistep problem. Let $b$ = the total amount budgeted for food, clothing, and entertainment and let $r$ = the income remaining after these allotments.

*Translate.*

| Food and clothing budget | plus | Entertainment budget | is | Total of these allotments |
|---|---|---|---|---|
| ↓ | ↓ | ↓ | ↓ | ↓ |
| 7825 | + | 2860 | = | $b$ |

| Food, clothing, and entertainment allotments | plus | Remaining income | is | Total income |
|---|---|---|---|---|
| ↓ | ↓ | ↓ | ↓ | ↓ |
| $b$ | + | $r$ | = | 38,283 |

*Solve.* We add on the left side to solve the first equation.

$$7825 + 2860 = b$$
$$10,685 = b$$

Now we substitute 10,685 for $b$ in the second equation and solve for $r$.

$$b + r = 38,283$$
$$10,685 + r = 38,283$$
$$10,685 + r - 10,685 = 38,283 - 10,685$$
$$r = 27,598$$

*Check.* We repeat the calculations. The answer checks.

*State.* After the allotments for food, clothing, and entertainment, $27,598 remains.

**62.** $7 + (4 + 3)^2 = 7 + 7^2$
$$= 7 + 49$$
$$= 56$$

Answer B is correct.

**63.** $7 + 4^2 + 3^2 = 7 + 16 + 9$
$$= 23 + 9$$
$$= 32$$

Answer A is correct.

**64.**
$$\begin{array}{r} 9\,d \\ \times \ \ d\,2 \\ \hline 8\,0\,3\,6 \end{array}$$

By using rough estimates, we see that the factor $d2 \approx 8100 \div 90 = 90$ or $d2 \approx 8000 \div 100 = 80$. Since $99 \times 92 = 9108$ and $98 \times 82 = 8036$, we have $d = 8$.

**65.**
$$2\,b\,1\,\overline{\smash{)}\,2\,3\,6{,}4\,2\,1}\ \ ^{9\,a\,1}$$

Since $250 \times 1000 = 250{,}000 \approx 236{,}421$ we deduce that $2b1 \approx 250$ and $9a1 \approx 1000$. By trial we find that $a = 8$ and $b = 4$.

**66.** At the beginning of each day the tunnel reaches $500\text{ ft} - 200\text{ ft}$, or $300\text{ ft}$, farther into the mountain than it did the day before. We calculate how far the tunnel reaches into the mountain at the beginning of each day, starting with Day 2.

Day 2: 300 ft

Day 3: 300 ft + 300 ft = 600 ft

Day 4: 600 ft + 300 ft = 900 ft

Day 5: 900 ft + 300 ft = 1200 ft

Day 6: 1200 ft + 300 ft = 1500 ft

We see that the tunnel reaches 1500 ft into the mountain at the beginning of Day 6. On Day 6 the crew tunnels an additional 500 ft, so the tunnel reaches 1500 ft + 500 ft, or 2000 ft, into the mountain. Thus, it takes 6 days to reach the copper deposit.

## Chapter 1 Discussion and Writing Exercises

**1.** No; if subtraction were associative, then $a - (b - c) = (a - b) - c$ for any $a$, $b$, and $c$. But, for example,
$$12 - (8 - 4) = 12 - 4 = 8,$$
whereas
$$(12 - 8) - 4 = 4 - 4 = 0.$$

Since $8 \neq 0$, this example shows that subtraction is not associative.

**2.** By rounding prices and estimating their sum a shopper can estimate the total grocery bill while shopping. This is particularly useful if the shopper wants to spend no more than a certain amount.

**3.** Answers will vary. Anthony is driving from Kansas City to Minneapolis, a distance of 512 miles. He stops for gas after driving 183 miles. How much farther must he drive?

**4.** The parentheses are not necessary in the expression $9 - (4 \cdot 2)$. Using the rules for order of operations, the multiplication would be performed before the subtraction even if the parentheses were not present.

The parentheses are necessary in the expression $(3 \cdot 4)^2$; $(3 \cdot 4)^2 = 12^2 = 144$, but $3 \cdot 4^2 = 3 \cdot 16 = 48$.

## Chapter 1 Test

**1.** $\boxed{5}\,4\,6{,}7\,8\,9$

The digit 5 tells the number of hundred thousands.

**2.** $8843 = 8$ thousands $+ 8$ hundreds $+ 4$ tens $+ 3$ ones

**3.**

$$\underbrace{38}{,}\underbrace{403}{,}\underbrace{277}$$

Thirty-eight million, ⎤
four hundred three thousand, ⎤
two hundred seventy-seven

**4.**
$$\begin{array}{r} 6\,8\,1\,1 \\ +\ 3\,1\,7\,8 \\ \hline 9\,9\,8\,9 \end{array}$$
Add ones, add tens, add hundreds, and then add thousands.

**5.**
$$\begin{array}{r} {}^{1\ \ 1\ \ \ 1} \\ 4\,5{,}8\,8\,9 \\ +\ 1\,7{,}9\,0\,2 \\ \hline 6\,3{,}7\,9\,1 \end{array}$$

**6.**
$$\begin{array}{r} {}^{2\ 1\ 1} \\ 1\,2\,3\,9 \\ 8\,4\,3 \\ 3\,0\,1 \\ +\ \ 7\,8\,2 \\ \hline 3\,1\,6\,5 \end{array}$$

**7.**
$$\begin{array}{r} 6\,2\,0\,3 \\ +\ 4\,3\,1\,2 \\ \hline 1\,0{,}5\,1\,5 \end{array}$$

**8.**
$$\begin{array}{r} 7\,9\,8\,3 \\ -\ 4\,3\,5\,3 \\ \hline 3\,6\,3\,0 \end{array}$$
Subtract ones, subtract tens, subtract hundreds, and then subtract thousands.

**9.**
$$\begin{array}{r} {}^{6\ 14} \\ 2\,9\,7\,\not{4} \\ -\ 1\,9\,3\,5 \\ \hline 1\,0\,3\,9 \end{array}$$

**10.**
$$\begin{array}{r} {}^{8\ 9\ 17} \\ 8\,\not{9}\,\not{0}\,7 \\ -\ 2\,0\,5\,9 \\ \hline 6\,8\,4\,8 \end{array}$$

**11.**
$$\begin{array}{r} \overset{12}{\underset{}{}} \\ {}^{1}\,\overset{2}{}\,\overset{9}{}\,16 \\ 2\,3,0\,6\,7 \\ -\,1\,7,8\,9\,2 \\ \hline 5\,1\,7\,5 \end{array}$$

**12.**
$$\begin{array}{r} {}^{5}\,{}^{6}\,{}^{7} \\ 4\,5\,6\,8 \\ \times\qquad 9 \\ \hline 4\,1,1\,1\,2 \end{array}$$

**13.**
$$\begin{array}{r} {}^{5}\,{}^{4}\,{}^{3} \\ 8\,8\,7\,6 \\ \times\,6\,0\,0 \\ \hline 5,3\,2\,5,6\,0\,0 \end{array}$$  Multiply by 6 hundreds (We write 00 and then multiply 8876 by 6.)

**14.**
$$\begin{array}{r} 6\,5 \\ \times\,3\,7 \\ \hline 4\,5\,5 \\ 1\,9\,5\,0 \\ \hline 2\,4\,0\,5 \end{array}$$  Multiplying by 7
Multiplying by 30
Adding

**15.**
$$\begin{array}{r} 6\,7\,8 \\ \times\,7\,8\,8 \\ \hline 5\,4\,2\,4 \\ 5\,4\,2\,4\,0 \\ 4\,7\,4\,6\,0\,0 \\ \hline 5\,3\,4,2\,6\,4 \end{array}$$

**16.**
$$\begin{array}{r} 3 \\ 4\,\overline{)1\,5} \\ 1\,2 \\ \hline 3 \end{array}$$
The answer is 3 R 3.

**17.**
$$\begin{array}{r} 7\,0 \\ 6\,\overline{)4\,2\,0} \\ 4\,2 \\ \hline 0 \\ 0 \\ \hline 0 \end{array}$$
The answer is 70.

**18.**
$$\begin{array}{r} 9\,7 \\ 89\,\overline{)8\,6\,3\,3} \\ 8\,0\,1 \\ \hline 6\,2\,3 \\ 6\,2\,3 \\ \hline 0 \end{array}$$
The answer is 97.

**19.**
$$\begin{array}{r} 8\,0\,5 \\ 44\,\overline{)3\,5,4\,2\,8} \\ 3\,5\,2 \\ \hline 2\,2\,8 \\ 2\,2\,0 \\ \hline 8 \end{array}$$
The answer is 805 R 8.

**20.** Round 34,528 to the nearest thousand.

$$3\,4,\boxed{5}\,2\,8$$
$\uparrow$

The digit 4 is in the thousands place. Consider the next digit to the right, 5. Since 5 is 5 or higher, round 4 thousands up to 5 thousands. Then change all digits to the right of thousands to zeros.

The answer is 35,000.

**21.** Round 34,528 to the nearest ten.

$$3\,4,5\,2\,\boxed{8}$$
$\uparrow$

The digit 2 is in the tens place. Consider the next digit to the right, 8. Since 8 is 5 or higher, round 2 tens up to 3 tens. Then change the digit to the right of tens to zero.

The answer is 34,530.

**22.** Round 34,528 to the nearest hundred.

$$3\,4,5\,\boxed{2}\,8$$
$\uparrow$

The digit 5 is in the hundreds place. Consider the next digit to the right, 2. Since 2 is 4 or lower, round down, meaning that 5 hundreds stays as 5 hundreds. Then change all digits to the right of hundreds to zero.

The answer is 34,500.

**23.**
$$\begin{array}{cc} & \text{Rounded to} \\ & \text{the nearest hundred} \\ 2\,3,6\,4\,9 & 2\,3,6\,0\,0 \\ +\,5\,4,7\,4\,6 & +\,5\,4,7\,0\,0 \\ \hline & 7\,8,3\,0\,0 \leftarrow \text{Estimated answer} \end{array}$$

**24.**
$$\begin{array}{cc} & \text{Rounded to} \\ & \text{the nearest hundred} \\ 5\,4,7\,5\,1 & 5\,4,8\,0\,0 \\ -\,2\,3,6\,4\,9 & -\,2\,3,6\,0\,0 \\ \hline & 3\,1,2\,0\,0 \leftarrow \text{Estimated answer} \end{array}$$

**25.**
$$\begin{array}{cc} & \text{Rounded to} \\ & \text{the nearest hundred} \\ 8\,2\,4 & 8\,0\,0 \\ \times\,4\,8\,9 & \times\,5\,0\,0 \\ \hline & 4\,0\,0,0\,0\,0 \leftarrow \text{Estimated answer} \end{array}$$

**26.** Since 34 is to the right of 17 on the number line, $34 > 17$.

**27.** Since 117 is to the left of 157 on the number line, $117 < 157$.

**28.**
$$28 + x = 74$$
$$28 + x - 28 = 74 - 28 \quad \text{Subtracting 28 on both sides}$$
$$x = 46$$

Check:
$$\begin{array}{c} 28 + x = 74 \\ \hline 28 + 46 \; ? \; 74 \\ \hline 74 \;\mid\; \text{TRUE} \end{array}$$

The solution is 46.

**29.** $169 \div 13 = n$

We carry out the division.

$$
\begin{array}{r}
1\,3 \\
13\overline{)1\,6\,9} \\
\underline{1\,3} \\
3\,9 \\
\underline{3\,9} \\
0
\end{array}
$$

The solution is 13.

**30.**     $38 \cdot y = 532$

$\dfrac{38 \cdot y}{38} = \dfrac{532}{38}$     Dividing by 38 on both sides

$y = 14$

Check:  $\begin{array}{c} 38 \cdot y = 532 \\ \hline 38 \cdot 14 \ ? \ 532 \\ 532 \ \Big| \qquad \text{TRUE} \end{array}$

The solution is 14.

**31.**     $381 = 0 + a$

$381 = a$     Adding on the right side

The solution is 381.

**32. Familiarize.** Let $s =$ the number of calories in an 8-oz serving of skim milk.

**Translate.**

| Number of calories in skim milk | plus | How many more calories | is | Number of calories in whole milk |
|:---:|:---:|:---:|:---:|:---:|
| ↓ | ↓ | ↓ | ↓ | ↓ |
| $s$ | $+$ | $63$ | $=$ | $146$ |

**Solve.** We subtract 63 on both sides of the equation.

$s + 63 = 146$

$s + 63 - 63 = 146 - 63$

$s = 83$

**Check.** Since 63 calories more than 83 calories is $83 + 63$, or 146 calories, the answer checks.

**State.** An 8-oz serving of skim milk contains 83 calories.

**33. Familiarize.** Let $s =$ the number of staplers that can be filled. We can think of this as repeated subtraction, taking successive sets of 250 staples and putting them into $s$ staplers.

**Translate.**

| Number of staples | divided by | Number in each stapler | is | Number of staplers filled |
|:---:|:---:|:---:|:---:|:---:|
| ↓ | ↓ | ↓ | ↓ | ↓ |
| $5000$ | $\div$ | $250$ | $=$ | $s$ |

**Solve.** We carry out the division.

$$
\begin{array}{r}
2\,0 \\
250\overline{)5\,0\,0\,0} \\
\underline{5\,0\,0} \\
0 \\
\underline{0} \\
0
\end{array}
$$

Then $20 = s$.

**Check.** We can multiply the number of staplers filled by the number of staples in each one.

$20 \cdot 250 = 5000$

The answer checks.

**State.** 20 staplers can be filled from a box of 5000 staples.

**34. Familiarize.** Let $a =$ the total land area of the five largest states, in square meters. Since we are combining the areas of the states, we can add.

**Translate.**

$571,951 + 261,797 + 155,959 + 145,552 + 121,356 = a$

**Solve.** We carry out the addition.

$$
\begin{array}{r}
\scriptstyle 2\ 1\ \ 3\ \ 3\ 2 \\
5\,7\,1\,,9\,5\,1 \\
2\,6\,1\,,7\,9\,7 \\
1\,5\,5\,,9\,5\,9 \\
1\,4\,5\,,5\,5\,2 \\
+\ 1\,2\,1\,,3\,5\,6 \\
\hline
1\,,2\,5\,6\,,6\,1\,5
\end{array}
$$

Then $1,256,615 = a$.

**Check.** We can repeat the calculation. We can also estimate the result by rounding. We will round to the nearest ten thousand.

$571,951 + 261,797 + 155,959 + 145,552 + 121,356$

$\approx 570,000 + 260,000 + 160,000 + 150,000 + 120,000$

$= 1,260,000$

Since $1,260,000 \approx 1,256,615$, we have a partial check.

**State.** The total land area of Alaska, Texas, California, Montana, and New Mexico is 1,256,615 sq mi.

**35. Familiarize.** Let $n =$ the number of 12-packs that can be filled. We can think of this as repeated subtraction, taking successive sets of 12 snack cakes and putting them into $n$ packages.

**Translate.**

| Number of cakes | divided by | Number in each package | is | Number of 12-packs |
|:---:|:---:|:---:|:---:|:---:|
| ↓ | ↓ | ↓ | ↓ | ↓ |
| $22,231$ | $\div$ | $12$ | $=$ | $n$ |

**Solve**. We carry out the division.

```
        1 8 5 2
1 2 [ 2 2, 2 3 1
      1 2
      ‾‾‾‾
      1 0 2
        9 6
        ‾‾‾
        6 3
        6 0
        ‾‾‾
          3 1
          2 4
          ‾‾‾
            7
```

Then 1852 R 7 = $n$.

**Check**. We multiply the number of packages by 12 and then add the remainder, 7.

$$12 \cdot 1852 = 22,224$$

$$22,224 + 7 = 22,231$$

The answer checks.

**State**. 1852 twelve-packs can be filled. There will be 7 cakes left over.

**36.** a) We will use the formula Perimeter $= 2 \cdot$ length $+ 2 \cdot$ width to find the perimeter of each pool table in inches. We will use the formula Area $=$ length $\cdot$ width to find the area of each pool table, in sq in.

For the 50 in. by 100 in. table:

$$\text{Perimeter} = 2 \cdot 100 \text{ in.} + 2 \cdot 50 \text{ in.}$$
$$= 200 \text{ in.} + 100 \text{ in.}$$
$$= 300 \text{ in.}$$
$$\text{Area} = 100 \text{ in.} \cdot 50 \text{ in.} = 5000 \text{ sq in.}$$

For the 44 in. by 88 in. table:

$$\text{Perimeter} = 2 \cdot 88 \text{ in.} + 2 \cdot 44 \text{ in.}$$
$$= 176 \text{ in.} + 88 \text{ in.}$$
$$= 264 \text{ in.}$$
$$\text{Area} = 88 \text{ in.} \cdot 44 \text{ in.} = 3872 \text{ sq in.}$$

For the 38 in. by 76 in. table:

$$\text{Perimeter} = 2 \cdot 76 \text{ in.} + 2 \cdot 38 \text{ in.}$$
$$= 152 \text{ in.} + 76 \text{ in.}$$
$$= 228 \text{ in.}$$
$$\text{Area} = 76 \text{ in.} \cdot 38 \text{ in.} = 2888 \text{ sq in.}$$

b) Let $a =$ the number of square inches by which the area of the largest table exceeds the area of the smallest table. We subtract to find $a$.

$$a = 5000 \text{ sq in.} - 2888 \text{ sq in.} = 2112 \text{ sq in.}$$

**37. Familiarize**. This a multistep problem. Let $b =$ the total cost of the black cartridges, $p =$ the total cost of the photo cartridges, and $t =$ the total cost of the entire purchase.

**Translate**.

For the black ink cartridges:

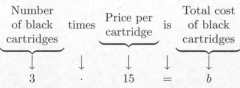

For the photo cartridges:

| Number of photo cartridges | times | Price per cartridge | is | Total cost of photo cartridges |
|---|---|---|---|---|
| ↓ | ↓ | ↓ | ↓ | ↓ |
| 2 | · | 25 | = | $p$ |

For the total cost of the order:

| Cost of black cartridges | plus | Cost of photo cartridges | is | Total cost of purchase |
|---|---|---|---|---|
| ↓ | ↓ | ↓ | ↓ | ↓ |
| $b$ | + | $p$ | = | $t$ |

**Solve**. We solve the first two equations and then add the solutions.

$$3 \cdot 15 = b$$
$$45 = b$$
$$2 \cdot 25 = p$$
$$50 = p$$
$$b + p = t$$
$$45 + 50 = t$$
$$95 = t$$

**Check**. We repeat the calculations. The answer checks.

**State**. The total cost of the purchase was $95.

**38.** Exponential notation for $12 \cdot 12 \cdot 12 \cdot 12$ is $12^4$.

**39.** $7^3 = 7 \cdot 7 \cdot 7 = 343$

**40.** $10^5 = 10 \cdot 10 \cdot 10 \cdot 10 \cdot 10 = 100,000$

**41.**
$$35 - 1 \cdot 28 \div 4 + 3$$
$$= 35 - 28 \div 4 + 3 \quad \text{Doing all multiplications and}$$
$$= 35 - 7 + 3 \quad \text{divisions in order from left to right}$$
$$= 28 + 3 \quad \text{Doing all additions and subtractions}$$
$$= 31 \quad \text{in order from left to right}$$

**42.**
$$10^2 - 2^2 \div 2$$
$$= 100 - 4 \div 2 \quad \text{Evaluating the exponential expressions}$$
$$= 100 - 2 \quad \text{Dividing}$$
$$= 98 \quad \text{Subtracting}$$

**43.**
$$(25 - 15) \div 5$$
$$= 10 \div 5 \quad \text{Doing the calculation inside the parentheses}$$
$$= 2 \quad \text{Dividing}$$

**44.**
$$2^4 + 24 \div 12$$
$$= 16 + 24 \div 12 \quad \text{Evaluating the exponential expression}$$
$$= 16 + 2 \quad \text{Dividing}$$
$$= 18 \quad \text{Adding}$$

**45.**    $8 \times \{(20-11) \cdot [(12+48) \div 6 - (9-2)]\}$
$= 8 \times \{9 \cdot [60 \div 6 - 7]\}$
$= 8 \times \{9 \cdot [10 - 7]\}$
$= 8 \times \{9 \cdot 3\}$
$= 8 \times 27$
$= 216$

**46.** We add the numbers and then divide by the number of addends.
$$\frac{97+99+87+89}{4} = \frac{372}{4} = 93$$
Answer A is correct.

**47.** **Familiarize**. We make a drawing.

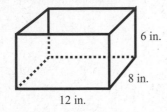

Observe that the dimensions of two sides of the container are 8 in. by 6 in. The area of each is 8 in. · 6 in. and their total area is $2 \cdot 8$ in. · 6 in. The dimensions of the other two sides are 12 in. by 6 in. The area of each is 12 in. · 6 in. and their total area is $2 \cdot 12$ in. · 6 in. The dimensions of the bottom of the box are 12 in. by 8 in. and its area is 12 in. · 8 in. Let $c$ = the number of square inches of cardboard that are used for the container.

**Translate**. We add the areas of the sides and the bottom of the container.
$$2 \cdot 8 \text{ in.} \cdot 6 \text{ in.} + 2 \cdot 12 \text{ in.} \cdot 6 \text{ in.} + 12 \text{ in.} \cdot 8 \text{ in.} = c$$

**Solve**. We carry out the calculation.
$$2 \cdot 8 \text{ in.} \cdot 6 \text{ in.} + 2 \cdot 12 \text{ in.} \cdot 6 \text{ in.} + 12 \text{ in.} \cdot 8 \text{ in.} = c$$
$$96 \text{ sq in.} + 144 \text{ sq in.} + 96 \text{ sq in.} = c$$
$$336 \text{ sq in.} = c$$

**Check**. We can repeat the calculations. The answer checks.

**State**. 336 sq in. of cardboard are used for the container.

**48.** We can reduce the number of trials required by simplifying the expression on the left side of the equation and then using the addition principle.
$$359 - 46 + a \div 3 \times 25 - 7^2 = 339$$
$$359 - 46 + a \div 3 \times 25 - 49 = 339$$
$$359 - 46 + \frac{a}{3} \times 25 - 49 = 339$$
$$359 - 46 + \frac{25 \cdot a}{3} - 49 = 339$$
$$313 + \frac{25 \cdot a}{3} - 49 = 339$$
$$264 + \frac{25 \cdot a}{3} = 339$$
$$264 + \frac{25 \cdot a}{3} - 264 = 339 - 264$$
$$\frac{25 \cdot a}{3} = 75$$

We see that when we multiply $a$ by 25 and divide by 3, the result is 75. By trial, we find that $\frac{25 \cdot 9}{3} = \frac{225}{3} = 75$, so $a = 9$. We could also reason that since $75 = 25 \cdot 3$ and $9/3 = 3$, we have $a = 9$.

**49.** **Familiarize**. First observe that a 10-yr loan with monthly payments has a total of $10 \cdot 12$, or 120, payments. Let $m$ = the number of monthly payments represented by $9160 and let $p$ = the number of payments remaining after $9160 has been repaid.

**Translate**. First we will translate to an equation that can be used to find $m$. Then we will write an equation that can be used to find $p$.

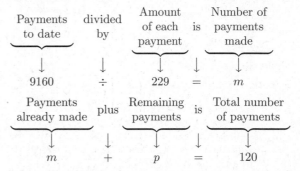

**Solve**. To solve the first equation we carry out the division.

$$\begin{array}{r} 40 \\ 229\overline{)9160} \\ \underline{916} \phantom{0} \\ 0 \\ 0 \\ \underline{0} \end{array}$$

Thus, $m = 40$.

Now we solve the second equation.
$$m + p = 120$$
$$40 + p = 120 \qquad \text{Substituting 40 for } m$$
$$40 + p - 40 = 120 - 40$$
$$p = 80$$

**Check**. We can approach the problem in a different way to check the answer. In 10 years, Cara's loan payments will total $120 \cdot \$229$, or $27,480. If $9160 has already been paid, then $27,480 - \$9160$, or $18,320, remains to be paid. Since $80 \cdot \$229 = \$18,320$, the answer checks.

**State**. 80 payments remain on the loan.

# Chapter 2

# Integers

## Exercise Set 2.1

1. The integer $-282$ corresponds to 282 ft below sea level.

3. The integer 24 corresponds to 24° above zero; the integer $-2$ corresponds to 2° below zero.

5. The integer 950,000,000 corresponds to a temperature of 950,000,000°F above zero; the integer $-460$ corresponds to a temperature of 460°F below zero.

7. The integer $-34$ corresponds to team A being 34 pins behind team B; the integer 15 corresponds to team B being 15 pins ahead of team C.

9. The integer $-39,868$ corresponds to a decrease of 39,868 residents.

11.

13.

15. Since 8 is to the right of 0, we have $8 > 0$.

17. Since $-8$ is to the left of 3, we have $-8 < 3$.

19. Since 8 is to the right of $-8$, we have $8 > -8$.

21. Since $-10$ is to the left of $-5$, we have $-10 < -5$.

23. Since $-5$ is to the right of $-11$, we have $-5 > -11$.

25. Since $-6$ is to the left of $-5$, we have $-6 < -5$.

27. Since $-7$ is to the left of 0, we have $-7 < 0$.

29. Since 1 is to the right of $-15$, we have $1 > -15$.

31. The distance of $-3$ from 0 is 3, so $|-3| = 3$.

33. The distance of 18 from 0 is 18, so $|18| = 18$.

35. The distance of 325 from 0 is 325, so $|325| = 325$.

37. The distance of $-29$ from 0 is 29, so $|-29| = 29$.

39. The distance of $-300$ from 0 is 300, so $|-300| = 300$.

41. The distance of 53 from 0 is 53, so $|53| = 53$.

43.
$$\begin{array}{r} \overset{1}{\phantom{+}9\,1\,8\,2} \\ +\,4\,3\,6\,7 \\ \hline 1\,3,5\,4\,9 \end{array}$$

45.
$$\begin{array}{r} \overset{1}{\phantom{+}}\overset{1}{3}\,2,0\,4\,7 \\ +\,1\,8,5\,6\,2 \\ \hline 5\,0,6\,0\,9 \end{array}$$

47.
$$\begin{array}{r} \overset{11}{\phantom{-}}\overset{\phantom{1}9\;17}{\cancel{1}\,\cancel{2}\cancel{0}\,7} \\ -\;\;9\,4\,8 \\ \hline 2\,5\,9 \end{array}$$

49. $|-5| = 5$ and $|-2| = 2$. Since 5 is to the right of 2, we have $|-5| > |-2|$.

51. $|-8| = 8$ and $|8| = 8$, so $|-8| = |8|$.

## Exercise Set 2.2

1. $-9 + 2$  The absolute values are 9 and 2. The difference is $9 - 2$, or 7. The negative number has the larger absolute value, so the answer is negative.  $-9 + 2 = -7$

3. $-10 + 6$  The absolute values are 10 and 6. The difference is $10 - 6$, or 4. The negative number has the larger absolute value, so the answer is negative.  $-10 + 6 = -4$

5. $-8 + 8$  A positive and a negative number. The numbers have the same absolute value. The sum is 0.  $-8 + 8 = 0$

7. $-3 + (-5)$  Two negatives. Add the absolute values, 3 and 5, getting 8. Make the answer negative.  $-3 + (-5) = -8$

9. $-7 + 0$  One number is 0. The answer is the other number. $-7 + 0 = -7$

11. $0 + (-27)$  One number is 0. The answer is the other number.  $0 + (-27) = -27$

13. $17 + (-17)$  A positive and a negative number. The numbers have the same absolute value. The sum is 0. $17 + (-17) = 0$

15. $-17 + (-25)$  Two negatives. Add the absolute values, 17 and 25, getting 42. Make the answer negative.  $-17 + (-25) = -42$

17. $18 + (-18)$  A positive and a negative number. The numbers have the same absolute value. The sum is 0. $18 + (-18) = 0$

19. $-18 + 18$  A positive and a negative number. The numbers have the same absolute value. The sum is 0.  $-18 + 18 = 0$

21. $8 + (-5)$  The absolute values are 8 and 5. The difference is $8 - 5$, or 3. The positive number has the larger absolute value, so the answer is positive.  $8 + (-5) = 3$

23. $-4 + (-5)$  Two negatives. Add the absolute values, 4 and 5, getting 9. Make the answer negative.  $-4 + (-5) = -9$

25. $13 + (-6)$  The absolute values are 13 and 6. The difference is $13 - 6$, or 7. The positive number has the larger absolute value, so the answer is positive.  $13 + (-6) = 7$

**27.** $-25 + 25$ A positive and a negative number. The numbers have the same absolute value. The sum is 0.   $-25 + 25 = 0$

**29.** $63 + (-18)$    The absolute values are 63 and 18. The difference is $63 - 18$, or 45. The positive number has the larger absolute value, so the answer is positive. $63 + (-18) = 45$

**31.** $-6 + 4$ The absolute values are 6 and 4. The difference is $6 - 4$, or 2. The negative number has the larger absolute value, so the answer is negative. $-6 + 4 = -2$

**33.** $-2 + (-5)$ Two negatives. Add the absolute values, 2 and 5, getting 7. Make the answer negative. $-2 + (-5) = -7$

**35.** $-22 + 3$ The absolute values are 22 and 3. The difference is $22 - 3$, or 19. The negative number has the larger absolute value, so the answer is negative.   $-22 + 3 = -19$

**37.** $-5 + (-7) + 6 = -12 + 6$    Adding the negative numbers

$= -6$          Adding the results

**39.**     $-4 + 7 + (-4)$

     $= 3 + (-4)$     Adding from left

     $= -1$           to right

**41.** $75 + (-14) + (-17) + (-5)$

a) $-14 + (-17) + (-5) = -36$   Adding the negative numbers

b) $75 + (-36) = 39$    Adding the results

**43.** $-44 + (-3) + 95 + (-5)$

a) $-44 + (-3) + (-5) = -52$   Adding the negative numbers

b) $-52 + 95 = 43$    Adding the results

**45.** $98 + (-54) + 113 + (-998) + 44 + (-612) + (-18) + 334$

a) $98 + 113 + 44 + 334 = 589$ Adding the positive numbers

b) $-54 + (-998) + (-612) + (-18) = -1682$ Adding the negative numbers

c) $589 + (-1682) = -1093$ Adding the results

**47.** The additive inverse of 24 is $-24$ because $24 + (-24) = 0$.

**49.** The additive inverse of $-26$ is 26 because $-26 + 26 = 0$.

**51.** The additive inverse of 103 is $-103$ because $103 + (-103) = 0$.

**53.** The additive inverse of 46 is $-46$ because $46 + (-46) = 0$.

**55.** If $x = 9$ then $-x = -(9) = -9$. (The additive inverse of 9 is $-9$.)

**57.** If $x = -14$, then $-x = -(-14) = 14$. (The additive inverse of $-14$ is 14.

**59.** If $x = -65$ then $-(-x) = -[-(-65)] = -65$. (The opposite of the opposite of $-65$ is $-65$.)

**61.** If $x = 5$, then $-(-x) = -(-5) = 5$. (The opposite of the opposite of 5 is 5.)

**63.** $-(-14) = 14$

**65.** $-(10) = -10$

**67.**
$$
\begin{array}{r}
\overset{7\ \ 1}{1\ 9\ 2} \\
\times\ \ \ 1\ 8 \\
\hline
1\ 5\ 3\ 6 \\
1\ 9\ 2\ 0 \\
\hline
3\ 4\ 5\ 6
\end{array}
$$
   Multiplying by 8   (row 1536)
   Multiplying by 10   (row 1920)
   Adding   (row 3456)

**69.**
$$
\begin{array}{r}
\overset{\ \ \ \ 2\ \ 2}{\overset{\ \ \ \ 3\ \ 2}{6\ 4\ 0\ 3}} \\
\times\ \ \ 7\ 0\ 8 \\
\hline
5\ 1\ 2\ 2\ 4 \\
4\ 4\ 8\ 2\ 1\ 0\ 0 \\
\hline
4,5\ 3\ 3,3\ 2\ 4
\end{array}
$$
   Multiplying by 8   (row 51224)
   Multiplying by 7 hundreds   (row 4482100)

**71.**
$$
\begin{array}{r}
1\ 2\ 7 \\
54\overline{)6\ 9\ 0\ 4} \\
5\ 4\ 0\ 0 \\
\hline
1\ 5\ 0\ 4 \\
1\ 0\ 8\ 0 \\
\hline
4\ 2\ 4 \\
3\ 7\ 8 \\
\hline
4\ 6
\end{array}
$$

The answer is 127 R 46.

**73.** Round 641,539 to the nearest ten.

$$6\ 4\ 1,5\ 3\ \boxed{9}$$
$$\uparrow$$

The digit 3 is in the tens place. Consider the next digit to the right. Since the digit, 9, is 5 or higher, round 3 tens up to 4 tens. Then change the digit to the right of the tens digit to zero.

The answer is 641,540.

**75.** Round 641,539 to the nearest thousand.

$$6\ 4\ 1,\ \boxed{5}\ 3\ 9$$
$$\uparrow$$

The digit 1 is in the thousands place. Consider the next digit to the right. Since the digit, 5, is 5 or higher, round 1 thousand up to 2 thousands. Then change all digits to the right of the thousands digit to zeros.

The answer is 642,000.

**77.** When $x$ is positive, the opposite of $x$, $-x$ is negative.

**79.** We use a calculator.

$-345,882 + (-295,097) = -640,979$

**81.** If $n$ is positive, $-n$ is negative. Then $-n + m$, the sum of two negative numbers, is negative.

## Exercise Set 2.3

**1.** $3 - 7 = 3 + (-7) = -4$

**3.** $0 - 7 = 0 + (-7) = -7$

**5.** $-8 - (-2) = -8 + 2 = -6$

**7.** $-10 - (-10) = -10 + 10 = 0$

**9.** $12 - 16 = 12 + (-16) = -4$

**11.** $20 - 27 = 20 + (-27) = -7$

**13.** $-9 - (-3) = -9 + 3 = -6$

**15.** $-11 - (-11) = -11 + 11 = 0$

**17.** $8 - (-3) = 8 + 3 = 11$

**19.** $-6 - 8 = -6 + (-8) = -14$

**21.** $-4 - (-9) = -4 + 9 = 5$

**23.** $2 - 9 = 2 + (-9) = -7$

**25.** $0 - 5 = 0 + (-5) = -5$

**27.** $-5 - (-2) = -5 + 2 = -3$

**29.** $2 - 25 = 2 + (-25) = -23$

**31.** $-42 - 26 = -42 + (-26) = -68$

**33.** $-71 - 2 = -71 + (-2) = -73$

**35.** $24 - (-92) = 24 + 92 = 116$

**37.** $-2 - 0 = -2 + 0 = -2$

**39.** $3 - 8 = 3 + (-8) = -5$

**41.** $2 - 3 = 2 + (-3) = -1$

**43.** $-3 - 2 = -3 + (-2) = -5$

**45.** $13 - (-9) = 13 + 9 = 22$

**47.** $6 - (-13) = 6 + 13 = 19$

**49.** $-14 - 6 = -14 + (-6) = -20$

**51.** $1 - 9 = 1 + (-9) = -8$

**53.** $11 - 21 = 11 + (-21) = -10$

**55.** $7 - 10 = 7 + (-10) = -3$

**57.** $16 - 23 = 16 + (-23) = -7$

**59.** $-47 - (-17) = -47 + 17 = -30$

**61.** $-9 - (-9) = -9 + 9 = 0$

**63.** $122 - 123 = 122 + (-123) = -1$

**65.** $18 - (-15) - 3 - (-5) + 2 = 18 + 15 + (-3) + 5 + 2 = 37$

**67.** $-31 + (-28) - (-14) - 17 = (-31) + (-28) + 14 + (-17) = -62$

**69.** $-93 - (-84) - 41 - (-56) = (-93) + 84 + (-41) + 56 = 6$

**71.** $-5 - (-30) + 30 + 40 - (-12) = (-5) + 30 + 30 + 40 + 12 = 107$

**73.** $132 - (-21) + 45 - (-21) = 132 + 21 + 45 + 21 = 219$

**75.** Let $D$ = the difference in elevation, in feet.

| Difference in elevation | is | Highest elevation | minus | Lowest elevation |
|---|---|---|---|---|
| ↓ | ↓ | ↓ | ↓ | ↓ |
| $D$ | $=$ | $20,320$ | $-$ | $(-282)$ |

We carry out the subtraction.

$$D = 20,320 - (-282) = 20,320 + 282 = 20,602$$

The difference in elevation is 20,602 ft.

**77.** Let $F$ = the final temperature.

| Final temperature | = | Beginning temperature | + | Rise in temperature | + | Fall in temperature |
|---|---|---|---|---|---|---|
| ↓ | ↓ | ↓ | ↓ | ↓ | ↓ | ↓ |
| $F$ | $=$ | $32$ | $+$ | $15$ | $+$ | $(-50)$ |

We carry out the addition.

$$F = 32 + 15 + (-50) = -3$$

The final temperature was $-3°$.

**79.** Let $C$ = the total change in the level of the lake, in feet.

| Change in level | = | First change | + | Second change | + | Third change | + | Fourth change |
|---|---|---|---|---|---|---|---|---|
| ↓ | ↓ | ↓ | ↓ | ↓ | ↓ | ↓ | ↓ | ↓ |
| $C$ | $=$ | $-2$ | $+$ | $1$ | $+$ | $(-5)$ | $+$ | $3$ |

We carry out the addition.

$$C = -2 + 1 + (-5) + 3 = -3$$

The lake level had gone down 3 feet at the end of four months.

**81.** Let $B$ = the final balance.

| Final balance | = | Original balance | − | Amount of first check | + | Deposit | − | Amount of second check |
|---|---|---|---|---|---|---|---|---|
| ↓ | ↓ | ↓ | ↓ | ↓ | ↓ | ↓ | ↓ | ↓ |
| $B$ | $=$ | $460$ | $-$ | $530$ | $+$ | $75$ | $-$ | $90$ |

We carry out the computation.

$$B = 460 - 530 + 75 - 90$$
$$= -70 + 75 - 90$$
$$= 5 - 90$$
$$= -85$$

The balance in the account is $-\$85$. (That is, the account is \$85 overdrawn.)

**83.** Let $T$ = the amount by which the temperature dropped, in degrees Fahrenheit.

| Temperature drop | is | Higher temperature | minus | Lower temperature |
|---|---|---|---|---|
| ↓ | ↓ | ↓ | ↓ | ↓ |
| $T$ | $=$ | $44$ | $-$ | $(-56)$ |

We carry out the subtraction.

$$T = 44 - (-56) = 44 + 56 = 100$$

The temperature dropped 100°F.

**85.** Let $E$ = the difference between the elevations, in feet.

| Difference in elevations | is | Higher elevation | minus | Lower elevation |
|---|---|---|---|---|
| ↓ | ↓ | ↓ | ↓ | ↓ |
| $E$ | $=$ | $5672$ | $-$ | $(-4)$ |

We carry out the subtraction.

$$E = 5672 - (-4) = 5672 + 4 = 5676$$

The difference between the elevations is 5676 ft.

**87.** $4^3 = 4 \cdot 4 \cdot 4 = 64$

**89.** $5 \cdot 4 + 9 = 20 + 9 = 29$

**91.** $2 + (5 + 3)^2 = 2 + 8^2 = 2 + 64 = 66$

**93.** *Familiarize.* Let $n$ = the number of 12-oz cans that can be filled. We think of an array consisting of 96 oz with 12 oz in each row.

The number $n$ corresponds to the number of rows in the array.

*Translate and Solve.* We translate to an equation and solve it.

$$96 \div 12 = n \qquad \begin{array}{r} 8 \\ 12\overline{\smash{)}96} \\ \underline{96} \\ 0 \end{array}$$

*Check.* We multiply the number of cans by 12: $8 \cdot 12 = 96$. The result checks.

*State.* Eight 12-oz cans can be filled.

**95.** True

**97.** True

**99.** True by the definition of opposites.

**101.** True

## Chapter 2 Mid-Chapter Review

**1.** The statement is true. See page 594 in the text.

**2.** If $a > b$, then $a$ lies to the right of $b$ on the number line. Thus, the given statement is false.

**3.** The absolute value of a number is its distance from zero on the number line. Since distance is always nonnegative, the absolute value of a number is always nonnegative. The given statement is true.

**4.** $-x = -(-4) = 4$

$-(-x) = -(-(-4)) = -(4) = -4$

**5.** $5 - 13 = 5 + (-13) = -8$

**6.** $-6 - 7 = -6 + (-7) = -13$

**7.** The integer 450 corresponds to a \$450 deposit; the integer $-79$ corresponds to writing a check for \$79.

**8.** The integer 20 corresponds to a 20° increase in temperature; the integer $-23$ corresponds to a 23° drop in temperature.

**9.** We locate the point $-3$ on the number line and mark it with a dot.

**10.** We locate the point 0 on the number line and mark it with a dot.

**11.** Since $-6$ is to the left of 6, we have $-6 < 6$.

**12.** Since $-5$ is to the left of $-3$, we have $-5 < -3$.

**13.** Since $-9$ is to the right of $-10$, we have $-9 > -10$.

**14.** Since 5 is to the right of 0, we have $5 > 0$.

**15.** The distance of 15 from 0 is 15, so $|15| = 15$.

**16.** The distance of $-18$ from 0 is 18, so $|-18| = 18$.

**17.** The distance of 0 from 0 is 0, so $|0| = 0$.

**18.** The distance of $-12$ from 0 is 12, so $|-12| = 12$.

**19.** The additive inverse of $-5$ is 5 because $-5 + 5 = 0$.

**20.** The additive inverse of 7 is $-7$ because $7 + (-7) = 0$.

**21.** The additive inverse of 0 is 0 because $0 + 0 = 0$.

**22.** The additive inverse of $-49$ is 49 because $-49 + 49 = 0$.

**23.** If $x = -19$, then $-x = -(-19) = 19$.

**24.** If $x = 2$, then $-(-x) = -(-2) = 2$.

**25.** $7 + (-9)$   The absolute values are 7 and 9. The difference is $9 - 7$, or 2. The negative number has the larger absolute value, so the answer is negative.   $7 + (-9) = -2$

**26.** $-3 + 1$   The absolute values are 3 and 1. The difference is $3 - 1$, or 2. The negative number has the larger absolute value, so the answer is negative.   $-3 + 1 = -2$

**27.** $3 + (-3)$   A positive and a negative number. The numbers have the same absolute value. The sum is 0.   $3 + (-3) = 0$

**28.** $-8 + (-9)$   Two negative numbers. Add the absolute values, 8 and 9, getting 17. Make the answer negative. $-8 + (-9) = -17$

**29.** $2 + (-12)$   The absolute values are 2 and 12. The difference is $12 - 2$, or 10. The negative number has the larger absolute value, so the answer is negative. $2 + (-12) = -10$

**30.** $-4 + (-3)$   Two negative numbers. Add the absolute values, 4 and 3, getting 7. Make the answer negative. $-4 + (-3) = -7$

**31.** $-14+5$   The absolute values are 14 and 5. The difference is $14-5$, or 9. The negative number has the larger absolute value, so the answer is negative.   $-14+5 = -9$

**32.** $19+(-21)$   The absolute values are 19 and 21. The difference is $21-19$, or 2. The negative number has the larger absolute value, so the answer is negative.
$19+(-21) = -2$

**33.** $-4-6 = -4+(-6) = -10$

**34.** $5-(-11) = 5+11 = 16$

**35.** $-1-(-3) = -1+3 = 2$

**36.** $12-24 = 12+(-24) = -12$

**37.** $-8-(-4) = -8+4 = -4$

**38.** $-1-5 = -1+(-5) = -6$

**39.** $12-14 = 12+(-14) = -2$

**40.** $6-(-7) = 6+7 = 13$

**41.** $16-(-9)-20-(-4) = 16+9+(-20)+4 = 9$

**42.** $-4+(-10)-(-3)-12 = -4+(-10)+3+(-12) = -23$

**43.** $17-(-25)+15-(-18) = 17+25+15+18 = 75$

**44.** $-9+(-3)+16-(-10) = -9+(-3)+16+10 = 14$

**45.** Let $T$ = the difference in the temperatures, in degrees Celsius.

| Difference in temperatures | is | Higher temperature | minus | Lower temperature |
|---|---|---|---|---|
| ↓ | ↓ | ↓ | ↓ | ↓ |
| $T$ | $=$ | $25$ | $-$ | $(-8)$ |

We carry out the subtraction.
$$T = 25-(-8) = 25+8 = 33$$
The difference in the two temperature is 33°C.

**46.** Let $S$ = the final value of the stock.

| Final value | = | Beginning price | + | First change | + | Second change | + | Third change |
|---|---|---|---|---|---|---|---|---|
| ↓ | ↓ | ↓ | ↓ | ↓ | ↓ | ↓ | ↓ | ↓ |
| $S$ | $=$ | $56$ | $+$ | $(-3)$ | $+$ | $1$ | $+$ | $(-6)$ |

We carry out the addition.
$$S = 56+(-3)+1+(-6) = 48$$
The final value of the stock was $48.

**47.** Answers will vary.

**48.** The absolute value of a number is its distance from 0, and distance is always nonnegative.

**49.** Answers may vary. If we think of the addition on the number line, we start at a negative number and move to the left. This always brings us to a point on the negative portion of the number line.

**50.** Yes; consider $m-(-n)$ where both $m$ and $n$ are positive. Then $m-(-n) = m+n$. Now $m+n$, the sum of two positive numbers, is positive.

**Exercise Set 2.4**

**1.** $-16$

**3.** $-24$

**5.** $-72$

**7.** $16$

**9.** $42$

**11.** $-120$

**13.** $-238$

**15.** $1200$

**17.** $84$

**19.** $-12$

**21.** $24$

**23.** $21$

**25.** $-69$

**27.** $27$

**29.** $-18$

**31.** $-45$

**33.** $7 \cdot (-4) \cdot (-3) \cdot 5 = 7 \cdot 12 \cdot 5 = 7 \cdot 60 = 420$

**35.** $-3 \cdot 2 \cdot (-6) = -6 \cdot (-6) = 36$

**37.** $-3 \cdot (-4) \cdot (-5) = 12 \cdot (-5) = -60$

**39.** $-2 \cdot (-5) \cdot (-3) \cdot (-5) = 10 \cdot 15 = 150$

**41.** $-90$

**43.** $-7 \cdot (-21) \cdot 13 = 147 \cdot 13 = 1911$

**45.** $-4 \cdot (-2) \cdot 7 = 8 \cdot 7 = 56$

**47.** $-3(-2)(5) = 6(5) = 30$

**49.** $4 \cdot (-4) \cdot (-5) \cdot (-12) = -16 \cdot (60) = -960$

**51.** $7 \cdot (-7) \cdot 6 \cdot (-6) = -49 \cdot (-36) = 1764$

**53.** $(-5)(8)(-3)(-2) = -40(6) = -240$

**55.** $(-14) \cdot (-27) \cdot (-2) = 378 \cdot (-2) = -756$

**57.** $(-8)(-9)(-10) = 72(-10) = -720$

**59.** $(-6)(-7)(-8)(-9)(-10) = 42 \cdot 72 \cdot (-10) = 3024 \cdot (-10) = -30,240$

**61.** The <u>average</u> of a set of numbers is the sum of the numbers divided by the number of addends.

**63.** The statement $5 \cdot 4 = 4 \cdot 5$ illustrates the <u>commutative</u> law of multiplication.

**65.** The <u>absolute value</u> of a number is its distance from zero on the number line.

**67.** The <u>difference</u> $a - b$ is the number $c$ for which $a = b + c$.

**69.** a) $a$ and $b$ have different signs;

b) either $a$ or $b$ is zero or both are zero;

c) $a$ and $b$ have the same sign

---

## Exercise Set 2.5

---

**1.** $36 \div (-6) = -6$  Check: $-6 \cdot (-6) = 36$

**3.** $26 \div (-2) = -13$  Check: $-13 \cdot (-2) = 26$

**5.** $-16 \div 8 = -2$  Check: $-2 \cdot 8 = -16$

**7.** $-48 \div (-12) = 4$  Check: $4(-12) = -48$

**9.** $-72 \div 9 = -8$  Check: $-8 \cdot 9 = -72$

**11.** $-100 \div (-50) = 2$  Check: $2(-50) = -100$

**13.** $-108 \div 9 = -12$  Check: $9(-12) = -108$

**15.** $200 \div (-25) = -8$  Check: $-8(-25) = 200$

**17.** Not defined

**19.** $81 \div (-9) = -9$  Check: $-9 \cdot (-9) = 81$

**21.** First we multiply to find the change in temperature $t$ in the 18 minutes from 11:00 AM to 11:18 AM:

$t = 3 \cdot 18 = 54$

The temperature dropped 54°C.

Now we subtract to find the temperature $T$ at 11:18 AM:

$T = 0 - 54 = -54$

At 11:18 AM the temperature was −54°C.

**23.** First we multiply to find the amount $d$ by which the price per share dropped in 3 hr:

$d = 2 \cdot 3 = 6$

The price per share dropped $6 in 3 hr.

Now we subtract to find the price $p$ of the stock after 3 hr:

$p = 32 - 6 = 26$

The price per share was $26 after 3 hr.

**25.** First we multiply to find the number of meters $m$ that the diver rises in 9 min:

$m = 7 \cdot 9 = 63$

The diver rises 63 m in 9 min.

Now we subtract to find the diver's distance $d$ from the surface, in meters:

$d = 95 - 63 = 32$

The diver is 32 m below the surface.

**27.** $8 - 2 \cdot 3 - 9 = 8 - 6 - 9$  Multiplying

$\phantom{8 - 2 \cdot 3 - 9} = 2 - 9$  Doing all additions and subtractions in order

$\phantom{8 - 2 \cdot 3 - 9} = -7$  from left to right

**29.** $(8 - 2 \cdot 3) - 9 = (8 - 6) - 9$  Multiplying inside parentheses

$\phantom{(8 - 2 \cdot 3) - 9} = 2 - 9$  Subtracting inside parentheses

$\phantom{(8 - 2 \cdot 3) - 9} = -7$  Subtracting

**31.** $16 \cdot (-24) + 50 = -384 + 50$  Multiplying

$\phantom{16 \cdot (-24) + 50} = -334$  Adding

**33.** $2^4 + 2^3 - 10 = 16 + 8 - 10$  Evaluating exponential expressions

$\phantom{2^4 + 2^3 - 10} = 24 - 10$  Adding and subtracting in order

$\phantom{2^4 + 2^3 - 10} = 14$  from left to right

**35.** $5^3 + 26 \cdot 71 - (16 + 25 \cdot 3)$

$= 5^3 + 26 \cdot 71 - (16 + 75)$  Multiplying inside parentheses

$= 5^3 + 26 \cdot 71 - 91$  Adding inside parentheses

$= 125 + 26 \cdot 71 - 91$  Evaluating the exponential expression

$= 125 + 1846 - 91$  Multiplying

$= 1971 - 91$  Adding and subtracting in order from left

$= 1880$  to right

**37.** $4 \cdot 5 - 2 \cdot 6 + 4 = 20 - 12 + 4$  Multiplying

$\phantom{4 \cdot 5 - 2 \cdot 6 + 4} = 8 + 4$

$\phantom{4 \cdot 5 - 2 \cdot 6 + 4} = 12$

**39.** $4^3 \div 8 = 64 \div 8$  Evaluating the exponential expression

$\phantom{4^3 \div 8} = 8$  Dividing

**41.** $8(-7) + 6(-5) = -56 - 30$  Multiplying

$\phantom{8(-7) + 6(-5)} = -86$

**43.** $19 - 5(-3) + 3 = 19 + 15 + 3$  Multiplying

$\phantom{19 - 5(-3) + 3} = 34 + 3$

$\phantom{19 - 5(-3) + 3} = 37$

**45.** $9 \div (-3) + 16 \div 8 = -3 + 2$  Dividing

$\phantom{9 \div (-3) + 16 \div 8} = -1$

**47.** $-4^2 + 6 = -16 + 6$

$\phantom{-4^2 + 6} = -10$

**49.** $-8^2 - 3 = -64 - 3$

$\phantom{-8^2 - 3} = -67$

**51.** $12 - 20^3 = 12 - 8000$

$\phantom{12 - 20^3} = -7988$

**53.** $2 \times 10^3 - 5000 = 2 \times 1000 - 5000$

$\phantom{2 \times 10^3 - 5000} = 2000 - 5000$

$\phantom{2 \times 10^3 - 5000} = -3000$

**55.** $6[9 - (3 - 4)] = 6[9 - (-1)]$  Subtracting inside the innermost parentheses

$\phantom{6[9 - (3 - 4)]} = 6[9 + 1]$

$\phantom{6[9 - (3 - 4)]} = 6[10]$

$\phantom{6[9 - (3 - 4)]} = 60$

**57.** $-1000 \div (-100) \div 10 = 10 \div 10$  Doing the divisions in order from left to right

$\phantom{-1000 \div (-100) \div 10} = 1$

**59.** $8 - (7 - 9) = 8 - (-2)$
$= 8 + 2$
$= 10$

**61.** $(10 - 6^2) \div (3^2 + 2^2)$
$= (10 - 36) \div (9 + 4)$   Evaluating the exponential expressions
$= -26 \div 13$   Subtracting and adding
$= -2$   Dividing

**63.** $[20(8 - 3) - 4(10 - 3)] \div [10(2 - 6) + 2(7 + 4)]$
$= [20(5) - 4(7)] \div [10(-4) + 2(11)]$
Doing the calculations in parentheses
$= [100 - 28] \div [-40 + 22]$   Multiplying
$= 72 \div (-18)$   Subtracting and adding
$= -4$   Dividing

**65.** $4, 6\,7\,\boxed{8}, 9\,5\,2$

The digit 8 means 8 thousands.

**67.** $7\,1\,4\,\boxed{8}$

The digit 8 means 8 ones.

**69.**
$\begin{array}{r} {\scriptstyle 8\ \ 9\ \ 9\ \ 11} \\ \cancel{9\,0\,0\,1} \\ -\ 6\,7\,9\,8 \\ \hline 2\,2\,0\,3 \end{array}$

**71.**
$\begin{array}{r} {\scriptstyle 16\ \ 10\ \ 10} \\ {\scriptstyle 5\ \ \cancel{6}\ \ \cancel{0}\ \ \cancel{0}\ \ 13} \\ \cancel{6}\,7, \cancel{1}\,\cancel{1}\,\cancel{3} \\ -\ 2\,9,\,8\,7\,4 \\ \hline 3\,7,\,2\,3\,9 \end{array}$

**73. Familiarize.** We make a drawing. We let $A =$ the area.

64 ft

78 ft

**Translate.** Using the formula for area, we have
$A = l \cdot w = 78 \cdot 64.$

Using the formula for perimeter, we have
$P = 2l + 2w = 2 \cdot 78 + 2 \cdot 64.$

**Solve.** We carry out the computations.
$\begin{array}{r} 7\,8 \\ \times\ \ 6\,4 \\ \hline 3\,1\,2 \\ 4\,6\,8\,0 \\ \hline 4\,9\,9\,2 \end{array}$

Thus, $A = 4992.$

$P = 2 \cdot 78 + 2 \cdot 64 = 156 + 128 = 284$

**Check.** We repeat the calculations. The answers check.

**State.** The area is 4992 sq ft. The perimeter is 284 ft.

**75.** Use a calculator.
$(19 - 17^2) \div (13^2 - 34)$
$= (19 - 289) \div (169 - 34)$
$= -270 \div 135$
$= -2$

**77.** $-n$ and $m$ are both negative, so $-n \div m$ is the quotient of two negative numbers and, thus, is positive.

**79.** $-n \div m$ is positive (see Exercise 79), so $-(-n \div m)$ is the opposite of a positive number and, thus, is negative.

**81.** $-n$ is negative and $-m$ is positive, so $-n \div (-m)$ is the quotient of a negative and a positive number and, thus, is negative. Then $-[-n \div (-m)]$ is the opposite of a negative number and, thus, is positive.

## Chapter 2 Concept Reinforcement

**1.** False; see page 95 in the text.

**2.** True; see pages 96 and 97 in the text.

**3.** True; see page 109 in the text.

**4.** For a number $n$, $-(-n) = n \neq \dfrac{1}{n}$. The given statement is false.

## Chapter 2 Important Concepts

**1.** Locate the point 4 on the number line and mark it with a dot.

$\xleftarrow[\ -6\,-5\,-4\,-3\,-2\,-1\ \ 0\ \ 1\ \ 2\ \ 3\ \ 4\ \ 5\ \ 6\ ]{\overset{4}{\bullet}}\rightarrow$

**2.** Since $-5$ is to the left of $-4$ on the number line, we have $-5 < -4$.

**3.** a) The number is negative, so we make it positive.
$|-17| = 17$
   b) The number is positive, so the absolute value is the same as the number. $|14| = 14$

**4.** $6 + (-9)$   The absolute values are 6 and 9. The difference is $9 - 6$, or 3. The negative number has the larger absolute value, so the answer is negative.   $6 + (-9) = -3$

**5.** $-5 + (-3)$   Two negative numbers. We add the absolute values, 5 and 3, getting 8. Make the answer negative. $-5 + (-3) = -8$

**6.** $6 - (-8) = 6 + 8 = 14$

**7.** $-9(-8) = 72$

**8.** $6(-15) = -90$

**9.** $-32 \div (-8) = 4$   Check: $4(-8) = -32$

**10.** $48 \div (-12) = -4$   Check: $-4(-12) = 48$

**11.**  $4 - 8^2 \div (10 - 6) = 4 - 8^2 \div 4$
$$= 4 - 64 \div 4$$
$$= 4 - 16$$
$$= -12$$

## Chapter 2 Review Exercises

1. The integer $-45$ corresponds to a debt of \$45; the integer 72 corresponds to having \$72 in a savings account.

2. The distance of $-38$ from 0 is 38, so $|-38| = 38$.

3. The distance of 7 from 0 is 7, so $|7| = 7$.

4. The distance of 0 from 0 is 0, so $|0| = 0$.

5. The distance of $-2$ from 0 is 2, so $|-2| = 2$. Then $-|-2| = -(2) = -2$.

6. Since $-3$ is to the left of 10, we have $-3 < 10$.

7. Since $-1$ is to the right of $-6$, we have $-1 > -6$.

8. Since 11 is to the right of $-12$, we have $11 > -12$.

9. Since $-2$ is to the left of $-1$, we have $-2 < -1$.

10.

11.

12. The opposite of 8 is $-8$ because $8 + (-8) = 0$.

13. The opposite of $-14$ is 14 because $-14 + 14 = 0$.

14. If $x = -34$, then $-x = -(-34) = 34$.

15. If $x = 5$, then $-(-x) = -(-5) = 5$.

16. $4 + (-7)$

The absolute values are 4 and 7. The difference is $7 - 4$, or 3. The negative number has the larger absolute value, so the answer is negative.  $4 + (-7) = -3$

17. $-8 + 1$

The absolute values are 8 and 1. The difference is $8 - 1$, or 7. The negative number has the larger absolute value, so the answer is negative.  $-8 + 1 = -7$

18. $6 + (-9) + (-8) + 7$
  a) Add the negative numbers: $-9 + (-8) = -17$
  b) Add the positive numbers: $6 + 7 = 13$
  c) Add the results: $-17 + 13 = -4$

19. $-4 + 5 + (-12) + (-4) + 10$
  a) Add the negative numbers: $-4 + (-12) + (-4) = -20$
  b) Add the positive numbers: $5 + 10 = 15$
  c) Add the results: $-20 + 15 = -5$

20. $-3 - (-7) = -3 + 7 = 4$

21. $-9 - 5 = -9 + (-5) = -14$

22. $-4 - 4 = -4 + (-4) = -8$

23. $-9 \cdot (-6) = 54$

24. $-3(13) = -39$

25. $7 \cdot (-8) = -56$

26. $3 \cdot (-7) \cdot (-2) \cdot (-5) = -21 \cdot 10 = -210$

27. $35 \div (-5) = -7$     Check: $-7 \cdot (-5) = 35$

28. $-51 \div 17 = -3$     Check: $-3 \cdot (17) = -51$

29. $-42 \div (-7) = 6$     Check: $6 \cdot (-7) = -42$

30. $(-3 - 12) - 8(-7) = -15 - 8(-7)$
$$= -15 + 56$$
$$= 41$$

31.  $[-12(-3) - 2^3] - (-9)(-10)$
$$= [-12(-3) - 8] - (-9)(-10)$$
$$= [36 - 8] - (-9)(-10)$$
$$= 28 - (-9)(-10)$$
$$= 28 - 90$$
$$= -62$$

32. $625 \div (-25) \div 5 = -25 \div 5 = -5$

33.  $-16 \div 4 - 30 \div (-5) = -4 - (-6)$
$$= -4 + 6$$
$$= 2$$

34. $9[(7 - 14) - 13] = 9[-7 - 13] = 9[-20] = -180$

35. Let $t$ = the total gain or loss. We represent the gains as positive numbers and the loss as a negative number. We add the gains and the loss to find $t$.
$$t = 5 + (-12) + 15 = -7 + 15 = 8$$
There is a total gain of 8 yd.

36. Let $a$ = Kaleb's total assets after he borrows \$300.

| Total assets | is | Initial assets | minus | Amount of loan |
|---|---|---|---|---|
| $\downarrow$ | $\downarrow$ | $\downarrow$ | $\downarrow$ | $\downarrow$ |
| $a$ | $=$ | 170 | $-$ | 300 |

We carry out the subtraction.
$$a = 170 - 300 = -130$$
Kaleb's total assets were $-\$130$.

37. First we multiply to find the total drop $d$ in the price:
$$d = 4(-\$2) = -\$8$$
Now we add this number to the opening price to find the price $p$ after 4 hr:
$$p = \$18 + (-\$8) = \$10$$
After 4 hr the price of the stock was \$10 per share.

**38.** Let $p =$ the price of each DVD.

$$\underbrace{\text{Original balance}}_{\downarrow} \ \underbrace{\text{minus}}_{\downarrow} \ \underbrace{7}_{\downarrow} \ \underbrace{\text{times}}_{\downarrow} \ \underbrace{\text{price of each DVD}}_{\downarrow} \ \underbrace{\text{is}}_{\downarrow} \ \underbrace{\text{New balance}}_{\downarrow}$$
$$68 \quad - \quad 7 \quad \cdot \quad p \quad = \quad -65$$

We solve the equation.

$$68 - 7p = -65$$
$$68 - 7p - 68 = -65 - 68$$
$$-7p = -133$$
$$\frac{-7p}{-7} = \frac{-133}{-7}$$
$$p = 19$$

Each DVD cost $19.

**39.** $8 - (-5) - 7 - (-9) = 8 + 5 + (-7) + 9$
$$= 13 + (-7) + 9$$
$$= 6 + 9$$
$$= 15$$

Answer C is correct.

**40.** $-3 \cdot 4 - 12 \div 4 = -12 - 3 = -12 + (-3) = -15$

Answer B is correct.

**41.** a) $-7 + (-6) + (-5) + (-4) + (-3) + (-2) + (-1) + 0 + 1 + 2 + 3 + 4 + 5 + 6 + 7 + 8$

b) Since one of the factors is 0, the product is 0.

**42.** $9 - (3 - 4) + 5 = 15$

**43.** $-|8 - (-4 \div 2) - 3 \cdot 5| = -|8 - (-2) - 3 \cdot 5|$
$$= -|8 + 2 - 3 \cdot 5|$$
$$= -|8 + 2 - 15|$$
$$= -|10 - 15|$$
$$= -|-5|$$
$$= -5$$

**44.** $(|-6 - 3| + 3^2 - |-3|) \div (-3)$
$$= (|-6 - 3| + 9 - |-3|) \div (-3)$$
$$= (|-9| + 9 - |-3|) \div (-3)$$
$$= (9 + 9 - 3) \div (-3)$$
$$= (18 - 3) \div (-3)$$
$$= 15 \div (-3)$$
$$= -5$$

## Chapter 2 Discussion and Writing Exercises

**1.** If the negative integer has the larger absolute value, the answer is negative.

**2.** We know that $a + (-a) = 0$, so the opposite of $-a$ is $a$. That is, $-(-a) = a$.

**3.** At 4 p.m. the temperature in Circle City was 23°. By 11 p.m. the temperature had dropped 32 °. What was the temperature at 11 p.m.?

**4.** We know that the product of an even number of negative numbers is positive, and the product of an odd number of negative numbers is negative. Since $(-7)^8$ is equivalent to the product of eight negative numbers, it will be a positive number. Similarly, since $(-7)^{11}$ is equivalent to the product of eleven negative numbers, it will be a negative number.

**5.** Jake is expecting the multiplication to be performed before the division.

**6.** Consider $\frac{a}{b} = q$ where $a$ and $b$ are both negative integers. Then $q \cdot b = a$, so $q$ must be a positive number.

## Chapter 2 Test

**1.** Since $-4$ is to the left of 0 on the number line, we have $-4 < 0$.

**2.** Since $-3$ is to the right of $-8$ on the number line, we have $-3 > -8$.

**3.** Since $-7$ is to the right of $-8$ on the number line, we have $-7 > -8$.

**4.** Since $-1$ is to the left of 1 on the number line, we have $-1 < 1$.

**5.** The distance of $-7$ from 0 is 7, so $|-7| = 7$.

**6.** The distance of 94 from 0 is 94, so $|94| = 94$.

**7.** The distance of $-27$ from 0 is 27, so $|-27| = 27$. Then $-|-27| = -27$.

**8.** The opposite of 23 is $-23$ because $23 + (-23) = 0$.

**9.** The opposite of $-14$ is 14 because $-14 + 14 = 0$.

**10.** If $x = -8$, then $-x = -(-8) = 8$.

**11.** 

**12.** $31 - (-47) = 31 + 47 = 78$

**13.** $-8 + 4 + (-7) + 3 = -4 + (-7) + 3$
$$= -11 + 3$$
$$= -8$$

**14.** $-13 + 15 = 2$

**15.** $2 - (-8) = 2 + 8 = 10$

**16.** $32 - 57 = 32 + (-57) = -25$

**17.** $18 + (-3) = 15$

**18.** $4 \cdot (-12) = -48$

**19.** $-8 \cdot (-3) = 24$

**20.** $-45 \div 5 = -9$     Check: $-9 \cdot 5 = -45$

**21.** $-63 \div (-7) = 9$     Check: $9 \cdot (-7) = -63$

**22.** $64 \div (-16) = -4$     Check: $-4 \cdot (-16) = 64$

**23.**
$$-2(16) - [2(-8) - 5^3] = -2(16) - [2(-8) - 125]$$
$$= -2(16) - [-16 - 125]$$
$$= -2(16) - [-141]$$
$$= -2(16) + 141$$
$$= -32 + 141$$
$$= 109$$

**24.** Let $D$ = the difference in the temperatures.

| Difference in temperature | is | Higher temperature | minus | Lower temperature |
|---|---|---|---|---|
| ↓ | ↓ | ↓ | ↓ | ↓ |
| $D$ | = | $-67$ | $-$ | $(-81)$ |

We carry out the subtraction.
$$D = -67 - (-81) = -67 + 81 = 14$$

The average high temperature is 14°F higher than the average low temperature.

**25.** Let $P$ = the number of points by which the market has changed over the five week period.

| Total change | = | Week 1 change | + | Week 2 change | + | Week 3 change | + |
|---|---|---|---|---|---|---|---|
| ↓ | | ↓ | | ↓ | | ↓ | |
| $P$ | = | $-13$ | + | $(-16)$ | + | $36$ | + |

| Week 4 change | + | Week 5 change |
|---|---|---|
| ↓ | | ↓ |
| $(-11)$ | + | $19$ |

We carry out the computation.
$$P = -13 + (-16) + 36 + (-11) + 19$$
$$= -29 + 36 + (-11) + 19$$
$$= 7 + (-11) + 19$$
$$= -4 + 19$$
$$= 15$$

The market rose 15 points.

**26.** First we multiply to find the total decrease $d$ in the population.
$$d = 6 \cdot 420 = 2520$$

The population decreased by 2520 over the six year period.

Now we subtract to find the new population $p$.
$$18,600 - 2520 = 16,080$$

After 6 yr the population was 16,080.

**27.** First we subtract to find the total drop in temperature $t$.
$$t = 17°C - (-17°C) = 17°C + 17°C = 34°C$$

Then we divide to find by how many degrees $d$ the temperature dropped each minute in the 17 minutes from 11:08 A.M. to 11:25 A.M.
$$d = 34 \div 17 = 2$$

The temperature dropped 2°C each minute.

**28.** If $x = 14$, then $-(-x) = -(-14) = 14$. (The opposite of the opposite of 14 is 14.)

Answer D is correct.

**29.**
$$|-27 - 3(4)| - |-36| + |-12|$$
$$= |-27 - 12| - |-36| + |-12|$$
$$= |-39| - |-36| + |-12|$$
$$= 39 - 36 + 12$$
$$= 3 + 12$$
$$= 15$$

**30.** Let $d$ = the difference in the depths. We represent the depth of the Marianas Trench as $-11,033$ m and the depth of the Puerto Rico Trench are $-8648$ m.

| Difference in depths | is | Higher depth | minus | Lower depth |
|---|---|---|---|---|
| ↓ | ↓ | ↓ | ↓ | ↓ |
| $d$ | = | $-8648$ | $-$ | $(-11,033)$ |

We carry out the subtraction.
$$d = -8648 - (-11,033) = -8648 + 11,033 = 2385$$

The Puerto Rico Trench is 2385 m higher than the Marianas Trench.

**31.** a) 6, 5, 3, 0, ___, ___, ___

Observe that $5 = 6 - \boxed{1}$, $3 = 5 - \boxed{2}$, and $0 = 3 - \boxed{3}$.

To find the next three numbers in the sequence we subtract 4, 5, and 6, in order, from the preceding number. We have
$$0 - 4 = -4,$$
$$-4 - 5 = -9,$$
$$-9 - 6 = -15.$$

b) 14, 10, 6, 2, ___, ___, ___

Observe that each number is 4 less than the one that precedes it. Then we find the next three numbers as follows:
$$2 - 4 = -2,$$
$$-2 - 4 = -6,$$
$$-6 - 4 = -10.$$

c) $-4, -6, -9, -13,$ ___, ___, ___

Observe that $-6 = -4 - \boxed{2}$, $-9 = -6 - \boxed{3}$, and $-13 = -9 - \boxed{4}$. To find the next three numbers in the sequence we subtract 5, 6, and 7, in order, from the preceding number. We have
$$-13 - 5 = -18,$$
$$-18 - 6 = -24,$$
$$-24 - 7 = -31.$$

d) 64, −32, 16, −8, ___, ___, ___

Observe that we find each number by dividing the preceding number by −2. Then we find the next three numbers as follows:

$$\frac{-8}{-2} = 4,$$

$$\frac{4}{-2} = -2,$$

$$\frac{-2}{-2} = 1.$$

# Chapter 3

# Fraction Notation: Multiplication and Division

## Exercise Set 3.1

**1.** We divide the first number by the second.

$$
\begin{array}{r}
3 \\
14\overline{)52} \\
42 \\
\hline
10
\end{array}
$$

The remainder is not 0, so 14 is not a factor of 52.

**3.** We divide the first number by the second.

$$
\begin{array}{r}
25 \\
25\overline{)625} \\
500 \\
\hline
125 \\
125 \\
\hline
0
\end{array}
$$

The remainder is 0, so 25 is a factor of 625.

**5.** We find as many two-factor factorizations as we can:

$18 = 1 \cdot 18 \qquad 18 = 3 \cdot 6$
$18 = 2 \cdot 9$

Factors: 1, 2, 3, 6, 9, 18

**7.** We find as many two-factor factorizations as we can:

$54 = 1 \cdot 54 \qquad 54 = 3 \cdot 18$
$54 = 2 \cdot 27 \qquad 54 = 6 \cdot 9$

Factors: 1, 2, 3, 6, 9, 18, 27, 54

**9.** We find as many two-factor factorizations as we can:

$4 = 1 \cdot 4 \qquad 4 = 2 \cdot 2$

Factors: 1, 2, 4

**11.** The only factorization is $1 = 1 \cdot 1$.

Factor: 1

**13.** We find as many two-factor factorizations as we can:

$98 = 1 \cdot 98 \qquad 98 = 7 \cdot 14$
$98 = 2 \cdot 49$

Factors: 1, 2, 7, 14, 49, 98

**15.** We find as many two-factor factorizations as we can:

$255 = 1 \cdot 255 \qquad 255 = 5 \cdot 51$
$255 = 3 \cdot 85 \qquad 255 = 15 \cdot 17$

Factors: 1, 3, 5, 15, 17, 51, 85, 255

**17.**

| | |
|---|---|
| $1 \cdot 4 = 4$ | $6 \cdot 4 = 24$ |
| $2 \cdot 4 = 8$ | $7 \cdot 4 = 28$ |
| $3 \cdot 4 = 12$ | $8 \cdot 4 = 32$ |
| $4 \cdot 4 = 16$ | $9 \cdot 4 = 36$ |
| $5 \cdot 4 = 20$ | $10 \cdot 4 = 40$ |

**19.**

| | |
|---|---|
| $1 \cdot 20 = 20$ | $6 \cdot 20 = 120$ |
| $2 \cdot 20 = 40$ | $7 \cdot 20 = 140$ |
| $3 \cdot 20 = 60$ | $8 \cdot 20 = 160$ |
| $4 \cdot 20 = 80$ | $9 \cdot 20 = 180$ |
| $5 \cdot 20 = 100$ | $10 \cdot 20 = 200$ |

**21.**

| | |
|---|---|
| $1 \cdot 3 = 3$ | $6 \cdot 3 = 18$ |
| $2 \cdot 3 = 6$ | $7 \cdot 3 = 21$ |
| $3 \cdot 3 = 9$ | $8 \cdot 3 = 24$ |
| $4 \cdot 3 = 12$ | $9 \cdot 3 = 27$ |
| $5 \cdot 3 = 15$ | $10 \cdot 3 = 30$ |

**23.**

| | |
|---|---|
| $1 \cdot 12 = 12$ | $6 \cdot 12 = 72$ |
| $2 \cdot 12 = 24$ | $7 \cdot 12 = 84$ |
| $3 \cdot 12 = 36$ | $8 \cdot 12 = 96$ |
| $4 \cdot 12 = 48$ | $9 \cdot 12 = 108$ |
| $5 \cdot 12 = 60$ | $10 \cdot 12 = 120$ |

**25.**

| | |
|---|---|
| $1 \cdot 10 = 10$ | $6 \cdot 10 = 60$ |
| $2 \cdot 10 = 20$ | $7 \cdot 10 = 70$ |
| $3 \cdot 10 = 30$ | $8 \cdot 10 = 80$ |
| $4 \cdot 10 = 40$ | $9 \cdot 10 = 90$ |
| $5 \cdot 10 = 50$ | $10 \cdot 10 = 100$ |

**27.**

| | |
|---|---|
| $1 \cdot 9 = 9$ | $6 \cdot 9 = 54$ |
| $2 \cdot 9 = 18$ | $7 \cdot 9 = 63$ |
| $3 \cdot 9 = 27$ | $8 \cdot 9 = 72$ |
| $4 \cdot 9 = 36$ | $9 \cdot 9 = 81$ |
| $5 \cdot 9 = 45$ | $10 \cdot 9 = 90$ |

**29.** We divide 26 by 6.

$$
\begin{array}{r}
4 \\
6\overline{)26} \\
24 \\
\hline
2
\end{array}
$$

Since the remainder is not 0, 26 is not divisible by 6.

**31.** We divide 1880 by 8.

$$
\begin{array}{r}
235 \\
8\overline{)1880} \\
1600 \\
\hline
280 \\
240 \\
\hline
40 \\
40 \\
\hline
0
\end{array}
$$

Since the remainder is 0, 1880 is divisible by 8.

**33.** We divide 256 by 16.

$$
\begin{array}{r}
1\ 6 \\
16\,\overline{\smash{)}\,2\ 5\ 6} \\
1\ 6\ 0 \\
\hline
9\ 6 \\
9\ 6 \\
\hline
0
\end{array}
$$

Since the remainder is 0, 256 is divisible by 16.

**35.** We divide 4227 by 9.

$$
\begin{array}{r}
4\ 6\ 9 \\
9\,\overline{\smash{)}\,4\ 2\ 2\ 7} \\
3\ 6\ 0\ 0 \\
\hline
6\ 2\ 7 \\
5\ 4\ 0 \\
\hline
8\ 7 \\
8\ 1 \\
\hline
6
\end{array}
$$

Since the remainder is not 0, 4227 is not divisible by 9.

**37.** We divide 8650 by 16.

$$
\begin{array}{r}
5\ 4\ 0 \\
16\,\overline{\smash{)}\,8\ 6\ 5\ 0} \\
8\ 0\ 0\ 0 \\
\hline
6\ 5\ 0 \\
6\ 4\ 0 \\
\hline
1\ 0
\end{array}
$$

Since the remainder is not 0, 8650 is not divisible by 16.

**39.** 1 is neither prime nor composite.

**41.** The number 9 has factors 1, 3, and 9.

Since 9 is not 1 and not prime, it is composite.

**43.** The number 11 is prime. It has only the factors 1 and 11.

**45.** The number 29 is prime. It has only the factors 1 and 29.

**47.**
$$
\begin{array}{r}
2 \quad \leftarrow \ 2 \text{ is prime.} \\
2\,\overline{\smash{)}\,4} \\
2\,\overline{\smash{)}\,8}
\end{array}
$$
$8 = 2 \cdot 2 \cdot 2$

**49.**
$$
\begin{array}{r}
7 \quad \leftarrow \ 7 \text{ is prime.} \\
2\,\overline{\smash{)}\,1\ 4}
\end{array}
$$
$14 = 2 \cdot 7$

**51.**
$$
\begin{array}{r}
7 \quad \leftarrow \ 7 \text{ is prime.} \\
3\,\overline{\smash{)}\,2\ 1} \\
2\,\overline{\smash{)}\,4\ 2}
\end{array}
$$
$42 = 2 \cdot 3 \cdot 7$

**53.**
$$
\begin{array}{r}
5 \quad \leftarrow \ 5 \text{ is prime.} \\
5\,\overline{\smash{)}\,2\ 5}
\end{array}
$$
(25 is not divisible by 2 or 3. We move to 5.)

$25 = 5 \cdot 5$

**55.**
$$
\begin{array}{r}
5 \quad \leftarrow \ 5 \text{ is prime.} \\
5\,\overline{\smash{)}\,2\ 5} \\
2\,\overline{\smash{)}\,5\ 0}
\end{array}
$$
(25 is not divisible by 2 or 3. We move to 5.)

$50 = 2 \cdot 5 \cdot 5$

**57.**
$$
\begin{array}{r}
1\ 3 \quad \leftarrow \ 13 \text{ is prime.} \\
13\,\overline{\smash{)}\,1\ 6\ 9}
\end{array}
$$
(169 is not divisible by 2, 3, 5, 7 or 11. We move to 13.)

$169 = 13 \cdot 13$

**59.**
$$
\begin{array}{r}
5 \quad \leftarrow \ 5 \text{ is prime.} \\
5\,\overline{\smash{)}\,2\ 5} \\
2\,\overline{\smash{)}\,5\ 0} \\
2\,\overline{\smash{)}\,1\ 0\ 0}
\end{array}
$$
(25 is not divisible by 2 or 3. We move to 5.)

$100 = 2 \cdot 2 \cdot 5 \cdot 5$

We can also use a factor tree.

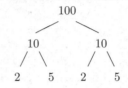

**61.**
$$
\begin{array}{r}
7 \quad \leftarrow \ 7 \text{ is prime.} \\
5\,\overline{\smash{)}\,3\ 5}
\end{array}
$$
(35 is not divisible by 2 or 3. We move to 5.)

$35 = 5 \cdot 7$

**63.**
$$
\begin{array}{r}
3 \quad \leftarrow \ 3 \text{ is prime.} \\
3\,\overline{\smash{)}\,9} \\
2\,\overline{\smash{)}\,1\ 8} \\
2\,\overline{\smash{)}\,3\ 6} \\
2\,\overline{\smash{)}\,7\ 2}
\end{array}
$$
(9 is not divisible by 2. We move to 3.)

$72 = 2 \cdot 2 \cdot 2 \cdot 3 \cdot 3$

We can also use a factor tree, as shown in Example 11 in the text.

**65.**
$$
\begin{array}{r}
1\ 1 \quad \leftarrow \ 11 \text{ is prime.} \\
7\,\overline{\smash{)}\,7\ 7}
\end{array}
$$
(77 is not divisible by 2, 3, or 5. We move to 7.)

$77 = 7 \cdot 11$

**67.**
$$
\begin{array}{r}
1\ 0\ 3 \quad \leftarrow \ 103 \text{ is prime.} \\
7\,\overline{\smash{)}\,7\ 2\ 1} \\
2\,\overline{\smash{)}\,1\ 4\ 4\ 2} \\
2\,\overline{\smash{)}\,2\ 8\ 8\ 4}
\end{array}
$$

$2884 = 2 \cdot 2 \cdot 7 \cdot 103$

We can also use a factor tree.

```
              2884
             /    \
            4      721
           / \    /   \
          2   2  7    103
```

**69.**
$$
\begin{array}{r}
1\ 7 \quad \leftarrow \ 17 \text{ is prime.} \\
3\,\overline{\smash{)}\,5\ 1}
\end{array}
$$
(51 is not divisible by 2. We move to 3.)

$51 = 3 \cdot 17$

**71.**
$$\begin{array}{r} 5 \\ 5\overline{)25} \\ 3\overline{)75} \\ 2\overline{)150} \\ 2\overline{)300} \\ 2\overline{)600} \\ 2\overline{)1200} \end{array} \quad \leftarrow 5 \text{ is prime}$$

$1200 = 2 \cdot 2 \cdot 2 \cdot 2 \cdot 3 \cdot 5 \cdot 5$

**73.**
$$\begin{array}{r} 13 \\ 7\overline{)91} \\ 3\overline{)273} \end{array} \quad \leftarrow 13 \text{ is prime}$$

$273 = 3 \cdot 7 \cdot 13$

**75.**
$$\begin{array}{r} 17 \\ 11\overline{)187} \\ 3\overline{)561} \\ 2\overline{)1122} \end{array} \quad \leftarrow 17 \text{ is prime}$$

$1122 = 2 \cdot 3 \cdot 11 \cdot 17$

**77.**
$$\begin{array}{r} 13 \\ \times \ 2 \\ \hline 26 \end{array}$$

**79.**
$$\begin{array}{r} \overset{3}{2}5 \\ \times \ 17 \\ \hline 175 \\ 250 \\ \hline 425 \end{array}$$
      Multiplying by 7
      Multiplying by 10
      Adding

**81.** Zero divided by any nonzero number is 0. Thus, $0 \div 22 = 0$.

**83.** Any nonzero number divided by itself is 1. Thus, $22 \div 22 = 1$.

**85.** *Familiarize*. Let $t$ = the total amount Kate took in.

*Translate*. We write an equation.

| Number of pounds | times | Price per pound | is | Total amount taken in |
|---|---|---|---|---|
| ↓ | ↓ | ↓ | ↓ | ↓ |
| 43 | · | 22 | = | $t$ |

*Solve*. We carry out the multiplication.

$$\begin{array}{r} 43 \\ \times \ 22 \\ \hline 86 \\ 860 \\ \hline 946 \end{array}$$

Thus, $946 = t$.

*Check*. We can repeat the calculation. The answer checks.

*State*. Kate took in a total of $946.

**87.** Row 1: 48, 90, 432, 63; row 2: 7, 2, 2, 10, 8, 6, 21, 10; row 3: 9, 18, 36, 14, 12, 11, 21; row 4: 29, 19, 42

---

## Exercise Set 3.2

**1.** A number is divisible by 2 if its <u>ones digit</u> is even.

4<u>6</u> is divisible by 2 because <u>6</u> is even.
22<u>4</u> is divisible by 2 because <u>4</u> is even.
1<u>9</u> is not divisible by 2 because <u>9</u> is not even.
55<u>5</u> is not divisible by 2 because <u>5</u> is not even.
30<u>0</u> is divisible by 2 because <u>0</u> is even.
3<u>6</u> is divisible by 2 because <u>6</u> is even.
45,27<u>0</u> is divisible by 2 because <u>0</u> is even.
444<u>4</u> is divisible by 2 because <u>4</u> is even.
8<u>5</u> is not divisible by 2 because <u>5</u> is not even.
71<u>1</u> is not divisible by 2 because <u>1</u> is not even.
13,25<u>1</u> is not divisible by 2 because <u>1</u> is not even.
254,76<u>5</u> is not divisible by 2 because <u>5</u> is not even.
25<u>6</u> is divisible by 2 because <u>6</u> is even.
806<u>4</u> is divisible by 2 because <u>4</u> is even.
186<u>7</u> is not divisible by 2 because <u>7</u> is not even.
21,56<u>8</u> is divisible by 2 because <u>8</u> is even.

**3.** A number is divisible by 4 if the <u>number</u> named by the last <u>two</u> digits is divisible by 4.

<u>46</u> is not divisible by 4 because <u>46</u> is not divisible by 4.
2<u>24</u> is divisible by 4 because <u>24</u> is divisible by 4.
<u>19</u> is not divisible by 4 because <u>19</u> is not divisible by 4.
5<u>55</u> is not divisible by 4 because <u>55</u> is not divisible by 4.
3<u>00</u> is divisible by 4 because <u>00</u> is divisible by 4.
<u>36</u> is divisible by 4 because <u>36</u> is divisible by 4.
45,2<u>70</u> is not divisible by 4 because <u>70</u> is not divisible by 4.
44<u>44</u> is divisible by 4 because <u>44</u> is divisible by 4.
<u>85</u> is not divisible by 4 because <u>85</u> is not divisible by 4.
7<u>11</u> is not divisible by 4 because <u>11</u> is not divisible by 4.
13,2<u>51</u> is not divisible by 4 because <u>51</u> is not divisible by 4.
254,7<u>65</u> is not divisible by 4 because <u>65</u> is not divisible by 4.
2<u>56</u> is divisible by 4 because <u>56</u> is divisible by 4.
80<u>64</u> is divisible by 4 because <u>64</u> is divisible by 4.
18<u>67</u> is not divisible by 4 because <u>67</u> is not divisible by 4.
21,5<u>68</u> is divisible by 4 because <u>68</u> is divisible by 4.

**5.** For a number to be divisible by 6, the sum of the digits must be divisible by 3 and the ones digit must be 0, 2, 4, 6 or 8 (even). It is most efficient to determine if the ones digit is even first and then, if so, to determine if the sum of the digits is divisible by 3.

46 is not divisible by 6 because 46 is not divisible by 3.

$$\underset{\uparrow}{4 + 6 = 10}$$
Not divisible by 3

224 is not divisible by 6 because 224 is not divisible by 3.

$$2 + 2 + 4 = 8$$
$$\uparrow$$
Not divisible by 3

19 is not divisible by 6 because 19 is not even.

19
$\uparrow$
Not even

555 is not divisible by 6 because 555 is not even.

555
$\uparrow$
Not even

300 is divisible by 6.

300          $3 + 0 + 0 = 3$
$\uparrow$          $\uparrow$
Even          Divisible by 3

36 is divisible by 6.

36          $3 + 6 = 9$
$\uparrow$          $\uparrow$
Even          Divisible by 3

45,270 is divisible by 6.

45,270          $4 + 5 + 2 + 7 + 0 = 18$
$\uparrow$          $\uparrow$
Even          Divisible by 3

4444 is not divisible by 6 because 4444 is not divisible by 3.

$$4 + 4 + 4 + 4 = 16$$
$$\uparrow$$
Not divisible by 3

85 is not divisible by 6 because 85 is not even.

85
$\uparrow$
Not even

711 is not divisible by 6 because 711 is not even.

711
$\uparrow$
Not even

13,251 is not divisible by 6 because 13,251 is not even.

13,251
$\uparrow$
Not even

254,765 is not divisible by 6 because 254,765 is not even.

254,765
$\uparrow$
Not even

256 is not divisible by 6 because 256 is not divisible by 3.

$$2 + 5 + 6 = 13$$
$$\uparrow$$
Not divisible by 3

8064 is divisible by 6.

8064          $8 + 0 + 6 + 4 = 18$
$\uparrow$          $\uparrow$
Even          Divisible by 3

1867 is not divisible by 6 because 1867 is not even.

1867
$\uparrow$
Not even

21,568 is not divisible by 6 because 21,568 is not divisible by 3.

$$2 + 1 + 5 + 6 + 8 = 22$$
$$\uparrow$$
Not divisible by 3

**7.** A number is divisible by 9 if the sum of the digits is divisible by 9.

46 is not divisible by 9 because $4 + 6 = 10$ and 10 is not divisible by 9.

224 is not divisible by 9 because $2 + 2 + 4 = 8$ and 8 is not divisible by 9.

19 is not divisible by 9 because $1 + 9 = 10$ and 10 is not divisible by 9.

555 is not divisible by 9 because $5 + 5 + 5 = 15$ and 15 is not divisible by 9.

300 is not divisible by 9 because $3 + 0 + 0 = 3$ and 3 is not divisible by 9.

36 is divisible by 9 because $3 + 6 = 9$ and 9 is divisible by 9.

45,270 is divisible by 9 because $4 + 5 + 2 + 7 + 0 = 18$ and 18 is divisible by 9.

4444 is not divisible by 9 because $4 + 4 + 4 + 4 = 16$ and 16 is not divisible by 9.

85 is not divisible by 9 because $8 + 5 = 13$ and 13 is not divisible by 9.

711 is divisible by 9 because $7 + 1 + 1 = 9$ and 9 is divisible by 9.

13,251 is not divisible by 9 because $1 + 3 + 2 + 5 + 1 = 12$ and 12 is not divisible by 9.

254,765 is not divisible by 9 because $2 + 5 + 4 + 7 + 6 + 5 = 29$ and 29 is not divisible by 9.

256 is not divisible by 9 because $2 + 5 + 6 = 13$ and 13 is not divisible by 9.

8064 is divisible by 9 because $8 + 0 + 6 + 4 = 18$ and 18 is divisible by 9.

1867 is not divisible by 9 because $1 + 8 + 6 + 7 = 22$ and 22 is not divisible by 9.

21,568 is not divisible by 9 because $2 + 1 + 5 + 6 + 8 = 22$ and 22 is not divisible by 9.

**9.** A number is divisible by 3 if the sum of the digits is divisible by 3.

56 is not divisible by 3 because $5 + 6 = 11$ and 11 is not divisible by 3.

324 is divisible by 3 because $3 + 2 + 4 = 9$ and 9 is divisible by 3.

784 is not divisible by 3 because $7 + 8 + 4 = 19$ and 19 is not divisible by 3.

55,555 is not divisible by 3 because $5 + 5 + 5 + 5 + 5 = 25$ and 25 is not divisible by 3.

200 is not divisible by 3 because $2+0+0=2$ and 2 is not divisible by 3.

42 is divisible by 3 because $4+2=6$ and 6 is divisible by 3.

501 is divisible by 3 because $5+0+1=6$ and 6 is divisible by 3.

3009 is divisible by 3 because $3+0+0+9=12$ and 12 is divisible by 3.

75 is divisible by 3 because $7+5=12$ and 12 is divisible by 3.

812 is not divisible by 3 because $8+1+2=11$ and 11 is not divisible by 3.

2345 is not divisible by 3 because $2+3+4+5=14$ and 14 is not divisible by 3.

2001 is divisible by 3 because $2+0+0+1=3$ and 3 is divisible by 3.

35 is not divisible by 3 because $3+5=8$ and 8 is not divisible by 3.

402 is divisible by 3 because $4+0+2=6$ and 6 is divisible by 3.

111,111 is divisible by 3 because $1+1+1+1+1+1=6$ and 6 is divisible by 3.

1005 is divisible by 3 because $1+0+0+5=6$ and 6 is divisible by 3.

**11.** A number is divisible by 5 if the ones digit is 0 or 5.

5<u>6</u> is not divisible by 5 because the ones digit (6) is not 0 or 5.

32<u>4</u> is not divisible by 5 because the ones digit (4) is not 0 or 5.

78<u>4</u> is not divisible by 5 because the ones digit (4) is not 0 or 5.

55,55<u>5</u> is divisible by 5 because the ones digit is 5.

20<u>0</u> is divisible by 5 because the ones digit is 0.

4<u>2</u> is not divisible by 5 because the ones digit (2) is not 0 or 5.

50<u>1</u> is not divisible by 5 because the ones digit (1) is not 0 or 5.

300<u>9</u> is not divisible by 5 because the ones digit (9) is not 0 or 5.

7<u>5</u> is divisible by 5 because the ones digit is 5.

81<u>2</u> is not divisible by 5 because the ones digit (2) is not 0 or 5.

234<u>5</u> is divisible by 5 because the ones digit is 5.

200<u>1</u> is not divisible by 5 because the ones digit (1) is not 0 or 5.

3<u>5</u> is divisible by 5 because the ones digit is 5.

40<u>2</u> is not divisible by 5 because the ones digit (2) is not 0 or 5.

111,11<u>1</u> is not divisible by 5 because the ones digit (1) is not 0 or 5.

100<u>5</u> is divisible by 5 because the ones digit is 5.

**13.** A number is divisible by 8 if the <u>number</u> named by the last <u>three</u> digits is divisible by 8.

<u>56</u> is divisible by 8.

<u>324</u> is not divisible by 8.

<u>784</u> is divisible by 8.

55,<u>555</u> is not divisible by 8 because 555 is not divisible by 8.

<u>200</u> is divisible by 8.

<u>42</u> is not divisible by 8.

<u>501</u> is not divisible by 8.

3<u>009</u> is not divisible by 8 because 9 is not divisible by 8.

<u>75</u> is not divisible by 8.

<u>812</u> is not divisible by 8.

2<u>345</u> is not divisible by 8 because 345 is not divisible by 8.

2<u>001</u> is not divisible by 8 because 1 is not divisible by 8.

<u>35</u> is not divisible by 8.

<u>402</u> is not divisible by 8.

111,<u>111</u> is not divisible by 8 because 111 is not divisible by 8.

1<u>005</u> is not divisible by 8 because 5 is not divisible by 8.

**15.** A number is divisible by 10 if the ones digit is 0.

Of the numbers under consideration, the only one whose ones digit is 0 is 200. Therefore, 200 is divisible by 10. None of the other numbers is divisible by 10.

**17.** A number is divisible by 2 if its ones digit is even. The numbers whose ones digits are even are 313,332, 7624, 111,126, 876, 1110, 5128, 64,000, and 9990.

**19.** A number is divisible by 6 if its one digit is even and the sum of its digits is divisible by 3. The numbers whose ones digit are even are given in Exercise 17 above. We find the sum of the digits of each one.

$3+1+3+3+3+2=15$; 15 is divisible by 3, so 313,332 is divisible by 6.

$7+6+2+4=19$; 19 is not divisible by 3, so 7624 is not divisible by 6.

$1+1+1+1+2+6=12$; 12 is divisible by 3, so 111,126 is divisible by 6.

$8+7+6=21$; 21 is divisible by 3, so 876 is divisible by 6.

$1+1+1+0=3$; 3 is divisible by 3, so 1110 is divisible by 6.

$5+1+2+8=16$; 16 is not divisible by 3, so 5128 is not divisible by 6.

$6+4+0+0+0=10$; 10 is not divisible by 3, so 64,000 is not divisible by 6.

$9+9+9+0=27$; 27 is divisible by 3, so 9990 is divisible by 6.

**21.** A number is divisible by 9 if the sum of its digits is divisible by 9.

$3+0+5=8$; 8 is not divisible by 9, so 305 is not divisible by 9.

$1 + 1 + 0 + 1 = 3$; 3 is not divisible by 9, so 1101 is not divisible by 9.

$1 + 3 + 0 + 2 + 5 = 11$; 11 is not divisible by 9, so 13,025 is not divisible by 9.

$3 + 1 + 3 + 3 + 3 + 2 = 15$; 15 is not divisible by 9, so 313,332 is not divisible by 9.

$7 + 6 + 2 + 4 = 19$; 19 is not divisible by 9, so 7624 is not divisible by 9.

$1 + 1 + 1 + 1 + 2 + 6 = 12$; 12 is not divisible by 9, so 111,126 is not divisible by 9.

$8 + 7 + 6 = 21$; 21 is not divisible by 9, so 876 is not divisible by 9.

$1 + 1 + 1 + 0 = 3$; 3 is not divisible by 9, so 1110 is not divisible by 9.

$5 + 1 + 2 + 8 = 16$; 16 is not divisible by 9, so 5128 is not divisible by 9.

$6 + 4 + 0 + 0 + 0 = 10$; 10 is not divisible by 9, so 64,000 is not divisible by 9.

$9 + 9 + 9 + 0 = 27$; 27 is divisible by 9, so 9990 is divisible by 9.

$1 + 2 + 6 + 1 + 1 + 1 = 12$; 12 is not divisible by 9, so 126,111 is not divisible by 9.

**23.** A number is divisible by 10 if its ones digit is 0. Then the numbers 1110, 64,000, and 9990 are divisible by 10.

**25.**
$$56 + x = 194$$
$$56 + x - 56 = 194 - 56 \quad \text{Subtracting 56 on both sides}$$
$$x = 138$$

The solution is 138.

**27.**
$$3008 = x + 2134$$
$$3008 - 2134 = x + 2134 - 2134 \quad \text{Subtracting 2134}$$
$$\text{on both sides}$$
$$874 = x$$

The solution is 874.

**29.**
$$24 \cdot m = 624$$
$$\frac{24 \cdot m}{24} = \frac{624}{24} \quad \text{Dividing by 24 on both sides}$$
$$m = 26$$

The solution is 26.

**31.**
```
     2 3 4
 9 | 2 1 0 6
     1 8 0 0
     -------
       3 0 6
       2 7 0
       -----
         3 6
         3 6
         ---
           0
```

The answer is 234.

**33.** *Familiarize.* We visualize the situation. Let $g =$ the number of gallons of gasoline the automobile will use to travel 1485 mi.

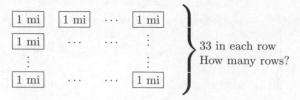

} 33 in each row
How many rows?

*Translate.* We translate to an equation.

| Number of miles | divided by | Miles per gallon | is | Number of gallons |
|---|---|---|---|---|
| ↓ | ↓ | ↓ | ↓ | ↓ |
| 1485 | ÷ | 33 | = | g |

*Solve.* We carry out the division.

```
        4 5
 3 3 | 1 4 8 5
      1 3 2 0
      -------
        1 6 5
        1 6 5
        -----
            0
```

Thus $45 = g$, or $g = 45$.

*Check.* We can repeat the calculation. The answer checks.

*State.* The automobile will use 45 gallons of gasoline to travel 1485 mi.

**35.** 78$\underline{0}$0 is divisible by 2 because the ones digit (0) is even.

$7800 \div 2 = 3900$ so $7800 = 2 \cdot 3900$.

390$\underline{0}$ is divisible by 2 because the ones digit (0) is even.

$3900 \div 2 = 1950$ so $3900 = 2 \cdot 1950$ and $7800 = 2 \cdot 2 \cdot 1950$.

195$\underline{0}$ is divisible by 2 because the ones digit (0) is even.

$1950 \div 2 = 975$ so $1950 = 2 \cdot 975$ and $7800 = 2 \cdot 2 \cdot 2 \cdot 975$.

97$\underline{5}$ is not divisible by 2 because the ones digit (5) is not even. Move on to 3.

975 is divisible by 3 because the sum of the digits ($9 + 7 + 5 = 21$) is divisible by 3.

$975 \div 3 = 325$ so $975 = 3 \cdot 325$ and $7800 = 2 \cdot 2 \cdot 2 \cdot 3 \cdot 325$.

Since 975 is not divisible by 2, none of its factors is divisible by 2. Therefore, we no longer need to check for divisibility by 2.

325 is not divisible by 3 because the sum of the digits ($3 + 2 + 5 = 10$) is not divisible by 3. Move on to 5.

32$\underline{5}$ is divisible by 5 because the ones digit is 5.

$325 \div 5 = 65$ so $325 = 5 \cdot 65$ and $7800 = 2 \cdot 2 \cdot 2 \cdot 3 \cdot 5 \cdot 65$.

Since 325 is not divisible by 3, none of its factors is divisible by 3. Therefore, we no longer need to check for divisibility by 3.

6$\underline{5}$ is divisible by 5 because the ones digit is 5.

$65 \div 5 = 13$ so $65 = 5 \cdot 13$ and $7800 = 2 \cdot 2 \cdot 2 \cdot 3 \cdot 5 \cdot 5 \cdot 13$.

13 is prime so the prime factorization of 7800 is $2 \cdot 2 \cdot 2 \cdot 3 \cdot 5 \cdot 5 \cdot 13$.

**37.** 277$\underline{2}$ is divisible by 2 because the ones digit (2) is even.

$2772 \div 2 = 1386$ so $2772 = 2 \cdot 1386$.

138$\underline{6}$ is divisible by 2 because the ones digit (6) is even.

$1386 \div 2 = 693$ so $1386 = 2 \cdot 693$ and $2772 = 2 \cdot 2 \cdot 693$.

69<u>3</u> is not divisible by 2 because the ones digit (3) is not even. We move to 3.

693 is divisible by 3 because the sum of the digits ($6 + 9 + 3 = 18$) is divisible by 3.

$693 \div 3 = 231$ so $693 = 3 \cdot 231$ and $2772 = 2 \cdot 2 \cdot 3 \cdot 231$.

Since 693 is not divisible by 2, none of its factors is divisible by 2. Therefore, we no longer need to check divisibility by 2.

231 is divisible by 3 because the sum of the digits ($2 + 3 + 1 = 6$) is divisible by 3.

$231 \div 3 = 77$ so $231 = 3 \cdot 77$ and $2772 = 2 \cdot 2 \cdot 3 \cdot 3 \cdot 77$.

77 is not divisible by 3 since the sum of the digits ($7 + 7 = 14$) is not divisible by 3. We move to 5.

7<u>7</u> is not divisible by 5 because the ones digit (7) is not 0 or 5. We move to 7.

We have not stated a test for divisibility by 7 so we will just try dividing by 7.

$$\begin{array}{r} 1\;1 \\ 7\overline{)\,7\;7} \end{array} \leftarrow 11 \text{ is prime}$$

$77 \div 7 = 11$ so $77 = 7 \cdot 11$ and the prime factorization of 2772 is $2 \cdot 2 \cdot 3 \cdot 3 \cdot 7 \cdot 11$.

**39.** The sum of the given digits is $9 + 5 + 8$, or 22. If the number is divisible by 99, it is also divisible by 9 since 99 is divisible by 9. The smallest number that is divisible by 9 and also greater than 22 is 27. Then the sum of the two missing digits must be at least $27 - 22$, or 5. We try various combinations of two digits whose sum is 5, using a calculator to divide the resulting number by 99:

  95,058 is not divisible by 99.

  95,148 is not divisible by 99.

  95,238 is divisible by 99.

Thus, the missing digits are 2 and 3 and the number is 95,238.

## Exercise Set 3.3

**1.** The top number is the numerator, and the bottom number is the denominator.

$$\frac{3}{4} \begin{array}{l} \leftarrow \text{Numerator} \\ \leftarrow \text{Denominator} \end{array}$$

**3.** The top number is the numerator, and the bottom number is the denominator.

$$\frac{11}{2} \begin{array}{l} \leftarrow \text{Numerator} \\ \leftarrow \text{Denominator} \end{array}$$

**5.** The top number is the numerator, and the bottom number is the denominator.

$$\frac{0}{7} \begin{array}{l} \leftarrow \text{Numerator} \\ \leftarrow \text{Denominator} \end{array}$$

**7.** The acre is divided into 12 equal parts. The unit is $\frac{1}{12}$. The denominator is 12. We have 6 parts shaded. This tells us that the numerator is 6. Thus, $\frac{6}{12}$ is shaded.

**9.** The yard is divided into 8 equal parts. The unit is $\frac{1}{8}$. The denominator is 8. We have 1 part shaded. This tells us that the numerator is 1. Thus, $\frac{1}{8}$ is shaded.

**11.** Each rectangle is divided into 3 equal parts. The unit is $\frac{1}{3}$. The denominator is 3. We see that 4 of the parts are shaded, so the numerator is 4. Thus, $\frac{4}{3}$ is shaded.

**13.** Each rectangle is divided into 8 equal parts. The unit is $\frac{1}{8}$. The denominator is 8. We see that 9 of the parts are shaded, so the numerator is 9. Thus, $\frac{9}{8}$ is shaded.

**15.** Each rectangle is divided into 6 equal parts. The unit is $\frac{1}{6}$. The denominator is 6. We see that 7 of the parts are shaded, so the numerator is 7. Thus, $\frac{7}{6}$ is shaded.

**17.** The triangle is divided into 4 equal parts. The unit is $\frac{1}{4}$. The denominator is 4. We have 3 parts shaded. This tells us that the numerator is 3. Thus, $\frac{3}{4}$ is shaded.

**19.** The rectangle is divided into 12 equal parts. The unit is $\frac{1}{12}$. The denominator is 12. All 12 parts are shaded. This tells us that the numerator is 12. Thus, $\frac{12}{12}$ is shaded.

**21.** The pie is divided into 8 equal parts. The unit is $\frac{1}{8}$. The denominator is 8. We have 4 parts shaded. This tells us that the numerator is 4. Thus, $\frac{4}{8}$ is shaded.

**23.** There are 8 circles, and 5 are shaded. Thus, $\frac{5}{8}$ of the circles are shaded.

**25.** There are 7 objects, and 4 are shaded. Thus, $\frac{4}{7}$ of the objects are shaded.

**27.** Each inch on the ruler is divided into 16 equal parts. The shading extends to the 12th mark, so $\frac{12}{16}$ is shaded.

**29.** Each inch on the ruler is divided into 16 equal parts. The shading extends to the 38th mark, so $\frac{38}{16}$ is shaded.

**31.** The gas gauge is divided into 8 equal parts.

  a) The needle is 2 marks from the E (empty) mark, so the amount of gas in the tank is $\frac{2}{8}$ of a full tank.

  b) The needle is 6 marks from the F (full) mark, so $\frac{6}{8}$ of a full tank of gas has been burned.

**33.** The gas gauge is divided into 8 equal parts.

a) The needle is 3 marks from the E (empty) mark, so the amount of gas in the tank is $\frac{3}{8}$ of a full tank.

b) The needle is 5 marks from the F (full) mark, so $\frac{5}{8}$ of a full tank of gas has been burned.

**35.** a) There are 8 people in the set and 5 are women, so the desired ratio is $\frac{5}{6}$.

b) There are 5 women and 3 men, so the ratio of women to men is $\frac{5}{3}$.

c) There are 8 people in the set and 3 are men, so the desired ratio is $\frac{3}{8}$.

d) There are 3 men and 5 women, so the ratio of men to women is $\frac{3}{5}$.

**37.** a) In Minnesota there are 1014 registered nurses per 100,000 residents, so the ratio is $\frac{1014}{100,000}$.

b) In Hawaii there are 750 registered nurses per 100,000 residents, so the ratio is $\frac{750}{100,000}$.

c) In Florida there are 812 registered nurses per 100,000 residents, so the ratio is $\frac{812}{100,000}$.

d) In Kentucky there are 922 registered nurses per 100,000 residents, so the ratio is $\frac{922}{100,000}$.

e) In New York there are 865 registered nurses per 100,000 residents, so the ratio is $\frac{865}{100,000}$.

f) In the District of Columbia there are 1379 registered nurses per 100,000 residents, so the ratio is $\frac{1379}{100,000}$.

**39.** a) The ratio is $\frac{4}{15}$.

b) The number of orders not delivered is $15 - 4$, or 11. The ratio is $\frac{4}{11}$.

c) From part (b) we know that the number of orders not delivered is 11. The ratio is $\frac{11}{15}$.

**41.** The ratio of voters in the 18-24 age group in the 2004 presidential election to all voters in the election is $\frac{16}{100}$.

The ratio of voters in the 18-24 age group in the 2008 presidential election to all voters in the election is $\frac{18.5}{100}$.

**43.** Remember: $\frac{0}{n} = 0$, for any whole number $n$ that is not 0.

$$\frac{0}{8} = 0$$

Think of dividing an object into 8 parts and taking none of them. We get 0.

**45.** $\frac{8-1}{9-8} = \frac{7}{1}$     Remember: $\frac{n}{1} = n$.

$$\frac{7}{1} = 7$$

Think of taking 7 objects and dividing each into 1 part. (We do not divide them.) We have 7 objects.

**47.** Remember: $\frac{n}{n} = 1$, for any integer $n$ that is not 0.

$$\frac{-20}{-20} = 1$$

**49.** Remember: $\frac{n}{n} = 1$, for any integer $n$ that is not 0.

$$\frac{45}{45} = 1$$

If we divide an object into 45 parts and take 45 of them, we get all of the object (1 whole object).

**51.** Remember: $\frac{0}{n} = 0$, for any integer $n$ that is not 0.

$$\frac{0}{238} = 0$$

Think of dividing an object into 238 parts and taking none of them. We get 0.

**53.** Remember: $\frac{n}{n} = 1$, for any integer $n$ that is not 0.

$$\frac{238}{238} = 1$$

If we divide an object into 238 parts and take 238 of them, we get all of the object (1 whole object).

**55.** Remember: $\frac{n}{n} = 1$, for any integer $n$ that is not 0.

$$\frac{3}{3} = 1$$

If we divide an object into 3 parts and take 3 of them, we get all of the object (1 whole object).

**57.** Remember: $\frac{0}{n} = 0$, for any integer $n$ that is not 0.

$$\frac{0}{-87} = 0$$

**59.** Remember: $\frac{n}{n} = 1$, for any integer $n$ that is not 0.

$$\frac{18}{18} = 1$$

**61.** Remember: $\frac{n}{1} = n$

$$\frac{-18}{1} = -18$$

**63.** Remember: $\frac{n}{0}$ is not defined for any integer $n$.

$\frac{729}{0}$ is not defined.

**65.** $\frac{5}{6-6} = \frac{5}{0}$

Remember: $\frac{n}{0}$ is not defined for any integer $n$. Thus, $\frac{5}{6-6}$ is not defined.

**67.** Round 3 4,5 6 $\boxed{2}$ to the nearest ten.

The digit 6 is in the tens place. Consider the next digit to the right. Since the digit, 2, is 4 or lower, round down, meaning that 6 tens stays as 6 tens. Then change the digit to the right of the tens digit to zero.

The answer is 34,560.

**69.** Round 3 4, $\boxed{5}$ 6 2 to the nearest thousand.

The digit 4 is in the thousands place. Consider the next digit to the right. Since the digit, 5, is 5 or higher, round 4 thousands up to 5 thousands. Then change all digits to the right of the thousands digit to zeros.

The answer is 35,000.

**71.** *Familiarize*. Let $m =$ the number by which the membership in 2008 exceeded the membership in 1983. We visualize the situation.

| Members in 1983 | Additional members in 2008 |
|---|---|
| 22,190 | $m$ |
| Members in 2008 | |
| 46,655 | |

*Translate*.

Members in 1983 $+$ Additional members is Members in 2008

$$22{,}190 \quad + \quad m \quad = \quad 46{,}655$$

*Solve*. We subtract 22,190 on both sides of the equation.

$$22{,}190 + m = 46{,}655$$
$$22{,}190 + m - 22{,}190 = 46{,}655 - 22{,}190$$
$$m = 24{,}465$$

*Check*. We can add the difference, 24,465, to the number of members in 1983, 22,190: $22{,}190 + 24{,}465 = 46{,}655$. We get the number of members in 2008, so the answer checks.

*State*. In 2008 the New York Road Runners had 24,465 more members than in 1983.

**73.**
$$\begin{array}{r} {\scriptstyle 3\ 9\ \overset{12}{\cancel{2}}\ 15} \\ \cancel{4\ 0\ \cancel{3}\ \cancel{5}} \\ -\ 2\ 3\ 6\ 9 \\ \hline 1\ 6\ 6\ 6 \end{array}$$

**75.** $7 + (-14)$ The absolute values are 7 and 14. The difference is $14 - 7$, or 7. The negative number has the larger absolute value, so the answer is negative. $7 + (-14) = -7$

**77.** We can think of the object as being divided into 6 sections, each the size of the area shaded. Thus, $\frac{1}{6}$ of the object is shaded.

**79.** We can think of the object as being divided into 16 sections, each the size of one of the shaded sections. Since 2 sections are shaded, $\frac{2}{16}$ of the object is shaded. We could also express this as $\frac{1}{8}$.

**81.** The set contains 5 objects, so we shade 3 of them.

**83.** The figure has 5 rows, so we shade 3 of them.

## Exercise Set 3.4

**1.** $\dfrac{2}{5} \cdot \dfrac{2}{3} = \dfrac{2 \cdot 2}{5 \cdot 3} = \dfrac{4}{15}$

**3.** $10 \cdot \dfrac{7}{9} = \dfrac{10 \cdot 7}{9} = \dfrac{70}{9}$

**5.** $-\dfrac{2}{3} \times \dfrac{1}{5} = -\dfrac{2 \times 1}{3 \times 5} = -\dfrac{2}{15}$

**7.** $-\dfrac{2}{11} \cdot 4 = -\dfrac{2 \cdot 4}{11} = -\dfrac{8}{11}$

**9.** $\dfrac{7}{8} \cdot \dfrac{7}{8} = \dfrac{7 \cdot 7}{8 \cdot 8} = \dfrac{49}{64}$

**11.** $5 \times \dfrac{1}{8} = \dfrac{5 \times 1}{8} = \dfrac{5}{8}$

**13.** $-\dfrac{2}{3} \cdot \dfrac{7}{13} = -\dfrac{2 \cdot 7}{3 \cdot 13} = -\dfrac{14}{39}$

**15.** $\dfrac{1}{2} \cdot \dfrac{1}{3} = \dfrac{1 \cdot 1}{2 \cdot 3} = \dfrac{1}{6}$

**17.** $\dfrac{2}{5} \cdot (-3) = -\dfrac{2 \cdot 3}{5} = -\dfrac{6}{5}$

**19.** $3 \cdot \dfrac{1}{5} = \dfrac{3 \cdot 1}{5} = \dfrac{3}{5}$

**21.** $7 \cdot \dfrac{3}{4} = \dfrac{7 \cdot 3}{4} = \dfrac{21}{4}$

**23.** $\dfrac{3}{4} \cdot \dfrac{3}{4} = \dfrac{3 \cdot 3}{4 \cdot 4} = \dfrac{9}{16}$

**25.** $\dfrac{1}{10} \cdot \left(-\dfrac{1}{100}\right) = -\dfrac{1 \cdot 1}{10 \cdot 100} = -\dfrac{1}{1000}$

**27.** $-\dfrac{2}{5} \cdot (-1) = \dfrac{2 \cdot 1}{5} = \dfrac{2}{5}$

**29.** $\dfrac{6}{5} \cdot \dfrac{6}{5} = \dfrac{6 \cdot 6}{5 \cdot 5} = \dfrac{36}{25}$

**31.** $\dfrac{2}{11} \cdot 4 = \dfrac{2 \cdot 4}{11} = \dfrac{8}{11}$

**33.** $\dfrac{14}{15} \cdot \dfrac{13}{19} = \dfrac{14 \cdot 13}{15 \cdot 19} = \dfrac{182}{285}$

**35. Familiarize.** Let $r$ = the number of yards of ribbon needed to make 8 bows.

**Translate.**

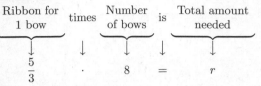

$$\frac{5}{3} \cdot 8 = r$$

**Solve.** We multiply.

$$\frac{5}{3} \cdot 8 = \frac{5 \cdot 8}{3} = \frac{40}{3}$$

Thus, $\frac{40}{3} = r$.

**Check.** We repeat the calculation. The answer checks.

**State.** $\frac{40}{3}$ yd of ribbon is needed to make 8 bows.

**37. Familiarize.** Let $b$ = the fractional part of high school basketball players who play professional basketball.

**Translate.** We are finding $\frac{1}{75}$ of $\frac{1}{35}$, so the multiplication sentence $\frac{1}{75} \cdot \frac{1}{35} = b$ corresponds to this situation.

**Solve.** We multiply.

$$\frac{1}{75} \cdot \frac{1}{35} = \frac{1 \cdot 1}{75 \cdot 35} = \frac{1}{2625}$$

Thus, $\frac{1}{2625} = b$.

**Check.** We repeat the calculation. The answer checks.

**State.** $\frac{1}{2625}$ of high school basketball players play professional basketball.

**39. Familiarize.** Let $c$ = the portion of a cheesecake that corresponds to $\frac{1}{2}$ piece.

**Translate.** We are finding $\frac{1}{2}$ of $\frac{1}{12}$, so the multiplication sentence $\frac{1}{2} \cdot \frac{1}{12} = c$ corresponds to this situation.

**Solve.** We multiply.

$$\frac{1}{2} \cdot \frac{1}{12} = \frac{1 \cdot 1}{2 \cdot 12} = \frac{1}{24}$$

Thus, $\frac{1}{24} = c$.

**Check.** We repeat the calculation. The answer checks.

**State.** $\frac{1}{2}$ piece is $\frac{1}{24}$ of a cheesecake.

**41. Familiarize.** Let $f$ = the fraction of the floor that has been tiled.

**Translate.** We can multiply the fraction of the length that has been covered by the fraction of the width that has been covered to find the fraction of the area of the floor that has been tiled. We have $f = \frac{3}{5} \cdot \frac{3}{4}$.

**Solve.** We multiply.

$$f = \frac{3}{5} \cdot \frac{3}{4} = \frac{3 \cdot 3}{5 \cdot 4} = \frac{9}{20}$$

**Check.** We repeat the calculation. The answer checks.

**State.** $\frac{9}{20}$ of the floor has been tiled.

**43.**

$$9 \overline{)\begin{array}{l} 3\ 0\ 0\ 1 \\ 2\ 7,0\ 0\ 9 \end{array}}$$
$$\underline{2\ 7\ 0\ 0\ 0}$$
$$\phantom{2\ 7\ 0\ 0}9$$
$$\phantom{2\ 7\ 0\ 0}\underline{9}$$
$$\phantom{2\ 7\ 0\ 0}0$$

The answer is 3001.

**45.** $7140 \div (-35) = -204$     Check: $-204 \cdot (-35) = 7140$

**47.** $4,\ 6\ 7\ \boxed{8},\ 9\ 5\ 2$

The digit 8 means 8 thousands.

**49.** $7\ 1\ 4\ \boxed{8}$

The digit 8 means 8 ones.

**51.** $17 - 4^2 = 17 - 16$     Evaluating the exponential
                                                    expression
$\phantom{17 - 4^2} = 1$          Subtracting

**53.** $\phantom{=} -144 \div 12 \div (-4)$
$= -12 \div (-4)$          Dividing in order from
$= 3$                              left to right

**55.** Use a calculator.
$$\frac{341}{517} \cdot \frac{209}{349} = \frac{341 \cdot 209}{517 \cdot 349} = \frac{71,269}{180,433}$$

**57.** $\left(\frac{2}{5}\right)^3 \left(-\frac{7}{9}\right) = \frac{8}{125}\left(-\frac{7}{9}\right)$     Evaluating the exponential
                                                                                                            expression
$$= -\frac{8 \cdot 7}{125 \cdot 9}$$
$$= -\frac{56}{1125}$$

## Exercise Set 3.5

**1.** Since $2 \cdot 5 = 10$, we multiply by $\frac{5}{5}$.

$$\frac{1}{2} = \frac{1}{2} \cdot \frac{5}{5} = \frac{1 \cdot 5}{2 \cdot 5} = \frac{5}{10}$$

**3.** Since $8 \cdot 4 = 32$, we multiply by $\frac{4}{4}$.

$$\frac{5}{8} = \frac{5}{8} \cdot \frac{4}{4} = \frac{5 \cdot 4}{8 \cdot 4} = \frac{20}{32}$$

**5.** Since $10 \cdot 3 = 30$, we multiply by $\frac{3}{3}$.

$$-\frac{9}{10} = -\frac{9}{10} \cdot \frac{3}{3} = -\frac{9 \cdot 3}{10 \cdot 3} = -\frac{27}{30}$$

**7.** Since $8 \cdot 4 = 32$, we multiply by $\frac{4}{4}$.

$$\frac{7}{8} = \frac{7}{8} \cdot \frac{4}{4} = \frac{7 \cdot 4}{8 \cdot 4} = \frac{28}{32}$$

**9.** Since $12 \cdot 4 = 48$, we multiply by $\frac{4}{4}$.

$$\frac{5}{12} = \frac{5}{12} \cdot \frac{4}{4} = \frac{5 \cdot 4}{12 \cdot 4} = \frac{20}{48}$$

**11.** Since $18 \cdot (-3) = -54$, we multiply by $\dfrac{-3}{-3}$.

$$\frac{17}{18} = \frac{17}{18} \cdot \left(\frac{-3}{-3}\right) = \frac{17 \cdot (-3)}{18 \cdot (-3)} = \frac{-51}{-54}$$

**13.** Since $3 \cdot 15 = 45$, we multiply by $\dfrac{15}{15}$.

$$\frac{5}{3} = \frac{5}{3} \cdot \frac{15}{15} = \frac{5 \cdot 15}{3 \cdot 15} = \frac{75}{45}$$

**15.** Since $22 \cdot 6 = 132$, we multiply by $\dfrac{6}{6}$.

$$\frac{7}{22} = \frac{7}{22} \cdot \frac{6}{6} = \frac{7 \cdot 6}{22 \cdot 6} = \frac{42}{132}$$

**17.**
$$\frac{2}{4} = \frac{1 \cdot 2}{2 \cdot 2} \quad \longleftarrow \text{Factor the numerator}$$
$$\phantom{\frac{2}{4}} \quad \longleftarrow \text{Factor the denominator}$$
$$= \frac{1}{2} \cdot \frac{2}{2} \quad \longleftarrow \text{Factor the fraction}$$
$$= \frac{1}{2} \cdot 1 \quad \longleftarrow \frac{2}{2} = 1$$
$$= \frac{1}{2} \quad \longleftarrow \text{Removing a factor of 1}$$

**19.**
$$-\frac{6}{8} = -\frac{3 \cdot 2}{4 \cdot 2} \quad \longleftarrow \text{Factor the numerator}$$
$$\phantom{-\frac{6}{8}} \quad \longleftarrow \text{Factor the denominator}$$
$$= -\frac{3}{4} \cdot \frac{2}{2} \quad \longleftarrow \text{Factor the fraction}$$
$$= -\frac{3}{4} \cdot 1 \quad \longleftarrow \frac{2}{2} = 1$$
$$= -\frac{3}{4} \quad \longleftarrow \text{Removing a factor of 1}$$

**21.**
$$\frac{2}{15} = \frac{1 \cdot 3}{5 \cdot 3} \quad \longleftarrow \text{Factor the numerator}$$
$$\phantom{\frac{2}{15}} \quad \longleftarrow \text{Factor the denominator}$$
$$= \frac{1}{5} \cdot \frac{3}{3} \quad \longleftarrow \text{Factor the fraction}$$
$$= \frac{1}{5} \cdot 1 \quad \longleftarrow \frac{3}{3} = 1$$
$$= \frac{1}{5} \quad \longleftarrow \text{Removing a factor of 1}$$

**23.** $\dfrac{-24}{8} = \dfrac{-3 \cdot 8}{1 \cdot 8} = \dfrac{-3}{1} \cdot \dfrac{8}{8} = \dfrac{-3}{1} \cdot 1 = \dfrac{-3}{1} = -3$

**25.** $\dfrac{18}{24} = \dfrac{3 \cdot 6}{4 \cdot 6} = \dfrac{3}{4} \cdot \dfrac{6}{6} = \dfrac{3}{4} \cdot 1 = \dfrac{3}{4}$

**27.** $\dfrac{14}{16} = \dfrac{7 \cdot 2}{8 \cdot 2} = \dfrac{7}{8} \cdot \dfrac{2}{2} = \dfrac{7}{8} \cdot 1 = \dfrac{7}{8}$

**29.** $\dfrac{12}{10} = \dfrac{6 \cdot 2}{5 \cdot 2} = \dfrac{6}{5} \cdot \dfrac{2}{2} = \dfrac{6}{5} \cdot 1 = \dfrac{6}{5}$

**31.** $\dfrac{16}{48} = \dfrac{1 \cdot 16}{3 \cdot 16} = \dfrac{1}{3} \cdot \dfrac{16}{16} = \dfrac{1}{3} \cdot 1 = \dfrac{1}{3}$

**33.** $\dfrac{150}{-25} = \dfrac{6 \cdot 25}{-1 \cdot 25} = \dfrac{6}{-1} \cdot \dfrac{25}{25} = \dfrac{6}{-1} \cdot 1 = \dfrac{6}{-1} = -6$

We could also simplify $\dfrac{150}{-25}$ by doing the division

$150 \div (-25)$. That is, $\dfrac{150}{-25} = 150 \div (-25) = -6$.

**35.** $-\dfrac{17}{51} = -\dfrac{1 \cdot 17}{3 \cdot 17} = -\dfrac{1}{3} \cdot \dfrac{17}{17} = -\dfrac{1}{3} \cdot 1 = -\dfrac{1}{3}$

**37.** We use the tests for divisibility to factor the numerator and the denominator.

$$\frac{390}{1410}$$
$$= \frac{3 \cdot 10 \cdot 13}{3 \cdot 10 \cdot 47} \quad \text{390 and 140 are divisible by 3 and by 10}$$
$$= \frac{3 \cdot 10}{3 \cdot 10} \cdot \frac{13}{47}$$
$$= \frac{13}{47}$$

**39.** We use the tests for divisibility to factor the numerator and the denominator.

$$-\frac{1080}{2688}$$
$$= -\frac{3 \cdot 8 \cdot 45}{3 \cdot 8 \cdot 112} \quad \text{1080 and 2688 are divisible by 3 and by 8}$$
$$= \frac{3 \cdot 8}{3 \cdot 8} \cdot \left(-\frac{45}{112}\right)$$
$$= -\frac{45}{112}$$

**41.** We multiply these two numbers: We multiply these two numbers:

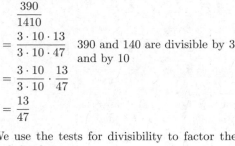

$3 \cdot 12 = 36 \qquad 4 \cdot 9 = 36$

Since $36 = 36$, $\dfrac{3}{4} = \dfrac{9}{12}$.

**43.** We multiply these two numbers: We multiply these two numbers:

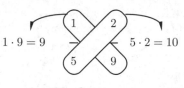

$1 \cdot 9 = 9 \qquad 5 \cdot 2 = 10$

Since $9 \neq 10$, $\dfrac{1}{5} \neq \dfrac{2}{9}$.

**45.** We multiply these two numbers: We multiply these two numbers:

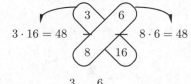

$3 \cdot 16 = 48 \qquad 8 \cdot 6 = 48$

Since $48 = 48$, $\dfrac{3}{8} = \dfrac{6}{16}$.

**47.** We multiply these     We multiply these
two numbers:               two numbers:

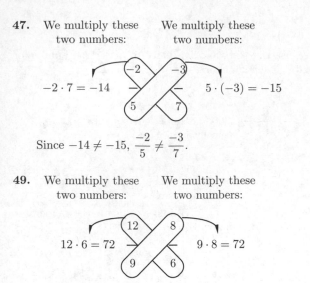

$-2 \cdot 7 = -14$          $5 \cdot (-3) = -15$

Since $-14 \neq -15$, $\dfrac{-2}{5} \neq \dfrac{-3}{7}$.

**49.** We multiply these     We multiply these
two numbers:               two numbers:

$12 \cdot 6 = 72$          $9 \cdot 8 = 72$

Since $72 = 72$, $\dfrac{12}{9} = \dfrac{8}{6}$.

**51.** First we write $-\dfrac{5}{2}$ as $\dfrac{-5}{2}$.

We multiply these     We multiply these
two numbers:               two numbers:

$-5 \cdot 7 = -35$          $2 \cdot (-17) = -34$

Since $-35 \neq -34$, $\dfrac{-5}{2} \neq \dfrac{-17}{7}$, or $-\dfrac{5}{2} \neq \dfrac{-17}{2}$.

**53.** First we write $-\dfrac{30}{100}$ as $\dfrac{30}{-100}$.

We multiply these     We multiply these
two numbers:               two numbers:

$3 \cdot (-100) = -300$          $-10 \cdot 30 = -300$

Since $-300 = -300$, $\dfrac{3}{-10} = \dfrac{30}{-100}$, or $\dfrac{3}{-10} = -\dfrac{30}{100}$.

**55.** We multiply these          We multiply these
two numbers:                    two numbers:

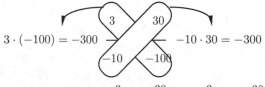

$5 \cdot 1000 = 5000$          $10 \cdot 520 = 5200$

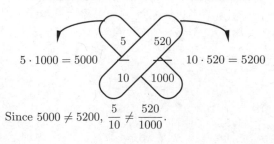

Since $5000 \neq 5200$, $\dfrac{5}{10} \neq \dfrac{520}{1000}$.

**57.** $\dfrac{28}{100} = \dfrac{2 \cdot 2 \cdot 7}{2 \cdot 2 \cdot 5 \cdot 5} = \dfrac{2 \cdot 2}{2 \cdot 2} \cdot \dfrac{7}{5 \cdot 5} = \dfrac{7}{25}$

**59.** $\dfrac{25}{100} = \dfrac{1 \cdot 25}{4 \cdot 25} = \dfrac{1}{4} \cdot \dfrac{25}{25} = \dfrac{1}{4}$

**61.** *Familiarize*. We make a drawing. We let $A =$ the area.

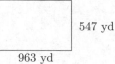
547 yd

963 yd

*Translate*. Using the formula for area, we have
$$A = l \cdot w = 963 \cdot 547.$$

Using the formula for perimeter, we have
$$P = 2l + 2w = 2 \cdot 963 + 2 \cdot 547$$

*Solve*. We carry out the computations.

$$
\begin{array}{r}
9\,6\,3 \\
\times\,5\,4\,7 \\
\hline
6\,7\,4\,1 \\
3\,8\,5\,2\,0\phantom{0} \\
4\,8\,1\,5\,0\,0\phantom{00} \\
\hline
5\,2\,6,7\,6\,1
\end{array}
$$

Thus, $A = 526,761$.

$P = 2 \cdot 963 + 2 \cdot 547 = 1926 + 1094 = 3020$

*Check*. We repeat the calculations. The answers check.

*State*. The area is 526,761 yd². The perimeter is 3020 yd.

**63.**
$$
\begin{array}{r}
{\scriptstyle 7\,\,9\,\,13} \\
\cancel{8\,0\,3} \\
-\,6\,1\,7 \\
\hline
1\,8\,6
\end{array}
$$

**65.**      $5280 = 1760 + t$

$5280 - 1760 = 1760 + t - 1760$   Subtracting 1760
on both sides

$3520 = t$

The solution is 3520.

**67.** $\dfrac{63}{82} \neq \dfrac{77}{100}$, because $63 \cdot 100 \neq 82 \cdot 77$. Thus, the student did not get the same portion of each test correct.

## Chapter 3 Mid-Chapter Review

**1.** The statement is true. See page 129 in the text.

**2.** The statement is false. For example 15 is not divisible by 6 but it is divisible by 3.

**3.**   We multiply these          We multiply these
two numbers:                    two numbers:

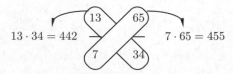

$13 \cdot 34 = 442$          $7 \cdot 65 = 455$

Since $442 \neq 455$, $\dfrac{13}{7} \neq \dfrac{65}{34}$. The given statement is false.

**4.** The statement is true. See page 130 in the text.

**5.** $\frac{n}{n} = 1$ for any whole number $n$ that is not 0, so $\frac{25}{25} = 1$.

**6.** $\frac{0}{n} = 0$ for any whole number $n$ that is not 0, so $\frac{0}{9} = 0$.

**7.** $\frac{n}{1} = n$ for any whole number $n$, so $\frac{-8}{1} = -8$.

**8.** $\frac{6}{13} = \frac{6}{13} \cdot \frac{3}{3} = \frac{18}{39}$

**9.** $\frac{70}{225} = \frac{2 \cdot 5 \cdot 7}{3 \cdot 3 \cdot 5 \cdot 5}$

$= \frac{5}{5} \cdot \frac{2 \cdot 7}{3 \cdot 3 \cdot 5}$

$= 1 \cdot \frac{14}{45}$

$= \frac{14}{45}$

**10.** A number is divisible by 2 if it has an even ones digit. A number is divisible by 10 if its ones digit is 0. Thus, we find the numbers with a ones digit of 2, 4, 6, or 8. Those numbers are 84, 17,576, 224, 132, 594, 504, and 1632. These are the numbers divisible by 2 but not by 10.

**11.** A number is divisible by 4 if the number named by its last two digits is divisible by 4. A number is divisible by 8 if the number named by its last three digits is divisible by 8. The numbers that are divisible by 8 are 17,576, 224, 120, 504, and 1632. Of the remaining numbers, those divisible by 4 are 84, 300, 132, 500, and 180. These are the numbers that are divisible by 4 but not by 8.

**12.** A number is divisible by 4 if the number named by its last two digits is divisible by 4. A number is divisible by 6 if its ones digit is even and the sum of its digits is divisible by 3. The numbers divisible by 4 are 84, 300, 17,576, 224, 132, 500, 180, 120, 504, and 1632. Of these, the numbers for which the sum of the digits is not divisible by 3 are 17,576, 224, and 500. These the numbers that are divisible by 4 but not by 6.

**13.** A number is divisible by 3 if the sum of its digits is divisible by 3. A number is divisible by 9 if the sum of its digits is divisible by 9. The numbers divisible by 3 are 84, 300, 132, 180, 351, 594, 120, 1125, 495, 14,850, 504, and 1632. Of these, the numbers that are not divisible by 9 are 84, 300, 132, 120, and 1632. These are the numbers that are divisible by 3 but not by 9.

**14.** A number is divisible by 4 if the number named by its last two digits is divisible by 4. A number is divisible by 5 if its ones digit is 0 or 5. A number is divisible by 6 if its ones digit is even and the sum of its digits is divisible by 3. From Exercise 12, we know that the numbers that are divisible by 4 are 84, 300, 17,576, 224, 132, 500, 180, 120, 504, and 1632. Of these, 300, 500, 180, and 120 have a ones digit of 0 or 5. Of these, 300, 180, and 120 are divisible by 6. These are the numbers that are divisible by 4, 5, and 6.

**15.** The number 61 is prime. It has only the factors 1 and 61.

**16.** The number 2 is prime. It has only the factors 1 and 2.

**17.** The number 91 has factors 1, 7, 13, and 91. Since 91 is not 1 and not prime, it is composite.

**18.** The number 1 is neither prime nor composite.

**19.** To find all the factors of 160, we find all the two-factor factorizations.

$160 = 1 \cdot 160 \qquad 160 = 5 \cdot 32$

$160 = 2 \cdot 80 \qquad 160 = 8 \cdot 20$

$160 = 4 \cdot 40 \qquad 160 = 10 \cdot 16$

Factors: 1, 2, 4, 5, 8, 10, 16, 20, 32, 40, 80, 160

Now we find the prime factorization of 160.

```
        5   ← 5 is prime
    2 ⌐1 0
    2 ⌐2 0
    2 ⌐4 0
    2 ⌐8 0
    2 ⌐1 6 0
```

$160 = 2 \cdot 2 \cdot 2 \cdot 2 \cdot 2 \cdot 5$

**20.** To find all the factors of 222, we find all the two-factor factorizations.

$222 = 1 \cdot 222 \qquad 222 = 3 \cdot 74$

$222 = 2 \cdot 111 \qquad 222 = 6 \cdot 37$

Factors: 1, 2, 3, 6, 37, 74, 111, 222

Now we find the prime factorization.

```
        3 7   ← 37 is prime
    3 ⌐1 1 1
    2 ⌐2 2 2
```

$222 = 2 \cdot 3 \cdot 37$

**21.** To find all the factors of 98, we find all the two-factor factorizations.

$98 = 1 \cdot 98 \qquad 98 = 7 \cdot 14$

$98 = 2 \cdot 49$

Factors: 1, 2, 7, 14, 49, 98

Now we find the prime factorization.

```
        7   ← 7 is prime
    7 ⌐4 9
    2 ⌐9 8
```

$98 = 2 \cdot 7 \cdot 7$

**22.** To find all the factorizations of 315, we find all the two-factor factorizations.

$315 = 1 \cdot 315 \qquad 315 = 7 \cdot 45$

$315 = 3 \cdot 105 \qquad 315 = 9 \cdot 35$

$315 = 5 \cdot 63 \qquad 315 = 15 \cdot 21$

Factors: 1, 3, 5, 7, 9, 15, 21, 35, 45, 63, 105, 315

Now we find the prime factorization.

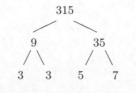

$315 = 3 \cdot 3 \cdot 5 \cdot 7$

23. The rectangle is divided into 24 equal parts. The unit is $\frac{1}{24}$. The denominator is 24. We have 8 parts shaded. This tells us that the numerator is 8. Thus, $\frac{8}{24}$ is shaded. We can simplify $\frac{8}{24}$:

$$\frac{8}{24} = \frac{8 \cdot 1}{8 \cdot 3} = \frac{8}{8} \cdot \frac{1}{3} = \frac{1}{3}.$$

24. Each rectangle is divided into 6 equal parts. The unit is $\frac{1}{6}$. The denominator is 6. We see that 8 of the parts are shaded, so the numerator is 8. Thus, $\frac{8}{6}$ is shaded. We can simplify $\frac{8}{6}$:

$$\frac{8}{6} = \frac{2 \cdot 4}{2 \cdot 3} = \frac{2}{2} \cdot \frac{4}{3} = \frac{4}{3}$$

25. $7 \cdot \frac{1}{9} = \frac{7 \cdot 1}{9} = \frac{7}{9}$

26. $\frac{4}{15} \cdot \frac{2}{3} = \frac{4 \cdot 2}{15 \cdot 3} = \frac{8}{45}$

27. $\frac{5}{11} \cdot (-8) = -\frac{5 \cdot 8}{11} = -\frac{40}{11}$

28. $-\frac{2}{9} \cdot \frac{4}{5} = -\frac{2 \cdot 4}{9 \cdot 5} = -\frac{8}{45}$

29. $-\frac{3}{4} \cdot \left(-\frac{7}{8}\right) = \frac{3 \cdot 7}{4 \cdot 8} = \frac{21}{32}$

30. $-5 \cdot \frac{2}{3} = -\frac{5 \cdot 2}{3} = -\frac{10}{3}$

31. $\frac{1}{8} \cdot \frac{5}{2} = \frac{1 \cdot 5}{8 \cdot 2} = \frac{5}{16}$

32. $\frac{3}{7} \cdot \left(-\frac{9}{4}\right) = -\frac{3 \cdot 9}{7 \cdot 4} = -\frac{27}{28}$

33. $\frac{24}{60} = \frac{2 \cdot 2 \cdot 2 \cdot 3}{2 \cdot 2 \cdot 3 \cdot 5} = \frac{2 \cdot 2 \cdot 3}{2 \cdot 2 \cdot 3} \cdot \frac{2}{5} = \frac{2}{5}$

34. $\frac{220}{60} = \frac{2 \cdot 10 \cdot 11}{2 \cdot 3 \cdot 10} = \frac{2 \cdot 10}{2 \cdot 10} \cdot \frac{11}{3} = \frac{11}{3}$

35. $\frac{17}{17} = 1$

36. $\frac{0}{23} = 0$

37. $\frac{315}{435} = \frac{3 \cdot 3 \cdot 5 \cdot 7}{3 \cdot 5 \cdot 29} = \frac{3 \cdot 5}{3 \cdot 5} \cdot \frac{3 \cdot 7}{29} = \frac{21}{29}$

38. $\frac{14}{0}$ is not defined.

39.    We multiply these       We multiply these
         two numbers:            two numbers:

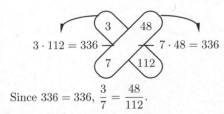

$3 \cdot 112 = 336$          $7 \cdot 48 = 336$

Since $336 = 336$, $\frac{3}{7} = \frac{48}{112}$.

40.    We multiply these       We multiply these
         two numbers:            two numbers:

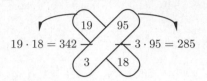

$19 \cdot 18 = 342$          $3 \cdot 95 = 285$

Since $342 \ne 285$, $\frac{19}{3} \ne \frac{95}{18}$.

41. The ratio is $\frac{60}{500}$. This can be simplified:

$$\frac{60}{500} = \frac{2 \cdot 3 \cdot 10}{2 \cdot 5 \cdot 5 \cdot 10} = \frac{2 \cdot 10}{2 \cdot 10} \cdot \frac{3}{5 \cdot 5} = \frac{3}{25}.$$

42. **Familiarize**. We will use the formula $A = l \times w$.

   **Translate**. We substitute in the formula.
$$A = l \times w = \frac{7}{100} \times \frac{3}{100}$$

   **Solve**. We carry out the multiplication.
$$A = \frac{7}{100} \times \frac{3}{100} = \frac{7 \times 3}{100 \times 100} = \frac{21}{10,000}.$$

   **Check**. We repeat the calculation. The answer checks.

   **State**. The area of the rink is $\frac{21}{10,000}$ m$^2$.

43. Find the product of two prime numbers.

44. Using the divisibility tests, it is quickly clear that none of the even-numbered years is a prime number. In addition, the divisibility tests for 5 and 3 show that 2000, 2001, 2005, 2007, 2013, 2015, and 2019 are not prime numbers. Then the years 2003, 2009, 2011, and 2017 can be divided by prime numbers to determine if they are prime. When we do this, we find that 2011 and 2017 are prime numbers.

   If the divisibility tests are not used, each of the numbers from 2000 to 2020 can be divided by prime numbers to determine if they are prime.

45. It is possible to cancel only when identical *factors* appear in the numerator and denominator of a fraction. Situations in which it is not possible to cancel include the occurrence of identical *terms* or *digits* in the numerator and denominator.

46. No; since the only factors of a prime number are the number itself and 1, two different prime numbers cannot contain a common factor (other than 1).

## Exercise Set 3.6

1. $\frac{2}{3} \cdot \frac{1}{2} = \frac{2 \cdot 1}{3 \cdot 2} = \frac{2}{2} \cdot \frac{1}{3} = 1 \cdot \frac{1}{3} = \frac{1}{3}$

3. $\frac{7}{8} \cdot \frac{1}{7} = \frac{7 \cdot 1}{8 \cdot 7} = \frac{7}{7} \cdot \frac{1}{8} = 1 \cdot \frac{1}{8} = \frac{1}{8}$

5. $-\frac{1}{8} \cdot \frac{4}{5} = -\frac{1 \cdot 4}{8 \cdot 5} = -\frac{1 \cdot 4}{2 \cdot 4 \cdot 5} = \frac{4}{4} \cdot \left(-\frac{1}{2 \cdot 5}\right) =$
$-\frac{1}{2 \cdot 5} = -\frac{1}{10}$

**7.** $\dfrac{1}{4} \cdot \dfrac{2}{3} = \dfrac{1 \cdot 2}{4 \cdot 3} = \dfrac{1 \cdot 2}{2 \cdot 2 \cdot 3} = \dfrac{2}{2} \cdot \dfrac{1}{2 \cdot 3} = \dfrac{1}{2 \cdot 3} = \dfrac{1}{6}$

**9.** $\dfrac{12}{5} \cdot \dfrac{9}{8} = \dfrac{12 \cdot 9}{5 \cdot 8} = \dfrac{4 \cdot 3 \cdot 9}{5 \cdot 2 \cdot 4} = \dfrac{4}{4} \cdot \dfrac{3 \cdot 9}{5 \cdot 2} = \dfrac{3 \cdot 9}{5 \cdot 2} = \dfrac{27}{10}$

**11.** $\dfrac{10}{9} \cdot \left(-\dfrac{7}{5}\right) = -\dfrac{10 \cdot 7}{9 \cdot 5} = -\dfrac{5 \cdot 2 \cdot 7}{9 \cdot 5} = \dfrac{5}{5} \cdot \left(-\dfrac{2 \cdot 7}{9}\right) =$
$-\dfrac{2 \cdot 7}{9} = -\dfrac{14}{9}$

**13.** $9 \cdot \dfrac{1}{9} = \dfrac{9 \cdot 1}{9} = \dfrac{9 \cdot 1}{9 \cdot 1} = 1$

**15.** $\dfrac{1}{3} \cdot 3 = \dfrac{1 \cdot 3}{3} = \dfrac{1 \cdot 3}{1 \cdot 3} = 1$

**17.** $\left(-\dfrac{7}{10}\right) \cdot \left(-\dfrac{10}{7}\right) = \dfrac{7 \cdot 10}{10 \cdot 7} = \dfrac{7 \cdot 10}{7 \cdot 10} = 1$

**19.** $\dfrac{7}{5} \cdot \dfrac{5}{7} = \dfrac{7 \cdot 5}{5 \cdot 7} = \dfrac{7 \cdot 5}{7 \cdot 5} = 1$

**21.** $\dfrac{1}{4} \cdot 8 = \dfrac{1 \cdot 8}{4} = \dfrac{8}{4} = \dfrac{4 \cdot 2}{4 \cdot 1} = \dfrac{4}{4} \cdot \dfrac{2}{1} = \dfrac{2}{1} = 2$

**23.** $-24 \cdot \dfrac{1}{6} = -\dfrac{24 \cdot 1}{6} = -\dfrac{24}{6} = -\dfrac{4 \cdot 6}{1 \cdot 6} = -\dfrac{4}{1} \cdot \dfrac{6}{6} = -\dfrac{4}{1} = -4$

**25.** $12 \cdot \dfrac{3}{4} = \dfrac{12 \cdot 3}{4} = \dfrac{4 \cdot 3 \cdot 3}{4 \cdot 1} = \dfrac{4}{4} \cdot \dfrac{3 \cdot 3}{1} = \dfrac{3 \cdot 3}{1} = 9$

**27.** $-\dfrac{3}{8} \cdot 24 = -\dfrac{3 \cdot 24}{8} = -\dfrac{3 \cdot 3 \cdot 8}{1 \cdot 8} = \dfrac{8}{8} \cdot \left(-\dfrac{3 \cdot 3}{1}\right) =$
$-\dfrac{3 \cdot 3}{1} = -9$

**29.** $-13 \cdot \left(-\dfrac{2}{5}\right) = \dfrac{13 \cdot 2}{5} = \dfrac{26}{5}$

**31.** $\dfrac{7}{10} \cdot 28 = \dfrac{7 \cdot 28}{10} = \dfrac{7 \cdot 2 \cdot 14}{2 \cdot 5} = \dfrac{2}{2} \cdot \dfrac{7 \cdot 14}{5} = \dfrac{7 \cdot 14}{5} = \dfrac{98}{5}$

**33.** $\dfrac{1}{6} \cdot 360 = \dfrac{1 \cdot 360}{6} = \dfrac{360}{6} = \dfrac{6 \cdot 60}{6 \cdot 1} = \dfrac{6}{6} \cdot \dfrac{60}{1} = \dfrac{60}{1} = 60$

**35.** $240 \cdot \left(-\dfrac{1}{8}\right) = -\dfrac{240 \cdot 1}{8} = -\dfrac{240}{8} = -\dfrac{8 \cdot 30}{8 \cdot 1} = \dfrac{8}{8} \cdot \left(-\dfrac{30}{1}\right) =$
$-\dfrac{30}{1} = -30$

**37.** $\dfrac{4}{10} \cdot \dfrac{5}{10} = \dfrac{4 \cdot 5}{10 \cdot 10} = \dfrac{2 \cdot 2 \cdot 5 \cdot 1}{2 \cdot 5 \cdot 2 \cdot 5} = \dfrac{2 \cdot 2 \cdot 5}{2 \cdot 2 \cdot 5} \cdot \dfrac{1}{5} = \dfrac{1}{5}$

**39.** $-\dfrac{8}{10} \cdot \dfrac{45}{100} = -\dfrac{8 \cdot 45}{10 \cdot 100} = -\dfrac{2 \cdot 2 \cdot 2 \cdot 5 \cdot 9}{2 \cdot 5 \cdot 2 \cdot 5 \cdot 2 \cdot 5}$
$= \dfrac{2 \cdot 2 \cdot 2 \cdot 5}{2 \cdot 2 \cdot 2 \cdot 5} \cdot \left(-\dfrac{9}{5 \cdot 5}\right) = -\dfrac{9}{5 \cdot 5} = -\dfrac{9}{25}$

**41.** $\dfrac{11}{24} \cdot \dfrac{3}{5} = \dfrac{11 \cdot 3}{24 \cdot 5} = \dfrac{11 \cdot 3}{3 \cdot 8 \cdot 5} = \dfrac{3}{3} \cdot \dfrac{11}{8 \cdot 5} = \dfrac{11}{8 \cdot 5} = \dfrac{11}{40}$

**43.** $-\dfrac{10}{21} \cdot \left(-\dfrac{3}{4}\right) = \dfrac{10 \cdot 3}{21 \cdot 4} = \dfrac{2 \cdot 5 \cdot 3}{3 \cdot 7 \cdot 2 \cdot 2}$
$= \dfrac{2 \cdot 3}{2 \cdot 3} \cdot \dfrac{5}{7 \cdot 2} = \dfrac{5}{7 \cdot 2} = \dfrac{5}{14}$

**45. Familiarize.** Let $n =$ the number of inches the screw will go into the piece of oak when it is turned 10 complete rotations.

**Translate.** We write an equation.

| Total distance | is | Distance for one revolution | times | Number of revolutions |
|---|---|---|---|---|
| ↓ | ↓ | ↓ | ↓ | ↓ |
| $n$ | $=$ | $\dfrac{1}{16}$ | $\cdot$ | 10 |

**Solve.** We carry out the multiplication.
$$n = \dfrac{1}{16} \cdot 10 = \dfrac{1 \cdot 10}{16}$$
$$= \dfrac{1 \cdot 2 \cdot 5}{2 \cdot 8} = \dfrac{2}{2} \cdot \dfrac{1 \cdot 5}{8}$$
$$= \dfrac{5}{8}$$

**Check.** We can repeat the calculation. We can also determine that the answer seems reasonable since we multiplied 10 by a number less than 10 and the result is less than 10. The answer checks.

**State.** The screw will go $\dfrac{5}{8}$ in. into the piece of oak when it is turned 10 completed rotations.

**47. Familiarize.** Let $a =$ the median income of people with associate's degrees.

**Translate.**

| Median income with associate's degree | is | $\dfrac{2}{3}$ | of | Median income with bachelor's degree |
|---|---|---|---|---|
| ↓ | ↓ | ↓ | ↓ | ↓ |
| $a$ | $=$ | $\dfrac{2}{3}$ | $\cdot$ | 72,420 |

**Solve.** We carry out the multiplication.
$$a = \dfrac{2}{3} \cdot 72,420 = \dfrac{2 \cdot 72,420}{3} = \dfrac{2 \cdot 3 \cdot 24,140}{3 \cdot 1}$$
$$= \dfrac{3}{3} \cdot \dfrac{2 \cdot 24,140}{1} = 48,280$$

**Check.** We can repeat the calculation. We can also observe that the answer seems reasonable since we multiplied 72,420 by a number less than 1 and the result is less than 72,420. The answer checks.

**State.** The median income of people with associate's degrees is $48,280.

**49. Familiarize.** We visualize the situation. We let $n =$ the number of addresses that will be incorrect after one year.

| Mailing list 2500 addresses | | | |
|---|---|---|---|
| 1/4 of the addresses $n$ | | | |

*Translate*.

$$\underbrace{\text{Number incorrect}}_{\downarrow} \;\; \underset{\downarrow}{\text{is}} \;\; \underset{\downarrow}{\frac{1}{4}} \;\; \underset{\downarrow}{\text{of}} \;\; \underbrace{\text{Number of addresses}}_{\downarrow}$$

$$n \qquad\qquad = \qquad \frac{1}{4} \;\; \cdot \qquad\qquad 2500$$

*Solve*. We carry out the multiplication.

$$n = \frac{1}{4} \cdot 2500 = \frac{1 \cdot 2500}{4} = \frac{2500}{4}$$

$$= \frac{4 \cdot 625}{4 \cdot 1} = \frac{4}{4} \cdot \frac{625}{1}$$

$$= 625$$

*Check*. We can repeat the calculation. We can also determine that the answer seems reasonable since we multiplied 2500 by a number less than 1 and the result is less than 2500. The answer checks.

*State*. After one year 625 addresses will be incorrect.

**51.** *Familiarize*. We draw a picture.

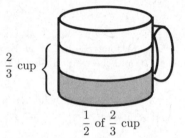

$\frac{2}{3}$ cup

$\frac{1}{2}$ of $\frac{2}{3}$ cup

We let $n =$ the amount of flour the chef should use.

*Translate*. The multiplication sentence

$$\frac{1}{2} \cdot \frac{2}{3} = n$$

corresponds to the situation.

*Solve*. We multiply and simplify:

$$n = \frac{1}{2} \cdot \frac{2}{3} = \frac{1 \cdot 2}{2 \cdot 3} = \frac{2}{2} \cdot \frac{1}{3} = \frac{1}{3}$$

*Check*. We can repeat the calculation. We can also determine that the answer seems reasonable since we multiplied $\frac{2}{3}$ by a number less than 1 and the result is less than $\frac{2}{3}$. The answer checks.

*State*. The chef should use $\frac{1}{3}$ cup of flour.

**53.** *Familiarize*. We visualize the situation. Let $a =$ the assessed value of the house.

| Value of house $154,000 | |
|---|---|
| 3/4 of the value $a | |

*Translate*. We write an equation.

$$\underbrace{\text{Assessed value}}_{\downarrow} \;\; \underset{\downarrow}{\text{is}} \;\; \underset{\downarrow}{\frac{3}{4}} \;\; \underset{\downarrow}{\text{of}} \;\; \underbrace{\text{the value of the house}}_{\downarrow}$$

$$a \qquad\qquad = \qquad \frac{3}{4} \;\; \cdot \qquad\qquad 154,000$$

*Solve*. We carry out the multiplication.

$$a = \frac{3}{4} \cdot 154,000 = \frac{3 \cdot 154,000}{4}$$

$$= \frac{3 \cdot 4 \cdot 38,500}{4 \cdot 1} = \frac{4}{4} \cdot \frac{3 \cdot 38,500}{1}$$

$$= 115,500$$

*Check*. We can repeat the calculation. We can also determine that the answer seems reasonable since we multiplied 154,000 by a number less than 1 and the result is less than 154,000. The answer checks.

*State*. The assessed value of the house is $115,500.

**55.** *Familiarize*. We draw a picture.

$\frac{2}{3}$ in.

1 in.
240 miles

We let $n =$ the number of miles represented by $\frac{2}{3}$ in.

*Translate*. The multiplication sentence

$$n = \frac{2}{3} \cdot 240$$

corresponds to the situation.

*Solve*. We multiply and simplify:

$$n = \frac{2}{3} \cdot 240 = \frac{2 \cdot 240}{3} = \frac{2 \cdot 3 \cdot 80}{1 \cdot 3}$$

$$= \frac{3}{3} \cdot \frac{2 \cdot 80}{1} = \frac{2 \cdot 80}{1}$$

$$= 160$$

*Check*. We can repeat the calculation. We can also determine that the answer seems reasonable since we multiplied 240 by a number less than 1 and the result is less than 240.

*State*. $\frac{2}{3}$ in. on the map represents 160 miles.

**57.** *Familiarize*. This is a multistep problem. First we find the amount of each of the given expenses. Then we find the total of these expenses and take it away from the annual income to find how much is spent for other expenses.

We let $f$, $h$, $c$, $s$, and $t$ represent the amounts spent on food, housing, clothing, savings, and taxes, respectively.

*Translate*. The following multiplication sentences correspond to the situation.

$$\frac{1}{5} \cdot 42,000 = f \qquad \frac{1}{14} \cdot 42,000 = s$$

$$\frac{1}{4} \cdot 42,000 = h \qquad \frac{1}{5} \cdot 42,000 = t$$

$$\frac{1}{10} \cdot 42,000 = c$$

*Solve*. We multiply and simplify.

$$f = \frac{1}{5} \cdot 42,000 = \frac{42,000}{5} = \frac{5 \cdot 8400}{5 \cdot 1} = \frac{5}{5} \cdot \frac{8400}{1} = 8400$$

$h = \dfrac{1}{4} \cdot 42{,}000 = \dfrac{42{,}000}{4} = \dfrac{4 \cdot 10{,}500}{4 \cdot 1} = \dfrac{4}{4} \cdot \dfrac{10{,}500}{1} = 10{,}500$

$c = \dfrac{1}{10} \cdot 42{,}000 = \dfrac{42{,}000}{10} = \dfrac{10 \cdot 4200}{10 \cdot 1} = \dfrac{10}{10} \cdot \dfrac{4200}{1} = 4200$

$s = \dfrac{1}{14} \cdot 42{,}000 = \dfrac{42{,}000}{14} = \dfrac{14 \cdot 3000}{14 \cdot 1} = \dfrac{14}{14} \cdot \dfrac{3000}{1} = 3000$

$t = \dfrac{1}{5} \cdot 42{,}000$; this is the same computation we did to find $f$ above. Thus, $t = 8400$.

We add to find the total of these expenses.

$$
\begin{array}{r}
\$\ 8\ 4\ 0\ 0 \\
1\ 0{,}5\ 0\ 0 \\
4\ 2\ 0\ 0 \\
3\ 0\ 0\ 0 \\
8\ 4\ 0\ 0 \\
\hline
\$\ 3\ 4{,}5\ 0\ 0
\end{array}
$$

We let $m =$ the amount spent on other expenses and subtract to find this amount.

| Annual income | minus | Total of itemized expenses | is | Total spent on other expenses |
|---|---|---|---|---|
| ↓ | ↓ | ↓ | ↓ | ↓ |
| $42,000 | − | $34,500 | = | $m$ |
|  |  | $7500 | = | $m$   Subtracting |

**Check.** We repeat the calculations. The results check.

**State.** $8400 is spent for food, $10,500 for housing, $4200 for clothing, $3000 for savings, $8400 for taxes, and $7500 for other expenses.

**59.**  $48 \cdot t = 1680$

$\dfrac{48 \cdot t}{48} = \dfrac{1680}{48}$

$t = 35$

The solution is 35.

**61.**  $3125 = 25 \cdot t$

$\dfrac{3125}{25} = \dfrac{25 \cdot t}{25}$   Dividing by 25 on both sides

$125 = t$

The solution is 125.

**63.**  $t + 28 = 5017$

$t = 5017 - 28$

$t = 4989$

The solution is 4989.

**65.**      $8797 = y + 2299$

$8797 - 2299 = y + 2299 - 2299$   Subtracting 2299 on both sides

$6498 = y$

The solution is 6498.

**67.**  $-14 - 2 = -14 + (-2) = -16$

**69.**  $8 - 12 = 8 + (-12) = -4$

**71.** Use a calculator and the table of prime numbers on page 131 of the text to find factors that are common to the numerator and denominator of the product.

$\dfrac{201}{535} \cdot \dfrac{4601}{6499} = \dfrac{201 \cdot 4601}{535 \cdot 6499}$

$= \dfrac{3 \cdot 67 \cdot 43 \cdot 107}{5 \cdot 107 \cdot 67 \cdot 97}$

$= \dfrac{67 \cdot 107}{67 \cdot 107} \cdot \dfrac{3 \cdot 43}{5 \cdot 97}$

$= \dfrac{3 \cdot 43}{5 \cdot 97}$

$= \dfrac{129}{485}$

**73. Familiarize.** We are told that $\dfrac{2}{3}$ of $\dfrac{7}{8}$ of the students are high school graduates who are older than 20, and $\dfrac{1}{7}$ of this fraction are left-handed. Thus, we want to find $\dfrac{1}{7}$ of $\dfrac{2}{3}$ of $\dfrac{7}{8}$. We let $f$ represent this fraction.

**Translate.** The multiplication sentence

$f = \dfrac{1}{7} \cdot \dfrac{2}{3} \cdot \dfrac{7}{8}$

corresponds to this situation.

**Solve.** We multiply and simplify.

$f = \dfrac{1}{7} \cdot \dfrac{2}{3} \cdot \dfrac{7}{8} = \dfrac{1 \cdot 2}{7 \cdot 3} \cdot \dfrac{7}{8} = \dfrac{1 \cdot 2 \cdot 7}{7 \cdot 3 \cdot 8} = \dfrac{1 \cdot 2 \cdot 7}{7 \cdot 3 \cdot 2 \cdot 4} = \dfrac{2 \cdot 7}{2 \cdot 7} \cdot \dfrac{1}{3 \cdot 4} = \dfrac{1}{3 \cdot 4} = \dfrac{1}{12}$

**Check.** We repeat the calculation. The result checks.

**State.** $\dfrac{1}{12}$ of the students are left-handed high school graduates over the age of 20.

**75. Familiarize.** If we divide the group of entering students into 8 equal parts and take 7 of them, we have the fractional part of the students that completed high school. Then the 1 part remaining, or $\dfrac{1}{8}$ of the students, did not graduate from high school. Similarly, if we divide the group of entering students into 3 equal parts and take 2 of them, we have the fractional part of the students that is older than 20. Then the 1 part remaining, or $\dfrac{1}{3}$ of the students, are 20 years old or younger. From Exercise 75 we know that $\dfrac{1}{7}$ of the students are left-handed. Thus, we want to find $\dfrac{1}{7}$ of $\dfrac{1}{3}$ of $\dfrac{1}{8}$. We let $f =$ this fraction.

**Translate.** The multiplication sentence

$f = \dfrac{1}{7} \cdot \dfrac{1}{3} \cdot \dfrac{1}{8}$

corresponds to this situation.

**Solve.** We multiply.

$f = \dfrac{1}{7} \cdot \dfrac{1}{3} \cdot \dfrac{1}{8} = \dfrac{1 \cdot 1 \cdot 1}{7 \cdot 3 \cdot 8} = \dfrac{1}{168}$

**Check.** We repeat the calculation. The result checks.

**State.** $\frac{1}{168}$ of the students did not graduate from high school, are 20 years old or younger, and are left-handed.

## Exercise Set 3.7

**1.** $\frac{5}{6}$  Interchange the numerator and denominator.

The reciprocal of $\frac{5}{6}$ is $\frac{6}{5}$. $\quad \left( \frac{5}{6} \cdot \frac{6}{5} = \frac{30}{30} = 1 \right)$

**3.** Think of 6 as $\frac{6}{1}$.

$\frac{6}{1}$  Interchange the numerator and denominator.

The reciprocal of $\frac{6}{1}$ is $\frac{1}{6}$. $\quad \left( \frac{6}{1} \cdot \frac{1}{6} = \frac{6}{6} = 1 \right)$

**5.** $\frac{1}{6}$  Interchange the numerator and denominator.

The reciprocal of $\frac{1}{6}$ is 6. $\quad \left( \frac{6}{1} = 6; \frac{1}{6} \cdot \frac{6}{1} = \frac{6}{6} = 1 \right)$

(Note that we also found that 6 and $\frac{1}{6}$ are reciprocals in Exercise 3.)

**7.** $-\frac{10}{3}$  Interchange the numerator and denominator.

The reciprocal of $-\frac{10}{3}$ is $-\frac{3}{10}$.

$\left[ -\frac{10}{3} \cdot \left( -\frac{3}{10} \right) = \frac{30}{30} = 1 \right]$

**9.** $\frac{3}{5} \div \frac{3}{4} = \frac{3}{5} \cdot \frac{4}{3}$ $\quad$ Multiplying the dividend $\left( \frac{3}{5} \right)$ by the reciprocal of the divisor $\left( \text{The reciprocal of } \frac{3}{4} \text{ is } \frac{4}{3}. \right)$

$= \frac{3 \cdot 4}{5 \cdot 3}$ $\quad$ Multiplying numerators and denominators

$= \frac{3}{3} \cdot \frac{4}{5} = \frac{4}{5}$ $\quad$ Simplifying

**11.** $-\frac{3}{5} \div \frac{9}{4} = -\frac{3}{5} \cdot \frac{4}{9}$ $\quad$ Multiplying the dividend $\left( \frac{3}{5} \right)$ by the reciprocal of the divisor $\left( \text{The reciprocal of } \frac{9}{4} \text{ is } \frac{4}{9}. \right)$

$= -\frac{3 \cdot 4}{5 \cdot 9}$ $\quad$ Multiplying numerators and denominators

$= -\frac{3 \cdot 4}{5 \cdot 3 \cdot 3}$

$= \frac{3}{3} \cdot \left( -\frac{4}{5 \cdot 3} \right)$ $\quad$ Simplifying

$= -\frac{4}{5 \cdot 3} = -\frac{4}{15}$

**13.** $\frac{4}{3} \div \frac{1}{3} = \frac{4}{3} \cdot 3 = \frac{4 \cdot 3}{3} = \frac{3}{3} \cdot 4 = 4$

**15.** $-\frac{1}{3} \div \left( -\frac{1}{6} \right) = -\frac{1}{3} \cdot (-6) = \frac{1 \cdot 6}{3} = \frac{1 \cdot 2 \cdot 3}{1 \cdot 3} = \frac{1 \cdot 3}{1 \cdot 3} \cdot 2 = 2$

**17.** $\frac{3}{8} \div 3 = \frac{3}{8} \cdot \frac{1}{3} = \frac{3 \cdot 1}{8 \cdot 3} = \frac{3}{3} \cdot \frac{1}{8} = \frac{1}{8}$

**19.** $\frac{12}{7} \div 4 = \frac{12}{7} \cdot \frac{1}{4} = \frac{12 \cdot 1}{7 \cdot 4} = \frac{4 \cdot 3 \cdot 1}{7 \cdot 4} = \frac{4}{4} \cdot \frac{3 \cdot 1}{7} = \frac{3 \cdot 1}{7} = \frac{3}{7}$

**21.** $12 \div \left( -\frac{3}{2} \right) = 12 \cdot \left( -\frac{2}{3} \right) = -\frac{12 \cdot 2}{3} = -\frac{3 \cdot 4 \cdot 2}{3 \cdot 1} = \frac{3}{3} \cdot \left( -\frac{4 \cdot 2}{1} \right) = -\frac{4 \cdot 2}{1} = -\frac{8}{1} = -8$

**23.** $28 \div \frac{4}{5} = 28 \cdot \frac{5}{4} = \frac{28 \cdot 5}{4} = \frac{4 \cdot 7 \cdot 5}{4 \cdot 1} = \frac{4}{4} \cdot \frac{7 \cdot 5}{1} = \frac{7 \cdot 5}{1} = 35$

**25.** $-\frac{5}{8} \div \frac{5}{8} = -\frac{5}{8} \cdot \frac{8}{5} = -\frac{5 \cdot 8}{8 \cdot 5} = -\frac{5 \cdot 8}{5 \cdot 8} = -1$

**27.** $\frac{8}{15} \div \left( -\frac{4}{5} \right) = \frac{8}{15} \cdot \left( -\frac{5}{4} \right) = -\frac{8 \cdot 5}{15 \cdot 4} = -\frac{2 \cdot 4 \cdot 5}{3 \cdot 5 \cdot 4} = \frac{4 \cdot 5}{4 \cdot 5} \cdot \left( -\frac{2}{3} \right) = -\frac{2}{3}$

**29.** $-\frac{9}{5} \div \left( -\frac{4}{5} \right) = -\frac{9}{5} \cdot \left( -\frac{5}{4} \right) = \frac{9 \cdot 5}{5 \cdot 4} = \frac{5}{5} \cdot \frac{9}{4} = \frac{9}{4}$

**31.** $120 \div \frac{5}{6} = 120 \cdot \frac{6}{5} = \frac{120 \cdot 6}{5} = \frac{5 \cdot 24 \cdot 6}{5 \cdot 1} = \frac{5}{5} \cdot \frac{24 \cdot 6}{1} = \frac{24 \cdot 6}{1} = 144$

**33.** $\frac{4}{5} \cdot x = 60$

$x = 60 \div \frac{4}{5}$ $\quad$ Dividing on both sides by $\frac{4}{5}$

$x = 60 \cdot \frac{5}{4}$ $\quad$ Multiplying by the reciprocal

$= \frac{60 \cdot 5}{4} = \frac{4 \cdot 15 \cdot 5}{4 \cdot 1} = \frac{4}{4} \cdot \frac{15 \cdot 5}{1} = \frac{15 \cdot 5}{1} = 75$

**35.** $-\frac{5}{3} \cdot y = \frac{10}{3}$

$y = \frac{10}{3} \div \left( -\frac{5}{3} \right)$ $\quad$ Dividing on both sides by $-\frac{5}{3}$

$y = \frac{10}{3} \cdot \left( -\frac{3}{5} \right)$ $\quad$ Multiplying by the reciprocal

$y = -\frac{10 \cdot 3}{3 \cdot 5} = -\frac{2 \cdot 5 \cdot 3}{3 \cdot 5 \cdot 1} = \frac{5 \cdot 3}{5 \cdot 3} \cdot \left( -\frac{2}{1} \right) = -\frac{2}{1} = -2$

**37.** $x \cdot \dfrac{25}{36} = \dfrac{5}{12}$

$$x = \dfrac{5}{12} \div \dfrac{25}{36} = \dfrac{5}{12} \cdot \dfrac{36}{25} = \dfrac{5 \cdot 36}{12 \cdot 25} = \dfrac{5 \cdot 3 \cdot 12}{12 \cdot 5 \cdot 5}$$

$$= \dfrac{5 \cdot 12}{5 \cdot 12} \cdot \dfrac{3}{5} = \dfrac{3}{5}$$

**39.** $n \cdot \dfrac{8}{7} = -360$

$$n = -360 \div \dfrac{8}{7} = -360 \cdot \dfrac{7}{8} = -\dfrac{360 \cdot 7}{8} = -\dfrac{8 \cdot 45 \cdot 7}{8 \cdot 1}$$

$$= \dfrac{8}{8} \cdot \left(-\dfrac{45 \cdot 7}{1}\right) = -\dfrac{45 \cdot 7}{1} = -315$$

**41. *Familiarize*.** We draw a picture. Let $c = $ the number of extension cords that can be made from 2240 ft of cable.

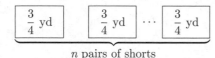

$c$ cords

***Translate*.** The multiplication that corresponds to the situation is
$$\dfrac{7}{3} \cdot c = 2240.$$

***Solve*.** We solve the equation by dividing on both sides by $\dfrac{7}{3}$ and carrying out the division:

$$c = 2240 \div \dfrac{7}{3} = 2240 \cdot \dfrac{3}{7} = \dfrac{2240 \cdot 3}{7} = \dfrac{7 \cdot 320 \cdot 3}{7 \cdot 1}$$

$$= \dfrac{7}{7} \cdot \dfrac{320 \cdot 3}{1} = \dfrac{320 \cdot 3}{1} = 960$$

***Check*.** We repeat the calculation. The answer checks.

***State*.** 960 $\dfrac{7}{3}$-ft extension cords can be made from 2240 ft of cable.

**43. *Familiarize*.** We draw a picture. We let $n = $ the number of pairs of basketball shorts that can be made.

$$\boxed{\dfrac{3}{4} \text{ yd}} \quad \boxed{\dfrac{3}{4} \text{ yd}} \quad \cdots \quad \boxed{\dfrac{3}{4} \text{ yd}}$$

$n$ pairs of shorts

***Translate*.** The multiplication that corresponds to the situation is
$$\dfrac{3}{4} \cdot n = 24.$$

***Solve*.** We solve the equation by dividing on both sides by $\dfrac{3}{4}$ and carrying out the division:

$$n = 24 \div \dfrac{3}{4} = 24 \cdot \dfrac{4}{3} = \dfrac{24 \cdot 4}{3} = \dfrac{3 \cdot 8 \cdot 4}{3 \cdot 1} = \dfrac{3}{3} \cdot \dfrac{8 \cdot 4}{1}$$

$$= \dfrac{8 \cdot 4}{1} = 32$$

***Check*.** We repeat the calculation. The answer checks.

***State*.** 32 pairs of basketball shorts can be made from 24 yd of fabric.

**45. *Familiarize*.** We draw a picture. We let $n = $ the number of sugar bowls that can be filled.

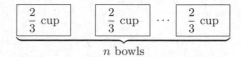

$n$ bowls

***Translate*.** We write a multiplication sentence:
$$\dfrac{2}{3} \cdot n = 16$$

***Solve*.** Solve the equation as follows:

$$\dfrac{2}{3} \cdot n = 16$$

$$n = 16 \div \dfrac{2}{3} = 16 \cdot \dfrac{3}{2} = \dfrac{16 \cdot 3}{2} = \dfrac{2 \cdot 8 \cdot 3}{2 \cdot 1}$$

$$= \dfrac{2}{2} \cdot \dfrac{8 \cdot 3}{1} = \dfrac{8 \cdot 3}{1} = 24$$

***Check*.** We repeat the calculation. The answer checks.

***State*.** 24 sugar bowls can be filled.

**47. *Familiarize*.** We draw a picture. We let $n = $ the amount the bucket could hold.

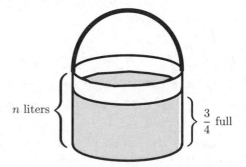

$n$ liters      $\dfrac{3}{4}$ full

***Translate*.** We write a multiplication sentence:
$$\dfrac{3}{4} \cdot n = 12$$

***Solve*.** Solve the equation as follows:

$$\dfrac{3}{4} \cdot n = 12$$

$$n = 12 \div \dfrac{3}{4} = 12 \cdot \dfrac{4}{3} = \dfrac{12 \cdot 4}{3} = \dfrac{3 \cdot 4 \cdot 4}{3 \cdot 1}$$

$$= \dfrac{3}{3} \cdot \dfrac{4 \cdot 4}{1} = \dfrac{4 \cdot 4}{1} = 16$$

***Check*.** We repeat the calculation. The answer checks.

***State*.** The bucket could hold 16 L.

**49. *Familiarize*.** This is a multistep problem. First we find the length of the total trip. Then we find how many kilometers were left to drive. We draw a picture. We let $n = $ the length of the total trip.

$\dfrac{5}{8}$ of the trip

180 km

$n$ km

*Translate*. We translate to an equation.

| Fraction of trip completed | times | Total length of trip | is | Amount already traveled |
|:---:|:---:|:---:|:---:|:---:|
| ↓ | ↓ | ↓ | ↓ | ↓ |
| $\frac{5}{8}$ | $\cdot$ | $n$ | $=$ | $180$ |

*Solve*. We solve the equation as follows:

$$\frac{5}{8} \cdot n = 180$$

$$n = 180 \div \frac{5}{8} = 180 \cdot \frac{8}{5} = \frac{5 \cdot 36 \cdot 8}{5 \cdot 1}$$

$$= \frac{5}{5} \cdot \frac{36 \cdot 8}{1} = \frac{36 \cdot 8}{1} = 288$$

The total trip was 288 km.

Now we find how many kilometers were left to travel. Let $t$ = this number.

| Length of total trip | minus | Distance traveled | is | Distance left to travel |
|:---:|:---:|:---:|:---:|:---:|
| ↓ | ↓ | ↓ | ↓ | ↓ |
| $288$ | $-$ | $180$ | $=$ | $t$ |

We carry out the subtraction:

$$288 - 180 = t$$
$$108 = t$$

*Check*. We repeat the calculation. The results check.

*State*. The total trip was 288 km. There were 108 km left to travel.

**51.** *Familiarize*. Let $p$ = the pitch of the screw, in inches. The distance the screw has traveled into the wallboard is found by multiplying the pitch by the number of complete rotations.

*Translate*. We translate to an equation.

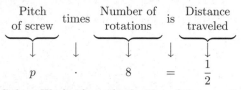

| Pitch of screw | times | Number of rotations | is | Distance traveled |
|:---:|:---:|:---:|:---:|:---:|
| ↓ | ↓ | ↓ | ↓ | ↓ |
| $p$ | $\cdot$ | $8$ | $=$ | $\frac{1}{2}$ |

*Solve*. We divide on both sides of the equation by 8 and carry out the division.

$$p = \frac{1}{2} \div 8 = \frac{1}{2} \cdot \frac{1}{8} = \frac{1 \cdot 1}{2 \cdot 8} = \frac{1}{16}$$

*Check*. We repeat the calculation. The answer checks.

*State*. The pitch of the screw is $\frac{1}{16}$ in.

**53.** The equation $14 + (2 + 30) = (14 + 2) + 30$ illustrates the <u>associative</u> law of addition.

**55.** A natural number that has exactly two different factors, only itself and 1, is called a <u>prime</u> number.

**57.** For any number $a$, $a + 0 = a$. The number 0 is the <u>additive</u> identity.

**59.** The set of <u>whole numbers</u> is 0, 1, 2, 3, 4, . . . .

**61.** Use a calculator.

$$\frac{711}{1957} \div \frac{10,033}{13,081} = \frac{711}{1957} \cdot \frac{13,081}{10,033}$$

$$= \frac{711 \cdot 13,081}{1957 \cdot 10,033}$$

$$= \frac{3 \cdot 3 \cdot 79 \cdot 103 \cdot 127}{19 \cdot 103 \cdot 79 \cdot 127}$$

$$= \frac{79 \cdot 103 \cdot 127}{79 \cdot 103 \cdot 127} \cdot \frac{3 \cdot 3}{19}$$

$$= \frac{9}{19}$$

**63.** $\left[\dfrac{9}{10} \div \left(-\dfrac{2}{5}\right) \div \dfrac{3}{8}\right]^2 = \left[\dfrac{9}{10} \cdot \left(-\dfrac{5}{2}\right) \div \dfrac{3}{8}\right]^2$

$$= \left(-\frac{9 \cdot 5}{10 \cdot 2} \div \frac{3}{8}\right)^2$$

$$= \left(-\frac{9 \cdot 5}{2 \cdot 5 \cdot 2} \div \frac{3}{8}\right)^2$$

$$= \left(-\frac{9}{2 \cdot 2} \div \frac{3}{8}\right)^2$$

$$= \left(-\frac{9}{2 \cdot 2} \cdot \frac{8}{3}\right)^2$$

$$= \left(-\frac{9 \cdot 8}{2 \cdot 2 \cdot 3}\right)^2$$

$$= \left(-\frac{3 \cdot 3 \cdot 2 \cdot 2 \cdot 2}{2 \cdot 2 \cdot 3 \cdot 1}\right)^2$$

$$= \left(-\frac{3 \cdot 2}{1}\right)^2 = (-6)^2 = 36$$

**65.** Let $n$ = the number.

$$\frac{1}{3} \cdot n = \frac{1}{4}$$

$$n = \frac{1}{4} \div \frac{1}{3} = \frac{1}{4} \cdot \frac{3}{1} = \frac{1 \cdot 3}{4 \cdot 1} = \frac{3}{4}$$

The number is $\frac{3}{4}$. Now we find $\frac{1}{2}$ of $\frac{3}{4}$.

$$\frac{1}{2} \cdot \frac{3}{4} = \frac{1 \cdot 3}{2 \cdot 4} = \frac{3}{8}$$

One-half of the number is $\frac{3}{8}$.

## Chapter 3 Concept Reinforcement

**1.** For any natural number $n$, $\dfrac{n}{n} = 1$ and $\dfrac{0}{n} = 0$. Since $1 > 0$, the statement is true.

**2.** A number is divisible by 10 only if its ones digit is 0. The statement is false.

**3.** Since 3 is a factor of 9, any number that is divisible by 9 is also divisible by 3. The statement is true.

**4.** $\frac{66}{187} = \frac{6 \cdot 11}{17 \cdot 11} = \frac{6}{17}$, so it is true that $\frac{6}{17}$ is equal to $\frac{66}{187}$.

## Chapter 3 Important Concepts

**1.** We find as many two-factor factorizations as we can.

$104 = 1 \cdot 104 \qquad 104 = 4 \cdot 26$
$104 = 2 \cdot 52 \qquad 104 = 8 \cdot 13$

Factors: 1, 2, 4, 8, 13, 26, 52, 104

**2.**
$$\begin{array}{r} 1\ 3 \leftarrow 13 \text{ is prime} \\ 2\overline{\smash{)}2\ 6} \\ 2\overline{\smash{)}5\ 2} \\ 2\overline{\smash{)}1\ 0\ 4} \end{array}$$

$104 = 2 \cdot 2 \cdot 2 \cdot 13$

**3.** $\frac{0}{18} = 0$; $\frac{-18}{-18} = 1$; $\frac{18}{1} = 18$

**4.** Since $12 \cdot 8 = 96$, we multiply by $\frac{8}{8}$.

$$\frac{7}{12} = \frac{7}{12} \cdot \frac{8}{8} = \frac{7 \cdot 8}{12 \cdot 8} = \frac{56}{96}$$

**5.** $\frac{100}{280} = \frac{2 \cdot 5 \cdot 10}{2 \cdot 2 \cdot 7 \cdot 10} = \frac{2 \cdot 10}{2 \cdot 10} \cdot \frac{5}{2 \cdot 7} = 1 \cdot \frac{5}{2 \cdot 7} = \frac{5}{14}$

**6.** We multiply these two numbers: We multiply these two numbers:

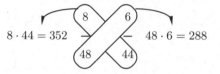

Since $352 \neq 288$, $\frac{8}{48} \neq \frac{6}{44}$.

**7.** $\frac{80}{3} \cdot \left(-\frac{21}{72}\right) = -\frac{80 \cdot 21}{3 \cdot 72} = -\frac{2 \cdot 5 \cdot 8 \cdot 3 \cdot 7}{3 \cdot 9 \cdot 8} =$

$\frac{3 \cdot 8}{3 \cdot 8} \cdot \left(-\frac{2 \cdot 5 \cdot 7}{9}\right) = 1 \cdot \left(-\frac{2 \cdot 5 \cdot 7}{9}\right) = -\frac{70}{9}$

**8.** $\frac{9}{4} \div \frac{45}{14} = \frac{9}{4} \cdot \frac{14}{45} = \frac{9 \cdot 14}{4 \cdot 45} = \frac{9 \cdot 2 \cdot 7}{2 \cdot 2 \cdot 9 \cdot 5} = \frac{9 \cdot 2}{9 \cdot 2} \cdot \frac{7}{2 \cdot 5} =$

$1 \cdot \frac{7}{2 \cdot 5} = \frac{7}{10}$

**9.** *Familiarize.* Let $w =$ the number of cups of water the vase can hold when full.

*Translate.*

$$\underbrace{\frac{7}{4} \text{ cups}}_{} \text{ is } \frac{3}{4} \text{ of } \underline{\text{full amount}}$$
$$\downarrow \quad \downarrow \quad \downarrow \quad \downarrow \qquad \downarrow$$
$$\frac{7}{4} \quad = \quad \frac{3}{4} \quad \cdot \qquad w$$

*Solve.* We divide by $\frac{3}{4}$ on both sides of the equation.

$$\frac{7}{4} \div \frac{3}{4} = \frac{7}{4} \cdot \frac{4}{3} = \frac{7 \cdot 4}{4 \cdot 3} = \frac{4}{4} \cdot \frac{7}{3} = \frac{7}{3}$$

Thus, $\frac{7}{3} = w$.

*Check.* We can repeat the calculation. We can also find $\frac{3}{4}$ of $\frac{7}{3}$: $\frac{3}{4} \cdot \frac{7}{3} = \frac{3 \cdot 7}{4 \cdot 3} = \frac{3}{3} \cdot \frac{7}{4} = \frac{7}{4}$. The answer checks.

*State.* The vase holds $\frac{7}{3}$ cups of water when full.

## Chapter 3 Review Exercises

**1.** We find as many two-factor factorizations as we can:

$60 = 1 \cdot 60 \qquad 60 = 4 \cdot 15$
$60 = 2 \cdot 30 \qquad 60 = 5 \cdot 12$
$60 = 3 \cdot 20 \qquad 60 = 6 \cdot 10$

Factors: 1, 2, 3, 4, 5, 6, 10, 12, 15, 20, 30, 60

**2.** We find as many two-factor factorizations as we can:

$176 = 1 \cdot 176 \qquad 176 = 8 \cdot 22$
$176 = 2 \cdot 88 \qquad 176 = 11 \cdot 16$
$176 = 4 \cdot 44$

Factors: 1, 2, 4, 8, 11, 16, 22, 44, 88, 176

**3.**
$1 \cdot 8 = 8 \qquad 6 \cdot 8 = 48$
$2 \cdot 8 = 16 \qquad 7 \cdot 8 = 56$
$3 \cdot 8 = 24 \qquad 8 \cdot 8 = 64$
$4 \cdot 8 = 32 \qquad 9 \cdot 8 = 72$
$5 \cdot 8 = 40 \qquad 10 \cdot 8 = 80$

**4.**
$$\begin{array}{r} 8\ 4 \\ 1\ 1\overline{\smash{)}9\ 2\ 4} \\ \underline{8\ 8\ 0} \\ 4\ 4 \\ \underline{4\ 4} \\ 0 \end{array}$$

Since the remainder is 0, 924 is divisible by 11.

**5.**
$$\begin{array}{r} 1\ 1\ 2 \\ 1\ 6\overline{\smash{)}1\ 8\ 0\ 0} \\ \underline{1\ 6\ 0\ 0} \\ 2\ 0\ 0 \\ \underline{1\ 6\ 0} \\ 4\ 0 \\ \underline{3\ 2} \\ 8 \end{array}$$

Since the remainder is not 0, 1800 is not divisible by 16.

**6.** The only factors of 37 are 1 and 37, so 37 is prime.

**7.** 1 is neither prime nor composite.

**8.** The number 91 has factors 1, 7, 13, and 91, so it is composite.

**9.**
$$\begin{array}{r} 7 \leftarrow 7 \text{ is prime.} \\ 5\overline{\smash{)}3\ 5} \\ 2\overline{\smash{)}7\ 0} \end{array}$$

$70 = 2 \cdot 5 \cdot 7$

**10.**
$$\begin{array}{r} 5 \leftarrow 5 \text{ is prime.} \\ 3\overline{\smash{)}1\ 5} \\ 2\overline{\smash{)}3\ 0} \end{array}$$

$30 = 2 \cdot 3 \cdot 5$

**11.**
$$\begin{array}{r} 5 \quad \leftarrow \text{ 5 is prime.} \\ 3\,\overline{\smash{\big)}\,1\,5} \\ 3\,\overline{\smash{\big)}\,4\,5} \end{array}$$
$$45 = 3 \cdot 3 \cdot 5$$

**12.**
$$\begin{array}{r} 5 \quad \leftarrow \text{ 5 is prime.} \\ 5\,\overline{\smash{\big)}\,2\,5} \\ 3\,\overline{\smash{\big)}\,7\,5} \\ 2\,\overline{\smash{\big)}\,1\,5\,0} \end{array}$$
$$150 = 2 \cdot 3 \cdot 5 \cdot 5$$

**13.**
$$\begin{array}{r} 3 \quad \leftarrow \text{ 3 is prime.} \\ 3\,\overline{\smash{\big)}\,9} \\ 3\,\overline{\smash{\big)}\,2\,7} \\ 3\,\overline{\smash{\big)}\,8\,1} \\ 2\,\overline{\smash{\big)}\,1\,6\,2} \\ 2\,\overline{\smash{\big)}\,3\,2\,4} \\ 2\,\overline{\smash{\big)}\,6\,4\,8} \end{array}$$
$$648 = 2 \cdot 2 \cdot 2 \cdot 3 \cdot 3 \cdot 3 \cdot 3$$

**14.**
$$\begin{array}{r} 7 \quad \leftarrow \text{ 7 is prime.} \\ 5\,\overline{\smash{\big)}\,3\,5} \\ 5\,\overline{\smash{\big)}\,1\,7\,5} \\ 5\,\overline{\smash{\big)}\,8\,7\,5} \\ 3\,\overline{\smash{\big)}\,2\,6\,2\,5} \\ 2\,\overline{\smash{\big)}\,5\,2\,5\,0} \end{array}$$
$$5250 = 2 \cdot 3 \cdot 5 \cdot 5 \cdot 5 \cdot 7$$

**15.** A number is divisible by 3 if the sum of its digits is divisible by 3. The numbers whose digits add to a multiple of 3 are 4344, 600, 93, 330, 255,555, 780, 2802, and 711.

**16.** A number is divisible by 2 if its ones digit is even. Thus, the numbers 140, 182, 716, 2432, 4344, 600, 330, 780, and 2802 are divisible by 2.

**17.** A number is divisible by 4 if the number named by its last two digits is divisible by 4. Thus, the numbers 140, 716, 2432, 4344, 600, and 780 are divisible by 4.

**18.** A number is divisible by 8 if the number named by its last three digits is divisible by 8. Thus, the numbers 2432, 4344, and 600 are divisible by 8.

**19.** A number is divisible by 5 if its ones digit is 0 or 5. Thus, the numbers 140, 95, 475, 600, 330, 255,555, and 780 are divisible by 5.

**20.** A number is divisible by 6 if its one digit is even and the sum of the digits is divisible by 3. The numbers whose ones digits are even are given in Exercise 16 above. Of these numbers, the ones whose digits add to a multiple of 3 are 4344, 600, 330, 780, and 2802.

**21.** A number is divisible by 9 if the sum of its digits is divisible by 9. The numbers whose digits add to a multiple of 9 are 255,555 and 711.

**22.** A number is divisible by 10 if its ones digit is 0. Thus, the numbers 140, 600, 330, and 780 are divisible by 10.

**23.** The top number is the numerator, and the bottom number is the denominator.
$$\begin{array}{l} 2 \quad \leftarrow \text{ Numerator} \\ \overline{7} \quad \leftarrow \text{ Denominator} \end{array}$$

**24.** The object is divided into 5 equal parts. The unit is $\frac{1}{5}$. The denominator is 5. We have 3 parts shaded. This tells us that the numerator is 3. Thus, $\frac{3}{5}$ is shaded.

**25.** Each rectangle is divided into 6 equal parts. The unit is $\frac{1}{6}$. The denominator is 6. We see that 7 of the parts are shaded, so the numerator is 7. Thus, $\frac{7}{6}$ is shaded.

**26.** There are 7 objects in the set, and 2 of the objects are shaded. Thus, $\frac{2}{7}$ of the set is shaded.

**27.** a) The ratio is $\frac{3}{5}$.

   b) The ratio is $\frac{5}{3}$.

   c) There are $3 + 5$, or 8, members of the committee. The desired ratio is $\frac{3}{8}$.

**28.** $\dfrac{12}{30} = \dfrac{2 \cdot 6}{5 \cdot 6} = \dfrac{2}{5} \cdot \dfrac{6}{6} = \dfrac{2}{5}$

**29.** $\dfrac{7}{28} = \dfrac{7 \cdot 1}{4 \cdot 7} = \dfrac{7}{7} \cdot \dfrac{1}{4} = \dfrac{1}{4}$

**30.** $\dfrac{n}{n} = 1$, for any integer $n$ that is not 0.

   $\dfrac{-23}{-23} = 1$

**31.** $\dfrac{0}{n} = 0$, for any integer $n$ that is not 0.

   $\dfrac{0}{-25} = 0$

**32.** $\dfrac{1170}{1200} = \dfrac{10 \cdot 117}{10 \cdot 120} = \dfrac{10}{10} \cdot \dfrac{117}{120} = \dfrac{117}{120} = \dfrac{3 \cdot 39}{3 \cdot 40} = \dfrac{3}{3} \cdot \dfrac{39}{40} = \dfrac{39}{40}$

**33.** $\dfrac{n}{1} = n$, for any integer $n$.

   $\dfrac{18}{1} = 18$

**34.** $-\dfrac{9}{27} = -\dfrac{1 \cdot 9}{3 \cdot 9} = -\dfrac{1}{3} \cdot \dfrac{9}{9} = -\dfrac{1}{3}$

**35.** $-\dfrac{88}{184} = -\dfrac{8 \cdot 11}{8 \cdot 23} = \dfrac{8}{8} \cdot \left(-\dfrac{11}{23}\right) = -\dfrac{11}{23}$

**36.** $\dfrac{n}{0}$ is not defined for any integer $n$.

   $\dfrac{18}{0}$ is not defined.

**37.** $\dfrac{48}{8} = \dfrac{6 \cdot 8}{1 \cdot 8} = \dfrac{6}{1} \cdot \dfrac{8}{8} = 6$

**38.** $\dfrac{140}{490} = \dfrac{10 \cdot 14}{10 \cdot 49} = \dfrac{10}{10} \cdot \dfrac{14}{49} = \dfrac{14}{49} = \dfrac{2 \cdot 7}{7 \cdot 7} = \dfrac{2}{7} \cdot \dfrac{7}{7} = \dfrac{2}{7}$

**39.** $-\dfrac{288}{2025} = -\dfrac{9 \cdot 32}{9 \cdot 225} = \dfrac{9}{9} \cdot \left(-\dfrac{32}{225}\right) = -\dfrac{32}{225}$

**40.** $\dfrac{15}{100} = \dfrac{3 \cdot 5}{20 \cdot 5} = \dfrac{3}{20} \cdot \dfrac{5}{5} = \dfrac{3}{20}$

$\dfrac{38}{100} = \dfrac{2 \cdot 19}{2 \cdot 50} = \dfrac{2}{2} \cdot \dfrac{19}{50} = \dfrac{19}{50}$

23 and 100 have no prime factors in common, so $\dfrac{23}{100}$ cannot be simplified.

$\dfrac{24}{100} = \dfrac{4 \cdot 6}{4 \cdot 25} = \dfrac{4}{4} \cdot \dfrac{6}{25} = \dfrac{6}{25}$

**41.**    We multiply these          We multiply these
      two numbers:               two numbers:

$3 \cdot 6 = 18$      $\begin{matrix} 3 & 4 \\ 5 & 6 \end{matrix}$      $5 \cdot 4 = 20$

Since $18 \neq 20$, $\dfrac{3}{5} \neq \dfrac{4}{6}$.

**42.** First we write $-\dfrac{8}{14}$ as $\dfrac{-8}{14}$.

    We multiply these          We multiply these
      two numbers:               two numbers:

$-4 \cdot 14 = -56$    $\begin{matrix} -4 & -8 \\ 7 & 14 \end{matrix}$    $7 \cdot (-8) = -56$

Since $-56 = -56$, $\dfrac{-4}{7} = \dfrac{-8}{14}$, or $\dfrac{-4}{7} = -\dfrac{8}{14}$.

**43.** First we write $-\dfrac{4}{5}$ as $\dfrac{-4}{5}$ and $-\dfrac{5}{6}$ as $\dfrac{-5}{6}$.

    We multiply these          We multiply these
      two numbers:               two numbers:

$-4 \cdot 6 = -24$    $\begin{matrix} -4 & -5 \\ 5 & 6 \end{matrix}$    $5 \cdot (-5) = -25$

Since $-24 \neq -25$, $\dfrac{-4}{5} \neq \dfrac{-5}{6}$, or $-\dfrac{4}{5} \neq -\dfrac{5}{6}$.

**44.**    We multiply these          We multiply these
      two numbers:               two numbers:

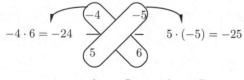

$4 \cdot 21 = 84$    $\begin{matrix} 4 & 28 \\ 3 & 21 \end{matrix}$    $3 \cdot 28 = 84$

Since $84 = 84$, $\dfrac{4}{3} = \dfrac{28}{21}$.

**45.** $4 \cdot \dfrac{3}{8} = \dfrac{4 \cdot 3}{8} = \dfrac{4 \cdot 3}{2 \cdot 4} = \dfrac{4}{4} \cdot \dfrac{3}{2} = \dfrac{3}{2}$

**46.** $\dfrac{7}{3} \cdot 24 = \dfrac{7 \cdot 24}{3} = \dfrac{7 \cdot 3 \cdot 8}{3 \cdot 1} = \dfrac{3}{3} = \dfrac{7 \cdot 8}{1} = \dfrac{7 \cdot 8}{1} = 56$

**47.** $-9 \cdot \dfrac{5}{18} = -\dfrac{9 \cdot 5}{18} = -\dfrac{9 \cdot 5}{2 \cdot 9} = \dfrac{9}{9} \cdot \left(-\dfrac{5}{2}\right) = -\dfrac{5}{2}$

**48.** $\dfrac{6}{5} \cdot (-20) = -\dfrac{6 \cdot 20}{5} = -\dfrac{6 \cdot 4 \cdot 5}{1 \cdot 5} = -\dfrac{6 \cdot 4}{1} \cdot \dfrac{5}{5} = -\dfrac{6 \cdot 4}{1} = -24$

**49.** $\dfrac{3}{4} \cdot \dfrac{8}{9} = \dfrac{3 \cdot 8}{4 \cdot 9} = \dfrac{3 \cdot 2 \cdot 4}{4 \cdot 3 \cdot 3} = \dfrac{3 \cdot 4}{3 \cdot 4} \cdot \dfrac{2}{3} = \dfrac{2}{3}$

**50.** $\dfrac{5}{7} \cdot \dfrac{1}{10} = \dfrac{5 \cdot 1}{7 \cdot 10} = \dfrac{5 \cdot 1}{7 \cdot 2 \cdot 5} = \dfrac{5}{5} \cdot \dfrac{1}{7 \cdot 2} = \dfrac{1}{7 \cdot 2} = \dfrac{1}{14}$

**51.** $-\dfrac{3}{7} \cdot \dfrac{14}{9} = -\dfrac{3 \cdot 14}{7 \cdot 9} = -\dfrac{3 \cdot 2 \cdot 7}{7 \cdot 3 \cdot 3} = \dfrac{3 \cdot 7}{3 \cdot 7} \cdot \left(-\dfrac{2}{3}\right) = -\dfrac{2}{3}$

**52.** $\dfrac{1}{4} \cdot \dfrac{2}{11} = \dfrac{1 \cdot 2}{4 \cdot 11} = \dfrac{1 \cdot 2}{2 \cdot 2 \cdot 11} = \dfrac{2}{2} \cdot \dfrac{1}{2 \cdot 11} = \dfrac{1}{2 \cdot 11} = \dfrac{1}{22}$

**53.** $\dfrac{4}{25} \cdot \dfrac{15}{16} = \dfrac{4 \cdot 15}{25 \cdot 16} = \dfrac{4 \cdot 3 \cdot 5}{5 \cdot 5 \cdot 4 \cdot 4} = \dfrac{4 \cdot 5}{4 \cdot 5} \cdot \dfrac{3}{5 \cdot 4} = \dfrac{3}{5 \cdot 4} = \dfrac{3}{20}$

**54.** $-\dfrac{11}{3} \cdot \left(-\dfrac{30}{77}\right) = \dfrac{11 \cdot 30}{3 \cdot 77} = \dfrac{11 \cdot 3 \cdot 10}{3 \cdot 11 \cdot 7} = \dfrac{3 \cdot 11}{3 \cdot 11} \cdot \dfrac{10}{7} = \dfrac{10}{7}$

**55.** Interchange the numerator and the denominator. The reciprocal of $\dfrac{4}{5}$ is $\dfrac{5}{4}$.

**56.** Think of $-3$ as $-\dfrac{3}{1}$ and interchange the numerator and the denominator. The reciprocal of $-3$ is $-\dfrac{1}{3}$.

**57.** Interchange the numerator and the denominator. The reciprocal of $\dfrac{1}{9}$ is $\dfrac{9}{1}$, or 9.

**58.** Interchange the numerator and the denominator. The reciprocal of $-\dfrac{47}{36}$ is $-\dfrac{36}{47}$.

**59.** $6 \div \dfrac{4}{3} = 6 \cdot \dfrac{3}{4} = \dfrac{6 \cdot 3}{4} = \dfrac{2 \cdot 3 \cdot 3}{2 \cdot 2} = \dfrac{2}{2} \cdot \dfrac{3 \cdot 3}{2} = \dfrac{3 \cdot 3}{2} = \dfrac{9}{2}$

**60.** $-\dfrac{5}{9} \div \dfrac{5}{18} = -\dfrac{5}{9} \cdot \dfrac{18}{5} = -\dfrac{5 \cdot 18}{9 \cdot 5} = -\dfrac{5 \cdot 2 \cdot 9}{9 \cdot 5 \cdot 1} = \dfrac{5 \cdot 9}{5 \cdot 9} \cdot \left(-\dfrac{2}{1}\right) = -\dfrac{2}{1} = -2$

**61.** $\dfrac{1}{6} \div \dfrac{1}{11} = \dfrac{1}{6} \cdot \dfrac{11}{1} = \dfrac{1 \cdot 11}{6 \cdot 1} = \dfrac{11}{6}$

**62.** $\dfrac{3}{14} \div \left(-\dfrac{6}{7}\right) = \dfrac{3}{14} \cdot \left(-\dfrac{7}{6}\right) = -\dfrac{3 \cdot 7}{14 \cdot 6} = -\dfrac{3 \cdot 7 \cdot 1}{2 \cdot 7 \cdot 2 \cdot 3} = \dfrac{3 \cdot 7}{3 \cdot 7} \cdot \left(-\dfrac{1}{2 \cdot 2}\right) = -\dfrac{1}{2 \cdot 2} = -\dfrac{1}{4}$

**63.** $-\dfrac{1}{4} \div \left(-\dfrac{1}{9}\right) = -\dfrac{1}{4} \cdot \left(-\dfrac{9}{1}\right) = \dfrac{1 \cdot 9}{4 \cdot 1} = \dfrac{9}{4}$

**64.** $180 \div \dfrac{3}{5} = 180 \cdot \dfrac{5}{3} = \dfrac{180 \cdot 5}{3} = \dfrac{3 \cdot 60 \cdot 5}{3 \cdot 1} = \dfrac{3}{3} \cdot \dfrac{60 \cdot 5}{1} = \dfrac{60 \cdot 5}{1} = 300$

**65.** $\dfrac{23}{25} \div \dfrac{23}{25} = \dfrac{23}{25} \cdot \dfrac{25}{23} = \dfrac{23 \cdot 25}{25 \cdot 23} = 1$

**66.** $-\dfrac{2}{3} \div \left(-\dfrac{3}{2}\right) = -\dfrac{2}{3} \cdot \left(-\dfrac{2}{3}\right) = \dfrac{2 \cdot 2}{3 \cdot 3} = \dfrac{4}{9}$

**67.** $\dfrac{5}{4} \cdot t = \dfrac{3}{8}$

$t = \dfrac{3}{8} \div \dfrac{5}{4}$     Dividing by $\dfrac{5}{4}$ on both sides

$t = \dfrac{3}{8} \cdot \dfrac{4}{5}$

$= \dfrac{3 \cdot 4}{8 \cdot 5} = \dfrac{3 \cdot 4}{2 \cdot 4 \cdot 5} = \dfrac{4}{4} \cdot \dfrac{3}{2 \cdot 5} = \dfrac{3}{2 \cdot 5} = \dfrac{3}{10}$

**68.** $x \cdot \dfrac{2}{3} = -160$

$x = -160 \div \dfrac{2}{3}$     Dividing by $\dfrac{2}{3}$ on both sides

$x = -160 \cdot \dfrac{3}{2}$

$= -\dfrac{160 \cdot 3}{2} = -\dfrac{2 \cdot 80 \cdot 3}{2 \cdot 1} = \dfrac{2}{2} \cdot \left(-\dfrac{80 \cdot 3}{1}\right)$

$= -\dfrac{80 \cdot 3}{1} = -240$

**69. Familiarize.** Let $d$ = the number of days it will take to repave the road.

**Translate.**

| Number of miles repaved each day | times | Number of days | is | Total number of miles repaved |
|:---:|:---:|:---:|:---:|:---:|
| $\dfrac{1}{12}$ | $\cdot$ | $d$ | $=$ | $\dfrac{3}{4}$ |

**Solve.** We divide by $\dfrac{1}{12}$ on both sides of the equation.

$d = \dfrac{3}{4} \div \dfrac{1}{12}$

$d = \dfrac{3}{4} \cdot \dfrac{12}{1} = \dfrac{3 \cdot 12}{4 \cdot 1} = \dfrac{3 \cdot 3 \cdot 4}{4 \cdot 1}$

$= \dfrac{4}{4} \cdot \dfrac{3 \cdot 3}{1} = \dfrac{3 \cdot 3}{1} = 9$

**Check.** We repeat the calculation. The answer checks.

**State.** It will take 9 days to repave the road.

**70. Familiarize.** Let $c$ = the number of metric tons of cotton produced in the United States in 2005-2006.

**Translate.**

| U.S. cotton production | was | $\dfrac{1}{5}$ | of | World cotton production |
|:---:|:---:|:---:|:---:|:---:|
| $c$ | $=$ | $\dfrac{1}{5}$ | $\cdot$ | $114,000,000$ |

**Solve.** We carry out the multiplication.

$c = \dfrac{1}{5} \cdot 114,000,000 = \dfrac{1 \cdot 114,000,000}{5} =$

$\dfrac{1 \cdot 5 \cdot 22,800,000}{1 \cdot 5} = \dfrac{5}{5} \cdot \dfrac{22,800,000}{1} = 22,800,000$

**Check.** We can repeat the calculation. The answer checks.

**State.** The United States produced 22,800,000 metric tons of cotton in 2005-2006.

**71. Familiarize.** Let $t$ = the total length of the trip, in km.

**Translate.**

| Distance driven | is | $\dfrac{3}{5}$ | of | Total distance |
|:---:|:---:|:---:|:---:|:---:|
| $600$ | $=$ | $\dfrac{3}{5}$ | $\cdot$ | $t$ |

**Solve.** We divide by $\dfrac{3}{5}$ on both sides of the equation.

$t = 600 \div \dfrac{3}{5}$

$t = 600 \cdot \dfrac{5}{3} = \dfrac{600 \cdot 5}{3} = \dfrac{3 \cdot 200 \cdot 5}{3 \cdot 1}$

$= \dfrac{3}{3} \cdot \dfrac{200 \cdot 5}{1} = \dfrac{200 \cdot 5}{1} = 1000$

**Check.** We repeat the calculation. The answer checks.

**State.** The trip is 1000 km long.

**72. Familiarize.** Let $x$ = the number of cups of peppers needed for $\dfrac{1}{2}$ recipe and $y$ = the amount needed for 3 recipes.

**Translate.** For $\dfrac{1}{2}$ recipe we want to find $\dfrac{1}{2}$ of $\dfrac{2}{3}$ cup, so we have the multiplication sentence $x = \dfrac{1}{2} \cdot \dfrac{2}{3}$. For 3 recipes we want to find 3 times $\dfrac{2}{3}$, so we have $y = 3 \cdot \dfrac{2}{3}$.

**Solve.** We carry out the multiplication.

$x = \dfrac{1}{2} \cdot \dfrac{2}{3} = \dfrac{1 \cdot 2}{2 \cdot 3} = \dfrac{2}{2} \cdot \dfrac{1}{3} = \dfrac{1}{3}$

$y = 3 \cdot \dfrac{2}{3} = \dfrac{3 \cdot 2}{3} = \dfrac{3 \cdot 2}{3 \cdot 1} = \dfrac{3}{3} \cdot \dfrac{2}{1} = \dfrac{2}{1} = 2$

**Check.** We repeat the calculations. The answer checks.

**State.** For $\dfrac{1}{2}$ recipe, $\dfrac{1}{3}$ cup of peppers are needed; 2 cups are needed for 3 recipes.

**73. Familiarize.** Let $w$ = the amount Bernardo earns for working $\dfrac{1}{7}$ of a day.

**Translate.** We want to find $\dfrac{1}{7}$ of $\$105$, so we have $w = \dfrac{1}{7} \cdot 105$.

**Solve.** We carry out the multiplication.

$w = \dfrac{1}{7} \cdot 105 = \dfrac{1 \cdot 105}{7} = \dfrac{1 \cdot 7 \cdot 15}{7 \cdot 1} = \dfrac{7}{7} \cdot \dfrac{1 \cdot 15}{1} = \dfrac{1 \cdot 15}{1} = 15$

**Check.** We repeat the calculation. The answer checks.

**State.** Bernardo earns $\$15$ for working $\dfrac{1}{7}$ of a day.

**74. Familiarize.** Let $b$ = the number of bags that can be made from 48 yd of fabric.

*Translate*.

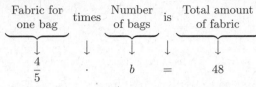

$$\frac{4}{5} \cdot b = 48$$

*Solve*. We divide by $\frac{4}{5}$ on both sides of the equation.

$$\frac{4}{5} \cdot b = 48$$

$$b = 48 \div \frac{4}{5}$$

$$b = 48 \cdot \frac{5}{4} = \frac{48 \cdot 5}{4} = \frac{4 \cdot 12 \cdot 5}{4 \cdot 1}$$

$$= \frac{4}{4} \cdot \frac{12 \cdot 5}{1} = \frac{12 \cdot 5}{1} = 60$$

*Check*. Since $\frac{4}{5} \cdot 60 = \frac{4 \cdot 60}{5} = \frac{4 \cdot 5 \cdot 12}{5 \cdot 1} = \frac{5}{5} \cdot \frac{4 \cdot 12}{1} = 48$, the answer checks.

*State*. 60 book bags can be made from 48 yd of fabric.

**75.** $\frac{2}{13} \cdot x = -\frac{1}{2}$

We divide by $\frac{2}{13}$ on both sides and carry out the division.

$$x = -\frac{1}{2} \div \frac{2}{13} = -\frac{1}{2} \cdot \frac{13}{2} = -\frac{1 \cdot 13}{2 \cdot 2} = -\frac{13}{4}$$

Answer D is correct.

**76.** $\frac{15}{26} \cdot \frac{13}{90} = \frac{15 \cdot 13}{26 \cdot 90} = \frac{3 \cdot 5 \cdot 13 \cdot 1}{2 \cdot 13 \cdot 2 \cdot 3 \cdot 3 \cdot 5}$

$$= \frac{3 \cdot 5 \cdot 13}{3 \cdot 5 \cdot 13} = \frac{1}{2 \cdot 2 \cdot 3} = \frac{1}{12}$$

Answer B is correct.

**77.** $\frac{19}{24} \div \frac{a}{b} = \frac{19}{24} \cdot \frac{b}{a} = \frac{19 \cdot b}{24 \cdot a} = \frac{187,853}{268,224}$

Then, assuming the quotient has not been simplified, we have

$$19 \cdot b = 187,853 \quad \text{and} \quad 24 \cdot a = 268,224$$

$$b = \frac{187,853}{19} \quad \text{and} \qquad a = \frac{268,224}{24}$$

$$b = 9887 \quad \text{and} \qquad a = 11,176.$$

**78.** 13 and 31 are both prime numbers, so 13 is a palindrome prime.

91 is not prime ($91 = 7 \cdot 13$), so 91 is not a palindrome prime.

16 is not prime ($16 = 2 \cdot 8 = 4 \cdot 4$), so it is not a palindrome prime.

11 is prime and when its digits are reversed we have 11 again, so 11 is a palindrome prime.

15 is not prime ($15 = 3 \cdot 5$), so it is not a palindrome prime.

24 is not prime ($24 = 2 \cdot 12 = 3 \cdot 8 = 4 \cdot 6$), so it is not a palindrome prime.

29 is prime but 92 is not ($92 = 2 \cdot 46 = 4 \cdot 23$), so 29 is not a palindrome prime.

101 is prime and when its digits are reversed we get 101 again, so 101 is a palindrome prime.

201 is not prime ($201 = 3 \cdot 67$), so it is not a palindrome prime.

37 and 73 are both prime numbers, so 37 is a palindrome prime.

## Chapter 3 Discussion and Writing Exercises

1. The student is probably multiplying the divisor by the reciprocal of the dividend rather than multiplying the dividend by the reciprocal of the divisor.

2. $9432 = 9 \cdot 1000 + 4 \cdot 100 + 3 \cdot 10 + 2 \cdot 1 = 9(999 + 1) + 4(99 + 1) + 3(9 + 1) + 2 \cdot 1 = 9 \cdot 999 + 9 \cdot 1 + 4 \cdot 99 + 4 \cdot 1 + 3 \cdot 9 + 3 \cdot 1 + 2 \cdot 1$. Since 999, 99, and 9 are each a multiple of 9, $9 \cdot 999$, $4 \cdot 99$, and $3 \cdot 9$ are multiples of 9. This leaves $9 \cdot 1 + 4 \cdot 1 + 3 \cdot 1 + 2 \cdot 1$, or $9 + 4 + 3 + 2$. If $9 + 4 + 3 + 2$, the sum of the digits, is divisible by 9, then 9432 is divisible by 9.

3. Taking $\frac{1}{2}$ of a number is equivalent to multiplying the number by $\frac{1}{2}$. Dividing by $\frac{1}{2}$ is equivalent to multiplying by the reciprocal of $\frac{1}{2}$, or 2. Thus taking $\frac{1}{2}$ of a number is not the same as dividing by $\frac{1}{2}$.

4. We first consider some object and take $\frac{4}{7}$ of it. We divide it into 7 parts and take 4 of them as shown by the shading below.

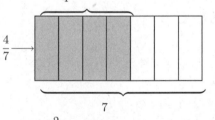

Next we take $\frac{2}{3}$ of the shaded area above. We divide it into 3 parts and take two of them as shown below.

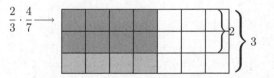

The entire object has been divided into 21 parts, 8 of which have been shaded. Thus, $\frac{2}{3} \cdot \frac{4}{7} = \frac{8}{21}$.

5. Since $\frac{1}{7}$ is a smaller number than $\frac{2}{3}$, there are more $\frac{1}{7}$'s in 5 than $\frac{2}{3}$'s. Thus, $5 \div \frac{1}{7}$ is a bigger number than $5 \div \frac{2}{3}$.

6.  No; in order to simplify a fraction, we must be able to remove a factor of the type $\frac{n}{n}, n \neq 0$, where $n$ is a factor that the numerator and denominator have in common.

---

## Chapter 3 Test

1.  We find as many "two-factor" factorizations of 300 as we can.

    $$1 \cdot 300$$
    $$2 \cdot 150$$
    $$3 \cdot 100$$
    $$4 \cdot 75$$
    $$5 \cdot 60$$
    $$6 \cdot 50$$
    $$10 \cdot 30$$
    $$12 \cdot 25$$
    $$15 \cdot 20$$

    If there are additional factors, they must be between 15 and 20. Since 16, 17, 18, and 19 are not factors of 300, we are finished. The factors of 300 are 1, 2, 3, 4, 5, 6, 10, 12, 15, 20, 25, 30, 50, 60, 75, 100, 150, and 300.

2.  The number 41 is prime. It has only the factors 41 and 1.

3.  The number 14 is composite. It has the factors 1, 2, 7, and 14.

4.  $\qquad \quad 3 \quad \leftarrow$ 3 is prime.
    $$3\ \overline{\smash{\big)}\ 9}$$
    $$2\ \overline{\smash{\big)}\ 1\,8}$$
    $$18 = 2 \cdot 3 \cdot 3$$

5.  We use a factor tree.

    $$60$$

    $$6 \qquad 10$$

    $$2 \quad 3 \quad 2 \quad 5$$

    $60 = 2 \cdot 3 \cdot 2 \cdot 5$, or $2 \cdot 2 \cdot 3 \cdot 5$

6.  1<u>784</u> is divisible by 8 because <u>784</u> is divisible by 8.

7.  $7 + 8 + 4 = 19$; since 19 is not divisible by 9, 784 is not divisible by 9.

8.  555<u>2</u> is not divisible by 5 because the ones digit (2) is not 0 or 5.

9.  The ones digit (2) is even; the sum of the digits $2+3+2+2$, or 9 is divisible by 3. Thus, 2322 is divisible by 6.

10.  $\quad 4 \quad \leftarrow$ Numerator
     $\quad \overline{5} \quad \leftarrow$ Denominator

11.  The figure is divided into 4 equal parts, so the unit is $\frac{1}{4}$ and the denominator is 4. Three of the units are shaded, so the numerator is 3. Thus, $\frac{3}{4}$ is shaded.

12.  There are 7 objects in the set, so the denominator is 7. Three of the objects are shaded, so the numerator is 3. Thus, $\frac{3}{7}$ of the set is shaded.

13.  a) The ratio of pass completions to attempts is $\frac{371}{555}$.

     b) The number of incomplete passes is $555 - 371$, or 184. Then the ratio of incomplete passes to attempts is $\frac{184}{555}$.

14.  $\frac{n}{1} = n$ for any integer $n$. Then $\frac{26}{1} = 26$.

15.  $\frac{n}{n} = 1$ for any integer $n$ that is not 0. Then $\frac{-12}{-12} = 1$.

16.  $\frac{0}{n} = 0$ for any integer $n$ that is not 0. Then $\frac{0}{16} = 0$.

17.  $-\frac{12}{24} = -\frac{1 \cdot 12}{2 \cdot 12} = -\frac{1}{2} \cdot \frac{12}{12} = -\frac{1}{2}$

18.  $\frac{42}{7} = \frac{6 \cdot 7}{1 \cdot 7} = \frac{6}{1} \cdot \frac{7}{7} = \frac{6}{1} = 6$

19.  $\frac{n}{0}$ is not defined for any integer $n$. Then $\frac{9}{0}$ is not defined.

20.  $\frac{7}{2-2} = \frac{7}{0}$

     $\frac{n}{0}$ is not defined for any integer $n$. Then $\frac{7}{2-2}$ is not defined.

21.  $-\frac{72}{108} = -\frac{2 \cdot 36}{3 \cdot 36} = -\frac{2}{3} \cdot \frac{36}{36} = -\frac{2}{3}$

22.  We multiply these     We multiply these
     two numbers:          two numbers:

     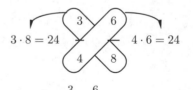

     $3 \cdot 8 = 24 \qquad\qquad\qquad 4 \cdot 6 = 24$

     Since $24 = 24$, $\frac{3}{4} = \frac{6}{8}$.

23.  We multiply these     We multiply these
     two numbers:          two numbers:

     $-5 \cdot 7 = -35 \qquad\qquad 4 \cdot (-9) = -36$

     Since $-35 \neq -36$, $\frac{-5}{4} \neq \frac{-9}{7}$.

24.  $\frac{4}{3} \cdot 24 = \frac{4 \cdot 24}{3} = \frac{4 \cdot 3 \cdot 8}{3 \cdot 1} = \frac{3}{3} \cdot \frac{4 \cdot 8}{1} = 1 \cdot \frac{4 \cdot 8}{1} = \frac{4 \cdot 8}{1} = 32$

25.  $-5 \cdot \frac{3}{10} = -\frac{5 \cdot 3}{10} = -\frac{5 \cdot 3}{2 \cdot 5} = \frac{5}{5} \cdot \left(-\frac{3}{2}\right) =$

     $1 \cdot \left(-\frac{3}{2}\right) = -\frac{3}{2}$

**26.** $\dfrac{2}{3} \cdot \dfrac{15}{4} = \dfrac{2 \cdot 15}{3 \cdot 4} = \dfrac{2 \cdot 3 \cdot 5}{3 \cdot 2 \cdot 2} = \dfrac{2 \cdot 3}{2 \cdot 3} \cdot \dfrac{5}{2} = 1 \cdot \dfrac{5}{2} = \dfrac{5}{2}$

**27.** $-\dfrac{22}{15} \cdot \left(-\dfrac{5}{33}\right) = \dfrac{22 \cdot 5}{15 \cdot 33} = \dfrac{2 \cdot 11 \cdot 5}{3 \cdot 5 \cdot 3 \cdot 11} = \dfrac{5 \cdot 11}{5 \cdot 11} \cdot \dfrac{2}{3 \cdot 3} =$

$1 \cdot \dfrac{2}{3 \cdot 3} = \dfrac{2}{3 \cdot 3} = \dfrac{2}{9}$

**28.** $\dfrac{5}{8}$    Interchange the numerator and denominator.

The reciprocal of $\dfrac{5}{8}$ is $\dfrac{8}{5}$.   $\left(\dfrac{5}{8} \cdot \dfrac{8}{5} = \dfrac{40}{40} = 1\right)$

**29.** $-\dfrac{1}{4}$    Interchange the numerator and denominator.

The reciprocal of $-\dfrac{1}{4}$ is $-\dfrac{4}{1}$, or $-4$.

$\left(-\dfrac{1}{4} \cdot (-4) = \dfrac{4}{4} = 1\right)$

**30.** Think of 18 as $\dfrac{18}{1}$.

$\dfrac{18}{1}$    Interchange the numerator and denominator.

The reciprocal of $\dfrac{18}{1}$ is $\dfrac{1}{18}$.   $\left(\dfrac{18}{1} \cdot \dfrac{1}{18} = \dfrac{18}{18} = 1\right)$

**31.** $\dfrac{1}{5} \div \dfrac{1}{8} = \dfrac{1}{5} \cdot \dfrac{8}{1} = \dfrac{1 \cdot 8}{5 \cdot 1} = \dfrac{8}{5}$

**32.** $12 \div \left(-\dfrac{2}{3}\right) = 12 \cdot \left(-\dfrac{3}{2}\right) = -\dfrac{12 \cdot 3}{2} = -\dfrac{2 \cdot 6 \cdot 3}{2 \cdot 1} =$

$\dfrac{2}{2} \cdot \left(-\dfrac{6 \cdot 3}{1}\right) = -\dfrac{6 \cdot 3}{1} = -18$

**33.** $-\dfrac{24}{5} \div \left(-\dfrac{28}{15}\right) = -\dfrac{24}{5} \cdot \left(-\dfrac{15}{28}\right) = \dfrac{24 \cdot 15}{5 \cdot 28} =$

$\dfrac{4 \cdot 6 \cdot 3 \cdot 5}{5 \cdot 4 \cdot 7} = \dfrac{4 \cdot 5}{4 \cdot 5} \cdot \dfrac{6 \cdot 3}{7} = \dfrac{6 \cdot 3}{7} = \dfrac{18}{7}$

**34.** $\dfrac{7}{8} \cdot x = -56$

$x = -56 \div \dfrac{7}{8}$   Dividing by $\dfrac{7}{8}$ on both sides

$x = -56 \cdot \dfrac{8}{7}$

$x = \dfrac{-56 \cdot 8}{7} = -\dfrac{7 \cdot 8 \cdot 8}{7 \cdot 1} = \dfrac{7}{7} \cdot \left(-\dfrac{8 \cdot 8}{1}\right) =$

$-\dfrac{8 \cdot 8}{1} = -64$

The solution is $-64$.

**35.** $t \cdot \dfrac{2}{5} = \dfrac{7}{10}$

$t = \dfrac{7}{10} \div \dfrac{2}{5}$   Dividing by $\dfrac{2}{5}$ on both sides

$t = \dfrac{7}{10} \cdot \dfrac{5}{2}$

$= \dfrac{7 \cdot 5}{10 \cdot 2} = \dfrac{7 \cdot 5}{2 \cdot 5 \cdot 2} = \dfrac{5}{5} \cdot \dfrac{7}{2 \cdot 2} = \dfrac{7}{2 \cdot 2} = \dfrac{7}{4}$

The solution is $\dfrac{7}{4}$.

**36.** *Familiarize.* Let $d$ = the number of students who live in dorms.

*Translate.* We translate to an equation.

$\underbrace{\text{How many students}}$   is   $\dfrac{5}{8}$   of   $\underbrace{\text{7000 students?}}$

$\qquad\quad \downarrow \qquad\qquad\quad \downarrow \;\; \downarrow \;\; \downarrow \qquad\qquad \downarrow$

$\qquad\quad d \qquad\qquad\quad = \dfrac{5}{8} \;\; \cdot \qquad\quad 7000$

*Solve.* We carry out the multiplication.

$d = \dfrac{5}{8} \cdot 7000 = \dfrac{5 \cdot 7000}{8} = \dfrac{5 \cdot 8 \cdot 875}{8 \cdot 1} =$

$\dfrac{8}{8} \cdot \dfrac{5 \cdot 875}{1} = 4375$

*Check.* We can check by repeating the calculation. The answer checks.

*State.* 4375 students live in dorms.

**37.** *Familiarize.* Let $l$ = the length of each piece of taffy, in meters.

*Translate.* We are dividing $\dfrac{9}{10}$ m into 12 equal pieces of length $l$. An equation that corresponds to the situation is $l = \dfrac{9}{10} \div 12$.

*Solve.* We carry out the division.

$l = \dfrac{9}{10} \div 12 = \dfrac{9}{10} \cdot \dfrac{1}{12} = \dfrac{9 \cdot 1}{10 \cdot 12} = \dfrac{3 \cdot 3 \cdot 1}{10 \cdot 3 \cdot 4} =$

$\dfrac{3}{3} \cdot \dfrac{3 \cdot 1}{10 \cdot 4} = \dfrac{3}{40}$

*Check.* The total length of 12 pieces of taffy, each of length $\dfrac{3}{40}$ m, is

$12 \cdot \dfrac{3}{40} = \dfrac{12 \cdot 3}{40} = \dfrac{3 \cdot 4 \cdot 3}{4 \cdot 10} = \dfrac{4}{4} \cdot \dfrac{3 \cdot 3}{10} = \dfrac{9}{10}$ m.

The answer checks.

*State.* The length of each piece of taffy will be $\dfrac{3}{40}$ m.

**38.** *Familiarize.* Let $t$ = the number of quarts of tea the thermos holds when it is full.

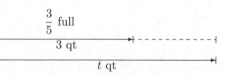

*Translate.* We translate to an equation.

| Fraction filled | of | Total capacity of thermos | is | Amount in thermos |
|---|---|---|---|---|
| $\downarrow$ | $\downarrow$ | $\downarrow$ | $\downarrow$ | $\downarrow$ |
| $\dfrac{3}{5}$ | $\cdot$ | $t$ | $=$ | $3$ |

*Solve.* We divide by $\dfrac{3}{5}$ on both sides and carry out the division.

$t = 3 \div \dfrac{3}{5} = 3 \cdot \dfrac{5}{3} = \dfrac{3 \cdot 5}{3} = \dfrac{3}{3} \cdot \dfrac{5}{1} = 5$

*Check.* Since $\dfrac{3}{5} \cdot 5 = \dfrac{3 \cdot 5}{5} = \dfrac{5}{5} \cdot \dfrac{3}{1} = 3$, the answer checks.

*State.* The thermos holds 5 qt of tea when it is full.

Copyright © 2011 Pearson Education, Inc. Publishing as Addison-Wesley.

**39. Familiarize.** Let $s =$ the number of inches the screw will go into the piece of walnut when it is turned 6 complete revolutions.

**Translate.** We translate to an equation.

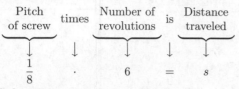

$$\frac{1}{8} \cdot 6 = s$$

**Solve.** We carry out the multiplication.

$$s = \frac{1}{8} \cdot 6 = \frac{1 \cdot 6}{8} = \frac{1 \cdot 2 \cdot 3}{2 \cdot 4} = \frac{2}{2} \cdot \frac{1 \cdot 3}{4} = \frac{3}{4}$$

**Check.** We repeat the calculation. The answer checks.

**State.** The screw will go $\frac{3}{4}$ in. into the piece of walnut.

**40.** Only the figures in C and D are divided into 6 equal parts, so in each of these the unit is $\frac{1}{6}$. The denominator is 6. In C, 7 of the units are shaded, so the numerator is 7 and $\frac{7}{6}$ is shaded. In D, 5 of the units are shaded, so the numerator is 5 and $\frac{5}{6}$ is shaded. We see that the correct answer is C.

**41. Familiarize.** This is a multistep problem. First we will find the number of acres Karl received. Then we will find how much of that land Eileen received. Let $k =$ the number of acres of land Karl received.

**Translate.** We translate to an equation.

$$k = \frac{7}{8} \cdot \frac{2}{3}$$

**Solve.** We carry out the multiplication.

$$k = \frac{7}{8} \cdot \frac{2}{3} = \frac{7 \cdot 2}{8 \cdot 3} = \frac{7 \cdot 2}{2 \cdot 4 \cdot 3} = \frac{2}{2} \cdot \frac{7}{4 \cdot 3} = \frac{7}{12}$$

Karl received $\frac{7}{12}$ acre of land. Let $a =$ the number of acres Eileen received. An equation that corresponds to this situation is

$$a = \frac{1}{4} \cdot \frac{7}{12}.$$

We solve the equation by carrying out the multiplication.

$$a = \frac{1}{4} \cdot \frac{7}{12} = \frac{1 \cdot 7}{4 \cdot 12} = \frac{7}{48}$$

**Check.** We repeat the calculations. The answer checks.

**State.** Eileen received $\frac{7}{48}$ acre of land.

**42.** First we will evaluate the exponential expression; then we will multiply and divide in order from left to right.

$$\left(\frac{3}{8}\right)^2 \div \frac{6}{7} \cdot \frac{2}{9} \div (-5) = \frac{9}{64} \div \frac{6}{7} \cdot \frac{2}{9} \div (-5)$$

$$= \frac{9}{64} \cdot \frac{7}{6} \cdot \frac{2}{9} \div (-5)$$

$$= \frac{9 \cdot 7}{64 \cdot 6} \cdot \frac{2}{9} \div (-5)$$

$$= \frac{9 \cdot 7 \cdot 2}{64 \cdot 6 \cdot 9} \div (-5)$$

$$= \frac{9 \cdot 7 \cdot 2}{64 \cdot 6 \cdot 9} \cdot \left(-\frac{1}{5}\right)$$

$$= -\frac{9 \cdot 7 \cdot 2 \cdot 1}{64 \cdot 6 \cdot 9 \cdot 5}$$

$$= -\frac{9 \cdot 7 \cdot 2 \cdot 1}{64 \cdot 2 \cdot 3 \cdot 9 \cdot 5}$$

$$= \frac{9 \cdot 2}{9 \cdot 2} \cdot \left(-\frac{7 \cdot 1}{64 \cdot 3 \cdot 5}\right)$$

$$= -\frac{7}{960}$$

## Cumulative Review Chapters 1 - 3

**1.** A word name for 7,453,062 is seven million, four hundred fifty-three thousand, sixty-two

**2.** 2 3, 7 $\boxed{4}$ 6, 5 9 1

The digit 4 means 4 ten thousands.

**3.** $\frac{n}{1} = n$, for any integer $n$.

$\frac{5}{1} = 5$

**4.** $\frac{n}{n} = 1$, for any integer $n$ that is not 0.

$\frac{-17}{-17} = 1$

**5.** $-\frac{56}{42} = -\frac{2 \cdot 2 \cdot 2 \cdot 7}{2 \cdot 3 \cdot 7} = -\frac{2 \cdot 2}{3} \cdot \frac{2 \cdot 7}{2 \cdot 7} = -\frac{2 \cdot 2}{3} = -\frac{4}{3}$

**6.** $\frac{n}{0}$ is not defined, for any integer $n$.

$\frac{32}{0}$ is not defined.

**7.** Think of 8 as $\frac{8}{1}$. Then interchange the numerator and the denominator.

The reciprocal of 8, or $\frac{8}{1}$, is $\frac{1}{8}$.

**8.** Interchange the numerator and denominator.

The reciprocal of $-\frac{3}{2}$ is $-\frac{2}{3}$.

9.
$$
\begin{array}{r}
{\scriptstyle 1\ \ 1} \\
4\,6\,5\,8 \\
+\ \ \ 7\,2\,9 \\
\hline
5\,3\,8\,7
\end{array}
$$

10. $-3 + (-9)$ Two negative numbers. Add the absolute values, 3 and 9, getting 12. Make the answer negative. $-3 + (-9) = -12$

11.
$$
\begin{array}{r}
{\scriptstyle \quad 13} \\
{\scriptstyle 4\ 9\ \not{3}\ 12} \\
\not{5}\,\not{0}\,\not{4}\,\not{2} \\
-\ 3\,6\,5\,8 \\
\hline
1\,3\,8\,4
\end{array}
$$

12. $-9 - 7 = -9 + (-7) = -16$

13.
$$
\begin{array}{r}
{\scriptstyle 1\ 2} \\
{\scriptstyle 3\ 4} \\
4\,5\,7 \\
\times\ \ \ 3\,6 \\
\hline
2\,7\,4\,2 \\
1\,3\,7\,1\,0 \\
\hline
1\,6,4\,5\,2
\end{array}
$$

14. $-8 \cdot 12 = -96$

15. $-\dfrac{5}{6} \cdot 15 = -\dfrac{5 \cdot 15}{6} = -\dfrac{5 \cdot 3 \cdot 5}{2 \cdot 3} = -\dfrac{5 \cdot 5}{2} \cdot \dfrac{3}{3} = -\dfrac{5 \cdot 5}{2} = -\dfrac{25}{2}$

16. $\dfrac{3}{7} \cdot \dfrac{14}{9} = \dfrac{3 \cdot 14}{7 \cdot 9} = \dfrac{3 \cdot 2 \cdot 7}{7 \cdot 3 \cdot 3} = \dfrac{3 \cdot 7}{3 \cdot 7} \cdot \dfrac{2}{3} = \dfrac{2}{3}$

17.
$$
\begin{array}{r}
4\,5\,1 \\
2\,4\,\overline{)\,1\,0,8\,4\,6} \\
9\,6\,0\,0 \\
\hline
1\,2\,4\,6 \\
1\,2\,0\,0 \\
\hline
4\,6 \\
2\,4 \\
\hline
2\,2
\end{array}
$$

The answer is 451 R 22.

18. $-56 \div (-7) = 8$      Check: $8(-7) = -56$

19. $-\dfrac{4}{7} \div 28 = -\dfrac{4}{7} \cdot \dfrac{1}{28} = -\dfrac{4 \cdot 1}{7 \cdot 28} = -\dfrac{4 \cdot 1}{7 \cdot 4 \cdot 7} = -\dfrac{1}{7 \cdot 7} \cdot \dfrac{4}{4} =$
$-\dfrac{1}{7 \cdot 7} = -\dfrac{1}{49}$

20. $\dfrac{3}{4} \div \dfrac{9}{16} = \dfrac{3}{4} \cdot \dfrac{16}{9} = \dfrac{3 \cdot 16}{4 \cdot 9} = \dfrac{3 \cdot 4 \cdot 4}{4 \cdot 3 \cdot 3} = \dfrac{3 \cdot 4}{3 \cdot 4} \cdot \dfrac{4}{3} = \dfrac{4}{3}$

21.
$$
\begin{aligned}
8^2 \div 8 \cdot 2 - (2 + 2 \cdot 7) \\
= 8^2 \div 8 \cdot 2 - (2 + 14) \\
= 8^2 \div 8 \cdot 2 - 16 \\
= 64 \div 8 \cdot 2 - 16 \\
= 8 \cdot 2 - 16 \\
= 16 - 16 \\
= 0
\end{aligned}
$$

22.
$$
\begin{aligned}
108 \div 9 - [3(18 - 5 \cdot 3)] \\
= 108 \div 9 - [3(18 - 15)] \\
= 108 \div 9 - [3(3)] \\
= 108 \div 9 - 9 \\
= 12 - 9 \\
= 3
\end{aligned}
$$

23.
$$
\begin{aligned}
-20 - 10 \div 5 + 2^3 &= -20 - 10 \div 5 + 8 \\
&= -20 - 2 + 8 \\
&= -22 + 8 \\
&= -14
\end{aligned}
$$

24.
$$
\begin{aligned}
(8 - 10^2) \div (5^2 - 2) &= (8 - 100) \div (25 - 2) \\
&= -92 \div 23 \\
&= -4
\end{aligned}
$$

25. We add the numbers and divide by the number of addends.
$$
\frac{85 + 91 + 80 + 88}{4} = \frac{344}{4} = 86
$$

26. Round 165,739 to the nearest thousand.

$$1\,6\,5, \boxed{7}\,3\,9$$
$\uparrow$

The digit 5 is in the thousands place. Consider the next digit to the right. Since the digit, 7, is 5 or higher, round 5 thousands up to 6 thousands. Then change all digits to the right of the thousands digit to zeros.

The answer is 166,000.

27. $9^2 = 9 \cdot 9 = 81$

28. $5^3 = 5 \cdot 5 \cdot 5 = 125$

29. $2^4 = 2 \cdot 2 \cdot 2 \cdot 2 = 16$

30. Since $-26$ is to the left of 2, we have $-26 < 2$.

31. Since 19 is to the right of 17, we have $19 > 17$.

32. The distance of 33 from 0 is 33, so $|33| = 33$.

33. The distance of $-86$ from 0 is 86, so $|-86| = 86$.

34. The distance of 0 from 0 is 0, so $|0| = 0$.

35. The opposite of 29 is $-29$ because $29 + (-29) = 0$.

36. The opposite of $-144$ is 144 because $-144 + 144 = 0$.

37. If $x = -7$, then $-(-x) = -[-(-7)] = -[7] = -7$.

38.
$$
\begin{array}{c}
\leftarrow\!+\!+\!+\!+\!+\!+\!+\!+\!+\!+\!\bullet\!+\!+\!\rightarrow \\
{\scriptstyle -2\ -1\ \ 0\ \ 1\ \ 2\ \ 3\ \ 4\ \ 5\ \ 6\ \ 7\ \ 8\ \ 9\ \ 10}
\end{array}
$$

39. The number 29 is prime. It has only the factors 1 and 29.

40. 1 is neither prime nor composite.

41. The number 24 has factors 1, 2, 3, 4, 6, 8, 12, and 24. Since 24 is not 1 and not prime, it is composite.

42.
$$
\begin{array}{r}
3 \quad\ \ \leftarrow\ \ 3 \text{ is prime.} \\
3\,\overline{)\,9} \\
2\,\overline{)\,1\,8} \\
2\,\overline{)\,3\,6}
\end{array}
$$
$36 = 2 \cdot 2 \cdot 3 \cdot 3$

43.

$135 = 3 \cdot 3 \cdot 3 \cdot 5$

**44.**
$$
\begin{array}{r}
5 \\
5\overline{)2\ 5} \\
3\overline{)7\ 5} \\
3\overline{)2\ 2\ 5} \\
3\overline{)6\ 7\ 5} \\
2\overline{)1\ 3\ 5\ 0}
\end{array}
\quad \leftarrow 5 \text{ is prime}
$$

$1350 = 2 \cdot 3 \cdot 3 \cdot 3 \cdot 5 \cdot 5$

**45.** 6330 is divisible by 2 because its one digit is even.

The sum of the digits, $6 + 3 + 3 + 0$, or 12, is divisible by 3, so 6330 is divisible by 3.

6330 is not divisible by 4 because the number named by its last two digits, 30, is not divisible by 4.

A number is divisible by 5 if its ones digit is 0 or 5. Thus, 6330 is divisible by 5.

As we found above, the ones digit of 6330 is even and the sum of the digits is divisible by 3. Thus, 6330 is divisible by 6.

6330 is not divisible by 8 because the number named by its last three digits, 330, is not divisible by 8.

The sum of the digits, 12, is not divisible by 9, so 6330 is not divisible by 9.

6330 is divisible by 10 because its ones digit is 0.

**46.**   We multiply these          We multiply these
two numbers:                two numbers:

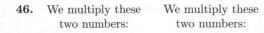

$5 \cdot 11 = 55$          $8 \cdot 7 = 56$

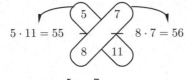

Since $55 \neq 56$, $\dfrac{5}{8} \neq \dfrac{7}{11}$.

**47.** We first rewrite $-\dfrac{5}{4}$ as $\dfrac{-5}{4}$ and $-\dfrac{20}{16}$ as $\dfrac{-20}{16}$.

We multiply these          We multiply these
two numbers:                two numbers:

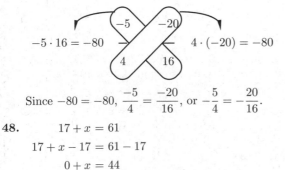

$-5 \cdot 16 = -80$          $4 \cdot (-20) = -80$

Since $-80 = -80$, $\dfrac{-5}{4} = \dfrac{-20}{16}$, or $-\dfrac{5}{4} = -\dfrac{20}{16}$.

**48.**
$$
\begin{aligned}
17 + x &= 61 \\
17 + x - 17 &= 61 - 17 \\
0 + x &= 44 \\
x &= 44
\end{aligned}
$$

The solution is 44.

**49.** $\dfrac{3}{5} \cdot y = -\dfrac{9}{25}$

$$
\begin{aligned}
y &= \frac{5}{3} \cdot \left(-\frac{9}{25}\right) \\
y &= -\frac{5 \cdot 9}{3 \cdot 25} = -\frac{5 \cdot 3 \cdot 3}{3 \cdot 5 \cdot 5} = -\frac{3}{5} \cdot \frac{5 \cdot 3}{5 \cdot 3} \\
y &= -\frac{3}{5}
\end{aligned}
$$

The solution is $-\dfrac{3}{5}$.

**50.** *Familiarize*. We visualize the situation. Let $y =$ the year in which Halley's Comet will appear again.

| Year of last appearance 1986 | Years between appearances 76 |
|---|---|
| Year of next appearance $y$ | |

*Translate*. We translate to an equation.

| Year of last appearance | plus | Years between appearances | is | Year of next appearance |
|---|---|---|---|---|
| ↓ | ↓ | ↓ | ↓ | ↓ |
| 1986 | + | 76 | = | $y$ |

*Solve*. We carry out the addition.

$$
\begin{array}{r}
{\scriptstyle 1\ 1\ 1} \\
1\ 9\ 8\ 6 \\
+ \quad 7\ 6 \\
\hline
2\ 0\ 6\ 2
\end{array}
$$

*Check*. We can estimate: $1986 + 76 \approx 1990 + 80 = 2070 \approx 2062$. The answer checks.

*State*. Halley's Comet will appear again in 2062.

**51.** *Familiarize*. This is a multistep problem. Let $p =$ the total cost of the wrapping paper, $c =$ the total cost of the candles, and $t =$ the total cost of the order.

*Translate*.

| Total cost of wrapping paper | is | Number of rolls | times | Price per roll |
|---|---|---|---|---|
| ↓ | ↓ | ↓ | ↓ | ↓ |
| $p$ | = | 4 | × | 5 |

| Total cost of candles | is | Number of candles | times | Price per candle |
|---|---|---|---|---|
| ↓ | ↓ | ↓ | ↓ | ↓ |
| $c$ | = | 2 | × | 8 |

| Total cost of order | is | Total cost of wrapping paper | plus | Total cost of candles |
|---|---|---|---|---|
| ↓ | ↓ | ↓ | ↓ | ↓ |
| $t$ | = | $p$ | + | $c$ |

*Solve*. We solve the equations.

$$p = 4 \times 5$$
$$p = 20$$

$$c = 2 \times 8$$
$$c = 16$$

$$t = p + c$$
$$t = 20 + 16$$
$$t = 36$$

*Check*. We can repeat the calculations. The answer checks.

*State*. The total cost of the order was $36.

**52.** First we multiply to find the change in temperature $t$ in the 11 minutes from 9:00 A.M. to 9:11 A.M.

$$t = 2 \cdot 11 = 22$$

The temperature dropped 22°C.

Now we subtract to find the temperature $T$ at 9:11 A.M.

$$T = 5 - 22 = -17.$$

At 9:11 A.M. the temperature was $-17$°C.

**53.** *Familiarize*. Let $s = $ the number of students with full-time jobs.

*Translate*. The number of students with full-time jobs is $\frac{2}{5}$ of 35, so we have

$$s = \frac{2}{5} \cdot 35.$$

*Solve*. We carry out the multiplication.

$$s = \frac{2}{5} \cdot 35 = \frac{2 \cdot 35}{5} = \frac{2 \cdot 5 \cdot 7}{5 \cdot 1} = \frac{5}{5} \cdot \frac{2 \cdot 7}{1} = 14$$

*Check*. We can repeat the calculation. The answer checks.

*State*. 14 students have full-time jobs.

**54.** a) $5 \cdot 20 = 100$ and $5 + 20 = 25$, so the numbers we want are 5 and 20.

b) $-4 \cdot (-25) = 100$ and $-4 + (-25) = -29$, so the numbers we want are $-4$ and $-25$.

c) $10 \cdot 10 = 100$ and $10 - 10 = 0$; also, $-10 \cdot (-10) = 100$ and $-10 - (-10) = 0$. The numbers we want are 10 and 10 or $-10$ and $-10$.

# Chapter 4

# Fraction Notation and Mixed Numerals

## Exercise Set 4.1

In this section we will find the LCM using the list of multiples method in Exercises 1 - 19 and the prime factorization method in Exercises 21 - 47.

**1.** a) 4 is a multiple of 2, so it is the LCM.

c) The LCM = 4.

**3.** a) 25 is not a multiple of 10.

b) Check multiples:

$2 \cdot 25 = 50$     A multiple of 10

c) The LCM = 50.

**5.** a) 40 is a multiple of 20, so it is the LCM.

c) The LCM = 40.

**7.** a) 27 is not a multiple of 18.

b) Check multiples:

$2 \cdot 27 = 54$     A multiple of 18

c) The LCM = 54.

**9.** a) 50 is not a multiple of 30.

b) Check multiples:

$2 \cdot 50 = 100$     Not a multiple of 30
$3 \cdot 50 = 150$     A multiple of 30

c) The LCM = 150.

**11.** a) 40 is not a multiple of 30.

b) Check multiples:

$2 \cdot 40 = 80$     Not a multiple of 30
$3 \cdot 40 = 120$     A multiple of 30

c) The LCM = 120.

**13.** a) 24 is not a multiple of 18.

b) Check multiples:

$2 \cdot 24 = 48$     Not a multiple of 18
$3 \cdot 24 = 72$     A multiple of 18

c) The LCM = 72.

**15.** a) 70 is not a multiple of 60.

b) Check multiples:

$2 \cdot 70 = 140$     Not a multiple of 60
$3 \cdot 70 = 210$     Not a multiple of 60
$4 \cdot 70 = 280$     Not a multiple of 60
$5 \cdot 70 = 350$     Not a multiple of 60
$6 \cdot 70 = 420$     A multiple of 60

c) The LCM = 420.

**17.** a) 36 is not a multiple of 16.

b) Check multiples:

$2 \cdot 36 = 72$     Not a multiple of 16
$3 \cdot 36 = 108$     Not a multiple of 16
$4 \cdot 36 = 144$     A multiple of 16

c) The LCM = 144.

**19.** a) 36 is not a multiple of 32.

b) Check multiples:

$2 \cdot 36 = 72$     Not a multiple of 32
$3 \cdot 36 = 108$     Not a multiple of 32
$4 \cdot 36 = 144$     Not a multiple of 32
$5 \cdot 36 = 180$     Not a multiple of 32
$6 \cdot 36 = 216$     Not a multiple of 32
$7 \cdot 36 = 252$     Not a multiple of 32
$8 \cdot 36 = 288$     A multiple of 32

c) The LCM = 288.

**21.** Note that each of the numbers 2, 3, and 5 is prime. They have no common prime factor. When this happens, the LCM is just the product of the numbers.

The LCM is $2 \cdot 3 \cdot 5$, or 30.

**23.** a) Find the prime factorization of each number.

$5 = 5$
$18 = 2 \cdot 3 \cdot 3$
$3 = 3$

b) Create a product by writing factors, using each the greatest number of times it occurs in any one factorization.

Consider the factor 2. The greatest number of times 2 occurs in any one factorization is one time. We write 2 as a factor one time.

$2 \cdot ?$

Consider the factor 3. The greatest number of times 3 occurs in any one factorization is two times. We write 3 as a factor two times.

$2 \cdot 3 \cdot 3 \cdot ?$

Consider the factor 5. The greatest number of times 5 occurs in any one factorization is one time. We write 5 as a factor one time.

$2 \cdot 3 \cdot 3 \cdot 5 \cdot ?$

Since there are no other prime factors in any of the factorizations, the LCM is $2 \cdot 3 \cdot 3 \cdot 5$, or 90.

**25.** a) Find the prime factorization of each number.

$24 = 2 \cdot 2 \cdot 2 \cdot 3$
$36 = 2 \cdot 2 \cdot 3 \cdot 3$
$12 = 2 \cdot 2 \cdot 3$

b) Create a product by writing factors, using each the greatest number of times it occurs in any one factorization.

Consider the factor 2. The greatest number of times 2 occurs in any one factorization is three. We write 2 as a factor three times.

$$2 \cdot 2 \cdot 2 \cdot \ ?$$

Consider the factor 3. The greatest number of times 3 occurs in any one factorization is two. We write 3 as a factor two times.

$$2 \cdot 2 \cdot 2 \cdot 3 \cdot 3 \cdot \ ?$$

Since there are no other prime factors in any of the factorizations, the LCM is $2 \cdot 2 \cdot 2 \cdot 3 \cdot 3$, or 72.

**27.** a) Find the prime factorization of each number.

$$\begin{aligned} 5 &= 5 \qquad \text{(5 is prime.)} \\ 12 &= 2 \cdot 2 \cdot 3 \\ 15 &= 3 \cdot 5 \end{aligned}$$

b) Create a product by writing each factor the greatest number of times it occurs in any one factorization.

The greatest number of times 2 occurs in any one factorization is two times.

The greatest number of times 3 occurs in any one factorization is one time.

The greatest number of times 5 occurs in any one factorization is one time.

Since there are no other prime factors in any of the factorizations, the LCM is $2 \cdot 2 \cdot 3 \cdot 5$, or 60.

**29.** a) Find the prime factorization of each number.

$$\begin{aligned} 9 &= 3 \cdot 3 \\ 12 &= 2 \cdot 2 \cdot 3 \\ 6 &= 2 \cdot 3 \end{aligned}$$

b) Create a product by writing each factor the greatest number of times it occurs in any one factorization.

The greatest number of times 2 occurs in any one factorization is two times.

The greatest number of times 3 occurs in any one factorization is two times.

Since there are no other prime factors in any of the factorizations, the LCM is $2 \cdot 2 \cdot 3 \cdot 3$, or 36.

**31.** a) Find the prime factorization of each number.

$$\begin{aligned} 180 &= 2 \cdot 2 \cdot 3 \cdot 3 \cdot 5 \\ 100 &= 2 \cdot 2 \cdot 5 \cdot 5 \\ 450 &= 2 \cdot 3 \cdot 3 \cdot 5 \cdot 5 \end{aligned}$$

b) Create a product by writing each factor the greatest number of times it occurs in any one factorization.

The greatest number of times 2 occurs in any one factorization is two times.

The greatest number of times 3 occurs in any one factorization is two times.

The greatest number of times 5 occurs in any one factorization is two times.

Since there are no other prime factors in any of the factorizations, the LCM is $2 \cdot 2 \cdot 3 \cdot 3 \cdot 5 \cdot 5$, or 900.

We can also find the LCM using exponents.

$$\begin{aligned} 180 &= 2^2 \cdot 3^2 \cdot 5^1 \\ 100 &= 2^2 \cdot 5^2 \\ 450 &= 2^1 \cdot 3^2 \cdot 5^2 \end{aligned}$$

The largest exponents of 2, 3, 5 in any of the factorizations are each 2. Thus, the LCM $= 2^2 \cdot 3^2 \cdot 5^2$, or 900.

**33.** Note that 8 is a factor of 48. If one number is a factor of another, the LCM is the greater number.

The LCM is 48.

The factorization method will also work here if you do not recognize at the outset that 8 is a factor of 48.

**35.** Note that 5 is a factor of 50. If one number is a factor of another, the LCM is the greater number.

The LCM is 50.

**37.** Note that 11 and 13 are prime. They have no common prime factor. When this happens, the LCM is just the product of the numbers.

The LCM is $11 \cdot 13$, or 143.

**39.** a) Find the prime factorization of each number.

$$\begin{aligned} 12 &= 2 \cdot 2 \cdot 3 \\ 35 &= 5 \cdot 7 \end{aligned}$$

b) Note that the two numbers have no common prime factor. When this happens, the LCM is just the product of the numbers.

The LCM is $12 \cdot 35$, or 420.

**41.** a) Find the prime factorization of each number.

$$\begin{aligned} 54 &= 3 \cdot 3 \cdot 3 \cdot 2 \\ 63 &= 3 \cdot 3 \cdot 7 \end{aligned}$$

b) Create a product by writing each factor the greatest number of times it occurs in any one factorization.

The greatest number of times 2 occurs in any one factorization is one time.

The greatest number of times 3 occurs in any one factorization is three times.

The greatest number of times 7 occurs in any one factorization is one time.

Since there are no other prime factors in any of the factorizations, the LCM is $2 \cdot 3 \cdot 3 \cdot 3 \cdot 7$, or 378.

**43.** a) Find the prime factorization of each number.

$$\begin{aligned} 81 &= 3 \cdot 3 \cdot 3 \cdot 3 \\ 90 &= 2 \cdot 3 \cdot 3 \cdot 5 \end{aligned}$$

b) Create a product by writing each factor the greatest number of times it occurs in any one factorization.

The greatest number of times 2 occurs in any one factorization is one time.

The greatest number of times 3 occurs in any one factorization is four times.

The greatest number of times 5 occurs in any one factorization is one time.

Since there are no other prime factors in any of the factorizations, the LCM is $2 \cdot 3 \cdot 3 \cdot 3 \cdot 3 \cdot 5$, or 810.

**45.** a) Find the prime factorization of each number.

$$36 = 2 \cdot 2 \cdot 3 \cdot 3$$
$$54 = 2 \cdot 3 \cdot 3 \cdot 3$$
$$80 = 2 \cdot 2 \cdot 2 \cdot 2 \cdot 5$$

b) Create a product by writing each factor the greatest number of times it occurs in any one factorization.

The greatest number of times 2 occurs in any one factorization is four times.

The greatest number of times 3 occurs in any one factorization is three times.

The greatest number of times 5 occurs in any one factorization is one time.

Since there are no other prime factors in any of the factorizations, the LCM is $2 \cdot 2 \cdot 2 \cdot 2 \cdot 3 \cdot 3 \cdot 3 \cdot 5$, or 2160.

**47.** a) Find the prime factorization of each number.

$$39 = 3 \cdot 13$$
$$91 = 7 \cdot 13$$
$$108 = 2 \cdot 2 \cdot 3 \cdot 3 \cdot 3$$
$$26 = 2 \cdot 13$$

b) Create a product by writing each factor the greatest number of times it occurs in any one factorization.

The greatest number of times 2 occurs in any one factorization is two times.

The greatest number of times 3 occurs in any one factorization is three times.

The greatest number of times 7 occurs in any one factorization is one time.

The greatest number of times 13 occurs in any one factorization is one time.

Since there are no other prime factors in any of the factorizations, the LCM is $2 \cdot 2 \cdot 3 \cdot 3 \cdot 3 \cdot 7 \cdot 13$, or 9828.

**49.** a) Find the prime factorization of each number.

$$2000 = 2 \cdot 2 \cdot 2 \cdot 2 \cdot 5 \cdot 5 \cdot 5$$
$$3000 = 2 \cdot 2 \cdot 2 \cdot 3 \cdot 5 \cdot 5 \cdot 5$$

b) Create a product by writing each factor the greatest number of times it occurs in any one factorization.

The greatest number of times 2 occurs in any one factorization is four times.

The greatest number of times 3 occurs in any one factorization is one time.

The greatest number of times 5 occurs in any one factorization is three times.

Since there are no other prime factors in any of the factorizations, the LCM is $2 \cdot 2 \cdot 2 \cdot 2 \cdot 3 \cdot 5 \cdot 5 \cdot 5$, or 6000.

**51.** We find the LCM of the number of years it takes Jupiter and Saturn to make a complete revolution around the sun.

Jupiter: $12 = 2 \cdot 2 \cdot 3$

Saturn: $30 = 2 \cdot 3 \cdot 5$

The LCM $= 2 \cdot 2 \cdot 3 \cdot 5$, or 60. Thus, Jupiter and Saturn will appear in the same direction in the night sky once every 60 years.

**53.** We find the LCM of the number of years it takes Saturn and Uranus to make a complete revolution around the sun.

Saturn: $30 = 2 \cdot 3 \cdot 5$

Uranus: $84 = 2 \cdot 2 \cdot 3 \cdot 7$

The LCM is $2 \cdot 2 \cdot 3 \cdot 5 \cdot 7$, or 420. Thus, Saturn and Uranus will appear in the same direction in the night sky once every 420 years.

**55.** *Familiarize*. Let $t =$ the number of tornadoes occurring from January through May in 2007.

*Translate*.

| January through May tornadoes in 2007 | plus | Increase in 2008 | is | January through May tornadoes in 2008 |
|---|---|---|---|---|
| ↓ | ↓ | ↓ | ↓ | ↓ |
| $t$ | $+$ | $348$ | $=$ | $1007$ |

*Solve*. We subtract 348 on both sides of the equation.

$$t + 348 = 1007$$
$$t + 348 - 348 = 1007 - 348$$
$$t = 659$$

*Check*. An increase of 348 over 659 is $659 + 348$, or 1007. The answer checks.

*State*. There were 659 tornadoes during the period from January through May in 2007.

**57.** $-\dfrac{4}{5} \div \dfrac{7}{10} = -\dfrac{4}{5} \cdot \dfrac{10}{7} = -\dfrac{4 \cdot 10}{5 \cdot 7} = -\dfrac{2 \cdot 2 \cdot 2 \cdot 5}{5 \cdot 7} =$

$-\dfrac{2 \cdot 2 \cdot 2}{7} \cdot \dfrac{5}{5} = -\dfrac{2 \cdot 2 \cdot 2 \cdot}{7} \cdot 1 = -\dfrac{8}{7}$

**59.** $\dfrac{4}{5} \cdot \dfrac{10}{12} = \dfrac{4 \cdot 10}{5 \cdot 12} = \dfrac{2 \cdot 2 \cdot 2 \cdot 5}{5 \cdot 2 \cdot 2 \cdot 3} = \dfrac{2}{3} \cdot \dfrac{2 \cdot 2 \cdot 5}{2 \cdot 2 \cdot 5} = \dfrac{2}{3} \cdot 1 = \dfrac{2}{3}$

**61.** The width of the carton will be the common width, 5 in. The length of the carton must be a multiple of both 6 and 8. The shortest length carton will be the least common multiple of 6 and 8.

$$6 = 2 \cdot 3$$
$$8 = 2 \cdot 2 \cdot 2$$

LCM is $2 \cdot 2 \cdot 2 \cdot 3$, or 24.

The shortest carton is 24 in. long.

## Exercise Set 4.2

**1.** $\dfrac{7}{8} + \dfrac{1}{8} = \dfrac{7+1}{8} = \dfrac{8}{8} = 1$

**3.** $\dfrac{1}{8} + \dfrac{5}{8} = \dfrac{1+5}{8} = \dfrac{6}{8} = \dfrac{3 \cdot 2}{4 \cdot 2} = \dfrac{3}{4} \cdot \dfrac{2}{2} = \dfrac{3}{4} \cdot 1 = \dfrac{3}{4}$

**5.**  $\dfrac{2}{3} + \dfrac{-5}{6}$   3 is a factor of 6, so the LCD is 6.

$= \dfrac{2}{3} \cdot \dfrac{2}{2} + \dfrac{-5}{6}$ ← This fraction already has the LCD as denominator.

Think: $3 \times \square = 6$. The answer is 2, so we multiply by 1, using $\dfrac{2}{2}$.

$= \dfrac{4}{6} + \dfrac{-5}{6}$

$= \dfrac{-1}{6}$, or $-\dfrac{1}{6}$

**7.**  $\dfrac{1}{8} + \dfrac{1}{6}$   $8 = 2 \cdot 2 \cdot 2$ and $6 = 2 \cdot 3$, so the LCD is $2 \cdot 2 \cdot 2 \cdot 3$, or 24

$= \dfrac{1}{8} \cdot \dfrac{3}{3} + \dfrac{1}{6} \cdot \dfrac{4}{4}$

Think: $6 \times \square = 24$. The answer is 4, so we multiply by 1, using $\dfrac{4}{4}$.

Think: $8 \times \square = 24$. The answer is 3, so we multiply by 1, using $\dfrac{3}{3}$.

$= \dfrac{3}{24} + \dfrac{4}{24}$

$= \dfrac{7}{24}$

**9.**  $\dfrac{-4}{5} + \dfrac{7}{10}$   5 is a factor of 10, so the LCD is 10.

$= \dfrac{-4}{5} \cdot \dfrac{2}{2} + \dfrac{7}{10}$ ← This fraction already has the LCD as denominator.

Think: $5 \times \square = 10$. The answer is 2, so we multiply by 1, using $\dfrac{2}{2}$.

$= \dfrac{-8}{10} + \dfrac{7}{10}$

$= \dfrac{-1}{10}$, or $-\dfrac{1}{10}$

**11.**  $\dfrac{5}{12} + \dfrac{3}{8}$   $12 = 2 \cdot 2 \cdot 3$ and $8 = 2 \cdot 2 \cdot 2$, so the LCD is $2 \cdot 2 \cdot 2 \cdot 3$, or 24.

$= \dfrac{5}{12} \cdot \dfrac{2}{2} + \dfrac{3}{8} \cdot \dfrac{3}{3}$

Think: $8 \times \square = 24$. The answer is 3, so we multiply by 1, using $\dfrac{3}{3}$.

Think: $12 \times \square = 24$. The answer is 2, so we multiply by 1, using $\dfrac{2}{2}$.

$= \dfrac{10}{24} + \dfrac{9}{24} = \dfrac{19}{24}$

**13.**  $\dfrac{3}{20} + \dfrac{3}{4}$   4 is a factor of 20, so the LCD is 20.

$= \dfrac{3}{20} + \dfrac{3}{4} \cdot \dfrac{5}{5}$   Multiplying by 1

$= \dfrac{3}{20} + \dfrac{15}{20} = \dfrac{18}{20} = \dfrac{9}{10}$

**15.**  $\dfrac{5}{6} + \dfrac{-7}{9}$   $6 = 2 \cdot 3$ and $9 = 3 \cdot 3$, so the LCD is $2 \cdot 3 \cdot 3$, or 18.

$= \dfrac{5}{6} \cdot \dfrac{3}{3} + \dfrac{-7}{9} \cdot \dfrac{2}{2}$   Multiplying by 1

$= \dfrac{15}{18} + \dfrac{-14}{18} = \dfrac{1}{18}$

**17.**  $\dfrac{3}{10} + \dfrac{1}{100}$   10 is a factor of 100, so the LCD is 100.

$= \dfrac{3}{10} \cdot \dfrac{10}{10} + \dfrac{1}{100}$

$= \dfrac{30}{100} + \dfrac{1}{100} = \dfrac{31}{100}$

**19.**  $\dfrac{5}{12} + \dfrac{4}{15}$   $12 = 2 \cdot 2 \cdot 3$ and $15 = 3 \cdot 5$, so the LCD is $2 \cdot 2 \cdot 3 \cdot 5$, or 60.

$= \dfrac{5}{12} \cdot \dfrac{5}{5} + \dfrac{4}{15} \cdot \dfrac{4}{4}$

$= \dfrac{25}{60} + \dfrac{16}{60} = \dfrac{41}{60}$

**21.**  $\dfrac{-9}{10} + \dfrac{99}{100}$   10 is a factor of 100, so the LCD is 100.

$= \dfrac{-9}{10} \cdot \dfrac{10}{10} + \dfrac{99}{100}$

$= \dfrac{-90}{100} + \dfrac{99}{100} = \dfrac{9}{100}$

**23.**  $\dfrac{7}{8} + \dfrac{0}{1}$   1 is a factor of 8, so the LCD is 8.

$= \dfrac{7}{8} + \dfrac{0}{1} \cdot \dfrac{8}{8}$

$= \dfrac{7}{8} + \dfrac{0}{8} = \dfrac{7}{8}$

Note that if we had observed at the outset that $\dfrac{0}{1} = 0$, the computation becomes $\dfrac{7}{8} + 0 = \dfrac{7}{8}$.

**25.**  $\dfrac{3}{8} + \dfrac{1}{6}$   $8 = 2 \cdot 2 \cdot 2$ and $6 = 2 \cdot 3$, so the LCD is $2 \cdot 2 \cdot 2 \cdot 3$, or 24.

$= \dfrac{3}{8} \cdot \dfrac{3}{3} + \dfrac{1}{6} \cdot \dfrac{4}{4}$

$= \dfrac{9}{24} + \dfrac{4}{24} = \dfrac{13}{24}$

**27.** $\dfrac{5}{12} + \dfrac{7}{24}$    12 is a factor of 24, so the LCD is 24.

$= \dfrac{5}{12} \cdot \dfrac{2}{2} + \dfrac{7}{24}$

$= \dfrac{10}{24} + \dfrac{7}{24} = \dfrac{17}{24}$

**29.** $\dfrac{3}{16} + \dfrac{5}{16} + \dfrac{4}{16} = \dfrac{3+5+4}{16} = \dfrac{12}{16} = \dfrac{3}{4}$

**31.** $\dfrac{8}{10} + \dfrac{7}{100} + \dfrac{4}{1000}$    10 and 100 are factors of 1000, so the LCD is 1000.

$= \dfrac{8}{10} \cdot \dfrac{100}{100} + \dfrac{7}{100} \cdot \dfrac{10}{10} + \dfrac{4}{1000}$

$= \dfrac{800}{1000} + \dfrac{70}{1000} + \dfrac{4}{1000} = \dfrac{874}{1000}$

$= \dfrac{437}{500}$

**33.** $\dfrac{3}{8} + \dfrac{-7}{12} + \dfrac{8}{15}$

$= \dfrac{3}{2 \cdot 2 \cdot 2} + \dfrac{-7}{2 \cdot 2 \cdot 3} + \dfrac{8}{3 \cdot 5}$    Factoring the denominators

The LCM is $2 \cdot 2 \cdot 2 \cdot 3 \cdot 5$, or 120.

$= \dfrac{3}{2 \cdot 2 \cdot 2} \cdot \dfrac{3 \cdot 5}{3 \cdot 5} + \dfrac{-7}{2 \cdot 2 \cdot 3} \cdot \dfrac{2 \cdot 5}{2 \cdot 5} + \dfrac{8}{3 \cdot 5} \cdot \dfrac{2 \cdot 2 \cdot 2}{2 \cdot 2 \cdot 2}$

In each case we multiply by 1 to obtain the LCD in the denominator.

$= \dfrac{3 \cdot 3 \cdot 5}{2 \cdot 2 \cdot 2 \cdot 3 \cdot 5} + \dfrac{-7 \cdot 2 \cdot 5}{2 \cdot 2 \cdot 3 \cdot 2 \cdot 5} + \dfrac{8 \cdot 2 \cdot 2}{3 \cdot 5 \cdot 2 \cdot 2 \cdot 2}$

$= \dfrac{45}{120} + \dfrac{-70}{120} + \dfrac{64}{120}$

$= \dfrac{39}{120} = \dfrac{13}{40}$

**35.** $\dfrac{15}{24} + \dfrac{7}{36} + \dfrac{91}{48}$

$= \dfrac{15}{2 \cdot 2 \cdot 2 \cdot 3} + \dfrac{7}{2 \cdot 2 \cdot 3 \cdot 3} + \dfrac{91}{2 \cdot 2 \cdot 2 \cdot 2 \cdot 3}$

Factoring the denominators.
The LCM is $2 \cdot 2 \cdot 2 \cdot 2 \cdot 3 \cdot 3$, or 144.

$= \dfrac{15}{2 \cdot 2 \cdot 2 \cdot 3} \cdot \dfrac{2 \cdot 3}{2 \cdot 3} + \dfrac{7}{2 \cdot 2 \cdot 3 \cdot 3} \cdot \dfrac{2 \cdot 2}{2 \cdot 2} +$

$\dfrac{91}{2 \cdot 2 \cdot 2 \cdot 3} \cdot \dfrac{3}{3}$

In each case we multiply by 1 to obtain the LCD in the denominator.

$= \dfrac{15 \cdot 2 \cdot 3}{2 \cdot 2 \cdot 2 \cdot 3 \cdot 2 \cdot 3} + \dfrac{7 \cdot 2 \cdot 2}{2 \cdot 2 \cdot 3 \cdot 3 \cdot 2 \cdot 2} + \dfrac{91 \cdot 3}{2 \cdot 2 \cdot 2 \cdot 2 \cdot 3 \cdot 3}$

$= \dfrac{90}{144} + \dfrac{28}{144} + \dfrac{273}{144} = \dfrac{391}{144}$

**37.** *Familiarize.* Let $d =$ the total distance Tate rode his Segway. This is the sum of the three distances he traveled.

*Translate.*

| Distance to library | plus | Distance to class | plus | Distance to work | is | Total distance |
|---|---|---|---|---|---|---|
| ↓ | ↓ | ↓ | ↓ | ↓ | ↓ | ↓ |
| $\dfrac{5}{6}$ | $+$ | $\dfrac{3}{4}$ | $+$ | $\dfrac{3}{2}$ | $=$ | $d$ |

*Solve.* We carry out the addition. The LCM of the denominators is 12.

$\dfrac{5}{6} \cdot \dfrac{2}{2} + \dfrac{3}{4} \cdot \dfrac{3}{3} + \dfrac{3}{2} \cdot \dfrac{6}{6} = d$

$\dfrac{10}{12} + \dfrac{9}{12} + \dfrac{18}{12} = d$

$\dfrac{37}{12} = d$

*Check.* We can repeat the calculation. Also note that the sum is larger than any of the individual distances, so the answer seems reasonable.

*State.* Tate rode his Segway $\dfrac{37}{12}$ mi.

**39.** *Familiarize.* We draw a picture. We let $p =$ the number of pounds of coffee beans in the mixture.

| $\dfrac{3}{16}$ lb | $\dfrac{5}{8}$ lb |
|---|---|
| $p$ | |

*Translate.*

| Pounds of decaffeinated beans | plus | Pounds of caffeinated beans | is | Total pounds of beans |
|---|---|---|---|---|
| ↓ | ↓ | ↓ | ↓ | ↓ |
| $\dfrac{3}{16}$ | $+$ | $\dfrac{5}{8}$ | $=$ | $p$ |

*Solve.* We carry out the addition. Since 16 is a multiple of 8, the LCM of the denominators is 16.

$\dfrac{3}{16} + \dfrac{5}{8} \cdot \dfrac{2}{2} = p$

$\dfrac{3}{16} + \dfrac{10}{16} = p$

$\dfrac{13}{16} = p$

*Check.* We check by repeating the calculation. We also note that the sum is larger than either of the individual weights, so the answer seems reasonable.

*State.* There was $\dfrac{13}{16}$ lb of coffee beans in the mixture.

**41.** *Familiarize.* First we find the number of quarts of liquid needed. Let $l =$ this amount. Then the liquid required if the recipe is doubled is $2 \cdot l$ and the amount required if the recipe is halved is $\dfrac{1}{2} \cdot l$.

*Translate.* We translate to an equation to find $l$.

$$\underbrace{\text{Ginger ale}}\quad \underbrace{\text{plus}}\quad \underbrace{\text{Strawberry soda}}\quad \underbrace{\text{is}}\quad \underbrace{\text{Total liquid}}$$
$$\downarrow\qquad \downarrow\qquad\quad \downarrow\qquad\quad \downarrow\qquad \downarrow$$
$$\frac{1}{5}\qquad + \qquad\quad \frac{3}{5}\qquad\quad = \qquad l$$

**Solve**. We carry out the addition.

$$\frac{1}{5} + \frac{3}{5} = l$$
$$\frac{4}{5} = l$$

Thus, we see that $\frac{4}{5}$ qt of liquid is needed for the recipe. If the recipe is doubled, the amount of liquid needed is $2 \cdot l = 2 \cdot \frac{4}{5} = \frac{8}{5}$ qt.

If the recipe is halved, the amount of liquid needed is
$\frac{1}{2} \cdot l = \frac{1}{2} \cdot \frac{4}{5} = \frac{1 \cdot 4}{2 \cdot 5} = \frac{1 \cdot 2 \cdot 2}{2 \cdot 5} = \frac{2}{2} \cdot \frac{1 \cdot 2}{5} = \frac{2}{5}$ qt.

**Check**.  We can repeat the calculations.  The answer checks.

**State**.  $\frac{4}{5}$ qt of liquid is needed for the recipe; $\frac{8}{5}$ qt is needed if the recipe is doubled; $\frac{2}{5}$ qt is needed if the recipe is halved.

**43. Familiarize**. Let $t$ = the thickness of the iced brownies, in inches. We see from the drawing in the text that the total thickness will be the sum of the thicknesses of the brownie and the icing.

**Translate**.

$$\underbrace{\text{Thickness}\atop\text{of brownie}}\ \underbrace{\text{plus}}\ \underbrace{\text{Thickness}\atop\text{of icing}}\ \underbrace{\text{is}}\ \underbrace{\text{Thickness of}\atop\text{iced brownie}}$$
$$\downarrow\qquad \downarrow\qquad \downarrow\qquad \downarrow\qquad \downarrow$$
$$\frac{11}{16}\qquad + \qquad \frac{5}{32}\qquad = \qquad t$$

**Solve**. We carry out the addition. Since 32 is a multiple of 16, the LCM of the denominators is 32.

$$\frac{11}{16} \cdot \frac{2}{2} + \frac{5}{32} = t$$
$$\frac{22}{32} + \frac{5}{32} = t$$
$$\frac{27}{32} = t$$

**Check**. We repeat the calculation. We also note that the sum is larger than either of the individual thicknesses, so the answer seems reasonable.

**State**. The thickness of the iced brownie is $\frac{27}{32}$ in.

**45. Familiarize**. Let $t$ = the total thickness, in inches. This total will be the sum of the thicknesses of the tile, the board, and the glue.

**Translate**.

$$\underbrace{\text{Thickness}\atop\text{of tile}}\ \underbrace{\text{plus}}\ \underbrace{\text{Thickness}\atop\text{of board}}\ \underbrace{\text{plus}}\ \underbrace{\text{Thickness}\atop\text{of glue}}\ \underbrace{\text{is}}\ \underbrace{\text{Total}\atop\text{thickness}}$$
$$\downarrow\quad \downarrow\quad \downarrow\quad \downarrow\quad \downarrow\quad \downarrow\quad \downarrow$$
$$\frac{5}{8}\quad + \quad \frac{7}{8}\quad + \quad \frac{3}{32}\quad = \quad t$$

**Solve**. We carry out the addition. The LCD is 32 since 8 is a factor of 32.

$$\frac{5}{8} \cdot \frac{4}{4} + \frac{7}{8} \cdot \frac{4}{4} + \frac{3}{32} = t$$
$$\frac{20}{32} + \frac{28}{32} + \frac{3}{32} = t$$
$$\frac{51}{32} = t$$

**Check**. We repeat the calculation. We also note that the sum is larger than any of the individual thicknesses, as expected.

**State**. The total thickness is $\frac{51}{32}$ in.

**47.**
$$\begin{array}{r} \overset{\scriptscriptstyle 1\ \ \overset{2}{4}}{5\ 1\ 6} \\ \times\quad 4\ 0\ 8 \\ \hline 4\ 1\ 2\ 8 \\ 2\ 0\ 6\ 4\ 0\ 0 \\ \hline 2\ 1\ 0,5\ 2\ 8 \end{array}$$
Multiplying by 8
Multiplying by 400
Adding

**49.** $-8 \cdot 7 = -56$

**51. Familiarize**. Let $p$ = the number of votes by which Gore's popular votes exceeded Bush's.

**Translate**.

$$\underbrace{\text{Bush's}\atop\text{votes}}\ \underbrace{\text{plus}}\ \underbrace{\text{Gore's}\atop\text{excess votes}}\ \underbrace{\text{is}}\ \underbrace{\text{Gore's}\atop\text{votes}}$$
$$\downarrow\qquad \downarrow\qquad \downarrow\qquad \downarrow\qquad \downarrow$$
$$50,459,211\ +\qquad p\qquad = 51,003,894$$

**Solve**. We subtract 50,459,211 on both sides of the equation.

$$50,459,211 + p = 51,003,894$$
$$50,459,211 + p - 50,459,211 = 51,003,894 - 50,459,211$$
$$p = 544,683$$

**Check**. Since $50,459,211 + 544,683 = 51,003,894$, the answer checks.

**State**. Albert A. Gore received 544,683 more popular votes than George W. Bush in 2000.

**53. Familiarize**. Let $v$ = the number of votes by which Kennedy's electoral votes exceeded Nixon's.

**Translate**. This is a "how much more" situation.

$$\underbrace{\text{Nixon's}\atop\text{votes}}\ \underbrace{\text{plus}}\ \underbrace{\text{Kennedy's}\atop\text{excess votes}}\ \underbrace{\text{is}}\ \underbrace{\text{Kennedy's}\atop\text{votes}}$$
$$\downarrow\qquad \downarrow\qquad \downarrow\qquad \downarrow\qquad \downarrow$$
$$219\qquad +\qquad v\qquad = \qquad 303$$

**Solve**. We subtract 219 on both sides of the equation.

$$219 + v = 303$$
$$219 + v - 219 = 303 - 219$$
$$v = 84$$

**Check**. Since $219 + 84 = 303$, the answer checks.

**State**. John F. Kennedy had 84 more electoral votes than Richard M. Nixon in 1960.

**55. Familiarize.** This is a multistep problem. Let $x$ = the total popular vote in 2000, $y$ = the total popular vote in 1976, and $n$ = the number of votes by which the 2000 total exceeded that in 1976.

**Translate.** First we add to find $x$ and $y$.

$$x = 50,459,211 + 51,003,894$$
$$y = 40,830,763 + 39,147,793$$

To find $n$, we write a third equation.

$$\underbrace{1976 \text{ votes}}_{\downarrow \atop y} \text{ plus } \underbrace{\text{excess 2000 votes}}_{\downarrow \atop n} \text{ is } \underbrace{2000 \text{ votes}}_{\downarrow \atop x}$$

$$y \quad + \quad n \quad = \quad x$$

**Solve.** We first carry out the additions to find $x$ and $y$.

$$x = 50,459,211 + 51,003,894$$
$$x = 101,463,105$$

$$y = 40,830,763 + 39,147,793$$
$$y - 79,978,556$$

Now we substitute 101,463,105 for $x$ and 79,978,556 for $y$ in the third equation and solve for $n$.

$$y + n = x$$
$$79,978,556 + n = 101,463,105$$
$$79,978,556 + n - 79,978,556 = 101,463,105 - 79,978,556$$
$$n = 21,484,549$$

**Check.** Since $79,978,556 + 21,484,549 = 101,463,105$, the answer checks.

**State.** The total popular vote in 2000 exceeded that in 1976 by 21,484,549 votes.

**57. Familiarize.** Let $h$ = the number of pounds by which honey produced in North Dakota exceeded honey production in California.

**Translate.**

$$\underbrace{\text{California production}} \text{ plus } \underbrace{\text{Excess North Dakota production}} \text{ is } \underbrace{\text{North Dakota production}}$$

$$19,760,000 \quad + \quad h \quad = \quad 25,900,000$$

**Solve.** We subtract 19,760,000 on both sides of the equation.

$$19,760,000 + h = 25,900,000$$
$$19,760,000 + h - 19,760,000 = 25,900,000 - 19,760,000$$
$$h = 6,140,000$$

**Check.** Since $19,760,000 + 6,140,000 = 25,900,000$, the answer checks.

**State.** Honey production in North Dakota exceeded honey production in California by 6,140,000 lb.

**59. Familiarize.** First we find the fractional part of the band's pay that the guitarist received. We let $f$ = this fraction.

**Translate.** We translate to an equation.

$$\underbrace{\text{One-third}}_{\downarrow \atop \frac{1}{3}} \underbrace{\text{of}}_{\downarrow \atop \cdot} \underbrace{\text{one-half}}_{\downarrow \atop \frac{1}{2}} \underbrace{\text{plus}}_{\downarrow \atop +} \underbrace{\text{one-fifth}}_{\downarrow \atop \frac{1}{5}} \underbrace{\text{of}}_{\downarrow \atop \cdot} \underbrace{\text{one-half}}_{\downarrow \atop \frac{1}{2}} \underbrace{\text{is}}_{\downarrow \atop =} \underbrace{\text{fractional part}}_{\downarrow \atop f}$$

**Solve.** We carry out the calculation.

$$\frac{1}{3} \cdot \frac{1}{2} + \frac{1}{5} \cdot \frac{1}{2} = f$$

$$\frac{1}{6} + \frac{1}{10} = f \qquad \text{LCD is 30.}$$

$$\frac{1}{6} \cdot \frac{5}{5} + \frac{1}{10} \cdot \frac{3}{3} = f$$

$$\frac{5}{30} + \frac{3}{30} = f$$

$$\frac{8}{30} = f$$

$$\frac{4}{15} = f$$

Now we find how much of the $1200 received by the band was paid to the guitarist. We let $p$ = the amount.

$$\underbrace{\text{Four-fifteenths}}_{\downarrow \atop \frac{4}{15}} \underbrace{\text{of}}_{\downarrow \atop \cdot} \underbrace{\$1200}_{\downarrow \atop 1200} \underbrace{=}_{\downarrow \atop =} \underbrace{\text{guitarist's pay}}_{\downarrow \atop p}$$

We solve the equation.

$$\frac{4}{15} \cdot 1200 = p$$

$$\frac{4 \cdot 1200}{15} = p$$

$$\frac{4 \cdot 3 \cdot 5 \cdot 80}{3 \cdot 5} = p$$

$$320 = p$$

**Check.** We repeat the calculations.

**State.** The guitarist received $\frac{4}{15}$ of the band's pay. This was $320.

---

## Exercise Set 4.3

**1.** When denominators are the same, subtract the numerators and keep the denominator.

$$\frac{5}{6} - \frac{1}{6} = \frac{5-1}{6} = \frac{4}{6} = \frac{2 \cdot 2}{2 \cdot 3} = \frac{2}{2} \cdot \frac{2}{3} = \frac{2}{3}$$

**3.** When denominators are the same, subtract the numerators and keep the denominator.

$$\frac{11}{12} - \frac{2}{12} = \frac{11-2}{12} = \frac{9}{12} = \frac{3 \cdot 3}{3 \cdot 4} = \frac{3}{3} \cdot \frac{3}{4} = \frac{3}{4}$$

**5.** The LCM of 8 and 4 is 8.

The first fraction already has the LCM as the denominator.

Think:  $4 \times \boxed{\phantom{0}} = 8$. The answer is 2, so we multiply $\frac{3}{4}$ by

1, using $\frac{2}{2}$.

$$\frac{1}{8} - \frac{3}{4} = \frac{1}{8} - \frac{3}{4} \cdot \frac{2}{2}$$

$$= \frac{1}{8} - \frac{6}{8} = -\frac{5}{8}$$

**7.** The LCM of 8 and 12 is 24.

$$\frac{1}{8} - \frac{1}{12} = \underbrace{\frac{1}{8} \cdot \frac{3}{3}} - \underbrace{\frac{1}{12} \cdot \frac{2}{2}}$$

Think:  $12 \times \boxed{\phantom{0}} = 24$. The answer is 2, so we multiply by 1, using $\frac{2}{2}$.

Think:  $8 \times \boxed{\phantom{0}} = 24$. The answer is 3, so we multiply by 1, using $\frac{3}{3}$.

$$= \frac{3}{24} - \frac{2}{24} = \frac{1}{24}$$

**9.** The LCM of 6 and 3 is 6.

$$\frac{5}{6} - \frac{4}{3} = \frac{5}{6} - \frac{4}{3} \cdot \frac{2}{2}$$

$$= \frac{5}{6} - \frac{8}{6} = -\frac{3}{6}$$

$$= -\frac{1 \cdot 3}{2 \cdot 3} = -\frac{1}{2} \cdot \frac{3}{3}$$

$$= -\frac{1}{2}$$

**11.** The LCM of 4 and 28 is 28.

$$\frac{3}{4} - \frac{3}{28} = \frac{3}{4} \cdot \frac{7}{7} - \frac{3}{28}$$

$$= \frac{21}{28} - \frac{3}{28}$$

$$= \frac{18}{28} = \frac{9 \cdot 2}{14 \cdot 2}$$

$$= \frac{9}{14} \cdot \frac{2}{2} = \frac{9}{14}$$

**13.** The LCM of 4 and 20 is 20.

$$\frac{3}{4} - \frac{3}{20} = \frac{3}{4} \cdot \frac{5}{5} - \frac{3}{20}$$

$$= \frac{15}{20} - \frac{3}{20} = \frac{12}{20}$$

$$= \frac{3 \cdot 4}{5 \cdot 4} = \frac{3}{5} \cdot \frac{4}{4}$$

$$= \frac{3}{5}$$

**15.** The LCM of 20 and 4 is 20.

$$\frac{1}{20} - \frac{3}{4} = \frac{1}{20} - \frac{3}{4} \cdot \frac{5}{5}$$

$$= \frac{1}{20} - \frac{15}{20} = -\frac{14}{20}$$

$$= -\frac{2 \cdot 7}{2 \cdot 10} = \frac{2}{2} \cdot \left(-\frac{7}{10}\right)$$

$$= -\frac{7}{10}$$

**17.** The LCM of 12 and 15 is 60.

$$\frac{5}{12} - \frac{2}{15} = \frac{5}{12} \cdot \frac{5}{5} - \frac{2}{15} \cdot \frac{4}{4}$$

$$= \frac{25}{60} - \frac{8}{60} = \frac{17}{60}$$

**19.** The LCM of 10 and 100 is 100.

$$\frac{6}{10} - \frac{7}{100} = \frac{6}{10} \cdot \frac{10}{10} - \frac{7}{100}$$

$$= \frac{60}{100} - \frac{7}{100} = \frac{53}{100}$$

**21.** The LCM of 15 and 25 is 75.

$$\frac{7}{15} - \frac{3}{25} = \frac{7}{15} \cdot \frac{5}{5} - \frac{3}{25} \cdot \frac{3}{3}$$

$$= \frac{35}{75} - \frac{9}{75} = \frac{26}{75}$$

**23.** The LCM of 10 and 100 is 100.

$$\frac{99}{100} - \frac{9}{10} = \frac{99}{100} - \frac{9}{10} \cdot \frac{10}{10}$$

$$= \frac{99}{100} - \frac{90}{100} = \frac{9}{100}$$

**25.** The LCM of 3 and 8 is 24.

$$-\frac{2}{3} - \frac{1}{8} = -\frac{2}{3} \cdot \frac{8}{8} - \frac{1}{8} \cdot \frac{3}{3}$$

$$= -\frac{16}{24} - \frac{3}{24}$$

$$= -\frac{19}{24}$$

**27.** The LCM of 5 and 2 is 10.

$$\frac{3}{5} - \frac{1}{2} = \frac{3}{5} \cdot \frac{2}{2} - \frac{1}{2} \cdot \frac{5}{5}$$

$$= \frac{6}{10} - \frac{5}{10}$$

$$= \frac{1}{10}$$

**29.** The LCM of 8 and 12 is 24.

$$\frac{3}{8}-\frac{5}{12}=\frac{3}{8}\cdot\frac{3}{3}-\frac{5}{12}\cdot\frac{2}{2}$$
$$=\frac{9}{24}-\frac{10}{24}$$
$$=-\frac{1}{24}$$

**31.** The LCM of 8 and 16 is 16.

$$\frac{7}{8}-\frac{1}{16}=\frac{7}{8}\cdot\frac{2}{2}-\frac{1}{16}$$
$$=\frac{14}{16}-\frac{1}{16}$$
$$=\frac{13}{16}$$

**33.** The LCM of 15 and 25 is 75.

$$\frac{4}{25}-\frac{17}{25}=\frac{4}{15}\cdot\frac{5}{5}-\frac{17}{25}\cdot\frac{3}{3}$$
$$=\frac{20}{75}-\frac{51}{73}$$
$$=-\frac{31}{75}$$

**35.** The LCM of 25 and 150 is 150.

$$\frac{23}{25}-\frac{112}{150}=\frac{23}{25}\cdot\frac{6}{6}-\frac{112}{150}$$
$$=\frac{138}{150}-\frac{112}{150}=\frac{26}{150}$$
$$=\frac{2\cdot13}{2\cdot75}=\frac{2}{2}\cdot\frac{13}{75}$$
$$=\frac{13}{75}$$

**37.** Since there is a common denominator, compare the numerators.

$$5<6,\text{ so }\frac{5}{8}<\frac{6}{8}.$$

**39.** The LCD is 12.

$$\frac{1}{3}\cdot\frac{4}{4}=\frac{4}{12}\quad\text{We multiply by 1 to get the LCD.}$$
$$\frac{1}{4}\cdot\frac{3}{3}=\frac{3}{12}\quad\text{We multiply by 1 to get the LCD.}$$

Since $4>3$, it follows that $\frac{4}{12}>\frac{3}{12}$, so $\frac{1}{3}>\frac{1}{4}$.

**41.** The LCD is 21.

$$\frac{-5}{7}\cdot\frac{3}{3}=\frac{-15}{21}$$
$$\frac{-2}{3}\cdot\frac{7}{7}=\frac{-14}{21}$$
Since $-15<-14$, it follows that $\frac{-15}{21}<\frac{-14}{20}$, so $\frac{-5}{7}<\frac{-2}{3}.$

**43.** The LCD is 30.

$$\frac{4}{5}\cdot\frac{6}{6}=\frac{24}{30}$$
$$\frac{5}{6}\cdot\frac{5}{5}=\frac{25}{30}$$
Since $24<25$, it follows that $\frac{24}{30}<\frac{25}{30}$, so $\frac{4}{5}<\frac{5}{6}.$

**45.** The LCD is 20.
$$\frac{-4}{5}\cdot\frac{4}{4}=\frac{-16}{20}$$
The denominator of $\frac{-19}{20}$ is the LCD.

Since $-16>-19$, it follows that $\frac{-16}{20}>\frac{-19}{20}$, so $\frac{-4}{5}>\frac{-19}{20}.$

**47.** The LCD is 20.

The denominator of $\frac{19}{20}$ is the LCD.
$$\frac{9}{10}\cdot\frac{2}{2}=\frac{18}{20}$$
Since $19>18$, it follows that $\frac{19}{20}>\frac{18}{20}$, so $\frac{19}{20}>\frac{9}{10}.$

**49.** The LCD is $13\cdot21$, or 273.

$$\frac{-41}{13}\cdot\frac{21}{21}=\frac{-861}{273}$$
$$\frac{-31}{21}\cdot\frac{13}{13}=\frac{-403}{273}$$

Since $-861<-403$, it follows that $\frac{-861}{273}<\frac{-403}{273}$, so $\frac{-41}{13}<\frac{-31}{21}.$

**51.**
$$x+\frac{1}{30}=\frac{1}{10}$$
$$x+\frac{1}{30}-\frac{1}{30}=\frac{1}{10}-\frac{1}{30}\quad\text{Subtracting }\frac{1}{30}\text{ on both sides}$$
$$x+0=\frac{1}{10}\cdot\frac{3}{3}-\frac{1}{30}\quad\text{The LCD is 30. We multiply by 1 to get the LCD.}$$
$$x=\frac{3}{30}-\frac{1}{30}=\frac{2}{30}$$
$$x=\frac{1\cdot2}{2\cdot15}=\frac{1}{15}\cdot\frac{2}{2}$$
$$x=\frac{1}{15}$$

The solution is $\frac{1}{15}.$

**53.** $$\frac{2}{3} + t = -\frac{4}{5}$$

$$\frac{2}{3} + t - \frac{2}{3} = -\frac{4}{5} - \frac{2}{3} \qquad \text{Subtracting } \frac{2}{3} \text{ on both sides.}$$

$$t + 0 = -\frac{4}{5} \cdot \frac{3}{3} - \frac{2}{3} \cdot \frac{5}{5} \qquad \text{The LCD is 15. We}$$
$$\text{multiply by 1 to get the LCD.}$$

$$t = -\frac{12}{15} - \frac{10}{15} = -\frac{22}{15}$$

The solution is $-\frac{22}{15}$.

**55.** $$x + \frac{1}{3} = \frac{5}{6}$$

$$x + \frac{1}{3} - \frac{1}{3} = \frac{5}{6} - \frac{1}{3}$$

$$x + 0 = \frac{5}{6} - \frac{1}{3} \cdot \frac{2}{2}$$

$$x = \frac{5}{6} - \frac{2}{6} = \frac{3}{6}$$

$$x = \frac{3 \cdot 1}{3 \cdot 2} = \frac{3}{3} \cdot \frac{1}{2}$$

$$x = \frac{1}{2}$$

The solution is $\frac{1}{2}$.

**57. Familiarize.** We visualize the situation. Let $t =$ the number of hours by which the time Kaitlyn spent on google.com exceeded the time she spent on chacha.com.

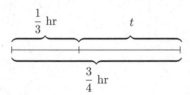

**Translate.**

| Time spent on chacha.com | plus | Excess time spent on google.com | is | Time spent on google.com |
|---|---|---|---|---|
| ↓ | ↓ | ↓ | ↓ | ↓ |
| $\frac{1}{3}$ | $+$ | $t$ | $=$ | $\frac{3}{4}$ |

**Solve.** We subtract $\frac{1}{3}$ on both sides of the equation.

$$\frac{1}{3} + t - \frac{1}{3} = \frac{3}{4} - \frac{1}{3}$$

$$t + 0 = \frac{3}{4} \cdot \frac{3}{3} - \frac{1}{3} \cdot \frac{4}{4} \qquad \text{The LCD is 12. We mul-}$$
$$\text{tiply by 1 to get the LCD.}$$

$$t = \frac{9}{12} - \frac{4}{12} = \frac{5}{12}$$

**Check.** We return to the original problem and add.

$$\frac{1}{3} + \frac{5}{12} = \frac{1}{3} \cdot \frac{4}{4} + \frac{5}{12} = \frac{4}{12} + \frac{5}{12} = \frac{9}{12} = \frac{3}{3} \cdot \frac{3}{4} = \frac{3}{4}$$

The answer checks.

**State.** Kaitlyn spent $\frac{5}{12}$ hr more on google.com than on chacha.com.

**59. Familiarize.** Let $d =$ the number of inches by which the depth of the long-life tread exceeds the depth of the more typical tread.

**Translate.**

| Typical tread depth | plus | Excess depth of long-life tire | is | Depth of long-life tire |
|---|---|---|---|---|
| ↓ | ↓ | ↓ | ↓ | ↓ |
| $\frac{11}{32}$ | $+$ | $d$ | $=$ | $\frac{3}{8}$ |

**Solve.** We subtract $\frac{11}{32}$ on both sides of the equation.

$$\frac{11}{32} + d = \frac{3}{8}$$

$$\frac{11}{32} + d - \frac{11}{32} = \frac{3}{8} - \frac{11}{32}$$

$$d + 0 = \frac{3}{8} \cdot \frac{4}{4} - \frac{11}{32} \qquad \text{The LCD is 32.}$$

$$d = \frac{12}{32} - \frac{11}{32}$$

$$d = \frac{1}{32}$$

**Check.** We return to the original problem and add.

$$\frac{11}{32} + \frac{1}{32} = \frac{12}{32} = \frac{4 \cdot 3}{4 \cdot 8} = \frac{4}{4} \cdot \frac{3}{8} = \frac{3}{8}$$

The answer checks.

**State.** The long-life tire tread is $\frac{1}{32}$ in. deeper than the more typical $\frac{11}{32}$ in. tread.

**61. Familiarize.** We visualize the situation. Let $x =$ the portion of the tub of popcorn that Cole ate.

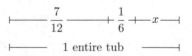

**Translate.**

| Ashley's portion | plus | Lauren's portion | plus | Cole's portion | is | Entire tub |
|---|---|---|---|---|---|---|
| ↓ | ↓ | ↓ | ↓ | ↓ | ↓ | ↓ |
| $\frac{7}{12}$ | $+$ | $\frac{1}{6}$ | $+$ | $x$ | $=$ | $1$ |

**Solve.** We begin by adding the fractions on the left side of the equation.

$$\frac{7}{12} + \frac{1}{6} \cdot \frac{2}{2} + x = 1 \qquad \text{The LCD is 12.}$$

$$\frac{7}{12} + \frac{2}{12} + x = 1$$

$$\frac{9}{12} + x = 1$$

$$\frac{3}{4} + x = 1 \qquad \text{Simplifying } \frac{9}{12}$$

$$\frac{3}{4} + x - \frac{3}{4} = 1 - \frac{3}{4} \qquad \text{Subtracting } \frac{3}{4} \text{ on both sides}$$

$$x + 0 = 1 \cdot \frac{4}{4} - \frac{3}{4} \qquad \text{The LCD is 4.}$$

$$x = \frac{4}{4} - \frac{3}{4}$$

$$x = \frac{1}{4}$$

**Check**. We return to the original problem and add.

$$\frac{7}{12} + \frac{1}{6} + \frac{1}{4} = \frac{7}{12} + \frac{1}{6} \cdot \frac{2}{2} + \frac{1}{4} \cdot \frac{3}{3} =$$

$$\frac{7}{12} + \frac{2}{12} + \frac{3}{12} = \frac{12}{12} = 1$$

**State**. Cole ate $\frac{1}{4}$ of the tub of popcorn.

**63. Familiarize**. We visualize the situation. Let $c =$ the amount of cheese remaining, in pounds.

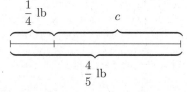

**Translate**. This is a "how much more" situation.

| Amount served | plus | Amount remaining | is | Original amount |
|---|---|---|---|---|
| ↓ | ↓ | ↓ | ↓ | ↓ |
| $\frac{1}{4}$ | $+$ | $c$ | $=$ | $\frac{4}{5}$ |

**Solve**. We subtract $\frac{1}{4}$ on both sides of the equation.

$$\frac{1}{4} + c - \frac{1}{4} = \frac{4}{5} - \frac{1}{4}$$

$$c + 0 = \frac{4}{5} \cdot \frac{4}{4} - \frac{1}{4} \cdot \frac{5}{5} \qquad \text{The LCD is 20.}$$

$$c = \frac{16}{20} - \frac{5}{20}$$

$$c = \frac{11}{20}$$

**Check**. Since $\frac{1}{4} + \frac{11}{20} = \frac{1}{4} \cdot \frac{5}{5} + \frac{11}{20} = \frac{5}{20} + \frac{11}{20} = \frac{16}{20} = \frac{4 \cdot 4}{4 \cdot 5} = \frac{4}{4} \cdot \frac{4}{5} = \frac{4}{5}$, the answer checks.

**State**. $\frac{11}{20}$ lb of cheese remains on the wheel.

**65.** Remember: $\frac{n}{n} = 1$, for any integer $n$ that is not 0.

$$\frac{-38}{-38} = 1$$

**67.** Remember: $\frac{n}{0}$ is not defined for any integer $n$.

$$\frac{124}{0} \text{ is not defined.}$$

**69.** $\frac{3}{7} \div \left(-\frac{9}{4}\right) = \frac{3}{7} \cdot \left(-\frac{4}{9}\right)$

$$= -\frac{3 \cdot 4}{7 \cdot 9}$$

$$= -\frac{3 \cdot 4}{7 \cdot 3 \cdot 3} = -\frac{4}{7 \cdot 3} \cdot \frac{3}{3}$$

$$= -\frac{4}{7 \cdot 3} = -\frac{4}{21}$$

**71.** $7 \div \frac{1}{3} = 7 \cdot \frac{3}{1} = \frac{7 \cdot 3}{1}$

$$= \frac{21}{1} = 21$$

**73. Familiarize**. Let $c =$ the number of crayons in the Crayola 64 box that were sold from 1958 to 2008.

**Translate**.

| Number of crayons in box | times | Number of boxes sold | is | Total number of crayons sold |
|---|---|---|---|---|
| ↓ | ↓ | ↓ | ↓ | ↓ |
| 64 | $\cdot$ | 200,000,000 | $=$ | $c$ |

**Solve**. We carry out the multiplication.

$$
\begin{array}{r}
2\,0\,0,0\,0\,0,0\,0\,0 \\
\times \qquad\qquad 6\,4 \\
\hline
8\,0\,0\,0\,0\,0\,0\,0\,0 \\
1\,2\,0\,0\,0\,0\,0\,0\,0\,0 \\
\hline
1\,2,8\,0\,0,0\,0\,0,0\,0\,0
\end{array}
$$

Thus, $12,800,000,000 = c$.

**Check**. We repeat the calculation. The answer checks.

**State**. About 12,800,000,000, or 12.8 billion, crayons in the 64 box were sold from 1958 to 2008.

**75.** Use a calculator.

$$x + \frac{16}{323} = \frac{10}{187}$$

$$x + \frac{16}{323} - \frac{16}{323} = \frac{10}{187} - \frac{16}{323}$$

$$x + 0 = \frac{10}{11 \cdot 17} - \frac{16}{17 \cdot 19}$$

$$x = \frac{10}{11 \cdot 17} \cdot \frac{19}{19} - \frac{16}{17 \cdot 19} \cdot \frac{11}{11} \qquad \text{The LCD is } 11 \cdot 17 \cdot 19.$$

$$x = \frac{190}{11 \cdot 17 \cdot 19} - \frac{176}{17 \cdot 19 \cdot 11}$$

$$x = \frac{14}{11 \cdot 17 \cdot 19}$$

$$x = \frac{14}{3553}$$

The solution is $\frac{14}{3553}$.

**77.** *Familiarize.* First we find how far the athlete swam. We let $s$ = this distance. We visualize the situation.

$$\underbrace{\longrightarrow \longrightarrow \cdots \longrightarrow}_{} \Big\} \ 10 \text{ laps}$$

$$\underbrace{\qquad\qquad}_{\frac{3}{80} \text{ km in each lap}}$$

*Translate.* We translate to the following equation:

$$s = 10 \cdot \frac{3}{80}$$

*Solve.* We carry out the multiplication.

$$s = 10 \cdot \frac{3}{80} = \frac{10 \cdot 3}{80}$$

$$s = \frac{10 \cdot 3}{10 \cdot 8} = \frac{10}{10} \cdot \frac{3}{8}$$

$$s = \frac{3}{8}$$

Now we find the distance the athlete must walk. We let $w$ = the distance.

$$\underbrace{\text{Distance}}_{\substack{\downarrow \\ \frac{3}{8}}} \underset{\underset{+}{\downarrow}}{\text{plus}} \underbrace{\text{Distance}}_{\substack{\downarrow \\ w}} \underset{\underset{=}{\downarrow}}{\text{is}} \underbrace{\frac{9}{10} \text{ km}}_{\substack{\downarrow \\ \frac{9}{10}}}$$

We solve the equation.

$$\frac{3}{8} + w = \frac{9}{10}$$

$$\frac{3}{8} + w - \frac{3}{8} = \frac{9}{10} - \frac{3}{8}$$

$$w + 0 = \frac{9}{10} \cdot \frac{4}{4} - \frac{3}{8} \cdot \frac{5}{5} \quad \text{The LCD is 40.}$$

$$w = \frac{36}{40} - \frac{15}{40}$$

$$w = \frac{21}{40}$$

*Check.* We add the distance swum and the distance walked:

$$\frac{3}{8} + \frac{21}{40} = \frac{3}{8} \cdot \frac{5}{5} + \frac{21}{40} = \frac{15}{40} + \frac{21}{40} = \frac{36}{40} = \frac{9 \cdot 4}{10 \cdot 4} =$$

$$\frac{9}{10} \cdot \frac{4}{4} = \frac{9}{10}$$

*State.* The athlete must walk $\frac{21}{40}$ km after swimming 10 laps.

**79.** $\dfrac{7}{8} - \dfrac{1}{10} \times \dfrac{5}{6} = \dfrac{7}{8} - \dfrac{1 \times 5}{10 \times 6} = \dfrac{7}{8} - \dfrac{1 \times 5}{2 \times 5 \times 6}$

$$= \frac{7}{8} - \frac{5}{5} \times \frac{1}{2 \times 6} = \frac{7}{8} - \frac{1}{2 \times 6} = \frac{7}{8} - \frac{1}{12}$$

$$= \frac{7}{8} \cdot \frac{3}{3} - \frac{1}{12} \cdot \frac{2}{2} = \frac{21}{24} - \frac{2}{24} = \frac{19}{24}$$

**81.** $\left(\dfrac{2}{3}\right)^2 - \left(\dfrac{3}{4}\right)^2 = \dfrac{4}{9} - \dfrac{9}{16} = \dfrac{4}{9} \cdot \dfrac{16}{16} - \dfrac{9}{16} \cdot \dfrac{9}{9} =$

$$\frac{64}{144} - \frac{81}{144} = -\frac{17}{144}$$

**83.** Add on the left side.

$$\frac{37}{157} + \frac{19}{107} = \frac{37}{157} \cdot \frac{107}{107} + \frac{19}{107} \cdot \frac{157}{157} =$$

$$\frac{3959}{16,799} + \frac{2983}{16,799} = \frac{6942}{16,799}$$

Then $6942 > 6941$, so $\dfrac{6942}{16,799} > \dfrac{6941}{16,799}$ and

$$\frac{37}{157} + \frac{19}{107} > \frac{6941}{16,799}.$$

**85.** Use the two cuts to cut the bar into three pieces as follows: one piece is $\frac{1}{7}$ of the bar, one is $\frac{2}{7}$ of the bar, and then the remaining piece is $\frac{4}{7}$ of the bar. On Day 1, give the contractor $\frac{1}{7}$ of the bar. On Day 2, have him/her return the $\frac{1}{7}$ and give him/her $\frac{2}{7}$ of the bar. On Day 3, add $\frac{1}{7}$ to what the contractor already has, making $\frac{3}{7}$ of the bar. On Day 4, have the contractor return the $\frac{1}{7}$ and $\frac{2}{7}$ pieces and give him/her the $\frac{4}{7}$ piece. On Day 5, add the $\frac{1}{7}$ piece to what the contractor already has, making $\frac{5}{7}$ of the bar. On Day 6, have the contractor return the $\frac{1}{7}$ piece and give him/her the $\frac{2}{7}$ to go with the $\frac{4}{7}$ piece he/she also has, making $\frac{6}{7}$ of the bar. On Day 7, give him/her the $\frac{1}{7}$ piece again. Now the contractor has all three pieces, or the entire bar. This assumes that he/she does not spend any part of the gold during the week.

## Exercise Set 4.4

**1.** [b]  [a] Multiply: $2 \cdot 32 = 64$.

$$32\frac{1}{2} = \frac{65}{2} \quad \text{[b] Add: } 64 + 1 = 65.$$

[a]  [c] Keep the denominator.

[b]  [a] Multiply: $6 \cdot 20 = 120$.

$$20\frac{5}{6} = \frac{125}{6} \quad \text{[b] Add: } 120 + 5 = 125.$$

[a]  [c] Keep the denominator.

[b]  [a] Multiply: $4 \cdot 11 = 44$.

$$11\frac{3}{4} = \frac{47}{4} \quad \text{[b] Add: } 44 + 3 = 47.$$

[a]  [c] Keep the denominator.

**3.** To convert $\frac{17}{4}$ to a mixed numeral, we divide.

$$4\overline{\smash)17} \quad \frac{17}{4} = 4\frac{1}{4}$$
$$\underline{16}$$
$$1$$

To convert $\frac{10}{3}$ to a mixed numeral, we divide.

$$3\overline{\smash)10} \quad \frac{10}{3} = 3\frac{1}{3}$$
$$\underline{9}$$
$$1$$

To convert $\frac{9}{8}$ to a mixed numeral, we divide.

$$8\overline{\smash)9} \quad \frac{9}{8} = 1\frac{1}{8}$$
$$\underline{8}$$
$$1$$

**5.** $\boxed{b}$   $\boxed{a}$ Multiply: $5 \cdot 3 = 15$.

$5\frac{2}{3} = \frac{17}{3}$   $\boxed{b}$ Add: $15 + 2 = 17$.

$\boxed{a}$   $\boxed{c}$ Keep the denominator.

**7.** $\boxed{b}$   $\boxed{a}$ Multiply: $3 \cdot 4 = 12$.

$3\frac{1}{4} = \frac{13}{4}$   $\boxed{b}$ Add: $12 + 1 = 13$.

$\boxed{a}$   $\boxed{c}$ Keep the denominator.

**9.** First consider $10\frac{1}{8}$.

$10\frac{1}{8} = \frac{81}{8}$   $(10 \cdot 8 = 80,\ 80 + 1 = 81)$

Then $-10\frac{1}{8} = -\frac{81}{8}$.

**11.** $5\frac{1}{10} = \frac{51}{10}$   $(5 \cdot 10 = 50,\ 50 + 1 = 51)$

**13.** $20\frac{3}{5} = \frac{103}{5}$   $(20 \cdot 5 = 100,\ 100 + 3 = 103)$

**15.** First consider $9\frac{5}{6}$.

$9\frac{5}{6} = \frac{59}{6}$   $(9 \cdot 6 = 54,\ 54 + 5 = 59)$

Then $-9\frac{5}{6} = -\frac{59}{6}$.

**17.** $7\frac{3}{10} = \frac{73}{10}$   $(7 \cdot 10 = 70,\ 70 + 3 = 73)$

**19.** $1\frac{5}{8} = \frac{13}{8}$   $(1 \cdot 8 = 8,\ 8 + 5 = 13)$

**21.** First consider $12\frac{3}{4}$.

$12\frac{3}{4} = \frac{51}{4}$   $(12 \cdot 4 = 48,\ 48 + 3 = 51)$

Then $-12\frac{3}{4} = -\frac{51}{4}$.

**23.** First consider $4\frac{3}{10}$.

$4\frac{3}{10} = \frac{43}{10}$   $(4 \cdot 10 = 40,\ 40 + 3 = 43)$

Then $-4\frac{3}{10} = -\frac{43}{10}$.

**25.** $2\frac{3}{100} = \frac{203}{100}$   $(2 \cdot 100 = 200,\ 200 + 3 = 203)$

**27.** $66\frac{2}{3} = \frac{200}{3}$   $(66 \cdot 3 = 198,\ 198 + 2 = 200)$

**29.** First consider $5\frac{29}{50}$.

$5\frac{29}{50} = \frac{279}{50}$   $(5 \cdot 50 = 250,\ 250 + 29 = 279)$

Then $-5\frac{29}{50} = -\frac{279}{50}$.

**31.** $101\frac{5}{16} = \frac{1621}{16}$   $(16 \cdot 101 = 1616,\ 1616 + 5 = 1621)$

**33.** To convert $\frac{18}{5}$ to a mixed numeral, we divide.

$$5\overline{\smash)18} \quad \frac{18}{5} = 3\frac{3}{5}$$
$$\underline{15}$$
$$3$$

**35.** To convert $\frac{14}{3}$ to a mixed numeral, we divide.

$$3\overline{\smash)14} \quad \frac{14}{3} = 4\frac{2}{3}$$
$$\underline{12}$$
$$2$$

**37.** First consider $\frac{27}{6}$.

$$6\overline{\smash)27} \quad \frac{27}{6} = 4\frac{3}{6} = 4\frac{1}{2}$$
$$\underline{24}$$
$$3$$

Then $-\frac{27}{6} = -4\frac{1}{2}$.

**39.** $$10\overline{\smash)57} \qquad \frac{57}{10} = 5\frac{7}{10}$$
$$\underline{50}$$
$$7$$

**41.** First consider $\dfrac{53}{7}$.

$$\begin{array}{r} 7 \\ 7\,\overline{)\,5\,3} \\ \underline{4\,9} \\ 4 \end{array} \qquad \dfrac{53}{7} = 7\dfrac{4}{7}$$

Then $-\dfrac{53}{7} = -7\dfrac{4}{7}$.

**43.**
$$\begin{array}{r} 7 \\ 6\,\overline{)\,4\,5} \\ \underline{4\,2} \\ 3 \end{array} \qquad \dfrac{45}{6} = 7\dfrac{3}{6} = 7\dfrac{1}{2}$$

**45.**
$$\begin{array}{r} 1\,1 \\ 4\,\overline{)\,4\,6} \\ \underline{4} \\ 6 \\ \underline{4} \\ 2 \end{array} \qquad \dfrac{46}{4} = 11\dfrac{2}{4} = 11\dfrac{1}{2}$$

**47.** First consider $\dfrac{12}{8}$.

$$\begin{array}{r} 1 \\ 8\,\overline{)\,1\,2} \\ \underline{8} \\ 4 \end{array} \qquad \dfrac{12}{8} = 1\dfrac{4}{8} = 1\dfrac{1}{2}$$

Then $-\dfrac{12}{8} = -1\dfrac{1}{2}$.

**49.**
$$\begin{array}{r} 7 \\ 100\,\overline{)\,7\,5\,7} \\ \underline{7\,0\,0} \\ 5\,7 \end{array} \qquad \dfrac{757}{100} = 7\dfrac{57}{100}$$

**51.** First consider $\dfrac{345}{8}$.

$$\begin{array}{r} 4\,3 \\ 8\,\overline{)\,3\,4\,5} \\ \underline{3\,2} \\ 2\,5 \\ \underline{2\,4} \\ 1 \end{array} \qquad \dfrac{345}{8} = 43\dfrac{1}{8}$$

Then $-\dfrac{345}{8} = -43\dfrac{1}{8}$.

**53.** We first divide as usual.

$$\begin{array}{r} 1\,0\,8 \\ 8\,\overline{)\,8\,6\,9} \\ \underline{8} \\ 6\,9 \\ \underline{6\,4} \\ 5 \end{array}$$

The answer is 108 R 5. We write a mixed numeral for the quotient as follows: $108\dfrac{5}{8}$.

**55.** We first divide as usual.

$$\begin{array}{r} 6\,1\,8 \\ 5\,\overline{)\,3\,0\,9\,1} \\ \underline{3\,0} \\ 9 \\ \underline{5} \\ 4\,1 \\ \underline{4\,0} \\ 1 \end{array}$$

The answer is 618 R 1. We write a mixed numeral for the quotient as follows: $618\dfrac{1}{5}$.

**57.**
$$\begin{array}{r} 4\,0 \\ 2\,1\,\overline{)\,8\,5\,2} \\ \underline{8\,4} \\ 1\,2 \end{array}$$

We get $40\dfrac{12}{21}$. This simplifies as $40\dfrac{4}{7}$.

**59.**
$$\begin{array}{r} 5\,5 \\ 1\,0\,2\,\overline{)\,5\,6\,1\,2} \\ \underline{5\,1\,0} \\ 5\,1\,2 \\ \underline{5\,1\,0} \\ 2 \end{array}$$

We get $55\dfrac{2}{102}$. This simplifies as $55\dfrac{1}{51}$.

**61.**
$$\begin{array}{r} 2\,2\,9\,2 \\ 3\,5\,\overline{)\,8\,0,2\,4\,3} \\ \underline{7\,0} \\ 1\,0\,2 \\ \underline{7\,0} \\ 3\,2\,4 \\ \underline{3\,1\,5} \\ 9\,3 \\ \underline{7\,0} \\ 2\,3 \end{array}$$

We get $2292\dfrac{23}{35}$.

**63.** Round 45,765 to the nearest hundred.

$$4\,5,7\,\boxed{6}\,5$$
$$\uparrow$$

The digit 7 is in the hundreds place. Consider the next digit to the right. Since the digit, 6, is 5 or higher, round 7 hundreds up to 8 hundreds. Then change all digits to the right of the hundreds digit to zero.

The answer is 45,800.

**65.** $\dfrac{200}{375} = \dfrac{8 \cdot 25}{15 \cdot 25} = \dfrac{8}{15} \cdot \dfrac{25}{25} = \dfrac{8}{15}$

**67.** $-\dfrac{160}{270} = -\dfrac{16 \cdot 10}{27 \cdot 10} = -\dfrac{16}{27} \cdot \dfrac{10}{10} = -\dfrac{16}{27}$

**69.** $\dfrac{6}{5} \cdot 15 = \dfrac{6 \cdot 15}{5} = \dfrac{6 \cdot 3 \cdot 5}{5 \cdot 1} = \dfrac{5}{5} \cdot \dfrac{6 \cdot 3}{1} = \dfrac{6 \cdot 3}{1} = \dfrac{18}{1} = 18$

**71.** $-\dfrac{7}{10} \cdot \left(-\dfrac{5}{14}\right) = \dfrac{7 \cdot 5}{10 \cdot 14} = \dfrac{7 \cdot 5 \cdot 1}{2 \cdot 5 \cdot 2 \cdot 7} = \dfrac{7 \cdot 5}{7 \cdot 5} \cdot \dfrac{1}{2 \cdot 2} =$

$\dfrac{1}{2 \cdot 2} = \dfrac{1}{4}$

**73.** $-\dfrac{2}{3} \div \dfrac{1}{36} = -\dfrac{2}{3} \cdot \dfrac{36}{1} = -\dfrac{2 \cdot 36}{3 \cdot 1} = -\dfrac{2 \cdot 3 \cdot 12}{3 \cdot 1} =$

$\dfrac{3}{3} \cdot \left(-\dfrac{2 \cdot 12}{1}\right) = -\dfrac{2 \cdot 12}{1} = -\dfrac{24}{1} = -24$

**75.** $200 \div \dfrac{15}{64} = 200 \cdot \dfrac{64}{15} = \dfrac{200 \cdot 64}{15} = \dfrac{5 \cdot 40 \cdot 64}{5 \cdot 3} =$

$\dfrac{5}{5} \cdot \dfrac{40 \cdot 64}{3} = \dfrac{40 \cdot 64}{3} = \dfrac{2560}{3}$

**77.** Use a calculator.

$\dfrac{128,236}{541} = 237\dfrac{19}{541}$

**79.** $\dfrac{56}{7} + \dfrac{2}{3} = 8 + \dfrac{2}{3} \qquad (56 \div 7 - 8)$

$= 8\dfrac{2}{3}$

**81.** 
$$
\begin{array}{r}
5\,2 \\
7\,\overline{)\,3\,6\,6} \\
3\,5 \\
\hline
1\,6 \\
1\,4 \\
\hline
2
\end{array}
\qquad \dfrac{366}{7} = 52\dfrac{2}{7}
$$

## Chapter 4 Mid-Chapter Review

**1.** The statement is true. To determine which of two numbers is greater when there is a common denominator, we compare the numerators. (This assumes that the fractions have the same sign.)

**2.** The statement is true. All positive mixed numerals have a whole number part and a fractional part, so all mixed numerals must be larger than the smallest non-zero number, 1.

**3.** The statement is false. The least common multiple of two natural numbers is the smallest number that is a multiple of both numbers.

**4.** The statement is false. To add fractions when denominators are the same, add the numerators and keep the denominator.

**5.** $\dfrac{11}{42} - \dfrac{3}{35} = \dfrac{11}{2 \cdot 3 \cdot 7} - \dfrac{3}{5 \cdot 7}$

$= \dfrac{11}{2 \cdot 3 \cdot 7} \cdot \dfrac{5}{5} - \dfrac{3}{5 \cdot 7} \cdot \dfrac{2 \cdot 3}{2 \cdot 3}$

$= \dfrac{11 \cdot 5}{2 \cdot 3 \cdot 7 \cdot 5} - \dfrac{3 \cdot 2 \cdot 3}{5 \cdot 7 \cdot 2 \cdot 3}$

$= \dfrac{55}{2 \cdot 3 \cdot 5 \cdot 7} - \dfrac{18}{2 \cdot 3 \cdot 5 \cdot 7}$

$= \dfrac{55 - 18}{2 \cdot 3 \cdot 5 \cdot 7} = \dfrac{37}{210}$

**6.** $x + \dfrac{1}{8} = \dfrac{2}{3}$

$x + \dfrac{1}{8} - \dfrac{1}{8} = \dfrac{2}{3} - \dfrac{1}{8}$

$x + 0 = \dfrac{2}{3} \cdot \dfrac{8}{8} - \dfrac{1}{8} \cdot \dfrac{3}{3}$

$x = \dfrac{16}{24} - \dfrac{3}{24}$

$x = \dfrac{13}{24}$

The solution is $\dfrac{13}{24}$.

**7.** For 45 and 50:

$45 = 3 \cdot 3 \cdot 5$

$50 = 2 \cdot 5 \cdot 5$

The LCM is $2 \cdot 3 \cdot 3 \cdot 5 \cdot 5$, or 450.

For 50 and 80:

$50 = 2 \cdot 5 \cdot 5$

$80 = 2 \cdot 2 \cdot 2 \cdot 2 \cdot 5$

The LCM is $2 \cdot 2 \cdot 2 \cdot 2 \cdot 5 \cdot 5$, or 400.

For 30 and 24:

$30 = 2 \cdot 3 \cdot 5$

$24 = 2 \cdot 2 \cdot 2 \cdot 3$

The LCM is $2 \cdot 2 \cdot 2 \cdot 3 \cdot 5$, or 120.

For 18, 24, and 80:

$18 = 2 \cdot 3 \cdot 3$

$24 = 2 \cdot 2 \cdot 2 \cdot 3$

$80 = 2 \cdot 2 \cdot 2 \cdot 2 \cdot 5$

The LCM is $2 \cdot 2 \cdot 2 \cdot 2 \cdot 3 \cdot 3 \cdot 5$, or 720.

For 30, 45, and 50:

$30 = 2 \cdot 3 \cdot 5$

$45 = 3 \cdot 3 \cdot 5$

$50 = 2 \cdot 5 \cdot 5$

The LCM is $2 \cdot 3 \cdot 3 \cdot 5 \cdot 5$, or 450.

**8.** $\dfrac{1}{5} + \dfrac{7}{45} = \dfrac{1}{5} \cdot \dfrac{9}{9} + \dfrac{7}{45} = \dfrac{9}{45} + \dfrac{7}{45} = \dfrac{16}{45}$

**9.** $\dfrac{5}{6} + \dfrac{2}{3} + \dfrac{7}{12} = \dfrac{5}{6} \cdot \dfrac{2}{2} + \dfrac{2}{3} \cdot \dfrac{4}{4} + \dfrac{7}{12} = \dfrac{10}{12} + \dfrac{8}{12} + \dfrac{7}{12} = \dfrac{25}{12}$

**10.** $\dfrac{1}{6} - \dfrac{2}{9} = \dfrac{1}{6} \cdot \dfrac{3}{3} - \dfrac{2}{9} \cdot \dfrac{2}{2} = \dfrac{3}{18} - \dfrac{4}{18} = -\dfrac{1}{18}$

**11.** $\dfrac{-5}{18} + \dfrac{1}{15} = \dfrac{-5}{2 \cdot 3 \cdot 3} + \dfrac{1}{3 \cdot 5}$

$= \dfrac{-5}{2 \cdot 3 \cdot 3} \cdot \dfrac{5}{5} + \dfrac{1}{3 \cdot 5} \cdot \dfrac{2 \cdot 3}{2 \cdot 3}$

$= \dfrac{-25}{2 \cdot 3 \cdot 3 \cdot 5} + \dfrac{6}{3 \cdot 5 \cdot 2 \cdot 3}$

$= \dfrac{-25 + 6}{2 \cdot 3 \cdot 3 \cdot 5} = \dfrac{-19}{90}$, or $-\dfrac{19}{90}$

**12.**  $\dfrac{19}{48} - \dfrac{11}{30} = \dfrac{19}{8 \cdot 6} - \dfrac{11}{5 \cdot 6}$

$\qquad\qquad = \dfrac{19}{8 \cdot 6} \cdot \dfrac{5}{5} - \dfrac{11}{5 \cdot 6} \cdot \dfrac{8}{8}$

$\qquad\qquad = \dfrac{95}{8 \cdot 6 \cdot 5} - \dfrac{88}{5 \cdot 6 \cdot 8}$

$\qquad\qquad = \dfrac{95 - 88}{5 \cdot 6 \cdot 8} = \dfrac{7}{240}$

**13.**  $\dfrac{3}{7} + \dfrac{15}{17} = \dfrac{3}{7} \cdot \dfrac{17}{17} + \dfrac{15}{17} \cdot \dfrac{7}{7} = \dfrac{51}{119} + \dfrac{105}{119} = \dfrac{156}{119}$

**14.**  $\dfrac{229}{720} - \dfrac{5}{24} = \dfrac{229}{720} - \dfrac{5}{24} \cdot \dfrac{30}{30} = \dfrac{229}{720} - \dfrac{150}{720} = \dfrac{79}{720}$

**15.**  $\dfrac{2}{35} - \dfrac{8}{65} = \dfrac{2}{5 \cdot 7} - \dfrac{8}{5 \cdot 13}$

$\qquad\qquad = \dfrac{2}{5 \cdot 7} \cdot \dfrac{13}{13} - \dfrac{8}{5 \cdot 13} \cdot \dfrac{7}{7}$

$\qquad\qquad = \dfrac{26}{5 \cdot 7 \cdot 13} - \dfrac{56}{5 \cdot 13 \cdot 7}$

$\qquad\qquad = -\dfrac{30}{5 \cdot 7 \cdot 13} = -\dfrac{5 \cdot 6}{5 \cdot 7 \cdot 13}$

$\qquad\qquad = \dfrac{-6}{7 \cdot 13} \cdot \dfrac{5}{5} = -\dfrac{6}{91}$

**16. Familiarize.** Let $d =$ the total distance Miguel jogs.

**Translate.**

| First distance | plus | Second distance | is | Total distance |
|---|---|---|---|---|
| ↓ | ↓ | ↓ | ↓ | ↓ |
| $\dfrac{4}{5}$ | $+$ | $\dfrac{2}{3}$ | $=$ | $d$ |

**Solve.** We carry out the addition.

$\dfrac{4}{5} + \dfrac{2}{3} = \dfrac{4}{5} \cdot \dfrac{3}{3} + \dfrac{2}{3} \cdot \dfrac{5}{5} = \dfrac{12}{15} + \dfrac{10}{15} = \dfrac{22}{15}$

Thus, $\dfrac{22}{15} = d$.

**Check.** We can repeat the calculation. Also note that the result is greater than either of the individual distances, so the answer seems reasonable.

**State.** Miguel jogs $\dfrac{22}{15}$ mi in all.

**17. Familiarize.** Let $t =$ the number of hours Kirby spent playing Brain Challenge.

**Translate.**

| Brain Challenge time | plus | Scrabble time | is | Total time |
|---|---|---|---|---|
| ↓ | ↓ | ↓ | ↓ | ↓ |
| $t$ | $+$ | $\dfrac{11}{4}$ | $=$ | $\dfrac{39}{5}$ |

**Solve.** We subtract $\dfrac{11}{4}$ on both sides of the equation.

$$t + \dfrac{11}{4} = \dfrac{39}{5}$$

$$t + \dfrac{11}{4} - \dfrac{11}{4} = \dfrac{39}{5} - \dfrac{11}{4}$$

$$t + 0 = \dfrac{39}{5} \cdot \dfrac{4}{4} - \dfrac{11}{4} \cdot \dfrac{5}{5}$$

$$t = \dfrac{156}{20} - \dfrac{55}{20}$$

$$t = \dfrac{101}{20}$$

**Check.** We can add to perform a check.

$\dfrac{101}{20} + \dfrac{11}{4} = \dfrac{101}{20} + \dfrac{11}{4} \cdot \dfrac{5}{5} = \dfrac{101}{20} + \dfrac{55}{20} = \dfrac{156}{20} =$

$\dfrac{4 \cdot 39}{4 \cdot 5} = \dfrac{4}{4} \cdot \dfrac{39}{5} = \dfrac{39}{5}$

The answer checks.

**State.** Kirby played Brain Challenge for $\dfrac{101}{20}$ hr.

**18.**  $\dfrac{4}{9} = \dfrac{4}{3 \cdot 3}; \dfrac{3}{10} = \dfrac{3}{2 \cdot 5}; \dfrac{2}{7}; \dfrac{1}{5}$

The LCD is $2 \cdot 3 \cdot 3 \cdot 5 \cdot 7$, or 630. We write each fraction with this denominator and then compare numerators.

$\dfrac{4}{9} = \dfrac{4}{3 \cdot 3} \cdot \dfrac{2 \cdot 5 \cdot 7}{2 \cdot 5 \cdot 7} = \dfrac{280}{630}$

$\dfrac{3}{10} = \dfrac{3}{2 \cdot 5} \cdot \dfrac{3 \cdot 3 \cdot 7}{3 \cdot 3 \cdot 7} = \dfrac{189}{630}$

$\dfrac{2}{7} = \dfrac{2}{7} \cdot \dfrac{2 \cdot 3 \cdot 3 \cdot 5}{2 \cdot 3 \cdot 3 \cdot 5} = \dfrac{180}{630}$

$\dfrac{1}{5} = \dfrac{1}{5} \cdot \dfrac{2 \cdot 3 \cdot 3 \cdot 7}{2 \cdot 3 \cdot 3 \cdot 7} = \dfrac{126}{630}$

Arranging the numbers in order from smallest to largest, we have

$\dfrac{1}{5}, \dfrac{2}{7}, \dfrac{3}{10}, \dfrac{4}{9}.$

**19.**  $\dfrac{2}{5} + x = \dfrac{9}{16}$

$\dfrac{2}{5} + x - \dfrac{2}{5} = \dfrac{9}{16} - \dfrac{2}{5}$      Subtracting $\dfrac{2}{5}$

$x + 0 = \dfrac{9}{16} \cdot \dfrac{5}{5} - \dfrac{2}{5} \cdot \dfrac{16}{16}$      The LCD is 16·5, or 80

$x = \dfrac{45}{80} - \dfrac{32}{80}$

$x = \dfrac{13}{80}$

The solution is $\dfrac{13}{80}$.

**20.**

```
        1 7
  1 5 ⟌ 2 6 3
        1 5
        ─────
        1 1 3
        1 0 5
        ─────
            8
```

The answer is 17 R 8. A mixed numeral for the answer is written $17\dfrac{8}{15}$.

**21.** $9\frac{3}{8} = \frac{75}{8}$  $(8 \cdot 9 = 72, 72 + 3 = 75)$

Answer C is correct.

**22.** $\begin{array}{r} 9 \\ 4\,\overline{)\,3\,9} \\ 3\,6 \\ \hline 3 \end{array}$

$\frac{39}{4} = 9\frac{3}{4}$

Answer C is correct.

**23.** No; if one number is a multiple of the other, for example, the LCM is the larger of the numbers.

**24.** We multiply by 1, using the notation $n/n$, to express each fraction in terms of the least common denominator.

**25.** Write $\frac{8}{5}$ as $\frac{16}{10}$ and $\frac{8}{2}$ as $\frac{40}{10}$ and use a drawing to show that $\frac{16}{10} - \frac{40}{10} \neq \frac{8}{3}$. You could also find the sum $\frac{8}{3} + \frac{8}{2}$ and show that it is not $\frac{8}{5}$.

**26.** No; $2\frac{1}{3} = \frac{7}{3}$ but $2 \cdot \frac{1}{3} = \frac{2}{3}$.

## Exercise Set 4.5

**1.**
$$\begin{array}{r} 20 \\ +\ 8\frac{3}{4} \\ \hline 28\frac{3}{4} \end{array}$$

**3.**
$$\begin{array}{r} 129\frac{7}{8} \\ +\ 56 \\ \hline 185\frac{7}{8} \end{array}$$

**5.**
$$\begin{array}{r} 2\frac{7}{8} \\ +3\frac{5}{8} \\ \hline 5\frac{12}{8} = 5 + \frac{12}{8} \\ = 5 + 1\frac{1}{2} \\ = 6\frac{1}{2} \end{array}$$

To find a mixed numeral for $\frac{12}{8}$ we divide:

$$\begin{array}{r} 1 \\ 8\,\overline{)\,1\,2} \\ 8 \\ \hline 4 \end{array} \qquad \frac{12}{8} = 1\frac{4}{8} = 1\frac{1}{2}$$

**7.** The LCD is 12.
$$\begin{array}{r} 1\ \boxed{\dfrac{1}{4} \cdot \dfrac{3}{3}} = \ 1\frac{3}{12} \\ +1\ \boxed{\dfrac{2}{3} \cdot \dfrac{4}{4}} = +1\frac{8}{12} \\ \hline 2\frac{11}{12} \end{array}$$

**9.** The LCD is 12.
$$\begin{array}{r} 8\ \boxed{\dfrac{3}{4} \cdot \dfrac{3}{3}} = \ 8\frac{9}{12} \\ +5\ \boxed{\dfrac{5}{6} \cdot \dfrac{2}{2}} = +5\frac{10}{12} \\ \hline 13\frac{19}{12} = 13 + \frac{19}{12} \\ = 13 + 1\frac{7}{12} \\ = 14\frac{7}{12} \end{array}$$

**11.** The LCD is 10.
$$\begin{array}{r} 3\ \boxed{\dfrac{2}{5} \cdot \dfrac{2}{2}} = \ 3\frac{4}{10} \\ +8\frac{7}{10} = +8\frac{7}{10} \\ \hline 11\frac{11}{10} = 11 + \frac{11}{10} \\ = 11 + 1\frac{1}{10} \\ = 12\frac{1}{10} \end{array}$$

**13.** The LCD is 24.
$$\begin{array}{r} 5\ \boxed{\dfrac{3}{8} \cdot \dfrac{3}{3}} = \ 5\frac{9}{24} \\ +10\ \boxed{\dfrac{5}{6} \cdot \dfrac{4}{4}} = +10\frac{20}{24} \\ \hline 15\frac{29}{24} = 15 + \frac{29}{24} \\ = 15 + 1\frac{5}{24} \\ = 16\frac{5}{24} \end{array}$$

**15.** The LCD is 10.
$$\begin{array}{r} 12\ \boxed{\dfrac{4}{5} \cdot \dfrac{2}{2}} = \ 12\frac{8}{10} \\ +8\frac{7}{10} = +8\frac{7}{10} \\ \hline 20\frac{15}{10} = 20 + \frac{15}{10} \\ = 20 + 1\frac{5}{10} \\ = 21\frac{5}{10} \\ = 21\frac{1}{2} \end{array}$$

**17.** The LCD is 8.

$$14\frac{5}{8} = 14\frac{5}{8}$$
$$+13\ \boxed{\frac{1}{4}\cdot\frac{2}{2}} = +13\frac{2}{8}$$
$$\overline{\qquad\qquad\qquad 27\frac{7}{8}}$$

**19.** The LCD is 24.

$$7\ \boxed{\frac{1}{8}\cdot\frac{3}{3}} = 7\frac{3}{24}$$
$$9\ \boxed{\frac{2}{3}\cdot\frac{8}{8}} = 9\frac{16}{24}$$
$$+10\ \boxed{\frac{3}{4}\cdot\frac{6}{6}} = +10\frac{18}{24}$$
$$\overline{\qquad\qquad\qquad 26\frac{37}{24}} = 26+\frac{37}{24}$$
$$= 26+1\frac{13}{24}$$
$$= 27\frac{13}{24}$$

**21.**
$$4\frac{1}{5} = 3\frac{6}{5}$$
$$-2\frac{3}{5} = -2\frac{3}{5}$$
$$\overline{\qquad\qquad 1\frac{3}{5}}$$

> Since $\frac{1}{5}$ is smaller than $\frac{3}{5}$, we cannot subtract until we borrow:
> $$4\frac{1}{5} = 3+\frac{5}{5}+\frac{1}{5} = 3+\frac{6}{5} = 3\frac{6}{5}$$

**23.** The LCD is 10.

$$6\ \boxed{\frac{3}{5}\cdot\frac{2}{2}} = 6\frac{6}{10}$$
$$-2\ \boxed{\frac{1}{2}\cdot\frac{5}{5}} = -2\frac{5}{10}$$
$$\overline{\qquad\qquad\qquad 4\frac{1}{10}}$$

**25.** The LCD is 24.

$$34\ \boxed{\frac{1}{3}\cdot\frac{8}{8}} = 34\frac{8}{24} = 33\frac{32}{24}$$
$$-12\ \boxed{\frac{5}{8}\cdot\frac{3}{3}} = -12\frac{15}{24} = -12\frac{15}{24}$$
$$\overline{\qquad\qquad\qquad\qquad\qquad 21\frac{17}{24}}$$

> $\Big($Since $\frac{8}{24}$ is smaller than $\frac{15}{24}$, we cannot subtract until we borrow: $34\frac{8}{24} = 33+\frac{24}{24}+\frac{8}{24} = 33+\frac{32}{24} = 33\frac{32}{24}.\Big)$

**27.**
$$21 = 20\frac{4}{4} \quad \Big(21 = 20+1 = 20+\frac{4}{4} = 20\frac{4}{4}\Big)$$
$$-8\frac{3}{4} = -8\frac{3}{4}$$
$$\overline{\qquad\qquad 12\frac{1}{4}}$$

**29.**
$$34 = 33\frac{8}{8} \quad \Big(34 = 33+1 = 33+\frac{8}{8} = 33\frac{8}{8}\Big)$$
$$-18\frac{5}{8} = -18\frac{5}{8}$$
$$\overline{\qquad\qquad 15\frac{3}{8}}$$

**31.** The LCD is 12.

$$21\ \boxed{\frac{1}{6}\cdot\frac{2}{2}} = 21\frac{2}{12} = 20\frac{14}{12}$$
$$-13\ \boxed{\frac{3}{4}\cdot\frac{3}{3}} = -13\frac{9}{12} = -13\frac{9}{12}$$
$$\overline{\qquad\qquad\qquad\qquad\qquad 7\frac{5}{12}}$$

> $\Big($Since $\frac{2}{12}$ is smaller than $\frac{9}{12}$, we cannot subtract until we borrow: $21\frac{2}{12} = 20+\frac{12}{12}+\frac{2}{12} = 20+\frac{14}{12} = 20\frac{14}{12}.\Big)$

**33.** The LCD is 8.

$$14\frac{1}{8} = 14\frac{1}{8} = 13\frac{9}{8}$$
$$-\boxed{\frac{3}{4}\cdot\frac{2}{2}} = -\frac{6}{8} = -\frac{6}{8}$$
$$\overline{\qquad\qquad\qquad\qquad 13\frac{3}{8}}$$

> $\Big($Since $\frac{1}{8}$ is smaller than $\frac{6}{8}$, we cannot subtract until we borrow: $14\frac{1}{8} = 13+\frac{8}{8}+\frac{1}{8} = 13+\frac{9}{8} = 13\frac{9}{8}.\Big)$

**35.** The LCD is 18.

$$25\ \boxed{\frac{1}{9}\cdot\frac{2}{2}} = 25\frac{2}{18} = 24\frac{20}{18}$$
$$-13\ \boxed{\frac{5}{6}\cdot\frac{3}{3}} = -13\frac{15}{18} = -13\frac{15}{18}$$
$$\overline{\qquad\qquad\qquad\qquad\qquad 11\frac{5}{18}}$$

> $\Big($Since $\frac{2}{18}$ is smaller than $\frac{15}{18}$, we cannot subtract until we borrow: $25\frac{2}{18} = 24+\frac{18}{18}+\frac{2}{18} = 24+\frac{20}{18} = 24\frac{20}{18}.\Big)$

**37. *Familiarize.*** Let $f$ = the total number of flats planted.

***Translate.*** We write an equation.

| Flats of impatiens | + | Flats of snapdragons | + | Flats of phlox | is | Total flats |
|---|---|---|---|---|---|---|
| ↓ | ↓ | ↓ | ↓ | ↓ | ↓ | ↓ |
| $4\frac{1}{2}$ | + | $6\frac{2}{3}$ | + | $3\frac{3}{8}$ | = | $f$ |

*Solve*. We add. The LCD is 24.

$$4 \,\boxed{\dfrac{1}{2}\cdot\dfrac{12}{12}} = \; 4\,\dfrac{12}{24}$$

$$6 \,\boxed{\dfrac{2}{3}\cdot\dfrac{8}{8}} = \; 6\,\dfrac{16}{24}$$

$$+3\,\boxed{\dfrac{3}{8}\cdot\dfrac{3}{3}} = +3\,\dfrac{9}{24}$$

$$\overline{\qquad\qquad\quad 13\,\dfrac{37}{24} = 13+\dfrac{37}{24}}$$

$$= 13 + 1\dfrac{13}{24}$$

$$= 14\dfrac{13}{24}$$

**Check**. We can repeat the calculation. Also note that the answer is reasonable since it is larger than any of the individual number of flats.

**State**. The landscaper planted $14\dfrac{13}{24}$ flats of flowers.

**39. Familiarize**. We let $w =$ the total weight of the meat.

**Translate**. We write an equation.

| Weight of one package | plus | Weight of second package | is | Total weight |
|---|---|---|---|---|
| $\downarrow$ | $\downarrow$ | $\downarrow$ | $\downarrow$ | $\downarrow$ |
| $1\dfrac{2}{3}$ | $+$ | $5\dfrac{3}{4}$ | $=$ | $w$ |

*Solve*. We carry out the addition. The LCD is 12.

$$1\,\boxed{\dfrac{2}{3}\cdot\dfrac{4}{4}} = \; 1\,\dfrac{8}{12}$$

$$+5\,\boxed{\dfrac{3}{4}\cdot\dfrac{3}{3}} = +5\,\dfrac{9}{12}$$

$$\overline{\qquad\qquad\quad 6\,\dfrac{17}{12} = 6+\dfrac{17}{12}}$$

$$= 6 + 1\dfrac{5}{12}$$

$$= 7\dfrac{5}{12}$$

**Check**. We repeat the calculation. We also note that the answer is larger than either of the individual weights, so the answer seems reasonable.

**State**. The total weight of the meat was $7\dfrac{5}{12}$ lb.

**41. Familiarize**. We let $h =$ Tara's excess height.

**Translate**. We have a "how much more" situation.

| Tom's height | plus | How much more height | is | Tara's height |
|---|---|---|---|---|
| $\downarrow$ | $\downarrow$ | $\downarrow$ | $\downarrow$ | $\downarrow$ |
| $59\dfrac{7}{12}$ | $+$ | $h$ | $=$ | $66$ |

*Solve*. We solve the equation as follows:

$$h = 66 - 59\dfrac{7}{12}$$

$$66 \;=\; 65\dfrac{12}{12}$$

$$-\,59\dfrac{7}{12} = -\,59\dfrac{7}{12}$$

$$\overline{\qquad\qquad\qquad\quad 6\dfrac{5}{12}}$$

**Check**. We add Tara's excess height to Tom's height:

$$6\dfrac{5}{12} + 59\dfrac{7}{12} = 65\dfrac{12}{12} = 66$$

The answer checks.

**State**. Tara is $6\dfrac{5}{12}$ in. taller.

**43. Familiarize**. Let $f =$ the total number of yards of upholstery fabric Art bought.

**Translate**.

| Amount of one fabric | plus | Amount of second fabric | is | Total amount of fabric |
|---|---|---|---|---|
| $\downarrow$ | $\downarrow$ | $\downarrow$ | $\downarrow$ | $\downarrow$ |
| $9\dfrac{1}{4}$ | $+$ | $10\dfrac{5}{6}$ | $=$ | $f$ |

*Solve*. We add. The LCD is 12.

$$9 \,\boxed{\dfrac{1}{4}\cdot\dfrac{3}{3}} = \; 9\,\dfrac{3}{12}$$

$$+10\,\boxed{\dfrac{5}{6}\cdot\dfrac{2}{2}} = +10\,\dfrac{10}{12}$$

$$\overline{\qquad\qquad\quad 19\,\dfrac{13}{12} = 19+\dfrac{13}{12}}$$

$$= 19 + 1\dfrac{1}{12}$$

$$= 20\dfrac{1}{12}$$

Thus, $20\dfrac{1}{12} = f$.

**Check**. We can repeat the calculation. We also observe that the result is larger than either of the individual amounts of fabric, so the answer seems reasonable.

**State**. Art bought $20\dfrac{1}{12}$ yd of fabric.

**45. Familiarize**. Let $d =$ the depth of the third stone, in inches. From the drawing in the text, we see that the sum of the depths of the bench and the three supporting stones must be 18 in.

**Translate.**

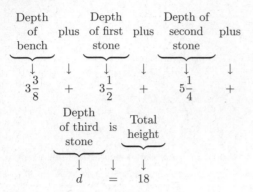

Depth of bench plus Depth of first stone plus Depth of second stone plus

$$3\frac{3}{8} + 3\frac{1}{2} + 5\frac{1}{4} +$$

Depth of third stone is Total height

$$d = 18$$

**Solve.** First we collect like terms on the left side of the equation. The LCD is 8.

$$3\frac{3}{8} + 3\frac{1}{2} + 5\frac{1}{4} + d = 18$$

$$3\frac{3}{8} + 3\frac{4}{8} + 5\frac{2}{8} + d = 18$$

$$11\frac{9}{8} + d = 18$$

$$12\frac{1}{8} + d = 18 \qquad \left(11\frac{9}{8} = 12\frac{1}{8}\right)$$

$$12\frac{1}{8} + d - 12\frac{1}{8} = 18 - 12\frac{1}{8}$$

$$d + 0 = 17\frac{8}{8} - 12\frac{1}{8}$$

$$d = 5\frac{7}{8}$$

**Check.** We can add the four depths.

$$3\frac{3}{8} + 3\frac{1}{2} + 5\frac{1}{4} + 5\frac{7}{8} = 3\frac{3}{8} + 3\frac{4}{8} + 5\frac{2}{8} + 5\frac{7}{8}$$

$$= 16\frac{16}{8} = 16 + \frac{16}{8}$$

$$= 16 + 2 = 18$$

Since we get the height of the bench, 18 in., the answer checks.

**State.** The depth of the third stone is $5\frac{7}{8}$ in.

**47. Familiarize.** We draw a picture. We let $D =$ the distance from Los Angeles at the end of the second day.

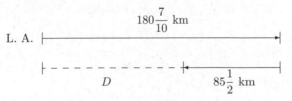

$180\frac{7}{10}$ km

L. A.

$D$        $85\frac{1}{2}$ km

**Translate.** We write an equation.

Distance away from Los Angeles minus Distance toward Los Angeles is Distance from Los Angeles

$$180\frac{7}{10} - 85\frac{1}{2} = D$$

**Solve.** To solve the equation we carry out the subtraction. The LCD is 10.

$$180\frac{7}{10} = 180\frac{7}{10}$$

$$- 85\boxed{\frac{1}{2}\cdot\frac{5}{5}} = - 85\frac{5}{10}$$

$$95\frac{2}{10} = 95\frac{1}{5}$$

**Check.** We add the distance from Los Angeles to the distance the person drove toward Los Angeles:

$$95\frac{1}{5} + 85\frac{1}{2} = 95\frac{2}{10} + 85\frac{5}{10} = 180\frac{7}{10}$$

This checks.

**State.** Kim Park was $95\frac{1}{5}$ mi from Los Angeles.

**49. Familiarize.** We let $P =$ the perimeter of the mirror.

**Translate.** We add the lengths of the four sides of the mirror.

$$P = 30\frac{1}{2} + 36\frac{5}{8} + 30\frac{1}{2} + 36\frac{5}{8}$$

**Solve.** We carry out the addition. The LCD is 8.

$$30\boxed{\frac{1}{2}\cdot\frac{4}{4}} = 30\frac{4}{8}$$

$$36\frac{5}{8} = 36\frac{5}{8}$$

$$30\boxed{\frac{1}{2}\cdot\frac{4}{4}} = 30\frac{4}{8}$$

$$+36\frac{5}{8} = + 36\frac{5}{8}$$

$$132\frac{18}{8} = 132 + \frac{18}{8}$$

$$= 132 + 2\frac{2}{8}$$

$$= 134\frac{2}{8}$$

$$= 134\frac{1}{4}$$

**Check.** We repeat the calculation. The answer checks.

**State.** The perimeter of the mirror is $134\frac{1}{4}$ in.

**51. Familiarize.** We make a drawing. Let $l =$ the length of the wood left over, in inches.

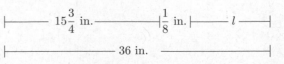

$15\frac{3}{4}$ in.   $\frac{1}{8}$ in.   $l$

36 in.

*Translate*.

$$\underbrace{\text{Length cut off}} + \underbrace{\text{Thickness of blade}} + \underbrace{\text{Length left over}} \;\text{is}\; \underbrace{\text{Original length}}$$

$$15\frac{3}{4} \;+\; \frac{1}{8} \;+\; l \;=\; 36$$

*Solve*. This is a two-step problem. First we add $15\frac{3}{4}$ and $\frac{1}{8}$. The LCD is 8.

$$
\begin{array}{r}
15\;\boxed{\dfrac{3}{4}\cdot\dfrac{2}{2}} = \quad 15\,\dfrac{6}{8} \\[2mm]
+ \quad \dfrac{1}{8} \quad\;\; = +\quad \dfrac{1}{8} \\[1mm]
\hline
15\,\dfrac{7}{8}
\end{array}
$$

Now we have $15\frac{7}{8} + l = 36$. We subtract $15\frac{7}{8}$ on both sides of the equation.

$$15\frac{7}{8} + l = 36$$
$$l = 36 - 15\frac{7}{8}$$

$$
\begin{array}{r}
36 \quad = \quad 35\,\dfrac{8}{8} \\[2mm]
- 15\,\dfrac{7}{8} = -15\,\dfrac{7}{8} \\[1mm]
\hline
20\,\dfrac{1}{8}
\end{array}
$$

*Check*. We repeat the calculations.

*State*. The piece of wood left over is $20\frac{1}{8}$ in. long.

**53. Familiarize**. We let $h =$ Rene's height.

*Translate*. We write an equation.

$$\underbrace{\text{Son's height}} + \underbrace{\text{Additional height of man}} = \underbrace{\text{Rene's height}}$$

$$72\frac{5}{6} \;+\; 5\frac{1}{4} \;=\; h$$

*Solve*. To solve we carry out the addition. The LCD is 12.

$$
\begin{array}{r}
72\;\boxed{\dfrac{5}{6}\cdot\dfrac{2}{2}} = \quad 72\,\dfrac{10}{12} \\[2mm]
+\;5\;\boxed{\dfrac{1}{4}\cdot\dfrac{3}{3}} = +\;5\,\dfrac{3}{12} \\[1mm]
\hline
77\,\dfrac{13}{12} = 78\,\dfrac{1}{12}
\end{array}
$$

*Check*. We repeat the calculation. We also note that the man's height is larger than his son's height, so the answer seems reasonable.

*State*. Rene is $78\frac{1}{12}$ in. tall.

**55. Familiarize**. We make a drawing. We let $t =$ the number of hours the designer worked on the third day.

$$\vdash 2\tfrac{1}{2}\text{ hr} \dashv \quad 4\tfrac{1}{5}\text{ hr} \quad \vdash t \dashv$$
$$\vdash\!\!\!\!\!\!\!\!\!\!\!\!\!\!\!\!\!\!\!\!\!\!\!\!\!\!\!\!\!\!\!\!\! 10\tfrac{1}{2}\text{ hr} \!\!\!\!\!\!\!\!\!\!\!\!\!\!\!\!\!\!\!\!\!\!\!\!\!\!\!\!\!\!\!\!\!\dashv$$

*Translate*. We write an addition sentence.

$$2\frac{1}{2} + 4\frac{1}{5} + t = 10\frac{1}{2}$$

*Solve*. This is a two-step problem.

First we add $2\frac{1}{2} + 4\frac{1}{5}$ to find the time worked on the first two days. The LCD is 10.

$$
\begin{array}{r}
2\;\boxed{\dfrac{1}{2}\cdot\dfrac{5}{5}} = \quad 2\,\dfrac{5}{10} \\[2mm]
+4\;\boxed{\dfrac{1}{5}\cdot\dfrac{2}{2}} = +4\,\dfrac{2}{10} \\[1mm]
\hline
6\,\dfrac{7}{10}
\end{array}
$$

Then we subtract $6\frac{7}{10}$ from $10\frac{1}{2}$ to find the time worked on the third day. The LCD is 10.

$$6\frac{7}{10} + t = 10\frac{1}{2}$$
$$t = 10\frac{1}{2} - 6\frac{7}{10}$$

$$
\begin{array}{r}
10\;\boxed{\dfrac{1}{2}\cdot\dfrac{5}{5}} = \quad 10\,\dfrac{5}{10} = \quad 9\,\dfrac{15}{10} \\[2mm]
-\;6\,\dfrac{7}{10} \quad = -\;6\,\dfrac{7}{10} = -6\,\dfrac{7}{10} \\[1mm]
\hline
3\,\dfrac{8}{10} = 3\dfrac{4}{5}
\end{array}
$$

*Check*. We repeat the calculations.

*State*. Sue worked $3\frac{4}{5}$ hr the third day.

**57.** The length of each of the five sides is $5\frac{3}{4}$ yd. We add to find the distance around the figure.

$$5\frac{3}{4} + 5\frac{3}{4} + 5\frac{3}{4} + 5\frac{3}{4} + 5\frac{3}{4} = 25\frac{15}{4} = 25 + 3\frac{3}{4} = 28\frac{3}{4}$$

The distance is $28\frac{3}{4}$ yd.

**59.** We see that $d$ and the two smallest distances combined are the same as the largest distance. We translate and solve.

$$2\frac{3}{4} + d + 2\frac{3}{4} = 12\frac{7}{8}$$

$$d = 12\frac{7}{8} - 2\frac{3}{4} - 2\frac{3}{4}$$

$$= 10\frac{1}{8} - 2\frac{3}{4} \quad \text{Subtracting } 2\frac{3}{4} \text{ from } 12\frac{7}{8}$$

$$= 7\frac{3}{8} \qquad \text{Subtracting } 2\frac{3}{4} \text{ from } 10\frac{1}{8}$$

The length of $d$ is $7\frac{3}{8}$ ft.

**61. Familiarize.** Let $f =$ the number of yards of fabric needed to make the outfit.

**Translate.** We write an equation.

$$\underbrace{\text{Fabric} \atop \text{for dress}} + \underbrace{\text{Fabric} \atop \text{for band}} + \underbrace{\text{Fabric} \atop \text{for jacket}} \text{ is } \underbrace{\text{Total} \atop \text{fabric}}$$

$$\downarrow \quad \downarrow \quad \downarrow \quad \downarrow \quad \downarrow \quad \downarrow \quad \downarrow$$

$$1\frac{3}{8} + \frac{5}{8} + 3\frac{3}{8} = f$$

**Solve.** We add.

$$
\begin{array}{r}
1\ \dfrac{3}{8} \\[2mm]
\dfrac{5}{8} \\[2mm]
+\ 3\ \dfrac{3}{8} \\[1mm]
\hline
4\ \dfrac{11}{8} = 4 + \dfrac{11}{8}
\end{array}
$$

$$= 4 + 1\frac{3}{8}$$

$$= 5\frac{3}{8}$$

**Check.** We can repeat the calculation. Also note that the answer is reasonable since it is larger than any of the individual amounts of fabric.

**State.** The outfit requires $5\frac{3}{8}$ yd of fabric.

**63. Familiarize.** The height of the window, $4\frac{5}{6}$ ft, will remain the same. The $8\frac{1}{4}$ ft will increase by $2\frac{1}{3}$ ft. Let $w =$ the new width of the window.

**Translate.**

$$\underbrace{\text{Original width}}_{} \text{ plus } \underbrace{\text{Added width}}_{} \text{ is } \underbrace{\text{New width}}_{}$$

$$\downarrow \quad \downarrow \quad \downarrow \quad \downarrow \quad \downarrow$$

$$8\frac{1}{4} + 2\frac{1}{3} = w$$

**Solve.** We carry out the addition. The LCD is 12.

$$
\begin{array}{r}
8\ \boxed{\dfrac{1}{4}\cdot\dfrac{3}{3}} = 8\ \dfrac{3}{12} \\[3mm]
+\ 2\ \boxed{\dfrac{1}{3}\cdot\dfrac{4}{4}} = +\ 2\ \dfrac{4}{12} \\[1mm]
\hline
10\ \dfrac{7}{12}
\end{array}
$$

**Check.** We repeat the calculation. The answer checks.

**State.** The new dimensions of the window are $4\frac{5}{6}$ ft $\times\ 10\frac{7}{12}$ ft.

**65. Familiarize.** We visualize the situation. Repeated subtraction, or division, works well here.

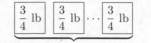

12 lb fills how many packages?

Let $n =$ the number of packages that can be made.

**Translate.** We translate to an equation.

$$n = 12 \div \frac{3}{4}$$

**Solve.** We carry out the division.

$$n = 12 \div \frac{3}{4} = 12 \cdot \frac{4}{3} = \frac{12 \cdot 4}{3}$$

$$= \frac{3 \cdot 4 \cdot 4}{3 \cdot 1} = \frac{3}{3} \cdot \frac{4 \cdot 4}{1}$$

$$= 16$$

**Check.** If each of 16 packages contains $\frac{3}{4}$ lb of cheese, a total of

$$16 \cdot \frac{3}{4} = \frac{16 \cdot 3}{4} = \frac{4 \cdot 4 \cdot 3}{4} = 4 \cdot 3,$$

or 12 lb of cheese is used. The answer checks.

**State.** 16 packages of cheese can be made from a 12-lb slab.

**67.** The sum of the digits is $9 + 9 + 9 + 3 = 30$. Since 30 is divisible by 3, then 9993 is divisible by 3.

**69.** The sum of the digits is $2 + 3 + 4 + 5 = 14$. Since 14 is not divisible by 9, then 2345 is not divisible by 9.

**71.** The ones digit of 2335 is not 0, so 2335 is not divisible by 10.

**73.** The last three digits of 18,888 are divisible by 8, so 18,888 is divisible by 8.

**75.**
$$\frac{15}{9} \cdot \frac{18}{39} = \frac{15 \cdot 18}{9 \cdot 39} = \frac{3 \cdot 5 \cdot 2 \cdot 3 \cdot 3}{3 \cdot 3 \cdot 3 \cdot 13}$$

$$= \frac{3 \cdot 3 \cdot 3}{3 \cdot 3 \cdot 3} = \frac{5 \cdot 2}{13}$$

$$= \frac{10}{13}$$

**77.** $-5\frac{1}{4} + \left(-3\frac{3}{8}\right) = -5\frac{2}{8} + \left(-3\frac{3}{8}\right) = -8\frac{5}{8}$

**79.** $-4\frac{5}{12} - 6\frac{5}{8} = -4\frac{5}{12} + \left(-6\frac{5}{8}\right)$

$$= -4\frac{10}{24} + \left(-6\frac{15}{24}\right)$$

$$= -10\frac{25}{24}$$

$$= -\left(10 + 1\frac{1}{24}\right)$$

$$= -11\frac{1}{24}$$

## Exercise Set 4.6

**1.** $8 \cdot 2\frac{5}{6}$

$= \frac{8}{1} \cdot \frac{17}{6}$    Writing fraction notation

$= \frac{8 \cdot 17}{1 \cdot 6} = \frac{2 \cdot 4 \cdot 17}{1 \cdot 2 \cdot 3} = \frac{2}{2} \cdot \frac{4 \cdot 17}{1 \cdot 3} = \frac{68}{3} = 22\frac{2}{3}$

**3.** $3\frac{5}{8} \cdot \left(-\frac{2}{3}\right)$

$= \frac{29}{8} \cdot \left(-\frac{2}{3}\right)$    Writing fraction notation

$= -\frac{29 \cdot 2}{8 \cdot 3} = -\frac{29 \cdot 2}{2 \cdot 4 \cdot 3} = \frac{2}{2} \cdot \left(-\frac{29}{4 \cdot 3}\right) = -\frac{29}{12} = -2\frac{5}{12}$

**5.** $3\frac{1}{2} \cdot 2\frac{1}{3} = \frac{7}{2} \cdot \frac{7}{3} = \frac{49}{6} = 8\frac{1}{6}$

**7.** $3\frac{2}{5} \cdot 2\frac{7}{8} = \frac{17}{5} \cdot \frac{23}{8} = \frac{391}{40} = 9\frac{31}{40}$

**9.** $-4\frac{7}{10} \cdot 5\frac{3}{10} = -\frac{47}{10} \cdot \frac{53}{10} = -\frac{2491}{100} = -24\frac{91}{100}$

**11.** $-20\frac{1}{2} \cdot \left(-4\frac{2}{3}\right) = -\frac{41}{2} \cdot \left(-\frac{14}{3}\right) = \frac{41 \cdot 14}{2 \cdot 3} =$

$\frac{41 \cdot 2 \cdot 7}{2 \cdot 3} = \frac{2}{2} \cdot \frac{41 \cdot 7}{3} = \frac{287}{3} = 95\frac{2}{3}$

**13.** $20 \div 3\frac{1}{5}$

$= 20 \div \frac{16}{5}$    Writing fractional notation

$= 20 \cdot \frac{5}{16}$    Multiplying by the reciprocal

$= \frac{20 \cdot 5}{16} = \frac{4 \cdot 5 \cdot 5}{4 \cdot 4} = \frac{4}{4} \cdot \frac{5 \cdot 5}{4} = \frac{25}{4} = 6\frac{1}{4}$

**15.** $8\frac{2}{5} \div 7$

$= \frac{42}{5} \div 7$    Writing fractional notation

$= \frac{42}{5} \cdot \frac{1}{7}$    Multiplying by the reciprocal

$= \frac{42 \cdot 1}{5 \cdot 7} = \frac{6 \cdot 7}{5 \cdot 7} = \frac{7}{7} \cdot \frac{6}{5} = \frac{6}{5} = 1\frac{1}{5}$

**17.** $-4\frac{3}{4} \div 1\frac{1}{3} = -\frac{19}{4} \div \frac{4}{3} = -\frac{19}{4} \cdot \frac{3}{4} = -\frac{19 \cdot 3}{4 \cdot 4} =$

$-\frac{57}{16} = -3\frac{9}{16}$

**19.** $1\frac{7}{8} \div 1\frac{2}{3} = \frac{15}{8} \div \frac{5}{3} = \frac{15}{8} \cdot \frac{3}{5} = \frac{15 \cdot 3}{8 \cdot 5} = \frac{5 \cdot 3 \cdot 3}{8 \cdot 5}$

$= \frac{5}{5} \cdot \frac{3 \cdot 3}{8} = \frac{3 \cdot 3}{8} = \frac{9}{8} = 1\frac{1}{8}$

**21.** $5\frac{1}{10} \div 4\frac{3}{10} = \frac{51}{10} \div \frac{43}{10} = \frac{51}{10} \cdot \frac{10}{43} = \frac{51 \cdot 10}{10 \cdot 43}$

$= \frac{10}{10} \cdot \frac{51}{43} = \frac{51}{43} = 1\frac{8}{43}$

**23.** $-20\frac{1}{4} \div (-90) = -\frac{81}{4} \div (-90) = -\frac{81}{4} \cdot \left(-\frac{1}{90}\right) =$

$\frac{81 \cdot 1}{4 \cdot 90} = \frac{9 \cdot 9 \cdot 1}{4 \cdot 9 \cdot 10} = \frac{9}{9} \cdot \frac{9 \cdot 1}{4 \cdot 10} = \frac{9}{40}$

**25. Familiarize.** Let $b =$ the number of beagles registered with The American Kennel Club.

**Translate.**

| $3\frac{4}{9}$ times | Number of beagles registered | is | Number of Labrador retrievers registered |
|---|---|---|---|
| ↓ ↓ | ↓ | ↓ | ↓ |
| $3\frac{4}{9}$ · | $b$ | = | $155,000$ |

**Solve.** We divide by $3\frac{4}{9}$ on both sides of the equation.

$b = 155,000 \div 3\frac{4}{9}$

$b = 155,000 \div \frac{31}{9}$

$b = 155,000 \times \frac{9}{31} = \frac{155,000 \cdot 9}{31}$

$b = \frac{31 \cdot 5000 \cdot 9}{31 \cdot 1} = \frac{31}{31} \cdot \frac{5000 \cdot 9}{1} = \frac{5000 \cdot 9}{1}$

$b = 45,000$

**Check.** Since $3\frac{4}{9} \cdot 45,000 = \frac{31}{9} \cdot 45,000 = 155,000,$ the answer checks.

**State.** There are 45,000 beagles registered with The American Kennel Club.

**27. Familiarize.** Let $p =$ the population of Alaska.

**Translate.**

| $6\frac{4}{5}$ times | Population of Alaska | is | Population of Alabama |
|---|---|---|---|
| ↓ ↓ | ↓ | ↓ | ↓ |
| $6\frac{4}{5}$ · | $p$ | = | $4,700,000$ |

**Solve.** We divide by $6\frac{4}{5}$ on both sides.

$6\frac{4}{5} \cdot p = 4,700,000$

$p = 4,700,000 \div 6\frac{4}{5}$

$p = 4,700,000 \div \frac{34}{5}$

$p = 4,700,000 \cdot \frac{5}{34}$

$p = \frac{4,700,000 \cdot 5}{34}$

$p \approx 691,176 \approx 690,000$

**Check.** We repeat the calculation. The answer checks.

**State.** The population of Alaska is about 690,000.

**29.** *Familiarize.* Let $A =$ the area of the mural, in square feet. Recall that the area of a rectangle is length times width.

*Translate.*

$$\begin{array}{ccccc} \text{Area} & = & \text{length} & \times & \text{width} \\ \downarrow & \downarrow & \downarrow & \downarrow & \downarrow \\ A & = & 9\frac{3}{8} & \times & 6\frac{2}{3} \end{array}$$

*Solve.* We carry out the multiplication.

$$A = 9\frac{3}{8} \times 6\frac{2}{3} = \frac{75}{8} \cdot \frac{20}{3} = \frac{75 \cdot 20}{8 \cdot 3}$$
$$= \frac{3 \cdot 25 \cdot 4 \cdot 5}{4 \cdot 2 \cdot 3} = \frac{3 \cdot 4}{3 \cdot 4} \cdot \frac{25 \cdot 5}{2}$$
$$= \frac{125}{2} = 62\frac{1}{2}$$

*Check.* We repeat the calculation. The answer checks.

*State.* The area of the mural is $62\frac{1}{2}$ ft$^2$.

**31.** *Familiarize.* Let $s =$ the number of teaspoons of sodium the average American woman consumes in ten days.

*Translate.* A multiplication corresponds to this situation.

$$s = 10 \cdot 1\frac{1}{3}$$

*Solve.* We carry out the multiplication.

$$s = 10 \cdot 1\frac{1}{3} = 10 \cdot \frac{4}{3} = \frac{10 \cdot 4}{3} = \frac{40}{3} = 13\frac{1}{3}$$

*Check.* We repeat the calculation. The answer checks.

*State.* In ten days the average American woman consumes $13\frac{1}{3}$ tsp of sodium.

**33.** *Familiarize.* We let $w =$ the weight of $5\frac{1}{2}$ cubic feet of water.

*Translate.* We write an equation.

$$\begin{array}{ccccc} \underbrace{\begin{array}{c}\text{Weight per} \\ \text{cubic foot}\end{array}} & \cdot & \underbrace{\begin{array}{c}\text{Number of} \\ \text{cubic feet}\end{array}} & = & \underbrace{\begin{array}{c}\text{Total} \\ \text{weight}\end{array}} \\ \downarrow & \downarrow & \downarrow & \downarrow & \downarrow \\ 62\frac{1}{2} & \cdot & 5\frac{1}{2} & = & w \end{array}$$

*Solve.* To solve the equation we carry out the multiplication.

$$w = 62\frac{1}{2} \cdot 5\frac{1}{2}$$
$$= \frac{125}{2} \cdot \frac{11}{2} = \frac{125 \cdot 11}{2 \cdot 2}$$
$$= \frac{1375}{4} = 343\frac{3}{4}$$

*Check.* We repeat the calculation. We also note that $62\frac{1}{2} \approx 60$ and $5\frac{1}{2} \approx 5$. Then the product is about 300. Our answer seems reasonable.

*State.* The weight of $5\frac{1}{2}$ cubic feet of water is $343\frac{3}{4}$ lb.

**35.** *Familiarize.* We let $t =$ the Fahrenheit temperature.

*Translate.*

$$\begin{array}{cccccc} \underbrace{\begin{array}{c}\text{Celsius} \\ \text{temperature}\end{array}} & \text{times } 1\frac{4}{5} & \text{plus} & 32° & \text{is} & \underbrace{\begin{array}{c}\text{Fahrenheit} \\ \text{temperature}\end{array}} \\ \downarrow & \downarrow \quad \downarrow & \downarrow & \downarrow & \downarrow & \downarrow \\ 20 & \cdot \quad 1\frac{4}{5} & + & 32 & = & t \end{array}$$

*Solve.* We multiply and then add, according to the rules for order of operations.

$$t = 20 \cdot 1\frac{4}{5} + 32 = \frac{20}{1} \cdot \frac{9}{5} + 32 = \frac{20 \cdot 9}{1 \cdot 5} + 32 =$$
$$\frac{4 \cdot 5 \cdot 9}{1 \cdot 5} + 32 = \frac{5}{5} \cdot \frac{4 \cdot 9}{1} + 32 = 36 + 32 = 68$$

*Check.* We repeat the calculation.

*State.* 68° Fahrenheit corresponds to 20° Celsius.

**37.** *Familiarize.* Let $c =$ the daily circulation of the *Wall Street Journal*.

*Translate.*

$$\begin{array}{ccccc} 5\frac{1}{4} & \text{times} & \underbrace{\begin{array}{c}\textit{Star Tribune} \\ \text{circulation}\end{array}} & \text{is} & \underbrace{\begin{array}{c}\textit{Wall Street} \\ \textit{Journal} \text{ circulation}\end{array}} \\ \downarrow & \downarrow & \downarrow & \downarrow & \downarrow \\ 5\frac{1}{4} & \cdot & 343,000 & = & c \end{array}$$

*Solve.* We carry out the multiplication.

$$5\frac{1}{4} \cdot 343,000 = \frac{21}{4} \cdot 343,000 = \frac{21 \cdot 343,000}{4} =$$
$$\frac{21 \cdot 4 \cdot 85,750}{4 \cdot 1} = \frac{4}{4} \cdot \frac{21 \cdot 85,750}{1} = \frac{21 \cdot 85,750}{1} =$$
$$1,800,750$$

*Check.* We repeat the calculation. The answer checks.

*State.* In 2002 the daily circulation of the *Wall Street Journal* was 1,800,750.

**39.** *Familiarize.* Let $p =$ the population of the United States in 2007.

*Translate.*

$$\begin{array}{ccccc} 3\frac{3}{4} & \text{times} & \underbrace{\begin{array}{c}\text{Population} \\ \text{of U.S.}\end{array}} & \text{is} & \underbrace{\begin{array}{c}\text{Population} \\ \text{of India}\end{array}} \\ \downarrow & \downarrow & \downarrow & \downarrow & \downarrow \\ 3\frac{3}{4} & \cdot & p & = & 1,130,100,000 \end{array}$$

*Solve.* We divide by $3\frac{3}{4}$ on both sides.

$$3\frac{3}{4} \cdot p = 1,130,100,000$$
$$p = 1,130,100,000 \div 3\frac{3}{4}$$
$$p = 1,130,100,000 \div \frac{15}{4}$$
$$p = 1,130,100,000 \cdot \frac{4}{15}$$
$$p = \frac{1,130,100,00 \cdot 4}{15} = \frac{\cancel{15} \cdot 75,340,000 \cdot 4}{\cancel{15} \cdot 1}$$
$$p = 301,360,000$$

**Check**. We can find $3\frac{3}{4}$ times 301,360,000.

$$3\frac{3}{4} \cdot 301,360,000 = \frac{15}{4} \cdot 301,360,000 =$$

$$\frac{15 \cdot 301,360,000}{4} = \frac{15 \cdot \cancel{4} \cdot 75,340,000}{\cancel{4} \cdot 1} =$$

$$1,130,100,000$$

This is the population of India, so the answer checks.

**State**. The population of the United States was 301,360,000 in 2007.

**41. Familiarize**. Let $f =$ the number of cups of flour and $s =$ the number of cups of sugar in the doubled recipe.

**Translate**. We write two equations. We multiply each of the original amounts by 2 to find the amounts in the doubled recipe.

$$f = 2 \cdot 2\frac{3}{4}$$
$$s = 2 \cdot 1\frac{1}{3}$$

**Solve**. We carry out the calculations.

$$f = 2 \cdot 2\frac{3}{4} = 2 \cdot \frac{11}{4} = \frac{2 \cdot 11}{4} = \frac{\cancel{2} \cdot 11}{\cancel{2} \cdot 2} = \frac{11}{2} = 5\frac{1}{2}$$

$$s = 2 \cdot 1\frac{1}{3} = 2 \cdot \frac{4}{3} = \frac{2 \cdot 4}{3} = \frac{8}{3} = 2\frac{2}{3}$$

**Check**. We repeat the calculations. The answer checks.

**State**. The chef will need $5\frac{1}{2}$ cups of flour and $2\frac{2}{3}$ cups of sugar for the doubled recipe.

**43. Familiarize**. We let $m =$ the number of miles per gallon the car got.

**Translate**. We write an equation.

| Total number of miles traveled | ÷ | Number of gallons of gas used | = | Miles per gallon |
|---|---|---|---|---|
| ↓ | ↓ | ↓ | ↓ | ↓ |
| 213 | ÷ | $14\frac{2}{10}$ | = | $m$ |

**Solve**. To solve the equation we carry out the division.

$$m = 213 \div 14\frac{2}{10} = 213 \div \frac{142}{10}$$

$$= 213 \cdot \frac{10}{142} = \frac{3 \cdot 71 \cdot 2 \cdot 5}{2 \cdot 71 \cdot 1}$$

$$= \frac{2 \cdot 71}{2 \cdot 71} \cdot \frac{3 \cdot 5}{1} = 15$$

**Check**. We repeat the calculation.

**State**. The car got 15 miles per gallon of gas.

**45. Familiarize**. We let $n =$ the number of cubic feet occupied by 25,000 lb of water.

**Translate**. We write an equation.

| Total weight | ÷ | Weight per cubic foot | = | Number of cubic feet |
|---|---|---|---|---|
| ↓ | ↓ | ↓ | ↓ | ↓ |
| 25,000 | ÷ | $62\frac{1}{2}$ | = | $n$ |

**Solve**. To solve the equation we carry out the division.

$$n = 25,000 \div 62\frac{1}{2} = 25,000 \div \frac{125}{2}$$

$$= 25,000 \cdot \frac{2}{125} = \frac{200 \cdot 125 \cdot 2}{125 \cdot 1}$$

$$= \frac{125}{125} \cdot \frac{200 \cdot 2}{1} = 400$$

**Check**. We repeat the calculation.

**State**. 400 cubic feet would be occupied.

**47. Familiarize**. We draw a picture.

| $\frac{1}{3}$ lb | $\frac{1}{3}$ lb | · · · · | $\frac{1}{3}$ lb |
|---|---|---|---|

$$\longleftarrow 5\frac{1}{2} \text{ lb} \longrightarrow$$

We let $s =$ the number of servings that can be prepared from $5\frac{1}{2}$ lb of flounder fillet.

**Translate**. The situation corresponds to a division sentence.

$$s = 5\frac{1}{2} \div \frac{1}{3}$$

**Solve**. We carry out the division.

$$s = 5\frac{1}{2} \div \frac{1}{3} = \frac{11}{2} \div \frac{1}{3}$$

$$= \frac{11}{2} \cdot \frac{3}{1} = \frac{33}{2}$$

$$= 16\frac{1}{2}$$

**Check**. We check by multiplying. If $16\frac{1}{2}$ servings are prepared, then

$$16\frac{1}{2} \cdot \frac{1}{3} = \frac{33}{2} \cdot \frac{1}{3} = \frac{3 \cdot 11 \cdot 1}{2 \cdot 3} = \frac{3}{3} \cdot \frac{11 \cdot 1}{2} = \frac{11}{2} = 5\frac{1}{2} \text{ lb}$$

of flounder is used. Our answer checks.

**State**. $16\frac{1}{2}$ servings can be prepared from $5\frac{1}{2}$ lb of flounder fillet.

**49. Familiarize**. The figure is composed of two rectangles. One has dimensions $s$ by $\frac{1}{2} \cdot s$, or $6\frac{7}{8}$ in. by $\frac{1}{2} \cdot 6\frac{7}{8}$ in. The other has dimensions $\frac{1}{2} \cdot s$ by $\frac{1}{2} \cdot s$, or $\frac{1}{2} \cdot 6\frac{7}{8}$ in. by $\frac{1}{2} \cdot 6\frac{7}{8}$ in. The total area is the sum of the areas of these two rectangles. We let $A =$ the total area.

**Translate**. We write an equation.

$$A = \left(6\frac{7}{8}\right) \cdot \left(\frac{1}{2} \cdot 6\frac{7}{8}\right) + \left(\frac{1}{2} \cdot 6\frac{7}{8}\right) \cdot \left(\frac{1}{2} \cdot 6\frac{7}{8}\right)$$

**Solve.** We carry out each multiplication and then add.

$$A = \left(6\frac{7}{8}\right) \cdot \left(\frac{1}{2} \cdot 6\frac{7}{8}\right) + \left(\frac{1}{2} \cdot 6\frac{7}{8}\right) \cdot \left(\frac{1}{2} \cdot 6\frac{7}{8}\right)$$

$$= \frac{55}{8} \cdot \left(\frac{1}{2} \cdot \frac{55}{8}\right) + \left(\frac{1}{2} \cdot \frac{55}{8}\right) \cdot \left(\frac{1}{2} \cdot \frac{55}{8}\right)$$

$$= \frac{55}{8} \cdot \frac{55}{16} + \frac{55}{16} \cdot \frac{55}{16}$$

$$= \frac{3025}{128} + \frac{3025}{256} = \frac{3025}{128} \cdot \frac{2}{2} + \frac{3025}{256}$$

$$= \frac{6050}{256} + \frac{3025}{256} = \frac{9075}{256}$$

$$= 35\frac{115}{256}$$

**Check.** We repeat the calculation.

**State.** The area is $35\frac{115}{256}$ sq in.

**51. Familiarize.** We make a drawing.

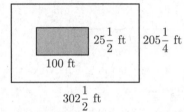

$302\frac{1}{2}$ ft

**Translate.** We let $A$ = the area of the lot not covered by the building.

$$\underbrace{\text{Area left over}}_{\downarrow} \quad \underset{\downarrow}{\text{is}} \quad \underbrace{\text{Area of lot}}_{\downarrow} \quad \underset{\downarrow}{\text{minus}} \quad \underbrace{\text{Area of building}}_{\downarrow}$$

$$A = \left(302\frac{1}{2}\right) \cdot \left(205\frac{1}{4}\right) - (100) \cdot \left(25\frac{1}{2}\right)$$

**Solve.** We do each multiplication and then find the difference.

$$A = \left(302\frac{1}{2}\right) \cdot \left(205\frac{1}{4}\right) - (100) \cdot \left(25\frac{1}{2}\right)$$

$$= \frac{605}{2} \cdot \frac{821}{4} - \frac{100}{1} \cdot \frac{51}{2}$$

$$= \frac{605 \cdot 821}{2 \cdot 4} - \frac{100 \cdot 51}{1 \cdot 2}$$

$$= \frac{605 \cdot 821}{2 \cdot 4} - \frac{2 \cdot 50 \cdot 51}{1 \cdot 2} = \frac{605 \cdot 821}{2 \cdot 4} - \frac{2}{2} \cdot \frac{50 \cdot 51}{1}$$

$$= \frac{496,705}{8} - 2550 = 62,088\frac{1}{8} - 2550$$

$$= 59,538\frac{1}{8}$$

**Check.** We repeat the calculation.

**State.** The area left over is $59,538\frac{1}{8}$ sq ft.

**53.** In the equation $420 \div 60 = 7$, 60 is called the <u>divisor</u>, 7 the <u>quotient</u>, and 420 the <u>dividend</u>.

**55.** The numbers 91, 95, and 111 are examples of <u>composite</u> numbers.

**57.** When simplifying $24 \div 4 + 4 \times 12 - 6 \div 2$, do all <u>multiplications</u> and <u>divisions</u> in order from left to right before doing all <u>additions</u> and <u>subtractions</u> in order from left to right.

**59.** In the expression $\frac{c}{d}$, we call $c$ the <u>numerator</u>.

**61.** Use a calculator.

$$15\frac{2}{11} \cdot 23\frac{31}{43} = \frac{167}{11} \cdot \frac{1020}{43} = \frac{167 \cdot 1020}{11 \cdot 43} =$$

$$\frac{170,340}{473} = 360\frac{60}{473}$$

**63.** $8 \div \frac{1}{2} + \frac{3}{4} - \left(5 - \frac{5}{8}\right)^2$

$$= 8 \div \frac{1}{2} + \frac{3}{4} - \left(\frac{40}{8} - \frac{5}{8}\right)^2$$

$$= 8 \div \frac{1}{2} + \frac{3}{4} - \left(\frac{35}{8}\right)^2$$

$$= 8 \div \frac{1}{2} + \frac{3}{4} - \frac{1225}{64}$$

$$= 8 \cdot 2 + \frac{3}{4} - \frac{1225}{64}$$

$$= 16 + \frac{3}{4} - \frac{1225}{64}$$

$$= \frac{1024}{64} + \frac{48}{64} - \frac{1225}{64}$$

$$= -\frac{153}{64} = -2\frac{25}{64}$$

**65.** $\frac{1}{3} \div \left(\frac{1}{2} - \frac{1}{5}\right) \times \frac{1}{4} + \frac{1}{6}$

$$= \frac{1}{3} \div \left(\frac{5}{10} - \frac{2}{10}\right) \times \frac{1}{4} + \frac{1}{6}$$

$$= \frac{1}{3} \div \frac{3}{10} \times \frac{1}{4} + \frac{1}{6}$$

$$= \frac{1}{3} \times \frac{10}{3} \times \frac{1}{4} + \frac{1}{6}$$

$$= \frac{10}{9} \times \frac{1}{4} + \frac{1}{6}$$

$$= \frac{2 \times 5 \times 1}{9 \times 2 \times 2} + \frac{1}{6} = \frac{2}{2} \times \frac{5 \times 1}{9 \times 2} + \frac{1}{6}$$

$$= \frac{5}{18} + \frac{1}{6} = \frac{5}{18} + \frac{3}{18} = \frac{8}{18} = \frac{4}{9}$$

**67.** $4\frac{1}{2} \div 2\frac{1}{2} + 8 - 4 \div \frac{1}{2}$

$= \frac{9}{2} \div \frac{5}{2} + 8 - 4 \div \frac{1}{2}$

$= \frac{9}{2} \cdot \frac{2}{5} + 8 - 4 \div \frac{1}{2}$

$= \frac{2}{2} \cdot \frac{9}{5} + 8 - 4 \div \frac{1}{2}$

$= \frac{9}{5} + 8 - 4 \div \frac{1}{2}$

$= \frac{9}{5} + 8 - 4 \cdot 2$

$= \frac{9}{5} + 8 - 8$

$= \frac{9}{5} + \frac{40}{5} - \frac{40}{5} = \frac{9}{5} = 1\frac{4}{5}$

## Exercise Set 4.7

**1.** $\frac{1}{2} \cdot \frac{1}{3} \cdot \frac{1}{4}$

$= \frac{1}{6} \cdot \frac{1}{4}$   Doing the multiplications in

$= \frac{1}{24}$   order from left to right

**3.** $6 \div 3 \div 5$

$= 2 \div 5$   Doing the divisions in

$= \frac{2}{5}$   order from left to right

**5.** $\frac{2}{3} \div \left(-\frac{4}{3}\right) \div \frac{7}{8}$

$= \frac{2}{3} \cdot \left(-\frac{3}{4}\right) \div \frac{7}{8}$   Doing the first division; multiplying by the reciprocal of $\frac{4}{3}$

$= -\frac{2 \cdot 3 \cdot 1}{3 \cdot 2 \cdot 2} \div \frac{7}{8}$

$= -\frac{1}{2} \div \frac{7}{8}$   Removing a factor of 1

$= -\frac{1}{2} \cdot \frac{8}{7}$   Dividing; multiplying by the reciprocal of $\frac{7}{8}$

$= -\frac{1 \cdot 2 \cdot 4}{2 \cdot 7}$

$= -\frac{4}{7}$   Removing a factor of 1

**7.** $\frac{5}{8} \div \frac{1}{4} - \frac{2}{3} \cdot \frac{4}{5}$

$= \frac{5}{8} \cdot \frac{4}{1} - \frac{2}{3} \cdot \frac{4}{5}$   Dividing

$= \frac{5 \cdot 4}{2 \cdot 4 \cdot 1} - \frac{2}{3} \cdot \frac{4}{5}$

$= \frac{5}{2} - \frac{2}{3} \cdot \frac{4}{5}$   Removing a factor of 1

$= \frac{5}{2} - \frac{2 \cdot 4}{3 \cdot 5}$   Multiplying

$= \frac{5}{2} - \frac{8}{15}$

$= \frac{5}{2} \cdot \frac{15}{15} - \frac{8}{15} \cdot \frac{2}{2}$   Multiplying by 1 to obtain the LCD

$= \frac{75}{30} - \frac{16}{30}$

$= \frac{59}{30}$, or $1\frac{29}{30}$   Subtracting

**9.** $\frac{3}{4} - \frac{2}{3} \cdot \left(\frac{1}{2} + \frac{2}{5}\right)$

$= \frac{3}{4} - \frac{2}{3} \cdot \left(\frac{5}{10} + \frac{4}{10}\right)$   Adding inside

$= \frac{3}{4} - \frac{2}{3} \cdot \frac{9}{10}$   the parentheses

$= \frac{3}{4} - \frac{2 \cdot 9}{3 \cdot 10}$   Multiplying

$= \frac{3}{4} - \frac{2 \cdot 3 \cdot 3}{3 \cdot 2 \cdot 5}$

$= \frac{3}{4} - \frac{2 \cdot 3}{2 \cdot 3} \cdot \frac{3}{5}$

$= \frac{3}{4} - \frac{3}{5}$

$= \frac{15}{20} - \frac{12}{20}$

$= \frac{3}{20}$   Subtracting

**11.** $28\frac{1}{8} - 5\frac{1}{4} + 3\frac{1}{2}$

$= 28\frac{1}{8} - 5\frac{2}{8} + 3\frac{1}{2}$   Doing the additions and

$= 27\frac{9}{8} - 5\frac{2}{8} + 3\frac{1}{2}$   subtractions in order

$= 22\frac{7}{8} + 3\frac{1}{2}$   from left to right

$= 22\frac{7}{8} + 3\frac{4}{8}$

$= 25\frac{11}{8}$

$= 26\frac{3}{8}$, or $\frac{211}{8}$

**13.** $\dfrac{7}{8} \div \dfrac{1}{2} \cdot \dfrac{1}{4}$

$= \dfrac{7}{8} \cdot \dfrac{2}{1} \cdot \dfrac{1}{4}$     Dividing

$= \dfrac{7 \cdot \cancel{2}}{\cancel{2} \cdot 4 \cdot 1} \cdot \dfrac{1}{4}$

$= \dfrac{7}{4} \cdot \dfrac{1}{4}$     Removing a factor of 1

$= \dfrac{7}{16}$     Multiplying

**15.** $\left(\dfrac{2}{3}\right)^2 - \dfrac{1}{3} \cdot 1\dfrac{1}{4}$

$= \dfrac{4}{9} - \dfrac{1}{3} \cdot 1\dfrac{1}{4}$    Evaluating the exponental expression

$= \dfrac{4}{9} - \dfrac{1}{3} \cdot \dfrac{5}{4}$

$= \dfrac{4}{9} - \dfrac{5}{12}$     Multiplying

$= \dfrac{4}{9} \cdot \dfrac{4}{4} - \dfrac{5}{12} \cdot \dfrac{3}{3}$

$= \dfrac{16}{36} - \dfrac{15}{36}$

$= \dfrac{1}{36}$     Subtracting

**17.** $-\dfrac{1}{2} - \left(\dfrac{1}{2}\right)^2 + \left(\dfrac{1}{2}\right)^3$

$= -\dfrac{1}{2} - \dfrac{1}{4} + \dfrac{1}{8}$    Evaluating the exponental expressions

$= -\dfrac{2}{4} - \dfrac{1}{4} + \dfrac{1}{8}$    Doing the additions and

$= -\dfrac{3}{4} + \dfrac{1}{8}$    subtractions in order

$= -\dfrac{6}{8} + \dfrac{1}{8}$    from left to right

$= -\dfrac{5}{8}$

**19.** Add the numbers and divide by the number of addends.

$\dfrac{\dfrac{2}{3} + \dfrac{7}{8}}{2}$

$= \dfrac{\dfrac{16}{24} + \dfrac{21}{24}}{2}$     The LCD is 24.

$= \dfrac{\dfrac{37}{24}}{2}$

$= \dfrac{37}{24} \cdot \dfrac{1}{2}$     Dividing

$= \dfrac{37}{48}$

**21.** Add the numbers and divide by the number of addends.

$\dfrac{\dfrac{1}{6} + \dfrac{1}{8} + \dfrac{3}{4}}{3}$

$= \dfrac{\dfrac{4}{24} + \dfrac{3}{24} + \dfrac{18}{24}}{3}$

$= \dfrac{\dfrac{25}{24}}{3}$

$= \dfrac{25}{24} \cdot \dfrac{1}{3}$

$= \dfrac{25}{72}$

**23.** Add the numbers and divide by the number of addends.

$\dfrac{3\dfrac{1}{2} + 9\dfrac{3}{8}}{2}$

$= \dfrac{3\dfrac{4}{8} + 9\dfrac{3}{8}}{2}$

$= \dfrac{12\dfrac{7}{8}}{2}$

$= \dfrac{\dfrac{103}{8}}{2}$

$= \dfrac{103}{8} \cdot \dfrac{1}{2}$

$= \dfrac{103}{16}$, or $6\dfrac{7}{16}$

**25.** We add the numbers and divide by the number of addends.

$\dfrac{15\dfrac{5}{32} + 20\dfrac{3}{16} + 12\dfrac{7}{8}}{3}$

$= \dfrac{15\dfrac{5}{32} + 20\dfrac{6}{32} + 12\dfrac{28}{32}}{3}$

$= \dfrac{47\dfrac{39}{32}}{3} = \dfrac{48\dfrac{7}{32}}{3}$

$= \dfrac{\dfrac{1543}{32}}{3} = \dfrac{1543}{32} \cdot \dfrac{1}{3}$

$= \dfrac{1543}{96} = 16\dfrac{7}{96}$

The average of the distances was $16\dfrac{7}{96}$ mi.

**27.** We add the numbers and divide by the number of addends.

$$\frac{7\frac{1}{2} + 8 + 9\frac{1}{2} + 10\frac{5}{8} + 11\frac{3}{4}}{5}$$

$$= \frac{7\frac{4}{8} + 8 + 9\frac{4}{8} + 10\frac{5}{8} + 11\frac{6}{8}}{5}$$

$$= \frac{45\frac{19}{8}}{5} = \frac{47\frac{3}{8}}{5}$$

$$= \frac{\frac{379}{8}}{5} = \frac{379}{8} \cdot \frac{1}{5}$$

$$= \frac{379}{40} = 9\frac{19}{40}$$

The average weight was $9\frac{19}{40}$ lb.

**29.** $\frac{2}{47}$

Because 2 is very small compared to 47, $\frac{2}{47} \approx 0$.

**31.** $\frac{7}{8}$

Because 7 is very close to 8, $\frac{7}{8} \approx 1$.

**33.** $\frac{6}{11}$

Because $2 \cdot 6 = 12$ and 12 is close to 11, the denominator is about twice the numerator. Thus, $\frac{6}{11} \approx \frac{1}{2}$.

**35.** $\frac{7}{15}$

Because $2 \cdot 7 = 14$ and 14 is close to 15, the denominator is about twice the numerator. Thus, $\frac{7}{15} \approx \frac{1}{2}$.

**37.** $\frac{7}{100}$

Because 7 is very small compared to 100, $\frac{7}{100} \approx 0$.

**39.** $\frac{19}{20}$

Because 19 is very close to 20, $\frac{19}{20} \approx 1$.

**41.** $2\frac{7}{8}$

Since $\frac{7}{8} \approx 1$, we have $2\frac{7}{8} = 2 + \frac{7}{8} \approx 2 + 1$, or 3.

**43.** $12\frac{5}{6}$

Since $\frac{5}{6} \approx 1$, we have $12\frac{5}{6} = 12 + \frac{5}{6} \approx 12 + 1$, or 13.

**45.** $\frac{4}{5} + \frac{7}{8} \approx 1 + 1 = 2$

**47.** $\frac{3}{8} + \frac{7}{13} + \frac{5}{9} \approx \frac{1}{2} + \frac{1}{2} + \frac{1}{2} = \frac{3}{2} = 1\frac{1}{2}$

**49.** $\frac{43}{100} + \frac{1}{10} - \frac{11}{1000} \approx \frac{1}{2} + 0 - 0 = \frac{1}{2}$

**51.** $7\frac{29}{80} + 10\frac{12}{13} \cdot 24\frac{2}{17} \approx 7\frac{1}{2} + 11 \cdot 24 =$

$7\frac{1}{2} + 264 = 271\frac{1}{2}$

**53.** $24 \div 7\frac{8}{9} \approx 24 \div 8 = 3$

**55.** $76\frac{3}{14} + 23\frac{19}{20} \approx 76 + 24 = 100$

**57.** $16\frac{1}{5} \div 2\frac{1}{11} + 25\frac{9}{10} - 4\frac{11}{23} \approx$

$16 \div 2 + 26 - 4\frac{1}{2} = 8 + 26 - 4\frac{1}{2} =$

$34 - 4\frac{1}{2} = 29\frac{1}{2}$

**59.** $-\frac{4}{5} \div \left(-\frac{3}{10}\right) = -\frac{4}{5} \cdot \left(-\frac{10}{3}\right) = \frac{4 \cdot 10}{5 \cdot 3} = \frac{4 \cdot 2 \cdot 5}{5 \cdot 3} =$

$\frac{5}{5} \cdot \frac{4 \cdot 2}{3} = \frac{8}{3}$

**61.** 1 is neither prime nor composite.

The only factors of 5 are 1 and 5, so 5 is prime.

The only factors of 7 are 1 and 7, so 7 is prime.

$9 = 3 \cdot 3$, so 9 is composite.

$14 = 2 \cdot 7$, so 14 is composite.

The only factors of 23 are 1 and 23, so 23 is prime.

The only factors of 43 are 1 and 43, so 43 is prime.

**63.** *Familiarize*. We make a drawing.

$\underbrace{\bigcirc \ \bigcirc \ \bigcirc \ \cdots \ \bigcirc}$ } 6 lb feeds how many people?
$\frac{3}{8}$ lb per person

We let $p =$ the number of people who can attend the luncheon.

*Translate*. The problem translates to the following equation:

$$p = 6 \div \frac{3}{8}$$

*Solve*. We carry out the division.

$$p = 6 \div \frac{3}{8}$$

$$= 6 \cdot \frac{8}{3} = \frac{6 \cdot 8}{3}$$

$$= \frac{2 \cdot 3 \cdot 8}{3 \cdot 1} = \frac{3}{3} \cdot \frac{2 \cdot 8}{1}$$

$$= 16$$

*Check*. If each of 16 people is allotted $\frac{3}{8}$ lb of cold cuts, a total of

$$16 \cdot \frac{3}{8} = \frac{16 \cdot 3}{8} = \frac{2 \cdot 8 \cdot 3}{8} = 2 \cdot 3,$$

or 6 lb of cold cuts are used. Our answer checks.

*State*. 16 people can attend the luncheon.

**65.**
$$\frac{a}{17} + \frac{1b}{23} = \frac{35a}{391}$$

$$\frac{a}{17} \cdot \frac{23}{23} + \frac{1b}{23} \cdot \frac{17}{17} = \frac{35a}{391}$$

$$\frac{a \cdot 23 + 1b \cdot 17}{391} = \frac{35a}{391}$$

Equating numerators, we have $a \cdot 23 + 1b \cdot 17 = 35a$. Try $a = 1$. We have:

$$1 \cdot 23 + 1b \cdot 17 = 351$$

$$23 + 1b \cdot 17 = 351$$

$$1b \cdot 17 = 351 - 23$$

$$1b \cdot 17 = 328$$

$$1b = \frac{328}{17} = 19\frac{5}{17}$$

Since $328/17$ is not a whole number, $a \neq 1$. Try $a = 2$. We have:

$$2 \cdot 23 + 1b \cdot 17 = 352$$

$$46 + 1b \cdot 17 = 352$$

$$1b \cdot 17 = 352 - 46$$

$$1b \cdot 17 = 306$$

$$1b = \frac{306}{17}$$

$$1b = 18$$

Thus, $a = 2$ and $b = 8$.

**67.** The largest sum will occur when the largest numbers, 4 and 5, are used for the numerators. Since $\frac{4}{3} + \frac{5}{2} > \frac{4}{2} + \frac{5}{3}$, the largest possible sum is $\frac{4}{3} + \frac{5}{2} = \frac{23}{6}$.

## Chapter 4 Concept Reinforcement

**1.** The statement is true; $5\frac{2}{3} = 5 + \frac{2}{3} = 5 \cdot 1 + \frac{2}{3} = 5 \cdot \frac{3}{3} + \frac{2}{3}$.

**2.** The statement is true. If one number is a multiple of the other, the LCM is the larger number. If one number is not a multiple of the other, then the LCM contains each prime factor that appears in either number the greatest number of times it occurs in any one factorization and, thus, is larger than both numbers.

**3.** The statement is false. If $\frac{a}{b} < 1$, then $a < b$.

**4.** The statement is true. Any positive mixed numeral is greater than 1, so the product of any two mixed numerals is greater than 1.

## Chapter 4 Important Concepts

**1.** $52 = 2 \cdot 2 \cdot 13$
$78 = 2 \cdot 3 \cdot 13$
The LCM is $2 \cdot 2 \cdot 3 \cdot 13$, or 156.

**2.**
$$\frac{19}{60} + \frac{11}{36} = \frac{19}{2 \cdot 2 \cdot 3 \cdot 5} + \frac{11}{2 \cdot 2 \cdot 3 \cdot 3}$$

$$= \frac{19}{2 \cdot 2 \cdot 3 \cdot 5} \cdot \frac{3}{3} + \frac{11}{2 \cdot 2 \cdot 3 \cdot 3} \cdot \frac{5}{5}$$

$$= \frac{57}{180} + \frac{55}{180} = \frac{112}{180}$$

$$= \frac{4 \cdot 28}{4 \cdot 45} = \frac{4}{4} \cdot \frac{28}{45} = \frac{28}{45}$$

**3.** $\frac{29}{35} - \frac{5}{7} = \frac{29}{35} - \frac{5}{7} \cdot \frac{5}{5} = \frac{29}{35} - \frac{25}{35} = \frac{4}{35}$

**4.** The LCD is $13 \cdot 12$, or 156.

$$\frac{3}{13} = \frac{3}{13} \cdot \frac{12}{12} = \frac{36}{156}$$

$$\frac{5}{12} = \frac{5}{12} \cdot \frac{13}{13} = \frac{65}{156}$$

Since $36 < 65$, $\frac{36}{156} < \frac{65}{156}$, and thus $\frac{3}{13} < \frac{5}{12}$.

**5.**
$$\frac{2}{9} + x = -\frac{9}{11}$$

$$\frac{2}{9} + x - \frac{2}{9} = -\frac{9}{11} - \frac{2}{9}$$

$$x = -\frac{9}{11} \cdot \frac{9}{9} - \frac{2}{9} \cdot \frac{11}{11}$$

$$x = -\frac{81}{99} - \frac{22}{99} = -\frac{103}{99}$$

The solution is $-\frac{103}{99}$.

**6.** $8\frac{2}{3} = \frac{26}{3} \quad (3 \cdot 8 = 24, 24 + 2 = 26)$

**7.** First we consider $\frac{47}{6}$.

$$\begin{array}{r} 7 \\ 6\overline{\smash{)}4\ 7} \\ 4\ 2 \\ \hline 5 \end{array} \qquad \frac{47}{6} = 7\frac{5}{6}$$

We have $\frac{47}{6} = 7\frac{5}{6}$, so $-\frac{47}{6} = -7\frac{5}{6}$.

**8.**
$$\begin{aligned} 10\frac{5}{7} &= 10\frac{20}{28} = 9\frac{48}{28} \\ -2\frac{3}{4} &= -2\frac{21}{28} = -2\frac{21}{28} \\ \hline & \qquad\qquad\quad 7\frac{27}{28} \end{aligned}$$

**9.** $-4\frac{1}{5} \cdot 3\frac{7}{15} = -\frac{21}{5} \cdot \frac{52}{15} = -\frac{21 \cdot 52}{5 \cdot 15} = -\frac{3 \cdot 7 \cdot 52}{5 \cdot 3 \cdot 5} =$

$-\frac{7 \cdot 52}{3 \cdot 5} \cdot \frac{3}{3} = -\frac{364}{25} = -14\frac{14}{25}$

**10.** *Familiarize.* Let $p =$ the population of Louisiana.

*Translate.*

$2\frac{1}{2}$ times $\underbrace{\text{Population of West Virginia}}$ is $\underbrace{\text{Population of Louisiana}}$

$\downarrow \quad \downarrow \qquad\qquad \downarrow \qquad \downarrow \qquad\qquad \downarrow$

$2\frac{1}{2} \quad \cdot \qquad\quad 1,800,000 \qquad = \qquad\quad p$

*Solve.* We carry out the multiplication.

$2\frac{1}{2} \cdot 1,800,000 = \frac{5}{2} \cdot 1,800,000 = \frac{5 \cdot 1,800,000}{2} =$

$\frac{5 \cdot 2 \cdot 900,000}{2 \cdot 1} = \frac{2}{2} \cdot \frac{5 \cdot 900,000}{1} = \frac{5 \cdot 900,000}{1} =$

$4,500,000$

Thus, $4,500,000 = p$.

*Check.* We repeat the calculation. The answer checks.

*State.* The population of Louisiana is 4,500,000.

**11.** $\frac{3}{2} \cdot 1\frac{1}{3} \div \left(\frac{2}{3}\right)^2 = \frac{3}{2} \cdot 1\frac{1}{3} \div \frac{4}{9}$

$= \frac{3}{2} \cdot \frac{4}{3} \div \frac{4}{9}$

$= \frac{3 \cdot 4}{2 \cdot 3} \div \frac{4}{9}$

$= \frac{3 \cdot 4}{2 \cdot 3} \cdot \frac{9}{4}$

$= \frac{3 \cdot 4 \cdot 9}{2 \cdot 3 \cdot 4} = \frac{3 \cdot 4}{3 \cdot 4} \cdot \frac{9}{2}$

$= \frac{9}{2}$, or $4\frac{1}{2}$

**12.** $1\frac{19}{20} + 3\frac{1}{8} - \frac{8}{17} \approx 2 + 3 - \frac{1}{2} = 5 - \frac{1}{2} = \frac{10}{2} - \frac{1}{2} = \frac{9}{2}$, or $4\frac{1}{2}$

---

## Chapter 4 Review Exercises

---

**1.** a) 18 is not a multiple of 12.

   b) Check multiples:

   $2 \cdot 18 = 36 \qquad$ A multiple of 12

   c) The LCM is 36.

**2.** a) 45 is not a multiple of 18.

   b) Check multiples:

   $2 \cdot 45 = 90 \qquad$ A multiple of 18

   c) The LCM is 90.

**3.** Note that 3 and 6 are factors of 30. Since the largest number, 30, has the other two numbers as factors, it is the LCM.

**4.** a) Find the prime factorization of each number.

   $26 = 2 \cdot 13$
   $36 = 2 \cdot 2 \cdot 3 \cdot 3$
   $54 = 2 \cdot 3 \cdot 3 \cdot 3$

b) Create a product by writing each factor the greatest number of times it occurs in any one factorization.

The greatest number of times 2 occurs in any one factorization is two times.

The greatest number of times 3 occurs in any one factorization is three times.

The greatest number of times 13 occurs in any one factorization is one time.

Since there are no other prime factors in any of the factorizations, the LCM is $2 \cdot 2 \cdot 3 \cdot 3 \cdot 3 \cdot 13$, or 1404.

**5.** The LCM of 5 and 8 is 40.

$\frac{6}{5} + \frac{3}{8} = \frac{6}{5} \cdot \frac{8}{8} + \frac{3}{8} \cdot \frac{5}{5}$

$= \frac{48}{40} + \frac{15}{40}$

$= \frac{63}{40}$

**6.** The LCM of 16 and 12 is 48.

$\frac{-5}{16} + \frac{1}{12} = \frac{-5}{16} \cdot \frac{3}{3} + \frac{1}{12} \cdot \frac{4}{4}$

$= \frac{-15}{48} + \frac{4}{48}$

$= \frac{-11}{48}$, or $-\frac{11}{48}$

**7.** The LCM of 5, 15, and 20 is 60.

$\frac{6}{5} + \frac{11}{15} + \frac{3}{20} = \frac{6}{5} \cdot \frac{12}{12} + \frac{11}{15} \cdot \frac{4}{4} + \frac{3}{20} \cdot \frac{3}{3}$

$= \frac{72}{60} + \frac{44}{60} + \frac{9}{60}$

$= \frac{125}{60} = \frac{5 \cdot 25}{5 \cdot 12}$

$= \frac{5}{5} \cdot \frac{25}{12} = \frac{25}{12}$

**8.** The LCM of 1000, 100, and 10 is 1000.

$\frac{1}{1000} + \frac{19}{100} + \frac{7}{10} = \frac{1}{1000} + \frac{19}{100} \cdot \frac{10}{10} + \frac{7}{10} \cdot \frac{100}{100}$

$= \frac{1}{1000} + \frac{190}{1000} + \frac{700}{1000}$

$= \frac{891}{1000}$

**9.** $\frac{5}{9} - \frac{2}{9} = \frac{3}{9} = \frac{3 \cdot 1}{3 \cdot 3} = \frac{3}{3} \cdot \frac{1}{3} = \frac{1}{3}$

**10.** The LCM of 8 and 4 is 8.

$\frac{7}{8} - \frac{3}{4} = \frac{7}{8} - \frac{3}{4} \cdot \frac{2}{2}$

$= \frac{7}{8} - \frac{6}{8} = \frac{1}{8}$

**11.** The LCM of 9 and 27 is 27.

$\frac{2}{9} - \frac{11}{27} = \frac{2}{9} \cdot \frac{3}{3} - \frac{11}{27}$

$= \frac{6}{27} - \frac{11}{27} = -\frac{5}{27}$

**12.** The LCM of 6 and 9 is 18.

$$-\frac{5}{6} - \frac{2}{9} = -\frac{5}{6} \cdot \frac{3}{3} - \frac{2}{9} \cdot \frac{2}{2}$$
$$= -\frac{15}{18} - \frac{4}{18} = \frac{-19}{18}, \text{ or } -\frac{19}{18}$$

**13.** The LCD is $7 \cdot 9$, or 63.

$$\frac{4}{7} \cdot \frac{9}{9} = \frac{36}{63}$$
$$\frac{5}{9} \cdot \frac{7}{7} = \frac{35}{63}$$

Since $36 > 35$, it follows that $\frac{36}{63} > \frac{35}{63}$, so $\frac{4}{7} > \frac{5}{9}$.

**14.** The LCD is $13 \cdot 9$, or 117.

$$\frac{-11}{13} \cdot \frac{9}{9} = \frac{-99}{117}$$
$$\frac{-8}{9} \cdot \frac{13}{13} = \frac{-104}{117}$$

Since $-99 > -104$, it follows that $\frac{-99}{117} > \frac{-104}{117}$, so $\frac{-11}{13} > \frac{-8}{9}$.

**15.**
$$x + \frac{2}{5} = \frac{7}{8}$$
$$x + \frac{2}{5} - \frac{2}{5} = \frac{7}{8} - \frac{2}{5}$$
$$x + 0 = \frac{7}{8} \cdot \frac{5}{5} - \frac{2}{5} \cdot \frac{8}{8}$$
$$x = \frac{35}{40} - \frac{16}{40}$$
$$x = \frac{19}{40}$$

The solution is $\frac{19}{40}$.

**16.**
$$\frac{1}{2} + y = -\frac{9}{10}$$
$$\frac{1}{2} + y - \frac{1}{2} = -\frac{9}{10} - \frac{1}{2}$$
$$y + 0 = -\frac{9}{10} - \frac{1}{2} \cdot \frac{5}{5}$$
$$y = -\frac{9}{10} - \frac{5}{10}$$
$$y = -\frac{14}{10} = -\frac{2 \cdot 7}{2 \cdot 5} = \frac{2}{2} \cdot \left(-\frac{7}{5}\right)$$
$$y = -\frac{7}{5}$$

The solution is $-\frac{7}{5}$.

**17.** $7\frac{1}{2} = \frac{15}{2}$   $(7 \cdot 2 = 14, 14 + 1 = 15)$

**18.** $8\frac{3}{8} = \frac{67}{8}$   $(8 \cdot 8 = 64, 64 + 3 = 67)$

**19.** $4\frac{1}{3} = \frac{13}{3}$   $(4 \cdot 3 = 12, 12 + 1 = 13)$

**20.** First we consider $10\frac{5}{7}$.

$$10\frac{5}{7} = \frac{75}{7}$$   $(10 \cdot 7 = 70, 70 + 5 = 75)$

Then $-10\frac{5}{7} = -\frac{75}{7}$.

**21.** To convert $\frac{7}{8}$ to a mixed numeral, we divide.

$$\begin{array}{r} 2 \\ 3\overline{)7} \\ 6 \\ \hline 1 \end{array}$$

$$\frac{7}{3} = 2\frac{1}{3}$$

**22.** To convert $\frac{27}{4}$ to a mixed numeral, we divide.

$$\begin{array}{r} 6 \\ 4\overline{)27} \\ 24 \\ \hline 3 \end{array}$$

$$\frac{27}{4} = 6\frac{3}{4}$$

**23.** First we consider $\frac{63}{5}$. To convert $\frac{63}{5}$ to a mixed numeral, we divide.

$$\begin{array}{r} 12 \\ 5\overline{)63} \\ 5 \\ \hline 13 \\ 10 \\ \hline 3 \end{array}$$

$\frac{63}{5} = 12\frac{3}{5}$, so $-\frac{63}{5} = -12\frac{3}{5}$.

**24.** To convert $\frac{7}{2}$ to a mixed numeral, we divide.

$$\begin{array}{r} 3 \\ 2\overline{)7} \\ 6 \\ \hline 1 \end{array}$$

$$\frac{7}{2} = 3\frac{1}{2}$$

**25.**
$$\begin{array}{r} 877 \\ 9\overline{)7896} \\ 72 \\ \hline 69 \\ 63 \\ \hline 66 \\ 63 \\ \hline 3 \end{array}$$

The answer is 877 R 3. Writing this as a mixed numeral, we have $877\frac{3}{9} = 877\frac{1}{3}$.

**26.**
$$\begin{array}{r} 456 \\ 23\overline{)10,493} \\ 92 \\ \hline 129 \\ 115 \\ \hline 143 \\ 138 \\ \hline 5 \end{array}$$

The answer is 456 R 5. Writing this as a mixed numeral, we have $456\frac{5}{23}$.

**27.** 
$$5\frac{3}{5}$$
$$+4\frac{4}{5}$$
$$9\frac{7}{5} = 9 + \frac{7}{5}$$
$$= 9 + 1\frac{2}{5}$$
$$= 10\frac{2}{5}$$

**28.** 
$$8\boxed{\frac{1}{3} \cdot \frac{5}{5}} = 8\frac{5}{15}$$
$$+3\boxed{\frac{2}{5} \cdot \frac{3}{3}} = +3\frac{6}{15}$$
$$11\frac{11}{15}$$

**29.** 
$$5\frac{5}{6}$$
$$+4\frac{5}{6}$$
$$9\frac{10}{6} = 9 + \frac{10}{6}$$
$$= 9 + 1\frac{4}{6}$$
$$= 10\frac{4}{6}$$
$$= 10\frac{2}{3}$$

**30.** 
$$2\quad\frac{3}{4} = 2\frac{3}{4}$$
$$+5\boxed{\frac{1}{2} \cdot \frac{2}{2}} = +5\frac{2}{4}$$
$$7\frac{5}{4} = 7 + \frac{5}{4}$$
$$= 7 + 1\frac{1}{4}$$
$$= 8\frac{1}{4}$$

**31.** 
$$12 = 11\frac{9}{9}$$
$$-4\frac{2}{9} = -4\frac{2}{9}$$
$$7\frac{7}{9}$$

**32.** 
$$9\boxed{\frac{3}{5} \cdot \frac{3}{3}} = 9\frac{9}{15} = 8\frac{24}{15}$$
$$-4\frac{13}{15} = -4\frac{13}{15} = -4\frac{13}{15}$$
$$4\frac{11}{15}$$

**33.** 
$$10\boxed{\frac{1}{4} \cdot \frac{5}{5}} = 10\frac{5}{20}$$
$$-6\boxed{\frac{1}{10} \cdot \frac{2}{2}} = -6\frac{2}{20}$$
$$4\frac{3}{20}$$

**34.** 
$$20\quad\frac{7}{24} = 20\frac{7}{24} = 19\frac{31}{24}$$
$$-6\boxed{\frac{11}{12} \cdot \frac{2}{2}} = -6\frac{22}{24} = -6\frac{22}{24}$$
$$13\frac{9}{24} = 13\frac{3}{8}$$

**35.** $6 \cdot 2\frac{2}{3} = 6 \cdot \frac{8}{3} = \frac{6 \cdot 8}{3} = \frac{2 \cdot 3 \cdot 8}{3 \cdot 1} = \frac{3}{3} \cdot \frac{2 \cdot 8}{1} = 16$

**36.** $5\frac{1}{4} \cdot \left(-\frac{2}{3}\right) = \frac{21}{4} \cdot \left(-\frac{2}{3}\right) = -\frac{21 \cdot 2}{4 \cdot 3} = -\frac{3 \cdot 7 \cdot 2}{2 \cdot 2 \cdot 3} =$
$$\frac{2 \cdot 3}{2 \cdot 3} \cdot \left(-\frac{7}{2}\right) = -\frac{7}{2} = -3\frac{1}{2}$$

**37.** $2\frac{1}{5} \cdot 1\frac{1}{10} = \frac{11}{5} \cdot \frac{11}{10} = \frac{11 \cdot 11}{5 \cdot 10} = \frac{121}{50} = 2\frac{21}{50}$

**38.** $-2\frac{2}{5} \cdot 2\frac{1}{2} = -\frac{12}{5} \cdot \frac{5}{2} = -\frac{12 \cdot 5}{5 \cdot 2} = -\frac{2 \cdot 6 \cdot 5}{5 \cdot 2 \cdot 1} =$
$$\frac{2 \cdot 5}{2 \cdot 5} \cdot \left(-\frac{6}{1}\right) = -6$$

**39.** $-27 \div 2\frac{1}{4} = -27 \div \frac{9}{4} = -27 \cdot \frac{4}{9} = -\frac{27 \cdot 4}{9} = -\frac{3 \cdot 9 \cdot 4}{9 \cdot 1} =$
$$\frac{9}{9} \cdot \left(-\frac{3 \cdot 4}{1}\right) = -12$$

**40.** $2\frac{2}{5} \div 1\frac{7}{10} = \frac{12}{5} \div \frac{17}{10} = \frac{12}{5} \cdot \frac{10}{17} = \frac{12 \cdot 10}{5 \cdot 17} = \frac{12 \cdot 2 \cdot 5}{5 \cdot 17} =$
$$\frac{5}{5} \cdot \frac{12 \cdot 2}{17} = \frac{24}{17} = 1\frac{7}{17}$$

**41.** $3\frac{1}{4} \div 26 = \frac{13}{4} \div 26 = \frac{13}{4} \cdot \frac{1}{26} = \frac{13 \cdot 1}{4 \cdot 26} = \frac{13 \cdot 1}{4 \cdot 2 \cdot 13} =$
$$\frac{13}{13} \cdot \frac{1}{4 \cdot 2} = \frac{1}{8}$$

**42.** $-4\frac{1}{5} \div \left(-4\frac{2}{3}\right) = -\frac{21}{5} \div \left(-\frac{14}{3}\right) = -\frac{21}{5} \cdot \left(-\frac{3}{14}\right) =$
$$\frac{21 \cdot 3}{5 \cdot 14} = \frac{3 \cdot 7 \cdot 3}{5 \cdot 2 \cdot 7} = \frac{7}{7} \cdot \frac{3 \cdot 3}{5 \cdot 2} = \frac{9}{10}$$

**43.** *Familiarize.* Let $f$ = the number of yards of fabric Gloria needs.

*Translate.*

| Fabric for dress | plus | Fabric for jacket | is | Total fabric needed |
|:---:|:---:|:---:|:---:|:---:|
| ↓ | ↓ | ↓ | ↓ | ↓ |
| $1\frac{5}{8}$ | $+$ | $2\frac{5}{8}$ | $=$ | $f$ |

**Solve**. We carry out the addition.

$$1\frac{5}{8}$$
$$+2\frac{5}{8}$$
$$\overline{\phantom{0}}$$
$$3\frac{10}{8} = 3 + \frac{10}{8}$$
$$= 3 + 1\frac{2}{8}$$
$$= 4\frac{2}{8} = 4\frac{1}{4}$$

**Check**. We repeat the calculation. The answer checks.

**State**. Gloria needs $4\frac{1}{4}$ yd of fabric.

**44.** We find the area of each rectangle and then add to find the total area. Recall that the area of a rectangle is length × width.

Area of rectangle A:

$$12 \times 9\frac{1}{2} = 12 \times \frac{19}{2} = \frac{12 \times 19}{2} = \frac{2 \cdot 6 \cdot 19}{2 \cdot 1} = \frac{2}{2} \cdot \frac{6 \cdot 19}{1} =$$
$$114 \text{ in}^2$$

Area of rectangle B:

$$8\frac{1}{2} \times 7\frac{1}{2} = \frac{17}{2} \times \frac{15}{2} = \frac{17 \times 15}{2 \times 2} = \frac{255}{4} = 63\frac{3}{4} \text{ in}^2$$

Sum of the areas:

$$114 \text{ in}^2 + 63\frac{3}{4} \text{ in}^2 = 177\frac{3}{4} \text{ in}^2$$

**45.** We subtract the area of rectangle B from the area of rectangle A.

$$114 = 113\frac{4}{4}$$
$$-63\frac{3}{4} = -63\frac{3}{4}$$
$$\overline{\phantom{0}}$$
$$50\frac{1}{4}$$

The area of rectangle A is $50\frac{1}{4}$ in$^2$ greater than the area of rectangle B.

**46. Familiarize**. We draw a picture. We let $t =$ the total thickness.

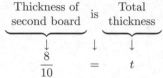

**Translate**. We translate to an equation.

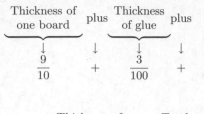

**Solve**. We carry out the addition. The LCD is 100 since 10 is a factor of 100.

$$\frac{9}{10} + \frac{3}{100} + \frac{8}{10} = t$$
$$\frac{9}{10} \cdot \frac{10}{10} + \frac{3}{100} + \frac{8}{10} \cdot \frac{10}{10} = t$$
$$\frac{90}{100} + \frac{3}{100} + \frac{80}{100} = t$$
$$\frac{173}{100} = t, \text{ or}$$
$$1\frac{73}{100} = t$$

**Check**. We repeat the calculation. We also note that the sum is larger than any of the individual thicknesses, as expected.

**State**. The result is $\frac{173}{100}$ in., or $1\frac{73}{100}$ in. thick.

**47. Familiarize**. Let $t =$ the number of pounds of turkey needed for 32 servings.

**Translate**.

**Solve**. We divide by $1\frac{1}{3}$ on both sides of the equation.

$$t = 32 \div 1\frac{1}{3}$$
$$t = 32 \div \frac{4}{3}$$
$$t = 32 \cdot \frac{3}{4} = \frac{32 \cdot 3}{4}$$
$$t = \frac{4 \cdot 8 \cdot 3}{4 \cdot 1} = \frac{4}{4} \cdot \frac{8 \cdot 3}{1}$$
$$t = 24$$

**Check**. Since $1\frac{1}{3} \cdot 24 = \frac{4}{3} \cdot 24 = \frac{4 \cdot 24}{3} = \frac{4 \cdot 3 \cdot 8}{3 \cdot 1} = \frac{3}{3} \cdot \frac{4 \cdot 8}{1} = 32$, the answer checks.

**State**. 24 pounds of turkey are needed for 32 servings.

**48.** *Familiarize*. This is a multistep problem. First we find the perimeter of the room, in feet. Then we find what the artist charges to paint that many feet. Let $P$ = the perimeter, in feet, and C the amount the artist charges.

*Translate*. To find the perimeter, we add the lengths of the four sides of the room.

$$P = 11\frac{3}{4} + 9\frac{1}{2} + 11\frac{3}{4} + 9\frac{1}{2}$$

We multiply to find the cost.

| Number of feet | times | Cost per foot | is | Total cost |
|---|---|---|---|---|
| ↓ | ↓ | ↓ | ↓ | ↓ |
| $P$ | · | 20 | = | $C$ |

*Solve*. First we find P.

$$P = 11\frac{3}{4} + 9\frac{1}{2} + 11\frac{3}{4} + 9\frac{1}{2} = 11\frac{3}{4} + 9\frac{2}{4} + 11\frac{3}{4} + 9\frac{2}{4} =$$

$$40\frac{10}{4} = 40 + \frac{10}{4} = 40 + 2\frac{2}{4} = 42\frac{2}{4} = 42\frac{1}{2}$$

Now we substitute $42\frac{1}{2}$ for P in the second equation and find C.

$$P \cdot 20 = C$$
$$42\frac{1}{2} \cdot 20 = C$$
$$\frac{85}{2} \cdot 20 = C$$
$$\frac{85 \cdot 20}{2} = C$$
$$\frac{85 \cdot \cancel{2} \cdot 10}{\cancel{2} \cdot 1} = C$$
$$850 = C$$

*Check*. We repeat the calculations. The answer checks.

*State*. The project will cost $850.

**49.** *Familiarize*. Let $s$ = the number of cups of shortening in the lower calorie cake.

*Translate*.

| New amount of shortening | plus | Amount of prune puree | is | Original amount of shortening |
|---|---|---|---|---|
| ↓ | ↓ | ↓ | ↓ | ↓ |
| $s$ | + | $3\frac{5}{8}$ | = | 12 |

*Solve*. We subtract $3\frac{5}{8}$ on both sides of the equation.

$$\begin{array}{r} 12 = 11\frac{8}{8} \\ -3\frac{5}{8} = -3\frac{5}{8} \\ \hline 8\frac{3}{8} \end{array}$$

Thus, $s = 8\frac{3}{8}$.

*Check*. $8\frac{3}{8} + 3\frac{5}{8} = 11\frac{8}{8} = 12$, so the answer checks.

*State*. The lower calorie recipe uses $8\frac{3}{8}$ cups of shortening.

**50.** For the $13\frac{1}{4}$ in. $\times$ $13\frac{1}{4}$ in. side:

Perimeter $= 13\frac{1}{4} + 13\frac{1}{4} + 13\frac{1}{4} + 13\frac{1}{4} = 52\frac{4}{4} = 52 + \frac{4}{4} =$ $52 + 1 = 53$ in.

Area $= 13\frac{1}{4} \times 13\frac{1}{4} = \frac{53}{4} \times \frac{53}{4} = \frac{53 \times 53}{4 \times 4} = \frac{2809}{16} = 175\frac{9}{16}$ in$^2$

For the $13\frac{1}{4}$ in. $\times$ $3\frac{1}{4}$ in. side:

Perimeter $= 13\frac{1}{4} + 3\frac{1}{4} + 13\frac{1}{4} + 3\frac{1}{4} = 32\frac{4}{4} = 32 + \frac{4}{4} =$ $32 + 1 = 33$ in.

Area $= 13\frac{1}{4} \times 3\frac{1}{4} = \frac{53}{4} \times \frac{13}{4} = \frac{53 \times 13}{4 \times 4} = \frac{689}{16} = 43\frac{1}{16}$ in$^2$

**51.** *Familiarize*. Let $s$ = the number of pies sold and let $l$ = the number of pies left over.

*Translate*.

| Number of pies sold | times | Number of pieces per pie | is | Number of pieces sold |
|---|---|---|---|---|
| ↓ | ↓ | ↓ | ↓ | ↓ |
| $s$ | · | 6 | = | 382 |

| Number of pies sold | plus | Number left over | is | Number of pies donated |
|---|---|---|---|---|
| ↓ | ↓ | ↓ | ↓ | ↓ |
| $s$ | + | $l$ | = | 83 |

*Solve*. To solve the first equation we divide by 6 on both sides.

$$s \cdot 6 = 382$$
$$s = \frac{382}{6} = 63\frac{2}{3}$$

Now we substitute $63\frac{2}{3}$ for $s$ in the second equation and solve for $l$.

$$s + l = 83$$
$$63\frac{2}{3} + l = 83$$
$$l = 83 - 63\frac{2}{3}$$
$$l = 19\frac{1}{3}$$

*Check*. We repeat the calculations. The answer checks.

*State*. $63\frac{2}{3}$ pies were sold; $19\frac{1}{3}$ pies were left over.

**52.** $\frac{1}{8} \div \frac{1}{4} + \frac{1}{2} = \frac{1}{8} \cdot \frac{4}{1} + \frac{1}{2}$
$$= \frac{4}{8} + \frac{1}{2}$$
$$= \frac{1}{2} + \frac{1}{2}$$
$$= 1$$

**53.**
$$\frac{4}{5} - \frac{1}{2} \cdot \left(1 + \frac{1}{4}\right) = \frac{4}{5} - \frac{1}{2} \cdot 1\frac{1}{4}$$
$$= \frac{4}{5} - \frac{1}{2} \cdot \frac{5}{4}$$
$$= \frac{4}{5} - \frac{5}{8}$$
$$= \frac{4}{5} \cdot \frac{8}{8} - \frac{5}{8} \cdot \frac{5}{5}$$
$$= \frac{32}{40} - \frac{25}{40}$$
$$= \frac{7}{40}$$

**54.**
$$20\frac{3}{4} - 1\frac{1}{2} \times 12 + \left(\frac{1}{2}\right)^2 = 20\frac{3}{4} - 1\frac{1}{2} \times 12 + \frac{1}{4}$$
$$= 20\frac{3}{4} - \frac{3}{2} \times 12 + \frac{1}{4}$$
$$= 20\frac{3}{4} - \frac{36}{2} + \frac{1}{4}$$
$$= 20\frac{3}{4} - 18 + \frac{1}{4}$$
$$= 2\frac{3}{4} + \frac{1}{4}$$
$$= 2\frac{4}{4} = 3$$

**55.**
$$\frac{\frac{1}{2} + \frac{1}{4} + \frac{1}{3} + \frac{1}{5}}{4} = \frac{\frac{30}{60} + \frac{15}{60} + \frac{20}{60} + \frac{12}{60}}{4}$$
$$= \frac{\frac{77}{60}}{4}$$
$$= \frac{77}{60} \cdot \frac{1}{4}$$
$$= \frac{77}{240}$$

**56.** Because $2 \cdot 29 = 58$ and $58$ is close to $59$, the denominator is about twice the numerator. Thus, $\frac{29}{59} \approx \frac{1}{2}$.

**57.** Because $2$ is very small compared to $59$, $\frac{2}{59} \approx 0$.

**58.** Because $61$ is very close to $59$, $\frac{61}{59} \approx 1$.

**59.** Since $\frac{7}{8} \approx 1$, we have $6\frac{7}{8} = 6 + \frac{7}{8} \approx 6 + 1$, or $7$.

**60.** Since $\frac{2}{17} \approx 0$, we have $10\frac{2}{17} = 10 + \frac{2}{17} \approx 10 + 0$, or $10$.

**61.** $\frac{11}{12} \cdot 5\frac{6}{13} \approx 1 \cdot 5\frac{1}{2} = 5\frac{1}{2}$

**62.** $\frac{1}{15} \cdot \frac{2}{3} \approx 0 \cdot 1 = 0$

**63.** $\frac{6}{11} + \frac{5}{6} + \frac{31}{29} \approx \frac{1}{2} + 1 + 1 = 2\frac{1}{2}$.

**64.**
$$32\frac{14}{15} + 27\frac{3}{4} - 4\frac{25}{28} \cdot 6\frac{37}{76}$$
$$\approx 33 + 28 - 5 \cdot 6\frac{1}{2}$$
$$= 33 + 28 - 5 \cdot \frac{13}{2}$$
$$= 33 + 28 - \frac{65}{2} = 33 + 28 - 32\frac{1}{2}$$
$$= 61 - 32\frac{1}{2}$$
$$= 28\frac{1}{2}$$

**65.**
$$\frac{1}{4} + \frac{2}{5} \div 5^2 = \frac{1}{4} + \frac{2}{5} \div 25$$
$$= \frac{1}{4} + \frac{2}{5} \cdot \frac{1}{25}$$
$$= \frac{1}{4} + \frac{2}{125}$$
$$= \frac{1}{4} \cdot \frac{125}{125} + \frac{2}{125} \cdot \frac{4}{4}$$
$$= \frac{125}{4 \cdot 125} + \frac{8}{4 \cdot 125}$$
$$= \frac{133}{500}$$
Answer A is correct.

**66.**
$$x + \frac{2}{3} = 5$$
$$x + \frac{2}{3} - \frac{2}{3} = 5 - \frac{2}{3}$$
$$x + 0 = 5 \cdot \frac{3}{3} - \frac{2}{3}$$
$$x = \frac{15}{3} - \frac{2}{3} = \frac{15 - 2}{3}$$
$$x = \frac{13}{3}, \text{ or } 4\frac{1}{3}$$
The solution is $\frac{13}{3}$, or $4\frac{1}{3}$. Answer D is correct.

**67.** The length of the act is the LCM of 6 min and 4 min.
$$6 = 2 \cdot 3$$
$$4 = 2 \cdot 2$$
Then the LCM $= 2 \cdot 2 \cdot 3$, or 12 min.

**68.** Since the largest fraction we can form is $\frac{6}{3}$, or 2, and $3\frac{1}{4} - 2 = \frac{5}{4}$, we know that both fractions must be greater than 1. By trial, we find true equation $\frac{6}{3} + \frac{5}{4} = 3\frac{1}{4}$.

---

## Chapter 4 Discussion and Writing Exercises

**1.** No; if the sum of the fractional parts of the mixed numerals is $\frac{n}{n}$, then the sum of the mixed numerals is an integer. For example, $1\frac{1}{5} + 6\frac{4}{5} = 7\frac{5}{5} = 8$.

**2.** A wheel makes $33\frac{1}{3}$ revolutions per minute. It rotates for $4\frac{1}{2}$ min. How many revolutions does it make?

**3.** The student is multiplying the whole numbers to get the whole number portion of the answer and multiplying fractions to get the fractional part of the answer. The student should have converted each mixed numeral to fractional notation, multiplied, simplified, and then converted back to a mixed numeral. The correct answer is $4\frac{6}{7}$.

**4.** It might be necessary to find the least common denominator before adding or subtracting. The least common denominator is the least common multiple of the denominators.

**5.** Suppose that a room has dimensions $15\frac{3}{4}$ ft by $28\frac{5}{8}$ ft. The equation $2 \cdot 15\frac{3}{4} + 2 \cdot 28\frac{5}{8} = 88\frac{3}{4}$ gives the perimeter of the room, in feet. Answers may vary.

**6.** The products $5 \cdot 3$ and $5 \cdot \frac{2}{7}$ should be added rather than multiplied together. The student could also have converted $3\frac{2}{7}$ to fractional notation, multiplied, simplified, and converted back to a mixed numeral. The correct answer is $16\frac{3}{7}$.

## Chapter 4 Test

**1.** We find the LCM using a list of multiples.

 a) 16 is not a multiple of 12.

 b) Check multiples of 16:

 $1 \cdot 16 = 16$    Not a multiple of 12
 $2 \cdot 16 = 32$    Not a multiple of 12
 $3 \cdot 16 = 48$    A multiple of 12

 The LCM $= 48$.

**2.** We will find the LCM using prime factorizations.

 a) Find the prime factorization of each number.

 $15 = 3 \cdot 5$
 $40 = 2 \cdot 2 \cdot 2 \cdot 5$
 $50 = 2 \cdot 5 \cdot 5$

 b) Create a product by writing factors that appear in the factorizations of 15, 40, and 50, using each the greatest number of times it occurs in any one factorization.

 The LCM is $2 \cdot 2 \cdot 2 \cdot 3 \cdot 5 \cdot 5$, or 600.

**3.** $\dfrac{1}{2} + \dfrac{5}{2} = \dfrac{1+5}{2} = \dfrac{6}{2} = 3$

**4.** $\dfrac{-7}{8} + \dfrac{2}{3}$    8 and 3 have no common factors, so the LCD is $8 \cdot 3$, or 24.

$= \dfrac{-7}{8} \cdot \dfrac{3}{3} + \dfrac{2}{3} \cdot \dfrac{8}{8}$

$= \dfrac{-21}{24} + \dfrac{16}{24}$

$= \dfrac{-5}{24}$, or $-\dfrac{5}{24}$

**5.** $\dfrac{7}{10} + \dfrac{19}{100} + \dfrac{31}{1000}$    10 and 100 are factors of 1000, so the LCD is 1000.

$= \dfrac{7}{10} \cdot \dfrac{100}{100} + \dfrac{19}{100} \cdot \dfrac{10}{10} + \dfrac{31}{1000}$

$= \dfrac{700}{1000} + \dfrac{190}{1000} + \dfrac{31}{1000} = \dfrac{921}{1000}$

**6.** $\dfrac{5}{6} - \dfrac{3}{6} = \dfrac{5-3}{6} = \dfrac{2}{6} = \dfrac{2 \cdot 1}{2 \cdot 3} = \dfrac{2}{2} \cdot \dfrac{1}{3} = \dfrac{1}{3}$

**7.** The LCM of 6 and 4 is 12.

$\dfrac{5}{6} - \dfrac{3}{4} = \dfrac{5}{6} \cdot \dfrac{2}{2} - \dfrac{3}{4} \cdot \dfrac{3}{3}$

$= \dfrac{10}{12} - \dfrac{9}{12} = \dfrac{1}{12}$

**8.** The LCM of 24 and 15 is 120.

$-\dfrac{17}{24} - \dfrac{1}{15} = -\dfrac{17}{24} \cdot \dfrac{5}{5} - \dfrac{1}{15} \cdot \dfrac{8}{8}$

$= -\dfrac{85}{120} - \dfrac{8}{120}$

$= -\dfrac{93}{120}$

$= -\dfrac{31 \cdot 3}{40 \cdot 3} = -\dfrac{31}{40} \cdot \dfrac{3}{3}$

$= -\dfrac{31}{40}$

**9.** $\dfrac{1}{4} + y = 4$

$\dfrac{1}{4} + y - \dfrac{1}{4} = 4 - \dfrac{1}{4}$    Subtracting $\dfrac{1}{4}$ on both sides

$y + 0 = 4 \cdot \dfrac{4}{4} - \dfrac{1}{4}$    The LCD is 4.

$y = \dfrac{16}{4} - \dfrac{1}{4}$

$y = \dfrac{15}{4}$

**10.** $x + \dfrac{2}{3} = \dfrac{11}{12}$

$x + \dfrac{2}{3} - \dfrac{2}{3} = \dfrac{11}{12} - \dfrac{2}{3}$    Subtracting $\dfrac{2}{3}$ on both sides

$x + 0 = \dfrac{11}{12} - \dfrac{2}{3} \cdot \dfrac{4}{4}$    The LCD is 12.

$x = \dfrac{11}{12} - \dfrac{8}{12} = \dfrac{3}{12}$

$x = \dfrac{3 \cdot 1}{3 \cdot 4} = \dfrac{3}{3} \cdot \dfrac{1}{4}$

$x = \dfrac{1}{4}$

**11.** The LCD is 175.

$$\frac{6}{7}\cdot\frac{25}{25}=\frac{150}{175}$$

$$\frac{21}{25}\cdot\frac{7}{7}=\frac{147}{175}$$

Since $150 > 147$, it follows that $\frac{150}{175} > \frac{147}{175}$, so $\frac{6}{7} > \frac{21}{25}$.

**12.** $3\frac{1}{2}=\frac{7}{2}$    $(3\cdot 2=6,\ 6+1=7)$

**13.** First consider $9\frac{7}{8}$.

$$9\frac{7}{8}=\frac{79}{8}\quad (9\cdot 8=72,\ 72+7=79)$$

Then $-9\frac{7}{8}=-\frac{79}{8}$.

**14.**
$$\begin{array}{r} 4 \\ 2\overline{)9} \\ 8 \\ \hline 1 \end{array}$$
    $\frac{9}{2}=4\frac{1}{2}$

**15.** First consider $\frac{74}{9}$.

$$\begin{array}{r} 8 \\ 9\overline{)74} \\ 72 \\ \hline 2 \end{array}$$
    $\frac{74}{9}=8\frac{2}{9}$

Then $-\frac{74}{9}=-8\frac{2}{9}$.

**16.**
$$\begin{array}{r} 162 \\ 11\overline{)1789} \\ 11 \\ \hline 68 \\ 66 \\ \hline 29 \\ 22 \\ \hline 7 \end{array}$$

The answer is $162\frac{7}{11}$.

**17.**
$$\begin{array}{r} 6\frac{2}{5} \\ +7\frac{4}{5} \\ \hline 13\frac{6}{5} \end{array} = 13+\frac{6}{5}$$
$$= 13+1\frac{1}{5}$$
$$= 14\frac{1}{5}$$

**18.** The LCD is 12.

$$\begin{array}{r} 9\boxed{\frac{1}{4}\cdot\frac{3}{3}}= 9\frac{3}{12} \\ +5\boxed{\frac{1}{6}\cdot\frac{2}{2}}= +5\frac{2}{12} \\ \hline 14\frac{5}{12} \end{array}$$

**19.** The LCD is 24.

$$\begin{array}{r} 10\boxed{\frac{1}{6}\cdot\frac{4}{4}} = 10\frac{4}{24} = 9\frac{28}{24} \\ -5\boxed{\frac{7}{8}\cdot\frac{3}{3}} = -5\frac{21}{24} = -5\frac{21}{24} \\ \hline 4\frac{7}{24} \end{array}$$

$\left(\text{Since } \frac{4}{24} \text{ is smaller than } \frac{21}{24}, \text{ we cannot subtract until we}\right.$
borrow: $\left.10\frac{4}{24}=9+\frac{24}{24}+\frac{4}{24}=9+\frac{28}{24}=9\frac{28}{24}.\right)$

**20.**
$$\begin{array}{r} 14 = 13\frac{6}{6} \\ -7\frac{5}{6} = -7\frac{5}{6} \\ \hline 6\frac{1}{6} \end{array}$$
$\left(14=13+1=13+\frac{6}{6}=13\frac{6}{6}\right)$

**21.** $9\cdot 4\frac{1}{3}=9\cdot\frac{13}{3}=\frac{9\cdot 13}{3}=\frac{3\cdot 3\cdot 13}{3\cdot 1}=\frac{3}{3}\cdot\frac{3\cdot 13}{1}=39$

**22.** $-6\frac{3}{4}\cdot\frac{2}{3}=-\frac{27}{4}\cdot\frac{2}{3}=-\frac{27\cdot 2}{4\cdot 3}=-\frac{3\cdot 9\cdot 2}{2\cdot 2\cdot 3}=$

$\frac{3\cdot 2}{3\cdot 2}\cdot\left(-\frac{9}{2}\right)=-\frac{9}{2}=-4\frac{1}{2}$

**23.** $2\frac{1}{3}\div 1\frac{1}{6}=\frac{7}{3}\div\frac{7}{6}=\frac{7}{3}\cdot\frac{6}{7}=\frac{7\cdot 6}{3\cdot 7}=\frac{7\cdot 2\cdot 3}{3\cdot 7\cdot 1}=$

$\frac{7\cdot 3}{7\cdot 3}\cdot\frac{2}{1}=2$

**24.** $2\frac{1}{12}\div(-75)=\frac{25}{12}\div(-75)=\frac{25}{12}\cdot\left(-\frac{1}{75}\right)=$

$-\frac{25\cdot 1}{12\cdot 75}=-\frac{25\cdot 1}{12\cdot 3\cdot 25}=\frac{25}{25}\cdot\left(-\frac{1}{12\cdot 3}\right)=-\frac{1}{36}$

**25.** *Familiarize.* Let $w=$ Rezazadeh's body weight, in kilograms.

*Translate.*

| Weight lifted | is | $2\frac{1}{2}$ | times | body weight |
|---|---|---|---|---|
| ↓ | ↓ | ↓ | ↓ | ↓ |
| 263 | = | $2\frac{1}{2}$ | · | $w$ |

*Solve.* We will divide by $2\frac{1}{2}$ on both sides of the equation.

$$263=2\frac{1}{2}w$$
$$263\div 2\frac{1}{2}=w$$
$$263\div\frac{5}{2}=w$$
$$263\cdot\frac{2}{5}=w$$
$$\frac{526}{5}=w$$
$$105\frac{1}{5}=w$$
$$105\approx w$$

**Check.** Since $2\frac{1}{2} \cdot 105 = \frac{5}{2} \cdot 105 = \frac{525}{2} = 262\frac{1}{2} \approx 263$, the answer checks.

**State.** Rezazadeh weighs about 105 kg.

**26. Familiarize.** Let $b$ = the number of books in the order.

**Translate.**

$$\underbrace{\begin{array}{c}\text{Weight of}\\\text{each book}\end{array}}_{} \quad \text{times} \quad \underbrace{\begin{array}{c}\text{Number}\\\text{of books}\end{array}}_{} \quad \text{is} \quad \underbrace{\begin{array}{c}\text{Total}\\\text{weight}\end{array}}_{}$$

$$2\frac{3}{4} \qquad \cdot \qquad b \qquad = \qquad 220$$

**Solve.** We will divide by $2\frac{3}{4}$ on both sides of the equation.

$$2\frac{3}{4} \cdot b = 220$$
$$b = 220 \div 2\frac{3}{4}$$
$$b = 220 \div \frac{11}{4}$$
$$b = 220 \cdot \frac{4}{11}$$
$$b = \frac{220 \cdot 4}{11} = \frac{11 \cdot 20 \cdot 4}{11 \cdot 1} = \frac{11}{11} \cdot \frac{20 \cdot 4}{1}$$
$$b = 80$$

**Check.** Since $80 \cdot 2\frac{3}{4} = 80 \cdot \frac{11}{4} = \frac{880}{4} = 220$, the answer checks.

**State.** There are 80 books in the order.

**27. Familiarize.** We add the three lengths across the top to find $a$ and the three lengths across the bottom to find $b$.

**Translate.**
$$a = 1\frac{1}{8} + \frac{3}{4} + 1\frac{1}{8}$$
$$b = \frac{3}{4} + 3 + \frac{3}{4}$$

**Solve.** We carry out the additions.
$$a = 1\frac{1}{8} + \frac{6}{8} + 1\frac{1}{8} = 2\frac{8}{8} = 2 + 1 = 3$$
$$b = \frac{3}{4} + 3 + \frac{3}{4} = 3\frac{6}{4} = 3 + 1\frac{2}{4} = 3 + 1\frac{1}{2} = 4\frac{1}{2}$$

**Check.** We can repeat the calculations. The answer checks.

**State.** a) The short length $a$ across the top is 3 in.

b) The length $b$ across the bottom is $4\frac{1}{2}$ in.

**28. Familiarize.** Let $t$ = the number of inches by which $\frac{3}{4}$ in. exceeds the actual thickness of the plywood.

**Translate.**

$$\underbrace{\text{Actual thickness}}_{} \quad \text{plus} \quad \underbrace{\text{Excess thickness}}_{} \quad \text{is} \quad \underbrace{\frac{3}{4}}_{} \text{ in.}$$

$$\frac{11}{16} \qquad + \qquad t \qquad = \qquad \frac{3}{4}$$

**Solve.** We will subtract $\frac{11}{16}$ on both sides of the equation.

$$\frac{11}{16} + t = \frac{3}{4}$$
$$\frac{11}{16} + t - \frac{11}{16} = \frac{3}{4} - \frac{11}{16}$$
$$t = \frac{3}{4} \cdot \frac{4}{4} - \frac{11}{16}$$
$$t = \frac{12}{16} - \frac{11}{16}$$
$$t = \frac{1}{16}$$

**Check.** Since $\frac{11}{16} + \frac{1}{16} = \frac{12}{16} = \frac{3}{4}$, the answer checks.

**State.** A $\frac{3}{4}$-in. piece of plywood is actually $\frac{1}{16}$ in. thinner than its name implies.

**29.** We add the heights and divide by the number of addends.

$$\frac{6\frac{5}{12} + 5\frac{11}{12} + 6\frac{7}{12}}{3} = \frac{17\frac{23}{12}}{3} = \frac{17 + 1\frac{11}{12}}{3} =$$

$$\frac{18\frac{11}{12}}{3} = \frac{227}{12} \div 3 = \frac{227}{12} \cdot \frac{1}{3} = \frac{227}{36} = 6\frac{11}{36}$$

The women's average height is $6\frac{11}{36}$ ft.

**30.** $\frac{2}{3} + 1\frac{1}{3} \cdot 2\frac{1}{8} = \frac{2}{3} + \frac{4}{3} \cdot \frac{17}{8} = \frac{2}{3} + \frac{4 \cdot 17}{3 \cdot 8} = \frac{2}{3} + \frac{4 \cdot 17}{3 \cdot 2 \cdot 4} =$

$\frac{2}{3} + \frac{4}{4} \cdot \frac{17}{3 \cdot 2} = \frac{2}{3} + \frac{17}{6} = \frac{2}{3} \cdot \frac{2}{2} + \frac{17}{6} = \frac{4}{6} + \frac{17}{6} = \frac{21}{6} =$

$3\frac{3}{6} = 3\frac{1}{2}$

**31.** $1\frac{1}{2} - \frac{1}{2}\left(\frac{1}{2} \div \frac{1}{4}\right) + \left(\frac{1}{2}\right)^2 = 1\frac{1}{2} - \frac{1}{2}\left(\frac{1}{2} \div \frac{1}{4}\right) + \frac{1}{4} =$

$1\frac{1}{2} - \frac{1}{2}\left(\frac{1}{2} \cdot \frac{4}{1}\right) + \frac{1}{4} = 1\frac{1}{2} - \frac{1}{2}\left(\frac{4}{2}\right) + \frac{1}{4} =$

$1\frac{1}{2} - \frac{4}{4} + \frac{1}{4} = 1\frac{1}{2} - 1 + \frac{1}{4} = \frac{1}{2} + \frac{1}{4} = \frac{1}{2} \cdot \frac{2}{2} + \frac{1}{4} =$

$\frac{2}{4} + \frac{1}{4} = \frac{3}{4}$

**32.** Because 3 is small in comparison to 82, $\frac{3}{82} \approx 0$.

**33.** Because 93 is nearly equal to 91, $\frac{93}{91} \approx 1$.

**34.** $256 \div 15\frac{19}{21} \approx 256 \div 16 = 16$

**35.**
$$43\frac{15}{31} \cdot 27\frac{5}{6} - 9\frac{15}{28} + 6\frac{5}{76}$$

$$\approx 43\frac{1}{2} \cdot 28 - 9\frac{1}{2} + 6$$

$$= \frac{87}{2} \cdot 28 - 9\frac{1}{2} + 6$$

$$= \frac{87 \cdot 28}{2} - 9\frac{1}{2} + 6$$

$$= \frac{87 \cdot \cancel{2} \cdot 14}{\cancel{2}} - 9\frac{1}{2} + 6$$

$$= 1218 - 9\frac{1}{2} + 6$$

$$= 1217\frac{2}{2} - 9\frac{1}{2} + 6$$

$$= 1208\frac{1}{2} + 6$$

$$= 1214\frac{1}{2}$$

**36.** a) Find the prime factorization of each number.

$$12 = 2 \cdot 2 \cdot 3$$
$$36 = 2 \cdot 2 \cdot 3 \cdot 3$$
$$60 = 2 \cdot 2 \cdot 3 \cdot 5$$

b) Create a product by writing factors that appear in the factorizations of 12, 36, and 60, using each factor the greatest number of times it appears in any one factorization.

The LCM is $2 \cdot 2 \cdot 3 \cdot 3 \cdot 5$, or 180.

Answer D is correct.

**37.** a) We find some common multiples of 8 and 6.

Multiples of 8:  8, 16, 24, 32, 40, 48, 56, 64, 72, . . .

Multiples of 6:  6, 12, 18, 24, 30, 36, 42, 48, 54, 60, 66, 72, . . .

Some common multiples are 24, 48, and 72. These are some class sizes for which study groups of 8 students or of 6 students can be organized with no students left out.

b) The smallest such class size is the least common multiple, 24.

**38.** *Familiarize.* First compare $\frac{1}{7}$ mi and $\frac{1}{8}$ mi. The LCD is 56.

$$\frac{1}{7} = \frac{1}{7} \cdot \frac{8}{8} = \frac{8}{56}$$
$$\frac{1}{8} = \frac{1}{8} \cdot \frac{7}{7} = \frac{7}{56}$$

Since $8 > 7$, then $\frac{8}{56} > \frac{7}{56}$ so $\frac{1}{7} > \frac{1}{8}$.

This tells us that Rebecca walks farther than Trent.

Next we will find how much farther Rebecca walks on each lap and then multiply by 17 to find how much farther she walks in 17 laps. Let $d$ represent how much farther Rebecca walks on each lap, in miles.

*Translate.* An equation that fits this situation is

$$\frac{1}{8} + d = \frac{1}{7}, \text{ or } \frac{7}{56} + d = \frac{8}{56}$$

*Solve.*

$$\frac{7}{56} + d = \frac{8}{56}$$

$$\frac{7}{56} + d - \frac{7}{56} = \frac{8}{56} - \frac{7}{56}$$

$$d = \frac{1}{56}$$

Now we multiply: $17 \cdot \frac{1}{56} = \frac{17}{56}$.

*Check.* We can think of the problem in a different way. In 17 laps Rebecca walks $17 \cdot \frac{1}{7}$, or $\frac{17}{7}$ mi, and Trent walks $17 \cdot \frac{1}{8}$, or $\frac{17}{8}$ mi. Then $\frac{17}{7} - \frac{17}{8} = \frac{17}{7} \cdot \frac{8}{8} - \frac{17}{8} \cdot \frac{7}{7} = \frac{136}{56} - \frac{119}{56} = \frac{17}{56}$, so Rebecca walks $\frac{17}{56}$ mi farther and our answer checks.

*State.* Rebecca walks $\frac{17}{56}$ mi farther than Trent.

## Cumulative Review Chapters 1 - 4

**1.** a) *Familiarize.* Let $t =$ the total number of miles David and Sally Jean skied.

*Translate.*

| First day's distance | plus | Second day's distance | plus | Third day's distance | is | Total distance skied |
|---|---|---|---|---|---|---|
| ↓ | ↓ | ↓ | ↓ | ↓ | ↓ | ↓ |
| $3\frac{2}{3}$ | $+$ | $6\frac{1}{8}$ | $+$ | $4\frac{3}{4}$ | $=$ | $t$ |

*Solve.* We carry out the addition.

$$3\boxed{\frac{2}{3} \cdot \frac{8}{8}} = 3\frac{16}{24}$$

$$6\boxed{\frac{1}{8} \cdot \frac{3}{3}} = 6\frac{3}{24}$$

$$+4\boxed{\frac{3}{4} \cdot \frac{6}{6}} = +4\frac{18}{24}$$

$$\overline{\qquad\qquad 13\frac{37}{24}} = 13 + \frac{37}{24}$$

$$= 13 + 1\frac{13}{24}$$

$$= 14\frac{13}{24}$$

*Check.* We repeat the calculation. The answer checks.

*State.* David and Sally Jean skied a total of $14\frac{13}{24}$ mi.

b) From part (a) we know that the sum of the three distances is $14\frac{13}{24}$. We divide this number by 3 to find the average number of miles skied per day.

$$\frac{14\frac{13}{24}}{3} = \frac{\frac{349}{24}}{3} = \frac{349}{24} \cdot \frac{1}{3} = \frac{349}{72} = 4\frac{61}{72}$$

An average of $4\frac{61}{72}$ mi was skied each day.

2. **Familiarize.** Let $p =$ the number of people who can get equal shares of the money.

**Translate.**

| Amount of each share | times | Number of people | is | Total amount divided |
|---|---|---|---|---|
| ↓ | ↓ | ↓ | ↓ | ↓ |
| 16 | · | $p$ | = | 496 |

**Solve.**

$$16 \cdot p = 496$$
$$p = \frac{496}{16}$$
$$p = 31$$

**Check.** $\$16 \cdot 31 = \$496$, so the answer checks.

**State.** 31 people can get equal \$16 shares from a total of \$496.

3. **Familiarize.** Let $x =$ the amount of salt required for $\frac{1}{2}$ recipe and $y =$ the amount required for 5 recipes.

**Translate.** We multiply by $\frac{1}{2}$ to find the amount of salt required for $\frac{1}{2}$ recipe and by 5 to find the amount for 5 recipes.

$$x = \frac{1}{2} \cdot \frac{4}{5}, \; y = 5 \cdot \frac{4}{5}$$

**Solve.** We carry out the multiplications.

$$x = \frac{1}{2} \cdot \frac{4}{5} = \frac{1 \cdot 4}{2 \cdot 5} = \frac{1 \cdot 2 \cdot 2}{2 \cdot 5} = \frac{2}{2} \cdot \frac{1 \cdot 2}{5} = \frac{2}{5}$$
$$y = 5 \cdot \frac{4}{5} = \frac{5 \cdot 4}{5} = \frac{5}{5} \cdot 4 = 4$$

**Check.** We repeat the calculations. The answer checks.

**State.** $\frac{2}{5}$ tsp of salt is required for $\frac{1}{2}$ recipe, and 4 tsp is required for 5 recipes.

4. **Familiarize.** Let $n =$ the number of $2\frac{3}{8}$-ft pieces that can be cut from a 38-ft wire.

**Translate.** We write a division sentence.

$$n = 38 \div 2\frac{3}{8}$$

**Solve.** We carry out the division.

$$n = 38 \div 2\frac{3}{8} = 38 \div \frac{19}{8} = 38 \cdot \frac{8}{19}$$
$$= \frac{38 \cdot 8}{19} = \frac{2 \cdot 19 \cdot 8}{19 \cdot 1} = \frac{19}{19} \cdot \frac{2 \cdot 8}{1}$$
$$= 16$$

**Check.** $16 \cdot 2\frac{3}{8} = 16 \cdot \frac{19}{8} = \frac{16 \cdot 19}{8} = \frac{2 \cdot 8 \cdot 19}{8 \cdot 1} =$

$\frac{8}{8} \cdot \frac{2 \cdot 19}{1} = 38$, so the answer checks.

**State.** 16 pieces can be cut from the wire.

5. **Familiarize.** Let $w =$ the total amount withdrawn for expenses and let $f =$ the amount left in the fund after the withdrawals.

**Translate.**

| First withdrawal | plus | Second withdrawal | is | Total withdrawals |
|---|---|---|---|---|
| ↓ | ↓ | ↓ | ↓ | ↓ |
| 148 | + | 167 | = | $w$ |

| Amount left in fund | is | Original amount | minus | Total withdrawals |
|---|---|---|---|---|
| ↓ | ↓ | ↓ | ↓ | ↓ |
| $f$ | = | 423 | − | $w$ |

**Solve.** We carry out the addition to solve the first equation.

$$148 + 167 = w$$
$$315 = w$$

Now we substitute 315 for $w$ in the second equation and carry out the subtraction.

$$f = 423 - w$$
$$f = 423 - 315$$
$$f = 108$$

**Check.** We repeat the calculations. The answer checks.

**State.** \$108 remains in the fund.

6. **Familiarize.** Let $w =$ the total number of miles Jermaine and Oleta walked.

**Translate.**

| Jermaine's distance | plus | Oleta's distance | is | Total distance |
|---|---|---|---|---|
| ↓ | ↓ | ↓ | ↓ | ↓ |
| $\frac{9}{10}$ | + | $\frac{3}{4}$ | = | $w$ |

**Solve.** We carry out the addition. The LCD is 20.

$$\frac{9}{10} + \frac{3}{4} = \frac{9}{10} \cdot \frac{2}{2} + \frac{3}{4} \cdot \frac{5}{5} = \frac{18}{20} + \frac{15}{20} = \frac{33}{20}$$

Thus, $w = \frac{33}{20}$.

**Check.** We repeat the calculations. The answer checks.

**State.** Jermaine and Oleta walked a total of $\frac{33}{20}$ mi.

7. We can think of the figure as being divided into 16 equal parts, each the size of the smaller area shaded. Then the larger shaded area is the equivalent of 4 smaller shaded areas. Thus, $\frac{5}{16}$ of the figure is shaded.

8. Each figure is divided into 3 equal parts. The unit is $\frac{1}{3}$, so the denominator is 3. We see that 4 of the parts are shaded. This means that the numerator is 4. Thus, $\frac{4}{3}$ is shaded.

**9.**

$$\begin{array}{r} \overset{1}{\phantom{+}}3\,7\,0\,4 \\ +5\,2\,7\,8 \\ \hline 8\,9\,8\,2 \end{array}$$

**10.**

$$\begin{array}{r} \overset{5\ \ 9\ 15}{7\,\cancel{6}\,\cancel{0}\,\cancel{5}} \\ -3\,0\,8\,7 \\ \hline 4\,5\,1\,8 \end{array}$$

**11.** $-27 + 12$

The absolute values are 27 and 12. The difference is $27 - 12$, or 15. The negative number has the larger absolute value, so the answer is negative.

$$-27 + 12 = -15$$

**12.** $\dfrac{3}{8} + \dfrac{1}{24} = \dfrac{3}{8} \cdot \dfrac{3}{3} + \dfrac{1}{24} = \dfrac{9}{24} + \dfrac{1}{24} = \dfrac{10}{24} = \dfrac{2 \cdot 5}{2 \cdot 12} =$

$\dfrac{2}{2} \cdot \dfrac{5}{12} = \dfrac{5}{12}$

**13.** $-20 - (-6) = -20 + 6 = -14$

**14.** $-\dfrac{3}{4} - \dfrac{1}{3} = -\dfrac{3}{4} \cdot \dfrac{3}{3} - \dfrac{1}{3} \cdot \dfrac{4}{4} = -\dfrac{9}{12} - \dfrac{4}{12} = -\dfrac{13}{12}$

**15.**

$$\begin{array}{r} 2 \quad \dfrac{3}{4} \quad = \quad 2\dfrac{3}{4} \\ +5\boxed{\dfrac{1}{2} \cdot \dfrac{2}{2}} = +5\dfrac{2}{4} \\ \hline 7\dfrac{5}{4} = 7 + \dfrac{5}{4} \\ = 7 + 1\dfrac{1}{4} \\ = 8\dfrac{1}{4} \end{array}$$

**16.**

$$\begin{array}{r} 2\boxed{\dfrac{1}{3} \cdot \dfrac{2}{2}} = \quad 2\dfrac{2}{6} \\ -1 \quad \dfrac{1}{6} \quad = -1\dfrac{1}{6} \\ \hline 1\dfrac{1}{6} \end{array}$$

**17.** $15 \cdot (-5) = -75$

**18.** $\dfrac{9}{10} \cdot \dfrac{5}{3} = \dfrac{9 \cdot 5}{10 \cdot 3} = \dfrac{3 \cdot 3 \cdot 5}{2 \cdot 5 \cdot 3} = \dfrac{3 \cdot 5}{3 \cdot 5} \cdot \dfrac{3}{2} = \dfrac{3}{2}$

**19.** $-18 \cdot \left(-\dfrac{5}{6}\right) = \dfrac{18 \cdot 5}{6} = \dfrac{3 \cdot 6 \cdot 5}{6 \cdot 1} = \dfrac{6}{6} \cdot \dfrac{3 \cdot 5}{1} = 15$

**20.** $2\dfrac{1}{3} \cdot 3\dfrac{1}{7} = \dfrac{7}{3} \cdot \dfrac{22}{7} = \dfrac{7 \cdot 22}{3 \cdot 7} = \dfrac{7}{7} \cdot \dfrac{22}{3} = \dfrac{22}{3} = 7\dfrac{1}{3}$

**21.**

$$\begin{array}{r} 7\,1\,5 \\ 6\overline{)4\,2\,9\,0} \\ \underline{4\,2} \\ 9 \\ \underline{6} \\ 3\,0 \\ \underline{3\,0} \\ 0 \end{array}$$

The answer is 715.

**22.**

$$\begin{array}{r} 5\,6 \\ 4\,5\overline{)2\,5\,3\,1} \\ \underline{2\,2\,5} \\ 2\,8\,1 \\ \underline{2\,7\,0} \\ 1\,1 \end{array}$$

The answer is 56 R 11.

**23.** The remainder is 11 and the divisor is 45, so a mixed numeral for the answer is $56\dfrac{11}{45}$.

**24.** $2\,7\,\boxed{5}\,3$

The digit 5 names the number of tens.

**25.** a) The total area is the sum of the areas of the two individual rectangles. We use the formula Area = length × width twice and add the results.

$$8\dfrac{1}{2} \cdot 11 + 6\dfrac{1}{2} \cdot 7\dfrac{1}{2}$$

$$= \dfrac{17}{2} \cdot 11 + \dfrac{13}{2} \cdot \dfrac{15}{2}$$

$$= \dfrac{187}{2} + \dfrac{195}{4} = \dfrac{187}{2} \cdot \dfrac{2}{2} + \dfrac{195}{4}$$

$$= \dfrac{374}{4} + \dfrac{195}{4} = \dfrac{569}{4}$$

$$= 142\dfrac{1}{4}$$

The area of the carpet is $142\dfrac{1}{4}$ ft$^2$.

b) The perimeter can be thought of as the sum of the perimeter of the larger rectangle and the two longer sides of the smaller rectangle. Thus, we have

$$\begin{array}{r} 8\,\dfrac{1}{2} \\ 11 \\ 8\,\dfrac{1}{2} \\ 11 \\ 7\,\dfrac{1}{2} \\ +\ 7\,\dfrac{1}{2} \\ \hline 52\,\dfrac{4}{2} = 52 + 2 = 54 \end{array}$$

The perimeter is 54 ft.

**26.** Round 38,478 to the nearest hundred.

$$3\,8,\,4\,\boxed{7}\,8$$
$$\uparrow$$

The digit 4 is in the hundreds place. Consider the next digit to the right. Since the digit, 7, is 5 or higher, round 4 hundreds up to 5 hundreds. Then change all digits to the right of the hundreds digit to zero.

The answer is 38,500.

**27.** $18 = 2 \cdot 3 \cdot 3$

$24 = 2 \cdot 2 \cdot 2 \cdot 3$

$\text{LCM} = 2 \cdot 2 \cdot 2 \cdot 3 \cdot 3 = 72$

**28.**
$$\left(\frac{1}{2}+\frac{2}{5}\right)^2 \div 3 + 6 \times \left(2+\frac{1}{4}\right)$$

$$= \left(\frac{1}{2}\cdot\frac{5}{5}+\frac{2}{5}\cdot\frac{2}{2}\right)^2 \div 3 + 6 \times \left(2\frac{1}{4}\right)$$

$$= \left(\frac{5}{10}+\frac{4}{10}\right)^2 \div 3 + 6 \times \frac{9}{4}$$

$$= \left(\frac{9}{10}\right)^2 \div 3 + 6 \times \frac{9}{4}$$

$$= \frac{81}{100} \div 3 + 6 \times \frac{9}{4}$$

$$= \frac{81}{100}\cdot\frac{1}{3} + 6 \times \frac{9}{4}$$

$$= \frac{81\cdot 1}{100\cdot 3} + \frac{6\cdot 9}{4}$$

$$= \frac{3\cdot 27\cdot 1}{100\cdot 3} + \frac{2\cdot 3\cdot 9}{2\cdot 2}$$

$$= \frac{3}{3}\cdot\frac{27\cdot 1}{100} + \frac{2}{2}\cdot\frac{3\cdot 9}{2}$$

$$= \frac{27}{100} + \frac{27}{2}$$

$$= \frac{27}{100} + \frac{27}{2}\cdot\frac{50}{50}$$

$$= \frac{27}{100} + \frac{1350}{100}$$

$$= \frac{1377}{100}, \text{ or } 13\frac{77}{100}$$

**29.** The LCD is 30.
$$\frac{4}{5}\cdot\frac{6}{6} = \frac{24}{30}$$
$$\frac{4}{6}\cdot\frac{5}{5} = \frac{20}{30}$$

Since $24 > 20$, it follows that $\frac{24}{30} > \frac{20}{30}$, so $\frac{4}{5} > \frac{4}{6}$.

**30.** The LCD is 39.
$$\frac{3}{13}\cdot\frac{3}{3} = \frac{9}{39}$$

The denominator of $\frac{9}{39}$ is 39.

Since $9 = 9$, it follows that $\frac{9}{39} = \frac{9}{39}$, so $\frac{3}{13} = \frac{9}{39}$.

(If we were simply testing these fractions for equality, we could have used cross products as in Section 3.5.)

**31.** The LCD is 84.
$$\frac{-3}{7}\cdot\frac{12}{12} = \frac{-36}{84}$$
$$\frac{-5}{12}\cdot\frac{7}{7} = \frac{-35}{84}$$

Since $-36 < -35$, it follows that $\frac{-36}{84} < \frac{-35}{84}$, so $\frac{-3}{7} < \frac{-5}{12}$.

**32.** Since 29 is very close to 30, $\frac{29}{30} \approx 1$.

**33.** Since $2\cdot 15 = 30$ and 30 is close to 29, the denominator is about twice the numerator. Thus, $\frac{15}{29} \approx \frac{1}{2}$.

**34.** Since 2 is very small compared to 43, $\frac{2}{43} \approx 0$.

**35.** $\frac{36}{45} = \frac{4\cdot 9}{5\cdot 9} = \frac{4}{5}\cdot\frac{9}{9} = \frac{4}{5}\cdot 1 = \frac{4}{5}$

**36.** Reminder: $\frac{0}{n} = 0$ for any integer $n$ that is not 0.
$$\frac{0}{-27} = 0$$

**37.** $\frac{320}{10} = \frac{32\cdot 10}{10\cdot 1} = \frac{10}{10}\cdot\frac{32}{1} = 1\cdot\frac{32}{1} = 32$

**38.** $4\frac{5}{8} = \frac{37}{8}$ $(4\cdot 8 = 32, \ 32+5 = 37)$

**39.**
$$3\overline{)17} \quad \begin{array}{r} 5 \\ \underline{15} \\ 2 \end{array}$$

We have $\frac{17}{3} = 5\frac{2}{3}$.

**40.**
$$x + 24 = 117$$
$$x + 24 - 24 = 117 - 24$$
$$x = 93$$
The solution is 93.

**41.**
$$x + \frac{7}{9} = \frac{4}{3}$$
$$x + \frac{7}{9} - \frac{7}{9} = \frac{4}{3} - \frac{7}{9}$$
$$x = \frac{4}{3}\cdot\frac{3}{3} - \frac{7}{9}$$
$$x = \frac{12}{9} - \frac{7}{9}$$
$$x = \frac{5}{9}$$
The solution is $\frac{5}{9}$.

**42.**
$$\frac{7}{9}\cdot t = -\frac{4}{3}$$
$$t = -\frac{4}{3} \div \frac{7}{9}$$
$$t = -\frac{4}{3}\cdot\frac{9}{7} = -\frac{4\cdot 9}{3\cdot 7} = -\frac{4\cdot 3\cdot 3}{3\cdot 7}$$
$$= \frac{3}{3}\cdot\left(-\frac{4\cdot 3}{7}\right) = -\frac{12}{7}$$
The solution is $-\frac{12}{7}$.

**43.** $y = 32,580 \div 36$

We carry out the division.

$$
\begin{array}{r}
905 \\
36\overline{)32,580} \\
\underline{324} \phantom{00} \\
180 \\
\underline{180} \\
0
\end{array}
$$

The solution is 905.

**44.** The factors of 68 are 1, 2, 4, 17, 34, 68.

A factorization of 68 is $2 \cdot 2 \cdot 17$ or $2 \cdot 34$.

The prime factorization of 68 is $2 \cdot 2 \cdot 17$.

The group of numbers divisible by 6 is 12, 54, 72, 300.

The group of numbers divisible by 8 is 8, 16, 24, 32, 40, 48, 64, 864.

The group of numbers divisible by 5 is 70, 95, 215.

The group of prime numbers is 2, 3, 17, 19, 23, 31, 47, 101.

**45.** 2001 is divisible by 3, so it is not prime. 2002 is divisible by 2, so it is not prime. The only factors of 2003 are 1 and 2003 itself, so 2003 is the smallest prime number larger than 2000.

# Chapter 5

# Decimal Notation

**1.** 486.34

   a) Write a word name for the whole number.    $\boxed{\text{Four hundred eighty-six}}$

   b) Write "and" for the decimal point.    Four hundred eighty-six $\boxed{\text{and}}$

   c) Write a word name for the number to the right of the decimal point, followed by the place value of the last digit.    Four hundred eighty-six and $\boxed{\text{thirty-four hundredths}}$

A word name for 486.34 is four hundred eighty-six and thirty-four hundredths.

**3.** 0.146

The whole number (the number to the left of the decimal point) is zero, so we write only a word name for the number to the right of the decimal point, followed by the place value of the last digit.

A word name for 0.146 is one hundred forty-six thousandths.

**5.** A word name for 249.89 is two hundred forty-nine and eighty-nine hundredths.

**7.** A word name for 3.785 is three and seven hundred eighty-five thousandths.

**9.** Negative thirty-four — and — eight hundred ninety-one thousandths —

$-34$ . $891$

**11.** 8.<u>3</u>    8.3.    $\dfrac{83}{10}$

1 place   Move 1 place.   1 zero

$8.3 = \dfrac{83}{10}$

**13.** 3.<u>56</u>    3.56.    $\dfrac{356}{100}$

2 places   Move 2 places.   2 zeros

$3.56 = \dfrac{356}{100}$

**15.** 46.<u>03</u>    46.03.    $\dfrac{4603}{100}$

2 places   Move 2 places.   2 zeros

$46.03 = \dfrac{4603}{100}$

**17.** $-0.\underline{00013}$    $-0.00013.$    $-\dfrac{13}{100,000}$

5 places   Move 5 places.   5 zeros

$-0.00013 = -\dfrac{13}{100,000}$

**19.** $-1.\underline{0008}$    $-1.0008.$    $-\dfrac{10,008}{10,000}$

4 places   Move 4 places.   4 zeros

$-1.0008 = -\dfrac{10,008}{10,000}$

**21.** 20.<u>003</u>    20.003.    $\dfrac{20,003}{1000}$

3 places   Move 3 places.   3 zeros

$20.003 = \dfrac{20,003}{1000}$

**23.** $\dfrac{8}{\underline{10}}$    0.8.

1 zero   Move 1 place.

$\dfrac{8}{10} = 0.8$

**25.** $\dfrac{889}{\underline{100}}$    8.89.

2 zeros   Move 2 places.

$\dfrac{889}{100} = 8.89$

**27.** $-\dfrac{3798}{\underline{1000}}$    $-3.798.$

3 zeros   Move 3 places.

$-\dfrac{3798}{1000} = -3.798$

**29.** $\dfrac{78}{\underline{10,000}}$    0.0078.

4 zeros   Move 4 places.

$\dfrac{78}{10,000} = 0.0078$

**31.** $\dfrac{19}{100,000}$    0.00019.

5 zeros    Move 5 places.

$\dfrac{19}{100,000} = 0.00019$

**33.** $-\dfrac{376,193}{1,000,000}$    −0.376193.

6 zeros    Move 6 places.

$-\dfrac{376,193}{1,000,000} = -0.376193$

**35.** $99\dfrac{44}{100} = 99 + \dfrac{44}{100} = 99$ and $\dfrac{44}{100} = 99.44$

**37.** $3\dfrac{798}{1000} = 3 + \dfrac{798}{1000} = 3$ and $\dfrac{798}{1000} = 3.798$

**39.** First consider $2\dfrac{1739}{10,000}$.

$2\dfrac{1739}{10,000} = 2 + \dfrac{1739}{10,000} = 2$ and $\dfrac{1739}{10,000} = 2.1739$

Then $-2\dfrac{1739}{10,000} = -2.1739$.

**41.** $8\dfrac{953,073}{1,000,000} = 8 + \dfrac{953,073}{1,000,000} =$

$8$ and $\dfrac{953,073}{1,000,000} = 8.953073$

**43.** To compare two positive numbers in decimal notation, start at the left and compare corresponding digits moving from left to right. When two digits differ, the number with the larger digit is the larger of the two numbers.

0.06

Different; 5 is larger than 0.

0.58

Thus, 0.58 is larger.

**45.** 0.905

Starting at the left, these digits are the first to differ; 1 is larger than 0.

0.91

Thus, 0.91 is larger.

**47.** To compare two negative numbers in decimal notation, start at the left and compare corresponding digits moving from left to right. When two digits differ, the number with the smaller digit is the larger of the two numbers.

−0.0009

Starting at the left, these digits are the first to differ, and 0 is smaller than 1.

−0.001

Thus, −0.0009 is larger.

**49.** 234.07

Starting at the left, these digits are the first to differ, and 5 is larger than 4.

235.07

Thus, 235.07 is larger.

**51.** $\dfrac{4}{100} = 0.04$ so we compare 0.004 and 0.04.

0.004

Starting at the left, these digits are the first to differ, and 4 is larger than 0.

0.04

Thus, 0.04 or $\dfrac{4}{100}$ is larger.

**53.** −0.4320

Starting at the left, these digits are the first to differ, and 0 is smaller than 5.

−0.4325

Thus, −0.432 is larger.

**55.** 0.1⬚1⬚   Hundredths digit is 4 or lower.
0.1   Round down.

**57.** −0.4⬚9⬚   Hundredths digit is 5 or higher.
−0.5   Round 4 up to 5.

**59.** 2.7⬚4⬚49   Hundredths digit is 4 or lower.
2.7   Round down.

**61.** −123.6⬚5⬚   Hundredths digit is 5 or higher.
−123.7   Round 6 up to 7.

**63.** 0.89⬚3⬚   Thousandths digit is 4 or lower.
0.89   Round down.

**65.** 0.66⬚6⬚6   Thousandths digit is 5 or higher.
0.67   Round up.

**67.** −0.99⬚5⬚   Thousandths digit is 5 or higher.
−1.00   Round 9 up.

(When we make the hundredths digit a 10, we carry 1 to the tenths place. This then requires us to carry 1 to the ones place.)

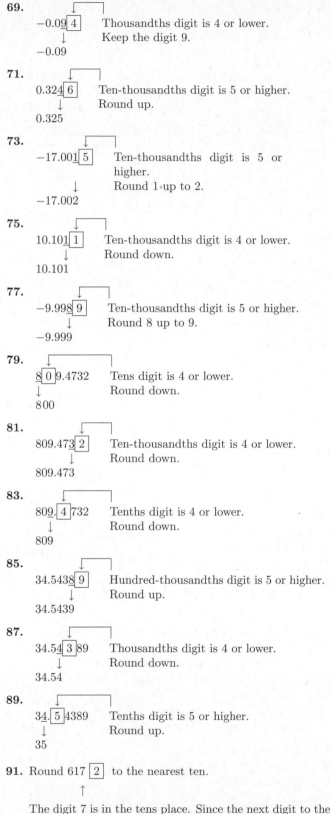

**69.**

$$-0.09\boxed{4}$$   Thousandths digit is 4 or lower.
Keep the digit 9.
$$-0.09$$

**71.**

$$0.324\boxed{6}$$   Ten-thousandths digit is 5 or higher.
Round up.
$$0.325$$

**73.**

$$-17.001\boxed{5}$$   Ten-thousandths digit is 5 or higher.
Round 1 up to 2.
$$-17.002$$

**75.**

$$10.101\boxed{1}$$   Ten-thousandths digit is 4 or lower.
Round down.
$$10.101$$

**77.**

$$-9.998\boxed{9}$$   Ten-thousandths digit is 5 or higher.
Round 8 up to 9.
$$-9.999$$

**79.**

$$\boxed{8\ 0}9.4732$$   Tens digit is 4 or lower.
Round down.
$$800$$

**81.**

$$809.473\boxed{2}$$   Ten-thousandths digit is 4 or lower.
Round down.
$$809.473$$

**83.**

$$809.\boxed{4}732$$   Tenths digit is 4 or lower.
Round down.
$$809$$

**85.**

$$34.5438\boxed{9}$$   Hundred-thousandths digit is 5 or higher.
Round up.
$$34.5439$$

**87.**

$$34.54\boxed{3}89$$   Thousandths digit is 4 or lower.
Round down.
$$34.54$$

**89.**

$$34.\boxed{5}4389$$   Tenths digit is 5 or higher.
Round up.
$$35$$

**91.** Round 617 $\boxed{2}$ to the nearest ten.
↑

The digit 7 is in the tens place. Since the next digit to the right, 2, is 4 or lower, round down, meaning that 7 tens stays as 7 tens. Then change the digit to the right of the tens digit to zero.

The answer is 6170.

**93.** Round 6 $\boxed{1}$ 72 to the nearest thousand.
↑

The digit 6 is in the thousands place. Since the next digit to the right, 1, is 4 or lower, round down, meaning that 6 thousands stays as 6 thousands. Then change all digits to the right of the thousands digit to zeros.

The answer is 6000.

**95.** We use a string of successive divisions.

$$
\begin{array}{r}
1\,7 \\
5\,\overline{)\ 8\ 5} \\
3\,\overline{)\ 2\ 5\ 5} \\
3\,\overline{)\ 7\ 6\ 5} \\
2\,\overline{)\ 1\ 5\ 3\ 0}
\end{array}
$$

$1530 = 2 \cdot 3 \cdot 3 \cdot 5 \cdot 17$, or $2 \cdot 3^2 \cdot 5 \cdot 17$

**97.** We use a string of successive divisions.

$$
\begin{array}{r}
1\,1 \\
7\,\overline{)\ 7\ 7} \\
7\,\overline{)\ 5\ 3\ 9} \\
2\,\overline{)\ 1\ 0\ 7\ 8} \\
2\,\overline{)\ 2\ 1\ 5\ 6} \\
2\,\overline{)\ 4\ 3\ 1\ 2}
\end{array}
$$

$4312 = 2 \cdot 2 \cdot 2 \cdot 7 \cdot 7 \cdot 11$, or $2^3 \cdot 7^2 \cdot 11$

**99.** The greatest number of decimals places occurring in any of the numbers is 6, so we add extra zeros to the first six numbers so that each number has 6 decimal places. Then we start at the left and compare corresponding digits, moving from left to right. The numbers, given from smallest to largest are $-2.109$, $-2.108$, $-2.1$, $-2.0302$, $-2.018$, $-2.0119$, $-2.000001$.

**101.** $6.78346\boxed{1902}$ ←Drop all decimal places
past the fifth place.
$$6.78346$$

**103.** $0.03030\boxed{3030303}$ ←Drop all decimal places
past the fifth place.
$$0.03030$$

---

## Exercise Set 5.2

**1.**
$$
\begin{array}{r}
\phantom{00}1\phantom{00} \\
3\,1\,6.2\,5 \\
+\ \ \ 1\,8.1\,2 \\
\hline
3\,3\,4.3\,7
\end{array}
$$
Add hundredths.
Add tenths.
Write a decimal point in the answer.
Add ones.
Add tens.
Add hundreds.

**3.**
$$
\begin{array}{r}
\phantom{0}1\ 1\phantom{00} \\
6\,5\,9.4\,0\,3 \\
+\ \ 9\,1\,6.8\,1\,2 \\
\hline
1\,5\,7\,6.2\,1\,5
\end{array}
$$
Add thousandths.
Add hundredths.
Add tenths.
Write a decimal point in the answer.
Add ones.
Add tens.
Add hundreds.

5.
```
    1   1
    9.1 0 4
+ 1 2 3.4 5 6
-------------
  1 3 2.5 6 0
```

7.
```
  2 0.0 1 2 4
+ 3 0.0 1 2 4
-------------
  5 0.0 2 4 8
```

9. Line up the decimal points.
```
      1
    3 9.0 0 0    Writing 2 extra zeros
  +  1.0 0 7
  -----------
    4 0.0 0 7
```

11.
```
  1 2 2 1
      4 7.8
    2 1 9.8 5 2
      4 3.5 9
  + 6 6 6.7 1 3
  -------------
    9 7 7.9 5 5
```

13. Line up the decimal points.
```
      1
      0.3 4 0    Writing an extra zero
      3.5 0 0    Writing 2 extra zeros
      0.1 2 7
  + 7 6 8.0 0 0    Writing in the decimal point
  -------------       and 3 extra zeros
    7 7 1.9 6 7    Adding
```

15.
```
  1 2 1   1
      9 9.6 0 0 1
    7 2 8 5.1 8 0 0
      5 0 0.0 4 2 0
  +   8 7 0.0 0 0 0
  -----------------
    8 7 5 4.8 2 2 1
```

17.
```
  4 11 2 11
  5 1.3 1     Borrow tenths to subtract hundredths.
-    2.2 9    Subtract hundredths.
-----------    Subtract tenths.
  4 9.0 2     Write a decimal point in the answer.
              Borrow tens to subtract ones.
              Subtract ones.
              Subtract tens.
```

19.
```
       11
   8  7 13
   9  2.3 4 1
 -    6.4 2
 -----------
   8 5.9 2 1
```

21.
```
    4 9 9 10
  2.5 0 0 0    Writing 3 extra zeros
- 0.0 0 2 5
-----------
  2.4 9 7 5
```

23.
```
    3 9 10
  3.4 0 0    Writing 2 extra zeros
- 0.0 0 3
---------
  3.3 9 7
```

25. Line up the decimal points. Write an extra zero if desired.
```
   17 11
  1 7 7 10
  2 8.7 0
- 1 9.3 5
---------
    8.8 5
```

27.
```
    3 10
  3 4.0 7
-  3 0.7
--------
    3.3 7
```

29.
```
    4 10
  8.4 5 0
- 7.4 0 5
---------
  1.0 4 5
```

31.
```
   5 10
  6.0 0 3
-   2.3
--------
  3.7 0 3
```

33.
```
    9 9 9 10
  1 0 0 0 0     Writing in the decimal point
- 0.0 0 9 8       and 4 extra zeros
-----------
  0.9 9 0 2     Subtracting
```

35.
```
    9 9 9 10
  1 0 0.0 0     Writing in the decimal point
-      0.3 4      and 4 extra zeros
-----------
   9 9.6 6
```

37.
```
   6 14
  7.4 8
- 2.6
------
  4.8 8
```

39.
```
   2 9 9 10
  3.0 0 0
- 2.0 0 6
---------
  0.9 9 4
```

41.
```
    8 9 9 10
  1 9.0 0 0
-   1.1 9 8
----------
  1 7.8 0 2
```

43.
```
   4 9 10
  6 5.0 0
- 1 3.8 7
---------
  5 1.1 3
```

45.
```
      8 17
  3 2.7 9 7 8
-    0.0 5 9 2
-------------
  3 2.7 3 8 6
```

47.
```
    6 9 10
  6.0 7 0 0
- 2.0 0 7 8
-----------
  4.0 6 2 2
```

49. −5.02 + 1.73    A positive and a negative number
```
    4 9 12
  5.0 2     Find the difference in
- 1.7 3     the absolute values
---------
  3.2 9
```
The negative number has the larger absolute value, so the answer is negative: −5.02 + 1.73 = −3.29.

51. 12.9 − 15.4 = 12.9 + (−15.4)
We add the opposite of 15.4. We have a positive and a negative number.
```
   4 14
  1 5.4     Finding the difference in
- 1 2.9     the absolute values
---------
    2.5
```
The negative number has the larger absolute value, so the answer is negative: 12.9 − 15.4 = −2.5.

**53.** $-2.9 + (-4.3)$   Two negative numbers

$$\begin{array}{r} \phantom{+}{}^{1}2\,.9 \\ +\,4\,.3 \\ \hline 7\,.2 \end{array}$$ Adding the absolute values

$-2.9 + (-4.3) = -7.2$   The sum of two negative numbers is negative.

**55.** $-4.301 + 7.68$   A negative and a positive number

$$\begin{array}{r} {}^{7}\,{}^{10}\\ 7.6\,\cancel{8}\,\cancel{0} \\ -\,4.3\,0\,1 \\ \hline 3.3\,7\,9 \end{array}$$ Finding the difference in the absolute values

The positive number has the larger absolute value, so the answer is positive: $-4.301 + 7.68 = 3.379$.

**57.**  $\phantom{=}-12.9 - 3.7$
$= -12.9 + (-3.7)$   Adding the opposite of 3.7
$= -16.6$   The sum of two negatives is negative.

**59.**  $\phantom{=}-2.1 - (-4.6)$
$= -2.1 + 4.6$   Adding the opposite of $-4.6$
$= 2.5$   Subtracting absolute values. Since 4.6 has the larger absolute value, the answer is positive.

**61.**  $\phantom{=}14.301 + (-17.82)$
$= -3.519$   Subtracting absolute values. Since $-17.82$ has the larger absolute value, the answer is negative.

**63.**  $\phantom{=}7.201 - (-2.4)$
$= 7.201 + 2.4$   Adding the opposite of $-2.4$
$= 9.601$   Adding

**65.**  $\phantom{=}23.9 + (-9.4)$
$= 75.5$   Subtracting absolute values. Since 96.9 has the larger absolute value, the answer is positive.

**67.**  $\phantom{=}-8.9 - (-12.7)$
$= -8.9 + 12.7$   Adding the opposite of $-12.7$
$= 3.8$   Subtracting absolute values. Since 12.7 has the larger absolute value, the answer is positive.

**69.**  $\phantom{=}-4.9 - 5.392$
$= -4.9 + (-5.392)$   Adding the opposite of 5.392
$= -10.292$   The sum of two negatives is negative.

**71.**  $\phantom{=}14.7 - 23.5$
$= 14.7 + (-23.5)$   Adding the opposite of 23.5
$= -8.8$   Subtracting absolute values. Since $-23.5$ has the larger absolute value, the answer is negative.

**73.**
$$x + 17.5 = 29.15$$
$$x + 17.5 - 17.5 = 29.15 - 17.5 \quad \text{Subtracting 17.5 on both sides}$$
$$x = 11.65$$

$$\begin{array}{r} {}^{8}\,{}^{11}\\ 2\,\cancel{9}.\,\cancel{1}\,5 \\ -\,1\,7.\,5 \\ \hline 1\,1.\,6\,5 \end{array}$$

**75.**
$$17.95 + p = 402.63$$
$$17.95 + p - 17.95 = 402.63 - 17.95$$
$$\text{Subtracting 17.95 on both sides}$$
$$p = 384.68$$

$$\begin{array}{r} {}^{3}\,{}^{9}\,{}^{11}\,{}^{15}\\ \cancel{4}\,\cancel{0}\,\cancel{2}.\,\cancel{6}\,\cancel{3} \\ -\,\phantom{0}1\,7.\,9\,5 \\ \hline 3\,8\,4.\,6\,8 \end{array}$$

**77.**
$$13,083.3 = x + 12,500.33$$
$$13,083.3 - 12,500.33 = x + 12,500.33 - 12,500.33$$
$$\text{Subtracting 12,500.33 on both sides}$$
$$582.97 = x$$

$$\begin{array}{r} {}^{2}\,{}^{10}\phantom{,}{}^{2}\,{}^{2}\,{}^{12}\,{}^{10}\\ 1\,\cancel{3},\,\cancel{0}\,8\,\cancel{3}.\,\cancel{3}\,\cancel{0} \\ -\,1\,2,\,5\,0\,0.\,3\,3 \\ \hline 5\,8\,2.\,9\,7 \end{array}$$

**79.**
$$x + 2349 = -17,684.3$$
$$x + 2349 - 2349 = -17,684.3 - 2349$$
$$\text{Subtracting 2349 on both sides}$$
$$x = -20,033.3 \quad \text{Adding absolute values and making the answer negative}$$

**81.** First we add the payments/debits:

$27.44 + 123.95 + 124.02 + 12.43 + 137.78 + 2800.00 = 3225.62$

Then we add the deposits/credits:

$1000.00 + 2500.00 + 18.88 = 3518.88$

We add the total of the deposits to the balance brought forward:

$9704.56 + 3518.88 = 13,223.44$

Now we subtract the debit total:

$13,223.44 - 3225.62 = 9997.82$

The result should be the ending balance, 10,483.66. Since $9997.82 \neq 10,483.66$, an error has been made. Now we successively add or subtract deposits/credits and payments/debits and check the result in the balance forward column.

$9704.56 - 27.44 = 9677.12$

This subtraction was done correctly.

$9677.12 + 1000.00 = 10,677.12$

This addition was done correctly.

$10,677.12 - 123.95 = 10,553.17$

This subtract was done correctly.

$10,553.17 - 124.02 = 10,429.15$

This subtraction was done incorrectly. It appears that 124.02 was added rather than subtracted. We correct the balance line and continue.

$$10,429.15 - 12.43 = 10,416.72$$

If the previous checkbook balance had been correct, this subtraction would have been correct. We work with the corrected balance and continue.

$$10,416.72 + 2500.00 = 12,916.72$$

$$12,916.72 - 137.78 = 12,778.94$$

$$12,778.94 + 18.88 = 12,797.82$$

$$12,797.82 - 2800.00 = 9997.82$$

The correct checkbook balance is $9997.82.

**83.**

$$34,\boxed{5}67 \quad \text{Hundreds digit is 5 or higher.}$$
$$\downarrow \qquad \text{Round up.}$$
$$35,000$$

**85.**
$$\frac{13}{24} - \frac{3}{8} = \frac{13}{24} - \frac{3}{8} \cdot \frac{3}{3}$$
$$= \frac{13}{24} - \frac{9}{24}$$
$$= \frac{13-9}{24} = \frac{4}{24}$$
$$= \frac{4 \cdot 1}{4 \cdot 6} = \frac{4}{4} \cdot \frac{1}{6}$$
$$= \frac{1}{6}$$

**87.**
$$\frac{1}{5} - \left(-\frac{1}{3}\right) = \frac{1}{5} + \frac{1}{3}$$
$$= \frac{1}{5} \cdot \frac{3}{3} + \frac{1}{3} \cdot \frac{5}{5}$$
$$= \frac{3}{15} + \frac{5}{15}$$
$$= \frac{8}{15}$$

**89. *Familiarize*.** We draw a picture.

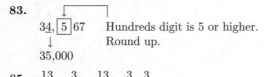

We let $s =$ the number of servings that can be prepared from $5\frac{1}{2}$ lb of flounder fillet.

***Translate*.** The situation corresponds to a division sentence.

$$s = 5\frac{1}{2} \div \frac{1}{3}$$

***Solve*.** We carry out the division.

$$s = 5\frac{1}{2} \div \frac{1}{3} = \frac{11}{2} \div \frac{1}{3}$$
$$= \frac{11}{2} \cdot \frac{3}{1} = \frac{33}{2}$$
$$= 16\frac{1}{2}$$

***Check*.** We check by multiplying. If $16\frac{1}{2}$ servings are prepared, then

$$16\frac{1}{2} \cdot \frac{1}{3} = \frac{33}{2} \cdot \frac{1}{3} = \frac{3 \cdot 11 \cdot 1}{2 \cdot 3} = \frac{3}{3} \cdot \frac{11 \cdot 1}{2} = \frac{11}{2} = 5\frac{1}{2} \text{ lb}$$

of flounder is used. Our answer checks.

***State*.** $16\frac{1}{2}$ servings can be prepared from $5\frac{1}{2}$ lb of flounder fillet.

**91.** First, "undo" the incorrect addition by subtracting 235.7 from the incorrect answer:

$$\begin{array}{r} 8\ 1\ 7.\ 2 \\ -\ 2\ 3\ 5.\ 7 \\ \hline 5\ 8\ 1.\ 5 \end{array}$$

The original minuend was 581.5. Now subtract 235.7 from this as the student originally intended:

$$\begin{array}{r} 5\ 8\ 1.\ 5 \\ -\ 2\ 3\ 5.\ 7 \\ \hline 3\ 4\ 5.\ 8 \end{array}$$

The correct answer is 345.8.

## Exercise Set 5.3

**1.**
$$\begin{array}{rl} 8.\ 6 & \text{(1 decimal place)} \\ \times\quad 7 & \text{(0 decimal places)} \\ \hline 6\ 0.\ 2 & \text{(1 decimal place)} \end{array}$$

**3.**
$$\begin{array}{rl} 0.\ 8\ 4 & \text{(2 decimal places)} \\ \times\qquad 8 & \text{(0 decimal places)} \\ \hline 6.\ 7\ 2 & \text{(2 decimal places)} \end{array}$$

**5.**
$$\begin{array}{rl} 6.\ 3 & \text{(1 decimal place)} \\ \times\ 0.\ 0\ 4 & \text{(2 decimal places)} \\ \hline 0.\ 2\ 5\ 2 & \text{(3 decimal places)} \end{array}$$

**7.**
$$\begin{array}{rl} 8\ 7 & \text{(0 decimal places)} \\ \times\ 0.\ 0\ 0\ 6 & \text{(3 decimal places)} \\ \hline 0.\ 5\ 2\ 2 & \text{(3 decimal places)} \end{array}$$

**9.** $\underline{10} \times 23.76 \qquad\qquad 23.7\underset{\underset{\textstyle\uparrow}{\rule{0.4em}{0pt}\llcorner}}{6}$

1 zero        Move 1 place to the right.

$10 \times 23.76 = 237.6$

**11.** First consider $1000 \times 583.686852$.

$\underline{1000} \times 583.686852 \qquad 583.686\underset{\underset{\textstyle\uparrow}{\rule{1.2em}{0pt}\llcorner}}{.852}$

3 zeros        Move 3 places to the right.

$1000 \times 583.686852 = 583,686.852$, so

$-1000 \times 583.686852 = -583,686.852$.

**13.** $-7.8 \times \underline{100} \qquad\qquad -7.8\underset{\underset{\textstyle\uparrow}{\rule{0.8em}{0pt}\llcorner}}{0}.$

2 zeros        Move 2 places to the right.

$-7.8 \times 100 = -780$

**15.** $0.\underline{1} \times 89.23 \qquad\qquad 8\underset{\underset{\textstyle\uparrow}{\rule{0.4em}{0pt}\lrcorner}}{.}9.23$

1 decimal place        Move 1 place to the left.

$0.1 \times 89.23 = 8.923$

**17.** $0.\underline{001} \times 97.68$      $0.097.68$

3 decimal places     Move 3 places to the left.

$0.001 \times 97.68 = 0.09768$

**19.** $-78.2 \times 0.\underline{01}$      $-0.78.2$

2 decimal places     Move 2 places to the left.

$-78.2 \times 0.01 = -0.782$

**21.**
```
    3 2. 6    (1 decimal place)
  ×    1 6    (0 decimal places)
  ---------
    1 9 5 6
    3 2 6 0
  ---------
    5 2 1. 6  (1 decimal place)
```

**23.**
```
     0. 9 8 4   (3 decimal places)
  ×       3. 3  (1 decimal place)
  -----------
     2 9 5 2
     2 9 5 2 0
  -----------
     3. 2 4 7 2 (4 decimal places)
```

**25.** $(374)(-2.4)$

First we multiply the absolute values.
```
     3 7 4    (0 decimal places)
  ×    2. 4   (1 decimal place)
  ---------
     1 4 9 6
     7 4 8 0
  ---------
     8 9 7. 6 (1 decimal place)
```
Since the product of a positive number and a negative number is negative, the answer is $-897.6$.

**27.** $-749(-0.43)$

We multiply the absolute values. Since the product of two negative numbers is positive, the answer will be positive.
```
       7 4 9    (0 decimal places)
  ×    0. 4 3   (2 decimal places)
  -----------
       2 2 4 7
     2 9 9 6 0
  -----------
     3 2 2. 0 7 (2 decimal places)
```

**29.**
```
     0. 8 7    (2 decimal places)
  ×      6 4   (0 decimal places)
  ---------
       3 4 8
     5 2 2 0
  ---------
     5 5. 6 8  (2 decimal places)
```

**31.**
```
      4 6. 5 0   (2 decimal places)
  ×         7 5  (0 decimal places)
  -------------
      2 3 2 5 0
    3 2 5 5 0 0
  -------------
    3 4 8 7. 5 0 (2 decimal places)
```
Since the last decimal place is 0, we could also write this answer as 3487.5.

**33.**
```
        8 1. 7    (1 decimal place)
  ×    0. 6 1 2   (3 decimal places)
  -------------
        1 6 3 4
        8 1 7 0
      4 9 0 2 0 0
  -------------
      5 0. 0 0 0 4 (4 decimal places)
```

**35.**
```
        1 0. 1 0 5   (3 decimal places)
  ×     1 1. 3 2 4   (3 decimal places)
  -----------------
        4 0 4 2 0
        2 0 2 1 0 0
      3 0 3 1 5 0 0
    1 0 1 0 5 0 0 0
  1 0 1 0 5 0 0 0 0
  -----------------
  1 1 4. 4 2 9 0 2 0 (6 decimal places)
```
or 114.42902

**37.**
```
      1 2. 3    (1 decimal place)
  ×    1. 0 8   (2 decimal places)
  -----------
        9 8 4
      1 2 3 0 0
  -----------
      1 3. 2 8 4 (3 decimal places)
```

**39.**
```
      3 2. 4    (1 decimal place)
  ×      2. 8   (1 decimal place)
  ---------
      2 5 9 2
      6 4 8 0
  ---------
      9 0. 7 2  (2 decimal places)
```

**41.**
```
     0. 0 0 3 4 2   (5 decimal places)
  ×         0. 8 4  (2 decimal places)
  -----------------
           1 3 6 8
         2 7 3 6 0
  -----------------
  0. 0 0 2 8 7 2 8  (7 decimal places)
```

**43.**
```
     0. 3 4 7   (3 decimal places)
  ×      2. 0 9 (2 decimal places)
  -----------
     3 1 2 3
     6 9 4 0 0
  -----------
  0. 7 2 5 2 3  (5 decimal places)
```

**45.** $3.005 \times (-0.623)$

First we multiply the absolute values.
```
       3. 0 0 5    (3 decimal places)
  ×    0. 6 2 3    (3 decimal places)
  --------------
       9 0 1 5
       6 0 1 0 0
     1 8 0 3 0 0 0
  --------------
     1. 8 7 2 1 1 5 (6 decimal places)
```
Since the product of a positive number and a negative number is negative, the answer is $-1.872115$.

**47.** $(-6.4)(-15.6)$

We multiply the absolute values. Since the product of two negative numbers is positive, the answer will be positive.
```
      1 5. 6    (1 decimal place)
  ×      6. 4   (1 decimal place)
  ---------
      6 2 4
      9 3 6 0
  ---------
      9 9. 8 4  (2 decimal places)
```

**49.** $\underline{1000} \times 45.678$      $45.678.$

3 zeros     Move 3 places to the right.

$1000 \times 45.678 = 45,678$

**51.** Move 2 places to the right.

$28.88 .¢

Change from $ sign in front to ¢ sign at end.

$28.88 = 2888¢

**53.** Move 2 places to the right.

$0.66 .¢

Change from $ sign in front to ¢ sign at end.

$0.66 = 66¢

**55.** Move 2 places to the left.

$0.34.¢

Change from ¢ sign at end to $ sign in front.

34¢ = $0.34

**57.** Move 2 places to the left.

$34.45.¢

Change from ¢ sign at end to $ sign in front.

3445¢ = $34.45

**59.** 47.3 billion = $47.3 \times 1,\underbrace{000,000,000}_{9 \text{ zeros}}$

47.300000000.

Move 9 places to the right.

47.3 billion = 47,300,000,000

**61.** 9.3 million = $9.3 \times 1,\underbrace{000,000}_{6 \text{ zeros}}$

9.300000.

Move 6 places to the right.

9.3 million = 9,300,000

**63.** 23.4 billion = $23.4 \times 1,\underbrace{000,000,000}_{9 \text{ zeros}}$

23.400000000.

Move 9 places to the right.

23.4 billion = 23,400,000,000

**65.** $2\frac{1}{3} \cdot 4\frac{4}{5} = \frac{7}{3} \cdot \frac{24}{5} = \frac{7 \cdot 3 \cdot 8}{3 \cdot 5}$

$$= \frac{3}{3} \cdot \frac{7 \cdot 8}{5} = \frac{56}{5}$$

$$= 11\frac{1}{5}$$

**67.**

$$4\frac{4}{5} = 4\frac{12}{15}$$
$$-2\frac{1}{3} = -2\frac{5}{15}$$
$$\overline{\phantom{-2}2\frac{7}{15}}$$

**69.**

```
        3 4 2
24 ⌈ 8 2 0 8
     7 2
     ---
     1 0 0
       9 6
       ---
         4 8
         4 8
         ---
           0
```

The answer is 342.

**71.**

```
       4 5 6 6
7 ⌈ 3 1, 9 6 2
    2 8
    ---
      3 9
      3 5
      ---
        4 6
        4 2
        ---
          4 2
          4 2
          ---
            0
```

The answer is 4566.

**73.** $49 \div (-7) = -7$

Check: $-7(-7) = 49$

**75.** (1 trillion) · (1 billion)

$= 1,\underbrace{000,000,000,000}_{12 \text{ zeros}} \times 1,\underbrace{000,000,000}_{9 \text{ zeros}}$

$= 1,\underbrace{000,000,000,000,000,000,000}_{21 \text{ zeros}}$

$= 10^{21} = 1$ sextillion

**77.** (1 trillion) · (1 trillion)

$= 1,\underbrace{000,000,000,000}_{12 \text{ zeros}} \times 1,\underbrace{000,000,000,000}_{12 \text{ zeros}}$

$= 1,\underbrace{000,000,000,000,000,000,000,000}_{24 \text{ zeros}}$

$= 10^{24} = 1$ septillion

## Exercise Set 5.4

**1.**
```
    2. 9 9
2 ) 5. 9 8
    4
    ─
    1 9
    1 8
    ──
      1 8
      1 8
      ──
        0
```
Divide as though dividing whole numbers. Place the decimal point directly above the decimal point in the dividend.

**3.**
```
    2 3. 7 8
4 ) 9 5. 1 2
    8
    ─
    1 5
    1 2
    ──
      3 1
      2 8
      ──
        3 2
        3 2
        ──
          0
```
Divide as though dividing whole numbers. Place the decimal point directly above the decimal point in the dividend.

**5.**
```
      7. 4 8
1 2 ) 8 9. 7 6
      8 4
      ──
      5 7
      4 8
      ──
        9 6
        9 6
        ──
          0
```

**7.**
```
      7. 2
3 3 ) 2 3 7. 6
      2 3 1
      ────
          6 6
          6 6
          ──
            0
```

**9.** First we consider $9.144 \div 8$.
```
    1. 1 4 3
8 ) 9. 1 4 4
    8
    ─
    1 1
     8
     ─
     3 4
     3 2
     ──
       2 4
       2 4
       ──
         0
```
Then $-9.144 \div 8 = -1.143$.

**11.** First we consider $12.123 \div 3$.
```
      4. 0 4 1
3 ) 1 2. 1 2 3
    1 2
    ──
      1 2
      1 2
      ──
          3
          3
          ─
          0
```
Then $12.123 \div (-3) = -4.041$.

**13.**
```
    0. 0 7
5 ) 0. 3 5
    3 5
    ──
     0
```

**15.**
```
        7 0.
0.1 2∧) 8.4 0∧
        8 4
        ──
         0
         0
         ─
         0
```
Multiply the divisor by 100 (move the decimal point 2 places). Multiply the same way in the dividend (move 2 places). Then divide.

**17.**
```
        2 0.
3.4∧) 6 8.0∧
      6 8
      ──
       0
       0
       ─
       0
```
Put a decimal point at the end of the whole number. Multiply the divisor by 10 (move the decimal point 1 place). Multiply the same way in the dividend (move 1 place), adding an extra 0. Then divide.

**19.** $-6 \div (-15)$

First we consider $6 \div 15$. The answer will be positive.
```
       0.4
1 5 ) 6.0
      6 0
      ──
       0
```
Put a decimal point at the end of the whole number. Write an extra 0 to the right of the decimal point. Then divide.

**21.**
```
        0.4 1
3 6 ) 1 4.7 6
      1 4 4
      ────
         3 6
         3 6
         ──
          0
```

**23.**
```
         8.5
3.2∧) 2 7.2∧0
      2 5 6
      ────
        1 6 0   Write an extra 0.
        1 6 0
        ────
            0
```

**25.**
```
        9.3
4.2∧) 3 9.0∧6
      3 7 8
      ────
        1 2 6
        1 2 6
        ────
            0
```

**27.** First consider $5 \div 8$.
```
    0.6 2 5
8 ) 5.0 0 0
    4 8
    ──
      2 0   Write an extra 0.
      1 6
      ──
        4 0   Write an extra 0.
        4 0
        ──
          0
```
Then $-5 \div 8 = -0.625$.

$$\begin{array}{r} 0.2\,6 \\ 0.4\,7_\wedge \overline{)\,0.1\,2_\wedge 2\,2} \\ 9\,4 \\ \hline 2\,8\,2 \\ 2\,8\,2 \\ \hline 0 \end{array}$$

**29.**

$$\begin{array}{r} 1\,5.6\,2\,5 \\ 4.8_\wedge \overline{)\,7\,5.0_\wedge 0\,0\,0} \\ 4\,8\,0 \\ \hline 2\,7\,0 \\ 2\,4\,0 \\ \hline 3\,0\ 0 \\ 2\,8\ 8 \\ \hline 1\,2\,0 \\ 9\,6 \\ \hline 2\,4\,0 \\ 2\,4\,0 \\ \hline 0 \end{array}$$

**31.**

$$\begin{array}{r} 2.3\,4 \\ 0.0\,3\,2_\wedge \overline{)\,0.0\,7\,4_\wedge 8\,8} \\ 6\,4 \\ \hline 1\,0\ 8 \\ 9\,6 \\ \hline 1\,2\,8 \\ 1\,2\,8 \\ \hline 0 \end{array}$$

**33.**

$$\begin{array}{r} 0.4\,7 \\ 8\,2\,\overline{)\,3\,8.5\,4} \\ 3\,2\,8 \\ \hline 5\,7\,4 \\ 5\,7\,4 \\ \hline 0 \end{array}$$

**35.**

**37.** $\dfrac{213.4567}{1000}$  $\quad$ 0.213.4567

3 zeros $\quad$ Move 3 places to the left.

$\dfrac{213.4567}{1000} = 0.2134567$

**39.** $\dfrac{-23.59}{10}$  $\quad$ $-23.5.9$

1 zero $\quad$ Move 1 place to the left.

$\dfrac{-23.59}{10} = -2.359$

**41.** $\dfrac{426.487}{100}$  $\quad$ 4.26.487

2 zeros $\quad$ Move 2 places to the left.

$\dfrac{426.487}{100} = 4.26487$

**43.** $\dfrac{-16.94}{0.1}$  $\quad$ $-16.9.4$

1 decimal place $\quad$ Move 1 place to the right.

$\dfrac{-16.94}{0.1} = -169.4$

**45.** $\dfrac{1.0237}{0.001}$  $\quad$ 1.023.7

3 decimal places $\quad$ Move 3 places to the right.

$\dfrac{1.0237}{0.001} = 1023.7$

**47.** $\dfrac{-42.561}{0.01}$  $\quad$ $-42.56.1$

2 decimal places $\quad$ Move 2 places to the right.

$\dfrac{-42.561}{0.01} = -4256.1$

**49.** $4.2 \cdot x = 39.06$

$\dfrac{4.2 \cdot x}{4.2} = \dfrac{39.06}{4.2}$ $\quad$ Dividing on both sides by 4.2

$x = 9.3$

$$\begin{array}{r} 0\,9.3 \\ 4.2\,_\wedge \overline{)\,3\,9.0_\wedge 6} \\ 3\,7\,8\ 0 \\ \hline 1\,2\,6 \\ 1\,2\,6 \\ \hline 0 \end{array}$$

The solution is 9.3.

**51.** $1000 \cdot y = -9.0678$

$\dfrac{1000 \cdot y}{1000} = \dfrac{-9.0678}{1000}$ $\quad$ Dividing on both sides by 1000

$y = -0.0090678$ $\quad$ Moving the decimal point 3 places to the left

The solution is $-0.0090678$.

**53.** $1048.8 = 23 \cdot t$

$\dfrac{1048.8}{23} = \dfrac{23 \cdot t}{23}$ $\quad$ Dividing on both sides by 23

$45.6 = t$

$$\begin{array}{r} 4\,5.6 \\ 2\,3\,\overline{)\,1\,0\,4\,8.8} \\ 9\,2\,0\,0 \\ \hline 1\,2\,8\,8 \\ 1\,1\,5\,0 \\ \hline 1\,3\,8 \\ 1\,3\,8 \\ \hline 0 \end{array}$$

The solution is 45.6.

**55.** $14 \times (82.6 + 67.9) = 14 \times (150.5)$ $\quad$ Doing the calculation inside the parentheses

$= 2107$ $\quad$ Multiplying

**57.** $0.003 - 3.03 \div 0.01 = 0.003 - 303$ $\quad$ Dividing first

$= -302.997$ $\quad$ Adding

**59.** $42 \times (10.6 + 0.024)$

$= 42 \times 10.624$ $\quad$ Doing the calculation inside the parentheses

$= 446.208$ $\quad$ Multiplying

**61.** $4.2 \times 5.7 + 0.7 \div 3.5$

$= 23.94 + 0.2$    Doing the multiplications and divisions in order from left to right

$= 24.14$    Adding

**63.** $-9.0072 + 0.04 \div 0.1^2$

$= -9.0072 + 0.04 \div 0.01$    Evaluating the exponential expression

$= -9.0072 + 4$    Dividing

$= -5.0072$    Adding

**65.** $(8 - 0.04)^2 \div 4 + 8.7 \times 0.4$

$= (7.96)^2 \div 4 + 8.7 \times 0.4$    Doing the calculation inside the parentheses

$= 63.3616 \div 4 + 8.7 \times 0.4$    Evaluating the exponential expression

$= 15.8404 + 3.48$    Doing the multiplications and divisions in order from left to right

$= 19.3204$    Adding

**67.** $86.7 + 4.22 \times (9.6 - 0.03)^2$

$= 86.7 + 4.22 \times (9.57)^2$    Doing the calculation inside the parentheses

$= 86.7 + 4.22 \times 91.5849$    Evaluating the exponential expression

$= 86.7 + 386.488278$    Multiplying

$= 473.188278$    Adding

**69.** $4 \div (-0.4) + 0.1 \times 5 - 0.1^2$

$= 4 \div (-0.4) + 0.1 \times 5 - 0.01$    Evaluating the exponential expression

$= -10 + 0.5 - 0.01$    Doing the multiplications and divisions in order from left to right

$= -9.51$    Adding and subtracting in order from left to right

**71.** $5.5^2 \times [(6 - 4.2) \div 0.06 + 0.12]$

$= 5.5^2 \times [1.8 \div 0.06 + 0.12]$    Doing the calculation in the innermost parentheses first

$= 5.5^2 \times [30 + 0.12]$    Doing the calculation inside the parentheses

$= 5.5^2 \times 30.12$

$= 30.25 \times 30.12$    Evaluating the exponential expression

$= 911.13$    Multiplying

**73.** $200 \times \{[(4 - 0.25) \div 2.5] - (4.5 - 4.025)\}$

$= 200 \times \{[3.75 \div 2.5] - 0.475\}$    Doing the calculations in the innermost parentheses first

$= 200 \times \{1.5 - 0.475\}$    Again, doing the calculations in the innermost parentheses

$= 200 \times 1.025$    Subtracting inside the parentheses

$= 205$    Multiplying

**75.** We add the numbers and then divide by the number of addends.

$(\$1276.59 + \$1350.49 + \$1123.78 + \$1402.56) \div 4$

$= \$5153.42 \div 4$

$= \$1288.355$

$\approx \$1288.36$

**77.** We add the amounts and divide by the number of addends, 5.

$$\frac{5.8 + 5.7 + 5.4 + 5.2 + 5.0}{5}$$

$$= \frac{27.1}{5} = 5.42$$

The average numbers of overnight camping stays in Park Service campgrounds from 2002 through 2006 was 5.42 million per year.

**79.** $\dfrac{36}{42} = \dfrac{6 \cdot 6}{6 \cdot 7} = \dfrac{6}{6} \cdot \dfrac{6}{7} = \dfrac{6}{7}$

**81.** $-\dfrac{38}{146} = -\dfrac{2 \cdot 19}{2 \cdot 73} = \dfrac{2}{2} \cdot \left(-\dfrac{19}{73}\right) = -\dfrac{19}{73}$

**83.**

$$
\begin{array}{r}
1\ 9 \\
3\overline{\smash{)}\ 5\ 7} \\
3\overline{\smash{)}\ 1\ 7\ 1} \\
2\overline{\smash{)}\ 3\ 4\ 2} \\
2\overline{\smash{)}\ 6\ 8\ 4}
\end{array}
$$

$684 = 2 \cdot 2 \cdot 3 \cdot 3 \cdot 19$, or $2^2 \cdot 3^2 \cdot 19$

**85.**

$$
\begin{array}{r}
2\ 2\ 3 \\
3\overline{\smash{)}\ 6\ 6\ 9} \\
3\overline{\smash{)}\ 2\ 0\ 0\ 7}
\end{array}
$$

$2007 = 3 \cdot 3 \cdot 223$, or $3^2 \cdot 223$

**87.** $10\dfrac{1}{2} + 4\dfrac{5}{8} = 10\dfrac{4}{8} + 4\dfrac{5}{8}$

$$= 14\dfrac{9}{8} = 15\dfrac{1}{8}$$

**89.** Use a calculator.

$9.0534 - 2.041^2 \times 0.731 \div 1.043^2$

$= 9.0534 - 4.165681 \times 0.731 \div 1.087849$    Evaluating the exponential expressions

$= 9.0534 - 3.045112811 \div 1.087849$    Multiplying and dividing in order from left to right

$= 9.0534 - 2.799205415$

$= 6.254194585$

**91.** $439.57 \times 0.01 \div 1000 \times \underline{\quad} = 4.3957$

$4.3957 \div 1000 \times \underline{\quad} = 4.3957$

$0.0043957 \times \underline{\quad} = 4.3957$

We need to multiply 0.0043957 by a number that moves the decimal point 3 places to the right. Thus, we need to multiply by 1000. This is the missing value.

**93.** $0.0329 \div 0.001 \times 10^4 \div \underline{\quad} = 3290$

$0.0329 \div 0.001 \times 10{,}000 \div \underline{\quad} = 3290$

$32.9 \times 10{,}000 \div \underline{\quad} = 3290$

$329{,}000 \div \underline{\quad} = 3290$

We need to divide 329,000 by a number that moves the decimal point 2 places to the left. Thus, we need to divide by 100. This is the missing value.

## Chapter 5 Mid-Chapter Review

1. In the number 308.00567, the number 6 names the ten-thousandths place. The statement is false.

2. The statement is true. See page 259 in the text.

3. The statement is true. See page 280 in the text.

4.
$$y + 12.8 = 23.35$$
$$y + 12.8 - 12.8 = 23.35 - 12.8$$
$$y + 0 = 10.55$$
$$y = 10.55$$

5. $5.6 + 4.3 \times (6.5 - 0.25)^2 = 5.6 + 4.3 \times (6.25)^2$
$$= 5.6 + 4.3 \times 39.0625$$
$$= 5.6 + 167.96875$$
$$= 173.56875$$

6. A word name for 9.69 is nine and sixty-nine hundredths.

7. $1.05$ million $= 1.05 \times 1$ million
$$= 1.05 \times 1,000,000$$
$$= 1,050,000$$

8.    4.<u>53</u>       4.53.      $\dfrac{453}{1\underline{00}}$

   2 places   Move 2 places.   2 zeros

$$4.53 = \frac{453}{100}$$

9.   $-0.\underline{287}$    $-0.287.$    $\dfrac{-287}{1\underline{000}}$

   3 places   Move 3 places.   3 zeros

$$-0.287 = \frac{-287}{1000}, \text{ or } -\frac{287}{1000}$$

10.  0.07

        Starting at the left, these digits are the first to differ; 1 is larger than 0.

  0.13

  Thus, 0.13 is larger.

11.  $-5.2$

        Starting at the left, these digits are the first to differ; 0 is smaller than 2.

  $-5.09$

  Thus, $-5.09$ is larger.

12.  $\dfrac{7}{1\underline{0}}$     0.7.

  1 zero   Move 1 place.

$$\frac{7}{10} = 0.7$$

13.  $\dfrac{-639}{1\underline{00}}$    $-6.39.$

  2 zeros   Move 2 places.

$$\frac{-639}{100} = -6.39$$

14. $35\dfrac{67}{100} = 35 + \dfrac{67}{100} = 35 \text{ and } \dfrac{67}{100} = 35.67$

15. $8\dfrac{2}{1000} = 8 + \dfrac{2}{1000} = 8 \text{ and } \dfrac{2}{1000} = 8.002$

16.
28.461|5|     Ten-thousandths digit is 5 or higher.
                    Round up.
28.462

17.
28.46|1|5     Thousandths digit is 4 or lower.
                    Round down.
28.46

18.
28.4|6|15     Hundredths digit is 5 or higher.
                    Round up.
28.5

19.
28.|4|615     Tenths digit is 4 or lower.
                    Round down.
28

20.
```
    1 1   1
  4 7.6 3 8
+   2.4 5 7
  5 0.0 9 5
```

21.
```
      1 1 1 1 1
    1 5.6 0 0     Writing 2 extra zeros
  2 3 4.7 2 9
      3.0 8 0     Writing an extra zero
+ 9 6 1.4 5 3
1 2 1 4.8 6 2
```

22. $-4.5 + 0.728$   A negative and a positive number

```
         14
   3  4 9 10
   4. 5 0 0     Finding the difference in
 - 0. 7 2 8     the absolute values
   3. 7 7 2
```

The negative number has the larger absolute value, so the answer is negative: $-4.5 + 0.728 = -3.772$.

23. Line up the decimal points.

```
     1
  1 6.00     Writing in the decimal point and 2
   0.34      extra zeros
+  1.90      Writing an extra zero
  1 8.24
```

24.
```
         11
   2  7 11 4 17
   3  2 1.5 7
 -   4 9.3 8
   2 7 2.1 9
```

25.
$$\overset{5\ \ 9\ \ 10}{5.\overset{}{6}\overset{}{0}\overset{}{0}} \quad \text{Writing 2 extra zeros}$$
$$\underline{-\ 0.0\ 0\ 7}$$
$$5.5\ 9\ 3$$

26. Line up the decimal points. Write an extra zero if desired.

$$\overset{\overset{13\ 12}{2\ \ \cancel{3}\ \ \cancel{2}\ 10}}{\cancel{3}\ 4.\cancel{3}\ \cancel{0}}$$
$$\underline{-\ 1\ 8.7\ 5}$$
$$1\ 5.5\ 5$$

27.
$$-49.07 - 9.7$$
$$= -49.07 + (-9.7) \quad \text{Adding the opposite of 9.7}$$
$$= -58.77 \qquad\qquad \text{The sum of two negative}$$
$$\qquad\qquad\qquad\qquad\quad \text{numbers is negative.}$$

28.
$$\begin{array}{rl} 4.\,6 & \text{(1 decimal place)} \\ \times\ \ 0.\,9 & \text{(1 decimal place)} \\ \hline 4.1\ 4 & \text{(2 decimal places)} \end{array}$$

29.
$$\begin{array}{rl} 1\ 5.\,3 & \text{(1 decimal place)} \\ \times\ 6.\,0\ 7 & \text{(2 decimal places)} \\ \hline 1\ 0\ 7\ 1 & \\ 9\ 1\ 8\ 0\ 0 & \\ \hline 9\ 2.\,8\ 7\ 1 & \text{(3 decimal places)} \end{array}$$

30. First consider $100 \times 81.236$.

$$\underline{100} \times 81.236 \qquad\qquad 81.23.6$$

2 zeros        Move 2 places to the right.

$$100 \times 81.236 = 8123.6, \text{ so } -100 \times 81.236 = -8123.6.$$

31. $0.\underline{1} \times 29.37$        $2.9.37$

1 decimal place     Move 1 place to the left.

$$0.1 \times 29.37 = 2.937$$

32.
$$\begin{array}{r} 5.0\ 6 \\ 4\ \overline{)\ 2\ 0.2\ 4} \\ \underline{2\ 0}\phantom{.2\ 4} \\ 2\ 4 \\ \underline{2\ 4} \\ 0 \end{array}$$

33. First we consider $21.76 \div 6.8$.

$$\begin{array}{r} 3.2 \\ 6.8_{\wedge}\overline{)\ 2\ 1.7_{\wedge}6} \\ \underline{2\ 0\ 4}\phantom{6} \\ 1\ 3\ 6 \\ \underline{1\ 3\ 6} \\ 0 \end{array}$$

Since a negative number divided by a positive number is negative, the answer is $-3.2$.

34.
$$\frac{76.34}{0.\underline{1}} \qquad\qquad 76.3.4$$

1 decimal place     Move 1 place to the right.

$$76.34 \div 0.1 = 763.4$$

35.
$$\frac{914.036}{\underline{1000}} \qquad\qquad 0.914.036$$

3 zeros        Move 3 places to the left.

$$914.036 \div 1000 = 0.914036$$

36. Move 2 places to the right.

$$\$20.45.\cancel{c}$$

Change from \$ sign in front to $\cancel{c}$ sign at end.

$$\$20.45 = 2045\cancel{c}$$

37. Move 2 places to the left.

$$\$1.47.\cancel{c}$$

Change from $\cancel{c}$ sign at end to \$ sign in front.

$$147\cancel{c} = \$1.47$$

38.
$$46.3 + x = -59$$
$$46.3 + x - 46.3 = -59 - 46.3 \quad \text{Subtracting 46.3}$$
$$\qquad\qquad\qquad\qquad\qquad\qquad\quad \text{on both sides}$$
$$x = -105.3$$

The solution is $-105.3$.

39.
$$42.84 = 5.1 \cdot y$$
$$\frac{42.84}{5.1} = \frac{5.1 \cdot y}{5.1} \quad \text{Dividing by 5.1 on both sides}$$
$$8.4 = y$$

The solution is $8.4$.

40.
$$6.594 + 0.5318 \div 0.01$$
$$= 6.594 + 53.18 \qquad\quad \text{Dividing}$$
$$= 59.774 \qquad\qquad\quad\ \text{Adding}$$

41.
$$7.3 \times 4.6 - 0.8 \div (-3.2)$$
$$= 33.58 + 0.25 \qquad\quad \text{Multiplying and dividing}$$
$$= 33.83 \qquad\qquad\qquad \text{Adding}$$

42. The student probably rounded over successively from the thousandths place as follows: $236.448 \approx 236.45 \approx 236.5 \approx 237$. The student should have considered only the tenths place and rounded down.

43. The decimal points were not lined up before the subtraction was carried out.

44. $10 \div 0.2 = \dfrac{10}{0.2} = \dfrac{10}{0.2} \cdot \dfrac{10}{10} = \dfrac{100}{2}$, or $100 \div 2$

45. $0.247 \div 0.1 = \dfrac{247}{1000} \div \dfrac{1}{10} = \dfrac{247}{1000} \cdot \dfrac{10}{1} = \dfrac{247 \cdot 10}{10 \cdot 100} = \dfrac{247}{100} = 2.47 \neq 0.0247$;

$0.247 \div 10 = \dfrac{247}{1000} \div 10 = \dfrac{247}{1000} \cdot \dfrac{1}{10} = \dfrac{247}{10,000} = 0.0247 \neq 2.47$

## Exercise Set 5.5

**1.** $\dfrac{23}{100} = 0.23$

**3.** $\dfrac{3}{5} = \dfrac{3}{5} \cdot \dfrac{2}{2}$    We use $\dfrac{2}{2}$ for 1 to get a
denominator of 10.

   $= \dfrac{6}{10} = 0.6$

**5.** $-\dfrac{13}{40} = -\dfrac{13}{40} \cdot \dfrac{25}{25}$    We use $\dfrac{25}{25}$ for 1 to get
a denominator of 1000.

   $= -\dfrac{325}{1000} = -0.325$

**7.** $\dfrac{1}{5} = \dfrac{1}{5} \cdot \dfrac{2}{2} = \dfrac{2}{10} = 0.2$

**9.** $-\dfrac{17}{20} = -\dfrac{17}{20} \cdot \dfrac{5}{5} = -\dfrac{85}{100} = -0.85$

**11.** $\dfrac{3}{8} = 3 \div 8$

$$
\begin{array}{r}
0.3\,7\,5 \\
8\,\overline{\smash{\big)}\,3.0\,0\,0} \\
\underline{2\,4}\phantom{00} \\
6\,0\phantom{0} \\
\underline{5\,6}\phantom{0} \\
4\,0 \\
\underline{4\,0} \\
0
\end{array}
$$

   $\dfrac{3}{8} = 0.375$

**13.** $-\dfrac{39}{40} = -\dfrac{39}{40} \cdot \dfrac{25}{25} = -\dfrac{975}{1000} = -0.975$

**15.** $\dfrac{13}{25} = \dfrac{13}{25} \cdot \dfrac{4}{4} = \dfrac{52}{100} = 0.52$

**17.** $-\dfrac{2502}{125} = -\dfrac{2502}{125} \cdot \dfrac{8}{8} = -\dfrac{20,016}{1000} = -20.016$

**19.** $\dfrac{1}{4} = \dfrac{1}{4} \cdot \dfrac{25}{25} = \dfrac{25}{100} = 0.25$

**21.** $-\dfrac{29}{25} = -\dfrac{29}{25} \cdot \dfrac{4}{4} = -\dfrac{116}{100} = -1.16$

**23.** $\dfrac{19}{16} = \dfrac{19}{16} \cdot \dfrac{625}{625} = \dfrac{11,875}{10,000} = 1.1875$

**25.** $\dfrac{4}{15} = 4 \div 15$

$$
\begin{array}{r}
0.\,2\,6\,6 \\
1\,5\,\overline{\smash{\big)}\,4.\,0\,0\,0} \\
\underline{3\,0}\phantom{00} \\
1\,0\,0\phantom{0} \\
\underline{9\,0}\phantom{0} \\
1\,0\,0 \\
\underline{9\,0} \\
1\,0
\end{array}
$$

Since 10 keeps reappearing as a remainder, the digits repeat and

$\dfrac{4}{15} = 0.2666\ldots$ or $0.2\overline{6}$.

**27.** $\dfrac{1}{3} = 1 \div 3$

$$
\begin{array}{r}
0.\,3\,3\,3 \\
3\,\overline{\smash{\big)}\,1.\,0\,0\,0} \\
\underline{9}\phantom{00} \\
1\,0\phantom{0} \\
\underline{9}\phantom{0} \\
1\,0 \\
\underline{9} \\
1
\end{array}
$$

Since 1 keeps reappearing as a remainder, the digits repeat and

$\dfrac{1}{3} = 0.333\ldots$ or $0.\overline{3}$.

**29.** First we consider $\dfrac{4}{3}$, or $4 \div 3$.

$$
\begin{array}{r}
1.\,3\,3 \\
3\,\overline{\smash{\big)}\,4.\,0\,0} \\
\underline{3}\phantom{00} \\
1\,0\phantom{0} \\
\underline{9}\phantom{0} \\
1\,0 \\
\underline{9} \\
1
\end{array}
$$

Since 1 keeps reappearing as a remainder, the digits repeat and

$\dfrac{4}{3} = 1.333\ldots$ or $1.\overline{3}$, so we have $-\dfrac{4}{3} = -1.\overline{3}$.

**31.** First we consider $\dfrac{7}{6}$, or $7 \div 6$.

$$
\begin{array}{r}
1.\,1\,6\,6 \\
6\,\overline{\smash{\big)}\,7.\,0\,0\,0} \\
\underline{6}\phantom{000} \\
1\,0\phantom{00} \\
\underline{6}\phantom{00} \\
4\,0\phantom{0} \\
\underline{3\,6}\phantom{0} \\
4\,0 \\
\underline{3\,6} \\
4
\end{array}
$$

Since 4 keeps reappearing as a remainder, the digits repeat and $\dfrac{7}{6} = 1.166\ldots$ or $1.1\overline{6}$, so we have $-\dfrac{7}{6} = 1.1\overline{6}$.

**33.** $\dfrac{4}{7} = 4 \div 7$

$$
\begin{array}{r}
0.\,5\;7\;1\;4\;2\;8\;5 \\
7\,\overline{\smash{)}\,4.\,0\;0\;0\;0\;0\;0\;0} \\
\underline{3\;5}\phantom{00000} \\
5\;0\phantom{0000} \\
\underline{4\;9}\phantom{0000} \\
1\;0\phantom{000} \\
\underline{7}\phantom{000} \\
3\;0\phantom{00} \\
\underline{2\;8}\phantom{00} \\
2\;0\phantom{0} \\
\underline{1\;4}\phantom{0} \\
6\;0 \\
\underline{5\;6} \\
4\;0 \\
\underline{3\;5} \\
5
\end{array}
$$

Since 5 reappears as a remainder, the sequence repeats and $\dfrac{4}{7} = 0.571428571428\ldots$ or $0.\overline{571428}$.

**35.** First we consider $\dfrac{11}{12}$, or $11 \div 12$.

$$
\begin{array}{r}
0.\,9\;1\;6\;6 \\
12\,\overline{\smash{)}\,1\,1.\,0\;0\;0\;0} \\
\underline{1\;0\;8}\phantom{000} \\
2\;0\phantom{00} \\
\underline{1\;2}\phantom{00} \\
8\;0\phantom{0} \\
\underline{7\;2}\phantom{0} \\
8\;0 \\
\underline{7\;2} \\
8
\end{array}
$$

Since 8 keeps reappearing as a remainder, the digits repeat and $\dfrac{11}{12} = 0.91666\ldots$ or $0.91\overline{6}$, so we have $-\dfrac{11}{12} = -0.91\overline{6}$.

**37.** Round $0.\,2\;\boxed{6}\;6\;6\ldots$ to the nearest tenth.

       Hundredths digit is 5 or higher.

   $0.\,3$         Round up.

Round $0.\,2\;6\;\boxed{6}\;6\ldots$ to the nearest hundredth.

       Thousandths digit is 5 or higher.

   $0.\,2\;7$       Round up.

Round $0.\,2\;6\;6\;\boxed{6}\;\ldots$ to the nearest thousandth.

       Ten-thousandths digit is 5 or higher.

   $0.\,2\;6\;7$     Round up.

**39.** Round $0.\,3\;\boxed{3}\;3\;3\ldots$ to the nearest tenth.

       Hundredths digit is 4 or lower.

   $0.\,3$         Round down.

Round $0.\,3\;3\;\boxed{3}\;3\ldots$ to the nearest hundredth.

       Thousandths digit is 4 or lower.

   $0.\,3\;3$       Round down.

Round $0.\,3\;3\;3\;\boxed{3}\;\ldots$ to the nearest thousandth.

       Ten-thousandths digit is 4 or lower.

   $0.\,3\;3\;3$     Round down.

**41.** Round $-1.\,3\;\boxed{3}\;3\;3\ldots$ to the nearest tenth.

       Hundredths digit is 4 or lower.

   $-1.\,3$      Keep the digit 3.

Round $-1.\,3\;3\;\boxed{3}\;3\ldots$ to the nearest hundredth.

       Thousandths digit is 4 or lower.

   $-1.\,3\;3$     Keep the digit 3.

Round $-1.\,3\;3\;3\;\boxed{3}\;\ldots$ to the nearest thousandth.

       Ten-thousandths digit is 4 or lower.

   $-1.\,3\;3\;3$    Keep the digit 3.

**43.** Round $-1.\,1\;\boxed{6}\;6\;6\ldots$ to the nearest tenth.

       Hundredths digit is 5 or higher.

   $-1.\,2$      Round 1 up to 2.

Round $-1.\,1\;6\;\boxed{6}\;6\ldots$ to the nearest hundredth.

       Thousandths digit is 5 or higher.

   $-1.\,1\;7$     Round 6 up to 7.

Round $-1.\,1\;6\;6\;\boxed{6}\;\ldots$ to the nearest thousandth.

       Ten-thousandths digit is 5 or higher.

   $-1.\,1\;6\;7$    Round 6 up to 7.

**45.** $0.\overline{571428}$

Round to the nearest tenth.

   $0.5\,\boxed{7}\,1428571428\ldots$

       Hundredths digit is 5 or higher.

   $0.6$         Round up.

Round to the nearest hundredth.

   $0.57\,\boxed{1}\,428571428\ldots$

       Thousandths digit is 4 or lower.

   $0.57$       Round down.

Round to the nearest thousandth.

   $0.571\,\boxed{4}\,28571428\ldots$

       Ten-thousandths digit is 4 or lower.

   $0.571$     Round down.

**47.** Round $-0.\,9\;\boxed{1}\;6\;6\ldots$ to the nearest tenth.

       Hundredths digit is 4 or lower.

   $-0.\,9$      Keep the digit 9.

Round $-0.\,9\;1\;\boxed{6}\;6\ldots$ to the nearest hundredth.

       Thousandths digit is 5 or higher.

   $-0.\,9\;2$     Round 1 up to 2.

Round $-0.9\,1\,6\,\boxed{6}\,\ldots$ to the nearest thousandth.

↑— Ten-thousandths digit is 5 or higher.

$-0.9\,1\,7$          Round 6 up to 7.

**49.** Round $0.1\,\boxed{8}\,1\,8\ldots$ to the nearest tenth.

↑— Hundredths digit is 5 or higher.

$0.2$          Round up.

Round $0.1\,8\,\boxed{1}\,8\ldots$ to the nearest hundredth.

↑— Thousandths digit is 4 or lower.

$0.1\,8$          Round down.

Round $0.1\,8\,1\,\boxed{8}\ldots$ to the nearest thousandth.

↑— Ten-thousandths digit is 5 or higher.

$0.1\,8\,2$          Round up.

**51.** Round $-0.2\,\boxed{7}\,7\,7\ldots$ to the nearest tenth.

↑— Hundredths digit is 5 or higher.

$-0.3$          Round 2 up to 3.

Round $-0.2\,7\,\boxed{7}\,7\ldots$ to the nearest hundredth.

↑— Thousandths digit is 5 or higher.

$-0.2\,8$          Round 7 up to 8.

Round $-0.2\,7\,7\,\boxed{7}\ldots$ to the nearest thousandth.

↑— Ten-thousandths digit is 5 or higher.

$-0.2\,7\,8$          Round 7 up to 8.

**53.** Note that there are 3 women and 4 men, so there are $3+4$, or 7, people.

(a) $\dfrac{\text{Women}}{\text{Number of people}} = \dfrac{3}{7} = 0.\overline{428571} \approx 0.429$

(b) $\dfrac{\text{Women}}{\text{Men}} = \dfrac{3}{4} = 0.75$

(c) $\dfrac{\text{Men}}{\text{Number of people}} = \dfrac{4}{7} = 0.\overline{571428} \approx 0.571$

(d) $\dfrac{\text{Men}}{\text{Women}} = \dfrac{4}{3} = 1.\overline{3} \approx 1.333$

**55.** $\dfrac{\text{Miles driven}}{\text{Gasoline used}} = \dfrac{285}{18} = 15.833\ldots \approx 15.8$

The gasoline mileage was about 15.8 miles per gallon.

**57.** $\dfrac{\text{Miles driven}}{\text{Gasoline used}} = \dfrac{324.8}{18.2} \approx 17.8$

The gasoline mileage was about 17.8 miles per gallon.

**59.** We add the wind speeds and divide by the number of addends, 6.

$$\frac{35.3 + 12.5 + 11.3 + 10.7 + 10.7 + 10.4}{6}$$

$$= \frac{90.9}{6} = 15.15 \approx 15.2$$

The average of the wind speeds is about 15.2 mph.

**61.** $29\dfrac{9}{16} = 29 + \dfrac{9}{16}$

We convert $\dfrac{9}{16}$ to decimal notation.

```
      0. 5 6 2 5
1 6 ⟌ 9. 0 0 0 0
      8 0
      1 0 0
        9 6
          4 0
          3 2
            8 0
            8 0
              0
```

We have $\dfrac{9}{16} = 0.5625$, so $\$29\dfrac{9}{16} = \$29.5625 \approx \$29.56$.

**63.** $27\dfrac{7}{8} = 27 + \dfrac{7}{8}$

In Example 6 we found that $\dfrac{7}{8} = 0.875$. Thus,

$\$27\dfrac{7}{8} = \$27.875 \approx \$27.88$.

**65.** $32\dfrac{31}{64} = 32 + \dfrac{31}{64}$

We convert $\dfrac{31}{64}$ to decimal notation.

```
        0. 4 8 4 3 7 5
6 4 ⟌ 3 1. 0 0 0 0 0 0
      2 5 6
        5 4 0
        5 1 2
          2 8 0
          2 5 6
            2 4 0
            1 9 2
              4 8 0
              4 4 8
                3 2 0
                3 2 0
                  0
```

We have $\dfrac{31}{64} = 0.484375$, so $\$32\dfrac{31}{64} = \$32.484375 \approx \$32.48$.

**67.** We will use the second method discussed in the text.

$\dfrac{7}{8} \times 12.64 = \dfrac{7}{8} \times \dfrac{1264}{100} = \dfrac{7 \cdot 1264}{8 \cdot 100}$

$= \dfrac{7 \cdot 2 \cdot 2 \cdot 2 \cdot 2 \cdot 79}{2 \cdot 2 \cdot 2 \cdot 2 \cdot 2 \cdot 5 \cdot 5}$

$= \dfrac{2 \cdot 2 \cdot 2 \cdot 2}{2 \cdot 2 \cdot 2 \cdot 2} \cdot \dfrac{7 \cdot 79}{2 \cdot 5 \cdot 5}$

$= 1 \cdot \dfrac{7 \cdot 79}{2 \cdot 5 \cdot 5}$

$= \dfrac{7 \cdot 79}{2 \cdot 5 \cdot 5} = \dfrac{553}{50}$, or 11.06

**69.** $2\dfrac{3}{4} + 5.65 = 2.75 + 5.65$   Writing $2\dfrac{3}{4}$ using decimal notation

$= 8.4$          Adding

**71.** We will use the first method discussed in the text.

$$\frac{47}{9} \times (-79.95) = 5.\overline{2} \times (-79.95)$$
$$\approx 5.222 \times (-79.95) = -417.4989$$

Note that this answer is not as accurate as those found using either of the other methods, due to rounding. The result using the other methods is $-417.51\overline{6}$.

**73.** $\frac{1}{2} - 0.5 = 0.5 - 0.5$   Writing $\frac{1}{2}$ using decimal notation
$$= 0$$

**75.** $4.875 - 2\frac{1}{16} = 4.875 - 2.0625$   Writing $2\frac{1}{16}$ using decimal notation
$$= 2.8125$$

**77.** We will use the third method discussed in the text.

$$\frac{5}{6} \times 0.0765 - \frac{5}{4} \times 0.1124 = \frac{5}{6} \times \frac{0.0765}{1} - \frac{5}{4} \times \frac{0.1124}{1}$$
$$= \frac{5 \times 0.0765}{6 \times 1} - \frac{5 \times 0.1124}{4 \times 1}$$
$$= \frac{0.3825}{6} - \frac{0.562}{4}$$
$$= 0.06375 - 0.1405$$
$$= -0.07675$$

**79.** We use the rules for order of operations, doing the multiplication first and then the division. Then we add.

$$\frac{4}{5} \times 384.8 + 24.8 \div \frac{8}{3} = 307.84 + 24.8 \cdot \frac{3}{8}$$
$$= 307.84 + 9.3$$
$$= 317.14$$

**81.** We do the multiplications in order from left to right. Then we subtract.

$$\frac{7}{8} \times 0.86 - 0.76 \times \frac{3}{4} = 0.7525 - 0.76 \times \frac{3}{4}$$
$$= 0.7525 - 0.57$$
$$= 0.1825$$

**83.** $3.375 \times 5\frac{1}{3} = 3.375 \times \frac{16}{3}$   Writing $5\frac{1}{3}$ using fractional notation
$$= 18$$   Multiplying

**85.** $6.84 \div \left(-2\frac{1}{2}\right)$
$$= 6.84 \div -2.5$$   Writing $-2\frac{1}{2}$ using decimal notation
$$= -2.736$$   Dividing

**87.** $9 \cdot 2\frac{1}{3} = \frac{9}{1} \cdot \frac{7}{3} = \frac{9 \cdot 7}{1 \cdot 3} = \frac{3 \cdot 3 \cdot 7}{1 \cdot 3} = \frac{3}{3} \cdot \frac{3 \cdot 7}{1} = 21$

**89.** $-84 \div 8\frac{2}{5} = -84 \div \frac{42}{5} = -\frac{84}{1} \cdot \frac{5}{42} = -\frac{84 \cdot 5}{42} =$
$$-\frac{42 \cdot 2 \cdot 5}{42 \cdot 1} = -\frac{2 \cdot 5}{1} \cdot \frac{42}{42} = -10$$

**91.** $17\frac{5}{6} + 32\frac{3}{8} = 17\frac{20}{24} + 32\frac{9}{24} = 49\frac{29}{24} = 50\frac{5}{24}$

**93.** $16\frac{1}{10} - 14\frac{3}{5} = 16\frac{1}{10} - 14\frac{6}{10} = 15\frac{11}{10} - 14\frac{6}{10} =$
$$1\frac{5}{10} = 1\frac{1}{2}$$

**95.** **Familiarize**. We draw a picture and let $c =$ the total number of cups of liquid ingredients.

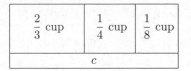

**Translate**. The problem can be translated to an equation as follows:

| Amount of water | plus | Amount of milk | plus | Amount of oil | is | Amount of liquid |
|:---:|:---:|:---:|:---:|:---:|:---:|:---:|
| ↓ | ↓ | ↓ | ↓ | ↓ | ↓ | ↓ |
| $\frac{2}{3}$ | $+$ | $\frac{1}{4}$ | $+$ | $\frac{1}{8}$ | $=$ | $c$ |

**Solve**. We carry out the addition. Since $3 = 3$, $4 = 2 \cdot 2$, and $8 = 2 \cdot 2 \cdot 2$, the LCM of the denominators is $3 \cdot 2 \cdot 2 \cdot 2$, or 24.

$$\frac{2}{3} + \frac{1}{4} + \frac{1}{8} = c$$
$$\frac{2}{3} \cdot \frac{8}{8} + \frac{1}{4} \cdot \frac{6}{6} + \frac{1}{8} \cdot \frac{3}{3} = c$$
$$\frac{16}{24} + \frac{6}{24} + \frac{3}{24} = c$$
$$\frac{25}{24} = c$$

**Check**. We repeat the calculation. We also note that the sum is larger than any of the individual amounts, as expected.

**State**. The recipe calls for $\frac{25}{24}$ cups, or $1\frac{1}{24}$ cups, of liquid ingredients.

**97.** Using a calculator we find that
$$\frac{1}{7} = 1 \div 7 = 0.\overline{142857}.$$

**99.** Using a calculator we find that
$$\frac{3}{7} = 3 \div 7 = 0.\overline{428571}.$$

**101.** Using a calculator we find that
$$\frac{5}{7} = 5 \div 7 = 0.\overline{714285}.$$

**103.** Using a calculator we find that
$$\frac{1}{9} = 1 \div 9 = 0.\overline{1}.$$

**105.** Using a calculator we find that
$$\frac{1}{999} = 0.\overline{001}.$$

## Exercise Set 5.6

**1.** We are estimating the sum

$$\$279.89 + \$149.99.$$

We round both numbers to the nearest ten. The estimate is
$$\$280 + \$150 = \$430.$$

Answer (d) is correct.

**3.** We are estimating the difference

$$\$279.89 - \$149.99.$$

We round both numbers to the nearest ten. The estimate is
$$\$280 - \$150 = \$130.$$

Answer (c) is correct.

**5.** We are estimating the product

$$6 \times \$79.99.$$

We round $\$79.99$ to the nearest ten. The estimate is

$$6 \times \$80 = \$480.$$

Answer (a) is correct.

**7.** We are estimating the quotient

$$\$830 \div \$79.99.$$

We round $\$830$ to the nearest hundred and $\$79.99$ to the nearest ten. The estimate is

$$\$800 \div \$80 = 10.$$

Answer (c) is correct.

**9.** This is about $0.0 + 1.3 + 0.3$, so the answer is about 1.6.

**11.** This is about $6 + 0 + 0$, so the answer is about 6.

**13.** This is about $52 + 1 + 7$, so the answer is about 60.

**15.** This is about $2.7 - 0.4$, so the answer is about 2.3.

**17.** This is about $200 - 20$, so the answer is about 180.

**19.** This is about $50 \times 8$, rounding 49 to the nearest ten and 7.89 to the nearest one, so the answer is about 400. Answer (a) is correct.

**21.** This is about $100 \times 0.08$, rounding 98.4 to the nearest ten and 0.083 to the nearest hundredth, so the answer is about 8. Answer (c) is correct.

**23.** This is about $4 \div 4$, so the answer is about 1. Answer (b) is correct.

**25.** This is about $75 \div 25$, so the answer is about 3. Answer (b) is correct.

**27.** We estimate the quotient $1760 \div 8.625$.

$$1800 \div 9 = 200$$

We estimate that 200 posts will be needed. Answers may vary depending on how the rounding was done.

**29.** We estimate the product $\$1.89 \times 12$.

$$\$2 \cdot 12 = \$24$$

We estimate the cost to be $\$24$. Answers may vary depending on how the rounding was done.

**31.** The decimal $0.57\overline{3}$ is an example of a <u>repeating</u> decimal.

**33.** The sentence $5(3+8) = 5\cdot3 + 5\cdot8$ illustrates the <u>distributive</u> law.

**35.** The number 1 is the <u>multiplicative</u> identity.

**37.** The least common <u>denominator</u> of two or more fractions is the least common <u>multiple</u> of their denominators.

**39.** We round each factor to the nearest ten. The estimate is $180 \times 60 = 10,800$. The estimate is close to the result given, so the decimal point was placed correctly.

**41.** We round each number on the left to the nearest one. The estimate is $20 - 1 \times 4 = 20 - 4 = 16$. The estimate is not close to the result given, so the decimal point was not placed correctly.

**43.** a) Observe that $2^{13} = 8192 \approx 8000$, $156,876.8 \approx 160,000$, and $8000 \times 20 = 160,000$. Thus, we want to find the product of $2^{13}$ and a number that is approximately 20. Since $0.37 + 18.78 = 19.15 \approx 20$, we add inside the parentheses and then multiply:

$$(0.37 + 18.78) \times 2^{13} = 156,876.8$$

We can use a calculator to confirm this result.

    b) Observe that $312.84 \approx 6 \cdot 50$. We start by multiplying 6.4 and 51.2, getting 327.68. Then we can use a calculator to find that if we add 2.56 to this product and then subtract 17.4, we have the desired result. Thus, we have

$$2.56 + 6.4 \times 51.2 - 17.4 = 312.84.$$

## Exercise Set 5.7

**1.** *Familiarize.* We let $a =$ the amount by which the amount of damage from Hurricane Katrina exceeded the amount of damage from Hurricane Andrew, in billions of dollars.

*Translate.*

| Andrew damage | plus | Excess Katrina damage | is | Katrina damage |
|:---:|:---:|:---:|:---:|:---:|
| ↓ | ↓ | ↓ | ↓ | ↓ |
| 38.1 | + | $a$ | = | 81.2 |

*Solve.* We subtract 38.1 on both sides of the equation.

$$38.1 + a = 81.2$$
$$38.1 + a - 38.1 = 81.2 - 38.1$$
$$a = 43.1$$

*Check.* We can check by adding 43.1 to 38.1 to get 81.2. The answer checks.

*State*. Hurricane Katrina was $43.1 billion more costly than Hurricane Andrew.

**3.** *Familiarize*. Let $p$ = the total number of passengers carried on the top two routes, in millions.

*Translate*.

| Passengers on New York City-Chicago route | plus | Passengers on Washington-Chicago route | is | Total passengers on top two routes |
|---|---|---|---|---|
| ↓ | ↓ | ↓ | ↓ | ↓ |
| 3.47 | + | 2.82 | = | $p$ |

*Solve*. We carry out the addition.

$$\begin{array}{r} \overset{1}{3.4}\,7 \\ +\,2.8\,2 \\ \hline 6.2\,9 \end{array}$$

*Check*. We can repeat the calculation. The answer checks.

*State*. 6.29 million passengers were carried on the top two routes.

**5.** *Familiarize*. Let $s$ = the amount the movie *Spider-Man 3* took in on its opening weekend, in millions of dollars.

*Translate*.

| Amount Spider-Man 3 took in | plus | Excess Dark Knight amount | is | Amount Dark Knight took in |
|---|---|---|---|---|
| ↓ | ↓ | ↓ | ↓ | ↓ |
| $s$ | + | 4.24 | = | 155.34 |

*Solve*. We subtract 4.24 on both sides of the equation.

$$s + 4.24 = 155.34$$
$$s + 4.24 - 4.24 = 155.34 - 4.24$$
$$s = 151.1$$

*Check*. We can add 151.1 and 4.24 to get 155.34. The answer checks.

*State*. *Spider-Man 3* took in $151.1 million on its opening weekend.

**7.** *Familiarize*. Let $t$ = the cost of drinking tap water for a year.

*Translate*.

| Cost of drinking tap water | plus | Excess cost of bottled water | is | Cost of drinking bottled water |
|---|---|---|---|---|
| ↓ | ↓ | ↓ | ↓ | ↓ |
| $t$ | + | 918.31 | = | 918.82 |

*Solve*. We subtract 918.31 on both sides of the equation.

$$t + 918.31 = 918.82$$
$$t + 918.31 - 918.31 = 918.82 - 918.31$$
$$t = 0.51$$

*Check*. We add 0.51 and 918.31 to get 918.82. The answer checks.

*State*. It costs $0.51 to drink tap water for a year.

**9.** *Familiarize*. We visualize the situation. We let $n$ = the new temperature.

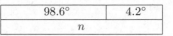

*Translate*. We are combining amounts.

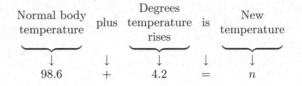

| Normal body temperature | plus | Degrees temperature rises | is | New temperature |
|---|---|---|---|---|
| ↓ | ↓ | ↓ | ↓ | ↓ |
| 98.6 | + | 4.2 | = | $n$ |

*Solve*. To solve the equation we carry out the addition.

$$\begin{array}{r} \overset{1}{9}\,8.6 \\ +\ \ 4.2 \\ \hline 1\,0\,2.8 \end{array}$$

Thus, $n = 102.8$.

*Check*. We can check by repeating the addition. We can also check by rounding:

$$98.6 + 4.2 \approx 99 + 4 = 103 \approx 102.8$$

*State*. The new temperature was 102.8°F.

**11.** *Familiarize*. We visualize the situation. Let $w$ = each winner's share.

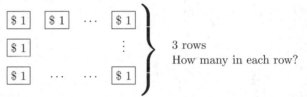

3 rows
How many in each row?

*Translate*.

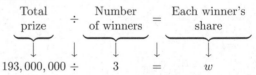

| Total prize | ÷ | Number of winners | = | Each winner's share |
|---|---|---|---|---|
| ↓ | ↓ | ↓ | ↓ | ↓ |
| 193,000,000 | ÷ | 3 | = | $w$ |

*Solve*. We carry out the division. We carry out the division, obtaining $64,333,333.\overline{3}$. Rounding to the nearest cent, or hundredth, we get $w = 64,333,333.33$.

*Check*. We can repeat the calculation. The answer checks.

*State*. Each winner's share is $64,333,333.33.

**13.** *Familiarize*. Let $A$ = the area, in sq cm, and $P$ = the perimeter, in cm.

*Translate*. We use the formulas $A = l \cdot w$ and $P = l + w + l + w$ and substitute 3.25 for $l$ and 2.5 for $w$.

$$A = l \cdot w = (3.25) \cdot (2.5)$$
$$P = l + w + l + w = 3.25 + 2.5 + 3.25 + 2.5$$

*Solve*. To find the area we carry out the multiplication.

$$\begin{array}{r} 3.\,2\,5 \\ \times\ 2.5 \\ \hline 1\ 6\ 2\ 5 \\ 6\ 5\ 0\ 0 \\ \hline 8.\,1\,2\,5 \end{array}$$

Thus, $A = 8.125$

To find the perimeter we carry out the addition.

$$\begin{array}{r} 3.\,2\,5 \\ 2.5 \\ 3.\,2\,5 \\ +\ 2.5 \\ \hline 1\,1.5\,0 \end{array}$$

Then $P = 11.5$.

**Check**. We can obtain partial checks by estimating.

$(3.25) \times (2.5) \approx 3 \times 3 \approx 9 \approx 8.125$

$3.25 + 2.5 + 3.25 + 2.5 \approx 3 + 3 + 3 + 3 = 12 \approx 11.5$

The answers check.

**State**. The area of the stamp is 8.125 sq cm, and the perimeter is 11.5 cm.

**15. Familiarize**. We visualize the situation. We let $m =$ the odometer reading at the end of the trip.

| 22,456.8 mi | 234.7 mi |
|:---:|:---:|
| $m$ | |

**Translate**. We are combining amounts.

$$\underbrace{\text{Reading}\atop\text{before trip}}_{22,456.8} \underset{\text{plus}}{+} \underbrace{\text{Miles}\atop\text{driven}}_{234.7} \underset{\text{is}}{=} \underbrace{\text{Reading at}\atop\text{end of trip}}_{m}$$

**Solve**. To solve the equation we carry out the addition.

$$\begin{array}{r} {\scriptstyle 1\ \ 1} \\ 2\,2,4\,5\,6.8 \\ +\quad 2\,3\,4.7 \\ \hline 2\,2,6\,9\,1.5 \end{array}$$

Thus, $m = 22,691.5$.

**Check**. We can check by repeating the addition. We can also check by rounding:

$22,456.8 + 234.7 \approx 22,460 + 230 = 22,690 \approx 22,691.5$

**State**. The odometer reading at the end of the trip was 22,691.5.

**17. Familiarize**. This is a two-step problem. First, we find the number of miles that have been driven between fillups. This is a "how-much-more" situation. We let $n =$ the number of miles driven.

**Translate and Solve**.

$$\underbrace{\text{First}\atop{\text{odometer}\atop\text{reading}}}_{26,342.8} \underset{\text{plus}}{+} \underbrace{\text{Number}\atop{\text{of miles}\atop\text{driven}}}_{n} \underset{\text{is}}{=} \underbrace{\text{Second}\atop{\text{odometer}\atop\text{reading}}}_{26,736.7}$$

To solve the equation we subtract 26,342.8 on both sides.

$n = 26,736.7 - 26,342.8$
$n = 393.9$

$$\begin{array}{r} 2\,6,7\,3\,6.7 \\ -\ 2\,6,3\,4\,2.8 \\ \hline 3\,9\,3.9 \end{array}$$

Second, we divide the total number of miles driven by the number of gallons. This gives us $m =$ the number of miles per gallon.

$393.9 \div 19.5 = m$

To find the number $m$, we divide.

$$\begin{array}{r} 2\,0.2\phantom{0} \\ 1\,9.5_\wedge\!\overline{\smash{)}\,3\,9\,3.\,9_\wedge 0} \\ \underline{3\,9\,0\,0}\phantom{0} \\ 3\,9\,0 \\ \underline{3\,9\,0} \\ 0 \end{array}$$

Thus, $m = 20.2$.

**Check**. To check, we first multiply the number of miles per gallon times the number of gallons:

$19.5 \times 20.2 = 393.9$

Then we add 393.9 to 26,342.8:

$26,342.8 + 393.9 = 26,736.7$

The number 20.2 checks.

**State**. The driver gets 20.2 miles per gallon.

**19. Familiarize**. We visualize a rectangular array consisting of 748.45 objects with 62.5 objects in each row. We want to find $n$, the number of rows.

**Translate**. We think (Total number of pounds) $\div$ (Pounds per cubic foot) = (Number of cubic feet).

$748.45 \div 62.5 = n$

**Solve**. We carry out the division.

$$\begin{array}{r} 1\,1.9\,7\,5\,2 \\ 6\,2.5_\wedge\!\overline{\smash{)}\,7\,4\,8.\,4_\wedge 5\,0\,0\,0} \\ \underline{6\,2\,5}\phantom{00000} \\ 1\,2\,3\,4\phantom{000} \\ \underline{6\,2\,5}\phantom{000} \\ 6\,0\,9\,5\phantom{00} \\ \underline{5\,6\,2\,5}\phantom{00} \\ 4\,7\,0\,0\phantom{0} \\ \underline{4\,3\,7\,5}\phantom{0} \\ 3\,2\,5\,0 \\ \underline{3\,1\,2\,5} \\ 1\,2\,5\,0 \\ \underline{1\,2\,5\,0} \\ 0 \end{array}$$

Thus, $n = 11.9752$.

**Check**. We obtain a partial check by rounding and estimating:

$748.45 \div 62.5 \approx 700 \div 70 = 10 \approx 11.9752$

**State**. The tank holds 11.9752 cubic feet of water.

**21. Familiarize**. We let $d =$ the distance around the figure, in cm.

**Translate**. We are combining lengths.

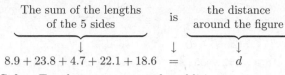

The sum of the lengths of the 5 sides **is** the distance around the figure

$$8.9 + 23.8 + 4.7 + 22.1 + 18.6 = d$$

**Solve**. To solve we carry out the addition.

$$
\begin{array}{r}
\overset{2\;\;3}{} \\
8.9 \\
2\,3.8 \\
4.7 \\
2\,2.1 \\
+\;1\,8.6 \\
\hline
7\,8.1
\end{array}
$$

Thus, $d = 78.1$.

**Check**. To check we can repeat the addition. We can also check by rounding:

$$8.9 + 23.8 + 4.7 + 22.1 + 18.6 \approx 9 + 24 + 5 + 22 + 19 = 79 \approx 78.1$$

**State**. The distance around the figure is 78.1 cm.

**23. Familiarize**. Let $d =$ the distance around the figure, in cm. The figure consists of 6 vertical sides, each with length 2.5 cm, and 6 horizontal sides, each with length 2.25 cm.

**Translate**.

Vertical distances **plus** Horizontal distances **is** Distance around figure

$$6 \times (2.5) + 6 \times (2.25) = d$$

**Solve**. We carry out the computation.

$$6 \times (2.5) + 6 \times (2.25) = 15 + 13.5 = 28.5$$

Thus, $d = 28.5$.

**Check**. We can obtain a partial check by estimating:

$$6 \times (2.5) + 6 \times (2.25) \approx 6 \times 3 + 6 \times 2 \approx 18 + 12 \approx 30 \approx 28.5$$

The answer checks.

**State**. The perimeter of the figure is 28.5 cm.

**25. Familiarize**. This is a two-step problem. First we find the total cost of the purchase. Then we find the amount of change Andrew received. Let $t =$ the total cost of the purchase, in dollars.

**Translate and Solve**.

Purchase price **plus** Sales tax **is** Total cost

$$23.99 + 1.68 = t$$

To solve the equation we carry out the addition.

$$
\begin{array}{r}
\overset{1\;\;1}{} \\
2\,3.9\,9 \\
+\;\;\;1.6\,8 \\
\hline
2\,5.6\,7
\end{array}
$$

Thus, $t = 25.67$.

Now we find the amount of the change.

We visualize the situation. We let $c =$ the amount of change.

| $50 | |
|---|---|
| $25.67 | $c$ |

Amount paid **minus** Amount of purchase **is** Amount of change

$$\$50 - \$25.67 = c$$

To solve the equation we carry out the subtraction.

$$
\begin{array}{r}
\overset{4\;\;9\;\;9\;\;10}{\cancel{5\,0.0\,0}} \\
-\;2\,5.6\,7 \\
\hline
2\,4.3\,3
\end{array}
$$

Thus, $c = \$24.33$.

**Check**. We check by adding $24.33 to $25.67 to get $50. This checks.

**State**. The change was $24.33.

**27. Familiarize**. This is a multistep problem. First we find the sum $s$ of the two 0.8 cm segments. Then we use this length to find $d$.

**Translate and Solve**.

Length of one small segment **plus** Length of other small segment **is** Total length

$$0.8 + 0.8 = s$$

To solve we carry out the addition.

$$
\begin{array}{r}
\overset{1}{} \\
0.8 \\
+\;0.8 \\
\hline
1.6
\end{array}
$$

Thus, $s = 1.6$.

Now we find $d$.

Total length of smaller segments **plus** length of $d$ **is** 3.91 cm

$$1.6 + d = 3.91$$

To solve we subtract 1.6 on both sides of the equation.

$$d = 3.91 - 1.6$$
$$d = 2.31$$

$$
\begin{array}{r}
3.9\,1 \\
-\;1.6\,0 \\
\hline
2.3\,1
\end{array}
$$

**Check**. We repeat the calculations.

**State**. The length $d$ is 2.31 cm.

**29. Familiarize**. This is a two-step problem. First, we find how many minutes there are in 2 hr. We let $m$ represent this number. Repeated addition fits this situation (Remember that 1 hr = 60 min.)

**Translate and Solve**.

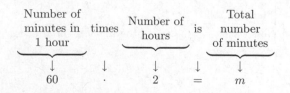

Number of minutes in 1 hour **times** Number of hours **is** Total number of minutes

$$60 \cdot 2 = m$$

To solve the equation we carry out the multiplication.

$$\begin{array}{r} 6\ 0 \\ \times\quad 2 \\ \hline 1\ 2\ 0 \end{array}$$

Thus, $m = 120$.

Next, we find how many calories are burned in 120 minutes. We let $t$ represent this number. Repeated addition fits this situation also.

| Number of calories burned in 1 minute | times | Number of minutes | is | Total number of calories burned |
|---|---|---|---|---|
| ↓ | ↓ | ↓ | ↓ | ↓ |
| 7.3 | × | 120 | = | $t$ |

To solve the equation we carry out the multiplication.

$$\begin{array}{r} 1\ 2\ 0 \\ \times\quad 7.\ 3 \\ \hline 3\ 6\ 0 \\ 8\ 4\ 0\ 0 \\ \hline 8\ 7\ 6.\ 0 \end{array}$$

Thus, $t = 876$.

**Check.** To check, we first divide the total number of calories by the number of calories burned in one minute to find the total number of minutes the person mowed:

$$876 \div 7.3 = 120$$

Then we divide 120 by 60 to find the number of hours:

$$120 \div 60 = 2$$

The number 876 checks.

**State.** In 2 hr of mowing, 876 calories would be burned.

31. **Familiarize.** This is a multistep problem. We will first find the total amount of the debits. Then we will find how much is left in the account after the debits are deducted. Finally, we will use this amount and the amount of the deposit to find the balance in the account after all the changes. We will let $d =$ the total amount of the debits.

**Translate and Solve.** We are combining amounts.

| First debit | plus | Second debit | plus | Third debit | is | Total amount of debits |
|---|---|---|---|---|---|---|
| ↓ | ↓ | ↓ | ↓ | ↓ | ↓ | ↓ |
| 23.82 | + | 507.88 | + | 98.32 | = | $d$ |

To solve the equation we carry out the addition.

$$\begin{array}{r} {\scriptstyle 1\ 2\ \ 2\ 1} \\ 2\ 3.8\ 2 \\ 5\ 0\ 7.8\ 8 \\ +\quad 9\ 8.3\ 2 \\ \hline 6\ 3\ 0.0\ 2 \end{array}$$

Thus, $d = 630.02$.

Now we let $a =$ the amount in the account after the debits are deducted.

| Original amount | less | Debit amount | is | New amount |
|---|---|---|---|---|
| ↓ | ↓ | ↓ | ↓ | ↓ |
| 1123.56 | − | 630.02 | = | $a$ |

To solve the equation we carry out the subtraction.

$$\begin{array}{r} {\scriptstyle 10} \\ {\scriptstyle \not 0\ \ 12} \\ \not 1\ \not 1\ \not 2\ 3.5\ 6 \\ -\quad 6\ 3\ 0.0\ 2 \\ \hline 4\ 9\ 3.5\ 4 \end{array}$$

Thus, $a = 493.54$.

Finally, we let $f =$ the amount in the account after the check is deposited.

| Amount after debits | plus | Amount of deposit | is | Final amount |
|---|---|---|---|---|
| ↓ | ↓ | ↓ | ↓ | ↓ |
| 493.54 | + | 678.20 | = | $f$ |

We carry out the addition.

$$\begin{array}{r} {\scriptstyle 1\ 1} \\ 4\ 9\ 3.5\ 4 \\ +\ 6\ 7\ 8.2\ 0 \\ \hline 1\ 1\ 7\ 1.7\ 4 \end{array}$$

Thus, $f = 1171.74$.

**Check.** We repeat the calculations.

**State.** There is $1171.74 in the account after the changes.

33. **Familiarize.** We make and label a drawing. The question deals with a rectangle and a square, so we also list the relevant area formulas. We let $g =$ the area covered by grass.

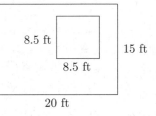

Area of a rectangle with length $l$ and width $w$:
$A = l \times w$

Area of a square with side $s$:  $A = s^2$

**Translate.** We subtract the area of the square from the area of the rectangle.

| Area of rectangle | minus | Area of square | is | Area covered by grass |
|---|---|---|---|---|
| ↓ | ↓ | ↓ | ↓ | ↓ |
| 20 × 15 | − | $(8.5)^2$ | = | $g$ |

**Solve.** We carry out the computations.

$$\begin{aligned} 20 \times 15 - (8.5)^2 &= g \\ 20 \times 15 - 72.25 &= g \\ 300 - 72.25 &= g \\ 227.75 &= g \end{aligned}$$

**Check.** We can repeat the calculations. Also note that 227.75 is less than the area of the yard but more than the area of the flower garden. This agrees with the impression given by our drawing.

***State.*** Grass covers 227.75 ft$^2$ of the yard.

**35. *Familiarize.*** The part of the at-bats that were hits is a fraction whose numerator is the number of hits and whose denominator is the number of at-bats. We let $a$ = the part of the at-bats that were hits.

***Translate.*** We think (Number of hits) $\div$ (Number of at-bats) = (Part of at-bats that were hits).

$$160 \div 439 = a$$

***Solve.*** We carry out the division.

```
        0.3 6 4 4
439 |1 6 0.0 0 0 0
     1 3 1 7
       2 8 3 0
       2 6 3 4
         1 9 6 0
         1 7 5 6
           2 0 4 0
           1 7 5 6
             2 8 4
```

We stop dividing at this point, because we will round to the nearest thousandth. Thus, $a \approx 0.364$.

***Check.*** We repeat the calculation. The answer checks.

***State.*** 0.364 of the at-bats were hits.

**37. *Familiarize.*** This is a three-step problem. We will find the area $S$ of a standard soccer field and the area $F$ of a standard football field using the formula Area = $l \cdot w$. Then we will find $E$, the amount by which the area of a soccer field exceeds the area of a football field.

***Translate and Solve.***

$$S = l \cdot w = 114.9 \times 74.4 = 8548.56$$
$$F = l \cdot w = 120 \times 53.3 = 6396$$

| Area of football field | plus | Excess area of soccer field | is | Area of soccer field |
|:---:|:---:|:---:|:---:|:---:|
| ↓ | ↓ | ↓ | ↓ | ↓ |
| 6396 | + | $E$ | = | 8548.56 |

To solve the equation we subtract 6396 on both sides.

$E = 8548.56 - 6396$
$E = 2152.56$

```
        4  14
    8 5̸ 4̸ 8.5 6
  − 6 3 9 6.0 0
    2 1 5 2.5 6
```

***Check.*** We can obtain a partial check by rounding and estimating:

$$114.9 \times 74.4 \approx 110 \times 75 = 8250 \approx 8548.56$$
$$120 \times 53.3 \approx 120 \times 50 = 6000 \approx 6396$$
$$8250 - 6000 = 2250 \approx 2152.56$$

***State.*** The area of a soccer field is 2152.56 sq yd greater than the area of a football field.

**39. *Familiarize.*** This is a two-step problem. First we find the number of eggs in 20 dozen (1 dozen = 12). We let $n$ represent this number.

***Translate and Solve.*** We think (Number of dozens) · (Number in a dozen) = (Number of eggs).

$$20 \cdot 12 = n$$
$$240 = n$$

Second, we find the cost $c$ of one egg. We think (Total cost) $\div$ (Number of eggs) = (Cost of one egg).

$$\$25.80 \div 240 = c$$

We carry out the division.

```
         0.1 0 7 5
240 |2 5.8 0 0 0
     2 4 0
       1 8 0 0
       1 6 8 0
         1 2 0 0
         1 2 0 0
               0
```

Thus, $c = 0.1075 \approx 0.108$ (rounded to the nearest tenth of a cent).

***Check.*** We repeat the calculations.

***State.*** Each egg cost about $0.108, or 10.8¢.

**41. *Familiarize.*** This is a multistep problem. First we find the number of overtime hours worked. Then we find the pay for the first 40 hours as well as the pay for the overtime hours. Finally we add these amounts to find the total pay. Let $h$ = the number of overtime hours worked.

***Translate and Solve.*** First we have a missing addend situation.

| First 40 hours | plus | Overtime hours | is | Total hours |
|:---:|:---:|:---:|:---:|:---:|
| ↓ | ↓ | ↓ | ↓ | ↓ |
| 40 | + | $h$ | = | 46 |

We subtract 40 on both sides of the equation.

$$40 + h = 46$$
$$40 + h - 40 = 46 - 40$$
$$h = 6$$

Now we multiply to find $p$, the amount of pay for the first 40 hours.

```
   1 8.5 0
 ×     4 0
 7 4 0.0 0
```

Thus, $p = 740$.

Next we multiply to find $a$, the additional pay for the overtime hours.

```
   2 7.7 5
 ×       6
 1 6 6.5 0
```

Then $a = 166.50$.

Finally we add to find $t$, the total pay.

```
   7 4 0.0 0
 + 1 6 6.5 0
   9 0 6.5 0
```

We have $t = 906.50$.

***Check.*** We repeat the calculations. The answer checks.

***State.*** The construction worker's pay was $906.50.

**43.** We add the population figures, keeping in mind that they are given in billions, and divide by the number of addends, 11.

$$(2.556 + 3.042 + 3.712 + 4.453 + 5.282 + 6.085 + 6.867 +$$

$$7.659 + 8.373 + 9.003 + 9.539) \div 11 = \frac{66.571}{11} \approx 6.052$$

The average population of the world for the years 1950 through 2050 will be about 6.052 billion.

**45. *Familiarize*.** We visualize the situation. We let $t =$ the number of degrees by which the temperature of the bath water exceeds normal body temperature.

| 98.6°F | $t$ |
|---|---|
| 100°F | |

***Translate*.** We have a missing addend situation.

$$\underbrace{\text{Normal body}}_{98.6} \underset{\text{plus}}{\text{temperature}} \quad \underbrace{\text{Additional}}_{t} \underset{\text{is}}{\text{degrees}} \quad \underbrace{\text{Bath water}}_{100} \text{temperature}$$

$$98.6 \quad + \quad t \quad = \quad 100$$

***Solve*.** To solve we subtract 98.6 on both sides of the equation.

$t = 100 - 98.6$
$t = 1.4$

$$\begin{array}{r} {\scriptstyle 9\ 9\ 10} \\ \cancel{1\,0\,0}.\cancel{0} \\ -\ \ 9\,8.\,6 \\ \hline 1.\,4 \end{array}$$

***Check*.** To check we add 1.4 to 98.6 to get 100. This checks.

***State*.** The temperature of the bath water is 1.4°F above normal body temperature.

**47. *Familiarize*.** We let $C =$ the cost of the home in Miami. Using the table in Example 8, find the indexes of Dallas and Miami.

***Translate*.** Using the formula given in Example 8, we translate to an equation.

$$C = \frac{\$125,000 \times 151}{72}$$

***Solve*.** We carry out the computation.

$$C = \frac{\$125,000 \times 151}{72}$$
$$= \frac{\$18,875,000}{72}$$
$$\approx \$262,153$$

***Check*.** We can repeat the computations. The answer checks.

***State*.** A home selling for $125,000 in Dallas would cost about $262,153 in Miami.

**49. *Familiarize*.** We let $C =$ the cost of the home in Juneau. Using the table in Example 8, find the indexes of Orlando and Juneau.

***Translate*.** Using the formula given in Example 8, we translate to an equation.

$$C = \frac{\$96,000 \times 96}{110}$$

***Solve*.** We carry out the computation.

$$C = \frac{\$96,000 \times 96}{110}$$
$$= \frac{\$9,216,000}{110}$$
$$\approx \$83,782$$

***Check*.** We can repeat the computations. The answer checks.

***State*.** A home selling for $96,000 in Orlando would cost about $83,782 in Juneau.

**51. *Familiarize*.** We let $C =$ the cost of the home in Boston. Using the table in Example 8, find the indexes of Louisville and Boston.

***Translate*.** Using the formula given in Example 8, we translate to an equation.

$$C = \frac{\$240,000 \times 327}{56}$$

***Solve*.** We carry out the computation.

$$C = \frac{\$240,000 \times 327}{56}$$
$$= \frac{\$78,480,000}{56}$$
$$\approx \$1,401,429$$

***Check*.** We can repeat the computations. The answer checks.

***State*.** A home selling for $240,000 in Louisville would cost about $1,401,429 in Boston.

**53. *Familiarize*.** We let $C =$ the cost of the home in Tulsa. Using the table in Example 8, find the indexes of Chicago and Tulsa.

***Translate*.** Using the formula given in Example 8, we translate to an equation.

$$C = \frac{\$140,000 \times 36}{173}$$

***Solve*.** We carry out the computation.

$$C = \frac{\$140,000 \times 36}{173}$$
$$= \frac{\$5,040,000}{173}$$
$$\approx \$29,133$$

***Check*.** We can repeat the computations. The answer checks.

***State*.** A house selling for $140,000 in Chicago would cost about $29,133 in Tulsa.

**55. *Familiarize*.** This is a multistep problem. First we find the number of minutes in excess of 450. Then we find the charge for the excess minutes. Finally we add this charge to the monthly charge for 450 minutes to find the total cost for the month. Let $m =$ the number of minutes in excess of 450.

**Translate and Solve**.

| First 450 minutes | plus | Excess minutes | is | Total minutes |
|:---:|:---:|:---:|:---:|:---:|
| ↓ | ↓ | ↓ | ↓ | ↓ |
| 450 | + | $m$ | = | 479 |

We subtract 450 on both sides of the equation.

$$450 + m = 479$$
$$450 + m - 450 = 479 - 450$$
$$m = 29$$

We see that 29 minutes are charged at the rate of $0.45 per minute. We multiply to find $c$, the cost of these minutes.

$$\begin{array}{r} 29 \\ \times\,0.4\,5 \\ \hline 1\,4\,5 \\ 1\,1\,6\,0 \\ \hline 1\,3.0\,5 \end{array}$$

Thus, $c = 13.05$.

Finally we add the cost of the first 450 minutes and the cost of the additional 29 minutes to find $t$, the total cost for the month.

$$\begin{array}{r} 3\,9.9\,9 \\ +\,1\,3.0\,5 \\ \hline 5\,3.0\,4 \end{array}$$

Thus, $t = 53.04$.

**Check**. We can repeat the calculations. The answer checks.

**State**. The total cost for the month was $53.04.

**57. Familiarize**. Let $d$ = the deduction, in dollars, that Sheila was allowed to take. We will express 50.5¢ as $0.505 and 58.5¢ as $0.585.

**Translate**.

| Miles driven Jan. 1-June 30 | times | Deduction per mile | plus |
|:---:|:---:|:---:|:---:|
| ↓ | ↓ | ↓ | ↓ |
| 3156 | × | 0.505 | + |

| Miles driven July 1-Dec. 31 | times | Deduction per mile | is | Total deduction |
|:---:|:---:|:---:|:---:|:---:|
| ↓ | ↓ | ↓ | ↓ | ↓ |
| 3678 | × | 0.585 | = | $d$ |

**Solve**. We carry out the computation using the rules for order of operations.

$$3156 \times 0.505 + 3678 \times 0.585$$
$$= 1593.78 + 2151.63$$
$$= 3745.41$$

Thus, $3745.41 = d$.

**Check**. We repeat the calculations. The answer checks.

**State**. Sheila was allowed to take a $3745.41 deduction.

**59. Familiarize**. This is a two-step problem. First we will find the number of 1000s in 165,500. Then we will find the amount paid in property taxes.

**Translate and Solve**. We divide to find the number of 1000s in 165,500.

$$\frac{165,500}{1000} = 165.5 \quad \text{(Moving the decimal pont 3 places to the right.)}$$

Now let $p$ = the amount paid in property tax.

| Number of $1000s of assessed value | times | Property taxes per $1000 | is | Property taxes paid |
|:---:|:---:|:---:|:---:|:---:|
| ↓ | ↓ | ↓ | ↓ | ↓ |
| 165.5 | × | 8.50 | = | $p$ |

We carry out the multiplication to solve the equation.

$$\begin{array}{r} 1\,6\,5.5 \\ \times\quad 8.\,5\,0 \\ \hline 8\,2\,7\,5\,0 \\ 1\,3\,2\,4\,0\,0\,0 \\ \hline 1\,4\,0\,6.7\,5\,0 \end{array}$$

Thus, $1406.75 = p$.

**Check**. We repeat the calculations. The answer checks.

**State**. The Brunners pay $1406.75 in property taxes each year.

**61.**
$$\begin{array}{r} {\scriptstyle 1\ 1\ 1} \\ 4\,5\,6\,9 \\ +\,1\,7\,6\,6 \\ \hline 6\,3\,3\,5 \end{array}$$

**63.**
$$4\,\boxed{\frac{1}{3} \cdot \frac{2}{2}} = \quad 4\frac{2}{6}$$
$$+\,2\,\boxed{\frac{1}{2} \cdot \frac{3}{3}} = +\,2\frac{3}{6}$$
$$\rule{3cm}{0.4pt}$$
$$6\frac{5}{6}$$

**65.**
$$\frac{2}{3} - \frac{5}{8} = \frac{2}{3} \cdot \frac{8}{8} - \frac{5}{8} \cdot \frac{3}{3}$$
$$= \frac{16}{24} - \frac{15}{24} = \frac{16-15}{24}$$
$$= \frac{1}{24}$$

**67.**
$$-\frac{5}{6} - \frac{7}{10} = -\frac{5}{6} \cdot \frac{5}{5} - \frac{7}{10} \cdot \frac{3}{3}$$
$$= -\frac{25}{30} - \frac{21}{30}$$
$$= -\frac{46}{30}$$
$$= -\frac{2 \cdot 23}{2 \cdot 15} = \frac{2}{2} \cdot \left(-\frac{23}{15}\right)$$
$$= -\frac{23}{15}$$

**69.**
$$\frac{3225}{6275} = \frac{25 \cdot 129}{25 \cdot 251} = \frac{25}{25} \cdot \frac{129}{251} = \frac{129}{251}$$

**71.**
$$-\frac{325}{625} = -\frac{25 \cdot 13}{25 \cdot 25} = \frac{25}{25} \cdot \left(-\frac{13}{25}\right) = -\frac{13}{25}$$

**73. Familiarize.** Visualize the situation as a rectangular array containing 469 revolutions with $16\frac{3}{4}$ revolutions in each row. We must determine how many rows the array has. (The last row may be incomplete.) We let $t$ = the time the wheel rotates.

**Translate.** The division that corresponds to the situation is

$$469 \div 16\frac{3}{4} = t.$$

**Solve.** We carry out the division.

$$t = 469 \div 16\frac{3}{4} = 469 \div \frac{67}{4} = 469 \cdot \frac{4}{67} =$$

$$\frac{67 \cdot 7 \cdot 4}{67 \cdot 1} = \frac{67}{67} \cdot \frac{7 \cdot 4}{1} = 28$$

**Check.** We check by multiplying the time by the number of revolutions per minute.

$$16\frac{3}{4} \cdot 28 = \frac{67}{4} \cdot 28 = \frac{67 \cdot 7 \cdot 4}{4 \cdot 1} = \frac{4}{4} \cdot \frac{67 \cdot 7}{1} = 469$$

The answer checks.

**State.** The water wheel rotated for 28 min.

**75. Familiarize.** Let $c$ = the number of calories by which a piece of pecan pie exceeds a piece of pumpkin pie.

**Translate.** This is a "how much more" situation.

$$\underbrace{\text{Pumpkin pie calories}}_{316} \quad \underbrace{\text{plus}}_{+} \quad \underbrace{\text{Excess pecan pie calories}}_{c} \quad \underbrace{\text{is}}_{=} \quad \underbrace{\text{Pecan pie calories}}_{502}$$

**Solve.** We subtract 316 on both sides of the equation.

$$316 + c = 502$$
$$316 + c - 316 = 502 - 316$$
$$c = 186$$

**Check.** $316 + 186 = 502$, so the answer checks.

**State.** A piece of pecan pie has 186 calories more than a piece of pumpkin pie.

**77. Familiarize.** This is a multistep problem. First we will find the total number of cards purchased. Then we will find the number of half-dozens in this number. Finally, we will find the purchase price. Let $t$ = the total number of cards, $h$ = the number of half-dozens, and $p$ = the purchase price. Recall that 1 dozen = 12, so $\frac{1}{2}$ dozen = $\frac{1}{2} \cdot 12 = 6$.

**Translate.** We write an equation to find the total number of cards purchased.

$$\underbrace{\text{Number of packs}}_{6} \quad \underbrace{\times}_{\times} \quad \underbrace{\text{Number per pack}}_{12} \quad \underbrace{=}_{=} \quad \underbrace{\text{Total number of cards}}_{t}$$

**Solve.** We carry out the multiplication.

$$t = 6 \times 12 = 72$$

Next we divide by 6 to find the number of half-dozens in 72:

$$h = 72 \div 6 = 12$$

Finally, we multiply the number of half-dozens by the price per half dozen to find the purchase price. Twelve dozen cents = $12 \cdot 12\cent = 144\cent = \$1.44$.

$$p = 12 \times \$1.44 = \$17.28$$

**Check.** We can repeat the calculations. The answer checks.

**State.** The purchase price of the cards is $17.28.

## Chapter 5 Concept Reinforcement

1. One thousand billions $= 1000 \times 1,000,000,000$
$$= 1,000,000,000,000$$
$$= \text{one trillion}$$
The statement is true.

2. The number of decimal places in the product of two number is the *sum* of the number of decimal places in the factors. The given statement is false.

3. Dividing a positive number by 0.1, 0.01, 0.001, and so on is equivalent to multiplying the number by 10, 100, 1000, and so on, so the quotient is larger than the dividend. The given statement is true.

4. A terminating decimal occurs only when the denominator of a fraction has only 2's or 5's, or both, as factors. The given statement is false.

5. An estimate found by rounding to the nearest ten uses numbers that are closer to the actual numbers in the calculation than one found by rounding to the nearest hundred, so an estimate found by rounding to the nearest ten is more accurate. The given statement is true.

## Chapter 5 Important Concepts

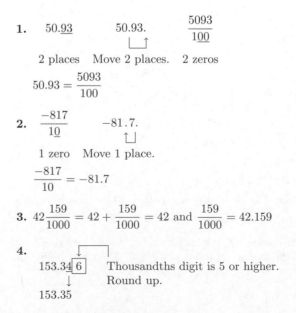

1. $50.9\underline{3}$ $\qquad$ $50.93.$ $\qquad$ $\dfrac{5093}{100}$

2 places $\quad$ Move 2 places. $\quad$ 2 zeros

$$50.93 = \frac{5093}{100}$$

2. $\dfrac{-817}{10}$ $\qquad$ $-81.7.$

1 zero $\quad$ Move 1 place.

$$\frac{-817}{10} = -81.7$$

3. $42\dfrac{159}{1000} = 42 + \dfrac{159}{1000} = 42 \text{ and } \dfrac{159}{1000} = 42.159$

4.

$$153.34\boxed{6} \qquad \text{Thousandths digit is 5 or higher.}$$
$$\downarrow \qquad\qquad \text{Round up.}$$
$$153.35$$

**5.** Line up the decimal points. Add an extra zero if desired.

$$
\begin{array}{r}
\overset{1}{\phantom{0}}\phantom{0} \\
5.5\,4\,0 \\
+\ 3\,3.0\,7\,1 \\
\hline
3\,8.6\,1\,1
\end{array}
$$

**6.** Line up the decimal points. Add an extra zero if desired.

$$
\begin{array}{r}
\phantom{00}{}^{10}\phantom{0}{}^{13} \\
{}^{1}\cancel{0}\ {}^{9}\cancel{3}\ 10 \\
2\,\cancel{2}\,\cancel{1}.\cancel{0}\,\cancel{4}\,\cancel{0} \\
-\ \ \ 1\,3.1\,9\,2 \\
\hline
2\,0\,7.8\,4\,8
\end{array}
$$

**7.**
$$
\begin{array}{r}
5.\,4\,6 \quad \text{(2 decimal places)}\\
\times\ \ \ \ \ 3.\,5 \quad \text{(1 decimal place)}\\
\hline
2\,7\,3\,0\\
1\,6\,3\,8\,0\\
\hline
1\,9.\,1\,1\,0
\end{array}
$$

**8.** $-17.6 \times 0.\underline{01}$ \qquad $-0.17.6$

2 decimal places \qquad Move 2 places to the left.

$-17.6 \times 0.01 = -0.176$

**9.** $\underline{1000} \times 60.437$ \qquad $60.437.$

3 zeros \qquad Move 3 places to the right.

$1000 \times 60.437 = 60,437$

**10.**
$$
\begin{array}{r}
7.4\phantom{0} \\
3.6_\wedge\overline{)\,2\,6\,6._\wedge4\,} \\
2\,5\,2\phantom{00} \\
\hline
1\,4\ \,4\phantom{0} \\
1\,4\ \,4\phantom{0} \\
\hline
0\phantom{0}
\end{array}
$$

**11.** $\dfrac{4.7}{\underline{100}}$ \qquad $0.04.7$

2 zeros \qquad Move 2 places to the left.

$\dfrac{4.7}{100} = 0.047$

**12.** $\dfrac{-156.9}{0.\underline{01}}$ \qquad $-156.90.$

2 decimal places \qquad Move 2 places to the right.

$\dfrac{-156.9}{0.01} = -15,690$

## Chapter 5 Review Exercises

**1.** 6.59 million $= 6.59 \times 1,\underbrace{000,000}_{6 \text{ zeros}}$

6.590000.

Move 6 places to the right.

6.59 million $= 6,590,000$

**2.** 3.1 billion $= 3.1 \times 1,\underbrace{000,000,000}_{9 \text{ zeros}}$

3.100000000.

Move 9 places to the right.

3.1 billion $= 3,100,000,000$

**3.** A word name for 3.47 is three and forty-seven hundredths.

**4.** A word name for 0.031 is thirty-one thousandths.

**5.** A word name for $-13.4$ is negative thirteen and four tenths.

**6.** A word name for 0.0007 is seven ten-thousandths.

**7.** $0.\underline{09}$ \qquad $0.09.$ \qquad $\dfrac{9}{100}$

2 places \quad Move 2 places. \quad 2 zeros

$0.09 = \dfrac{9}{100}$

**8.** $4.\underline{561}$ \qquad $4.561.$ \qquad $\dfrac{4561}{1000}$

3 places \quad Move 3 places. \quad 3 zeros

$4.561 = \dfrac{4561}{1000}$

**9.** $-0.\underline{089}$ \qquad $-0.089.$ \qquad $-\dfrac{89}{1000}$

3 places \quad Move 3 places. \quad 3 zeros

$-0.089 = -\dfrac{89}{1000}$

**10.** $-3.\underline{0227}$ \qquad $-3.0227.$ \qquad $\dfrac{-30,227}{10,000}$

4 places \quad Move 4 places. \quad 4 zeros

$-3.0227 = \dfrac{-30,227}{10,000}$, or $-\dfrac{30,227}{10,000}$

**11.** $\dfrac{34}{\underline{1000}}$ \qquad $0.034.$

3 zeros \quad Move 3 places.

$\dfrac{34}{1000} = 0.034$

**12.** $-\dfrac{42,603}{\underline{10,000}}$ \qquad $-4.2603.$

4 zeros \quad Move 4 places.

$-\dfrac{42,603}{10,000} = -4.2603$

**13.** $27\dfrac{91}{100} = 27 + \dfrac{91}{100} = 27$ and $\dfrac{91}{100} = 27.91$

**14.** First we consider $867\dfrac{6}{1000}$.

$867\dfrac{6}{1000} = 867 + \dfrac{6}{1000} = 867$ and $\dfrac{6}{1000} = 867.006$

Then $-867\dfrac{6}{1000} = -867.006$.

**15.** 0.034

Starting at the left, these digits are
the first to differ; 3 is larger than 1.

0.0185

Thus, 0.034 is larger.

**16.** 0.91

Starting at the left, these digits are
the first to differ; 9 is larger than 1.

0.19

Thus, 0.91 is larger.

**17.** 0.741

Starting at the left, these digits are
the first to differ; 7 is larger than 6.

0.6943

Thus, 0.741 is larger.

**18.** −1.038

Starting at the left, these digits are the
first to differ, and 3 is smaller than 4.

−1.041

Thus, −1.038 is larger.

**19.**

17.4⎡2⎤87    Hundredths digit is 4 or lower.
    ↓        Round down.
17.4

**20.**

17.42⎡8⎤7    Thousandths digit is 5 or higher.
    ↓        Round up.
17.43

**21.**

17.428⎡7⎤    Ten-thousandths digit is 5 or higher.
    ↓        Round up.
17.429

**22.**

17.⎡4⎤287    Tenths digit is 4 or lower.
   ↓         Round down.
17

**23.**
$$
\begin{array}{r}
\phantom{+}\overset{1}{\phantom{0}}\phantom{5}\overset{1}{\phantom{3}}\phantom{71} \\
2.0\,4\,8 \\
6\,5.3\,7\,1 \\
+\,5\,0\,7.1 \\
\hline
5\,7\,4.5\,1\,9
\end{array}
$$

**24.**
$$
\begin{array}{r}
\overset{1}{\phantom{0}} \\
0.6 \\
0.0\,0\,4 \\
0.0\,7 \\
+\,0.0\,0\,9\,8 \\
\hline
0.6\,8\,3\,8
\end{array}
$$

**25.** −219.3 + 2.8 + 7 = −216.5 + 7 = −209.5

**26.**
$$
\begin{array}{r}
0.4\,1\,0 \\
4.1\,0\,0 \\
4\,1.0\,0\,0 \\
+\phantom{4}0.0\,4\,1 \\
\hline
4\,5.5\,5\,1
\end{array}
$$

**27.**
$$
\begin{array}{r}
\overset{2}{3}\overset{9}{0}.\overset{9}{0}\,\overset{9}{0}\,\overset{9}{0}\,\overset{10}{\cancel{0}} \\
-\phantom{3}0.\,7\,9\,0\,8 \\
\hline
2\,9.\,2\,0\,9\,2
\end{array}
$$

**28.**
$$
\begin{array}{r}
\overset{7}{\cancel{8}}\,\overset{14}{\cancel{4}}\,\overset{4}{5}.\overset{9}{\cancel{0}}\,\overset{18}{\cancel{8}} \\
-\phantom{84}5\,4.\,7\,9 \\
\hline
7\,9\,0.\,2\,9
\end{array}
$$

**29.** 37.645 − (−8.497) = 37.645 + 8.497
$$
\begin{array}{r}
\overset{1}{\phantom{3}}\,\overset{1}{7}.\overset{1}{6}\,\overset{1}{4}\,5 \\
+\phantom{3}8.4\,9\,7 \\
\hline
4\,6.1\,4\,2
\end{array}
$$
The answer is 46.142.

**30.** −70.8 − 0.0109 = −70.8 + (−0.0109)

First we add absolute values.
$$
\begin{array}{r}
7\,0.8\,0\,0\,0 \quad \text{Adding 3 zeros} \\
+\phantom{70}0.0\,1\,0\,9 \\
\hline
7\,0.8\,1\,0\,9
\end{array}
$$

The sum of two negative numbers is negative, so the answer
is −70.8109.

**31.**
$$
\begin{array}{r}
4\,8 \\
\times\,0.\,2\,7 \\
\hline
3\,3\,6 \\
9\,6\,0 \\
\hline
1\,2.9\,6
\end{array}
$$

**32.** −0.174 · (−0.83)

We multiply the absolute values. The answer will be positive.
$$
\begin{array}{r}
0.\,1\,7\,4 \\
\times\,0.\,8\,3 \\
\hline
5\,2\,2 \\
1\,3\,9\,2\,0 \\
\hline
0.\,1\,4\,4\,4\,2
\end{array}
$$

**33.** 100 × 0.043

0.04.3
  ⌐→

2 zeros         Move 2 places to the right.

100 × 0.043 = 4.3

**34.** 0.001 × −24.68

−0.024.68
    ←⌐

3 decimal places    Move 3 places to the left.

0.001 × −24.68 = −0.02468

**35.**
$$
\begin{array}{r}
0.4\,5 \\
5\,2\,\overline{)\,2\,3.4\,0} \\
2\,0\,8\phantom{.0} \\
\hline
2\,6\,0\phantom{.} \\
2\,6\,0\phantom{.} \\
\hline
0\phantom{.}
\end{array}
$$

**36.**

$$
\begin{array}{r}
45.2 \\
2.6_\wedge \overline{)117.5_\wedge 2} \\
\underline{104\phantom{.5\wedge2}} \\
135 \\
\underline{130} \\
5\,2 \\
\underline{5\,2} \\
0
\end{array}
$$

**37.**

$$
\begin{array}{r}
1.022 \\
2.1\,4_\wedge \overline{)2.1\,8_\wedge 7\,0\,8} \\
\underline{2\,1\,4\phantom{08}} \\
4\,7\,0 \\
\underline{4\,2\,8} \\
4\,2\,8 \\
\underline{4\,2\,8} \\
0
\end{array}
$$

**38.** $-60 \div 8$

First we consider $60 \div 8$.

$$
\begin{array}{r}
7.5 \\
8\,\overline{)6\,0.0} \\
\underline{5\,6} \\
4\,0 \\
\underline{4\,0} \\
0
\end{array}
$$

$60 \div 8 = 7.5$, so $-60 \div 8 = -7.5$.

**39.** $\dfrac{276.3}{1000}$      $0.276.3$

3 zeros          Move 3 places to the left.

$$\frac{276.3}{1000} = 0.2763$$

**40.** $\dfrac{-13.892}{0.01}$      $-13.89.2$

2 decimal places     Move 2 places to the right.

$$\frac{-13.892}{0.01} = -1389.2$$

**41.**
$$x + 51.748 = 548.0275$$
$$x + 51.748 - 51.748 = 548.0275 - 51.748$$
$$x = 496.2795$$

The solution is 496.2795.

**42.**
$$3 \cdot x = -20.85$$
$$\frac{3 \cdot x}{3} = \frac{-20.85}{3}$$
$$x = -6.95$$

The solution is $-6.95$.

**43.**
$$10 \cdot y = 425.4$$
$$\frac{10 \cdot y}{10} = \frac{425.4}{10}$$
$$y = 42.54$$

The solution is 42.54.

**44.**
$$0.0089 + y = 5$$
$$0.0089 + y - 0.0089 = 5 - 0.0089$$
$$y = 4.9911$$

**45.** *Familiarize.* Let $h$ = Stacia's hourly wage.

*Translate.*

| Hourly wage | times | Number of hours worked | is | Total earnings |
|---|---|---|---|---|
| ↓ | ↓ | ↓ | ↓ | ↓ |
| $h$ | $\cdot$ | 40 | $=$ | 620.80 |

*Solve.*
$$h \cdot 40 = 620.80$$
$$\frac{h \cdot 40}{40} = \frac{620.80}{40}$$
$$h = 15.52$$

*Check.* $40 \cdot \$15.52 = \$620.80$, so the answer checks.

*State.* Stacia earns $15.52 per hour.

**46.** *Familiarize.* Let $d$ = the average consumption of fruits and vegetables per person in a day, in pounds.

*Translate.*

| Average consumption per day | times | Number of days in a year | is | Average consumption in a year |
|---|---|---|---|---|
| ↓ | ↓ | ↓ | ↓ | ↓ |
| $d$ | $\cdot$ | 365 | $=$ | 683.6 |

*Solve.*
$$d \cdot 365 = 683.6$$
$$\frac{d \cdot 365}{.365} = \frac{683.6}{365}$$
$$d \approx 1.9$$

*Check.* $1.9 \times 365 = 693.5 \approx 683.6$, so the answer checks. (Remember, we rounded the solution of the equation.)

*State.* The average person eats about 1.9 lb of fruits and vegetables in one day.

**47.** *Familiarize.* Let $a$ = the amount left in the account after the purchase was made.

*Translate.* We write a subtraction sentence.
$$a = 1034.46 - 249.99$$

*Solve.* We carry out the subtraction.

$$
\begin{array}{r}
\phantom{0}\,12\ 13\ 13 \\
9\ \not2\ \not3\ \not3\ 16 \\
\not1\not0\ \not3\,4.\,\not4\ \not6 \\
-\ \ 2\ 4\,9.\,9\ 9 \\
\hline
7\ 8\ 4.\,4\ 7
\end{array}
$$

Thus, $a = 784.47$.

*Check.* $\$784.47 + \$249.99 = \$1034.46$, so the answer checks.

*State.* There is $784.47 left in the account.

**48.** *Familiarize.* This is a multistep problem. First we find the number of minutes in excess of 700. Then we find the charge for the excess minutes. Finally we add this charge to the monthly charge for 700 minutes to find the total cost for the month. Let $m$ = the number of minutes in excess of 700.

**Translate and Solve.**

First 700 minutes  plus  Excess minutes  is  Total minutes

$$700 + m = 925$$

We subtract 700 on both sides of the equation.

$$700 + m = 925$$
$$700 + m - 700 = 925 - 700$$
$$m = 225$$

We see that 225 minutes are charged at the rate of $0.45 per minute. We multiply to find $c$, the cost of these minutes.

```
    2 2 5
  × 0. 4 5
  1 1 2 5
  9 0 0 0
1 0 1. 2 5
```

Thus, $c = 101.25$.

Finally we add the cost of the first 700 minutes and the cost of the additional 225 minutes to find $t$, the total cost for the month.

```
  6 9. 9 9
+ 1 0 1. 2 5
  1 7 1. 2 4
```

Thus, $t = 171.24$.

**Check.** We can repeat the calculations. The answer checks.

**State.** The total cost for the month was $171.24.

**49. Familiarize.** This is a two-step problem. First, we find the number of miles that have been driven between fillups. This is a "how-much-more" situation. We let $n =$ the number of miles driven.

**Translate and Solve.**

First odometer reading  plus  Number of miles driven  is  Second odometer reading

$$36,057.1 + n = 36,217.6$$

To solve the equation we subtract 36,057.1 on both sides.

$$n = 36,217.6 - 36,057.1$$
$$n = 160.5$$

```
  3 6, 2 1 7.6
− 3 6, 0 5 7.1
      1 6 0.5
```

Second, we divide the total number of miles driven by the number of gallons. This gives us $m =$ the number of miles per gallon.

$$160.5 \div 11.1 = m$$

To find the number $m$, we divide.

```
        1 4.4 5
11.1∧| 1 6 0. 5∧0 0
      1 1 1
        4 9 5
        4 4 4
          5 1 0
          4 4 4
            6 6 0
            5 5 5
            1 0 5
```

Thus, $m \approx 14.5$.

**Check.** To check, we first multiply the number of miles per gallon times the number of gallons:

$$11.1 \times 14.5 = 160.95$$

Then we add 160.95 to 36,057.1:

$$36,057.1 + 160.95 = 36,218.05 \approx 36,217.6$$

The number 14.5 checks.

**State.** The driver gets 14.5 miles per gallon.

**50.** a) **Familiarize.** Let $s =$ the total consumption of seafood per person, in pounds, for the seven given years.

**Translate.** We add the seven amounts shown in the graph in the text.

$$s = 12.4 + 14.9 + 14.8 + 15.2 + 16.3 + 16.5 + 16.1$$

**Solve.** We carry out the addition.

```
   3 3
   1 2. 4
   1 4. 9
   1 4. 8
   1 5. 2
   1 6. 3
   1 6. 5
 + 1 6. 1
 1 0 6. 2
```

**Check.** We repeat the calculation. The answer checks.

**State.** The total consumption of seafood per person for the seven given years was 106.2 lb.

b) We add the amounts and divide by the number of addends. From part (a) we know that the sum of the seven numbers is 106.2, so we have $106.2 \div 7$:

```
      1 5. 1 7
7 | 1 0 6. 2 0
    7
    3 6
    3 5
    1 2
      7
      5 0
      4 9
       1
```

Rounding to the nearest tenth, we find that the average annual seafood consumption per person was about 15.2 lb.

**51.** $7.82 \times 34.487 \approx 8 \times 34 = 272$

**52.** $219.875 - 4.478 \approx 220 - 4 = 216$

**53.** $\$45.78 + \$78.99 \approx \$46 + \$79 = \$125$

**54.** $\dfrac{13}{5} = \dfrac{13}{5} \cdot \dfrac{2}{2} = \dfrac{26}{10} = 2.6$

**55.** $-\dfrac{32}{25} = -\dfrac{32}{25} \cdot \dfrac{4}{4} = -\dfrac{128}{100} = -1.28$

**56.** $\dfrac{11}{4} = \dfrac{11}{4} \cdot \dfrac{25}{25} = \dfrac{275}{100} = 2.75$

**57.** First we consider $\dfrac{13}{4}$.

$$
\begin{array}{r}
3.25 \\
4\overline{\smash{)}13.00} \\
\underline{12}\phantom{.00} \\
10\phantom{0} \\
\underline{8}\phantom{0} \\
20 \\
\underline{20} \\
0
\end{array}
$$

$\dfrac{13}{4} = 3.25$, so $-\dfrac{13}{4} = -3.25$.

**58.**
$$
\begin{array}{r}
1.166 \\
6\overline{\smash{)}7.000} \\
\underline{6}\phantom{.000} \\
10 \\
\underline{6} \\
40 \\
\underline{36} \\
40 \\
\underline{36} \\
4
\end{array}
$$

Since 4 keeps reappearing as a remainder, the digits repeat and

$$\frac{7}{6} = 1.166\ldots, \text{ or } 1.1\overline{6}.$$

**59.**
$$
\begin{array}{r}
1.54 \\
11\overline{\smash{)}17.00} \\
\underline{11}\phantom{.00} \\
60 \\
\underline{55} \\
50 \\
\underline{44} \\
6
\end{array}
$$

Since 6 reappears as a remainder, the sequence repeats and

$$\frac{17}{11} = 1.5454\ldots, \text{ or } 1.\overline{54}.$$

**60.** Round 1. 5 $\underline{5}$ $\boxed{4}$ 5 4 $\ldots$ to the nearest tenth.

          Hundredths digit is 4 or lower.

    1. 5         Round down.

**61.** Round 1. 5 4 $\boxed{5}$ 4 $\ldots$ to the nearest hundredth.

          Thousandths digit is 5 or higher.

    1. 5 5         Round up.

**62.** Round 1. 5 4 5 $\underline{5}$ $\boxed{4}$ $\ldots$ to the nearest thousandth.

          Ten-thousandths digit is 4 or lower.

    1. 5 4 5         Round down.

**63.** Move 2 places to the left.

$\$82.73.\cancel{c}$

Change from $\cancel{c}$ sign at end to $ sign in front.

$8273\cancel{c} = \$82.73$

**64.** Move 2 places to the left.

$\$4.87.\cancel{c}$

Change from $\cancel{c}$ sign at end to $ sign in front.

$487\cancel{c} = \$4.87$

**65.** Move 2 places to the right.

$\$24.93.\cancel{c}$

Change from $ sign in front to $\cancel{c}$ sign at end.

$\$24.93 = 2493\cancel{c}$

**66.** Move 2 places to the right.

$\$9.86.\cancel{c}$

Change from $ sign in front to $\cancel{c}$ sign at end.

$\$9.86 = 986\cancel{c}$

**67.**
$$(8 - 1.23) \div (-4) + 5.6 \times 0.02$$
$$= 6.77 \div (-4) + 5.6 \times 0.02$$
$$= -1.6925 + 0.112$$
$$= -1.5805$$

**68.**
$$(1 + 0.07)^2 + 10^3 \div 10^2 + [4(10.1 - 5.6) + 8(11.3 - 7.8)]$$
$$= (1.07)^2 + 10^3 \div 10^2 + [4(4.5) + 8(3.5)]$$
$$= (1.07)^2 + 10^3 \div 10^2 + [18 + 28]$$
$$= (1.07)^2 + 10^3 \div 10^2 + 46$$
$$= 1.1449 + 1000 \div 100 + 46$$
$$= 1.1449 + 10 + 46$$
$$= 11.1449 + 46$$
$$= 57.1449$$

**69.** $\dfrac{3}{4} \times (-20.85) = 0.75 \times (-20.85) = -15.6375$

**70.** $\dfrac{346.295}{0.001}$　　　　　　　　346.295.

3 decimal places　　Move 3 places to the right.

$\dfrac{346.295}{0.001} = 346{,}295$

Answer D is correct.

**71.** $82.304 \div 17.287 \approx 80 \div 20 = 4$

Answer B is correct.

**72.** a) By trial we find the following true sentence:

$$2.56 \times 6.4 \div 51.2 - 17.4 + 89.7 = 72.62.$$

b) By trial we find the following true sentence:

$$(11.12 - 0.29) \times 3^4 = 877.23$$

**73.** $1 = 3 \cdot \dfrac{1}{3} = 3(0.3333\ldots) = 0.9999\cdots$, or $0.\overline{9}$

## Chapter 5 Discussion and Writing Exercises

**1.** Count the number of decimal places. Move the decimal point that many places to the right and write the result over a denominator of 1 followed by that many zeros.

**2.** $346.708 \times 0.1 = \dfrac{346{,}708}{1000} \times \dfrac{1}{10} = \dfrac{346{,}708}{10{,}000} =$

$34.6708 \neq 3467.08$

**3.** When the denominator of a fraction is a multiple of 10, long division is not the fastest way to convert the fraction to decimal notation. Many times when the denominator is a factor of some multiple of 10 this is also the case. The latter situation occurs when the denominator has only 2's or 5's or both as factors.

**4.** Multiply by 1 to get a denominator that is a power of 10:

$$\dfrac{44}{125} = \dfrac{44}{125} \cdot \dfrac{8}{8} = \dfrac{352}{1000} = 0.352.$$

We can also divide to find that $\dfrac{44}{125} = 0.352.$

## Chapter 5 Test

**1.**　　8.9 billion

$= 8.9 \times 1$ billion

$= 8.9 \times 1{,}000{,}000{,}000$　　9 zeros

$= 8{,}900{,}000{,}000$　　Moving the decimal point 9 places to the right

**2.**　　3.756 million

$= 3.756 \times 1$ million

$= 3.756 \times 1{,}000{,}000$　　6 zeros

$= 3{,}756{,}000$　　Moving the decimal point 6 places to the right

**3.** 2.34

a) Write a word name for the whole number.　　　$\boxed{\text{Two}}$

b) Write "and" for the decimal point.　　　Two

$\boxed{\text{and}}$

c) Write a word name for the number to the right　　Two
of the decimal point,　　　and
followed by the place　　$\boxed{\text{thirty-four hundredths}}$
value of the last digit.

A word name for 2.34 is two and thirty-four hundredths.

**4.** 105.0005

a) Write a word name for the whole number.　　$\boxed{\text{One hundred five}}$

b) Write "and" for the decimal point.　　One hundred five

$\boxed{\text{and}}$

c) Write a word name for the number to the right　　One hundred five
of the decimal point,　　　and
followed by the place　　$\boxed{\text{five ten-thousandths}}$
value of the last digit.

A word name for 105.0005 is one hundred five and five ten-thousandths.

**5.**　$0.\underline{91}$　　　$0.91.$　　　$\dfrac{91}{100}$

2 places　Move 2 places.　2 zeros

$0.91 = \dfrac{91}{100}$

**6.**　$-2.\underline{769}$　　$-2.769.$　　$-\dfrac{2769}{1000}$

3 places　Move 3 places.　3 zeros

$-2.769 = -\dfrac{2769}{1000}$

**7.**　$\dfrac{74}{1\underline{000}}$　　　$0.074.$

3 zeros　Move 3 places.

$\dfrac{74}{1000} = 0.074$

**8.**　$-\dfrac{37{,}047}{10{,}000}$　　$-3.7047.$

4 zeros　Move 4 places.

$-\dfrac{37{,}047}{10{,}000} = -3.7047$

**9.** $756\dfrac{9}{100} = 756 + \dfrac{9}{100} = 756 \text{ and } \dfrac{9}{100} = 756.09$

**10.** First consider $91\frac{703}{1000}$.

$$91\frac{703}{1000} = 91 + \frac{703}{1000} = 91 \text{ and } \frac{703}{1000} = 91.703$$

Then $-91\frac{703}{1000} = -91.703$.

**11.** To compare two numbers in decimal notation, start at the left and compare corresponding digits moving from left to right. When two digits differ, the number with the larger digit is the larger of the two numbers.

0.07

↑ Different; 1 is larger than 0.

0.162

Thus, 0.162 is larger.

**12.** −0.078

↑ Starting at the left, these digits are the first to differ; 6 is smaller than 7.

−0.06

Thus, −0.06 is larger.

**13.** 0.09

↑ Different; 9 is larger than 0.

0.9

Thus, 0.9 is larger.

**14.**

5. 6 783    Tenths digit is 5 or higher.
             Round up.
6

**15.**

5.67 8 3    Thousandths digit is 5 or higher.
             Round up.
5.68

**16.**

5.678 3    Ten-thousandths digit is 4 or lower.
            Round down.
5.678

**17.**

5.6 7 83    Hundredths digit is 5 or higher.
             Round up.
5.7

**18.**

$$
\begin{array}{ll}
0.7\,0\,0\,0 & \text{Writing 3 extra zeros} \\
0.0\,8\,0\,0 & \text{Writing 2 extra zeros} \\
0.0\,0\,9\,0 & \text{Writing an extra zero} \\
+\ 0.0\,0\,1\,2 & \\
\hline
0.7\,9\,0\,2 &
\end{array}
$$

**19.** $-102.4 + 6.1 + 78 = -96.3 + 78 = -18.3$

**20.** Line up the decimal points. We write in decimal points in the last two numbers and add extra zeros in the last three numbers.

$$
\begin{array}{r}
0.9\,3 \\
9.3\,0 \\
9\,3.0\,0 \\
+\ 9\,3\,0.0\,0 \\
\hline
1\,0\,3\,3.2\,3
\end{array}
$$

**21.**

$$
\begin{array}{r}
5\,2.6\,7\,8 \\
-\ 4.3\,2\,1 \\
\hline
4\,8.3\,5\,7
\end{array}
$$

**22.**

$$
\begin{array}{r}
2\,0.0\,0\,0\,0 \\
-0.9\,0\,9\,9 \\
\hline
1\,9.0\,9\,0\,1
\end{array}
$$
Writing 3 additional zeros

**23.** $-234.6788 - 81.7854$

We add the absolute values and then make the answer negative.

$$
\begin{array}{r}
2\,3\,4.6\,7\,8\,8 \\
+\ 8\,1.7\,8\,5\,4 \\
\hline
3\,1\,6.4\,6\,4\,2
\end{array}
$$

Thus $-234.6788 - 81.7854 = -316.4642$.

**24.**

$$
\begin{array}{rl}
0.1\,2\,5 & \text{(3 decimal places)} \\
\times\ \ 0.2\,4 & \text{(2 decimal places)} \\
\hline
5\,0\,0 & \\
2\,5\,0\,0 & \\
\hline
0.0\,3\,0\,0\,0 & \text{(5 decimal places)}
\end{array}
$$

**25.** $0.001 \times (-213.45)$

0.001 has 3 decimal places so we move the decimal point in −231.45 three places to the left.

$0.001 \times (-213.45) = -0.21345$

**26.** $1000 \times 73.962$        73.962.

3 zeros        Move 3 places to the right.

$1000 \times 73.962 = 73,962$

**27.** First we consider $19 \div 4$

$$
\begin{array}{r}
4.7\,5 \\
4\,\overline{)1\,9.0\,0} \\
1\,6 \\
\hline
3\,0 \\
2\,8 \\
\hline
2\,0 \\
2\,0 \\
\hline
0
\end{array}
$$

$19 \div 4 = 4.75$, so $-19 \div 4 = -4.75$.

**28.**

$$
\begin{array}{r}
3\,0.4 \\
3.3_\wedge\,\overline{)1\,0\,0.3_\wedge\,2} \\
9\,9 \\
\hline
1\,3\,2 \\
1\,3\,2 \\
\hline
0
\end{array}
$$

**29.**
$$82 \overline{)15.58} \quad \begin{array}{r} 0.19 \\ \hline \end{array}$$

$$
\begin{array}{r}
0.1\,9 \\
82\overline{)15.58} \\
\underline{82\phantom{.}} \\
73\,8 \\
\underline{73\,8} \\
0
\end{array}
$$

**30.** $\dfrac{-346.89}{1000}$ $\qquad$ $-0.346.89$ $\overset{\curvearrowleft}{\underset{\lfloor\_\rfloor}{}}$

3 zeros $\qquad$ Move 3 places to the left.

$\dfrac{-346.89}{1000} = -0.34689$

**31.** $\dfrac{346.89}{0.\underline{01}}$ $\qquad$ $346.89.$ $\underset{\lfloor\_\rceil}{}$

2 decimal places $\qquad$ Move 2 places to the right.

$\dfrac{346.89}{0.01} = 34{,}689$

**32.** $-4.8 \cdot y = 404.448$

$\dfrac{-4.8 \cdot y}{-4.8} = \dfrac{404.448}{-4.8}$ $\quad$ Dividing on both sides by $-4.8$

$y = -84.26$

The solution is $-84.26$.

**33.** $\qquad x + 0.018 = 9$

$x + 0.018 - 0.018 = 9 - 0.018$ $\quad$ Subtracting 0.018

$\qquad\qquad\qquad\qquad\qquad$ on both sides

$\qquad\qquad\qquad x = 8.982$

$$
\begin{array}{r}
{\scriptstyle 8\ \ 9\ \ 9\ 10} \\
\cancel{9}.\cancel{0}\,\cancel{0}\,\cancel{0} \\
-0.0\,1\,8 \\
\hline
8.9\,8\,2
\end{array}
$$

The solution is $8.982$.

**34.** *Familiarize*. This is a multistep problem. First we will find the number of minutes in excess of 1400. Then we will find the total cost of these minutes and, finally, we will find the total cell phone bill. Let $m$ = the number of minutes in excess of 1400.

*Translate and Solve*.

| First 1400 minutes | plus | Excess minutes | is | Total minutes |
|:---:|:---:|:---:|:---:|:---:|
| $\downarrow$ | $\downarrow$ | $\downarrow$ | $\downarrow$ | $\downarrow$ |
| 1400 | $+$ | $m$ | $=$ | 1510 |

To solve the equation, we subtract 1400 on both sides.

$\qquad m = 1510 - 1400 = 110$

Next we multiply by \$0.40 to find the cost $c$ of the 110 excess minutes.

$\qquad c = \$0.40 \cdot 110 = \$44$

Finally, we add the cost of the first 1400 minutes and the cost of the excess minutes to find the total charge, $t$.

$\qquad t = \$89.99 + \$44 = \$133.99$

*Check*. We can repeat the calculations. The answer checks.

*State*. The charge was \$133.99.

**35.** *Familiarize*. This is a two-step problem. First we will find the number of miles that are driven between fillups. Then we find the gas mileage. Let $n$ = the number of miles driven between fillups.

*Translate and Solve*.

| First odometer reading | plus | Number of miles driven | is | Second odometer reading |
|:---:|:---:|:---:|:---:|:---:|
| $\downarrow$ | $\downarrow$ | $\downarrow$ | $\downarrow$ | $\downarrow$ |
| 76,843 | $+$ | $n$ | $=$ | 77,310 |

To solve the equation, we subtract 76,843 on both sides.

$\qquad n = 77{,}310 - 76{,}843 = 467$

Now let $m$ = the number of miles driven per gallon.

| Number of miles per gallon | times | Number of gallons | is | Miles driven |
|:---:|:---:|:---:|:---:|:---:|
| $\downarrow$ | $\downarrow$ | $\downarrow$ | $\downarrow$ | $\downarrow$ |
| $m$ | $\cdot$ | 16.5 | $=$ | 467 |

We divide by 16.5 on both sides to find $m$.

$\qquad m = 467 \div 16.5$

$\qquad m = 28.\overline{30}$

$\qquad m \approx 28.3$ $\quad$ Rounding to the nearest tenth

*Check*. First we multiply the number of miles per gallon by the number of gallons to find the number of miles driven:

$\qquad 16.5 \cdot 28.3 = 466.95 \approx 467$

Then we add 467 mi to the first odometer reading:

$\qquad 76{,}843 + 467 = 77{,}310$

This is the second odometer reading, so the answer checks. .

*State*. The gas mileage is about 28.3 miles per gallon.

**36.** *Familiarize*. Let $b$ = the balance after the purchases are made.

*Translate*. We subtract the amounts of the three purchases from the original balance:

$\qquad b = 820 - 123.89 - 56.68 - 46.98$

*Solve*. We carry out the calculations.

$\qquad b = 820 - 123.89 - 56.68 - 46.98$

$\qquad\quad = 696.11 - 56.68 - 46.98$

$\qquad\quad = 639.43 - 46.98$

$\qquad\quad = 592.45$

*Check*. We can find the total amount of the purchases and then subtract to find the new balance.

$\qquad \$123.89 + \$56.68 + \$46.98 = \$227.55$

$\qquad \$820 - \$227.55 = \$592.45$

The answer checks.

*State*. After the purchases were made, the balance was \$592.45.

**37.** **Familiarize**. Let $c =$ the total cost of the copy paper.

**Translate**.

$$\underbrace{\text{Cost per case}}_{} \quad \text{times} \quad \underbrace{\text{Number of cases}}_{} \quad \text{is} \quad \underbrace{\text{Total cost}}_{}$$

| | | | | |
|---|---|---|---|---|
| ↓ | ↓ | ↓ | ↓ | ↓ |
| 41.99 | · | 7 | = | $c$ |

**Solve**. We carry out the multiplication.

$$\begin{array}{r} 4\,1.9\,9 \quad \text{(2 decimal places)} \\ \times \qquad 7 \\ \hline 2\,9\,3.9\,3 \quad \text{(2 decimal places)} \end{array}$$

Thus, $c = 293.93$.

**Check**. We can obtain a partial check by rounding and estimating:

$$41.99 \times 7 \approx 40 \times 7 = 280 \approx 293.93$$

**State**. The total cost of the copy paper is $293.93.

**38.** We add the numbers and divide by the number of addends.

$$\frac{89.4 + 76.2 + 61.9 + 59.8 + 49.9}{5} = \frac{337.2}{5} = 67.44$$

The average number of passengers is 67.44 million.

**39.** $8.91 \times 22.457 \approx 9 \times 22 = 198$

**40.** $78.2209 \div 16.09 \approx 80 \div 20 = 4$

**41.** $\dfrac{8}{5} = \dfrac{8}{5} \cdot \dfrac{2}{2} = \dfrac{16}{10} = 1.6$

**42.** $\dfrac{22}{25} = \dfrac{22}{25} \cdot \dfrac{4}{4} = \dfrac{88}{100} = 0.88$

**43.** $-\dfrac{21}{4} = -\dfrac{21}{4} \cdot \dfrac{25}{25} = -\dfrac{525}{100} = -5.25$

**44.** $\dfrac{3}{4} = 3 \div 4$

$$\begin{array}{r} 0.7\,5 \\ 4\,\overline{\smash{)}\,3.0\,0} \\ \underline{2\,8} \\ 2\,0 \\ \underline{2\,0} \\ 0 \end{array}$$

$\dfrac{3}{4} = 0.75$

**45.** First consider $\dfrac{11}{9}$.

$$\dfrac{11}{9} = 11 \div 9$$

$$\begin{array}{r} 1.2\,2 \\ 9\,\overline{\smash{)}\,1\,1.0\,0} \\ \underline{9} \\ 2\,0 \\ \underline{1\,8} \\ 2\,0 \\ \underline{1\,8} \\ 2 \end{array}$$

Since 2 keeps reappearing as a remainder, the digits repeat. We have $\dfrac{11}{9} = 1.222\ldots = 1.\overline{2}$, so $-\dfrac{11}{9} = -1.\overline{2}$.

**46.** $\dfrac{15}{7} = 15 \div 7$

$$\begin{array}{r} 2.1\,4\,2\,8\,5\,7 \\ 7\,\overline{\smash{)}\,1\,5.0\,0\,0\,0\,0\,0} \\ \underline{1\,4} \\ 1\,0 \\ \underline{7} \\ 3\,0 \\ \underline{2\,8} \\ 2\,0 \\ \underline{1\,4} \\ 6\,0 \\ \underline{5\,6} \\ 4\,0 \\ \underline{3\,5} \\ 5\,0 \\ \underline{4\,9} \\ 1 \end{array}$$

Since 1 reappears as a remainder, the sequence repeats and $\dfrac{15}{7} = 2.\overline{142857}$.

**47.** $2.1\boxed{4}2857\ldots$

Hundredths digit is 4 or lower.

$2.1$       Round down.

**48.** $2.14\boxed{2}857\ldots$

Thousandths digit is 4 or lower.

$2.14$       Round down.

**49.** $2.142\boxed{8}57\ldots$

Ten-thousandths digit is 5 or higher.

$2.143$       Round up.

**50.**
$$256 \div 3.2 \div 2 - 1.56 + 78.325 \times 0.02$$
$$= 80 \div 2 - 1.56 + 78.325 \times 0.02$$
$$= 40 - 1.56 + 78.325 \times 0.02$$
$$= 40 - 1.56 + 1.5665$$
$$= 38.44 + 1.5665$$
$$= 40.0065$$

**51.**
$$(1 - 0.08)^2 + 6[5(12.1 - 8.7) + 10(14.3 - 9.6)]$$
$$= (0.92)^2 + 6[5(3.4) + 10(4.7)]$$
$$= (0.92)^2 + 6[17 + 47]$$
$$= (0.92)^2 + 6[64]$$
$$= (0.92)^2 + 384$$
$$= 0.8464 + 384$$
$$= 384.8464$$

**52.**
$$-\dfrac{7}{8} \times (-345.6)$$
$$= -0.875 \times (-345.6) \quad \text{Writing } -\dfrac{7}{8} \text{ in decimal notation}$$
$$= 302.4$$

**53.** $\dfrac{2}{3} \times 79.95 - \dfrac{7}{9} \times 1.235 = \dfrac{2 \times 79.95}{3} - \dfrac{7 \times 1.235}{9}$

$$= \dfrac{159.9}{3} - \dfrac{8.645}{9}$$

$$= \dfrac{159.9}{3} \cdot \dfrac{3}{3} - \dfrac{8.645}{9}$$

$$= \dfrac{479.7}{9} - \dfrac{8.645}{9}$$

$$= \dfrac{471.055}{9}$$

$$= 52.339\overline{4}$$

**54.** Move 2 places to the left.

$\$9.49.\cancel{\phi}$

Change from $\phi$ sign at end to $ sign in front.

$949\phi = \$9.49$

Answer B is correct.

**55. Familiarize.** This is a two-step problem. First we will find the cost of membership for six months without the coupon. Let $c =$ this cost. Then we will find how much Allise will save if she uses the coupon.

**Translate and Solve.**

| Membership fee | plus | Monthly cost | times | Number of months | is | Cost without coupon |
|---|---|---|---|---|---|---|
| ↓ | ↓ | ↓ | ↓ | ↓ | ↓ | ↓ |
| 79 | + | 42.50 | · | 6 | = | c |

To solve the equation, we carry out the calculation.

$c = 79 + 42.50 \cdot 6 = 79 + 255 = 334$

Now let $s =$ the coupon savings.

| Cost with coupon | plus | Coupon savings | is | Cost without coupon |
|---|---|---|---|---|
| ↓ | ↓ | ↓ | ↓ | ↓ |
| 299 | + | s | = | 334 |

We subtract 299 on both sides.

$s = 334 - 299 = 35$

**Check.** We repeat the calculations. The answer checks.

**State.** Allise will save $35 if she uses the coupon.

**56.** First use a calculator to find decimal notation for each fraction.

$-\dfrac{2}{3} = -0.\overline{6}$

$-\dfrac{15}{19} \approx -0.789474$

$-\dfrac{11}{13} = -0.\overline{846153}$

$-\dfrac{5}{7} = -0.\overline{714285}$

$-\dfrac{13}{15} = -0.8\overline{6}$

$-\dfrac{17}{20} = -0.85$

Arranging these numbers from smallest to largest, we have
$-0.86, -0.85, -0.\overline{846153}, -0.789474, -0.\overline{714285}, -0.\overline{6}$.

Then, in fraction notation, the numbers from smallest to largest are

$$-\dfrac{13}{15}, -\dfrac{17}{20}, -\dfrac{11}{13}, -\dfrac{15}{19}, -\dfrac{5}{7}, -\dfrac{2}{3}.$$

## Cumulative Review Chapters 1 - 5

**1.** $2\dfrac{2}{9} = \dfrac{20}{9}$     $(9 \cdot 2 = 18,\ 18 + 2 = 20)$

**2.**   $-3.\underline{051}$      $-3.051.$      $\dfrac{-3051}{1000}$

   3 places   Move 3 places.   3 zeros

$-3.051 = \dfrac{-3051}{1000}$, or $-\dfrac{3051}{1000}$

**3.** $-\dfrac{7}{5} = -\dfrac{7}{5} \cdot \dfrac{2}{2} = -\dfrac{14}{10} = -1.4$

**4.** $\dfrac{6}{11} = 6 \div 11$

$$
\begin{array}{r}
0.5\,4\,5 \\
11\,\overline{)6.0\,0\,0} \\
5\,5\phantom{.000} \\
\hline
5\,0\phantom{0} \\
4\,4\phantom{0} \\
\hline
6\,0 \\
5\,5 \\
\hline
5
\end{array}
$$

Since 5 reappears as a remainder, the sequence repeats and $\dfrac{6}{11} = 0.5454\ldots$, or $0.\overline{54}$.

**5.** The number 43 has only the factors 43 and 1, so it is prime.

**6.** The number named by the last two digits of 2,053,752 is 52. Since 52 is divisible by 4, the number 2,053,752 is divisible by 4.

**7.**    $48 + 12 \div 4 - 10 \times 2 + 6892 \div 4$

| | |
|---|---|
| $= 48 + 3 - 20 + 1723$ | Multiplying and dividing in order from left to right |
| $= 51 - 20 + 1723$ | Adding and subtracting |
| $= 31 + 1723$ | in order from |
| $= 1754$ | left to right |

**8.**    $0.2 - \{0.1[1.2(3.95 - 1.65) + 1.5 \div 2.5]\}$

| | |
|---|---|
| $= 0.2 - \{0.1[1.2(2.3) + 1.5 \div 2.5]\}$ | Subtracting inside parentheses |
| $= 0.2 - \{0.1[2.76 + 0.6]\}$ | Multiplying and dividing inside brackets |
| $= 0.2 - \{0.1[3.36]\}$ | Adding inside brackets |
| $= 0.2 - 0.336$ | Multiplying |
| $= -0.136$ | Subtracting |

**9.**

584.97⌐3⌐   Thousandths digit is 4 or lower.
   ↓        Round down.
584.97

**10.**

$218.\overline{5} = 218.55\boxed{5}$   Thousandths digit is 5 or higher.
   ↓        Round up.
   218.56

**11.** $16.392 \times 9.715 \approx 16 \times 10 = 160$

**12.** $2.714 + 4.562 - 3.31 - 0.0023 \approx 2.7 + 4.6 - 3.3 - 0.0 = 4$

**13.** $6418 \times 1984 \approx 6400 \times 2000 = 12{,}800{,}000$

**14.** $717.832 \div 124.998 \approx 720 \div 120 = 6$

**15.**

$$2\frac{1}{4} = 2\frac{5}{20}$$
$$+3\frac{4}{5} = +3\frac{16}{20}$$
$$\overline{\phantom{+3}5\frac{21}{20}} = 5 + \frac{21}{20} = 5 + 1\frac{1}{20} = 6\frac{1}{20}$$

**16.**

$$
\begin{array}{r}
\overset{1\ \ 1}{3\,4,9\,2\,1} \\
9\,3,0\,9\,2 \\
+\,1\,1,1\,0\,3 \\
\hline
1\,3\,9,1\,1\,6
\end{array}
$$

**17.**

$$\frac{1}{6} + \frac{2}{3} + \frac{8}{9} = \frac{1}{6} \cdot \frac{3}{3} + \frac{2}{3} \cdot \frac{6}{6} + \frac{8}{9} \cdot \frac{2}{2}$$
$$= \frac{3}{18} + \frac{12}{18} + \frac{16}{18}$$
$$= \frac{3 + 12 + 16}{18} = \frac{31}{18}$$

**18.** $-143.9 + 2.053$

First we find the difference in absolute values.

$$
\begin{array}{r}
\overset{8\ \ 9\ 10}{1\,4\,3.9\cancel{0}\,\cancel{0}\,\cancel{0}} \\
-\ \ \ 2.0\,5\,3 \\
\hline
1\,4\,1.8\,4\,7
\end{array}
$$

The negative number has the greater absolute value, so the answer is negative: $-143.9 + 2.053 = -141.847$.

**19.**

$$
\begin{array}{r}
\overset{2\ 10\ 3\ 11}{7\,2\,\cancel{3},\cancel{0}\,\cancel{4}\,\cancel{1}} \\
-\ \ \ 1\,2,9\,0\,4 \\
\hline
7\,1\,0,1\,3\,7
\end{array}
$$

**20.** We add a decimal point and extra zeros.

$$
\begin{array}{r}
\overset{8\ \ 9\ \ 9\ 10}{1\,9.\cancel{0}\,\cancel{0}\,\cancel{0}} \\
-\ \ \ 5.9\,0\,3 \\
\hline
1\,3.0\,9\,7
\end{array}
$$

**21.**

$$5\frac{1}{7} = \phantom{-}4\frac{8}{7}$$
$$-4\frac{3}{7} = -4\frac{3}{7}$$
$$\overline{\phantom{-4\frac{3}{7} = }\ \frac{5}{7}}$$

**22.**

$$\frac{9}{10} - \frac{10}{11} = \frac{9}{10} \cdot \frac{11}{11} - \frac{10}{11} \cdot \frac{10}{10}$$
$$= \frac{99}{110} - \frac{100}{110}$$
$$= \frac{99 - 100}{110}$$
$$= \frac{-1}{110}, \text{ or } -\frac{1}{110}$$

**23.** $\dfrac{3}{8} \cdot \left(-\dfrac{4}{9}\right) = -\dfrac{3 \cdot 4}{8 \cdot 9} = -\dfrac{3 \cdot 4 \cdot 1}{2 \cdot 4 \cdot 3 \cdot 3} = -\dfrac{1}{2 \cdot 3} \cdot \dfrac{3 \cdot 4}{3 \cdot 4} = -\dfrac{1}{6}$

**24.**

$$
\begin{array}{r}
2\,5\,3\,2 \\
\times\ 2\,1\,0\,0 \\
\hline
2\,5\,3\,2\,0\,0 \\
5\,0\,6\,4\,0\,0\,0 \\
\hline
5,3\,1\,7,2\,0\,0
\end{array}
$$

**25.**

$$
\begin{array}{rl}
2\,3.9 & \text{(1 decimal place)} \\
\times\ \ \ 0.2 & \text{(1 decimal place)} \\
\hline
4.7\,8 & \text{(2 decimal places)}
\end{array}
$$

**26.**

$$
\begin{array}{rl}
2\,7.9\,4\,3\,1 & \text{(4 decimal places)} \\
\times\ \ \ \ \ \ 0.0\,0\,1 & \text{(3 decimal places)} \\
\hline
0.0\,2\,7\,9\,4\,3\,1 & \text{(7 decimal places)}
\end{array}
$$

**27.**

$$
\begin{array}{r}
2.1\,2\,2 \\
16.5_\wedge \overline{)\,3\,5.0_\wedge 1\,3\,0} \\
3\,3\,0 \\
\hline
2\,0\ 1 \\
1\,6\ 5 \\
\hline
3\ 6\,3 \\
3\ 3\,0 \\
\hline
3\,3\,0 \\
3\,3\,0 \\
\hline
0
\end{array}
$$

**28.**

$$
\begin{array}{r}
1\,8\,4\,3 \\
26\,\overline{)\,4\,7,9\,1\,8} \\
2\,6 \\
\hline
2\,1\ 9 \\
2\,0\ 8 \\
\hline
1\,1\,1 \\
1\,0\,4 \\
\hline
7\,8 \\
7\,8 \\
\hline
0
\end{array}
$$

**29.** $13.8621 \div 0.001 = \dfrac{13.8621}{0.001}$

$$\frac{13.8621}{0.001}\qquad\qquad 13.862.1$$

3 decimal places    Move 3 places to the right.

$13.8621 \div 0.001 = 13{,}862.1$

**30.** $-\dfrac{4}{9} \div \left(-\dfrac{8}{15}\right) = -\dfrac{4}{9} \cdot \left(-\dfrac{15}{8}\right) = \dfrac{4 \cdot 15}{9 \cdot 8} = \dfrac{4 \cdot 3 \cdot 5}{3 \cdot 3 \cdot 2 \cdot 4} =$

$\dfrac{4 \cdot 3}{4 \cdot 3} \cdot \dfrac{5}{3 \cdot 2} = \dfrac{5}{6}$

**31.** $\qquad 8.32 + x = 9.1$

$8.32 + x - 8.32 = 9.1 - 8.32 \quad$ Subtracting 8.32

$\qquad\qquad x = 0.78$

The solution is 0.78.

**32.** $\qquad 75 \cdot x = 2100$

$\qquad \dfrac{75 \cdot x}{75} = \dfrac{2100}{75} \quad$ Dividing by 75

$\qquad\qquad x = 28$

The solution is 28.

**33.** $\qquad y \cdot 9.47 = -81.6314$

$\qquad \dfrac{y \cdot 9.47}{9.47} = \dfrac{-81.6314}{9.47} \quad$ Dividing by 9.47

$\qquad\qquad y = -8.62$

The solution is $-8.62$.

**34.** $\qquad 1062 + y = 368,313$

$1062 + y - 1062 = 368,313 - 1062 \quad$ Subtracting 1062

$\qquad\qquad y = 367,251$

The solution is 367,251.

**35.** $\qquad t + \dfrac{5}{6} = \dfrac{8}{9}$

$t + \dfrac{5}{6} - \dfrac{5}{6} = \dfrac{8}{9} - \dfrac{5}{6} \qquad$ Subtracting $\dfrac{5}{6}$

$\qquad\qquad t = \dfrac{8}{9} \cdot \dfrac{2}{2} - \dfrac{5}{6} \cdot \dfrac{3}{3}$

$\qquad\qquad t = \dfrac{16}{18} - \dfrac{15}{18} = \dfrac{16 - 15}{18}$

$\qquad\qquad t = \dfrac{1}{18}$

The solution is $\dfrac{1}{18}$.

**36.** $\qquad -\dfrac{7}{8} \cdot t = \dfrac{7}{16}$

$\qquad \dfrac{-\dfrac{7}{8} \cdot t}{-\dfrac{7}{8}} = \dfrac{\dfrac{7}{16}}{-\dfrac{7}{8}} \qquad$ Dividing by $-\dfrac{7}{8}$

$\qquad t = \dfrac{7}{16} \cdot \left(-\dfrac{8}{7}\right) = -\dfrac{7 \cdot 8}{16 \cdot 7}$

$\qquad t = -\dfrac{7 \cdot 8 \cdot 1}{2 \cdot 8 \cdot 7} = -\dfrac{1}{2} \cdot \dfrac{7 \cdot 8}{7 \cdot 8}$

$\qquad t = -\dfrac{1}{2}$

The solution is $-\dfrac{1}{2}$.

**37.** *Familiarize*. Let $t$ = the total number of transplants.

*Translate*. We are combining the number of transplants, so we add.

$\qquad 16,646 + 6136 + 2147 = t$

*Solve*. We carry out the addition.

$$\begin{array}{r} \overset{1}{\phantom{0}}\overset{\phantom{0}1\ 1}{\phantom{0}} \\ 1\,6,6\,4\,6 \\ 6\,1\,3\,6 \\ +\quad 2\,1\,4\,7 \\ \hline 2\,4,9\,2\,9 \end{array}$$

Thus, $t = 24,929$.

*Check*. We can repeat the calculation. We can also estimate by rounding. We will round to the nearest hundred.

$\qquad 16,646 + 6136 + 2147 \approx 16,600 + 6100 + 2100$

$\qquad\qquad\qquad\qquad\qquad\quad = 24,800$

Since $24,800 \approx 24,929$, our answer seems reasonable.

*State*. There were 24,929 kidney, liver, and heart transplants performed.

**38.** *Familiarize*. Let $m$ = the revenue received from Medicare payments, in billions of dollars.

*Translate*.

| Medicare payments | $+$ | Excess private payments | is | Private insurance payments |
|---|---|---|---|---|
| $\downarrow$ | $\downarrow$ | $\downarrow$ | $\downarrow$ | $\downarrow$ |
| $m$ | $+$ | 88 | $=$ | 265 |

*Solve*. We subtract 88 on both sides of the equation.

$\qquad m + 88 = 265$

$m + 88 - 88 = 265 - 88$

$\qquad\qquad m = 177$

*Check*. We can add $177 + 88$ to get 265. The answer checks.

*State*. $177 billion was received from Medicare payments.

**39.** *Familiarize*. Let $s$ = the price of the sofa.

*Translate*.

| Down payment | is | $\dfrac{3}{10}$ | of | the price of the sofa |
|---|---|---|---|---|
| $\downarrow$ | $\downarrow$ | $\downarrow$ $\downarrow$ | | $\downarrow$ |
| 450 | $=$ | $\dfrac{3}{10}$ $\cdot$ | | $s$ |

**Solve**. We divide by $\dfrac{3}{10}$ on both sides of the equation.

$$450 = \frac{3}{10} \cdot s$$

$$\frac{450}{\frac{3}{10}} = \frac{\frac{3}{10} \cdot s}{\frac{3}{10}}$$

$$450 \cdot \frac{10}{3} = s$$

$$\frac{450 \cdot 10}{3} = s$$

$$\frac{\cancel{3} \cdot 150 \cdot 10}{\cancel{3} \cdot 1} = s$$

$$1500 = s$$

**Check**. We find $\dfrac{3}{10}$ of 1500:

$$\frac{3}{10} \cdot 1500 = \frac{3 \cdot 1500}{10} = \frac{3 \cdot 150 \cdot \cancel{10}}{\cancel{10} \cdot 1} = 450$$

This is the amount of the down payment, so the answer checks.

**State**. The sofa cost $1500.

**40. Familiarize**. Let $a =$ the amount of the loan.

**Translate**.

$$\underbrace{\text{Amount of loan}}_{\downarrow} \;\; \underset{\downarrow}{\text{is}} \;\; \underset{\downarrow}{\frac{2}{3}} \;\; \underset{\downarrow}{\text{of}} \;\; \underbrace{\text{Tuition}}_{\downarrow}$$

$$a \qquad\qquad = \quad \frac{2}{3} \;\; \cdot \quad 3600$$

**Solve**. We carry out the multiplication.

$$a = \frac{2}{3} \cdot 3600 = \frac{2 \cdot 3600}{3} = \frac{2 \cdot \cancel{3} \cdot 1200}{\cancel{3} \cdot 1} = 2400$$

**Check**.  Observe that $\dfrac{2400}{3600} = \dfrac{2 \cdot 1200}{3 \cdot 1200} = \dfrac{2}{3} \cdot \dfrac{1200}{1200} = \dfrac{2}{3}$.

Thus, 2400 represents $\dfrac{2}{3}$ of 3600. The answer checks.

**State**. The amount of the loan was $2400.

**41. Familiarize**. Let $b =$ the balance after the check is written.

**Translate**. We subtract to find the new balance.

$$b = 314.79 - 56.02$$

**Solve**. We carry out the subtraction.

$$\begin{array}{r} {\scriptstyle 10} \\[-2pt] {\scriptstyle 2\;\;\cancel{0}\;\;14} \\[-2pt] \cancel{3}\;\cancel{1}\;4.7\,9 \\ -\quad 5\;6.0\,2 \\ \hline 2\;5\;8.7\,7 \end{array}$$

Thus, $b = 258.77$.

**Check**. Since $258.77 + 56.02 = 314.79$, the answer checks.

**State**. The balance in the account is $258.77.

**42. Familiarize**. Let $m =$ the total number of pounds of meat sold.

**Translate**. We are combining amounts, so we add.

$$1\frac{1}{2} + 2\frac{3}{4} + 2\frac{1}{4} = m$$

**Solve**. We carry out the addition.

$$\begin{aligned} 1\frac{1}{2} &= \;\; 1\frac{2}{4} \\ 2\frac{3}{4} &= \;\; 2\frac{3}{4} \\ +2\frac{1}{4} &= +2\frac{1}{4} \\ \hline & 5\frac{6}{4} = 5 + \frac{6}{4} = 5 + 1\frac{2}{4} \\ &= 6\frac{2}{4} = 6\frac{1}{2} \end{aligned}$$

Thus, $m = 6\dfrac{1}{2}$.

**Check**. We repeat the calculation. The answer checks.

**State**. A total of $6\dfrac{1}{2}$ lb of meat was sold.

**43. Familiarize**. Let $s =$ the number of pounds of sugar used.

**Translate**. We are combining amounts, so we add.

$$\frac{1}{2} + \frac{2}{3} + \frac{5}{6} = s$$

**Solve**. We carry out the addition.

$$\begin{aligned} \frac{1}{2} + \frac{2}{3} + \frac{5}{6} &= \frac{1}{2} \cdot \frac{3}{3} + \frac{2}{3} \cdot \frac{2}{2} + \frac{5}{6} \\ &= \frac{3}{6} + \frac{4}{6} + \frac{5}{6} \\ &= \frac{3 + 4 + 5}{6} = \frac{12}{6} \\ &= 2 \end{aligned}$$

**Check**. We repeat the calculation. The answer checks.

**State**. The baker used 2 lb of sugar.

**44. Familiarize**. Let $A =$ the area.

**Translate**. We use the formula $A = l \times w$.

$$A = 23.6 \times 19.8$$

**Solve**. We carry out the multiplication.

$$\begin{array}{r} 2\,3.\,6 \\ \times\; 1\,9.\,8 \\ \hline 1\,8\,8\,8 \\ 2\,1\,2\,4\,0 \\ 2\,3\,6\,0\,0 \\ \hline 4\,6\,7.\,2\,8 \end{array}$$

Thus, $A = 467.28$.

**Check**. We can obtain a partial check by estimating the product.

$$23.6 \times 19.8 \approx 24 \times 20 = 480 \approx 467.28$$

The answer seems reasonable.

**State**. The area of the room is 467.28 sq ft.

**45.** $\left(\frac{3}{4}\right)^2 - \frac{1}{8} \cdot \left(3 - 1\frac{1}{2}\right)^2 = \frac{9}{16} - \frac{1}{8}\left(1\frac{1}{2}\right)^2$

$$= \frac{9}{16} - \frac{1}{8}\left(\frac{3}{2}\right)^2$$

$$= \frac{9}{16} - \frac{1}{8} \cdot \frac{9}{4}$$

$$= \frac{9}{16} - \frac{9}{32} = \frac{9}{16} \cdot \frac{2}{2} - \frac{9}{32}$$

$$= \frac{18}{32} - \frac{9}{32} = \frac{18 - 9}{32}$$

$$= \frac{9}{32}$$

**46.** $-1.2 \times 12.2 \div 0.1 \times 3.6 = -14.64 \div 0.1 \times 3.6$

$$= -146.4 \times 3.6$$

$$= -527.04$$

**47. *Familiarize*.** This is a two-step problem. First we will find the price of two cartons of juice. Then we will find the cost per carton with the coupon.

***Translate and Solve*.** Let $p$ = the price of two cartons of juice. We can multiply to find $p$.

$$p = 3.59 \times 2 = 7.18$$

Now let $c$ = the cost per carton with the coupon. We can divide to find $c$.

$$c = 7.18 \div 3 \approx 2.39$$

***Check*.** We can repeat the calculation. The answer checks.

***State*.** With the coupon, the cost per carton was $2.39.

**48. *Familiarize*.** First we convert $15\frac{3}{4}$ lb to ounces. (Instead, we could have converted $1\frac{3}{4}$ oz to pounds.)

$$15\frac{3}{4} \text{ lb} \cdot \frac{16 \text{ oz}}{1 \text{ lb}} = \frac{63}{4} \text{ lb} \cdot \frac{16 \text{ oz}}{1 \text{ lb}}$$

$$= \frac{63}{4} \cdot 16 \cdot \frac{\text{lb}}{\text{lb}} \cdot \text{oz}$$

$$= \frac{63 \cdot 16}{4} \cdot \text{oz} \qquad \left(\frac{\text{lb}}{\text{lb}} = 1\right)$$

$$= \frac{63 \cdot 4 \cdot \cancel{4}}{\cancel{4}} \text{ oz}$$

$$= 252 \text{ oz}$$

Now let $p$ = the number of packages in the carton.

***Translate*.** We divide to find $p$.

$$p = 252 \div 1\frac{3}{4}$$

***Solve*.** We carry out the division.

$$p = 252 \div 1\frac{3}{4} = 252 \div \frac{7}{4}$$

$$= 252 \cdot \frac{4}{7} = \frac{252 \cdot 4}{7} = \frac{7 \cdot 36 \cdot 4}{7 \cdot 1}$$

$$= \frac{7}{7} \cdot \frac{36 \cdot 4}{1} = 144$$

***Check*.** We have $144 \cdot 1\frac{3}{4} = 144 \cdot \frac{7}{4} = \frac{144 \cdot 7}{4} =$

$\frac{36 \cdot 4 \cdot 7}{4} = \frac{4}{4} \cdot \frac{36 \cdot 7}{1} = 252$. We get the weight of the carton, in ounces, so the answer checks.

***State*.** There are 144 packages in the carton.

# Chapter 6

# Ratio and Proportion

**1.** The ratio of 4 to 5 is $\dfrac{4}{5}$.

**3.** The ratio of 178 to 572 is $\dfrac{178}{572}$.

**5.** The ratio of 0.4 to 12 is $\dfrac{0.4}{12}$.

**7.** The ratio of 3.8 to 7.4 is $\dfrac{3.8}{7.4}$.

**9.** The ratio of 56.78 to 98.35 is $\dfrac{56.78}{98.35}$.

**11.** The ratio of $8\dfrac{3}{4}$ to $9\dfrac{5}{6}$ is $\dfrac{8\frac{3}{4}}{9\frac{5}{6}}$.

**13.** The ratio of the time of the current trip to the time of the trip on the space plane is $\dfrac{21}{4}$.

The ratio of the time of the trip on the space plane to the time of the current trip is $\dfrac{4}{21}$.

**15.** The ratio of all people to those covered by health insurance is $\dfrac{100}{84.7}$.

The ratio of those covered by health insurance to all people is $\dfrac{84.7}{100}$.

**17.** The ratio of the number of days worked to pay taxes in 2008 to the number of days in the year is $\dfrac{113}{366}$.

**19.** The ratio of all people to those in the 20-to-29 year age group is $\dfrac{1000}{126}$.

The ratio of those in the 20-to-29 year age group to all people is $\dfrac{126}{1000}$.

**21.** The ratio of width to length is $\dfrac{60}{100}$.

The ratio of length to width is $\dfrac{100}{60}$.

**23.** The ratio of 4 to 6 is $\dfrac{4}{6} = \dfrac{2 \cdot 2}{2 \cdot 3} = \dfrac{2}{2} \cdot \dfrac{2}{3} = \dfrac{2}{3}$.

**25.** The ratio of 18 to 24 is $\dfrac{18}{24} = \dfrac{3 \cdot 6}{4 \cdot 6} = \dfrac{3}{4} \cdot \dfrac{6}{6} = \dfrac{3}{4}$.

**27.** The ratio of 4.8 to 10 is $\dfrac{4.8}{10} = \dfrac{4.8}{10} \cdot \dfrac{10}{10} = \dfrac{48}{100} = \dfrac{4 \cdot 12}{4 \cdot 25} = \dfrac{4}{4} \cdot \dfrac{12}{25} = \dfrac{12}{25}$.

**29.** The ratio of 2.8 to 3.6 is $\dfrac{2.8}{3.6} = \dfrac{2.8}{3.6} \cdot \dfrac{10}{10} = \dfrac{28}{36} = \dfrac{4 \cdot 7}{4 \cdot 9} = \dfrac{4}{4} \cdot \dfrac{7}{9} = \dfrac{7}{9}$.

**31.** The ratio is $\dfrac{20}{30} = \dfrac{2 \cdot 10}{3 \cdot 10} = \dfrac{2}{3} \cdot \dfrac{10}{10} = \dfrac{2}{3}$.

**33.** The ratio is $\dfrac{56}{100} = \dfrac{4 \cdot 14}{4 \cdot 25} = \dfrac{4}{4} \cdot \dfrac{14}{25} = \dfrac{14}{25}$.

**35.** The ratio is $\dfrac{128}{256} = \dfrac{1 \cdot 128}{2 \cdot 128} = \dfrac{1}{2} \cdot \dfrac{128}{128} = \dfrac{1}{2}$.

**37.** The ratio is $\dfrac{0.48}{0.64} = \dfrac{0.48}{0.64} \cdot \dfrac{100}{100} = \dfrac{48}{64} = \dfrac{3 \cdot 16}{4 \cdot 16} = \dfrac{3}{4} \cdot \dfrac{16}{16} = \dfrac{3}{4}$.

**39.** The ratio of length to width is $\dfrac{478}{213}$.

The ratio of width to length is $\dfrac{213}{478}$.

**41.** We find the cross products:

$$12 \cdot 4 = 48 \qquad 8 \cdot 6 = 48$$

Since the cross products are equal, $\dfrac{12}{8} = \dfrac{6}{4}$.

**43.** We find the cross products:

$$-7 \cdot 9 = -63 \qquad 2 \cdot (-31) = -62$$

Since the cross products are not equal, $\dfrac{-7}{2} \neq \dfrac{-31}{9}$.

**45.**
```
      5 0
   4 ⟌ 2 0 0
      2 0
        0
        0
        0
```

The answer is 50.

**47.** First we consider $232 \div 16$.
```
         1 4. 5
   1 6 ⟌ 2 3 2. 0
         1 6 0
           7 2
           6 4
             8 0
             8 0
              0
```

$232 \div 16 = 14.5$, so $232 \div (-16) = -14.5$.

**49. *Familiarize*.** We let $h$ = Rocky's excess height.

***Translate*.** We have a "how much more" situation.

| Height of daughter | plus | How much more height | is | Rocky's height |
|:---:|:---:|:---:|:---:|:---:|
| ↓ | ↓ | ↓ | ↓ | ↓ |
| $180\frac{3}{4}$ | $+$ | $h$ | $=$ | $187\frac{1}{10}$ |

***Solve*.** We solve the equation as follows:

$$h = 187\frac{1}{10} - 180\frac{3}{4}$$

$$187 \boxed{\frac{1}{10} \cdot \frac{2}{2}} = 187\frac{2}{20}$$

$$180 \boxed{\frac{3}{4} \cdot \frac{5}{5}} = 180\frac{15}{20}$$

$$
\begin{array}{r}
187\frac{1}{10} = \quad 187\frac{2}{20} = \quad 186\frac{22}{20} \\
-180\frac{3}{4} = -180\frac{15}{20} = -180\frac{15}{20} \\
\hline
6\frac{7}{20}
\end{array}
$$

Thus, $h = 6\frac{7}{20}$.

***Check*.** We add Rocky's excess height to his daughter's height:

$$180\frac{3}{4} + 6\frac{7}{20} = 180\frac{15}{20} + 6\frac{7}{20} = 186\frac{22}{20} = 187\frac{2}{20} = 187\frac{1}{10}$$

The answer checks.

***State*.** Rocky is $6\frac{7}{20}$ cm taller.

**51.** $\dfrac{3\frac{3}{4}}{5\frac{7}{8}} = \dfrac{\frac{15}{4}}{\frac{47}{8}} = \dfrac{15}{4} \cdot \dfrac{8}{47} = \dfrac{15 \cdot 8}{4 \cdot 47} =$

$$\dfrac{15 \cdot 2 \cdot 4}{4 \cdot 47} = \dfrac{4}{4} \cdot \dfrac{15 \cdot 2}{47} = \dfrac{30}{47}$$

**53.** We divide each number in the ratio by 5. Since $5 \div 5 = 1$, $10 \div 5 = 2$, and $15 \div 5 = 3$, we have $1 : 2 : 3$.

---

## Exercise Set 6.2

**1.** $\dfrac{120 \text{ km}}{3 \text{ hr}} = 40 \dfrac{\text{km}}{\text{hr}}$

**3.** $\dfrac{217 \text{ mi}}{29 \text{ sec}} \approx 7.48 \dfrac{\text{mi}}{\text{sec}}$

**5.** $\dfrac{312.5 \text{ mi}}{12.5 \text{ gal}} = 25 \text{ mpg}$

**7.** $\dfrac{624 \text{ mi}}{19.5 \text{ gal}} = 32 \text{ mpg}$

**9.** $\dfrac{32{,}796 \text{ people}}{0.75 \text{ sq mi}} = 43{,}728 \text{ people/sq mi}$

**11.** $\dfrac{500 \text{ mi}}{20 \text{ hr}} = 25 \dfrac{\text{mi}}{\text{hr}}$

$\dfrac{20 \text{ hr}}{500 \text{ mi}} = 0.04 \dfrac{\text{hr}}{\text{mi}}$

**13.** $\dfrac{1254 \text{ points}}{51 \text{ games}} \approx 24.6 \dfrac{\text{points}}{\text{game}}$

**15.** $\dfrac{623 \text{ gal}}{1000 \text{ sq ft}} = 0.623 \text{ gal/ft}^2$

**17.** $\dfrac{186{,}000 \text{ mi}}{1 \text{ sec}} = 186{,}000 \dfrac{\text{mi}}{\text{sec}}$

**19.** $\dfrac{310 \text{ km}}{2.5 \text{ hr}} = 124 \dfrac{\text{km}}{\text{hr}}$

**21.** $\dfrac{1500 \text{ beats}}{60 \text{ min}} = 25 \dfrac{\text{beats}}{\text{min}}$

**23.** $\dfrac{\$4.19}{16 \text{ oz}} = \dfrac{419\cent}{16 \text{ oz}} \approx 26.188\cent/\text{oz}$

$\dfrac{\$5.29}{20 \text{ oz}} = \dfrac{529\cent}{20 \text{ oz}} = 26.450\cent/\text{oz}$

The 16-oz size is the better buy based on unit price alone.

**25.** $\dfrac{\$3.99}{45 \text{ oz}} = \dfrac{399\cent}{45 \text{ oz}} \approx 8.867\cent/\text{oz}$

$\dfrac{\$5.49}{75 \text{ oz}} = \dfrac{549\cent}{75 \text{ oz}} = 7.320\cent/\text{oz}$

The 75-oz size is the better buy based on unit price alone.

**27.** $\dfrac{\$2.50}{18 \text{ oz}} = \dfrac{250\cent}{18 \text{ oz}} \approx 13.889\cent/\text{oz}$

$\dfrac{\$4.89}{28 \text{ oz}} = \dfrac{489\cent}{28 \text{ oz}} \approx 17.464\cent/\text{oz}$

The 18-oz size is the better buy based on unit price alone.

**29.** $\dfrac{\$1.39}{10.7 \text{ oz}} = \dfrac{139\cent}{10.7 \text{ oz}} \approx 12.991\cent/\text{oz}$

$\dfrac{\$2.69}{26 \text{ oz}} = \dfrac{269\cent}{26 \text{ oz}} \approx 10.346\cent/\text{oz}$

The 26-oz size is the better buy based on unit price alone.

**31.** $\dfrac{\$5.99}{32 \text{ oz}} = \dfrac{599\cent}{32 \text{ oz}} \approx 18.719\cent/\text{oz}$

$\dfrac{\$6.99}{48 \text{ oz}} = \dfrac{699\cent}{48 \text{ oz}} \approx 14.563\cent/\text{oz}$

Brand E is the better buy based on unit price alone.

**33.** $\dfrac{\$2.49}{24 \text{ oz}} = \dfrac{249\cent}{24 \text{ oz}} = 10.375\cent/\text{oz}$

$\dfrac{\$3.29}{36 \text{ oz}} = \dfrac{329\cent}{36 \text{ oz}} \approx 9.139\cent/\text{oz}$

$\dfrac{\$3.69}{46 \text{ oz}} = \dfrac{369\cent}{46 \text{ oz}} \approx 8.022\cent/\text{oz}$

Brand H is the better buy based on unit price alone.

**35.** *Familiarize.* Let $t =$ the number of drive-in movie theaters in 1958.

*Translate.*

| Number of theaters in 2007 | plus | Excess number of theaters in 1958 | is | Number of theaters in 1958 |
|:---:|:---:|:---:|:---:|:---:|
| ↓ | ↓ | ↓ | ↓ | ↓ |
| 405 | + | 3658 | = | $t$ |

*Solve.* We carry out the addition.

$$\begin{array}{r} {\scriptstyle 1\ \ 1} \\ 4\,0\,5 \\ +\,3\,6\,5\,8 \\ \hline 4\,0\,6\,3 \end{array}$$

Thus, $4063 = t$.

*Check.* We can repeat the calculation. We can also do a partial check by rounding to the nearest hundred.

$$405 + 3658 \approx 400 + 3700 = 4100 \approx 4063.$$

Since 4100 is close to 4063, the answer is reasonable.

*State.* There were 4063 drive-in movie theaters in 1958.

**37.** *Familiarize.* Let $g =$ the amount by which gift-card sales rose from 2003 to 2007, in billions of dollars.

*Translate.*

| Gift-card sales in 2003 | plus | Rise in sales from 2003 to 2007 | is | Gift-card sales in 2007 |
|:---:|:---:|:---:|:---:|:---:|
| ↓ | ↓ | ↓ | ↓ | ↓ |
| 17.2 | + | $g$ | = | 26.3 |

*Solve.* We subtract 17.2 on both sides of the equation.

$$17.2 + g = 26.3$$
$$17.2 + g - 17.2 = 26.3 - 17.2$$
$$g = 9.1$$

*Check.* We add $17.2 + 9.1$ and get 26.3, so the answer checks.

*State.* Gift-card sales rose by $9.1 billion from 2003 to 2007.

**39.**
$$\begin{array}{r} 4\,5.\,6\,7 \\ \times\ \ \ \ 2.\,4 \\ \hline 1\,8\,2\,6\,8 \\ \cdot\,9\,1\,3\,4\,0 \\ \hline 1\,0\,9.\,6\,0\,8 \end{array}$$

**41.** First we consider $84.3 \times 69.2$.
$$\begin{array}{r} 6\,9.\,2 \\ \times\,8\,4.\,3 \\ \hline 2\,0\,7\,6 \\ 2\,7\,6\,8\,0 \\ 5\,5\,3\,6\,0\,0 \\ \hline 5\,8\,3\,3.\,5\,6 \end{array}$$
$84.3 \times 69.2 = 5833.56$, so $84.3 \times (-69.2) = -5833.56$.

**43.** For the 6-oz container: $\dfrac{65\cancel{c}}{6\ \text{oz}} \approx 10.833\cancel{c}/\text{oz}$

For the 5.5-oz container: $\dfrac{60\cancel{c}}{5.5\ \text{oz}} \approx 10.909\cancel{c}/\text{oz}$

---

**Exercise Set 6.3**

**1.** We can use cross products:

$$5 \cdot 9 = 45 \qquad \begin{matrix} 5 & 7 \\ 6 & 9 \end{matrix} \qquad 6 \cdot 7 = 42$$

Since the cross products are not the same, $45 \neq 42$, we know that the numbers are not proportional.

**3.** We can use cross products:

$$1 \cdot 20 = 20 \qquad \begin{matrix} 1 & 10 \\ 2 & 20 \end{matrix} \qquad 2 \cdot 10 = 20$$

Since the cross products are the same, $20 = 20$, we know that $\dfrac{1}{2} = \dfrac{10}{20}$, so the numbers are proportional.

**5.** We can use cross products:

$$2.4 \cdot 2.7 = 6.48 \qquad \begin{matrix} 2.4 & 1.8 \\ 3.6 & 2.7 \end{matrix} \qquad 3.6 \cdot 1.8 = 6.48$$

Since the cross products are the same, $6.48 = 6.48$, we know that $\dfrac{2.4}{3.6} = \dfrac{1.8}{2.7}$, so the numbers are proportional.

**7.** We can use cross products:

$$5\tfrac{1}{3} \cdot 9\tfrac{1}{2} = 50\tfrac{2}{3} \qquad \begin{matrix} 5\tfrac{1}{3} & 2\tfrac{1}{5} \\ 8\tfrac{1}{4} & 9\tfrac{1}{2} \end{matrix} \qquad 8\tfrac{1}{4} \cdot 2\tfrac{1}{5} = 18\tfrac{3}{20}$$

Since the cross products are not the same, $50\tfrac{2}{3} \neq 18\tfrac{3}{20}$, we know that the numbers are not proportional.

**9.**
$$\frac{18}{4} = \frac{x}{10}$$

$18 \cdot 10 = 4 \cdot x \qquad$ Equating cross products

$\dfrac{18 \cdot 10}{4} = \dfrac{4 \cdot x}{4} \qquad$ Dividing by 4

$\dfrac{18 \cdot 10}{4} = x$

$\dfrac{180}{4} = x \qquad$ Multiplying

$45 = x \qquad$ Dividing

**11.**
$$\frac{x}{8} = \frac{9}{6}$$

$6 \cdot x = 8 \cdot 9 \qquad$ Equating cross products

$\dfrac{6 \cdot x}{6} = \dfrac{8 \cdot 9}{6} \qquad$ Dividing by 6

$x = \dfrac{8 \cdot 9}{6}$

$x = \dfrac{72}{6} \qquad$ Multiplying

$x = 12 \qquad$ Dividing

**13.** $\dfrac{t}{12} = \dfrac{5}{6}$

$6 \cdot t = 12 \cdot 5$

$\dfrac{6 \cdot t}{6} = \dfrac{12 \cdot 5}{6}$

$t = \dfrac{12 \cdot 5}{6}$

$t = \dfrac{60}{6}$

$t = 10$

**15.** $\dfrac{2}{5} = \dfrac{8}{n}$

$2 \cdot n = 5 \cdot 8$

$\dfrac{2 \cdot n}{2} = \dfrac{5 \cdot 8}{2}$

$n = \dfrac{5 \cdot 8}{2}$

$n = \dfrac{40}{2}$

$n = 20$

**17.** $\dfrac{n}{15} = \dfrac{10}{30}$

$30 \cdot n = 15 \cdot 10$

$\dfrac{30 \cdot n}{30} = \dfrac{15 \cdot 10}{30}$

$n = \dfrac{15 \cdot 10}{30}$

$n = \dfrac{150}{30}$

$n = 5$

**19.** $\dfrac{16}{12} = \dfrac{24}{x}$

$16 \cdot x = 12 \cdot 24$

$\dfrac{16 \cdot x}{16} = \dfrac{12 \cdot 24}{16}$

$x = \dfrac{12 \cdot 24}{16}$

$x = \dfrac{288}{16}$

$x = 18$

**21.** $\dfrac{6}{11} = \dfrac{12}{x}$

$6 \cdot x = 11 \cdot 12$

$\dfrac{6 \cdot x}{6} = \dfrac{11 \cdot 12}{6}$

$x = \dfrac{11 \cdot 12}{6}$

$x = \dfrac{132}{6}$

$x = 22$

**23.** $\dfrac{20}{7} = \dfrac{80}{x}$

$20 \cdot x = 7 \cdot 80$

$\dfrac{20 \cdot x}{20} = \dfrac{7 \cdot 80}{20}$

$x = \dfrac{7 \cdot 80}{20}$

$x = \dfrac{560}{20}$

$x = 28$

**25.** $\dfrac{12}{9} = \dfrac{x}{7}$

$12 \cdot 7 = 9 \cdot x$

$\dfrac{12 \cdot 7}{9} = \dfrac{9 \cdot x}{9}$

$\dfrac{12 \cdot 7}{9} = x$

$\dfrac{84}{9} = x$

$\dfrac{28}{3} = x$     Simplifying

$9\dfrac{1}{3} = x$     Writing a mixed numeral

**27.** $\dfrac{x}{13} = \dfrac{2}{9}$

$9 \cdot x = 13 \cdot 2$

$\dfrac{9 \cdot x}{9} = \dfrac{13 \cdot 2}{9}$

$x = \dfrac{13 \cdot 2}{9}$

$x = \dfrac{26}{9}, \text{ or } 2\dfrac{8}{9}$

**29.** $\dfrac{100}{25} = \dfrac{20}{n}$

$100 \cdot n = 25 \cdot 20$

$\dfrac{100 \cdot n}{100} = \dfrac{25 \cdot 20}{100}$

$n = \dfrac{25 \cdot 20}{100}$

$n = \dfrac{500}{100}$

$n = 5$

**31.** $\dfrac{6}{y} = \dfrac{18}{15}$

$6 \cdot 15 = y \cdot 18$

$\dfrac{6 \cdot 15}{18} = \dfrac{y \cdot 18}{18}$

$\dfrac{6 \cdot 15}{18} = y$

$\dfrac{90}{18} = y$

$5 = y$

**33.** $\dfrac{x}{3} = \dfrac{0}{9}$

$x \cdot 9 = 3 \cdot 0$

$\dfrac{x \cdot 9}{9} = \dfrac{3 \cdot 0}{9}$

$x = \dfrac{0}{9}$

$x = 0$

**35.** $\dfrac{1}{2} = \dfrac{7}{x}$

$1 \cdot x = 2 \cdot 7$

$x = \dfrac{2 \cdot 7}{1}$

$x = 14$

**37.** $\dfrac{1.2}{4} = \dfrac{x}{9}$

$1.2 \cdot 9 = 4 \cdot x$

$\dfrac{1.2 \cdot 9}{4} = \dfrac{4 \cdot x}{4}$

$\dfrac{1.2 \cdot 9}{4} = x$

$\dfrac{10.8}{4} = x$

$2.7 = x$

**39.** $\dfrac{8}{2.4} = \dfrac{6}{y}$

$8 \cdot y = 2.4 \cdot 6$

$\dfrac{8 \cdot y}{8} = \dfrac{2.4 \cdot 6}{8}$

$y = \dfrac{2.4 \cdot 6}{8}$

$y = \dfrac{14.4}{8}$

$y = 1.8$

**41.** $\dfrac{t}{0.16} = \dfrac{0.15}{0.40}$

$0.40 \times t = 0.16 \times 0.15$

$\dfrac{0.40 \times t}{0.40} = \dfrac{0.16 \times 0.15}{0.40}$

$t = \dfrac{0.16 \times 0.15}{0.40}$

$t = \dfrac{0.024}{0.40}$

$t = 0.06$

**43.** $\dfrac{0.5}{n} = \dfrac{2.5}{3.5}$

$0.5 \cdot 3.5 = n \cdot 2.5$

$\dfrac{0.5 \cdot 3.5}{2.5} = \dfrac{n \cdot 2.5}{2.5}$

$\dfrac{0.5 \cdot 3.5}{2.5} = n$

$\dfrac{1.75}{2.5} = n$

$0.7 = n$

**45.** $\dfrac{1.28}{3.76} = \dfrac{4.28}{y}$

$1.28 \times y = 3.76 \times 4.28$

$\dfrac{1.28 \times y}{1.28} = \dfrac{3.76 \times 4.28}{1.28}$

$y = \dfrac{3.76 \times 4.28}{1.28}$

$y = \dfrac{16.0928}{1.28} = 12.5725$

**47.** $\dfrac{7}{\frac{1}{4}} = \dfrac{28}{x}$

$7 \cdot x = \dfrac{1}{4} \cdot 28$

$\dfrac{7 \cdot x}{7} = \dfrac{\frac{1}{4} \cdot 28}{7}$

$x = \dfrac{\frac{1}{4} \cdot 28}{7}$

$x = \dfrac{7}{7} = 1$

**49.** $\dfrac{\frac{1}{5}}{\frac{1}{10}} = \dfrac{\frac{1}{10}}{x}$

$\dfrac{1}{5} \cdot x = \dfrac{1}{10} \cdot \dfrac{1}{10}$

$x = \dfrac{5}{1} \cdot \dfrac{1}{10} \cdot \dfrac{1}{10}$  Dividing by $\dfrac{1}{5}$

$x = \dfrac{5 \cdot 1 \cdot 1}{1 \cdot 10 \cdot 10} = \dfrac{\cancel{5} \cdot 1}{2 \cdot \cancel{5} \cdot 10}$

$x = \dfrac{1}{20}$

**51.** $\dfrac{\frac{y}{3}}{\frac{5}{}} = \dfrac{\frac{7}{12}}{\frac{14}{15}}$

$y \cdot \dfrac{14}{15} = \dfrac{3}{5} \cdot \dfrac{7}{12}$

$y = \dfrac{15}{14} \cdot \dfrac{3}{5} \cdot \dfrac{7}{12}$  Dividing by $\dfrac{14}{15}$

$y = \dfrac{15 \cdot 3 \cdot 7}{14 \cdot 5 \cdot 12} = \dfrac{\cancel{3} \cdot \cancel{5} \cdot 3 \cdot 7}{2 \cdot 7 \cdot \cancel{5} \cdot \cancel{3} \cdot 4}$

$y = \dfrac{3}{8}$

**53.** $\dfrac{x}{1\frac{3}{5}} = \dfrac{2}{15}$

$x \cdot 15 = 1\dfrac{3}{5} \cdot 2$

$x \cdot 15 = \dfrac{8}{5} \cdot 2$

$x = \dfrac{1}{15} \cdot \dfrac{8}{5} \cdot 2$  Dividing by 15

$x = \dfrac{16}{75}$

**55.**
$$\frac{2\frac{1}{2}}{3\frac{1}{3}} = \frac{x}{4\frac{1}{4}}$$

$$2\frac{1}{2} \cdot 4\frac{1}{4} = 3\frac{1}{3} \cdot x$$

$$\frac{5}{2} \cdot \frac{17}{4} = \frac{10}{3} \cdot x$$

$$\frac{3}{10} \cdot \frac{5}{2} \cdot \frac{17}{4} = x \qquad \text{Dividing by } \frac{10}{3}$$

$$\frac{3}{\cancel{5} \cdot 2} \cdot \frac{\cancel{5}}{2} \cdot \frac{17}{4} = x$$

$$\frac{3 \cdot 17}{2 \cdot 2 \cdot 4} = x$$

$$\frac{51}{16} = x, \text{ or}$$

$$3\frac{3}{16} = x$$

**57.**
$$\frac{5\frac{1}{5}}{6\frac{1}{6}} = \frac{y}{3\frac{1}{2}}$$

$$5\frac{1}{5} \cdot 3\frac{1}{2} = 6\frac{1}{6} \cdot y$$

$$\frac{26}{5} \cdot \frac{7}{2} = \frac{37}{6} \cdot y$$

$$\frac{6}{37} \cdot \frac{26}{5} \cdot \frac{7}{2} = y \qquad \text{Dividing by } \frac{37}{6}$$

$$\frac{\cancel{2} \cdot 3 \cdot 26 \cdot 7}{37 \cdot 5 \cdot \cancel{2}} = y$$

$$\frac{546}{185} = y, \text{ or}$$

$$2\frac{176}{185} = y$$

**59.** A ratio is the <u>quotient</u> of two quantities.

**61.** To compute an <u>average</u> of a set of numbers, we add the numbers and then <u>divide</u> by the number of addends.

**63.** In the equation $103 - 13 = 90$, the <u>subtrahend</u> is 13.

**65.** The sentence $\frac{2}{5} \cdot \frac{4}{9} = \frac{4}{9} \cdot \frac{2}{5}$ illustrates the <u>commutative</u> law of multiplication.

**67.**
$$\frac{1728}{5643} = \frac{836.4}{x}$$

$$1728 \cdot x = 5643 \cdot 836.4$$

$$\frac{1728 \cdot x}{1728} = \frac{5643 \cdot 836.4}{1728}$$

$$x = \frac{5643 \cdot 836.4}{1728}$$

$$x \approx 2731.4 \qquad \text{Using a calculator to multiply and divide}$$

The solution is approximately 2731.4.

**69.** Babe Ruth:
$$\frac{1330 \text{ strikeouts}}{714 \text{ home runs}} \approx 1.863 \frac{\text{strikeouts}}{\text{home run}}$$
Mike Schmidt:
$$\frac{1883 \text{ strikeouts}}{548 \text{ home runs}} \approx 3.436 \frac{\text{strikeouts}}{\text{home run}}$$

## Chapter 6 Mid-Chapter Review

**1.** The statement is true. See page 340 in the text.

**2.** The statement is true. See page 347 in the text.

**3.** The statement is false. See Example 6 in Section 6.2, for example.

**4.** If $\frac{x}{t} = \frac{y}{s}$, then $x \cdot s = t \cdot y$. Thus, the given statement is false.

**5.** $\dfrac{120 \text{ mi}}{2 \text{ hr}} = \dfrac{120}{2} \dfrac{\text{mi}}{\text{hr}} = 60 \text{ mi/hr}$

**6.**
$$\frac{x}{4} = \frac{3}{6}$$
$$x \cdot 6 = 4 \cdot 3$$
$$\frac{x \cdot 6}{6} = \frac{4 \cdot 3}{6}$$
$$x = \frac{4 \cdot 3}{6}$$
$$x = 2$$

**7.** The ratio is $\dfrac{4}{7}$.

**8.** The ratio is $\dfrac{313}{199}$.

**9.** The ratio is $\dfrac{35}{17}$.

**10.** The ratio is $\dfrac{59}{101}$.

**11.** $\dfrac{8}{12} = \dfrac{2 \cdot 4}{3 \cdot 4} = \dfrac{2}{3} \cdot \dfrac{4}{4} = \dfrac{2}{3}$

**12.** $\dfrac{25}{75} = \dfrac{25 \cdot 1}{3 \cdot 25} = \dfrac{25}{25} \cdot \dfrac{1}{3} = \dfrac{1}{3}$

**13.** $\dfrac{32}{28} = \dfrac{4 \cdot 8}{4 \cdot 7} = \dfrac{4}{4} \cdot \dfrac{8}{7} = \dfrac{8}{7}$

**14.** $\dfrac{100}{76} = \dfrac{4 \cdot 25}{4 \cdot 19} = \dfrac{4}{4} \cdot \dfrac{25}{19} = \dfrac{25}{19}$

**15.** $\dfrac{112}{56} = \dfrac{2 \cdot 56}{56 \cdot 1} = \dfrac{2}{1} \cdot \dfrac{56}{56} = \dfrac{2}{1}$

**16.** $\dfrac{15}{3} = \dfrac{3 \cdot 5}{3 \cdot 1} = \dfrac{3}{3} \cdot \dfrac{5}{1} = \dfrac{5}{1}$

**17.** $\dfrac{2.4}{8.4} = \dfrac{2.4}{8.4} \cdot \dfrac{10}{10} = \dfrac{24}{84} = \dfrac{2 \cdot 12}{7 \cdot 12} = \dfrac{2}{7} \cdot \dfrac{12}{12} = \dfrac{2}{7}$

**18.** $\dfrac{0.27}{0.45} = \dfrac{0.27}{0.45} \cdot \dfrac{100}{100} = \dfrac{27}{45} = \dfrac{3 \cdot 9}{5 \cdot 9} = \dfrac{3}{5} \cdot \dfrac{9}{9} = \dfrac{3}{5}$

**19.** $\dfrac{243 \text{ mi}}{4 \text{ hr}} = \dfrac{243}{4} \dfrac{\text{mi}}{\text{hr}} = 60.75 \text{ mi/hr, or } 60.75 \text{ mph}$

**20.** $\dfrac{146 \text{ km}}{3 \text{ hr}} = \dfrac{146}{3} \cdot \dfrac{\text{km}}{\text{hr}} \approx 48.67 \text{ km/hr}$

**21.** $\dfrac{65 \text{ m}}{5 \text{ sec}} = \dfrac{65}{5} \dfrac{\text{m}}{\text{sec}} = 13 \text{ m/sec}$

**22.** $\dfrac{97 \text{ ft}}{6 \text{ sec}} = \dfrac{97}{6} \dfrac{\text{ft}}{\text{sec}} \approx 16.17 \text{ ft/sec}$

**23.** $\dfrac{189 \text{ in.}}{7 \text{ days}} = \dfrac{189}{7} \dfrac{\text{in.}}{\text{day}} = 27 \text{ in./day}$

**24.** $\dfrac{130 \text{ free throws made}}{140 \text{ attempts}} = \dfrac{130}{140} \dfrac{\text{free throws made}}{\text{attempt}} \approx$
0.929 free throws made/attempt

**25.** $\dfrac{\$2.09}{18 \text{ oz}} = \dfrac{209\cent}{18 \text{ oz}} \approx 11.611\cent/\text{oz}$

**26.** $\dfrac{\$5.99}{12 \text{ oz}} = \dfrac{599\cent}{12 \text{ oz}} \approx 49.917\cent/\text{oz}$

**27.** We can use cross products.

$3 \cdot 35 = 105 \qquad 7 \cdot 15 = 105$

(cross products diagram: 3, 15, 7, 35)

Since the cross products are the same, $105 = 105$, we know that $\dfrac{3}{7} = \dfrac{15}{35}$, so the numbers are proportional.

**28.** We can use cross products.

$9 \cdot 5 = 45 \qquad 7 \cdot 7 = 49$

(cross products diagram: 9, 7, 7, 5)

Since the cross products are not the same, $45 \neq 49$, we know that the numbers are not proportional.

**29.** We can use cross products.

$2.4 \cdot 2.8 = 6.72 \qquad 1.5 \cdot 3.2 = 4.8$

(cross products diagram: 2.4, 3.2, 1.5, 2.8)

Since the cross products are not the same, $6.72 \neq 4.8$, we know that the numbers are not proportional.

**30.** We can use cross products.

$1\frac{3}{4} \cdot 6\frac{2}{3} = 11\frac{2}{3} \qquad 1\frac{1}{3} \cdot 8\frac{3}{4} = 11\frac{2}{3}$

(cross products diagram: $1\frac{3}{4}$, $8\frac{3}{4}$, $1\frac{1}{3}$, $6\frac{2}{3}$)

Since the cross products are the same, $11\frac{2}{3} = 11\frac{2}{3}$, we know that the numbers are proportional.

**31.** $\dfrac{9}{15} = \dfrac{x}{20}$

$9 \cdot 20 = 15 \cdot x$

$\dfrac{9 \cdot 20}{15} = \dfrac{15 \cdot x}{15}$

$\dfrac{9 \cdot 20}{15} = x$

$\dfrac{180}{15} = x$

$12 = x$

**32.** $\dfrac{x}{24} = \dfrac{30}{18}$

$x \cdot 18 = 24 \cdot 30$

$\dfrac{x \cdot 18}{18} = \dfrac{24 \cdot 30}{18}$

$x = \dfrac{24 \cdot 30}{18}$

$x = \dfrac{720}{18}$

$x = 40$

**33.** $\dfrac{12}{y} = \dfrac{20}{15}$

$12 \cdot 15 = y \cdot 20$

$\dfrac{12 \cdot 15}{20} = \dfrac{y \cdot 20}{20}$

$\dfrac{12 \cdot 15}{20} = y$

$\dfrac{180}{20} = y$

$9 = y$

**34.** $\dfrac{2}{7} = \dfrac{10}{y}$

$2 \cdot y = 7 \cdot 10$

$\dfrac{2 \cdot y}{2} = \dfrac{7 \cdot 10}{2}$

$y = \dfrac{7 \cdot 10}{2}$

$y = \dfrac{70}{2}$

$y = 35$

**35.** $\dfrac{y}{1.2} = \dfrac{1.1}{0.6}$

$y \cdot 0.6 = 1.2 \cdot 1.1$

$\dfrac{y \cdot 0.6}{0.6} = \dfrac{1.2 \cdot 1.1}{0.6}$

$y = \dfrac{1.32}{0.6}$

$y = 2.2$

**36.**
$$\frac{0.24}{0.02} = \frac{y}{0.36}$$
$$0.24 \cdot 0.36 = 0.02 \cdot y$$
$$\frac{0.24 \cdot 0.36}{0.02} = \frac{0.02 \cdot y}{0.02}$$
$$\frac{0.24 \cdot 0.36}{0.02} = y$$
$$\frac{0.0864}{0.02} = y$$
$$4.32 = y$$

**37.**
$$\frac{\frac{1}{4}}{x} = \frac{\frac{1}{8}}{\frac{1}{4}}$$
$$\frac{1}{4} \cdot \frac{1}{4} = x \cdot \frac{1}{8}$$
$$\frac{1}{4} \cdot \frac{1}{4} \cdot \frac{8}{1} = x \qquad \text{Dividing by } \frac{1}{8}$$
$$\frac{8}{16} = x$$
$$\frac{1}{2} = x$$

**38.**
$$\frac{1\frac{1}{2}}{3\frac{1}{4}} = \frac{7\frac{1}{2}}{x}$$
$$1\frac{1}{2} \cdot x = 3\frac{1}{4} \cdot 7\frac{1}{2}$$
$$\frac{3}{2} \cdot x = \frac{13}{4} \cdot \frac{15}{2}$$
$$x = \frac{2}{3} \cdot \frac{13}{4} \cdot \frac{15}{2} \qquad \text{Dividing by } \frac{3}{2}$$
$$x = \frac{\cancel{2} \cdot 13 \cdot \cancel{3} \cdot 5}{\cancel{3} \cdot 4 \cdot \cancel{2}}$$
$$x = \frac{65}{4}, \text{ or } 16\frac{1}{4}$$

**39.** Yes; every ratio $\frac{a}{b}$ can be written as $\frac{\frac{a}{b}}{1}$.

**40.** By making some sketches, we see that the rectangle's length must be twice the width. We can show this algebraically as follows:
$$\frac{l}{2l + 2w} = \frac{1}{3}$$
$$3l = 2l + 2w$$
$$l = 2w$$

**41.** The student's approach is not necessarily a bad one. However, when we use the approach of equating cross-products we eliminate the need to find the least common denominator.

**42.** The longer a student studies, the higher his or her grade will be. Consider the proportion $\frac{96}{8} = \frac{78}{6\frac{1}{2}}$. This could represent the situation that one student gets a test grade of 96 after studying for 8 hr while another student gets a score of 78 after studying for $6\frac{1}{2}$ hr.

## Exercise Set 6.4

**1. Familiarize.** Let $h$ = the number of hours Lisa would have to study to receive a score of 92.

**Translate.** We translate to a proportion, keeping the number of hours in the numerators.
$$\begin{array}{r} \text{Hours} \rightarrow \\ \text{Score} \rightarrow \end{array} \frac{9}{75} = \frac{h}{92} \begin{array}{l} \leftarrow \text{Hours} \\ \leftarrow \text{Score} \end{array}$$

**Solve.** We solve the proportion.
$$9 \cdot 92 = 75 \cdot h \qquad \text{Equating cross products}$$
$$\frac{9 \cdot 92}{75} = \frac{75 \cdot h}{75}$$
$$\frac{9 \cdot 92}{75} = h$$
$$11.04 = h$$

**Check.** We substitute into the proportion and check cross products.
$$\frac{9}{75} = \frac{11.04}{92}$$
$$9 \cdot 92 = 828; \ 75 \cdot 11.04 = 828$$
The cross products are the same, so the answer checks.

**State.** Lisa would have to study 11.04 hr to get a score of 92.

**3. Familiarize.** Let $s$ = the number of students estimated to be in the class.

**Translate.** We translate to a proportion.
$$\begin{array}{r} \text{Class size} \rightarrow \\ \text{Lefties} \rightarrow \end{array} \frac{40}{6} = \frac{s}{9} \begin{array}{l} \leftarrow \text{Class size} \\ \leftarrow \text{Lefties} \end{array}$$

**Solve.**
$$40 \cdot 9 = 6 \cdot s$$
$$\frac{40 \cdot 9}{6} = s$$
$$\frac{2 \cdot 20 \cdot 3 \cdot 3}{2 \cdot 3} = s$$
$$20 \cdot 3 = s$$
$$60 = s$$

**Check.** We substitute in the proportion and check cross products.
$$\frac{40}{6} = \frac{60}{9}$$
$$40 \cdot 9 = 360; \ 6 \cdot 60 = 360$$
The cross products are the same, so the answer checks.

**State.** If a class includes 9 lefties, we estimate that there are 60 students in the class.

**5.** a) **Familiarize.** Let $g$ = the number of gallons of gasoline needed to drive 2690 mi.

**Translate.** We translate to a proportion.
$$\begin{array}{r} \text{Gallons} \rightarrow \\ \text{Miles} \rightarrow \end{array} \frac{15.5}{341} = \frac{g}{2690} \begin{array}{l} \leftarrow \text{Gallons} \\ \leftarrow \text{Miles} \end{array}$$

*Solve.*

$$15.5 \cdot 2690 = 341 \cdot g \quad \text{Equating cross products}$$

$$\frac{15.5 \cdot 2690}{341} = g$$

$$122 \approx g$$

**Check.** We find how far the car can be driven on 1 gallon of gasoline and then divide to find the number of gallons required for a 2690-mi trip.

$$341 \div 15.5 = 22 \text{ and } 2690 \div 22 \approx 122$$

The answer checks.

**State.** It will take about 122 gal of gasoline to drive 2690 mi.

b) **Familiarize.** Let $d =$ the number of miles the car can be driven on 140 gal of gasoline.

**Translate.** We translate to a proportion.

$$\begin{array}{l} \text{Gallons} \rightarrow \\ \text{Miles} \rightarrow \end{array} \frac{15.5}{341} = \frac{140}{d} \begin{array}{l} \leftarrow \text{Gallons} \\ \leftarrow \text{Miles} \end{array}$$

*Solve.*

$$15.5 \cdot d = 341 \cdot 140 \quad \text{Equating cross products}$$

$$d = \frac{341 \cdot 140}{15.5}$$

$$d = 3080$$

**Check.** From the check in part (a) we know that the car can be driven 22 mi on 1 gal of gasoline. We multiply to find how far it can be driven on 140 gal.

$$140 \cdot 22 = 3080$$

The answer checks.

**State.** The car can be driven 3080 mi on 140 gal of gasoline.

7. **Familiarize.** Let $a =$ the number of Americans who would be considered overweight or obese in 2015, in millions.

**Translate.** We translate to a proportion.

$$\begin{array}{l} \text{Overweight/obese} \rightarrow \\ \text{Total Americans} \rightarrow \end{array} \frac{66}{100} = \frac{a}{322} \begin{array}{l} \leftarrow \text{Overweight/obese} \\ \leftarrow \text{Total Americans} \end{array}$$

*Solve.*

$$66 \cdot 322 = 100 \cdot a \quad \text{Equating cross products}$$

$$\frac{66 \cdot 322}{100} = a$$

$$212.52 = a$$

**Check.** We substitute in the proportion and check cross products.

$$\frac{66}{100} = \frac{212.52}{322}$$

$$66 \cdot 322 = 21,252; \ 100 \cdot 212.52 = 21,252$$

The cross products are the same, so the answer checks.

**State.** At the given rate, 212.52 million, or 212,520,000 Americans would be overweight or obese in 2015.

9. **Familiarize.** Let $d =$ the number of defective bulbs in a lot of 2500.

**Translate.** We translate to a proportion.

$$\begin{array}{l} \text{Defective bulbs} \rightarrow \\ \text{Bulbs in lot} \rightarrow \end{array} \frac{7}{100} = \frac{d}{2500} \begin{array}{l} \leftarrow \text{Defective bulbs} \\ \leftarrow \text{Bulbs in lot} \end{array}$$

*Solve.*

$$7 \cdot 2500 = 100 \cdot d$$

$$\frac{7 \cdot 2500}{100} = d$$

$$\frac{7 \cdot 25 \cdot 100}{100} = d$$

$$7 \cdot 25 = d$$

$$175 = d$$

**Check.** We substitute in the proportion and check cross products.

$$\frac{7}{100} = \frac{175}{2500}$$

$$7 \cdot 2500 = 17,500; \ 100 \cdot 175 = 17,500$$

The cross products are the same, so the answer checks.

**State.** There will be 175 defective bulbs in a lot of 2500.

11. **Familiarize.** Let $s =$ the number of square feet of siding that Fred can paint with 7 gal of paint.

**Translate.** We translate to a proportion.

$$\begin{array}{l} \text{Gallons} \rightarrow \\ \text{Siding} \rightarrow \end{array} \frac{3}{1275} = \frac{7}{s} \begin{array}{l} \leftarrow \text{Gallons} \\ \leftarrow \text{Siding} \end{array}$$

*Solve.*

$$3 \cdot s = 1275 \cdot 7$$

$$s = \frac{1275 \cdot 7}{3}$$

$$s = \frac{3 \cdot 425 \cdot 7}{3}$$

$$s = 425 \cdot 7$$

$$s = 2975$$

**Check.** We find the number of square feet covered by 1 gallon of paint and then multiply that number by 7.

$$1275 \div 3 = 425 \text{ and } 425 \cdot 7 = 2975$$

The answer checks.

**State.** Fred can paint 2975 ft$^2$ of siding with 7 gal of paint.

13. **Familiarize.** Let $p =$ the number of published pages in a 540-page manuscript.

**Translate.** We translate to a proportion.

$$\begin{array}{l} \text{Published pages} \rightarrow \\ \text{Manuscript} \rightarrow \end{array} \frac{5}{6} = \frac{p}{540} \begin{array}{l} \leftarrow \text{Published pages} \\ \leftarrow \text{Manuscript} \end{array}$$

*Solve.*

$$5 \cdot 540 = 6 \cdot p$$

$$\frac{5 \cdot 540}{6} = p$$

$$\frac{5 \cdot 6 \cdot 90}{6} = p$$

$$5 \cdot 90 = p$$

$$450 = p$$

**Check.** We substitute in the proportion and check cross products.

$$\frac{5}{6} = \frac{450}{540}$$

$$5 \cdot 540 = 2700; \ 6 \cdot 450 = 2700$$

The cross products are the same, so the answer checks.

**State**. A 540-page manuscript will become 450 published pages.

**15.** a) **Familiarize**. Let $a =$ the number of British pounds equivalent to 50 U.S. dollars.

**Translate**. We translate to a proportion.

$$\text{U.S.} \;\rightarrow\; \frac{1}{0.5613} = \frac{50}{a} \;\begin{matrix}\leftarrow\; \text{U.S.}\\ \leftarrow\; \text{British}\end{matrix}$$
British $\rightarrow$

**Solve**.

$$1 \cdot a = 0.5613 \cdot 50 \quad \text{Equating cross products}$$
$$a = 28.065$$

**Check**. We substitute in the proportion and check cross products.

$$\frac{1}{0.5613} = \frac{50}{28.065}$$
$$1 \cdot 28.065 = 28.065; \; 0.5613 \cdot 50 = 28.065$$

The cross products are the same, so the answer checks.

**State**.  50 U.S. dollars would be worth 28.065 British pounds.

b) **Familiarize**. Let $c =$ the cost of the car in U.S. dollars.

**Translate**. We translate to a proportion.

$$\text{U.S.} \;\rightarrow\; \frac{1}{0.5613} = \frac{c}{8640} \;\begin{matrix}\leftarrow\; \text{U.S.}\\ \leftarrow\; \text{British}\end{matrix}$$
British $\rightarrow$

**Solve**.

$$1 \cdot 8640 = 0.5613 \cdot c \quad \text{Equating cross products}$$
$$\frac{1 \cdot 8640}{0.5613} = c$$
$$15{,}392.84 \approx c$$

**Check**. We substitute in the proportion and check cross products.

$$\frac{1}{0.5613} = \frac{15{,}392.84}{8640}$$
$$1 \cdot 8640 = 8640; \; 0.5613 \cdot 15{,}392.84 \approx 8640$$

The cross products are about the same. Remember that we rounded the value of $c$. The answer checks.

**State**. The car would cost $15,392.84 in U.S. dollars.

**17.** a) **Familiarize**. Let $a =$ the number of Japanese yen equivalent to 125 U.S. dollars.

**Translate**. We translate to a proportion.

$$\text{U.S.} \;\rightarrow\; \frac{1}{107.197} = \frac{125}{a} \;\begin{matrix}\leftarrow\; \text{U.S.}\\ \leftarrow\; \text{Japanese}\end{matrix}$$
Japanese $\rightarrow$

**Solve**.

$$1 \cdot a = 107.197 \cdot 125 \quad \text{Equating cross products}$$
$$a = 13{,}399.625$$

**Check**. We substitute in the proportion and check cross products.

$$\frac{1}{107.197} = \frac{125}{13{,}399.625}$$
$$1 \cdot 13{,}399.625 = 13{,}399.625; \; 107.197 \cdot 125 = 13{,}399.625$$

The cross products are the same, so the answer checks.

**State**. 125 U.S. dollars would be worth 13,399.625 Japanese yen.

b) **Familiarize**. Let $c =$ the cost of the vase in U.S. dollars.

**Translate**. We translate to a proportion.

$$\text{U.S.} \;\rightarrow\; \frac{1}{107.197} = \frac{c}{5120} \;\begin{matrix}\leftarrow\; \text{U.S.}\\ \leftarrow\; \text{Japanese}\end{matrix}$$
Japanese $\rightarrow$

**Solve**.

$$1 \cdot 5120 = 107.197 \cdot c \quad \text{Equating cross products}$$
$$\frac{1 \cdot 5120}{107.197} = c$$
$$47.76 \approx c$$

**Check**. We substitute in the proportion and check cross products.

$$\frac{1}{107.197} = \frac{47.76}{5120}$$
$$1 \cdot 5120 = 5120; \; 107.197 \cdot 47.76 = 5119.72872 \approx 5120$$

The cross products are about the same. Remember that we rounded the value of $c$. The answer checks.

**State**. The vase cost $47.76 in U.S. dollars.

**19.** **Familiarize**. Let $m =$ the number of miles the car will be driven in 1 year. Note that 1 year $=$ 12 months.

**Translate**.

$$\text{Months} \;\rightarrow\; \frac{8}{9000} = \frac{12}{m} \;\begin{matrix}\leftarrow\; \text{Months}\\ \leftarrow\; \text{Miles}\end{matrix}$$
Miles $\rightarrow$

**Solve**.

$$8 \cdot m = 9000 \cdot 12$$
$$m = \frac{9000 \cdot 12}{8}$$
$$m = \frac{2 \cdot 4500 \cdot 3 \cdot 4}{2 \cdot 4}$$
$$m = 4500 \cdot 3$$
$$m = 13{,}500$$

**Check**. We find the average number of miles driven in 1 month and then multiply to find the number of miles the car will be driven in 1 yr, or 12 months.

$$9000 \div 8 = 1125 \text{ and } 12 \cdot 1125 = 13{,}500$$

The answer checks.

**State**. At the given rate, the car will be driven 13,500 mi in one year.

**21.** **Familiarize**. Let $c =$ the number of calories in 6 cups of cereal.

**Translate**. We translate to a proportion, keeping the number of calories in the numerators.

$$\text{Calories} \;\rightarrow\; \frac{110}{3/4} = \frac{c}{6} \;\begin{matrix}\leftarrow\; \text{Calories}\\ \leftarrow\; \text{Cups}\end{matrix}$$
Cups $\rightarrow$

*Solve*. We solve the proportion.

$$110 \cdot 6 = \frac{3}{4} \cdot c \quad \text{Equating cross products}$$

$$\frac{110 \cdot 6}{3/4} = \frac{\frac{3}{4} \cdot c}{3/4}$$

$$\frac{110 \cdot 6}{3/4} = c$$

$$110 \cdot 6 \cdot \frac{4}{3} = c$$

$$880 = c$$

*Check*. We substitute into the proportion and check cross products.

$$\frac{110}{3/4} = \frac{880}{6}$$

$$110 \cdot 6 = 660; \quad \frac{3}{4} \cdot 880 = 660$$

The cross products are the same, so the answer checks.

*State*. There are 880 calories in 6 cups of cereal.

**23. *Familiarize***. Let $z$ = the number of pounds of zinc in the alloy.

*Translate*. We translate to a proportion.

$$\begin{array}{l} \text{Zinc} \rightarrow \\ \text{Copper} \rightarrow \end{array} \frac{3}{13} = \frac{z}{520} \begin{array}{l} \leftarrow \text{Zinc} \\ \leftarrow \text{Copper} \end{array}$$

*Solve*.

$$3 \cdot 520 = 13 \cdot z$$

$$\frac{3 \cdot 520}{13} = z$$

$$\frac{3 \cdot 13 \cdot 40}{13} = z$$

$$3 \cdot 40 = z$$

$$120 = z$$

*Check*. We substitute in the proportion and check cross products.

$$\frac{3}{13} = \frac{120}{520}$$

$$3 \cdot 520 = 1560; \quad 13 \cdot 120 = 1560$$

The cross products are the same, so the answer checks.

*State*. There are 120 lb of zinc in the alloy.

**25. *Familiarize***. Let $p$ = the number of gallons of paint Helen should buy.

*Translate*. We translate to a proportion.

$$\begin{array}{l} \text{Area} \rightarrow \\ \text{Paint} \rightarrow \end{array} \frac{950}{2} = \frac{30,000}{p} \begin{array}{l} \leftarrow \text{Area} \\ \leftarrow \text{Paint} \end{array}$$

*Solve*.

$$950 \cdot p = 2 \cdot 30,000$$

$$p = \frac{2 \cdot 30,000}{950}$$

$$p = \frac{2 \cdot 50 \cdot 600}{19 \cdot 50}$$

$$p = \frac{2 \cdot 600}{19}$$

$$p = \frac{1200}{19}, \text{ or } 63\frac{3}{19}$$

*Check*. We find the area covered by 1 gal of paint and then divide to find the number of gallons needed to paint a 30,000-ft$^2$ wall.

$$950 \div 2 = 475 \text{ and } 30,000 \div 475 = 63\frac{3}{19}$$

The answer checks.

*State*. Since Helen is buying paint in one gallon cans, she will have to buy 64 cans of paint.

**27. *Familiarize***. Let $s$ = the number of ounces of grass seed needed for 5000 ft$^2$ of lawn.

*Translate*. We translate to a proportion.

$$\begin{array}{l} \text{Seed} \rightarrow \\ \text{Area} \rightarrow \end{array} \frac{60}{3000} = \frac{s}{5000} \begin{array}{l} \leftarrow \text{Seed} \\ \leftarrow \text{Area} \end{array}$$

*Solve*.

$$60 \cdot 5000 = 3000 \cdot s$$

$$\frac{60 \cdot 5000}{3000} = s$$

$$100 = s$$

*Check*. We find the number of ounces of seed needed for 1 ft$^2$ of lawn and then multiply this number by 5000:

$$60 \div 3000 = 0.02 \text{ and } 5000(0.02) = 100$$

The answer checks.

*State*. 100 oz of grass seed would be needed to seed 5000 ft$^2$ of lawn.

**29. *Familiarize***. Let $D$ = the number of deer in the game preserve.

*Translate*. We translate to a proportion.

$$\begin{array}{l} \text{Deer tagged} \\ \text{originally} \end{array} \rightarrow \frac{318}{D} = \frac{56}{168} \begin{array}{l} \leftarrow \text{Tagged deer} \\ \text{caught later} \end{array}$$
$$\begin{array}{l} \text{Deer in game} \\ \text{preserve} \end{array} \rightarrow \qquad\qquad \begin{array}{l} \leftarrow \text{Deer caught} \\ \text{later} \end{array}$$

*Solve*.

$$318 \cdot 168 = 56 \cdot D$$

$$\frac{318 \cdot 168}{56} = D$$

$$954 = D$$

*Check*. We substitute in the proportion and check cross products.

$$\frac{318}{954} = \frac{56}{168}; \quad 318 \cdot 168 = 53,424; \quad 954 \cdot 56 = 53,424$$

Since the cross products are the same, the answer checks.

*State*. We estimate that there are 954 deer in the game preserve.

**31. *Familiarize***. Let $d$ = the actual distance between the cities.

*Translate*. We translate to a proportion.

$$\begin{array}{l} \text{Map distance} \rightarrow \\ \text{Actual distance} \rightarrow \end{array} \frac{1}{16.6} = \frac{3.5}{d} \begin{array}{l} \leftarrow \text{Map distance} \\ \leftarrow \text{Actual distance} \end{array}$$

*Solve*.

$$1 \cdot d = 16.6 \cdot 3.5$$

$$d = 58.1$$

*Check.* We use a different approach. Since 1 in. represents 16.6 mi, we multiply 16.6 by 3.5:

$$3.5(16.6) = 58.1$$

The answer checks.

*State.* The cities are 58.1 mi apart.

**33.** *Familiarize.* Let $g$ = the number of gallons of gasoline needed to travel 126 mi.

*Translate.* We translate to a proportion.

$$\begin{array}{cc} \text{Miles} \;\to\; \dfrac{84}{6.5} = \dfrac{126}{g} & \leftarrow \text{Miles} \\ \text{Gallons} \to & \leftarrow \text{Gallons} \end{array}$$

*Solve.*

$$84 \cdot g = 6.5 \cdot 126 \quad \text{Equating cross products}$$

$$g = \frac{6.5 \cdot 126}{84} \quad \text{Dividing by 84}$$

$$g = \frac{819}{84}$$

$$g = 9.75$$

*Check.* We substitute in the proportion and check cross products.

$$\frac{84}{6.5} = \frac{126}{9.75}$$

$$84 \cdot 9.75 = 819; \; 6.5 \cdot 126 = 819$$

The cross products are the same, so the answer checks.

*State.* 9.75 gallons of gasoline are needed to travel 126 mi.

**35.** a) *Familiarize.* Let $g$ = the number of games required for Ryan Howard to hit 50 home runs.

*Translate.* We translate to a proportion.

$$\begin{array}{cc} \text{Home runs} \to \dfrac{43}{147} = \dfrac{50}{g} & \leftarrow \text{Home runs} \\ \text{Games} \;\to & \leftarrow \;\; \text{Games} \end{array}$$

*Solve.*

$$43 \cdot g = 147 \cdot 50$$

$$g = \frac{147 \cdot 50}{43}$$

$$g \approx 171$$

*Check.*We substitute in the proportion and check cross products.

$$\frac{43}{147} = \frac{50}{171}$$

$$43 \cdot 171 = 7353, \; 147 \cdot 50 = 7350$$

Since we rounded the value for $g$ and $7353 \approx 7350$, we have a check.

*State.* At the given rate it would take 171 games for Ryan Howard to hit 50 home runs.

b) *Familiarize.* Let $h$ = the number of home runs Ryan Howard would hit in the 162-game baseball season.

*Translate.* We translate to a proportion.

$$\begin{array}{cc} \text{Home runs} \to \dfrac{43}{147} = \dfrac{h}{162} & \leftarrow \text{Home runs} \\ \text{Games} \;\to & \leftarrow \;\; \text{Games} \end{array}$$

*Solve.*

$$43 \cdot 162 = 147 \cdot h$$

$$\frac{43 \cdot 162}{147} = h$$

$$47 \approx h$$

*Check.*We substitute in the proportion and check cross products.

$$\frac{43}{147} = \frac{47}{162}$$

$$43 \cdot 162 = 6966; \; 147 \cdot 47 = 6909$$

Since we rounded the value for $h$ and $6966 \approx 6909$, we have a check.

*State.* At the given rate Ryan Howard would hit 47 home runs in the 162-game baseball season.

**37.** 1 is neither prime nor composite.

**39.** The number 28 has factors 1, 2, 4, 7, 14, and 28. It is composite.

**41.** The only factors of 47 are 47 itself and 1, so 47 is prime.

**43.**
$$\begin{array}{r} 7 \\ 2 \,\overline{\big|\, 1\,4\,} \\ 2 \,\overline{\big|\, 2\,8\,} \\ 2 \,\overline{\big|\, 5\,6\,} \end{array}$$

$$56 = 2 \cdot 2 \cdot 2 \cdot 7, \text{ or } 2^3 \cdot 7$$

**45.**
$$\begin{array}{r} 3\,1 \\ 3 \,\overline{\big|\, 9\,3\,} \end{array}$$

$$93 = 3 \cdot 31$$

**47.** *Familiarize.* Let $f$ = the number of faculty positions required to maintain the current student-to-faculty ratio after the university expands.

*Translate.* We translate to a proportion.

$$\begin{array}{cc} \text{Students} \to \dfrac{2700}{217} = \dfrac{2900}{f} & \leftarrow \text{Students} \\ \text{Faculty} \;\to & \leftarrow \text{Faculty} \end{array}$$

*Solve.*

$$2700 \cdot f = 217 \cdot 2900$$

$$f = \frac{217 \cdot 2900}{2700}$$

$$f = \frac{6293}{27}, \text{ or } 233\frac{2}{27}$$

Since it is impossible to create a fractional part of a position, we round up to the nearest whole position. Thus, 234 positions will be required after the university expands. We subtract to find how many new positions should be created:

$$234 - 217 = 17$$

*Check.* We substitute in the proportion and check cross products.

$$\frac{2700}{217} = \frac{2900}{6293/27}; \; 2700 \cdot \frac{6293}{27} = 629,300;$$

$$217 \cdot 2900 = 629,300$$

The cross products are the same, so the answer checks.

*State.* 17 new faculty positions should be created.

**49. Familiarize.** Let $r$ = the number of earned runs Cy Young gave up in his career.

**Translate.** We translate to a proportion.

$$\text{Runs} \rightarrow \frac{2.63}{9} = \frac{r}{7356} \leftarrow \text{Runs} \atop \leftarrow \text{Innings}$$

Runs $\rightarrow$ ... Innings $\rightarrow$

**Solve.**

$$2.63 \cdot 7356 = 9 \cdot r$$
$$\frac{2.63 \cdot 7356}{9} = r$$
$$2150 \approx r$$

**Check.** We substitute in the proportion and check cross products.

$$\frac{2.63}{9} = \frac{2150}{7356}$$

$2.63 \cdot 7356 = 19,346.28$ and $9 \cdot 2150 = 19,350 \approx 19,346.28$

The cross products are about the same, so the answer checks.

**State.** Cy Young gave up 2150 earned runs in his career.

**51.** $1 + 3 + 2 = 6$, and $\$900/6 = \$150$.

Then $1 \cdot \$150$, or $\$150$, would be spent on a CD player; $3 \cdot \$150$, or $\$450$, would be spent on a receiver; and $2 \cdot \$150$, or $\$300$, would be spent on speakers.

## Exercise Set 6.5

**1.** The ratio of $h$ to 5 is the same as the ratio of 45 to 9. We have the proportion

$$\frac{h}{5} = \frac{45}{9}.$$

Solve: $9 \cdot h = 5 \cdot 45$    Equating cross products

$\quad h = \dfrac{5 \cdot 45}{9}$    Dividing by 9 on both sides

$\quad h = 25$    Simplifying

The missing length $h$ is 25.

**3.** The ratio of $x$ to 2 is the same as the ratio of 2 to 3. We have the proportion

$$\frac{x}{2} = \frac{2}{3}.$$

Solve: $3 \cdot x = 2 \cdot 2$    Equating cross products

$\quad x = \dfrac{2 \cdot 2}{3}$    Dividing by 3 on both sides

$\quad x = \dfrac{4}{3}$, or $1\dfrac{1}{3}$

The missing length $x$ is $\dfrac{4}{3}$, or $1\dfrac{1}{3}$. We could also have used $\dfrac{x}{2} = \dfrac{1}{1\frac{1}{2}}$ to find $x$.

**5.** First we find $x$. The ratio of $x$ to 9 is the same as the ratio of 6 to 8. We have the proportion

$$\frac{x}{9} = \frac{6}{8}.$$

Solve: $8 \cdot x = 9 \cdot 6$

$\quad x = \dfrac{9 \cdot 6}{8}$

$\quad x = \dfrac{27}{4}$, or $6\dfrac{3}{4}$

The missing length $x$ is $\dfrac{27}{4}$, or $6\dfrac{3}{4}$.

Next we find $y$. The ratio of $y$ to 12 is the same as the ratio of 6 to 8. We have the proportion

$$\frac{y}{12} = \frac{6}{8}.$$

Solve: $8 \cdot y = 12 \cdot 6$

$\quad y = \dfrac{12 \cdot 6}{8}$

$\quad y = 9$

The missing length $y$ is 9.

**7.** First we find $x$. The ratio of $x$ to 2.5 is the same as the ratio of 2.1 to 0.7. We have the proportion

$$\frac{x}{2.5} = \frac{2.1}{0.7}.$$

Solve: $0.7 \cdot x = 2.5 \cdot 2.1$

$\quad x = \dfrac{2.5 \cdot 2.1}{0.7}$

$\quad x = 7.5$

The missing length $x$ is 7.5.

Next we find $y$. The ratio of $y$ to 2.4 is the same as the ratio of 2.1 to 0.7. We have the proportion

$$\frac{y}{2.4} = \frac{2.1}{0.7}.$$

Solve: $0.7 \cdot y = 2.4 \cdot 2.1$

$\quad y = \dfrac{2.4 \cdot 2.1}{0.7}$

$\quad y = 7.2$

The missing length $y$ is 7.2.

**9.** If we use the sun's rays to represent the third side of a triangle in a drawing of the situation, we see that we have similar triangles. We let $s$ = the length of a shadow cast by a person 2 m tall.

The ratio of $s$ to 5 is the same as the ratio of 2 to 8. We have the proportion

$$\frac{s}{5} = \frac{2}{8}.$$

Solve: $8 \cdot s = 5 \cdot 2$

$\quad s = \dfrac{5 \cdot 2}{8}$

$\quad s = \dfrac{5}{4}$, or 1.25

The length of a shadow cast by a person 2 m tall is 1.25 m.

**11.** If we use the sun's rays to represent the third side of a triangle in a drawing of the situation, we see that we have similar triangles. We let $h =$ the height of the tree.

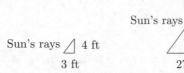

The ratio of $h$ to 4 is the same as the ratio of 27 to 3. We have the proportion
$$\frac{h}{4} = \frac{27}{3}.$$
Solve:  $3 \cdot h = 4 \cdot 27$
$$h = \frac{4 \cdot 27}{3}$$
$$h = 36$$
The tree is 36 ft tall.

**13.** The ratio of $h$ to 7 ft is the same as the ratio of 6 ft to 6 ft. We have the proportion
$$\frac{h}{7} = \frac{6}{6}.$$
Solve:  $6 \cdot h = 7 \cdot 6$
$$h = \frac{7 \cdot 6}{6}$$
$$h = 7$$
The wall is 7 ft high.

**15.** Since the ratio of $d$ to 25 ft is the same as the ratio of 40 ft to 10 ft, we have the proportion
$$\frac{d}{25} = \frac{40}{10}.$$
Solve:  $10 \cdot d = 25 \cdot 40$
$$d = \frac{25 \cdot 40}{10}$$
$$d = 100$$
The distance across the river is 100 ft.

**17.**   Width $\rightarrow$   $\dfrac{6}{9} = \dfrac{x}{6}$   $\leftarrow$ Width
Length $\rightarrow$                $\leftarrow$ Length

Solve: $\dfrac{2}{3} = \dfrac{x}{6}$      Rewriting $\dfrac{6}{9}$ as $\dfrac{2}{3}$
$2 \cdot 6 = 3 \cdot x$     Equating cross products
$$\frac{2 \cdot 6}{3} = x$$
$$\frac{2 \cdot 2 \cdot 3}{3} = x$$
$$2 \cdot 2 = x$$
$$4 = x$$
The missing length $x$ is 4.

**19.**   Width $\rightarrow$   $\dfrac{4}{7} = \dfrac{6}{x}$   $\leftarrow$ Width
Length $\rightarrow$                $\leftarrow$ Length

Solve:  $4 \cdot x = 7 \cdot 6$           Equating cross products
$$x = \frac{7 \cdot 6}{4}$$
$$x = \frac{7 \cdot 2 \cdot 3}{2 \cdot 2}$$
$$x = \frac{7 \cdot 3}{2}$$
$$x = \frac{21}{2}, \text{ or } 10\frac{1}{2}$$
The missing length $x$ is $10\frac{1}{2}$.

**21.** First we find $x$. The ratio of $x$ to 8 is the same as the ratio of 3 to 4. We have the proportion
$$\frac{x}{8} = \frac{3}{4}.$$
Solve:  $4 \cdot x = 8 \cdot 3$
$$x = \frac{8 \cdot 3}{4}$$
$$x = 6$$
The missing length $x$ is 6.

Next we find $y$. The ratio of $y$ to 7 is the same as the ratio of 3 to 4. We have the proportion
$$\frac{y}{7} = \frac{3}{4}.$$
Solve:  $4 \cdot y = 7 \cdot 3$
$$y = \frac{7 \cdot 3}{4}$$
$$y = \frac{21}{4}, \text{ or } 5\frac{1}{4}, \text{ or } 5.25$$
The missing length $y$ is $\frac{21}{4}$, or $5\frac{1}{4}$, or 5.25.

Finally we find $z$. The ratio of $z$ to 4 is the same as the ratio of 3 to 4. This statement tells us that $z$ must be 3. We could also calculate this using the proportion
$$\frac{z}{4} = \frac{3}{4}.$$
The missing length $z$ is 3.

**23.** First we find $x$. The ratio of $x$ to 8 is the same as the ratio of 2 to 3. We have the proportion
$$\frac{x}{8} = \frac{2}{3}.$$
Solve: $3 \cdot x = 8 \cdot 2$
$$x = \frac{8 \cdot 2}{3}$$
$$x = \frac{16}{3}, \text{ or } 5\frac{1}{3}$$
The missing length $x$ is $\frac{16}{3}$, or $5\frac{1}{3}$, or $5.\overline{3}$.

Next we find $y$. The ratio of $y$ to 7 is the same as the ratio of 2 to 3. We have the proportion
$$\frac{y}{7} = \frac{2}{3}.$$

Solve: $3 \cdot y = 7 \cdot 2$

$$y = \frac{7 \cdot 2}{3}$$

$$y = \frac{14}{3} = 4\frac{2}{3}$$

The missing length $y$ is $\frac{14}{3}$, or $4\frac{2}{3}$, or $4.\overline{6}$.

Finally we find $z$. The ratio of $z$ to 8 is the same as the ratio of 2 to 3. We have the proportion

$$\frac{z}{9} = \frac{2}{3}.$$

This is the same proportion we solved above when we found $x$. Then the missing length $z$ is $5\frac{1}{3}$, or $5.\overline{3}$.

**25.** Height $\to$ $\dfrac{h}{32} = \dfrac{5}{8}$ $\leftarrow$ Height
Width $\to$ $\phantom{\dfrac{h}{32}}$ $\phantom{\dfrac{5}{8}}$ $\leftarrow$ Width

Solve: $8 \cdot h = 32 \cdot 5$

$$h = \frac{32 \cdot 5}{8}$$

$$h = \frac{4 \cdot 8 \cdot 5}{8}$$

$$h = 4 \cdot 5$$

$$h = 20$$

The missing length is 20 ft.

**27.** The ratio of $h$ to 15 is the same as the ratio of 116 to 12. We have the proportion

$$\frac{h}{19} = \frac{120}{15}.$$

Solve: $15 \cdot h = 19 \cdot 120$

$$h = \frac{19 \cdot 120}{15}$$

$$h = 152$$

The addition will be 152 ft high.

**29. Familiarize.** This is a multistep problem.

First we find the total cost of the purchases. We let $c =$ this amount.

**Translate and Solve.**

| Price of book | plus | Price of CD | plus | Price of sweatshirt | is | Total cost |
|---|---|---|---|---|---|---|
| ↓ | ↓ | ↓ | ↓ | ↓ | ↓ | ↓ |
| $49.95 | + | $14.88 | + | $29.95 | = | c |

To solve the equation we carry out the addition.

$$\begin{array}{r} {\scriptstyle 2\ 2\ 1} \\ 4\,9.9\,5 \\ 1\,4.8\,8 \\ +\,2\,9.9\,5 \\ \hline 9\,4.7\,8 \end{array}$$

Thus, $c = \$94.78$.

Now we find how much more money the student needs to make these purchases. We let $m =$ this amount.

| Money student has | plus | How much more money | is | Total cost of purchases |
|---|---|---|---|---|
| ↓ | ↓ | ↓ | ↓ | ↓ |
| $34.97 | + | m | = | $94.78 |

To solve the equation we subtract 34.97 on both sides.

$$m = 94.78 - 34.97$$
$$m = 59.81$$

$$\begin{array}{r} {\scriptstyle 13} \\ {\scriptstyle 8\ \ \not{3}\ 17} \\ \not{9}\ \not{4}.7\,8 \\ -\,3\,4.9\,7 \\ \hline 5\,9.8\,1 \end{array}$$

**Check.** We repeat the calculations.

**State.** The student needs $59.81 more to make the purchases.

**31.**

$$\begin{array}{r} {\scriptstyle 7\ 7\ 1} \\ {\scriptstyle 3\ 3} \\ 8\,0.8\,9\,2 \\ \times\phantom{000}8.4 \\ \hline 3\,2\,3\,5\,6\,8 \\ 6\,4\,7\,1\,3\,6\,0 \\ \hline 6\,7\,9.4\,9\,2\,8 \end{array}$$

**33.** $100 \times 274.568 \qquad 274.56.8$

2 zeros      Move 2 places to the right.

$100 \times 274.568 = 27{,}456.8$

**35.** $\dfrac{17}{20} = \dfrac{17}{20} \cdot \dfrac{5}{5} = \dfrac{85}{100} = 0.85$

**37.**

$$\begin{array}{r} 0.9\,0\,9\,0 \\ 1\,1\,\overline{)\,1\,0.0\,0\,0} \\ 9\,9\phantom{000} \\ \hline 1\,0\,0\phantom{00} \\ 9\,9\phantom{00} \\ \hline 1\,0\phantom{0} \end{array}$$

Because we are rounding to the nearest thousandth, we stop here.

$$\frac{10}{11} \approx 0.909$$

**39.**

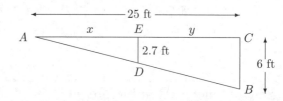

We note that triangle $ADE$ is similar to triangle $ABC$ and use this information to find the length $x$.

$$\frac{x}{25} = \frac{2.7}{6}$$

$$6 \cdot x = 25 \cdot 2.7$$

$$x = \frac{25 \cdot 2.7}{6}$$

$$x = 11.25$$

Thus the goalie should be 11.25 ft from point $A$. We subtract to find how far from the goal the goalie should be located.

$$25 - 11.25 = 13.75$$

The goalie should stand 13.75 ft from the goal.

**41.** From Exercise 27 we know that a height of 19 cm on the model corresponds to a height of 152 ft on the building. We let $h$ = the height of the model hoop. Then we translate to a proportion.

Model height $\rightarrow$ $\dfrac{19}{152} = \dfrac{h}{10}$ $\leftarrow$ Model height
Actual height $\rightarrow$ $\phantom{\dfrac{19}{152}}$ $\phantom{\dfrac{h}{10}}$ $\leftarrow$ Actual height

Solve: $19 \cdot 10 = 152 \cdot h$   Equating cross products

$$\frac{19 \cdot 10}{152} = h$$
$$1.25 = h$$

The model hoop should be 1.25 cm high.

**43.**
$$\frac{12.0078}{56.0115} = \frac{789.23}{y}$$
$$12.0078 \cdot y = 56.0115(789.23)$$
$$y = \frac{56.0115(789.23)}{12.0078}$$
$$y \approx 3681.437 \quad \text{Using a calculator}$$

**45.** First we find $x$. We see from the drawing that side $x$ is longer than side $y$, so the ratio of $x$ to 22.4 is the same as the ratio of 0.3 to 16.8. We have the proportion

$$\frac{x}{22.4} = \frac{0.3}{16.8}$$

Solve: $16.8 \cdot x = 22.4(0.3)$
$$x = \frac{22.4(0.3)}{16.8}$$
$$x = 0.4$$

The missing length $x$ is 0.4.

Now we find $y$. The ratio of $y$ to 19.7 is the same as the ratio of 0.3 to 16.8. We have the proportion

$$\frac{y}{19.7} = \frac{0.3}{16.8}.$$

Solve: $16.8 \cdot y = 19.7(0.3)$
$$y = \frac{19.7(0.3)}{16.8}$$
$$y \approx 0.35$$

The missing length $y$ is approximately 0.35.

## Chapter 6 Concept Reinforcement

**1.** The statement is true. See page 342 in the text.

**2.** Find cross products. For $\dfrac{a}{b} = \dfrac{c}{d}$, we have $ad = bc$. For $\dfrac{c}{a} = \dfrac{d}{b}$, we have $cb = ad$, or $bc = ad$, or $ad = bc$. The proportions have the same cross products, we see that $\dfrac{a}{b} = \dfrac{c}{d}$ can be written as $\dfrac{c}{a} = \dfrac{d}{b}$. The given statement is true.

**3.** Similar triangles have the same shape but not necessarily the same size. The given statement is false.

**4.** The statement is true. See page 374 in the text.

## Chapter 6 Important Concepts

**1.** The ratio is $\dfrac{17}{3}$.

**2.** $\dfrac{3.2}{2.8} = \dfrac{3.2}{2.8} \cdot \dfrac{10}{10} = \dfrac{32}{28} = \dfrac{4 \cdot 8}{4 \cdot 7} = \dfrac{4}{4} \cdot \dfrac{8}{7} = \dfrac{8}{7}$

**3.** $\dfrac{\$120}{16 \text{ hr}} = \dfrac{120}{16} \dfrac{\$}{\text{hr}} = \$7.5/\text{hr}$, or $\$7.50/\text{hr}$

**4.** $\dfrac{\$2.79}{28 \text{ oz}} = \dfrac{279\cent}{28 \text{ oz}} \approx 9.964\cent/\text{oz}$

$\dfrac{\$3.29}{32 \text{ oz}} = \dfrac{329\cent}{32 \text{ oz}} \approx 10.281\cent/\text{oz}$

Brand A is a better buy based on unit price alone.

**5.** We can use cross products.

$7 \cdot 27 = 189$   $\begin{matrix} 7 & 21 \\ 9 & 27 \end{matrix}$   $9 \cdot 21 = 189$

Since the cross products are the same, $189 = 189$, we know that the numbers are proportional.

**6.**
$$\frac{9}{x} = \frac{8}{3}$$
$$9 \cdot 3 = x \cdot 8 \quad \text{Equating cross products}$$
$$\frac{9 \cdot 3}{8} = \frac{x \cdot 8}{8}$$
$$\frac{9 \cdot 3}{8} = x$$
$$\frac{27}{8} = x$$

**7.** *Familiarize.* Let $d$ = the distance between the cities in reality, in miles.

*Translate.* We translate to a proportion.

Map distance $\rightarrow$ $\dfrac{\frac{1}{2}}{50} = \dfrac{1\frac{3}{4}}{d}$ $\leftarrow$ Map distance
Actual distance $\rightarrow$ $\phantom{xx}$ $\phantom{xx}$ $\leftarrow$ Actual distance

*Solve.*
$$\frac{1}{2} \cdot d = 50 \cdot 1\frac{3}{4}$$
$$\frac{1}{2} \cdot d = 50 \cdot \frac{7}{4}$$
$$d = \frac{2}{1} \cdot \frac{50}{1} \cdot \frac{7}{4} \quad \text{Dividing by } \frac{1}{2}$$
$$d = \frac{2 \cdot 2 \cdot 25 \cdot 7}{1 \cdot 1 \cdot 2 \cdot 2}$$
$$d = 175$$

*Check.* We substitute in the proportion and check cross products.
$$\frac{\frac{1}{2}}{50} = \frac{1\frac{3}{4}}{175}$$
$$\frac{1}{2} \cdot 175 = \frac{175}{2}; \quad 50 \cdot 1\frac{3}{4} = 50 \cdot \frac{7}{4} = \frac{350}{4} = \frac{175}{2}$$

The cross products are the same so the answer checks.

*State.* The cities are 175 mi apart in reality.

**8.** The ratio of 17.5 to $y$ is the same as the ratio of 15 to 18. We have the proportion

$$\frac{17.5}{y} = \frac{15}{18}.$$

Solve: $17.5 \cdot 18 = y \cdot 15$

$$\frac{17.5 \cdot 18}{15} = y$$

$$21 = y$$

We could also have used $\dfrac{17.5}{y} = \dfrac{20}{24}$ to find $y$.

## Chapter 6 Review Exercises

**1.** The ratio of 47 to 84 is $\dfrac{47}{84}$.

**2.** The ratio of 46 to 1.27 is $\dfrac{46}{1.27}$.

**3.** The ratio of 83 to 100 is $\dfrac{83}{100}$.

**4.** The ratio of 0.72 to 197 is $\dfrac{0.72}{197}$.

**5.** a) The ratio of 12,480 to 16,640 is $\dfrac{12,480}{16,640}$.

We can simplify this ratio as follows:

$$\frac{12,480}{16,640} = \frac{3 \cdot 5 \cdot 8 \cdot 8 \cdot 13}{4 \cdot 5 \cdot 8 \cdot 8 \cdot 13} = \frac{3}{4} \cdot \frac{5 \cdot 8 \cdot 8 \cdot 13}{5 \cdot 8 \cdot 8 \cdot 13} = \frac{3}{4}$$

b) The total of both kinds of fish sold is 12,480 lb + 16,640 lb, or 29,120 lb. Then the ratio of salmon sold to the total amount of both kinds of fish sold is $\dfrac{16,640}{29,120}$.

We can simplify this ratio as follows:

$$\frac{16,640}{29,120} = \frac{4 \cdot 5 \cdot 8 \cdot 8 \cdot 13}{5 \cdot 7 \cdot 8 \cdot 8 \cdot 13} = \frac{4}{7} \cdot \frac{5 \cdot 8 \cdot 8 \cdot 13}{5 \cdot 8 \cdot 8 \cdot 13} = \frac{4}{7}$$

**6.** $\dfrac{9}{12} = \dfrac{3 \cdot 3}{3 \cdot 4} = \dfrac{3}{3} \cdot \dfrac{3}{4} = \dfrac{3}{4}$

**7.** $\dfrac{3.6}{6.4} = \dfrac{3.6}{6.4} \cdot \dfrac{10}{10} = \dfrac{36}{64} = \dfrac{4 \cdot 9}{4 \cdot 16} = \dfrac{4}{4} \cdot \dfrac{9}{16} = \dfrac{9}{16}$

**8.** $\dfrac{377 \text{ mi}}{14.5 \text{ gal}} = 26 \dfrac{\text{mi}}{\text{gal}}$, or 26 mpg

**9.** $\dfrac{472,500 \text{ revolutions}}{75 \text{ min}} = 6300 \dfrac{\text{revolutions}}{\text{min}}$, or 6300 rpm

**10.** $\dfrac{319 \text{ gal}}{500 \text{ ft}^2} = 0.638 \text{ gal/ft}^2$

**11.** $\dfrac{\$18.99}{300 \text{ tablets}} = \dfrac{1899\cancel{c}}{300 \text{ tablets}} = 6.33\cancel{c}/\text{tablet}$

**12.** $\dfrac{\$2.69}{24 \text{ oz}} = \dfrac{269\cancel{c}}{24 \text{ oz}} \approx 11.208\cancel{c}/\text{oz}$

**13.** $\dfrac{\$4.79}{32 \text{ oz}} = \dfrac{479\cancel{c}}{32 \text{ oz}} \approx 14.969\cancel{c}/\text{oz}$

$\dfrac{\$5.99}{48 \text{ oz}} = \dfrac{599\cancel{c}}{48 \text{ oz}} \approx 12.479\cancel{c}/\text{oz}$

$\dfrac{\$9.99}{64 \text{ oz}} = \dfrac{999\cancel{c}}{64 \text{ oz}} \approx 15.609\cancel{c}/\text{oz}$

The 48 oz size has the lowest unit price.

**14.** We can use cross products:

$9 \cdot 60 = 540 \qquad 15 \cdot 36 = 540$

Since the cross products are the same, $540 = 540$, we know that the numbers are proportional.

**15.** We can use cross products:

$24 \cdot 46.25 = 1110 \qquad 37 \cdot 40 = 1480$

Since the cross products are not the same, $1110 \neq 1480$, we know that the numbers are not proportional.

**16.** $\dfrac{8}{9} = \dfrac{x}{36}$

$8 \cdot 36 = 9 \cdot x \qquad$ Equating cross products

$\dfrac{8 \cdot 36}{9} = \dfrac{9 \cdot x}{9}$

$\dfrac{288}{9} = x$

$32 = x$

**17.** $\dfrac{6}{x} = \dfrac{48}{56}$

$6 \cdot 56 = x \cdot 48$

$\dfrac{6 \cdot 56}{48} = \dfrac{x \cdot 48}{48}$

$\dfrac{336}{48} = x$

$7 = x$

**18.** $\dfrac{120}{\frac{3}{7}} = \dfrac{7}{x}$

$120 \cdot x = \dfrac{3}{7} \cdot 7$

$120 \cdot x = 3$

$\dfrac{120 \cdot x}{120} = \dfrac{3}{120}$

$x = \dfrac{1}{40}$

**19.** $\dfrac{4.5}{120} = \dfrac{0.9}{x}$

$4.5 \cdot x = 120 \cdot 0.9$

$\dfrac{4.5 \cdot x}{4.5} = \dfrac{120 \cdot 0.9}{4.5}$

$x = \dfrac{108}{4.5}$

$x = 24$

**20.** *Familiarize*. Let $d =$ the number of defective circuits in a lot of 585.

*Translate*. We translate to a proportion.

Defective $\quad\to\quad \dfrac{3}{65} = \dfrac{d}{585} \quad\leftarrow$ Defective

Total circuits $\to \quad 65 \qquad 585 \quad\leftarrow$ Total circuits

*Solve*. We solve the proportion.

$$\frac{3}{65} = \frac{d}{585}$$

$$3 \cdot 585 = 65 \cdot d$$

$$\frac{3 \cdot 585}{65} = d$$

$$27 = d$$

*Check*. We substitute in the proportion and check cross products.

$$\frac{3}{65} = \frac{27}{585}$$

$$3 \cdot 585 = 1755; \; 65 \cdot 27 = 1755$$

The cross products are the same, so the answer checks.

*State*. It would be expected that 27 defective circuits would occur in a lot of 585 circuits.

**21.** a) *Familiarize*. Let $c =$ the number of Canadian dollars equivalent to 250 U.S. dollars.

*Translate*. We translate to a proportion.

U.S. dollars $\quad\to\quad \dfrac{1}{1.068} = \dfrac{250}{c} \quad$ U.S. dollars

Canadian dollars$\to 1.068 \qquad c \;\leftarrow$Canadian dollars

*Solve*.

$$1 \cdot c = 1.068 \cdot 250 \quad \text{Equating cross products}$$

$$c = 267$$

*Check*. We substitute in the proportion and check cross products.

$$\frac{1}{1.068} = \frac{250}{267}$$

$$1 \cdot 267 = 267; \; 1.068 \cdot 250 = 267$$

The cross products are the same, so the answer checks.

*State*. 250 U.S. dollars would be worth 267 Canadian dollars.

b) *Familiarize*. Let $c =$ the cost of the sweatshirt in U.S. dollars.

*Translate*. We translate to a proportion.

U.S. dollars $\quad\to\quad \dfrac{1}{1.068} = \dfrac{c}{50} \;\leftarrow$ U.S. dollars

Canadian dollars$\to 1.068 \qquad 50 \;\leftarrow$Canadian dollars

*Solve*.

$$1 \cdot 50 = 1.068 \cdot c \quad \text{Equating cross products}$$

$$\frac{1 \cdot 50}{1.068} = c$$

$$46.82 \approx c$$

*Check*. We substitute in the proportion and check cross products.

$$\frac{1}{1.068} = \frac{46.82}{50}$$

$$1 \cdot 50 = 50; \; 1.068 \cdot 46.82 \approx 50.004$$

The cross products are about the same. Remember that we rounded the value of $c$. The answer checks.

*State*. The sweatshirt cost $46.82 in U.S. dollars.

**22.** *Familiarize*. Let $d =$ the number of miles the train will travel in 13 hr.

*Translate*. We translate to a proportion.

Miles $\quad\to\quad \dfrac{448}{7} = \dfrac{d}{13} \quad\leftarrow$ Miles

Hours $\quad\to\quad 7 \qquad 13 \quad\leftarrow$ Hours

*Solve*.

$$448 \cdot 13 = 7 \cdot d \quad \text{Equating cross products}$$

$$\frac{448 \cdot 13}{7} = \frac{7 \cdot d}{7}$$

$$832 = d$$

*Check*. We find how far the train travels in 1 hr and then multiply by 13:

$$448 \div 7 = 64 \text{ and } 64 \cdot 13 = 832$$

The answer checks.

*State*. The train will travel 832 mi in 13 hr.

**23.** *Familiarize*. Let $a =$ the number of acres required to produce 97.2 bushels of tomatoes.

*Translate*. We translate to a proportion.

Acres $\quad\to\quad \dfrac{15}{54} = \dfrac{a}{97.2}$

Bushels $\to \quad 54 \qquad 97.2$

*Solve*.

$$15 \cdot 97.2 = 54 \cdot a \quad \text{Equating cross products}$$

$$\frac{15 \cdot 97.2}{54} = \frac{54 \cdot a}{54}$$

$$27 = a$$

*Check*. We substitute in the proportion and check cross products.

$$\frac{15}{54} = \frac{27}{97.2}$$

$$15 \cdot 97.2 = 1458; \; 54 \cdot 27 = 1458$$

The answer checks.

*State*. 27 acres are required to produce 97.2 bushels of tomatoes.

**24.** *Familiarize*. Let $g =$ the number of pounds of trash produced in Austin, Texas in one day.

*Translate*. We translate to a proportion.

Trash $\quad\to\quad \dfrac{23}{5} = \dfrac{g}{743,074} \quad\leftarrow$ Trash

People $\quad\to\quad 5 \qquad 743,074 \quad\leftarrow$ People

*Solve*.

$$23 \cdot 743,074 = 5 \cdot g \quad \text{Equating cross products}$$

$$\frac{23 \cdot 743,074}{5} = \frac{5 \cdot g}{5}$$

$$3,418,140 \approx g$$

*Check*. We can divide to find the amount of garbage produced by one person and then multiply to find the amount produced by 743,074 people.

$23 \div 5 = 4.6$ and $4.6 \cdot 743,074 = 3,418,140.4 \approx 3,418,140.$
The answer checks.

**State.** About 3,418,140 lb of garbage is produced in Austin, Texas in one day.

**25. Familiarize.** Let $w$ = the number of inches of water to which $4\frac{1}{2}$ ft of snow melts.

**Translate.** We translate to a proportion.

$$\text{Snow} \rightarrow \frac{1\frac{1}{2}}{2} = \frac{4\frac{1}{2}}{w} \leftarrow \text{Snow} \\ \text{Water} \rightarrow \quad \leftarrow \text{Water}$$

**Solve.**

$$1\frac{1}{2} \cdot w = 2 \cdot 4\frac{1}{2} \quad \text{Equating cross products}$$

$$\frac{3}{2} \cdot w = 2 \cdot \frac{9}{2}$$

$$\frac{3}{2} \cdot w = 9$$

$$w = 9 \div \frac{3}{2} \quad \text{Dividing by } \frac{3}{2} \text{ on both sides}$$

$$w = 9 \cdot \frac{2}{3}$$

$$w = \frac{9 \cdot 2}{3}$$

$$w = 6$$

**Check.** We substitute in the proportion and check cross products.

$$\frac{1\frac{1}{2}}{2} = \frac{4\frac{1}{2}}{6}$$

$$1\frac{1}{2} \cdot 6 = \frac{3}{2} \cdot 6 = \frac{3 \cdot 6}{2} = 9; \quad 2 \cdot 4\frac{1}{2} = 2 \cdot \frac{9}{2} = \frac{2 \cdot 9}{2} = 9$$

The cross products are the same, so the answer checks.

**State.** $4\frac{1}{2}$ ft of snow will melt to 6 in. of water.

**26. Familiarize.** Let $l$ = the number of lawyers we would expect to find in Chicago.

**Translate.** We translate to a proportion.

$$\text{Lawyers} \rightarrow \frac{4.8}{1000} = \frac{l}{2,842,518} \leftarrow \text{Lawyers} \\ \text{Population} \rightarrow \quad \leftarrow \text{Population}$$

**Solve.**

$$4.8 \cdot 2,842,518 = 1000 \cdot l \quad \text{Equating cross products}$$

$$\frac{4.8 \cdot 2,842,518}{1000} = \frac{1000 \cdot l}{1000}$$

$$13,644 \approx l$$

**Check.** We substitute in the proportion and check cross products.

$$\frac{4.8}{1000} = \frac{13,644}{2,842,518}$$

$$4.8 \cdot 2,842,518 = 13,644,086.4;$$

$$1000 \cdot 13,644 = 13,644,000 \approx 13,644,086.4$$

The answer checks.

**State.** We would expect that there would be about 13,644 lawyers in Chicago.

**27.** The ratio of $x$ to 7 is the same as the ratio of 6 to 9.

$$\frac{x}{7} = \frac{6}{9}$$

$$x \cdot 9 = 7 \cdot 6$$

$$x = \frac{7 \cdot 6}{9} = \frac{7 \cdot 2 \cdot 3}{3 \cdot 3}$$

$$x = \frac{7 \cdot 2}{3} \cdot \frac{3}{3} = \frac{7 \cdot 2}{3}$$

$$x = \frac{14}{3}, \text{ or } 4\frac{2}{3}$$

We could also have used the proportion $\frac{x}{7} = \frac{2}{3}$ to find $x$.

**28.** The ratio of $x$ to 8 is the same as the ratio of 7 to 5.

$$\frac{x}{8} = \frac{7}{5}$$

$$x \cdot 5 = 8 \cdot 7$$

$$x = \frac{8 \cdot 7}{5}$$

$$x = \frac{56}{5}, \text{ or } 11\frac{1}{5}$$

The ratio of $y$ to 9 is the same as the ratio of 7 to 5.

$$\frac{y}{9} = \frac{7}{5}$$

$$y \cdot 5 = 9 \cdot 7$$

$$y = \frac{9 \cdot 7}{5}$$

$$y = \frac{63}{5}, \text{ or } 12\frac{3}{5}$$

**29.** If we use the sun's rays to represent the third side of a triangle in a drawing of the situation, we see that we have similar triangles. We let $h$ = the height of the billboard, in feet.

Sun's rays ◿ 8 ft
5 ft

Sun's rays ◹ $h$
25 ft

The ratio of $h$ to 8 is the same as the ratio of 25 to 5.

$$\frac{h}{8} = \frac{25}{5}$$

$$h \cdot 5 = 8 \cdot 25$$

$$h = \frac{8 \cdot 25}{5}$$

$$h = 40$$

The billboard is 40 ft high.

**30.** The ratio of $x$ to 2 is the same as the ratio of 9 to 6.

$$\frac{x}{2} = \frac{9}{6}$$

$$x \cdot 6 = 2 \cdot 9$$

$$x = \frac{2 \cdot 9}{6}$$

$$x = 3$$

The ratio of $y$ to 7 is the same as the ratio of 9 to 6.

$$\frac{y}{7} = \frac{9}{6}$$

$$y \cdot 6 = 7 \cdot 9$$

$$y = \frac{7 \cdot 9}{6}$$

$$y = \frac{21}{2}, \text{ or } 10\frac{1}{2}$$

The ratio of $z$ to 5 is the same as the ratio of 9 to 6.

$$\frac{z}{5} = \frac{9}{6}$$

$$z \cdot 6 = 5 \cdot 9$$

$$z = \frac{5 \cdot 9}{6} = \frac{5 \cdot 3 \cdot 3}{2 \cdot 3}$$

$$z = \frac{5 \cdot 3}{2} \cdot \frac{3}{3} = \frac{5 \cdot 3}{2}$$

$$z = \frac{15}{2}, \text{ or } 7\frac{1}{2}$$

**31.** $\dfrac{18 \text{ servings}}{25 \text{ lb}} = 0.72 \text{ serving/lb}$

Answer B is correct.

**32. Familiarize**. Let $p$ = the price of 5 dozen eggs.

**Translate**. We translate to a proportion.

$$\text{Eggs} \rightarrow \frac{3}{5.04} = \frac{5}{p} \leftarrow \text{Eggs}$$
$$\text{Price} \rightarrow \phantom{\frac{3}{5.04}} \phantom{=} \phantom{\frac{5}{p}} \leftarrow \text{Price}$$

**Solve**. We solve the proportion.

$$\frac{3}{5.04} = \frac{5}{p}$$

$$3 \cdot p = 5.04 \cdot 5$$

$$\frac{3 \cdot p}{3} = \frac{5.04 \cdot 5}{3}$$

$$p = 8.4$$

**Check**. We substitute in the proportion and check cross products.

$$\frac{3}{5.04} = \frac{5}{8.4}$$

$$3 \cdot 8.4 = 25.2; \; 5.04 \cdot 5 = 25.2$$

The cross products are the same, so the answer checks.

**State**. 5 dozen eggs would cost $8.40.

Answer C is correct.

**33.** In 8 rolls of towels with 60 sheets per roll there are $8 \cdot 60$, or 480 sheets.

$$\frac{\$6.38}{480 \text{ sheets}} = \frac{638\cancel{c}}{480 \text{ sheets}} \approx 1.329\cancel{c}/\text{sheet}$$

In 15 rolls of towels with 60 sheets per roll there are $15 \cdot 60$, or 900 sheets.

$$\frac{\$13.99}{900 \text{ sheets}} = \frac{1399\cancel{c}}{900 \text{ sheets}} \approx 1.554\cancel{c}/\text{sheet}$$

In 6 rolls of towels with 165 sheets per roll there are $6 \cdot 165$, or 990 sheets.

$$\frac{\$10.99}{990 \text{ sheets}} = \frac{1099\cancel{c}}{990 \text{ sheets}} \approx 1.110\cancel{c}/\text{sheet}$$

The package containing 6 big rolls with 165 sheets per roll is the best buy based on unit price alone.

**34.** The ratio of $x$ to 5678 is the same as the ratio of 2530.5 to 3374.

$$\frac{x}{5678} = \frac{2530.5}{3374}$$

$$x \cdot 3374 = 5678 \cdot 2530.5$$

$$x = \frac{5678 \cdot 2530.5}{3374}$$

$$x = 4258.5$$

The ratio of $z$ to 7570.7 is the same as the ratio of 3374 to 2530.5.

$$\frac{z}{7570.7} = \frac{3374}{2530.5}$$

$$z \cdot 2530.5 = 7570.7 \cdot 3374$$

$$z = \frac{7570.7 \cdot 3374}{2530.5}$$

$$z \approx 10,094.3$$

**38.** First we divide to find how many gallons of finishing paint are needed.

$$4950 \div 450 = 11 \text{ gal}$$

Next we write and solve a proportion to find how many gallons of primer are needed. Let $p$ = the amount of primer needed.

$$\text{Finishing paint} \rightarrow \frac{2}{3} = \frac{11}{p} \leftarrow \text{Finishing paint}$$
$$\text{Primer} \rightarrow \phantom{\frac{2}{3}} \phantom{=} \phantom{\frac{11}{p}} \leftarrow \text{Primer}$$

$$2 \cdot p = 3 \cdot 11$$

$$p = \frac{3 \cdot 11}{2}$$

$$p = \frac{33}{2}, \text{ or } 16.5$$

Thus, 11 gal of finishing paint and 16.5 gal of primer should be purchased.

---

## Chapter 6 Discussion and Writing Exercises

**1.** In terms of cost, a low faculty-to-student ratio is less expensive than a high faculty-to-student ratio. In terms of quality of education and student satisfaction, a high faculty-to-student ratio is more desirable. A college president must balance the cost and quality issues.

**2.** Unit prices can be used to solve proportions involving money. In Example 3, for instance, we would have divided $90 by the unit price, or the price per ticket, to find the number of tickets that could be purchased for $90.

**3.** Leslie used 4 gal of gasoline to drive 92 mile. At the same rate, how many gallons would be needed to travel 368 mi?

**4.** Yes; consider the following pair of triangles.

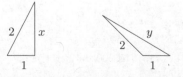

Two pairs of sides are proportional, but we can see that $x$ is shorter than $y$ so the ratio of $x$ to $y$ is clearly not the same as the ratio of 1 to 1 (or 2 to 2).

## Chapter 6 Test

**1.** The ratio of 85 to 97 is $\dfrac{85}{97}$.

**2.** The ratio of 0.34 to 124 is $\dfrac{0.34}{124}$.

**3.** $\dfrac{18}{20} = \dfrac{2 \cdot 9}{2 \cdot 10} = \dfrac{2}{2} \cdot \dfrac{9}{10} = \dfrac{9}{10}$

**4.** $\dfrac{0.75}{0.96} = \dfrac{0.75}{0.96} \cdot \dfrac{100}{100}$   Clearing the decimals

$\quad = \dfrac{75}{96}$

$\quad = \dfrac{3 \cdot 25}{3 \cdot 33} = \dfrac{3}{3} \cdot \dfrac{25}{33}$

$\quad = \dfrac{25}{33}$

**5.** $\dfrac{10 \text{ ft}}{16 \text{ sec}} = \dfrac{10}{16} \dfrac{\text{ft}}{\text{sec}} = 0.625 \text{ ft/sec}$

**6.** $\dfrac{16 \text{ servings}}{12 \text{ lb}} = \dfrac{16}{12} \dfrac{\text{servings}}{\text{lb}} = \dfrac{4}{3}$ servings/lb, or

$1\dfrac{1}{3}$ servings/lb

**7.** $\dfrac{464 \text{ mi}}{14.5 \text{ gal}} = \dfrac{464}{14.5} \dfrac{\text{mi}}{\text{gal}} = 32 \text{ mpg}$

**8.** $\dfrac{\$2.49}{16 \text{ oz}} = \dfrac{249 \text{ cents}}{16 \text{ oz}} = \dfrac{249}{16} \dfrac{\text{cents}}{\text{oz}} \approx 15.563 \text{ cents/oz}$

**9.** $\dfrac{\$6.59}{40 \text{ oz}} = \dfrac{659\cent}{40 \text{ oz}} = 16.475\cent/\text{oz}$

$\dfrac{\$6.99}{50 \text{ oz}} = \dfrac{699\cent}{50 \text{ oz}} = 13.980\cent/\text{oz}$

$\dfrac{\$11.49}{100 \text{ oz}} = \dfrac{1149\cent}{100 \text{ oz}} = 11.490\cent/\text{oz}$

$\dfrac{\$24.99}{150 \text{ oz}} = \dfrac{2499\cent}{150 \text{ oz}} = 16.660\cent/\text{oz}$

The 100-oz size has the lowest unit price.

**10.** We can use cross products:

$7 \cdot 72 = 504 \quad \dfrac{7}{8} \bowtie \dfrac{63}{72} \quad 8 \cdot 63 = 504$

Since the cross products are the same, $504 = 504$, we know that $\dfrac{7}{8} \; \dfrac{63}{72}$, so the numbers are proportional.

**11.** We can use cross products:

$1.3 \cdot 15.2 = 19.76 \quad \dfrac{1.3}{3.4} \bowtie \dfrac{5.6}{15.2} \quad 3.4 \cdot 5.6 = 19.04$

Since the cross products are not the same, $19.76 \neq 19.04$, we know that $\dfrac{1.3}{3.4} \neq \dfrac{5.6}{15.2}$, so the numbers are not proportional.

**12.** $\dfrac{9}{4} = \dfrac{27}{x}$

$9 \cdot x = 4 \cdot 27$   Equating cross products

$\dfrac{9 \cdot x}{9} = \dfrac{4 \cdot 27}{9}$

$x = \dfrac{4 \cdot \cancel{9} \cdot 3}{\cancel{9} \cdot 1}$

$x = 12$

**13.** $\dfrac{150}{2.5} = \dfrac{x}{6}$

$150 \cdot 6 = 2.5 \cdot x$   Equating cross products

$\dfrac{150 \cdot 6}{2.5} = \dfrac{2.5 \cdot x}{2.5}$

$\dfrac{900}{2.5} = x$

$360 = x$

**14.** $\dfrac{x}{100} = \dfrac{27}{64}$

$x \cdot 64 = 100 \cdot 27$   Equating cross products

$\dfrac{x \cdot 64}{64} = \dfrac{100 \cdot 27}{64}$

$x = \dfrac{2700}{64}$

$x = 42.1875$

**15.** $\dfrac{68}{y} = \dfrac{17}{25}$

$68 \cdot 25 = y \cdot 17$   Equating cross products

$\dfrac{68 \cdot 25}{17} = \dfrac{y \cdot 17}{17}$

$\dfrac{4 \cdot \cancel{17} \cdot 25}{\cancel{17} \cdot 1} = y$

$100 = y$

**16. *Familiarize*.** Let $d$ = the distance the boat would travel in 42 hr.

***Translate*.** We translate to a proportion.

$\begin{array}{ll} \text{Distance} \rightarrow & \dfrac{432}{12} = \dfrac{d}{42} \leftarrow \text{Distance} \\ \text{Time} \rightarrow & \phantom{\dfrac{432}{12} = \dfrac{d}{42}} \leftarrow \text{Time} \end{array}$

***Solve*.**

$432 \cdot 42 = 12 \cdot d$   Equating cross products

$\dfrac{432 \cdot 42}{12} = \dfrac{12 \cdot d}{12}$

$1512 = d$   Multiplying and dividing

***Check*.** We substitute in the proportion and check cross products.

$\dfrac{432}{12} = \dfrac{1512}{42}$

$432 \cdot 42 = 18,144; \; 12 \cdot 1512 = 18,144$

The cross products are the same, so the answer checks.

***State*.** The boat would travel 1512 km in 42 hr.

**17. Familiarize**. Let $m =$ the number of minutes the watch will lose in 24 hr.

**Translate**. We translate to a proportion.

$$\text{Minutes lost} \to \frac{2}{10} = \frac{m}{24} \leftarrow \text{Minutes lost}$$
$$\text{Hours} \to \qquad\qquad \leftarrow \text{Hours}$$

**Solve**.
$$2 \cdot 24 = 10 \cdot m \quad \text{Equating cross products}$$
$$\frac{2 \cdot 24}{10} = \frac{10 \cdot m}{10}$$
$$4.8 = m \qquad \text{Multiplying and dividing}$$

**Check**. We substitute in the proportion and check cross products.
$$\frac{2}{10} = \frac{4.8}{24}$$
$$2 \cdot 24 = 48; \ 10 \cdot 4.8 = 48$$

The cross products are the same, so the answer checks.

**State**. The watch will lose 4.8 min in 24 hr.

**18. Familiarize**. Let $d =$ the actual distance between the cities.

**Translate**. We translate to a proportion.
$$\text{Map distance} \to \frac{3}{225} = \frac{7}{d} \leftarrow \text{Map distance}$$
$$\text{Actual distance} \to \qquad\qquad \leftarrow \text{Actual distance}$$

**Solve**.
$$3 \cdot d = 225 \cdot 7 \quad \text{Equating cross products}$$
$$\frac{3 \cdot d}{3} = \frac{225 \cdot 7}{3}$$
$$d = 525 \qquad \text{Multiplying and dividing}$$

**Check**. We substitute in the proportion and check cross products.
$$\frac{3}{225} = \frac{7}{525}$$
$$3 \cdot 525 = 1575; \ 225 \cdot 7 = 1575$$

The cross products are the same, so the answer checks.

**State**. The cities are 525 mi apart.

**19.** If we use the sun's rays to represent the third side of the triangles in the drawing of the situation in the text, we see that we have similar triangles. The ratio of 3 to $h$ is the same as the ratio of 5 to 110. We have the proportion
$$\frac{3}{h} = \frac{5}{110}.$$
Solve: $3 \cdot 110 = h \cdot 5$
$$\frac{3 \cdot 110}{5} = \frac{h \cdot 5}{5}$$
$$66 = h$$
The tower is 66 m high.

**20. a) Familiarize**. Let $c =$ the value of 450 U.S. dollars in Hong Kong dollars.

**Translate**. We translate to a proportion.
$$\text{U.S. dollars} \to \frac{1}{7.781} = \frac{450}{c} \leftarrow \text{U.S. dollars}$$
$$\text{Hong Kong} \to \qquad\qquad \leftarrow \text{Hong Kong}$$
$$\text{dollars} \qquad\qquad\qquad \text{dollars}$$

**Solve**.
$$1 \cdot c = 7.781 \cdot 450 \quad \text{Equating cross products}$$
$$c = 3501.45$$

**Check**. We substitute in the proportion and check cross products.
$$\frac{1}{7.781} = \frac{450}{3501.45}$$
$$1 \cdot 3501.45 = 3501.45; \ 7.781 \cdot 450 = 3501.45$$

The cross products are the same, so the answer checks.

**State**. 450 U.S. dollars would be worth 3501.45 Hong Kong dollars.

**b) Familiarize**. Let $d =$ the price of the DVD player in U.S. dollars.

**Translate**. We translate to a proportion.
$$\text{U.S. dollars} \to \frac{1}{7.781} = \frac{d}{795} \leftarrow \text{U.S. dollars}$$
$$\text{Hong Kong} \to \qquad\qquad \leftarrow \text{Hong Kong}$$
$$\qquad\qquad\qquad\qquad \text{dollars}$$

**Solve**.
$$1 \cdot 795 = 7.781 \cdot d \quad \text{Equating cross products}$$
$$\frac{1 \cdot 795}{7.781} = \frac{7.781 \cdot d}{7.781}$$
$$102.17 \approx d$$

**Check**. We use a different approach. Since 1 U.S. dollar is worth 7.781 Hong Kong dollars, we multiply 102.17 by 7.781:
$$102.17(7.781) \approx 795.$$
This is the price in Hong Kong dollars, so the answer checks.

**State**. The DVD player would cost $102.17 in U.S. dollars.

**21. Familiarize**. Let $c =$ the cost of a turkey dinner for 14 people.

**Translate**. We translate to a proportion.
$$\text{People} \to \frac{8}{33.81} = \frac{14}{c} \leftarrow \text{People}$$
$$\text{Cost} \to \qquad\qquad \leftarrow \text{Cost}$$

**Solve**.
$$8 \cdot c = 33.81 \cdot 14$$
$$c = \frac{33.81 \cdot 14}{8}$$
$$c \approx 59.17$$

**Check**. We substitute in the proportion and check cross products.
$$\frac{8}{33.81} = \frac{14}{59.17}$$
$$8 \cdot 59.17 = 473.36; \ 33.81 \cdot 14 = 473.34$$
Since $473.36 \approx 473.34$, the answer checks.

**State**. It would cost about $59.17 to serve a turkey dinner for 14 people.

**22.** The ratio of 10 to $x$ is the same as the ratio of 5 to 4. We have the proportion
$$\frac{10}{x} = \frac{5}{4}.$$

Solve: $10 \cdot 4 = x \cdot 5$

$$\frac{10 \cdot 4}{5} = \frac{x \cdot 5}{5}$$

$$8 = x$$

The ratio of 11 to $y$ is the same as the ratio of 5 to 4. We have the proportion

$$\frac{11}{y} = \frac{5}{4}.$$

Solve: $11 \cdot 4 = y \cdot 5$

$$\frac{11 \cdot 4}{5} = \frac{y \cdot 5}{5}$$

$$8.8 = y$$

**23.** The ratio of 3 to $x$ is the same as the ratio of 5 to 8. We have the proportion

$$\frac{3}{x} = \frac{5}{8}.$$

Solve: $3 \cdot 8 = x \cdot 5$

$$\frac{3 \cdot 8}{5} = x$$

$$\frac{24}{5} = x, \text{ or}$$

$$4.8 = x$$

The ratio of 4 to $y$ is the same as the ratio of 5 to 8. We have the proportion.

$$\frac{4}{y} = \frac{5}{8}.$$

Solve: $4 \cdot 8 = y \cdot 5$

$$\frac{4 \cdot 8}{5} = y$$

$$\frac{32}{5} = y, \text{ or}$$

$$6.4 = y$$

The ratio of 7.5 to $z$ is the same as the ratio of 5 to 8. We have the proportion

$$\frac{7.5}{z} = \frac{5}{8}.$$

Solve: $7.5 \cdot 8 = z \cdot 5$

$$\frac{7.5 \cdot 8}{5} = \frac{z \cdot 5}{5}$$

$$12 = z$$

**24.** $\dfrac{4\frac{1}{2} \text{ mi}}{1\frac{1}{2}\text{hr}} = \dfrac{4\frac{1}{2}}{1\frac{1}{2}}\dfrac{\text{mi}}{\text{hr}} = \dfrac{\frac{9}{2}}{\frac{3}{2}}\dfrac{\text{mi}}{\text{hr}} = \dfrac{9}{2} \cdot \dfrac{2}{3}\dfrac{\text{mi}}{\text{hr}} = \dfrac{9 \cdot 2}{2 \cdot 3}\dfrac{\text{mi}}{\text{hr}} =$

$\dfrac{3 \cdot 3 \cdot 2}{2 \cdot 3 \cdot 1}\dfrac{\text{mi}}{\text{hr}} = \dfrac{3 \cdot 2}{3 \cdot 2} \cdot \dfrac{3}{1}\dfrac{\text{mi}}{\text{hr}} = 3\dfrac{\text{mi}}{\text{hr}}, \text{ or 3 mph}$

The correct answer is C.

**25.** **Familiarize.** Since there are 128 oz in a gallon, there are $8 \cdot 128$ oz, or 1024 oz, in 8 gallons. Let $m = $ Nancy's guess for the number of marbles in an 8-gal jar.

**Translate.** We translate to a proportion.

$$\begin{array}{c}\text{Ounces} \to \\ \text{Marbles} \to\end{array} \frac{8}{46} = \frac{1024}{m} \begin{array}{c}\leftarrow \text{Ounces} \\ \leftarrow \text{Marbles}\end{array}$$

**Solve.**

$8 \cdot m = 46 \cdot 1024$    Equating cross products

$$\frac{8 \cdot m}{8} = \frac{46 \cdot 1024}{8}$$

$$m = 5888$$

**Check.** We substitute in the proportion and check cross products.

$$\frac{8}{46} = \frac{1024}{5888}$$

$$8 \cdot 5888 = 47,104; \; 46 \cdot 1024 = 47,104$$

The cross products are the same, so the answer checks.

**State.** Nancy should guess that there are 5888 marbles in the jar.

---

# Cumulative Review Chapters 1 - 6

**1.**
$$\begin{array}{r} \overset{1\ 1\ \ \ 1\ 1}{2\ 7.\ 6\ 8\ 0} \\ 3.\ 0\ 1\ 9 \\ +\ 4\ 8\ 3.\ 2\ 9\ 7 \\ \hline 5\ 1\ 3.\ 9\ 9\ 6 \end{array}$$ Writing an extra zero

**2.** $2\boxed{\dfrac{1}{3} \cdot \dfrac{4}{4}} = 2\dfrac{4}{12}$

$\phantom{2}+4\,\dfrac{5}{12} \phantom{xx} = +4\,\dfrac{5}{12}$

$\phantom{xxxxxxxxxxxxx} 6\dfrac{9}{12} = 6\dfrac{3}{4}$

**3.** The LCD is 140.

$$-\frac{6}{35} + \left(-\frac{5}{28}\right) = -\frac{6}{35} \cdot \frac{4}{4} + \left(-\frac{5}{28}\right) \cdot \frac{5}{5}$$

$$= -\frac{24}{140} + \left(-\frac{25}{140}\right)$$

$$= -\frac{49}{140} = -\frac{7 \cdot 7}{7 \cdot 20}$$

$$= \frac{7}{7} \cdot \left(-\frac{7}{20}\right) = -\frac{7}{20}$$

**4.**
$$\begin{array}{r} \overset{\phantom{x}\overset{11}{\phantom{x}}}{\overset{3\ 9\ \cancel{1}\ 9\ 10}{4\,0.\,\cancel{2}\,\cancel{0}\,\cancel{0}}} \\ -\ 9.\ 7\ 0\ 9 \\ \hline 3\ 0.\ 4\ 9\ 1 \end{array}$$ Writing 2 extra zeros

**5.**
$$\begin{array}{r} \overset{2\ 18\ 1\ 10}{7\,\cancel{3}.\,\cancel{8}\,\cancel{2}\,\cancel{0}} \\ -\ 0.\ 9\ 0\ 8 \\ \hline 7\ 2.\ 9\ 1\ 2 \end{array}$$ Writing an extra zero

**6.** The LCD is 60.

$$-\frac{4}{15} - \frac{3}{20} = -\frac{4}{15} \cdot \frac{4}{4} - \frac{3}{20} \cdot \frac{3}{3}$$

$$= -\frac{16}{60} - \frac{9}{60}$$

$$= -\frac{25}{60} = -\frac{5 \cdot 5}{12 \cdot 5}$$

$$= -\frac{5}{12} \cdot \frac{5}{5} = -\frac{5}{12}$$

**7.**
$$
\begin{array}{r}
3\,7.6\,4 \\
\times\quad 5.9 \\
\hline
3\,3\,8\,7\,6 \\
1\,8\,8\,2\,0\,0 \\
\hline
2\,2\,2.0\,7\,6
\end{array}
$$

**8.** We move the decimal point 2 places to the right.
$$5.678 \times 100 = 567.8$$

**9.** $2\dfrac{1}{3} \cdot \left(-1\dfrac{2}{7}\right) = \dfrac{7}{3} \cdot \left(-\dfrac{9}{7}\right) = -\dfrac{7 \cdot 9}{3 \cdot 7} = -\dfrac{7 \cdot 3 \cdot 3}{3 \cdot 7 \cdot 1} =$
$\dfrac{7 \cdot 3}{7 \cdot 3} \cdot \left(-\dfrac{3}{1}\right) = -\dfrac{3}{1} = -3$

**10.**
$$
\begin{array}{r}
4\,3.\phantom{0} \\
2.3_\wedge\,\overline{)\,9\,8.9_\wedge} \\
9\,2\phantom{.0} \\
\hline
6\,9 \\
6\,9 \\
\hline
0
\end{array}
$$

**11.**
$$
\begin{array}{r}
8\,9\,9 \\
5\,4\,\overline{)\,4\,8,5\,4\,6} \\
4\,3\,2 \\
\hline
5\,3\,4 \\
4\,8\,6 \\
\hline
4\,8\,6 \\
4\,8\,6 \\
\hline
0
\end{array}
$$

**12.** $-\dfrac{7}{11} \div \dfrac{14}{33} = -\dfrac{7}{11} \cdot \dfrac{33}{14} = -\dfrac{7 \cdot 33}{11 \cdot 14} = -\dfrac{7 \cdot 3 \cdot 11}{11 \cdot 2 \cdot 7} =$
$\dfrac{7 \cdot 11}{7 \cdot 11} \cdot \left(-\dfrac{3}{2}\right) = -\dfrac{3}{2}$

**13.** $30{,}074 = 3$ ten thousands $+ 0$ thousands $+ 0$ hundreds $+$ 7 tens $+ 4$ ones, or 3 ten thousands $+ 7$ tens $+ 4$ ones

**14.** A word name for 120.07 is one hundred twenty and seven hundredths.

**15.** To compare two numbers in decimal notation, start at the left and compare corresponding digits moving from left to right. When two digits differ, the number with the larger digit is the larger of the two numbers.

0.7

$\downarrow$  Different; 7 is larger than 6.

0.698

Thus, 0.7 is larger.

**16.** To compare two negative numbers in decimal notation, start at the left and compare corresponding digits moving from left to right. When two digits differ, the number with the smaller digit is the larger of the two numbers.

$-0.799$

$\uparrow$  Different; 7 is smaller than 8.

$-0.8$

Thus, $-0.799$ is larger.

**17.**
$$
\begin{array}{r}
3 \quad \leftarrow 3 \text{ is prime.} \\
3\,\overline{)\,9} \\
2\,\overline{)\,1\,8} \\
2\,\overline{)\,3\,6} \\
2\,\overline{)\,7\,2} \\
2\,\overline{)\,1\,4\,4}
\end{array}
$$
Thus, $144 = 2 \cdot 2 \cdot 2 \cdot 2 \cdot 3 \cdot 3$, or $2^4 \cdot 3^2$.

**18.** $18 = 2 \cdot 3 \cdot 3$
$30 = 2 \cdot 3 \cdot 5$
The LCM is $2 \cdot 3 \cdot 3 \cdot 5$, or 90.

**19.** The rectangle is divided into 8 equal parts. The unit is $\dfrac{1}{8}$. The denominator is 8. We have 5 parts shaded. This tells us that the numerator is 5. Thus, $\dfrac{5}{8}$ is shaded.

**20.** $\dfrac{90}{144} = \dfrac{2 \cdot 3 \cdot 3 \cdot 5}{2 \cdot 2 \cdot 2 \cdot 2 \cdot 3 \cdot 3} = \dfrac{2 \cdot 3 \cdot 3}{2 \cdot 3 \cdot 3} \cdot \dfrac{5}{2 \cdot 2 \cdot 2} = \dfrac{5}{2 \cdot 2 \cdot 2} = \dfrac{5}{8}$

**21.** $\dfrac{3}{5} \times 9.53 = 0.6 \times 9.53 = 5.718$

**22.** $\dfrac{1}{3} \times 0.645 - \dfrac{3}{4} \times 0.048 = 0.215 - 0.036 = 0.179$

**23.** The ratio of 0.3 to 15 is $\dfrac{0.3}{15}$.

**24.** We can use cross products:

$3 \cdot 75 = 225$    $\dfrac{3}{9} \quad \dfrac{25}{75}$    $9 \cdot 25 = 225$

Since the cross products are the same, $225 = 225$, we know that $\dfrac{3}{9} = \dfrac{25}{75}$, so the numbers are proportional.

**25.** $\dfrac{660 \text{ meters}}{12 \text{ seconds}} = 55 \text{ m/sec}$

**26.** $\dfrac{\$0.99}{8 \text{ oz}} = \dfrac{99\cent}{8 \text{ oz}} = 12.375\cent/\text{oz}$
$\dfrac{\$3.29}{24.5 \text{ oz}} = \dfrac{329\cent}{24.5 \text{ oz}} \approx 13.429\cent/\text{oz}$
The 8-oz size has the lower unit price.

**27.** $\dfrac{14}{25} = \dfrac{x}{54}$
$14 \cdot 54 = 25 \cdot x$   Equating cross products
$\dfrac{14 \cdot 54}{25} = \dfrac{25 \cdot x}{25}$
$30.24 = x$
The solution is 30.24.

**28.** $423 = 16 \cdot t$
$\dfrac{423}{16} = \dfrac{16 \cdot t}{16}$
$26.4375 = t$
The solution is 26.4375.

**29.** $\dfrac{2}{3} \cdot y = \dfrac{16}{27}$

$$y = \dfrac{16}{27} \div \dfrac{2}{3}$$

$$y = \dfrac{16}{27} \cdot \dfrac{3}{2} = \dfrac{16 \cdot 3}{27 \cdot 2} = \dfrac{2 \cdot 8 \cdot 3}{3 \cdot 9 \cdot 2}$$

$$y = \dfrac{2 \cdot 3}{2 \cdot 3} \cdot \dfrac{8}{9} = \dfrac{8}{9}$$

The solution is $\dfrac{8}{9}$.

**30.** $\dfrac{7}{16} = \dfrac{56}{x}$

$7 \cdot x = 16 \cdot 56$     Equating cross products

$$\dfrac{7 \cdot x}{7} = \dfrac{16 \cdot 56}{7}$$

$$x = 128$$

The solution is 128.

**31.** $\qquad 34.56 + n = -67.9$

$34.56 + n - 34.56 = -67.9 - 34.56$

$$n = -102.46$$

The solution is $-102.46$.

**32.** $\qquad t + \dfrac{7}{25} = \dfrac{5}{7}$

$$t + \dfrac{7}{25} - \dfrac{7}{25} = \dfrac{5}{7} - \dfrac{7}{25}$$

$$t = \dfrac{5}{7} \cdot \dfrac{25}{25} - \dfrac{7}{25} \cdot \dfrac{7}{7}$$

$$t = \dfrac{125}{175} - \dfrac{49}{175}$$

$$t = \dfrac{76}{175}$$

**33. Familiarize.** Let $c =$ the number of calories in $\dfrac{3}{4}$ cup of fettuccini alfredo.

**Translate.** We write a multiplication sentence.

$$c = \dfrac{3}{4} \cdot 520$$

**Solve.** We carry out the multiplication.

$$c = \dfrac{3}{4} \cdot 520 = \dfrac{3 \cdot 520}{4} = \dfrac{3 \cdot 4 \cdot 130}{4 \cdot 1}$$

$$= \dfrac{4}{4} \cdot \dfrac{3 \cdot 130}{1} = \dfrac{3 \cdot 130}{1} = 390$$

**Check.** We repeat the calculation. The answer checks.

**State.** There are 390 calories in $\dfrac{3}{4}$ cup of fettuccini alfredo.

**34. Familiarize.** Let $t =$ the number of minutes required to stamp out 1295 washers.

**Translate.** We translate to a proportion.

$$\begin{array}{c} \text{Washers} \to \\ \text{Time} \quad \to \end{array} \dfrac{925}{5} = \dfrac{1295}{t} \begin{array}{c} \leftarrow \text{Washers} \\ \leftarrow \text{Time} \end{array}$$

**Solve**.

$925 \cdot t = 5 \cdot 1295$     Equating cross products

$$\dfrac{925 \cdot t}{925} = \dfrac{5 \cdot 1295}{925}$$

$$t = 7$$

**Check.** The number of washers that can be stamped out in 1 min is $925 \div 5$, or 185, so in 7 min $7 \cdot 185$, or 1295, washers can be stamped out. The answer checks.

**State.** It will take 7 min to stamp out 1295 washers.

**35. Familiarize.** Let $j =$ the number of cups of juice left over.

**Translate.** This is a "how much more" situation.

$$\underbrace{\text{Juice}}_{\displaystyle 3\frac{1}{2}} \underset{\text{plus}}{+} \underbrace{\text{Juice}}_{\displaystyle j} \underset{\text{is}}{=} \underbrace{\text{Amount of}}_{\displaystyle 5\frac{3}{4}}$$

(Juice used plus Juice left over is Amount of juice in can)

**Solve**.

$$3\dfrac{1}{2} + j = 5\dfrac{3}{4}$$

$$3\dfrac{1}{2} + j - 3\dfrac{1}{2} = 5\dfrac{3}{4} - 3\dfrac{1}{2}$$

$$j = 5\dfrac{3}{4} - 3\dfrac{2}{4} \qquad \left(\dfrac{1}{2} = \dfrac{2}{4}\right)$$

$$j = 2\dfrac{1}{4}$$

**Check.** $3\dfrac{1}{2} + 2\dfrac{1}{4} = 3\dfrac{2}{4} + 2\dfrac{1}{4} = 5\dfrac{3}{4}$, so the answer checks.

**State.** There are $2\dfrac{1}{4}$ cups of juice left over.

**36. Familiarize.** Let $d =$ the number of doors that can be hung in 8 hr.

**Translate**.

$$\underbrace{\text{Time to hang}}_{\displaystyle \frac{2}{3}} \underset{\text{times}}{\cdot} \underbrace{\text{Number}}_{\displaystyle d} \underset{\text{is}}{=} \underbrace{\text{Total}}_{\displaystyle 8}$$

(Time to hang one door times Number of doors is Total time)

**Solve**.

$$\dfrac{2}{3} \cdot d = 8$$

$$d = 8 \div \dfrac{2}{3}$$

$$d = 8 \cdot \dfrac{3}{2} = \dfrac{8 \cdot 3}{2} = \dfrac{2 \cdot 4 \cdot 3}{2 \cdot 1}$$

$$= \dfrac{2}{2} \cdot \dfrac{4 \cdot 3}{1} = \dfrac{4 \cdot 3}{1}$$

$$= 12$$

**Check.** $\dfrac{2}{3} \cdot 12 = \dfrac{2 \cdot 12}{3} = \dfrac{2 \cdot 3 \cdot 4}{3 \cdot 1} = \dfrac{3}{3} \cdot \dfrac{2 \cdot 4}{1} = 8$, so the answer checks.

**State.** 12 doors can be hung in 8 hr.

**37.** $\dfrac{337.62}{8 \text{ hr}} = 42.2025$ mi/hr, so the car travels 42.2025 mi in 1 hr.

**38.** *Familiarize.* Let $m =$ the number of orbits made during the mission.

*Translate.*

$$\underbrace{\text{Number of orbits per day}}_{\displaystyle\downarrow \atop 16} \underset{\underset{\cdot}{}}{\text{times}} \underbrace{\text{Number of days}}_{\displaystyle\downarrow \atop 8.25} \underset{\underset{=}{}}{\text{is}} \underbrace{\text{Total number of orbits}}_{\displaystyle\downarrow \atop m}$$

*Solve.* We multiply.

$$\begin{array}{r} 8.\,2\,5 \\ \times\,1\,6 \\ \hline 4\,9\,5\,0 \\ 8\,2\,5\,0 \\ \hline 1\,3\,2.\,0\,0 \end{array}$$

Thus, $m = 132$.

*Check.* We repeat the calculation. The answer checks.

*State.* 132 orbits were made during the mission.

**39.** 2 is the only even prime number, so answer D is correct.

**40.** If the mileage is 28.16 miles per gallon, then the rate of gallons per mile is $\dfrac{1 \text{ gallon}}{28.16 \text{ miles}}$:

$$\frac{1}{28.16} = \frac{1}{28.16} \cdot \frac{100}{100} = \frac{100}{2816} = \frac{4 \cdot 25}{4 \cdot 704} = \frac{4}{4} \cdot \frac{25}{704} = \frac{25}{704}$$

Answer B is correct.

**41.**

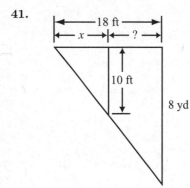

We express 8 yd as $8 \cdot 3$ ft, or 24 ft. Then the ratio of $x$ to 10 is the same as the ratio of 18 to 24.

$$\frac{x}{10} = \frac{18}{24}$$

$$x \cdot 24 = 10 \cdot 18$$

$$\frac{x \cdot 24}{24} = \frac{10 \cdot 18}{24}$$

$$x = 7.5$$

Now we subtract to find the length labeled "?" in the drawing.

$$18 - 7.5 = 10.5, \text{ or } 10\frac{1}{2}$$

The goalie should stand $10\frac{1}{2}$ ft in front of the goal.

# Chapter 7

# Percent Notation

---

## Exercise Set 7.1

**1.** $90\% = \dfrac{90}{100}$     A ratio of 90 to 100

$90\% = 90 \times \dfrac{1}{100}$     Replacing % with $\times \dfrac{1}{100}$

$90\% = 90 \times 0.01$     Replacing % with $\times 0.01$

**3.** $12.5\% = \dfrac{12.5}{100}$     A ratio of 12.5 to 100

$12.5\% = 12.5 \times \dfrac{1}{100}$     Replacing % with $\times \dfrac{1}{100}$

$12.5\% = 12.5 \times 0.01$     Replacing % with $\times 0.01$

**5.** 67%

a) Replace the percent symbol with ×0.01.

$67 \times 0.01$

b) Move the decimal point two places to the left.

0.67.

Thus, 67% = 0.67.

**7.** 45.6%

a) Replace the percent symbol with ×0.01.

$45.6 \times 0.01$

b) Move the decimal point two places to the left.

0.45.6

Thus, 45.6% = 0.456.

**9.** 59.01%

a) Replace the percent symbol with ×0.01.

$59.01 \times 0.01$

b) Move the decimal point two places to the left.

0.59.01

Thus, 59.01% = 0.5901.

**11.** 10%

a) Replace the percent symbol with ×0.01.

$10 \times 0.01$

b) Move the decimal point two places to the left.

0.10.

Thus, 10% = 0.1.

**13.** 1%

a) Replace the percent symbol with ×0.01.

$1 \times 0.01$

b) Move the decimal point two places to the left.

0.01.

Thus, 1% = 0.01.

**15.** 200%

a) Replace the percent symbol with ×0.01.

$200 \times 0.01$

b) Move the decimal point two places to the left.

2.00.

Thus, 200% = 2.

**17.** 0.1%

a) Replace the percent symbol with ×0.01.

$0.1 \times 0.01$

b) Move the decimal point two places to the left.

0.00.1

Thus, 0.1% = 0.001.

**19.** 0.09%

a) Replace the percent symbol with ×0.01.

$0.09 \times 0.01$

b) Move the decimal point two places to the left.

0.00.09

Thus, 0.09% = 0.0009.

**21.** 0.18%

a) Replace the percent symbol with ×0.01.

$0.18 \times 0.01$

b) Move the decimal point two places to the left.

0.00.18

Thus, 0.18% = 0.0018.

**23.** 23.19%

a) Replace the percent symbol with ×0.01.

$23.19 \times 0.01$

b) Move the decimal point two places to the left.

0.23.19

Thus, 23.19% = 0.2319.

**25.** $14\frac{7}{8}\%$

  a) Convert $14\frac{7}{8}$ to decimal notation and replace the percent symbol with $\times 0.01$.

    $14.875 \times 0.01$

  b) Move the decimal point two places to the left.

    0.14.875

    Thus, $14\frac{7}{8}\% = 0.14875$.

**27.** $56\frac{1}{2}\%$

  a) Convert $56\frac{1}{2}$ to decimal notation and replace the percent symbol with $\times 0.01$.

    $56.5 \times 0.01$

  b) Move the decimal point two places to the left.

    0.56.5

    Thus, $56\frac{1}{2}\% = 0.565$.

**29.** 97%

  a) Replace the percent symbol with $\times 0.01$.

    $97 \times 0.01$

  b) Move the decimal point two places to the left.

    0.97.

  Thus, $97\% = 0.97$.

58%

  a) Replace the percent symbol with $\times 0.01$.

    $58 \times 0.01$

  b) Move the decimal point two places to the left.

    0.58.

  Thus, $58\% = 0.58$.

**31.** 7%

  a) Replace the percent symbol with $\times 0.01$.

    $7 \times 0.01$

  b) Move the decimal point two places to the left.

    0.07.

  Thus, $7\% = 0.07$.

8%

  a) Replace the percent symbol with $\times 0.01$.

    $8 \times 0.01$

  b) Move the decimal point two places to the left.

    0.08.

  Thus, $8\% = 0.08$.

**33.** 54.8%

  a) Replace the percent symbol with $\times 0.01$.

    $54.8 \times 0.01$

  b) Move the decimal point two places to the left.

    0.54.8

  Thus, $54.8\% = 0.548$.

**35.** 0.47

  a) Move the decimal point two places to the right.

    0.47.

  b) Write a percent symbol: 47%

  Thus, $0.47 = 47\%$.

**37.** 0.03

  a) Move the decimal point two places to the right.

    0.03.

  b) Write a percent symbol: 3%

  Thus, $0.03 = 3\%$.

**39.** 8.7

  a) Move the decimal point two places to the right.

    8.70.

  b) Write a percent symbol: 870%

  Thus, $8.7 = 870\%$.

**41.** 0.334

  a) Move the decimal point two places to the right.

    0.33.4

  b) Write a percent symbol: 33.4%

  Thus, $0.334 = 33.4\%$.

**43.** 0.75

  a) Move the decimal point two places to the right.

    0.75.

  b) Write a percent symbol: 75%

  Thus, $0.75 = 75\%$.

**45.** 0.4

  a) Move the decimal point two places to the right.

    0.40.

  b) Write a percent symbol: 40%

  Thus, $0.4 = 40\%$.

**47.** 0.006

a) Move the decimal point two places to the right.

0.00.6

b) Write a percent symbol: 0.6%

Thus, 0.006 = 0.6%.

**49.** 0.017

a) Move the decimal point two places to the right.

0.01.7

b) Write a percent symbol: 1.7%

Thus, 0.017 = 1.7%.

**51.** 0.2718

a) Move the decimal point two places to the right.

0.27.18

b) Write a percent symbol: 27.18%

Thus, 0.2718 = 27.18%.

**53.** 0.0239

a) Move the decimal point two places to the right.

0.02.39

b) Write a percent symbol: 2.39%

Thus, 0.0239 = 2.39%.

**55.** 0.27

a) Move the decimal point two places to the right.

0.27.

b) Write a percent symbol: 27%

Thus, 0.27 = 27%.

**57.** 0.057

a) Move the decimal point two places to the right.

0.05.7

b) Write a percent symbol: 5.7%

Thus, 0.057 = 5.7%.

0.176

a) Move the decimal point two places to the right.

0.17.6

b) Write a percent symbol: 17.6%

Thus, 0.176 = 17.6%.

**59.** 0.906

a) Move the decimal point two places to the right.

0.90.6

b) Write a percent symbol: 90.6%

Thus, 0.906 = 90.6%.

0.88

a) Move the decimal point two places to the right.

0.88.

b) Write a percent symbol: 88%

Thus, 0.88 = 88%.

**61.** 64%

a) Replace the percent symbol with ×0.01.

$64 \times 0.01$

b) Move the decimal point two places to the left.

0.64.

Thus, 64% = 0.64.

30%

a) Replace the percent symbol with ×0.01.

$30 \times 0.01$

b) Move the decimal point two places to the left.

0.30.

Thus, 30% = 0.3.

4%

a) Replace the percent symbol with ×0.01.

$4 \times 0.01$

b) Move the decimal point two places to the left.

0.04.

Thus, 4% = 0.04.

2%

a) Replace the percent symbol with ×0.01.

$2 \times 0.01$

b) Move the decimal point two places to the left.

0.02.

Thus, 2% = 0.02.

**63.** To convert $\dfrac{100}{3}$ to a mixed numeral, we divide.

$$\begin{array}{r} 3\,3 \\ 3\,\overline{)1\,0\phantom{0}} \\ 9\,0 \\ \hline 1\,0 \\ 9 \\ \hline 1 \end{array} \qquad \dfrac{100}{3} = 33\tfrac{1}{3}$$

**65.** To convert $\dfrac{75}{8}$ to a mixed numeral, we divide.

$$\begin{array}{r} 9 \\ 8\,\overline{)7\,5} \\ 7\,2 \\ \hline 3 \end{array} \qquad \dfrac{75}{8} = 9\tfrac{3}{8}$$

**67.** First we consider $\dfrac{567}{98}$.

To convert $\dfrac{567}{98}$ to a mixed numeral, we divide.

$$\begin{array}{r} 5 \\ 9\,8\,\overline{)5\,6\,7} \\ 4\,9\,0 \\ \hline 7\,7 \end{array} \qquad \dfrac{567}{98} = 5\dfrac{77}{98} = 5\dfrac{11}{14}$$

$\dfrac{567}{98} = 5\dfrac{11}{14}$, so $-\dfrac{567}{98} = -5\dfrac{11}{14}$

**69.** To convert $\dfrac{2}{3}$ to decimal notation, we divide.

$$\begin{array}{r} 0.6\,6 \\ 3\,\overline{)2.0\,0} \\ 1\,8 \\ \hline 2\,0 \\ 1\,8 \\ \hline 2 \end{array}$$

Since 2 keeps reappearing as a remainder, the digits repeat and
$$\dfrac{2}{3} = 0.66\ldots \quad \text{or} \quad 0.\overline{6}.$$

**71.** First we consider $\dfrac{5}{6}$.

To convert $\dfrac{5}{6}$ to decimal notation, we divide.

$$\begin{array}{r} 0.8\,3 \\ 6\,\overline{)5.0\,0} \\ 4\,8 \\ \hline 2\,0 \\ 1\,8 \\ \hline 2 \end{array}$$

Since 2 keeps reappearing as a remainder, the digits repeat and
$$\dfrac{5}{6} = 0.833\ldots \quad \text{or} \quad 0.8\overline{3}, \text{ so } -\dfrac{5}{6} = -0.8\overline{3}.$$

**73.** To convert $\dfrac{8}{3}$ to decimal notation, we divide.

$$\begin{array}{r} 2.6\,6 \\ 3\,\overline{)8.0\,0} \\ 6 \\ \hline 2\,0 \\ 1\,8 \\ \hline 2\,0 \end{array}$$

Since 2 keeps reappearing as a remainder, the digits repeat and
$$\dfrac{8}{3} = 2.66\ldots \text{ or } 2.\overline{6}.$$

**75.** $\dfrac{1}{2} = \dfrac{1}{2} \cdot \dfrac{50}{50} = \dfrac{50}{100} = 50\%$

**77.** $\dfrac{7}{10} = \dfrac{7}{10} \cdot \dfrac{10}{10} = \dfrac{70}{100} = 70\%$

**79.** One of the five equal-sized portions of the figure is shaded, so $\dfrac{1}{5}$ is shaded. We find percent notation for $\dfrac{1}{5}$.

$$\dfrac{1}{5} = \dfrac{1}{5} \cdot \dfrac{20}{20} = \dfrac{20}{100} = 20\%$$

Thus, 20% is shaded.

## Exercise Set 7.2

**1.** We use the definition of percent as a ratio.

$$\dfrac{41}{100} = 41\%$$

**3.** We use the definition of percent as a ratio.

$$\dfrac{5}{100} = 5\%$$

**5.** We multiply by 1 to get 100 in the denominator.

$$\dfrac{2}{10} = \dfrac{2}{10} \cdot \dfrac{10}{10} = \dfrac{20}{100} = 20\%$$

**7.** We multiply by 1 to get 100 in the denominator.

$$\dfrac{3}{10} = \dfrac{3}{10} \cdot \dfrac{10}{10} = \dfrac{30}{100} = 30\%$$

**9.** $\dfrac{1}{2} = \dfrac{1}{2} \cdot \dfrac{50}{50} = \dfrac{50}{100} = 50\%$

**11.** Find decimal notation by division.

$$\begin{array}{r} 0.8\,7\,5 \\ 8\,\overline{)7.0\,0\,0} \\ 6\,4 \\ \hline 6\,0 \\ 5\,6 \\ \hline 4\,0 \\ 4\,0 \\ \hline 0 \end{array}$$

$$\dfrac{7}{8} = 0.875$$

Convert to percent notation.

$$0.87.5$$

$$\dfrac{7}{8} = 87.5\%, \text{ or } 87\tfrac{1}{2}\%$$

**13.** $\dfrac{4}{5} = \dfrac{4}{5} \cdot \dfrac{20}{20} = \dfrac{80}{100} = 80\%$

**15.** Find decimal notation by division.

$$
\begin{array}{r}
0.6\ 6\ 6 \\
3\,\overline{)\,2.0\ 0\ 0} \\
\underline{1\ 8} \\
2\ 0 \\
\underline{1\ 8} \\
2\ 0 \\
\underline{1\ 8} \\
2
\end{array}
$$

We get a repeating decimal: $\dfrac{2}{3} = 0.66\overline{6}$

Convert to percent notation.

0.66.$\overline{6}$
$\quad$└─↑

$\dfrac{2}{3} = 66.\overline{6}\%$, or $66\dfrac{2}{3}\%$

**17.**
$$
\begin{array}{r}
0.1\ 6\ 6 \\
6\,\overline{)\,1.0\ 0\ 0} \\
\underline{6} \\
4\ 0 \\
\underline{3\ 6} \\
4\ 0 \\
\underline{3\ 6} \\
4
\end{array}
$$

We get a repeating decimal: $\dfrac{1}{6} = 0.16\overline{6}$

Convert to percent notation.

0.16.$\overline{6}$
$\quad$└─↑

$\dfrac{1}{6} = 16.\overline{6}\%$, or $16\dfrac{2}{3}\%$

**19.**
$$
\begin{array}{r}
0.1\ 8\ 7\ 5 \\
16\,\overline{)\,3.0\ 0\ 0\ 0} \\
\underline{1\ 6} \\
1\ 4\ 0 \\
\underline{1\ 2\ 8} \\
1\ 2\ 0 \\
\underline{1\ 1\ 2} \\
8\ 0 \\
\underline{8\ 0} \\
0
\end{array}
$$

$\dfrac{3}{16} = 0.1875$

Convert to percent notation.

0.18.75
$\quad$└─↑

$\dfrac{3}{16} = 18.75\%$, or $18\dfrac{3}{4}\%$

**21.**
$$
\begin{array}{r}
0.8\ 1\ 2\ 5 \\
16\,\overline{)\,1\ 3.0\ 0\ 0\ 0} \\
\underline{1\ 2\ 8} \\
2\ 0 \\
\underline{1\ 6} \\
4\ 0 \\
\underline{3\ 2} \\
8\ 0 \\
\underline{8\ 0} \\
0
\end{array}
$$

$\dfrac{13}{16} = 0.8125$

Convert to percent notation.

0.81.25
$\quad$└─↑

$\dfrac{13}{16} = 81.25\%$, or $81\dfrac{1}{4}\%$

**23.** $\dfrac{4}{25} = \dfrac{4}{25} \cdot \dfrac{4}{4} = \dfrac{16}{100} = 16\%$

**25.** $\dfrac{1}{20} = \dfrac{1}{20} \cdot \dfrac{5}{5} = \dfrac{5}{100} = 5\%$

**27.** $\dfrac{17}{50} = \dfrac{17}{50} \cdot \dfrac{2}{2} = \dfrac{34}{100} = 34\%$

**29.** $\dfrac{2}{25} = \dfrac{2}{25} \cdot \dfrac{4}{4} = \dfrac{8}{100} = 8\%$

$\dfrac{59}{100} = 59\%$

**31.** $\dfrac{11}{50} = \dfrac{11}{50} \cdot \dfrac{2}{2} = \dfrac{22}{100} = 22\%$

**33.** $\dfrac{3}{25} = \dfrac{3}{25} \cdot \dfrac{4}{4} = \dfrac{12}{100} = 12\%$

**35.** $\dfrac{3}{20} = \dfrac{3}{20} \cdot \dfrac{5}{5} = \dfrac{15}{100} = 15\%$

**37.** $85\% = \dfrac{85}{100}$ $\qquad$ Definition of percent

$\qquad = \dfrac{5 \cdot 17}{5 \cdot 20}$ ⎫

$\qquad = \dfrac{5}{5} \cdot \dfrac{17}{20}$ ⎬ Simplifying

$\qquad = \dfrac{17}{20}$ ⎭

**39.** $62.5\% = \dfrac{62.5}{100}$ $\qquad$ Definition of percent

$\qquad = \dfrac{62.5}{100} \cdot \dfrac{10}{10}$ $\qquad$ Multiplying by 1 to eliminate the decimal point in the numerator

$\qquad = \dfrac{625}{1000}$

$\qquad = \dfrac{5 \cdot 125}{8 \cdot 125}$ ⎫

$\qquad = \dfrac{5}{8} \cdot \dfrac{125}{125}$ ⎬ Simplifying

$\qquad = \dfrac{5}{8}$ ⎭

**41.** $33\frac{1}{3}\% = \frac{100}{3}\%$     Converting from mixed numeral to fraction notation

$= \frac{100}{3} \times \frac{1}{100}$     Definition of percent

$= \frac{100 \cdot 1}{3 \cdot 100}$     Multiplying

$= \frac{1}{3} \cdot \frac{100}{100}$

$= \frac{1}{3}$     Simplifying

**43.** $16.\overline{6}\% = 16\frac{2}{3}\%$     $\left(16.\overline{6} = 16\frac{2}{3}\right)$

$= \frac{50}{3}\%$     Converting from mixed numeral to fractional notation

$= \frac{50}{3} \times \frac{1}{100}$     Definition of percent

$= \frac{50 \cdot 1}{3 \cdot 50 \cdot 2}$     Multiplying

$= \frac{1}{2 \cdot 3} \cdot \frac{50}{50}$

$= \frac{1}{6}$     Simplifying

**45.** $7.25\% = \frac{7.25}{100} = \frac{7.25}{100} \cdot \frac{100}{100}$

$= \frac{725}{10,000} = \frac{29 \cdot 25}{400 \cdot 25} = \frac{29}{400} \cdot \frac{25}{25}$

$= \frac{29}{400}$

**47.** $0.8\% = \frac{0.8}{100} = \frac{0.8}{100} \cdot \frac{10}{10}$

$= \frac{8}{1000} = \frac{1 \cdot 8}{125 \cdot 8} = \frac{1}{125} \cdot \frac{8}{8}$

$= \frac{1}{125}$

**49.** $25\frac{3}{8}\% = \frac{203}{8}\%$

$= \frac{203}{8} \times \frac{1}{100}$     Definition of percent

$= \frac{203}{800}$

**51.** $78\frac{2}{9}\% = \frac{704}{9}\%$

$= \frac{704}{9} \times \frac{1}{100}$     Definition of percent

$= \frac{4 \cdot 176 \cdot 1}{9 \cdot 4 \cdot 25}$

$= \frac{4}{4} \cdot \frac{176 \cdot 1}{9 \cdot 25}$

$= \frac{176}{225}$

**53.** $64\frac{7}{11}\% = \frac{711}{11}\%$

$= \frac{711}{11} \times \frac{1}{100}$

$= \frac{711}{1100}$

**55.** $150\% = \frac{150}{100} = \frac{3 \cdot 50}{2 \cdot 50} = \frac{3}{2} \cdot \frac{50}{50} = \frac{3}{2}$

**57.** $0.0325\% = \frac{0.0325}{100} = \frac{0.0325}{100} \cdot \frac{10,000}{10,000} = \frac{325}{1,000,000} = \frac{25 \cdot 13}{25 \cdot 40,000} = \frac{25}{25} \cdot \frac{13}{40,000} = \frac{13}{40,000}$

**59.** Note that $33.\overline{3}\% = 33\frac{1}{3}\%$ and proceed as in Exercise 41;

$33.\overline{3}\% = \frac{1}{3}$.

**61.** $6\% = \frac{6}{100}$

$= \frac{2 \cdot 3}{2 \cdot 50} = \frac{2}{2} \cdot \frac{3}{50}$

$= \frac{3}{50}$

**63.** $12\% = \frac{12}{100}$

$= \frac{4 \cdot 3}{4 \cdot 25} = \frac{4}{4} \cdot \frac{3}{25}$

$= \frac{3}{25}$

**65.** $75\% = \frac{75}{100}$

$= \frac{25 \cdot 3}{25 \cdot 4} = \frac{25}{25} \cdot \frac{3}{4}$

$= \frac{3}{4}$

**67.** $15\% = \frac{15}{100}$

$= \frac{5 \cdot 3}{5 \cdot 20} = \frac{5}{5} \cdot \frac{3}{20}$

$= \frac{3}{20}$

**69.** $20.9\% = \frac{20.9}{100}$

$= \frac{20.9}{100} \cdot \frac{10}{10}$

$= \frac{209}{1000}$

**71.** $\frac{1}{8} = 1 \div 8$

$$\begin{array}{r} 0.1\,2\,5 \\ 8\,\overline{)\,1.0\,0\,0} \\ \underline{8\phantom{.000}} \\ 2\,0\phantom{0} \\ \underline{1\,6\phantom{0}} \\ 4\,0 \\ \underline{4\,0} \\ 0 \end{array}$$

$\dfrac{1}{8} = 0.125 = 12\dfrac{1}{2}\%$, or $12.5\%$

$\dfrac{1}{6} = 1 \div 6$

$$\begin{array}{r} 0.1\ 6\ 6 \\ 6\,\overline{\big)\,1.0\ 0\ 0} \\ 6\phantom{.000} \\ \overline{\phantom{0}4\ 0} \\ 3\ 6\phantom{0} \\ \overline{\phantom{00}4\ 0} \\ 3\ 6 \\ \overline{\phantom{000}4} \end{array}$$

We get a repeating decimal: $0.1\overline{6}$

$0.16.\overline{6} \qquad\qquad 0.1\overline{6} = 16.\overline{6}\%$
  $\llcorner\uparrow$

$\dfrac{1}{6} = 0.1\overline{6} = 16.\overline{6}\%$, or $16\dfrac{2}{3}\%$

$20\% = \dfrac{20}{100} = \dfrac{1}{5} \cdot \dfrac{20}{20} = \dfrac{1}{5}$

$0.20. \qquad\qquad 20\% = 0.2$
  $\ulcorner\lrcorner$

$\dfrac{1}{5} = 0.2 = 20\%$

$0.25. \qquad\qquad 0.25 = 25\%$
  $\llcorner\uparrow$

$25\% = \dfrac{25}{100} = \dfrac{1}{4} \cdot \dfrac{25}{25} = \dfrac{1}{4}$

$\dfrac{1}{4} = 0.25 = 25\%$

$33\dfrac{1}{3}\% = \dfrac{\frac{100}{3}}{\phantom{x}}\% = \dfrac{100}{3} \times \dfrac{1}{100} = \dfrac{100}{300} = \dfrac{1}{3} \cdot \dfrac{100}{100} = \dfrac{1}{3}$

$0.33.\overline{3} \qquad\qquad 33.\overline{3}\% = 0.33\overline{3}$, or $0.\overline{3}$
  $\ulcorner\lrcorner$

$\dfrac{1}{3} = 0.\overline{3} = 33\dfrac{1}{3}\%$, or $33.\overline{3}\%$

$37.5\% = \dfrac{37.5}{100} = \dfrac{37.5}{100} \cdot \dfrac{10}{10} = \dfrac{375}{1000} = \dfrac{3}{8} \cdot \dfrac{125}{125} = \dfrac{3}{8}$

$0.37.5 \qquad\qquad 37.5\% = 0.375$
  $\ulcorner\lrcorner$

$\dfrac{3}{8} = 0.375 = 37\dfrac{1}{2}\%$, or $37.5\%$

$40\% = \dfrac{40}{100} = \dfrac{2}{5} \cdot \dfrac{20}{20} = \dfrac{2}{5}$

$0.40. \qquad\qquad 40\% = 0.4$
  $\ulcorner\lrcorner$

$\dfrac{2}{5} = 0.4 = 40\%$

$\dfrac{1}{2} = \dfrac{1}{2} \cdot \dfrac{5}{5} = \dfrac{5}{10} = 0.5$

$\dfrac{1}{2} = \dfrac{1}{2} \cdot \dfrac{50}{50} = \dfrac{50}{100} = 5\%$

$\dfrac{1}{2} = 0.5 = 50\%$

**73.**    $0.50. \qquad\qquad 0.5 = 50\%$
    $\llcorner\uparrow$

$50\% = \dfrac{50}{100} = \dfrac{1}{2} \cdot \dfrac{50}{50} = \dfrac{1}{2}$

$\dfrac{1}{2} = 0.5 = 50\%$

$\dfrac{1}{3} = 1 \div 3$

$$\begin{array}{r} 0.3 \\ 3\,\overline{\big)\,1.0} \\ 9 \\ \overline{\phantom{0}1} \end{array}$$

We get a repeating decimal: $0.\overline{3}$

$0.33.\overline{3} \qquad\qquad 0.\overline{3} = 33.\overline{3}\%$
    $\llcorner\uparrow$

$\dfrac{1}{3} = 0.\overline{3} = 33.\overline{3}\%$, or $33\dfrac{1}{3}\%$

$25\% = \dfrac{25}{100} = \dfrac{25}{25} \cdot \dfrac{1}{4} = \dfrac{1}{4}$

$0.25. \qquad\qquad 25\% = 0.25$
  $\ulcorner\lrcorner$

$\dfrac{1}{4} = 0.25 = 25\%$

$16\dfrac{2}{3}\% = \dfrac{\frac{50}{3}}{\phantom{x}}\% = \dfrac{50}{3} \times \dfrac{1}{100} = \dfrac{50 \cdot 1}{3 \cdot 2 \cdot 50} = \dfrac{50}{50} \cdot \dfrac{1}{6} = \dfrac{1}{6}$

$\dfrac{1}{6} = 1 \div 6$

$$\begin{array}{r} 0.1\ 6 \\ 6\,\overline{\big)\,1.0\ 0} \\ 6\phantom{.00} \\ \overline{\phantom{0}4\ 0} \\ 3\ 6 \\ \overline{\phantom{00}4} \end{array}$$

We get a repeating decimal: $0.1\overline{6}$

$\dfrac{1}{6} = 0.1\overline{6} = 16\dfrac{2}{3}\%$, or $16.\overline{6}\%$

0.12.5          $0.125 = 12.5\%$

$12.5\% = \dfrac{12.5}{100} = \dfrac{12.5}{100} \cdot \dfrac{10}{10} = \dfrac{125}{1000} = \dfrac{125}{125} \cdot \dfrac{1}{8} = \dfrac{1}{8}$

$\mathbf{\dfrac{1}{8} = 0.125 = 12.5\%, \ or \ 12\dfrac{1}{2}\%}$

$\dfrac{3}{4} = \dfrac{3}{4} \cdot \dfrac{25}{25} = \dfrac{75}{100} = 75\%$

0.75.          $75\% = 0.75$

$\mathbf{\dfrac{3}{4} = 0.75 = 75\%}$

$0.8\overline{3} = 0.83.\overline{3}$      $0.8\overline{3} = 83.\overline{3}\%$

$83.\overline{3}\% = 83\dfrac{1}{3}\% = \dfrac{250}{3}\% = \dfrac{250}{3} \times \dfrac{1}{100} = \dfrac{5 \cdot 50}{3 \cdot 2 \cdot 50} =$
$\dfrac{5}{6} \cdot \dfrac{50}{50} = \dfrac{5}{6}$

$\mathbf{\dfrac{5}{6} = 0.8\overline{3} = 83.\overline{3}\%, \ or \ 83\dfrac{1}{3}\%}$

$\dfrac{3}{8} = 3 \div 8$

```
      0.3 7 5
  8 | 3.0 0 0
      2 4
      ───
        6 0
        5 6
        ───
          4 0
          4 0
          ───
            0
```

$\dfrac{3}{8} = 0.375$

0.37.5          $0.375 = 37.5\%$

$\mathbf{\dfrac{3}{8} = 0.375 = 37.5\%, \ or \ 37\dfrac{1}{2}\%}$

**75.** $\quad 13 \cdot x = 910$

$\qquad \dfrac{13 \cdot x}{13} = \dfrac{910}{13}$

$\qquad\qquad x = 70$

**77.** $\quad 0.05 \times b = -20$

$\qquad \dfrac{0.05 \times b}{0.05} = \dfrac{-20}{0.05}$

$\qquad\qquad b = -400$

**79.** $\quad \dfrac{24}{37} = \dfrac{15}{x}$

$\qquad 24 \cdot x = 37 \cdot 15$     Equating cross products

$\qquad\qquad x = \dfrac{37 \cdot 15}{24}$

$\qquad\qquad x = 23.125$

**81.** $\quad \dfrac{9}{10} = \dfrac{x}{5}$

$\qquad 9 \cdot 5 = 10 \cdot x$

$\qquad \dfrac{9 \cdot 5}{10} = x$

$\qquad \dfrac{45}{10} = x$

$\qquad \dfrac{9}{2} = x, \ or$

$\qquad 4.5 = x$

**83.**
```
        3 3
    3 | 1 0 0
        9
        ───
        1 0
          9
        ───
          1
```

$\dfrac{100}{3} = 33\dfrac{1}{3}$

**85.**
```
        8 3
    3 | 2 5 0
        2 4
        ───
        1 0
          9
        ───
          1
```

$\dfrac{250}{3} = 83\dfrac{1}{3}$

**87.** First consider $\dfrac{345}{8}$.

```
        4 3
    8 | 3 4 5
        3 2 0
        ─────
          2 5
          2 4
          ───
            1
```

$\dfrac{345}{8} = 43\dfrac{1}{8}$, so $-\dfrac{345}{8} = -43\dfrac{1}{8}$.

**89.**
```
        1 8
    4 | 7 5
        4
        ───
        3 5
        3 2
        ───
          3
```

$\dfrac{75}{4} = 18\dfrac{3}{4}$

**91.** $1\dfrac{1}{17} = \dfrac{18}{17}$     $(1 \cdot 17 = 17, \ 17 + 1 = 18)$

**93.** First consider $101\dfrac{1}{2}$.

$101\dfrac{1}{2} = \dfrac{203}{2}$     $(101 \cdot 2 = 202, \ 202 + 1 = 203)$

Then $-101\dfrac{1}{2} = -\dfrac{203}{2}$.

**95.** Use a calculator.

$\dfrac{41}{369} = 0.11.\overline{1} = 11.\overline{1}\%$

**97.**  $2.5\overline{74631} = 2.57.\overline{46317} = 257.\overline{46317}\%$
        └─↑

**99.**  $\dfrac{14}{9}\% = \dfrac{14}{9} \times \dfrac{1}{100} = \dfrac{2 \cdot 7 \cdot 1}{9 \cdot 2 \cdot 50} = \dfrac{2}{2} \cdot \dfrac{7}{450} = \dfrac{7}{450}$

To find decimal notation for $\dfrac{7}{450}$ we divide.

```
              0.0 1 5 5
        4 5 0 |7.0 0 0 0
                4 5 0
                2 5 0 0
                2 2 5 0
                  2 5 0 0
                  2 2 5 0
                    2 5 0
```

We get a repeating decimal:  $\dfrac{14}{9}\% = 0.01\overline{5}$

**101.**  $\dfrac{729}{7}\% = \dfrac{729}{7} \times \dfrac{1}{100} = \dfrac{729}{700}$

To find decimal notation for $\dfrac{729}{700}$ we divide.

```
                1.0 4 1 4 2 8 5 7
        7 0 0 |7 2 9.0 0 0 0 0 0 0 0 0
                7 0 0
                2 9 0 0
                2 8 0 0
                  1 0 0 0
                    7 0 0
                    3 0 0 0
                    2 8 0 0
                      2 0 0 0
                      1 4 0 0
                        6 0 0 0
                        5 6 0 0
                          4 0 0 0
                          3 5 0 0
                            5 0 0 0
                            4 9 0 0
                              1 0 0
```

We get a repeating decimal:  $\dfrac{729}{7}\% = 1.04\overline{142857}.$

**103.** We will express each number in decimal notation.

$16\dfrac{1}{6}\% = 0.161\overline{6}$

$1.6$

$\dfrac{1}{6}\% = 0.001\overline{6}$

$\dfrac{1}{2} = 0.5$

$0.2$

$1.6\% = 0.016$

$1\dfrac{1}{6}\% = 0.011\overline{6}$

$0.5\% = 0.005$

$\dfrac{2}{7}\% = 0.00\overline{285714}$

$0.\overline{54} = 0.545454$

Arranging the numbers from smallest to largest, we have
$\dfrac{1}{6}\%,\ \dfrac{2}{7}\%,\ 0.5\%,\ 1\dfrac{1}{6}\%,\ 1.6\%,\ 16\dfrac{1}{6}\%,\ 0.2,\ \dfrac{1}{2},\ 0.\overline{54},\ 1.6.$

---

## Exercise Set 7.3

**1.** What is 32% of 78?
$$\begin{array}{ccccc} \downarrow & \downarrow & \downarrow & \downarrow & \downarrow \\ a & = & 32\% & \times & 78 \end{array}$$

**3.** 89 is what percent of 99?
$$\begin{array}{ccccc} \downarrow\ \downarrow & & \downarrow & & \downarrow\ \downarrow \\ 89\ = & & p & & \times\ 99 \end{array}$$

**5.** 13 is 25% of what?
$$\begin{array}{ccccc} \downarrow\ \downarrow & \downarrow & \downarrow & \downarrow & \downarrow \\ 13\ = & 25\% & \times & & b \end{array}$$

**7.** What is 85% of 276?

*Translate:*  $a = 85\% \cdot 276$

*Solve:*  The letter is by itself.  To solve the equation we convert 85% to decimal notation and multiply.

```
          2 7 6
        × 0. 8 5      (85% = 0.85)
        1 3 8 0
        2 2 0 8 0
    a = 2 3 4. 6 0
```

234.6 is 85% of 276.  The answer is 234.6.

**9.** 150% of 30 is what?

*Translate:*  $150\% \times 30 = a$

*Solve:*  Convert 150% to decimal notation and multiply.

```
          3 0
        × 1. 5        (150% = 1.5)
        1 5 0
        3 0 0
    a = 4 5. 0
```

150% of 30 is 45.  The answer is 45.

**11.** What is 6% of \$300?

*Translate:*  $a = 6\% \cdot \$300$

*Solve:*  Convert 6% to decimal notation and multiply.

```
        $ 3 0 0
        × 0. 0 6       (6% = 0.06)
    a = $ 1 8. 0 0
```

\$18 is 6% of \$300.  The answer is \$18.

**13.** 3.8% of 50 is what?

*Translate:*  $3.8\% \cdot 50 = a$

*Solve:*  Convert 3.8% to decimal notation and multiply.

```
          5 0
        × 0. 0 3 8     (3.8% = 0.038)
          4 0 0
        1 5 0 0
    a = 1. 9 0 0
```

3.8% of 50 is 1.9.  The answer is 1.9.

**15.** $39 is what percent of $50?

*Translate:*  $39 = n \times 50$

*Solve:*  To solve the equation we divide on both sides by 50 and convert the answer to percent notation.

$$n \cdot 50 = 39$$

$$\frac{n \cdot 50}{50} = \frac{39}{50}$$

$$n = 0.78 = 78\%$$

$39 is 78% of $50. The answer is 78%.

**17.** 20 is what percent of 10?

*Translate:*  $20 = n \times 10$

*Solve:*  To solve the equation we divide on both sides by 10 and convert the answer to percent notation.

$$n \cdot 10 = 20$$

$$\frac{n \cdot 10}{10} = \frac{20}{10}$$

$$n = 2 = 200\%$$

20 is 200% of 10. The answer is 200%.

**19.** What percent of $300 is $150?

*Translate:*  $n \times 300 = 150$

*Solve:*  $n \cdot 300 = 150$

$$\frac{n \cdot 300}{300} = \frac{150}{300}$$

$$n = 0.5 = 50\%$$

50% of $300 is $150. The answer is 50%.

**21.** What percent of 80 is 100?

*Translate:*  $n \times 80 = 100$

*Solve:*  $n \cdot 80 = 100$

$$\frac{n \cdot 80}{80} = \frac{100}{80}$$

$$n = 1.25 = 125\%$$

125% of 80 is 100. The answer is 125%.

**23.** 20 is 50% of what?

*Translate:*  $20 = 50\% \times b$

*Solve:*  To solve the equation we divide on both sides by 50%:

$$\frac{20}{50\%} = \frac{50\% \times b}{50\%}$$

$$\frac{20}{0.5} = b \quad (50\% = 0.5)$$

$$40 = b$$

```
            4 0 .
0. 5∧⌐2 0. 0 ∧
      2 0
        0
        0
        0
```

20 is 50% of 40. The answer is 40.

**25.** 40% of what is $16?

*Translate:*  $40\% \times b = 16$

*Solve:*  To solve the equation we divide on both sides by 40%:

$$\frac{40\% \times b}{40\%} = \frac{16}{40\%}$$

$$b = \frac{16}{0.4} \quad (40\% = 0.4)$$

$$b = 40$$

```
            4 0 .
0. 4∧⌐1 6. 0 ∧
      1 6
        0
        0
        0
```

40% of $40 is $16. The answer is $40.

**27.** 56.32 is 64% of what?

*Translate:*  $56.32 = 64\% \times b$

*Solve:*  $$\frac{56.32}{64\%} = \frac{64\% \times b}{64\%}$$

$$\frac{56.32}{0.64} = b$$

$$88 = b$$

```
              8 8 .
0. 6 4∧⌐ 5 6. 3 2 ∧
        5 1 2
        5 1 2
        5 1 2
            0
```

56.32 is 64% of 88. The answer is 88.

**29.** 70% of what is 14?

*Translate:*  $70\% \times b = 14$

*Solve:*  $$\frac{70\% \times b}{70\%} = \frac{14}{70\%}$$

$$b = \frac{14}{0.7}$$

$$b = 20$$

```
            2 0 .
0. 7 ∧⌐1 4. 0 ∧
      1 4
        0
        0
        0
```

70% of 20 is 14. The answer is 20.

**31.** What is $62\frac{1}{2}\%$ of 10?

*Translate:*  $a = 62\frac{1}{2}\% \times 10$

*Solve:*  $a - 0.625 \times 10 \quad (62\frac{1}{2}\% = 0.625)$

$$a = 6.25 \qquad \text{Multiplying}$$

6.25 is $62\frac{1}{2}\%$ of 10. The answer is 6.25.

**33.** What is 8.3% of $10,200?

*Translate:*  $a = 8.3\% \times 10,200$

*Solve:*  $a = 8.3\% \times 10,200$

$$a = 0.083 \times 10,200 \quad (8.3\% = 0.083)$$

$$a = 846.6 \qquad \text{Multiplying}$$

$846.60 is 8.3% of $10,200. The answer is $846.60.

**35.** 2.5% of what is 30.4?

*Translate:* $2.5\% \times b = 30.4$

*Solve:* $\dfrac{2.5\% \times b}{2.5\%} = \dfrac{30.4}{2.5\%}$

$b = \dfrac{30.4}{0.025}$

$b = 1216$

$$
\begin{array}{r}
1\,2\,1\,6\,. \\
0.0\,2\,5_\wedge \overline{)\,3\,0\,.\,4\,0\,0\,_\wedge\,} \\
\underline{2\,5\phantom{.0000}} \\
5\,4\phantom{.000} \\
\underline{5\,0\phantom{.000}} \\
4\,0\phantom{.00} \\
\underline{2\,5\phantom{.00}} \\
1\,5\,0 \\
\underline{1\,5\,0} \\
0
\end{array}
$$

2.5% of 1216 is 30.4. The answer is 1216.

**37.** $\underset{\substack{\text{2 decimal}\\\text{places}}}{0.\underline{09}} = \underset{\text{2 zeros}}{\dfrac{9}{100}}$

**39.** $\underset{\substack{\text{3 decimal}\\\text{places}}}{0.\underline{875}} = \underset{\text{3 zeros}}{\dfrac{875}{1000}}$

$\dfrac{875}{1000} = \dfrac{7 \cdot 125}{8 \cdot 125} = \dfrac{7}{8} \cdot \dfrac{125}{125} = \dfrac{7}{8}$

Thus, $0.875 = \dfrac{875}{1000}$, or $\dfrac{7}{8}$.

**41.** $\underset{\substack{\text{4 decimal}\\\text{places}}}{-0.\underline{9375}} = -\underset{\text{4 zeros}}{\dfrac{9375}{10,000}}$

$-\dfrac{9375}{10,000} = -\dfrac{15 \cdot 625}{16 \cdot 625} = -\dfrac{15}{16} \cdot \dfrac{625}{625} = -\dfrac{15}{16}$

Thus, $-0.9375 = -\dfrac{9375}{10,000}$, or $-\dfrac{15}{16}$.

**43.** $\dfrac{89}{100}$    0.89.

2 zeros    Move 2 places

$\dfrac{89}{100} = 0.89$

**45.** $-\dfrac{3}{10}$    $-0.3.$

1 zero    Move 1 place

$-\dfrac{3}{10} = -0.3$

**47.** Estimate: Round 7.75% to 8% and $10,880 to $11,000. Then translate:

  What is 8% of $11,000?

$a = 8\% \times 11,000$

We convert 8% to decimal notation and multiply.

$$
\begin{array}{r}
1\,1,0\,0\,0 \\
\times\phantom{0}0.0\,8 \\
\hline
8\,8\,0.0\,0
\end{array}
\quad (8\% = 0.08)
$$

$880 is about 7.75% of $10,880. (Answers may vary.)

Calculate: First we translate.

  What is 7.75% of $10,880?

$a = 7.75\% \times 10,880$

Use a calculator to multiply:

  $0.0775 \times 10,880 = 843.2$

$843.20 is 7.75% of $10,880.

**49.** Estimate: Round $2496 to $2500 and 24% to 25%. Then translate:

  $2500 is 25% of what?

$2500 = 25\% \times b$

We convert 25% to decimal notation and divide.

$\dfrac{2500}{0.25} = \dfrac{0.25 \times b}{0.25}$

$10,000 = b$

$2496 is 24% of about $10,000. (Answers may vary.)

Calculate: First we translate.

  $2496 is 24% of what?

$2496 = 0.24 \times b$

Use a calculator to divide:

$\dfrac{2496}{0.24} = 10,400$

$2496 is 24% of $10,400.

**51.** We translate:

  40% of $18\frac{3}{4}$% of $25,000 is what?

$40\% \times 18\frac{3}{4}\% \times 25,000 = a$

We convert 40% and $18\frac{3}{4}$% to decimal notation and multiply.

$0.4 \times 0.1875 \times 25,000 = a$

$$
\begin{array}{r}
0.1\,8\,7\,5 \\
\times\phantom{00}0.4 \\
\hline
0.0\,7\,5\,0\,0
\end{array}
$$

$$
\begin{array}{r}
2\,5,0\,0\,0 \\
\times\phantom{00}0.0\,7\,5 \\
\hline
1\,2\,5\,0\,0\,0 \\
1\,7\,5\,0\,0\,0\,0 \\
\hline
1\,8\,7\,5.0\,0\,0
\end{array}
$$

40% of $18\frac{3}{4}$% of $25,000 is $1875.

## Exercise Set 7.4

**1.** What is 37% of 74?

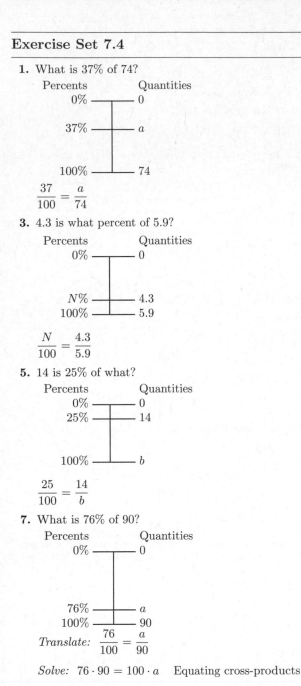

Percents        Quantities
0% ——— 0

37% ——— $a$

100% ——— 74

$$\frac{37}{100} = \frac{a}{74}$$

**3.** 4.3 is what percent of 5.9?

Percents        Quantities
0% ——— 0

$N\%$ ——— 4.3
100% ——— 5.9

$$\frac{N}{100} = \frac{4.3}{5.9}$$

**5.** 14 is 25% of what?

Percents        Quantities
0% ——— 0
25% ——— 14

100% ——— $b$

$$\frac{25}{100} = \frac{14}{b}$$

**7.** What is 76% of 90?

Percents        Quantities
0% ——— 0

76% ——— $a$
100% ——— 90

*Translate:* $\frac{76}{100} = \frac{a}{90}$

*Solve:* $76 \cdot 90 = 100 \cdot a$   Equating cross-products

$$\frac{76 \cdot 90}{100} = \frac{100 \cdot a}{100}$$   Dividing by 100

$$\frac{6840}{100} = a$$

$$68.4 = a$$   Simplifying

68.4 is 76% of 90. The answer is 68.4.

**9.** 70% of 660 is what?

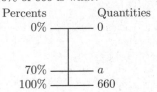

Percents        Quantities
0% ——— 0

70% ——— $a$
100% ——— 660

*Translate:* $\frac{70}{100} = \frac{a}{660}$

*Solve:* $70 \cdot 660 = 100 \cdot a$   Equating cross-products

$$\frac{70 \cdot 660}{100} = \frac{100 \cdot a}{100}$$   Dividing by 100

$$\frac{46,200}{100} = a$$

$$462 = a$$   Simplifying

70% of 660 is 462. The answer is 462.

**11.** What is 4% of 1000?

Percents        Quantities
0% ——— 0
4% ——— $a$

100% ——— 1000

*Translate:* $\frac{4}{100} = \frac{a}{1000}$

*Solve:* $4 \cdot 1000 = 100 \cdot a$

$$\frac{4 \cdot 1000}{100} = \frac{100 \cdot a}{100}$$

$$\frac{4000}{100} = a$$

$$40 = a$$

40 is 4% of 1000. The answer is 40.

**13.** 4.8% of 60 is what?

Percents        Quantities
0% ——— 0
4.8% ——— $a$

100% ——— 60

*Translate:* $\frac{4.8}{100} = \frac{a}{60}$

*Solve:* $4.8 \cdot 60 = 100 \cdot a$

$$\frac{4.8 \cdot 60}{100} = \frac{100 \cdot a}{100}$$

$$\frac{288}{100} = a$$

$$2.88 = a$$

4.8% of 60 is 2.88. The answer is 2.88.

**15.** $24 is what percent of $96?

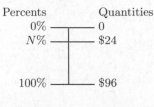

Percents        Quantities
0% ——— 0
$N\%$ ——— $24

100% ——— $96

*Translate:* $\dfrac{N}{100} = \dfrac{24}{96}$

*Solve:* $96 \cdot N = 100 \cdot 24$

$$\dfrac{96N}{96} = \dfrac{100 \cdot 24}{96}$$

$$N = \dfrac{100 \cdot 24}{96}$$

$$N = 25$$

$24 is 25% of $96. The answer is 25%.

**17.** 102 is what percent of 100?

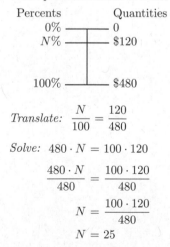

*Translate:* $\dfrac{N}{100} = \dfrac{102}{100}$

*Solve:* $100 \cdot N = 100 \cdot 102$

$$\dfrac{100 \cdot N}{100} = \dfrac{100 \cdot 102}{100}$$

$$N = \dfrac{100 \cdot 102}{100}$$

$$N = 102$$

102 is 102% of 100. The answer is 102%.

**19.** What percent of $480 is $120?

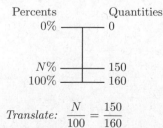

*Translate:* $\dfrac{N}{100} = \dfrac{120}{480}$

*Solve:* $480 \cdot N = 100 \cdot 120$

$$\dfrac{480 \cdot N}{480} = \dfrac{100 \cdot 120}{480}$$

$$N = \dfrac{100 \cdot 120}{480}$$

$$N = 25$$

25% of $480 is $120. The answer is 25%.

**21.** What percent of 160 is 150?

Percents          Quantities
  0% ——————— 0


 N% ————|———— 150
100% ————|———— 160

*Translate:* $\dfrac{N}{100} = \dfrac{150}{160}$

*Solve:* $160 \cdot N = 100 \cdot 150$

$$\dfrac{160 \cdot N}{160} = \dfrac{100 \cdot 150}{160}$$

$$N = \dfrac{100 \cdot 150}{160}$$

$$N = 93.75$$

93.75% of 160 is 150. The answer is 93.75%, or $93\dfrac{3}{4}\%$.

**23.** $18 is 25% of what?

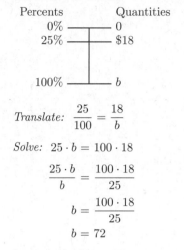

*Translate:* $\dfrac{25}{100} = \dfrac{18}{b}$

*Solve:* $25 \cdot b = 100 \cdot 18$

$$\dfrac{25 \cdot b}{b} = \dfrac{100 \cdot 18}{25}$$

$$b = \dfrac{100 \cdot 18}{25}$$

$$b = 72$$

$18 is 25% of $72. The answer is $72.

**25.** 60% of what is $54.

Percents          Quantities
  0% ——————— 0


 60% ————|———— 54
100% ————|———— b

*Translate:* $\dfrac{60}{100} = \dfrac{54}{b}$

*Solve:* $60 \cdot b = 100 \cdot 54$

$$\dfrac{60 \cdot b}{b} = \dfrac{100 \cdot 54}{60}$$

$$b = \dfrac{100 \cdot 54}{60}$$

$$b = 90$$

60% of 90 is 54. The answer is 90.

**27.** 65.12 is 74% of what?

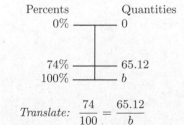

*Translate:* $\dfrac{74}{100} = \dfrac{65.12}{b}$

*Solve:* $74 \cdot b = 100 \cdot 65.12$

$$\frac{74 \cdot b}{74} = \frac{100 \cdot 65.12}{74}$$

$$b = \frac{100 \cdot 65.12}{74}$$

$$b = 88$$

65.12 is 74% of 88. The answer is 88.

**29.** 80% of what is 16?

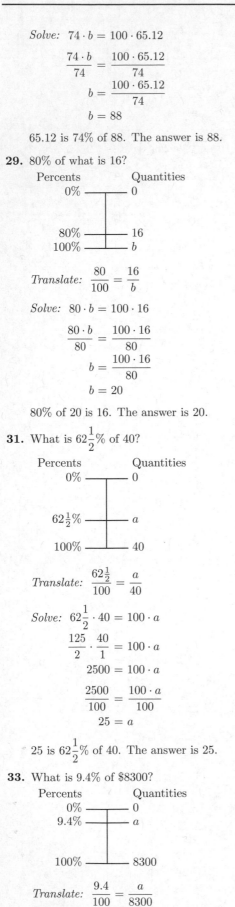

Percents      Quantities

0% ——— 0

80% ——— 16
100% ——— $b$

*Translate:* $\dfrac{80}{100} = \dfrac{16}{b}$

*Solve:* $80 \cdot b = 100 \cdot 16$

$$\frac{80 \cdot b}{80} = \frac{100 \cdot 16}{80}$$

$$b = \frac{100 \cdot 16}{80}$$

$$b = 20$$

80% of 20 is 16. The answer is 20.

**31.** What is $62\frac{1}{2}\%$ of 40?

Percents      Quantities

0% ——— 0

$62\frac{1}{2}\%$ ——— $a$

100% ——— 40

*Translate:* $\dfrac{62\frac{1}{2}}{100} = \dfrac{a}{40}$

*Solve:* $62\frac{1}{2} \cdot 40 = 100 \cdot a$

$$\frac{125}{2} \cdot \frac{40}{1} = 100 \cdot a$$

$$2500 = 100 \cdot a$$

$$\frac{2500}{100} = \frac{100 \cdot a}{100}$$

$$25 = a$$

25 is $62\frac{1}{2}\%$ of 40. The answer is 25.

**33.** What is 9.4% of $8300?

Percents      Quantities

0% ——— 0
9.4% ——— $a$

100% ——— 8300

*Translate:* $\dfrac{9.4}{100} = \dfrac{a}{8300}$

*Solve:* $9.4 \cdot 8300 = 100 \cdot a$

$$\frac{9.4 \cdot 8300}{100} = \frac{100 \cdot a}{100}$$

$$\frac{78,020}{100} = a$$

$$780.2 = a$$

$780.20 is 9.4% of $8300. The answer is $780.20.

**35.** 80.8 is $40\frac{2}{5}\%$ of what?

Percents      Quantities

0% ——— 0

$40\frac{2}{5}\%$ ——— 80.8

100% ——— $b$

*Translate:* $\dfrac{40\frac{2}{5}}{100} = \dfrac{80.8}{b}$

*Solve:* $40\frac{2}{5} \cdot b = 100 \cdot 80.8$

$$40.4 \cdot b = 100 \cdot 80.8 \qquad \left(40\frac{2}{5} = 40.4\right)$$

$$\frac{40.4 \cdot b}{40.4} = \frac{100 \cdot 80.8}{40.4}$$

$$b = \frac{100 \cdot 80.8}{40.4}$$

$$b = 200$$

80.8 is $40\frac{2}{5}\%$ of 200. The answer is 200.

**37.**
$$\frac{x}{188} = \frac{2}{47}$$

$$47 \cdot x = 188 \cdot 2$$

$$x = \frac{188 \cdot 2}{47}$$

$$x = \frac{4 \cdot 47 \cdot 2}{47}$$

$$x = 8$$

**39.**
$$\frac{4}{7} = \frac{x}{14}$$

$$4 \cdot 14 = 7 \cdot x$$

$$\frac{4 \cdot 14}{7} = x$$

$$\frac{4 \cdot 2 \cdot 7}{7} = x$$

$$8 = x$$

**41.**
$$\frac{5000}{t} = \frac{3000}{60}$$

$$5000 \cdot 60 = 3000 \cdot t$$

$$\frac{5000 \cdot 60}{3000} = t$$

$$\frac{5 \cdot 1000 \cdot 3 \cdot 20}{3 \cdot 1000} = t$$

$$100 = t$$

**43.** $\dfrac{x}{1.2} = \dfrac{36.2}{5.4}$

$5.4 \cdot x = 1.2(36.2)$

$x = \dfrac{1.2(36.2)}{5.4}$

$x = 8.0\overline{4}$

**45. *Familiarize*.** Let $q$ = the number of quarts of liquid ingredients the recipe calls for.

***Translate*.**

| Butter-milk | plus | Skim milk | plus | Oil | is | Total liquid ingredients |
|---|---|---|---|---|---|---|
| ↓ | ↓ | ↓ | ↓ | ↓ | ↓ | ↓ |
| $\frac{1}{2}$ | $+$ | $\frac{1}{3}$ | $+$ | $\frac{1}{16}$ | $=$ | $q$ |

***Solve*.** We carry out the addition. The LCM of the denominators is 48, so the LCD is 48.

$\dfrac{1}{2} \cdot \dfrac{24}{24} + \dfrac{1}{3} \cdot \dfrac{16}{16} + \dfrac{1}{16} \cdot \dfrac{3}{3} = q$

$\dfrac{24}{48} + \dfrac{16}{48} + \dfrac{3}{48} = q$

$\dfrac{43}{48} = q$

***Check*.** We repeat the calculation. The answer checks.

***State*.** The recipe calls for $\dfrac{43}{48}$ qt of liquid ingredients.

**47.** Estimate: Round 8.85% to 9%, and $12,640 to $12,600.

What is 9% of $12,600?

| Percents | Quantities |
|---|---|
| 0% | 0 |
| 9% | $a$ |
| 100% | $12,600 |

*Translate:* $\dfrac{9}{100} = \dfrac{a}{12,600}$

*Solve:* $9 \cdot 12,600 = 100 \cdot a$

$\dfrac{9 \cdot 12,600}{100} = \dfrac{100 \cdot a}{100}$

$\dfrac{113,400}{100} = a$

$1134 = a$

$1134 is about 8.85% of $12,640. (Answers may vary.)

Calculate:

What is 8.85% of $12,640?

| Percents | Quantities |
|---|---|
| 0% | 0 |
| 8.85% | $a$ |
| 100% | $12,640 |

*Translate:* $\dfrac{8.85}{100} = \dfrac{a}{12,640}$

*Solve:* $8.85 \cdot 12,640 = 100 \cdot a$

$\dfrac{8.85 \cdot 12,640}{100} = \dfrac{100 \cdot a}{100}$

$\dfrac{111,864}{100} = a$   Use a calculator to multiply and divide.

$1118.64 = a$

$1118.64 is 8.85% of $12,640.

---

## Chapter 7 Mid-Chapter Review

1. The statement is true. See page 395 in the text.

2. The statement is false. The symbol % is equivalent to $\times 0.01$.

3. We begin by writing decimal notation for each number.

$\dfrac{1}{10} = 0.1$

$1\% = 0.01$

$0.1\% = 0.001$

$10\% = 0.1$

$\dfrac{1}{100} = 0.01$

We see that the smallest number is 0.001, or 0.1%, so the given statement is true.

4. $\dfrac{1}{2}\% = \dfrac{1}{2} \cdot \dfrac{1}{100} = \dfrac{1}{200}$

5. $\dfrac{80}{1000} = \dfrac{8}{100} = 8\%$

6. $5.5\% = \dfrac{5.5}{100} = \dfrac{55}{1000} = \dfrac{11}{200}$

7. $0.375 = \dfrac{375}{1000} = \dfrac{37.5}{100} = 37.5\%$

8. $15 = p \times 80$

$\dfrac{15}{80} = \dfrac{p \times 80}{80}$

$\dfrac{15}{80} = p$

$0.1875 = p$

$18.75\% = p$

9. 28%

a) Replace the percent symbol with $\times 0.01$.

$28 \times 0.01$

b) Move the decimal point two places to the left.

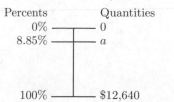

Thus, $28\% = 0.28$.

**10.** 0.15%

    a) Replace the percent symbol with ×0.01.

       0.15 × 0.01

    b) Move the decimal point two places to the left.

       0.00.15

    Thus, 0.15% = 0.0015.

**11.** $5\frac{3}{8}\% = 5.375\%$

    a) Replace the percent symbol with ×0.01.

       5.375 × 0.01

    b) Move the decimal point two places to the left.

       0.05.375

    Thus, $5\frac{3}{8}\% = 0.05375$.

**12.** 240%

    a) Replace the percent symbol with ×0.01.

       240 × 0.01

    b) Move the decimal point two places to the left.

       2.40.

    Thus, 240% = 2.4.

**13.** 0.71

    a) Move the decimal point two places to the right.

       0.71.

    b) Write a percent symbol: 71%

    Thus, 0.71 = 71%.

**14.** We use the definition of percent as a ratio.

$$\frac{9}{100} = 9\%$$

**15.** 0.3891

    a) Move the decimal point two places to the right.

       0.38.91

    b) Write a percent symbol: 38.91%

    Thus, 0.3891 = 38.91%.

**16.** Find decimal notation by division.

```
       0.1 8 7 5
  1 6 [ 3.0 0 0 0
       1 6
       1 4 0
       1 2 8
         1 2 0
         1 1 2
             8 0
             8 0
               0
```

$\frac{3}{16} = 0.1875$

Convert to percent notation.

    0.18.75

$\frac{3}{16} = 18.75\%$, or $18\frac{3}{4}\%$

**17.** 0.005

    a) Move the decimal point two places to the right.

       0.00.5

    b) Write a percent symbol: 0.5%

    Thus, 0.005 = 0.5%.

**18.** $\frac{37}{50} = \frac{37}{50} \cdot \frac{2}{2} = \frac{74}{100} = 74\%$

**19.** $6 = 6 \cdot \frac{100}{100} = \frac{600}{100} = 600\%$

**20.** Find decimal notation by division.

```
       0.8 3 3
  6 [ 5.0 0 0
       4 8
         2 0
         1 8
           2 0
           1 8
             2
```

We get a repeating decimal: $\frac{5}{6} = 0.83\overline{3}$

Convert to percent notation.

    0.83.$\overline{3}$

$\frac{5}{6} = 83.\overline{3}\%$, or $83\frac{1}{3}\%$

**21.** $85\% = \frac{85}{100} = \frac{5 \cdot 17}{5 \cdot 20} = \frac{5}{5} \cdot \frac{17}{20} = \frac{17}{20}$

**22.** $0.048\% = \frac{0.048}{100}$

$$= \frac{0.048}{100} \cdot \frac{1000}{1000}$$

$$= \frac{48}{100,000} = \frac{16 \cdot 3}{16 \cdot 6250}$$

$$= \frac{16}{16} \cdot \frac{3}{6250} = \frac{3}{6250}$$

**23.** $22\frac{3}{4}\% = 22.75\% = \frac{22.75}{100}$

$$= \frac{22.75}{100} \cdot \frac{100}{100} = \frac{2275}{10,000}$$

$$= \frac{25 \cdot 91}{25 \cdot 400} = \frac{25}{25} \cdot \frac{91}{400}$$

$$= \frac{91}{400}$$

**24.** $16.\overline{6} = 16\frac{2}{3}\% = \frac{50}{3}\%$

$= \frac{50}{3} \cdot \frac{1}{100} = \frac{50}{300}$

$= \frac{50 \cdot 1}{50 \cdot 6} = \frac{50}{50} \cdot \frac{1}{6}$

$= \frac{1}{6}$

**25.** Five of the eight equal parts are shaded, so the shaded area is $\frac{5}{8}$ of the figure. We convert $\frac{5}{8}$ to decimal notation and then to percent notation.

```
    0.6 2 5
8 ) 5.0 0 0
    4 8
    ─────
      2 0
      1 6
      ─────
        4 0
        4 0
        ───
          0
```

$\frac{5}{8} = 0.625 = 62.5\%$, or $62\frac{1}{2}\%$

**26.** Nine of the twenty equal parts are shaded, so $\frac{9}{20}$ of the area is shaded.

$\frac{9}{20} = \frac{9}{20} \cdot \frac{5}{5} = \frac{45}{100} = 45\%$

**27.** 25% of what is 14.5?

*Translate:* $25\% \times b = 14.5$

*Solve:* $\frac{25\% \times b}{25\%} = \frac{14.5}{25\%}$

$b = \frac{14.5}{0.25}$

$b = 58$

25% of 58 is 14.5. The answer is 58.

**28.** 220 is what percent of 1320?

*Translate:* $220 = p \times 1320$

*Solve:* $\frac{220}{1320} = \frac{p \times 1320}{1320}$

$0.16\overline{6} = p$

$16.\overline{6} = p$

220 is $16.\overline{6}$ of 1320. The answer is $16.\overline{6}\%$, or $16\frac{2}{3}\%$.

**29.** What is 3.2% of 80,000?

*Translate:* $a = 3.2\% \cdot 80,000$

*Solve:* Convert 3.2% to decimal notation and multiply.

```
        8 0,0 0 0
    ×      0 .0 3 2
    ─────────────
      1 6 0 0 0 0
    2 4 0 0 0 0
    ─────────────
a = 2 5 6 0.0 0 0
```

2560 is 3.2% of 80,000. The answer is 2560.

**30.** $17.50 is 35% of what?

| Percents | Quantities |
|---|---|
| 0% | 0 |
| 35% | $17.50 |
| 100% | b |

*Translate:* $\frac{35}{100} = \frac{\$17.50}{b}$

*Solve:* $35 \cdot b = 100 \cdot \$17.50$

$\frac{35 \cdot b}{35} = \frac{100 \cdot \$17.50}{35}$

$b = \frac{\$1750}{35}$

$b = \$50$

$17.50 is 35% of $50. The answer is $50,.

**31.** What percent of $800 is $160?

| Percents | Quantities |
|---|---|
| 0% | 0 |
| N% | $160 |
| 100% | $800 |

*Translate:* $\frac{N}{100} = \frac{\$160}{\$800}$

*Solve:* $N \cdot 800 = 100 \cdot 160$

$\frac{N \cdot 800}{800} = \frac{100 \cdot 160}{800}$

$N = \frac{100 \cdot 160}{800}$

$N = 20$

20% of $800 is $160. The answer is 20%.

**32.** 130% of $350 is what?

| Percents | Quantities |
|---|---|
| 0% | 0 |
| 100% | $350 |
| 130% | a |

*Translate:* $\frac{100}{130} = \frac{\$350}{a}$

*Solve:* $100 \cdot a = 130 \cdot \$350$

$\frac{100 \cdot a}{100} = \frac{130 \cdot \$350}{100}$

$a = \frac{130 \cdot \$350}{100}$

$a = \$455$

130% of $350 is $455. The answer is $455.

**33.** We begin by writing the numbers in decimal notation.

$\dfrac{1}{2}\% = 0.5\% = 0.005$

$5\% = 0.05$

$0.275$

$\dfrac{13}{100} = 0.13$

$1\% = 0.01$

$0.1\% = 0.001$

$0.05\% = 0.0005$

$\dfrac{3}{10} = 0.3$

$\dfrac{7}{20} = \dfrac{7}{20} \cdot \dfrac{5}{5} = \dfrac{35}{100} = 0.35$

$10\% = 0.1$

Arranging the numbers from smallest to largest, we have
$0.05\%,\ 0.1\%,\ \dfrac{1}{2}\%,\ 1\%,\ 5\%,\ 10\%,\ \dfrac{13}{100},\ 0.275,\ \dfrac{3}{10},\ \dfrac{7}{20}$.

**34.** 8.5 is $2\dfrac{1}{2}\%$ of what?

*Translate:* $\quad 8.5 = 2\dfrac{1}{2}\% \times b$

*Solve:* $\quad \dfrac{8.5}{2\frac{1}{2}\%} = \dfrac{2\frac{1}{2}\% \times b}{2\frac{1}{2}\%}$

$\qquad\qquad \dfrac{8.5}{0.025} = b$

$\qquad\qquad\quad 340 = b$

Answer D is correct.

**35.** \$102,000 is what percent of \$3.6 million?

*Translate:* $\quad 102,000 = p \times 3,600,000$

*Solve:* $\quad \dfrac{102,000}{3,600,000} = \dfrac{p \times 3,600,000}{3,600,000}$

$\qquad\qquad \dfrac{102,000}{3,600,000} = p$

$\qquad\qquad\quad 0.028\overline{3} = p$

$\qquad\qquad\quad 2.8\overline{3}\% = p,\ \text{or}$

$\qquad\qquad\quad 2\dfrac{5}{6}\% = p$

Answer B is correct.

**36.** Some will say that the conversion will be done most accurately by first finding decimal notation. Others will say that it is more efficient to become familiar with some or all of the fraction and percent equivalents that appear inside the back cover and to make the conversion by going directly from fraction notation to percent notation.

**37.** Since $40\% \div 10 = 4\%$, we can divide 36.8 by 10, obtaining 3.68. Since $400\% = 40\% \times 10$, we can multiply 36.8 by 10, obtaining 368.

**38.** Answers may vary. Some will say this is a good idea since it makes the computations in the solution easier. Others will say it is a poor idea since it adds an extra step to the solution.

**39.** They all represent the same number.

## Exercise Set 7.5

**1.** *Familiarize.* Let $s$ and $j$ represent the number of foreign students from South Korea and Japan, respectively.

*Translate.* We translate to percent equations.

$$\underbrace{\text{What number}}_{s} \text{ is } 10.70\% \text{ of } 583,000?$$
$$s = 10.70\% \cdot 583,000$$

$$\underbrace{\text{What number}}_{j} \text{ is } 6.05\% \text{ of } 583,000?$$
$$j = 6.05\% \cdot 583,000$$

*Solve.* We convert the percent notation to decimal notation and multiply.

$s = 0.1070 \cdot 583,000 = 62,381$

$j = 0.0605 \cdot 583,000 = 35,271.5 \approx 35,272$

*Check.* We can repeat the calculations. We can also do a partial check by estimating:

$10.70\% \cdot 583,000 \approx 10\% \cdot 600,000 = 60,000;$

$0.0605 \cdot 583,000 \approx 6\% \cdot 600,000 = 36,000.$

Since 60,000 is close to 62,381 and 36,000 is close to 35,272, our answer seems reasonable.

*State.* There were 62,381 students from South Korea and 35,272 students from Japan.

**3.** *Familiarize.* We note that the amount of the raise can be found and then added to the old salary. A drawing helps us visualize the situation.

| $43,200 | $ ? |
|----------|-----|
| 100% | 8% |

We let $x =$ the new salary.

*Translate.* We translate to a percent equation.

$$\text{What is } \underbrace{\text{the old salary}} \text{ plus } 8\% \text{ of } \underbrace{\text{the old salary?}}$$
$$x = 43,200 + 8\% \times 43,200$$

*Solve.* We convert 8% to a decimal and simplify.

$x = 43,200 + 0.08 \times 43,200$

$\quad = 43,200 + 3456 \qquad$ The raise is \$3456.

$\quad = 46,656$

*Check.* To check, we note that the new salary is 100% of the old salary plus 8% of the old salary, or 108% of the old salary. Since $1.08 \times 43,200 = 46,656$, our answer checks.

*State.* The new salary is \$46,656.

**5.** *Familiarize.* Let $a =$ the number of items on the test.

*Translate.* We translate to a proportion.

$$\dfrac{85}{100} = \dfrac{119}{a}$$

**Solve.**

$$\frac{85}{100} = \frac{119}{a}$$

$$85 \cdot a = 100 \cdot 119$$

$$\frac{85 \cdot a}{85} = \frac{100 \cdot 119}{85}$$

$$a = 140$$

**Check.** We can repeat the calculation. Also note that $\frac{119}{140} = 85\%$. The answer checks.

**State.** There were 140 items on the test.

7. **Familiarize.** Let $f =$ the total number of acres of farm land in the United States.

   **Translate.** We translate to a percent equation.

   $$47,000,000 \text{ is } 5\% \text{ of } \underbrace{\text{what number?}}$$
   $$\downarrow \qquad \downarrow \ \downarrow \ \downarrow \qquad \downarrow$$
   $$47,000,000 = 5\% \ \cdot \qquad f$$

   **Solve.**

   $$47,000,000 = 0.05 \cdot f \quad (5\% = 0.05)$$

   $$\frac{47,000,000}{0.05} = \frac{0.05 \cdot f}{0.05} \quad \text{Dividing by 0.05}$$

   $$940,000,000 = f$$

   **Check.** We find 5% of 940,000,000: $0.05 \cdot 940,000,000 = 47,000,000$. Since this is the number of acres of farm land in Kansas, the answer checks.

   **State.** There are 940,000,000 acres of farm land in the United States.

9. **Familiarize.** Since the car depreciates 25% in the first year, its value after the first year is $100\% - 25\%$, or 75%, of the original value. To find the decrease in value, we ask:

   $27,300 is 75% of what?

   Let $b =$ the original cost.

   **Translate.** We translate to an equation.

   $$\$27,300 \text{ is } 75\% \text{ of what?}$$
   $$\downarrow \qquad \downarrow \ \downarrow \ \downarrow \qquad \downarrow$$
   $$\$27,300 = 75\% \times \quad b$$

   **Solve.**

   $$27,300 = 75\% \times b$$

   $$\frac{27,300}{75\%} = \frac{75\% \times b}{75\%}$$

   $$\frac{27,300}{0.75} = b$$

   $$36,400 = b$$

   **Check.** We find 25% of 36,400 and then subtract this amount from 36,400:

   $$0.25 \times 36,400 = 9100 \text{ and}$$

   $$36,400 - 9100 = 27,300$$

   The answer checks.

   **State.** The original cost was $36,400.

11. **Familiarize.** First we find the number of items Pedro got correct. Let $b$ represent this number.

**Translate.** We translate to a percent equation.

$$\underbrace{\text{What number}} \text{ is } 95\% \text{ of } 80?$$
$$\downarrow \qquad\qquad \downarrow \ \downarrow \ \downarrow \ \downarrow$$
$$b \qquad\qquad = 95\% \ \cdot \ 80$$

**Solve.** We convert 95% to decimal notation and multiply.

$$b = 0.95 \cdot 80 = 76$$

We subtract to find the number of items Pedro got incorrect:

$$80 - 76 = 4$$

**Check.** We can repeat the calculations. The answer checks.

**State.** Pedro got 76 items correct and 4 items incorrect.

13. **Familiarize.** Let $x$ and $y$ represent the number of people under the age of 15 in Egypt and in the United States, respectively.

**Translate.** We translate to proportions.

$$\frac{32.6}{100} = \frac{x}{75,449,000} \text{ and } \frac{20.4}{100} = \frac{y}{305,468,000}$$

**Solve.**

$$\frac{32.6}{100} = \frac{x}{75,449,000}$$

$$32.6 \cdot 75,449,000 = 100 \cdot x$$

$$\frac{32.6 \cdot 75,449,000}{100} = \frac{100 \cdot x}{100}$$

$$24,596,374 = x$$

$$\frac{20.4}{100} = \frac{y}{305,468,000}$$

$$20.4 \cdot 305,468,000 = 100 \cdot y$$

$$\frac{20.4 \cdot 305,468,000}{100} = \frac{100 \cdot y}{100}$$

$$62,315,472 = y$$

**Check.** We can repeat the calculations. Also note that $\frac{24,596,374}{75,449,000} \approx \frac{25,000,000}{75,000,000} = 33\frac{1}{3}\% \approx 32.4\%$ and $\frac{62,315,472}{305,468,000} \approx \frac{60,000,000}{300,000,000} = 20\% \approx 20.4\%$. The answer checks.

**State.** In Egypt, 24,596,374 people are under the age of 15. There are 62,315,472 people in this age group in the United States.

15. **Familiarize.** Let $p =$ the percent of patients waiting for transplants who are waiting for a liver transplant.

**Translate.** We write a percent equation.

$$16,737 \text{ is } \underbrace{\text{what percent}} \text{ of } 96,749?$$
$$\downarrow \quad \downarrow \qquad \downarrow \qquad \downarrow \quad \downarrow$$
$$16,737 = \qquad p \qquad \cdot \ 96,749$$

**Solve.** We divide on both sides by 96,749 and express the result in percent notation.

$$16,737 = p \cdot 96,749$$

$$\frac{16,737}{96,749} = \frac{p \cdot 96,749}{96,749}$$

$$0.173 \approx p$$

$$17.3\% = p$$

**Check**. We find 17.3% of 96,749:

$17.3\% \cdot 96,749 = 0.173 \cdot 96,749 = 16,737.577 \approx 16,737$. The answer checks.

**State**. About 17.3% of patients waiting for a transplant are waiting for a liver transplant.

**17. Familiarize**. Let $t$ = the total cost of the meal, including the tip.

**Translate**.

$$\underbrace{\text{Total cost}}_{\downarrow} \underset{\downarrow}{\text{is}} \underbrace{\text{Food cost}}_{\downarrow} \underset{\downarrow}{\text{plus}} \underset{\downarrow}{18\%} \underset{\downarrow}{\text{of}} \underbrace{\text{Food cost}}_{\downarrow}$$
$$t \quad = \quad 195 \quad + \quad 18\% \cdot \quad 195$$

**Solve**. We convert 18% to decimal notation and simplify.

$t = 195 + 0.18 \cdot 195$

$t = 195 + 35.10$     The tip is $35.10.

$t = 230.10$

**Check**. To check, note that the total cost of the meal is 100% of the cost of the food plus 18% of the cost of the food, or 118% of the cost of the food. Since $1.18 \times \$195 = \$230.10$, the answer checks.

**State**. The total cost of the meal is $230.10.

**19. Familiarize**. Let $f$ = the number of fast-food cooks in the United States.

**Translate**. We translate to a percent equation.

$$392,850 \text{ is } 64.2\% \text{ of } \underbrace{\text{what number}}_{\downarrow}?$$
$$392,850 = 64.2\% \times \quad f$$

**Solve**.

$392,850 = 64.2\% \times f$

$392,850 = 0.642 \times f$     $(64.2\% = 0.642)$

$\dfrac{392,850}{0.642} = \dfrac{0.642 \times f}{0.642}$     Dividing by 0.642

$612,000 \approx f$

**Check**. We find 64.2% of 612,000:

$64.2\% \cdot 612,000 = 0.642 \times 612,000 = 392,904$. Since 392,904 is close to 392,850, the answer checks. (Remember, we rounded to get 612,000.)

**State**. There are about 612,000 fast-food cooks in the United States.

**21. Familiarize**. First we find the amount of the solution that is alcohol. We let $a$ = this amount, in mL.

**Translate**. We translate to a percent equation.

$$\text{What is } 8\% \text{ of } 540?$$
$$a \quad = 8\% \times 540$$

**Solve**. We convert 8% to decimal notation and multiply.

$8\% \times 540 = 0.08 \times 540 = 43.2$

Now we find the amount that is water. We let $w$ = this amount, in mL.

$$\underbrace{\text{Total amount}}_{\downarrow} \underset{\downarrow}{\text{minus}} \underbrace{\text{Amount of alcohol}}_{\downarrow} \underset{\downarrow}{\text{is}} \underbrace{\text{Amount of water}}_{\downarrow}$$
$$540 \quad - \quad 43.2 \quad = \quad w$$

To solve the equation we carry out the subtraction.

$w = 540 - 43.2 = 496.8$

**Check**. We can repeat the calculations. Also, observe that, since 8% of the solution is alcohol, 92% is water. Because $92\%$ of $540 = 0.92 \times 540 = 496.8$, our answer checks.

**State**. The solution contains 43.2 mL of alcohol and 496.8 mL of water.

**23. Familiarize**. Let $f$, $a$, $n$, and $m$ represent the percent of people in active military service in the Air Force, Army, Navy, and Marines, respectively.

**Translate**. We translate to proportions.

$$\frac{f}{100} = \frac{349,000}{1,385,000}; \quad \frac{a}{100} = \frac{505,000}{1,385,000};$$
$$\frac{n}{100} = \frac{350,000}{1,385,000}; \quad \frac{m}{100} = \frac{180,000}{1,385,000}$$

**Solve**. We solve each proportion, beginning by equating cross products.

$$f \cdot 1,385,000 = 100 \cdot 349,000$$
$$\frac{f \cdot 1,385,000}{1,385,000} = \frac{100 \cdot 349,000}{1,385,000}$$
$$f \approx 25.2$$

$$a \cdot 1,385,000 = 100 \cdot 505,000$$
$$\frac{a \cdot 1,385,000}{1,385,000} = \frac{100 \cdot 505,000}{1,385,000}$$
$$a \approx 36.5$$

$$n \cdot 1,385,000 = 100 \cdot 350,000$$
$$\frac{n \cdot 1,385,000}{1,385,000} = \frac{100 \cdot 350,000}{1,385,000}$$
$$n \approx 25.3$$

$$m \cdot 1,385,000 = 100 \cdot 180,000$$
$$\frac{m \cdot 1,385,000}{1,385,000} = \frac{100 \cdot 180,000}{1,385,000}$$
$$m \approx 13.0$$

**Check**. We can repeat the calculations. Also note that the sum of the percents is $25.2\% + 36.5\% + 25.3\% + 13.0\%$, or 100%, so the answer seems reasonable.

**State**. The percent of the people in active military service in the United States are as follows: Air Force: 25.2%, Army: 36.5%, Navy: 25.3%, Marines: 13.0%.

**25. Familiarize**. Use the drawing in the text to visualize the situation. Note that the increase in the amount was $42.

Let $n$ = the percent of increase.

**Translate**. We translate to a percent equation.

$$\$42 \text{ is } \underbrace{\text{what percent}}_{\downarrow} \text{ of } \$800?$$
$$42 = \quad n \quad \times \quad 800$$

*Solve*. We divide by 800 on both sides and convert the result to percent notation.

$$42 = n \times 800$$
$$\frac{42}{800} = \frac{n \times 800}{800}$$
$$0.05 = n$$
$$5\% = n$$

*Check*. Find 5% of 840: $5\% \times 840 = 0.05 \times 840 = 42$. Since this is the amount of the increase, the answer checks.

*State*. The percent of increase was 5%.

**27.** *Familiarize*. We use the drawing in the text to visualize the situation. Note that the weight loss is 24 lb.

We let $n$ = the percent of decrease.

*Translate*. We translate to a percent equation.

$24 is \underbrace{what\ percent}\ of\ \$160?$
$\downarrow \downarrow \qquad \downarrow \qquad \downarrow \downarrow$
$24 = \qquad n \quad \times \ 160$

*Solve*. To solve the equation, we divide on both sides by 160 and convert the result to percent notation.

$$\frac{24}{160} = \frac{n \times 160}{160}$$
$$0.15 = n$$
$$15\% = n$$

*Check*. We find 15% of 160: $15\% \times 160 = 0.15 \times 160 = 24$. Since this is the amount of weight lost, the answer checks.

*State*. The percent of decrease was 15%.

**29.** *Familiarize*. First we find the amount of decrease.

$$\begin{array}{r} \$\ 2\ 3.\ 4\ 3 \\ -\ 1\ 5.\ 3\ 1 \\ \hline \$\quad 8.\ 1\ 2 \end{array}$$

Let $N$ = the percent of decrease.

*Translate*. We translate to a proportion.

$$\frac{N}{100} = \frac{8.12}{23.43}$$

*Solve*.
$$\frac{N}{100} = \frac{8.12}{23.43}$$
$$N \cdot 23.43 = 100 \cdot 8.12$$
$$\frac{N \cdot 23.43}{23.43} = \frac{100 \cdot 8.12}{23.43}$$
$$N \approx 34.7$$

*Check*. We can repeat the calculations. Also note that 34.7% of \$23.43 ≈ 30% · \$25 = \$7.50 ≈ \$8.12. The answer checks.

*State*. The percent of decrease is about 34.7%.

**31.** *Familiarize*. First we find the amount of increase, in billions of rides.

$$\begin{array}{r} 2.\ 8 \\ -\ 2.\ 1 \\ \hline 0.\ 7 \end{array}$$

Let $p$ = the percent of increase.

*Translate*. We translate to a percent equation.

$0.7\ is\ \underbrace{what\ percent}\ of\ 2.1?$
$\downarrow \downarrow \qquad \downarrow \qquad \downarrow \downarrow$
$0.7 = \qquad p \quad \times\ 2.1$

*Solve*. We divide on both sides by 2.1 and convert the result to percent notation.

$$0.7 = p \times 2.1$$
$$\frac{0.7}{2.1} = \frac{p \times 2.1}{2.1}$$
$$0.\overline{3} = p$$
$$33.\overline{3}\% = p, \text{ or}$$
$$33\frac{1}{3}\% = p$$

*Check*. We find $33\frac{1}{3}$% of 2.1: $33\frac{1}{3}\% \times 2.1 = \frac{1}{3} \times 2.1 = 0.7$. This is the amount of the increase, so the answer checks.

*State*. The percent of increase was $33\frac{1}{3}$%.

**33.** *Familiarize*. First we find the amount of overdraft fees in 2001, in billions of dollars.

$$\begin{array}{r} 4\ 5.\ 6 \\ -\ 1\ 5.\ 1 \\ \hline 3\ 0.\ 5 \end{array}$$

Let $p$ = the percent of increase.

*Translate*. We translate to a percent equation.

$15.1\ is\ \underbrace{what\ percent}\ of\ 30.5?$
$\downarrow \downarrow \qquad \downarrow \qquad \downarrow \downarrow$
$15.1 = \qquad p \quad \times\ 30.5$

*Solve*. We divide on both sides by 30.5 and convert the result to percent notation.

$$15.1 = p \times 30.5$$
$$\frac{15.1}{30.5} = \frac{p \times 30.5}{30.5}$$
$$0.495 \approx p$$
$$49.5\% \approx p$$

*Check*. We find 49.5% of 30.5:
$$49.5\% \times 30.5 = 0.495 \times 30.5 = 15.0975 \approx 15.1.$$
This is the amount of the increase, so the answer checks.

*State*. The percent of increase is about 49.5%.

**35.** *Familiarize*. First we find the amount of decrease.

$$\begin{array}{r} 1\ 1\ 4,\ 4\ 6\ 9 \\ -\quad 4\ 6,\ 8\ 6\ 6 \\ \hline 6\ 7,\ 6\ 0\ 3 \end{array}$$

Let $N$ = the percent of decrease.

*Translate*. We translate to a proportion.

$$\frac{N}{100} = \frac{67,603}{114,469}$$

*Solve*.
$$\frac{N}{100} = \frac{67,603}{114,469}$$
$$N \cdot 114,469 = 100 \cdot 67,603$$
$$\frac{N \cdot 114,469}{114,469} = \frac{100 \cdot 67,603}{114,469}$$
$$N \approx 59$$

**Check**. If the percent of decrease is 59%, then the 2008 number is 100% − 59%, or 41% of the 2007 number. We find 41% of 114,469: $41\% \cdot 114,469 = 0.41 \cdot 114,469 \approx 46,932 \approx 46,866$. The answer checks.

**State**. The percent of decrease is about 59%.

**37.** **Familiarize**. First we find the amount of decrease.

$$\begin{array}{r} 1\,7\,3,7\,9\,4 \\ -\,1\,5\,7,2\,8\,4 \\ \hline 1\,6,5\,1\,0 \end{array}$$

Let $p$ = the percent of decrease.

**Translate**. We translate to a percent equation.

16, 510 is what percent of 173,794?

$$16,510 = \quad p \quad \cdot \quad 173,794$$

**Solve**. We divide on both sides by 173,794 and convert the result to percent notation.

$$16,510 = p \cdot 173,794$$
$$\frac{16,510}{173,794} = \frac{p \cdot 173,794}{173,794}$$
$$0.095 \approx p$$
$$9.5\% \approx p$$

**Check**. We find 9.5% of 173,794: $9.5\% \times 173,794 = 0.095 \times 173,794 = 16,510.43 \approx 16,510$. Since we get approximately the amount of decrease, the answer checks.

**State**. The percent of decrease is about 9.5%.

**39.** **Familiarize**. This is a multistep problem. First we find the area of a cross-section of a finished board and of a rough board using the formula $A = l \cdot w$. Then we find the amount of wood removed in planing and drying and finally we find the percent of wood removed. Let $f$ = the area of a cross-section of a finished board and let $r$ = the area of a cross-section of a rough board.

**Translate**. We find the areas.

$$f = 3\frac{1}{2} \cdot 1\frac{1}{2}$$
$$r = 4 \cdot 2$$

**Solve**. We carry out the multiplications.

$$f = 3\frac{1}{2} \cdot 1\frac{1}{2} = \frac{7}{2} \cdot \frac{3}{2} = \frac{21}{4}$$
$$r = 4 \cdot 2 = 8$$

Now we subtract to find the amount of wood removed in planing and drying.

$$8 - \frac{21}{4} = \frac{32}{4} - \frac{21}{4} = \frac{11}{4}$$

Finally we find $p$, the percent of wood removed in planing and drying.

$\frac{11}{4}$ is what percent of 8?

$$\frac{11}{4} = \quad p \quad \cdot \quad 8$$

We solve the equation.

$$\frac{11}{4} = p \cdot 8$$
$$\frac{1}{8} \cdot \frac{11}{4} = p$$
$$\frac{11}{32} = p$$
$$0.34375 = p$$
$$34.375\% = p, \text{ or}$$
$$34\frac{3}{8}\% = p$$

**Check**. We repeat the calculations. The answer checks.

**State**. 34.375%, or $34\frac{3}{8}\%$, of the wood is removed in planing and drying.

**41.** **Familiarize**. First we subtract to find the amount of change.

$$\begin{array}{r} 6\,2\,3,9\,0\,8 \\ -\,6\,0\,8,8\,2\,7 \\ \hline 1\,5,0\,8\,1 \end{array}$$

Now let $N$ = the percent of change.

**Translate**. We translate to a proportion.

$$\frac{N}{100} = \frac{15,081}{608,827}$$

**Solve**.

$$\frac{N}{100} = \frac{15,081}{608,827}$$
$$N \cdot 608,827 = 100 \cdot 15,081$$
$$N = \frac{100 \cdot 15,081}{608,827}$$
$$N \approx 2.5$$

**Check**. We can repeat the calculations. Also note that $102.5\% \cdot 608,827 \approx 624,028 \approx 623,908$. The answer checks.

**State**. The population of Vermont increased by 15,081. This was a 2.5% increase.

**43.** **Familiarize**. First we subtract to find the population in 2000.

$$\begin{array}{r} 6,1\,6\,6,3\,1\,8 \\ -\,1,0\,3\,5,6\,8\,6 \\ \hline 5,1\,3\,0,6\,3\,2 \end{array}$$

Now let $p$ = the percent of change.

**Translate**. We translate to an equation.

1,035,686 is what percent of 5,130,632?

$$1,035,686 = \quad p \quad \cdot \quad 5,130,632$$

**Solve**.

$$1,035,686 = p \cdot 5,130,632$$
$$\frac{1,035,686}{5,130,632} = p$$
$$0.202 \approx p$$
$$20.2\% \approx p$$

**Check**. We can repeat the calculations. Also note that $120.2\% \cdot 5,130,632 \approx 6,167,020 \approx 6,166,318$. The answer checks.

***State***. The population of Arizona was 5,130,632 in 2000. The population had increased by about 20.2% in 2006.

**45.** ***Familiarize***. First we add to find the population in 2006.

$$
\begin{array}{r}
1,2\,9\,3,9\,5\,3 \\
+\ \ \ 1\,7\,2,5\,1\,2 \\
\hline
1,4\,6\,6,4\,6\,5
\end{array}
$$

Now let $N$ = the percent of change.

***Translate***. We translate to a proportion.

$$\frac{N}{100} = \frac{172,512}{1,293,953}$$

***Solve***.

$$
\begin{aligned}
\frac{N}{100} &= \frac{172,512}{1,293,953} \\
N \cdot 1,293,953 &= 100 \cdot 172,512 \\
N &= \frac{100 \cdot 172,512}{1,293,953} \\
N &\approx 13.3
\end{aligned}
$$

***Check***. We can repeat the calculations. Also note that $113.2\% \cdot 1,293,953 \approx 1,464,755 \approx 1,466,465$. The answer checks.

***State***. The population of Idaho in 2006 was 1,466,465. The population had increased by about 13.2% in 2006.

**47.** ***Familiarize***. First we find the amount of decrease.

$$
\begin{array}{r}
6\,4\,2,2\,0\,0 \\
-\,6\,3\,5,8\,6\,7 \\
\hline
6\,3\,3\,3
\end{array}
$$

Let $p$ = the percent of decrease.

***Translate***. We translate to an equation.

6333 is what percent of 642,200?

$$6333 = p \cdot 642,200$$

***Solve***.

$$
\begin{aligned}
6333 &= p \cdot 642,200 \\
\frac{6333}{642,200} &= \frac{p \cdot 642,200}{642,200} \\
0.0010 &\approx p \\
1.0\% &\approx p
\end{aligned}
$$

***Check***. We can repeat the calculations. Also note that the population in 2006 was $100\% - 1\%$, or 99%, of the population in 2000. We find 99% of 642,200: $0.99 \times 642,200 = 635,778 \approx 635,867$. The answer checks.

***State***. The percent of decrease was about 1.0%.

**49.** $\dfrac{25}{11} = 25 \div 11$

$$
\begin{array}{r}
2.\,2\,7 \\
11\,\overline{)2\,5.\,0\,0} \\
\underline{2\,2}\ \ \ \\
3\,0\ \\
\underline{2\,2}\ \\
8\,0 \\
\underline{7\,7} \\
3
\end{array}
$$

Since the remainders begin to repeat, we have a repeating decimal.

$$\frac{25}{11} = 2.\overline{27}$$

**51.** First consider $\dfrac{27}{8}$.

$$\frac{27}{8} = 27 \div 8$$

$$
\begin{array}{r}
3.\,3\,7\,5 \\
8\,\overline{)2\,7.\,0\,0\,0} \\
\underline{2\,4}\ \ \ \ \\
3\,0\ \ \\
\underline{2\,4}\ \ \\
6\,0\ \\
\underline{5\,6}\ \\
4\,0 \\
\underline{4\,0} \\
0
\end{array}
$$

$\dfrac{27}{8} = 3.375$, so $-\dfrac{27}{8} = -3.375$.

We could also do this conversion as follows:

$$-\frac{27}{8} = -\frac{27}{8} \cdot \frac{125}{125} = -\frac{3375}{1000} = -3.375$$

**53.** $\dfrac{23}{25} = \dfrac{23}{25} \cdot \dfrac{4}{4} = \dfrac{92}{100} = 0.92$

**55.** $\dfrac{14}{32} = 14 \div 32$

$$
\begin{array}{r}
0.\,4\,3\,7\,5 \\
32\,\overline{)1\,4.\,0\,0\,0\,0} \\
\underline{1\,2\,8}\ \ \ \ \\
1\,2\,0\ \ \ \\
\underline{9\,6}\ \ \ \\
2\,4\,0\ \ \\
\underline{2\,2\,4}\ \ \\
1\,6\,0\ \\
\underline{1\,6\,0}\ \\
0
\end{array}
$$

$$\frac{14}{32} = 0.4375$$

(Note that we could have simplified the fraction first, getting $\dfrac{7}{16}$ and then found the quotient $7 \div 16$.)

**57.** Think of $-\dfrac{34,809}{10,000}$ as $\dfrac{-34,809}{10,000}$.

Since 10,000 has 4 zeros, we move the decimal point in the number in the numerator 4 places to the left.

$$\frac{-34,809}{10,000} = -3.4809$$

**59.** ***Familiarize***. Let $c$ = the cost of the dinner before the tip and without the coupon.

***Translate***.

| Cost of food | plus | 20% of | Cost of food | minus | \$10 | is | \$40.40 |

$$c + 20\% \cdot c - 10 = 40.40$$

*Solve.*

$$c + 20\% \cdot c - 10 = 40.40$$
$$1 \cdot c + 0.2 \cdot c - 10 = 40.40$$
$$1.2 \cdot c - 10 = 40.40$$
$$1.2 \cdot c - 10 + 10 = 40.40 + 10$$
$$1.2 \cdot c = 50.40$$
$$\frac{1.2 \cdot c}{1.2} = \frac{50.40}{1.2}$$
$$c = 42$$

**Check.** 20% of $42 is $0.2 \times \$42 = \$8.40$; $\$42 + \$8.40 = \$50.40$ and $\$50.40 - \$10 = \$40.40$. The answer checks.

**State.** Before the tip and without the coupon, the meal would cost $42.

## Exercise Set 7.6

**1.** The sales tax on an item costing $239 is

$$\underbrace{\text{Sales tax rate}}_{4\%} \times \underbrace{\text{Purchase price}}_{\$239},$$

or $0.04 \times 239$, or $9.56$. Thus the tax is $9.56.

**3.** The sales tax on an item costing $29.50 is

$$\underbrace{\text{Sales tax rate}}_{5.5\%} \times \underbrace{\text{Purchase price}}_{\$29.50},$$

or $0.055 \times 29.50$, or $1.62$ (rounded to the nearest cent). Thus the tax is $1.62.

**5.** a) We first find the cost of the pillows. It is
$$4 \times \$39.95 = \$159.80.$$

b) The sales tax on items costing $159.80 is

$$\underbrace{\text{Sales tax rate}}_{7.25\%} \times \underbrace{\text{Purchase price}}_{\$159.80},$$

or $0.0725 \times 159.80$, or about $11.59. Thus the tax is $11.59.

c) The total price is given by the purchase price plus the sales tax:
$$\$159.80 + \$11.59 = \$171.39.$$

To check, note that the total price is the purchase price plus 7.25% of the purchase price. Thus the total price is 107.25% of the purchase price. Since $1.0725 \times \$159.80 = \$171.3855 \approx \$171.39$, we have a check. The total price is $171.39.

**7.** *Rephrase:* Sales tax is what percent of purchase price?

*Translate:* $30 = r \times 750$

To solve the equation, we divide on both sides by 750.
$$\frac{30}{750} = \frac{r \times 750}{750}$$
$$0.04 = r$$
$$4\% = r$$

The sales tax rate is 4%.

**9.** *Rephrase:* Sales tax is what percent of purchase price?

*Translate:* $9.12 = r \times 456$

To solve the equation, we divide on both sides by 456.
$$\frac{9.12}{456} = \frac{r \times 456}{456}$$
$$0.02 = r$$
$$2\% = r$$

The sales tax rate is 2%.

**11.** *Rephrase:* Sales tax is 2% of what?

*Translate:* $112 = 2\% \times b$, or
$112 = 0.02 \times b$

To solve the equation, we divide on both sides by 0.02.
$$\frac{112}{0.02} = \frac{0.02 \times b}{0.02}$$
$$5600 = b$$

The purchase price is $5600.

**13.** a) We first find the cost of the chocolates. It is
$$6 \times \$17.95 = \$107.70.$$

b) The total tax rate is the city tax rate plus the state tax rate, or $4.375\% + 4\% = 8.375\%$. The sales tax paid on items costing $107.70 is

$$\underbrace{\text{Sales tax rate}}_{8.375\%} \times \underbrace{\text{Purchase price}}_{\$107.70},$$

or $0.08375 \times \$107.70$, or about $9.02. Thus the tax is $9.02.

c) The total price is given by the purchase price plus the sales tax:
$$\$107.70 + \$9.02 = \$116.72.$$

To check, note that the total price is the purchase price plus 8.375% of the purchase price. Thus the total price is 108.375% of the purchase price. Since $1.08375 \times 107.70 \approx 116.72$, we have a check. The total amount paid for the 6 boxes of chocolates is $116.72.

**15.** a) We first find the cost of the ceiling fans. It is
$$3 \times \$84.49 = \$253.47.$$

b) The total tax rate is the county tax rate plus the state tax rate, or $2.5\% + 6.5\% = 9\%$. The sales tax paid on items costing $253.47 is

Sales tax rate × Purchase price
↓          ↓          ↓
9%     ×    $253.47,

or 0.09 × $253.47, or about $22.81. Thus the tax is $22.81.

c) The total price is given by the purchase price plus the sales tax:

$253.47 + $22.81 = $276.28.

To check, note that the total price is the purchase price plus 9% of the purchase price. Thus the total price is 109% of the purchase price. Since 1.09 × 253.47 ≈ 276.28, we have a check. The total amount paid for the 3 ceiling fans is $276.28.

**17.** a) We first find the cost of the basketballs. It is

6 × $29.95 = $179.70.

b) The total tax rate is the city tax rate plus the county tax rate plus the state tax rate, or 1% + 3% + 4% = 8%. The sales tax paid on items costing $179.70 is

Sales tax rate × Purchase price
↓          ↓          ↓
8%     ×    $179.70,

or 0.08 × $179.70, or about $14.38. Thus the tax is $14.38.

c) The total price is given by the purchase price plus the sales tax:

$179.70 + $14.38 = $194.08.

To check, note that the total price is the purchase price plus 8% of the purchase price. Thus the total price is 108% of the purchase price. Since 1.08 × 179.70 ≈ 194.08, we have a check. The total amount paid for the 6 basketballs is $194.08.

**19.** Commission = Commission rate × Sales
$C$ = 21% × 12,500

This tells us what to do. We multiply.

```
    1 2, 5 0 0
×      0. 2 1      (21% = 0.21)
    1 2 5 0 0
  2 5 0 0 0 0
  2 6 2 5. 0 0
```

The commission is $2625.

**21.** Commission = Commission rate × Sales
408 = $r$ × 3400

To solve this equation we divide on both sides by 3400:

$$\frac{408}{3400} = \frac{r \times 3400}{3400}$$
0.12 = $r$
12% = $r$

The commission rate is 12%.

**23.** Commission = Commission rate × Sales
12,950 = 7% × $S$

To solve this equation we divide on both sides by 0.07:

$$\frac{12,950}{0.07} = \frac{0.07 \times S}{0.07}$$
185,000 = $S$

The home sold for $185,000.

**25.** Commission = Commission rate × Sales
$C$ = 8% × 68,000

This tells us what to do. We multiply.

```
    6 8, 0 0 0
×       0. 0 8      (8% = 0.08)
    5 4 4 0. 0 0
```

The commission is $5440.

**27.** Commission = Commission rate × Sales
1147.50 = $r$ × 7650

To solve this equation we divide on both sides by 7650.

$$\frac{1147.50}{7650} = \frac{r \times 7650}{7650}$$
0.15 = $r$
15% = $r$

The commission rate is 15%.

**29.** First we find the commission on the first $1000 of sales.
Commission = Commission rate × Sales
$C$ = 4% × 1000

This tells us what to do. We multiply.

```
    1 0 0 0
×     0. 0 4
    4 0. 0 0
```

The commission on the first $1000 of sales is $40.

Next we subtract to find the amount of sales over $1000.

$5500 − $1000 = $4500

Sabrina had $4500 in sales over $1000.

Then we find the commission on the sales over $1000.
Commission = Commission rate × Sales
$C$ = 7% × 4500

This tells us what to do. We multiply.

```
    4 5 0 0
×     0. 0 7
    3 1 5. 0 0
```

The commission on the sales over $1000 is $315.

Finally we add to find the total commission.

$40 + $315 = $355

The total commission is $355.

**31.** Discount = Rate of discount × Original price
$D$ = 10% × $300

Convert 10% to decimal notation and multiply.

```
    3 0 0
×     0. 1      (10% = 0.10 = 0.1)
    3 0. 0
```

The discount is $30.

Sale price = Original price − Discount
$$S \quad = \quad 300 \quad - \quad 30$$

We subtract:
$$\begin{array}{r} 3\,0\,0 \\ -\ 3\,0 \\ \hline 2\,7\,0 \end{array}$$

To check, note that the sale price is 90% of the original price: $0.9 \times 300 = 270$.

The sale price is $270.

**33.** Discount = Rate of discount × Original price
$$D \quad = \quad 15\% \quad \times \quad \$17$$

Convert 15% to decimal notation and multiply.
$$\begin{array}{r} 1\,7 \\ \times\ 0.\,1\,5 \\ \hline 8\,5 \\ 1\,7\,0 \\ \hline 2.\,5\,5 \end{array} \qquad (15\% = 0.15)$$

The discount is $2.55.

Sale price = Original price − Discount
$$S \quad = \quad 17 \quad - \quad 2.55$$

We subtract:
$$\begin{array}{r} 1\,7.\,0\,0 \\ -\ 2.\,5\,5 \\ \hline 1\,4.\,4\,5 \end{array}$$

To check, note that the sale price is 85% of the original price: $0.85 \times 17 = 14.45$.

The sale price is $14.45.

**35.** Discount = Rate of discount × Original price
$$12.50 \quad = \quad 10\% \quad \times \quad M$$

To solve the equation we divide on both sides by 0.1.
$$\frac{12.50}{0.1} = \frac{0.1 \times M}{0.1}$$
$$125 = M$$

The original price is $125.

Sale price = Original price − Discount
$$S \quad = \quad 125.00 \quad - \quad 12.50$$

We subtract:
$$\begin{array}{r} 1\,2\,5.\,0\,0 \\ -\ 1\,2.\,5\,0 \\ \hline 1\,1\,2.\,5\,0 \end{array}$$

To check, note that the sale price is 90% of the original price: $0.9 \times 125 = 112.50$.

The sale price is $112.50.

**37.** Discount = Rate of discount × Original price
$$240 \quad = \quad r \quad \times \quad 600$$

To solve the equation we divide on both sides by 600.
$$\frac{240}{600} = \frac{r \times 600}{600}$$

We can simplify by removing a factor of 1:
$$r = \frac{240}{600} = \frac{2}{5} \cdot \frac{120}{120} = \frac{2}{5} = 0.4 = 40\%$$

The rate of discount is 40%.

Sale price = Original price − Discount
$$S \quad = \quad 600 \quad - \quad 240$$

We subtract:
$$\begin{array}{r} 6\,0\,0 \\ -\ 2\,4\,0 \\ \hline 3\,6\,0 \end{array}$$

To check, note that a 40% discount rate means that 60% of the original price is paid. Since $\frac{360}{600} = 0.6$, or 60%, we have a check.

The sale price is $360.

**39.** Discount = Original price − Sale price
$$15 \quad = \quad M \quad - \quad 35$$

We add 35 on both sides of the equation:
$$15 + 35 = M$$
$$50 = M$$

The original price is $50.

Discount = Rate of discount × Original price
$$15 \quad = \quad R \quad \times \quad 50$$

To solve the equation we divide on both sides by 50.
$$\frac{15}{50} = \frac{R \times 50}{50}$$
$$0.3 = R$$
$$30\% = R$$

To check note that a discount rate of 30% means that 70% of the original price is paid: $0.7 \times 50 = 35$. Since this is the sale price, the answer checks.

The rate of discount is 30%.

**41.** Discount = Original price − Sale price
$$D \quad = \quad 3999 \quad - \quad 3150$$

We subtract:
$$\begin{array}{r} 3\,9\,9\,9 \\ -\ 3\,1\,5\,0 \\ \hline 8\,4\,9 \end{array}$$

The discount is $849.

Discount = Rate of discount × Original price
$$849 \quad = \quad R \quad \times \quad 3999$$

To solve the equation we divide on both sides by 3999.
$$\frac{849}{3999} = \frac{R \times 3999}{3999}$$
$$0.212 \approx R$$
$$21.2\% \approx R$$

To check, note that a discount rate of 21.2% means that 78.8% of the original price is paid: $0.788 \times 3999 = 3151.212 \approx 3150$. Since that is the sale price, the answer checks.

The rate of discount is 21.2%.

**43.**
$$\frac{x}{12} = \frac{24}{16}$$
$$16 \cdot x = 12 \cdot 24 \qquad \text{Equating cross-products}$$
$$x = \frac{12 \cdot 24}{16} \qquad \text{Dividing by 16 on both sides}$$
$$x = \frac{288}{16}$$
$$x = 18$$

The solution is 18.

**45.**  $0.64 \times x = 170$

$\dfrac{0.64 \cdot x}{0.64} = \dfrac{170}{0.64}$   Dividing by 0.64 on both sides

$x = 265.625$

The solution is 265.625.

**47.**  $\dfrac{5}{9} = 5 \div 9$

$$
\begin{array}{r}
0.5\,5 \\
9\,\overline{\smash{)}\,5.0\,0} \\
4\,5 \\ \hline
5\,0 \\
4\,5 \\ \hline
5
\end{array}
$$

We get a repeating decimal.

$\dfrac{5}{9} = 0.\overline{5}$

**49.**  $\dfrac{11}{12} = 11 \div 12$

$$
\begin{array}{r}
0.9\,1\,6\,6 \\
1\,2\,\overline{\smash{)}\,1\,1.0\,0\,0\,0} \\
1\,0\,8 \\ \hline
2\,0 \\
1\,2 \\ \hline
8\,0 \\
7\,2 \\ \hline
8\,0 \\
7\,2 \\ \hline
8
\end{array}
$$

We get a repeating decimal.

$\dfrac{11}{12} = 0.91\overline{6}$

**51.**  First consider $\dfrac{15}{7}$.

$\dfrac{15}{7} = 15 \div 7$

$$
\begin{array}{r}
2.1\,4\,2\,8\,5\,7 \\
7\,\overline{\smash{)}\,1\,5.0\,0\,0\,0\,0\,0} \\
1\,4 \\ \hline
1\,0 \\
7 \\ \hline
3\,0 \\
2\,8 \\ \hline
2\,0 \\
1\,4 \\ \hline
6\,0 \\
5\,6 \\ \hline
4\,0 \\
3\,5 \\ \hline
5\,0 \\
4\,9 \\ \hline
1
\end{array}
$$

We get a repeating decimal.

$\dfrac{15}{7} = 2.\overline{142857}$, so $-\dfrac{15}{7} = -2.\overline{142857}$.

**53.**  4.03 trillion $= 4.03 \times 1$ trillion

$= 4.03 \times 1,000,000,000,000$

$= 4,030,000,000,000$

**55.**  42.7 million $= 42.7 \times 1$ million

$= 42.7 \times 1,000,000$

$= 42,700,000$

**57.** *Familiarize*. The subscription price is $100\% - 69.94\%$, or $30.06\%$ of the cover price. Let $p =$ the cover price.

*Translate*.

$$
\underbrace{\text{Subscription price}}_{} \text{ is } 30.06\% \text{ of } \underbrace{\text{cover price}}_{}
$$

$$
\begin{array}{ccccc}
\downarrow & \downarrow & \downarrow & \downarrow & \downarrow \\
1.50 & = & 30.06\% & \times & p
\end{array}
$$

*Solve*.

$1.50 = 30.06\% \times p$

$1.50 = 0.3006 \times p$

$\dfrac{1.50}{0.3006} = \dfrac{0.3006 \times p}{0.3006}$

$4.99 \approx p$

*Check*. We find $69.94\%$ of 4.99 and then subtract this amount from 4.99:

$0.6994 \times 4.99 \approx 3.49$, $4.99 - 3.49 = 1.50$.

Since we get the subscription price, the answer checks.

*State*. The cover price was $4.99.

**59.**  First we find the commission on the first $5000 in sales.

$$
\begin{array}{ccc}
\text{Commission} = & \text{Commission rate} & \times \text{ Sales} \\
C \quad = & 10\% & \times \ 5000
\end{array}
$$

Using a calculator we find that $0.1 \times 5000 = 500$, so the commission on the first $5000 in sales was $500. We subtract to find the additional commission:

$\$2405 - \$500 = \$1905$

Now we find the amount of sales required to earn $1905 at a commission rate of $15\%$.

$$
\begin{array}{ccc}
\text{Commission} = & \text{Commission rate} & \times \text{ Sales} \\
1905 \quad = & 15\% & \times \quad S
\end{array}
$$

Using a calculator to divide 1905 by $15\%$, or 0.15, we get 12,700.

Finally we add to find the total sales:

$\$5000 + \$12,700 = \$17,700$

---

## Exercise Set 7.7

**1.**  $I = P \cdot r \cdot t$

$= \$200 \times 4\% \times 1$

$= \$200 \times 0.04$

$= \$8$

**3.**  $I = P \cdot r \cdot t$

$= \$4300 \times 10.56\% \times \dfrac{1}{4}$

$= \dfrac{\$4300 \times 0.1056}{4}$

$= \$113.52$

**5.**  $I = P \cdot r \cdot t$

$= \$20,000 \times 4\dfrac{5}{8}\% \times 1$

$= \$20,000 \times 0.04625$

$= \$925$

**7.** $I = P \cdot r \cdot t$

$= \$50,000 \times 5\frac{3}{8}\% \times \frac{1}{4}$

$= \dfrac{\$50,000 \times 0.05375}{4}$

$\approx \$671.88$

**9. a)** We express 60 days as a fractional part of a year and find the interest.

$I = P \cdot r \cdot t$

$= \$10,000 \times 9\% \times \dfrac{60}{365}$

$= \$10,000 \times 0.09 \times \dfrac{60}{365}$

$\approx \$147.95$   Using a calculator

The interest due for 60 days is \$147.95.

**b)** The total amount that must be paid after 60 days is the principal plus the interest.

$10,000 + 147.95 = 10,147.95$

The total amount due is \$10,147.95.

**11. a)** We express 90 days as a fractional part of a year and find the interest.

$I = P \cdot r \cdot t$

$= \$6500 \times 5\frac{1}{4}\% \times \dfrac{90}{365}$

$= \$6500 \times 0.0525 \times \dfrac{90}{365}$

$\approx \$84.14$   Using a calculator

The interest due for 90 days is \$84.14.

**b)** The total amount that must be paid after 90 days is the principal plus the interest.

$6500 + 84.14 = 6584.14$

The total amount due is \$6584.14.

**13. a)** We express 30 days as a fractional part of a year and find the interest.

$I = P \cdot r \cdot t$

$= \$5600 \times 10\% \times \dfrac{30}{365}$

$= \$5600 \times 0.1 \times \dfrac{30}{365}$

$\approx \$46.03$   Using a calculator

The interest due for 30 days is \$46.03.

**b)** The total amount that must be paid after 30 days is the principal plus the interest.

$5600 + 46.03 = 5646.03$

The total amount due is \$5646.03.

**15. a)** After 1 year, the account will contain 105% of \$400.

$1.05 \times \$400 = \$420$

$$
\begin{array}{r}
4\ 0\ 0 \\
\times\ \ 1.0\ 5 \\
\hline
2\ 0\ 0\ 0 \\
4\ 0\ 0\ 0\ 0 \\
\hline
4\ 2\ 0.0\ 0
\end{array}
$$

**b)** At the end of the second year, the account will contain 1.05% of \$420.

$1.05 \times \$420 = \$441$

$$
\begin{array}{r}
4\ 2\ 0 \\
\times\ \ 1.0\ 5 \\
\hline
2\ 1\ 0\ 0 \\
4\ 2\ 0\ 0\ 0 \\
\hline
4\ 4\ 1.0\ 0
\end{array}
$$

The amount in the account after 2 years is \$441.

(Note that we could have used the formula

$A = P \cdot \left(1 + \dfrac{r}{n}\right)^{n \cdot t}$, substituting \$400 for $P$, 5% for $r$,

1 for $n$, and 2 for $t$.)

**17.** We use the compound interest formula, substituting \$2000 for $P$, 8.8% for $r$, 1 for $n$, and 4 for $t$.

$A = P \cdot \left(1 + \dfrac{r}{n}\right)^{n \cdot t}$

$= \$2000 \cdot \left(1 + \dfrac{8.8\%}{1}\right)^{1 \cdot 4}$

$= \$2000 \cdot (1 + 0.088)^4$

$= \$2000 \cdot (1.088)^4$

$\approx \$2802.50$

The amount in the account after 4 years is \$2802.50.

**19.** We use the compound interest formula, substituting \$4300 for $P$, 10.56% for $r$, 1 for $n$, and 6 for $t$.

$A = P \cdot \left(1 + \dfrac{r}{n}\right)^{n \cdot t}$

$= \$4300 \cdot \left(1 + \dfrac{10.56\%}{1}\right)^{1 \cdot 6}$

$= \$4300 \cdot (1 + 0.1056)^6$

$= \$4300 \cdot (1.1056)^6$

$\approx \$7853.38$

The amount in the account after 6 years is \$7853.38.

**21.** We use the compound interest formula, substituting \$20,000 for $P$, $6\frac{5}{8}\%$ for $r$, 1 for $n$, and 25 for $t$.

$A = P \cdot \left(1 + \dfrac{r}{n}\right)^{n \cdot t}$

$= \$20,000 \cdot \left(1 + \dfrac{6\frac{5}{8}\%}{1}\right)^{1 \cdot 25}$

$= \$20,000 \cdot (1 + 0.06625)^{25}$

$= \$20,000 \cdot (1.06625)^{25}$

$\approx \$99,427.40$

The amount in the account after 25 years is \$99,427.40.

**23.** We use the compound interest formula, substituting $4000 for $P$, 6% for $r$, 2 for $n$, and 1 for $t$.

$$A = P \cdot \left(1 + \frac{r}{n}\right)^{n \cdot t}$$

$$= \$4000 \cdot \left(1 + \frac{6\%}{2}\right)^{2 \cdot 1}$$

$$= \$4000 \cdot \left(1 + \frac{0.06}{2}\right)^{2}$$

$$= \$4000 \cdot (1.03)^2$$

$$= \$4243.60$$

The amount in the account after 1 year is $4243.60.

**25.** We use the compound interest formula, substituting $20,000 for $P$, 8.8% for $r$, 2 for $n$, and 4 for $t$.

$$A = P \cdot \left(1 + \frac{r}{n}\right)^{n \cdot t}$$

$$= \$20,000 \cdot \left(1 + \frac{8.8\%}{2}\right)^{2 \cdot 4}$$

$$= \$20,000 \cdot \left(1 + \frac{0.088}{2}\right)^{8}$$

$$= \$20,000 \cdot (1.044)^8$$

$$\approx \$28,225.00$$

The amount in the account after 4 years is $28,225.00.

**27.** We use the compound interest formula, substituting $5000 for $P$, 10.56% for $r$, 2 for $n$, and 6 for $t$.

$$A = P \cdot \left(1 + \frac{r}{n}\right)^{n \cdot t}$$

$$= \$5000 \cdot \left(1 + \frac{10.56\%}{2}\right)^{2 \cdot 6}$$

$$= \$5000 \cdot \left(1 + \frac{0.1056}{2}\right)^{12}$$

$$= \$5000 \cdot (1.0528)^{12}$$

$$\approx \$9270.87$$

The amount in the account after 6 years is $9270.87.

**29.** We use the compound interest formula, substituting $20,000 for $P$, $7\frac{5}{8}$% for $r$, 2 for $n$, and 25 for $t$.

$$A = P \cdot \left(1 + \frac{r}{n}\right)^{n \cdot t}$$

$$= \$20,000 \cdot \left(1 + \frac{7\frac{5}{8}\%}{2}\right)^{2 \cdot 25}$$

$$= \$20,000 \cdot \left(1 + \frac{0.07625}{2}\right)^{50}$$

$$= \$20,000 \cdot (1.038125)^{50}$$

$$\approx \$129,871.09$$

The amount in the account after 25 years is $129,871.09.

**31.** We use the compound interest formula, substituting $4000 for $P$, 6% for $r$, 12 for $n$, and $\frac{5}{12}$ for $t$.

$$A = P \cdot \left(1 + \frac{r}{n}\right)^{n \cdot t}$$

$$= \$4000 \cdot \left(1 + \frac{6\%}{12}\right)^{12 \cdot \frac{5}{12}}$$

$$= \$4000 \cdot \left(1 + \frac{0.06}{12}\right)^{5}$$

$$= \$4000 \cdot (1.005)^5$$

$$\approx \$4101.01$$

The amount in the account after 5 months is $4101.01.

**33.** We use the compound interest formula, substituting $1200 for $P$, 10% for $r$, 4 for $n$, and 1 for $t$.

$$A = P \cdot \left(1 + \frac{r}{n}\right)^{n \cdot t}$$

$$= \$1200 \cdot \left(1 + \frac{10\%}{4}\right)^{4 \cdot 1}$$

$$= \$1200 \cdot \left(1 + \frac{0.1}{4}\right)^{4}$$

$$= \$1200 \cdot (1.025)^4$$

$$\approx \$1324.58$$

The amount in the account after 1 year is $1324.58.

**35.** First we find the amount of interest on $1278.56 at 19.6% for one month.

$$I = P \cdot r \cdot t$$

$$= \$1278.56 \times 0.196 \times \frac{1}{12}$$

$$\approx \$20.88$$

We subtract to find the amount applied to decrease the principal.

$$\$25.57 - \$20.88 = \$4.69$$

We also subtract to find the balance after the payment.

$$\$1278.56 - \$4.69 = \$1273.87$$

**37.** a) We multiply the balance by 2%:

$$0.02 \times \$4876.54 = \$97.5308.$$

Antonio's minimum payment, rounded to the nearest dollar, is $98.

b) We find the amount of interest on $4876.54 at 21.3% for one month.

$$I = P \cdot r \cdot t$$

$$= \$4876.54 \times 0.213 \times \frac{1}{12}$$

$$\approx \$86.56$$

We subtract to find the amount applied to decrease the principal in the first payment.

$$\$98 - \$86.56 = \$11.44$$

The principal is decreased by $11.44 with the first payment.

c) We find the amount of interest on $4876.54 at 12.6% for one month.

$$I = P \cdot r \cdot t$$

$$= \$4876.54 \times 0.126 \times \frac{1}{12}$$

$$\approx \$51.20$$

We subtract to find the amount applied to decrease the principal in the first payment.

$$\$98 - \$51.20 = \$46.80.$$

The principal is decreased by $46.80 with the first payment.

d) With the 12.6% rate the principal was decreased by $46.80 − $11.44, or $35.36 more than at the 21.3% rate. This also means that the interest at 12.6% is $35.36 less than at 21.3%.

**39.** If the product of two numbers is 1, they are <u>reciprocals</u> of each other.

**41.** The number 0 is the <u>additive</u> identity.

**43.** The distance around an object is its <u>perimeter</u>.

**45.** A natural number that has exactly two different factors, only itself and 1, is called a <u>prime</u> number.

**47.** For a principle $P$ invested at 9% compounded monthly, to find the amount in the account at the end of 1 year we would multiply $P$ by $(1 + 0.09/12)^{12}$. Since $(1 + 0.09/12)^{12} = 1.0075^{12} \approx 1.0938$, the effective yield is approximately 9.38%.

## Chapter 7 Concept Reinforcement

**1.** The statement is true.

**2.** We begin by writing each number in decimal notation.

$$0.5\% = 0.005$$

$$\frac{5}{1000} = 0.005$$

$$\frac{1}{2}\% = 0.5\% = 0.005$$

$$\frac{1}{5} = 0.2$$

$$0.\overline{1}$$

We see that $0.2 > 0.\overline{1}$, so $0.\overline{1}$ is not the largest number. The given statement is false.

**3.** Principal A grows to the amount

$$A\left(1 + \frac{0.04}{4}\right)^{4 \cdot 2} = A(1 + 0.01)^8 = A(1.01)^8.$$

Principal B grows to the amount

$$B\left(1 + \frac{0.02}{2}\right)^{2 \cdot 4} = B(1 + 0.01)^8 = B(1.01)^8.$$

Since $A = B$, the interest from the investments is the same. The given statement is true.

## Chapter 7 Important Concepts

**1.** $62\frac{5}{8}\% = 62.625\% = 62.625 \times 0.01 = 0.62625$

**2.** First we divide to find decimal notation.

$$\begin{array}{r} 0.6\,3\,6 \\ 1\,1\,\overline{)7.0\,0\,0} \\ \underline{6\,6} \\ 4\,0 \\ \underline{3\,3} \\ 7\,0 \\ \underline{6\,6} \\ 3 \end{array}$$

We get a repeating decimal, $0.63\overline{63}$.

$$\frac{7}{11} = 0.63\overline{63} = 63.\overline{63}\%, \text{ or } 63\frac{7}{11}\%$$

**3.** $6.8\% = \dfrac{6.8}{100} = \dfrac{6.8}{100} \cdot \dfrac{10}{10} = \dfrac{68}{1000} = \dfrac{4 \cdot 17}{4 \cdot 250} =$

$\dfrac{4}{4} \cdot \dfrac{17}{250} = \dfrac{17}{250}$

**4.** We translate to a percent equation.

$$12 = p \cdot 288$$

$$\frac{12}{288} = \frac{p \cdot 288}{288}$$

$$\frac{12}{288} = p$$

$$0.041\overline{6} = p$$

$$4.1\overline{6}\% = p, \text{ or}$$

$$4\frac{1}{6}\% = p$$

Thus, 12 is $4.1\overline{6}\%$, or $4\frac{1}{6}\%$, of 288.

**5.** We translate to a proportion.

$$\frac{3}{100} = \frac{300}{b}$$

$$3 \cdot b = 100 \cdot 300 \quad \text{Equating cross products}$$

$$\frac{3 \cdot b}{3} = \frac{100 \cdot 300}{3}$$

$$b = \frac{100 \cdot 300}{3}$$

$$b = 10,000$$

Thus, 3% of 10,000 is 300.

**6.** *Familiarize.* First we find the amount of increase.

$$\begin{array}{r} 4\,6.2\,0 \\ -\,4\,0.0\,7 \\ \hline 6.1\,3 \end{array}$$

Let $p =$ the percent of increase.

*Translate.*

6.13 is what percent of 40.07?

$$\begin{array}{ccccc} \downarrow & \downarrow & \downarrow & \downarrow & \downarrow \\ 6.13 & = & p & \times & 40.07 \end{array}$$

*Solve*.

$$6.13 = p \times 40.07$$

$$\frac{6.13}{40.07} = \frac{p \times 40.07}{40.07}$$

$$\frac{6.13}{40.07} = p$$

$$0.153 \approx p$$

$$15.3\% = p$$

**Check**. If the percent of increase is 15.3%, then the 2008 cost was the 2007 cost plus 15.3% of the 2007 cost, or 115.3% of the 2007 cost. Since 115.3% of $40.07 is 1.153 × $40.07 = $46.20071 ≈ $46.20, the answer checks.

**State**. The percent of increase was about 15.3%.

**7.** $\underbrace{\text{Sales tax}}$ is $\underbrace{\text{what percent}}$ of $\underbrace{\text{purchase price}}$

$$\begin{array}{ccccc} \downarrow & \downarrow & \downarrow & \downarrow & \downarrow \\ 1102.20 & = & r & \times & 18,370 \end{array}$$

$$\frac{1102.20}{18,370} = \frac{r \times 18,370}{18,370}$$

$$\frac{1102.20}{18,370} = r$$

$$0.06 = r$$

$$6\% = r$$

The sales tax rate is 6%.

**8.** Commission = Commission rate × Sales

$$12,950 \quad = \quad 7\% \quad \times \quad S$$

To solve this equation we divide on both sides by 0.07:

$$\frac{12,950}{0.07} = \frac{0.07 \times S}{0.07}$$

$$185,000 = S$$

The home sold for $185,000.

**9.** We express 60 days as a fractional part of a year.

$$I = P \cdot r \cdot t$$

$$= \$2500 \times 5\frac{1}{2}\% \times \frac{60}{365}$$

$$= \$2500 \times 0.055 \times \frac{60}{365}$$

$$\approx \$22.60$$

The interest is $22.60.

The total amount due is $2500 + $22.60 = $2522.60.

**10.** $A = P\left(1 + \dfrac{r}{n}\right)^{n \cdot t}$

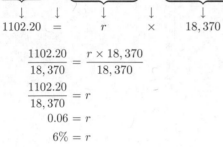

$$= \$6000\left(1 + \frac{4\frac{3}{4}\%}{4}\right)^{4 \cdot 2}$$

$$= \$6000\left(1 + \frac{0.0475}{4}\right)^{8}$$

$$= \$6000(1.011875)^{8}$$

$$= \$6594.26$$

---

## Chapter 7 Review Exercises

**1.** $4\% = 4 \times 0.01 = 0.04$

  $14.4\% = 14.4 \times 0.01 = 0.144$

**2.** $62.1\% = 62.1 \times 0.01 = 0.621$

  $84.2\% = 84.2 \times 0.01 = 0.842$

**3.** First we divide to find decimal notation.

$$\begin{array}{r} 0.3\,7\,5 \\ 8\,\overline{)\,3.0\,0\,0} \\ \underline{2\,4} \\ 6\,0 \\ \underline{5\,6} \\ 4\,0 \\ \underline{4\,0} \\ 0 \end{array}$$

$$\frac{3}{8} = 0.375$$

Now convert 0.375 to percent notation by moving the decimal point two places to the right and writing a percent symbol.

$$\frac{3}{8} = 37.5\%$$

**4.** First we divide to find decimal notation.

$$\begin{array}{r} 0.3\,3\,3 \\ 3\,\overline{)\,1.0\,0\,0} \\ \underline{9} \\ 1\,0 \\ \underline{9} \\ 1\,0 \\ \underline{9} \\ 1 \end{array}$$

We get a repeating decimal: $\dfrac{1}{3} = 0.33\overline{3}$. We convert $0.33\overline{3}$ to percent notation by moving the decimal point two places to the right and writing a percent symbol.

$$\frac{1}{3} = 33.\overline{3}\%, \text{ or } 33\frac{1}{3}\%$$

**5.** Move the decimal point two places to the right and write a percent symbol.

$$1.7 = 170\%$$

**6.** Move the decimal point two places to the right and write a percent symbol.

$$0.065 = 6.5\%$$

**7.** $24\% = \dfrac{24}{100} = \dfrac{4 \cdot 6}{4 \cdot 25} = \dfrac{4}{4} \cdot \dfrac{6}{25} = \dfrac{6}{25}$

**8.** $6.3\% = \dfrac{6.3}{100} = \dfrac{6.3}{100} \cdot \dfrac{10}{10} = \dfrac{63}{1000}$

**9.** *Translate*. $30.6 = p \times 90$

  *Solve*. We divide by 90 on both sides and convert to percent notation.

$$30.6 = p \times 90$$

$$\frac{30.6}{90} = \frac{p \times 90}{90}$$

$$0.34 = p$$

$$34\% = p$$

30.6 is 34% of 90.

**10.** *Translate.* $63 - 84\% \times b$

*Solve.* We divide by 84% on both sides.

$$63 = 84\% \times b$$

$$\frac{63}{84\%} = \frac{84\% \times b}{84\%}$$

$$\frac{63}{0.84} = b$$

$$75 = b$$

63 is 84% of 75.

**11.** *Translate.* $a = 38\frac{1}{2}\% \times 168$

*Solve.* Convert $38\frac{1}{2}\%$ to decimal notation and multiply.

$$
\begin{array}{r}
1\,6\,8 \\
\times\,0.\,3\,8\,5 \\
\hline
8\,4\,0 \\
1\,3\,4\,4\,0 \\
5\,0\,4\,0\,0 \\
\hline
6\,4.\,6\,8\,0
\end{array}
$$

64.68 is $38\frac{1}{2}\%$ of 168.

**12.** 24 percent of what is 16.8?

| Percents | Quantities |
|----------|-----------|
| 0% | 0 |
| 24% | 16.8 |
| 100% | b |

*Translate:* $\dfrac{24}{100} = \dfrac{16.8}{b}$

*Solve:* $24 \cdot b = 100 \cdot 16.8$

$$\frac{24 \cdot b}{24} = \frac{100 \cdot 16.8}{24}$$

$$b = \frac{100 \cdot 16.8}{24}$$

$$b = 70$$

24% of 70 is 16.8. The answer is 16.8.

**13.** 42 is what percent of 30?

| Percents | Quantities |
|----------|-----------|
| 0% | 0 |
| 100% | 30 |
| N% | 42 |

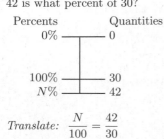

*Translate:* $\dfrac{N}{100} = \dfrac{42}{30}$

*Solve:* $30 \cdot N = 100 \cdot 42$

$$\frac{30 \cdot N}{30} = \frac{100 \cdot 42}{30}$$

$$N = \frac{4200}{30}$$

$$N = 140$$

42 is 140% of 30. The answer is 140%.

**14.** What is 10.5% of 84?

| Percents | Quantities |
|----------|-----------|
| 0% | 0 |
| 10.5% | a |
| 100% | 84 |

*Translate:* $\dfrac{10.5}{100} = \dfrac{a}{84}$

*Solve:* $10.5 \cdot 84 = 100 \cdot a$

$$\frac{10.5 \cdot 84}{100} = \frac{100 \cdot a}{100}$$

$$\frac{882}{100} = a$$

$$8.82 = a$$

8.82 is 10.5% of 84. The answer is 8.82.

**15.** **Familiarize.** Let $c =$ the number of students who would choose chocolate as their favorite ice cream and $b =$ the number who would choose butter pecan.

**Translate.** We translate to two equations.

$$\underbrace{\text{What number}}_{c} \underset{=}{\text{ is }} \underset{8.9\%}{8.9\%} \cdot \underset{2000}{2000}?$$

$$\underbrace{\text{What number}}_{b} \underset{=}{\text{ is }} \underset{4.2\%}{4.2\%} \cdot \underset{2000}{2000}?$$

**Solve.** We convert percent notation to decimal notation and multiply.

$$c = 0.089 \cdot 2000 = 178$$

$$b = 0.042 \cdot 2000 = 84$$

**Check.** We can repeat the calculation. We can also do partial checks by estimating.

$$8.9\% \cdot 2000 \approx 9\% \cdot 2000 = 180;$$

$$4.2\% \cdot 2000 \approx 4\% \cdot 2000 = 80$$

Since 180 is close to 178 and 80 is close to 84, our answers seem reasonable.

**State.** 178 students would choose chocolate as their favorite ice cream and 80 would choose butter pecan.

**16.** **Familiarize.** Let $p =$ the percent of people in the U.S. who take at least one kind of prescription drug per day.

**Translate.** We translate to a proportion.

$$\frac{p}{100} = \frac{140.3}{305}$$

*Solve*. We equate cross products.

$$p \cdot 305 = 100 \cdot 140.3$$
$$\frac{p \cdot 305}{305} = \frac{100 \cdot 140.3}{305}$$
$$p = 0.46$$
$$p = 46\%$$

*Check*. $46\% \cdot 305$ million $= 0.46 \cdot 305$ million $= 140.3$ million. The answer checks.

*State*. In the U.S. 46% of the people take at least one kind of prescription drug per day.

**17. Familiarize.** Let $w =$ the total output of water from the body per day.

*Translate*.

200 mL is 8% of what number?
$$200 = 8\% \cdot w$$

*Solve*.

$$200 = 8\% \cdot w$$
$$200 = 0.08 \cdot w$$
$$\frac{200}{0.08} = \frac{0.08 \cdot w}{0.08}$$
$$2500 = w$$

*Check*. $8\% \cdot 2500 = 0.08 \cdot 2500 = 200$, so the answer checks.

*State*. The total output of water from the body is 2500 mL per day.

**18. Familiarize.** First we subtract to find the amount of the increase.

$$\begin{array}{r} {\scriptstyle 7\ 14} \\ 8\ 4 \\ -\ 7\ 5 \\ \hline 9 \end{array}$$

Now let $p =$ the percent of increase.

*Translate*. We translate to a proportion.

$$\frac{p}{100} = \frac{9}{75}$$

*Solve*. We equate cross products.

$$p \cdot 75 = 100 \cdot 9$$
$$\frac{p \cdot 75}{75} = \frac{100 \cdot 9}{75}$$
$$p = 12$$

*Check*. $12\% \cdot 75 = 0.12 \cdot 75 = 9$, the amount of the increase, so the answer checks.

*State*. Jason's score increased 12%.

**19. Familiarize.** Let $s =$ the new score. Note that the new score is the original score plus 15% of the original score.

New score is Original score plus 15% of Original score
$$s = 80 + 15\% \cdot 80$$

*Solve*. We convert 15% to decimal notation and carry out the computation.

$$s = 80 + 0.15 \cdot 80 = 80 + 12 = 92$$

*Check*. We repeat the calculation. The answer checks.

*State*. Jenny's new score was 92.

**20.** The meals tax is

Meal tax rate × Cost of meal
$$7\frac{1}{2}\% \times \$320,$$

or $0.075 \times \$320$, or \$24.

**21.**

Sales tax is what percent of purchase price?
$$\$453.60 = r \times 7560$$

To solve the equation, we divide on both sides by 7560.

$$\frac{\$453.60}{7560} = \frac{r \times 7560}{7560}$$
$$0.06 = r$$
$$6\% = r$$

The sales tax rate is 6%.

**22.**

$$\text{Commission} = \text{Commission rate} \times \text{Sales}$$
$$753.50 = r \times 6850$$

To solve this equation, we divide on both sides by 6850.

$$\frac{753.50}{6850} = \frac{r \times 6850}{6850}$$
$$0.11 = r$$
$$11\% = r$$

The commission rate is 11%.

**23.**

$$\text{Discount} = \text{Rate of discount} \times \text{Original price}$$
$$D = 12\% \times \$350$$

Convert 12% to decimal notation and multiply.

$$\begin{array}{r} 3\ 5\ 0 \\ \times\ 0.1\ 2 \\ \hline 7\ 0\ 0 \\ 3\ 5\ 0\ 0 \\ \hline 4\ 2.0\ 0 \end{array}$$

The discount is \$42.

$$\text{Sale price} = \text{Original price} - \text{Discount}$$
$$S = \$350 - \$42$$

We subtract:

$$\begin{array}{r} {\scriptstyle 4\ 10} \\ 3\ 5\ 0 \\ -\ 4\ 2 \\ \hline 3\ 0\ 8 \end{array}$$

The sale price is \$308.

**24.** First we find the discount.

$$\$305 - \$262.30 = \$42.70$$

Now we find the rate of discount.

$$\text{Discount} = \text{Rate of discount} \times \text{Original price}$$
$$42.70 = r \times 305$$

$$\frac{42.70}{305} = \frac{r \times 305}{305}$$

$$\frac{42.70}{305} = r$$

$$0.14 = r$$

$$14\% = r$$

The rate of discount is 14%.

**25.**  Commission = Commission rate × Sales
$$C \quad = \quad 7\% \quad \times 42,000$$

We convert 7% to decimal notation and multiply.

$$\begin{array}{r} 4\,2,\,0\,0\,0 \\ \times \quad\quad 0.\,0\,7 \\ \hline 2\,9\,4\,0.\,0\,0 \end{array}$$

The commission is $2940.

**26.**  First we subtract to find the discount.

$$\$82 - \$67 = \$15$$

Discount = Rate of discount × Original price
$$15 \quad = \quad\quad r \quad\quad \times \quad\quad 82$$

We divide on both sides by 82.

$$\frac{15}{82} = \frac{r \times 82}{82}$$

$$0.183 \approx r$$

$$18.3\% \approx r$$

The rate of discount is about 18.3%.

**27.**  $I = P \cdot r \cdot t$

$$= \$1800 \times 6\% \times \frac{1}{3}$$

$$= \$1800 \times 0.06 \times \frac{1}{3}$$

$$= \$36$$

**28.**  a)  $I = P \cdot r \cdot t$

$$= \$24,000 \times 10\% \times \frac{60}{365}$$

$$= \$24,000 \times 0.1 \times \frac{60}{365}$$

$$\approx \$394.52$$

b)  $\$24,000 + \$394.52 = \$24,394.52$

**29.**  $I = P \cdot r \cdot t$

$$= \$2200 \times 5.5\% \times 1$$

$$= \$2200 \times 0.055 \times 1$$

$$= \$121$$

**30.**  $A = P \cdot \left(1 + \dfrac{r}{n}\right)^{n \cdot t}$

$$= \$7500 \cdot \left(1 + \frac{4\%}{12}\right)^{12 \cdot \frac{1}{4}}$$

$$= \$7500 \cdot \left(1 + \frac{0.04}{12}\right)^{3}$$

$$\approx \$7575.25$$

**31.**  $A = P \cdot \left(1 + \dfrac{r}{n}\right)^{n \cdot t}$

$$= \$8000 \cdot \left(1 + \frac{9\%}{1}\right)^{1 \cdot 2}$$

$$= \$8000 \cdot (1 + 0.09)^2$$

$$= \$8000 \cdot (1.09)^2$$

$$= \$9504.80$$

**32.**  a) 2% of $6428.74 = 0.02 \times \$6428.74 \approx \$129$

b)  $I = P \cdot r \cdot t$

$$= \$6428.74 \times 0.187 \times \frac{1}{12}$$

$$\approx \$100.18$$

The amount of interest is $100.18.

$129 - \$100.18 = \$28.82$, so the principal is reduced by $28.82.

c)  $I = P \cdot r \cdot t$

$$= \$6428.74 \times 0.132 \times \frac{1}{12}$$

$$\approx \$70.72$$

The amount of interest is $70.72.

$129 - \$70.72 = \$58.28$, so the principal is reduced by $58.28 with the lower interest rate.

d) With the 13.2% rate the principal was decreased by $58.28 - \$28.82$, or $29.46, more than at the 18.7% rate. This also means that the interest at 13.2% is $29.46 less than at 18.7%.

**33.**  First we find the discount.
Discount = Rate of discount × Original price
$$D \quad = \quad\quad 15\% \quad\quad \times \quad\quad 16,500$$

We multiply.

$$\begin{array}{r} 1\,6,\,5\,0\,0 \\ \times \quad\quad 0.\,1\,5 \\ \hline 8\,2\,5\,0\,0 \\ 1\,6\,5\,0\,0\,0 \\ \hline 2\,4\,7\,5.\,0\,0 \end{array}$$

The discount is $2475. Now we find the sale price.

Sale price = Original price − Discount
$$S \quad = \quad 16,500 \quad - \quad 2475$$
$$S \quad = \quad 14,025$$

The sale price is $14,025. Answer A is correct.

**34.**  $A = 10,500 \left(1 + \dfrac{6\%}{2}\right)^{2 \cdot 1\frac{1}{2}}$

$$= 10,500 \left(1 + \frac{0.06}{2}\right)^{2 \cdot \frac{3}{2}}$$

$$= 10,500(1 + 0.03)^3$$

$$= 10,500(1.03)^3$$

$$\approx 11,473.63$$

The amount in the account is $11,473.63. Answer C is correct.

**35. *Familiarize*.** Let $d =$ the original price of the dress. After the 40% discount, the sale price is 60% of $d$, or $0.6d$. Let $p =$ the percent by which the sale price must be increased to return to the original price.

***Translate*.**

| Sale price | plus | what percent | of | sale price | is | original price? |
|---|---|---|---|---|---|---|
| ↓ | ↓ | ↓ | ↓ | ↓ | ↓ | ↓ |
| $0.6d$ | $+$ | $p$ | $\cdot$ | $0.6d$ | $=$ | $d$ |

***Solve*.**

$$0.6d + p \cdot 0.6d = d$$

$$(1+p)(0.6d) = d \quad \text{Factoring on the left}$$

$$\frac{(1+p)(0.6d)}{0.6d} = \frac{d}{0.6d}$$

$$1 + p = \frac{1}{0.6} \cdot \frac{d}{d}$$

$$1 + p = 1.66\overline{6}$$

$$p = 1.66\overline{6} - 1$$

$$p = 0.66\overline{6}$$

$$p = 66.\overline{6}\%, \text{ or } 66\frac{2}{3}\%$$

***Check*.** Suppose the dress cost \$100. Then the sale price is 60% of \$100, or \$60. Now $66\frac{2}{3}\% \cdot \$60 = \$40$ and $\$60 + \$40 = \$100$, the original price. Since the answer checks for this specific price, it seems to be reasonable.

***State*.** The sale price must be increased $66\frac{2}{3}\%$ after the sale to return to the original price.

**36.** Let $S =$ the original salary. After a 3% raise, the salary becomes $103\% \cdot S$, or $1.03S$. After a 6% raise, the new salary is $1.06\% \cdot 1.03S$, or $1.06(1.03S)$. Finally, after a 9% raise, the salary is $109\% \cdot 1.06(1.03S)$, or $1.09(1.06)(1.03S)$. Multiplying, we get $1.09(1.06)(1.03S) = 1.190062S$. This is equivalent to $119.0062\% \cdot S$, so the original salary has increased about 19%.

---

# Chapter 7 Discussion and Writing Exercises

**1.** A 40% discount is better. When successive discounts are taken, each is based on the previous discounted price rather than on the original price. A 20% discount followed by a 22% discount is the same as a 37.6% discount off the original price.

**2.** Let $S =$ the original salary. After both raises have been given, the two situations yield the same salary: $1.05 \cdot 1.1S = 1.1 \cdot 1.05S$. However, the first situation is better for the wage earner, because $1.1S$ is earned the first year when a 10% raise is given while in the second situation $1.05S$ is earned that year.

**3.** No; the 10% discount was based on the original price rather than on the sale price.

**4.** For a number $n$, 40% of 50% of $n$ is $0.4(0.5n)$, or $0.2n$, or 20% of $n$. Thus, taking 40% of 50% of a number is the same as taking 20% of the number.

**5.** The interest due on the 30 day loan will be \$41.10 while that due on the 60 day loan will be \$131.51. This could be an argument in favor of the 30 day loan. On the other hand the 60 day loan puts twice as much cash at the firm's disposal for twice as long as the 30 day loan. This could be an argument in favor of the 60 day loan.

**6.** Answers will vary.

---

# Chapter 7 Test

**1.** 14.7%

a) Replace the percent symbol with $\times 0.01$.

$$14.7 \times 0.01$$

b) Move the decimal point two places to the left.

0.14.7

Thus, $14.7\% = 0.147$.

**2.** 0.38

a) Move the decimal point two places to the right.

0.38.

b) Write a percent symbol: 38%

Thus, $0.38 = 38\%$.

**3.**

```
      1.3 7 5
  8 ) 1 1.0 0 0
      8
      ‾‾‾
      3 0
      2 4
      ‾‾‾
        6 0
        5 6
        ‾‾‾
          4 0
          4 0
          ‾‾‾
            0
```

$$\frac{11}{8} = 1.375$$

Convert to percent notation.

1.37.5

$$\frac{11}{8} = 137.5\%, \text{ or } 137\frac{1}{2}\%$$

**4.** $65\% = \dfrac{65}{100}$      Definition of percent

$$= \frac{5 \cdot 13}{5 \cdot 20}$$

$$= \frac{5}{5} \cdot \frac{13}{20} \quad \left.\right\} \text{ Simplifying}$$

$$= \frac{13}{20}$$

**5.** Translate:  What is 40% of 55?

$$a = 40\% \cdot 55$$

Solve: We convert 40% to decimal notation and multiply.

$$a = 40\% \cdot 55$$
$$= 0.4 \cdot 55 = 22$$

The answer is 22.

**6.** What percent of 80 is 65?

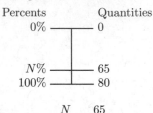

Percents        Quantities
0% ———— 0
N% ———— 65
100% ———— 80

Translate:  $\dfrac{N}{100} = \dfrac{65}{80}$

Solve:  $80 \cdot N = 100 \cdot 65$

$$\frac{80 \cdot N}{80} = \frac{100 \cdot 65}{80}$$
$$N = \frac{6500}{80}$$
$$N = 81.25$$

The answer is 81.25%.

**7. Familiarize.** Let $k$ = the number of kidney transplants, $l$ = the number of liver transplants, and $h$ = the number of heart transplants in 2006.

**Translate.** We translate to three equations.

What number is 59% of 28,291?

$$k = 59\% \cdot 28,291$$

What number is 22% of 28,291?

$$l = 22\% \cdot 28,291$$

What number is 8% of 28,291?

$$h = 8\% \cdot 28,291$$

**Solve.** To solve each equation we convert percent notation to decimal notation and multiply.

$$k = 59\% \cdot 28,291 = 0.59 \cdot 28,291 \approx 16,692$$
$$l = 22\% \cdot 28,291 = 0.22 \cdot 28,291 \approx 6224$$
$$h = 8\% \cdot 28,291 = 0.08 \cdot 28,291 \approx 2263$$

**Check.** We repeat the calculations. The answers check.

**State.** In 2006, there were 16,692 kidney transplants, 6224 liver transplants, and 2263 heart transplants.

**8. Familiarize.** Let $b$ = the number of at-bats.

**Translate.** We translate to a proportion. We are asking "175 is 28.64% of what?"

$$\frac{28.64}{100} = \frac{175}{b}$$

**Solve.**

$$28.64 \cdot b = 100 \cdot 175 \quad \text{Equating cross products}$$
$$\frac{28.64 \cdot b}{28.64} = \frac{100 \cdot 175}{28.64}$$
$$b \approx 611$$

**Check.** We can repeat the calculation. Also note that $\dfrac{175}{611} \approx \dfrac{150}{600} = \dfrac{1}{4} = 25\% \approx 28.64\%$. The answer checks.

**State.** Garrett Atkins had about 611 at-bats.

**9. Familiarize.** We first find the amount of decrease.

$$\begin{array}{r} \overset{1\ \ 10\ \ 5\ \ 17}{2\,0,\,6\,7\,9} \\ -\ 1\,9,\,2\,9\,2 \\ \hline 1\,3\,8\,7 \end{array}$$

Let $p$ = the percent of decrease.

**Translate.** We translate to an equation.

1387 is what percent of 20,679?

$$1387 = p \cdot 20,679$$

**Solve.**

$$1387 = p \cdot 20,679$$
$$\frac{1387}{20,679} = \frac{p \cdot 20,679}{20,679}$$
$$0.067 \approx p$$
$$6.7\% \approx p$$

**Check.** Note that $6.7\% \approx 7\%$. With a decrease of approximately 7%, the number of adoptions in 2007 should be about $100\% - 7\%$, or 93%, of the number of adoptions in 2006. Since $93\% \cdot 20,679 = 0.93 \cdot 20,679 \approx 19,231 \approx 19,292$, the answer checks.

**State.** The percent of decrease was about 6.7%.

**10. Familiarize.** Let $p$ = the percent of people who live in Asia.

**Translate.** We translate to an equation.

4,002,000,000 is what percent of 6,603,000,000?

$$4,002,000,000 = p \cdot 6,603,000,000$$

**Solve.**

$$4,002,000,000 = p \cdot 6,603,000,000$$
$$\frac{4,002,000,000}{6,603,000,000} = \frac{p \cdot 6,603,000,000}{6,603,000,000}$$
$$0.606 \approx p$$
$$60.6\% \approx p$$

**Check.** We find 60.6% of 6,603,000,000:

$$60.6\% \cdot 6,603,000,000 = 0.606 \cdot 6,603,000,000 \approx$$
$$4,001,418,000 \approx 4,002,000,000$$

The answer checks.

**State.** About 60.6% of people living in the world today live in Asia.

**11.** The sales tax on an item costing \$560 is

$$\underbrace{\text{Sales tax rate}}_{\downarrow} \times \underbrace{\text{Purchase price}}_{\downarrow}$$

$$\qquad 4.5\% \qquad\quad \times \qquad\quad \$560,$$

or $0.045 \times \$560$, or \$25.20. Thus the tax is \$25.20.

The total price is given by the purchase price plus the sales tax:

$$\$560 + \$25.20 = \$585.20$$

**12.** Commission = Commission rate × Sales

$$
\begin{aligned}
C &= \quad 15\% \quad\times 4200 \\
C &= \quad 0.15 \quad\times 4200 \\
C &= \quad 630
\end{aligned}
$$

The commission is \$630.

**13.** Discount = Rate of discount × Marked price

$$
\begin{aligned}
D &= \quad 20\% \quad\times \quad \$200
\end{aligned}
$$

Convert 20% to decimal notation and multiply.

$$
\begin{array}{r}
2\,0\,0 \\
\times \ \ 0.\,2 \\
\hline
4\,0.\,0
\end{array}
\qquad (20\% = 0.20 = 0.2)
$$

The discount is \$40.

Sale price = Marked price − Discount

$$
\begin{aligned}
S &= \quad 200 \quad - \quad 40
\end{aligned}
$$

We subtract:
$$
\begin{array}{r}
2\,0\,0 \\
-\ \ 4\,0 \\
\hline
1\,6\,0
\end{array}
$$

To check, note that the sale price is 80% of the marked price: $0.8 \times 200 = 160$.

The sale price is \$160.

**14.** $I = P \cdot r \cdot t = \$120 \times 7.1\% \times 1$

$$
\begin{aligned}
&= \$120 \times 0.071 \times 1 \\
&= \$8.52
\end{aligned}
$$

**15.** $I = P \cdot r \cdot t = \$5200 \times 6\% \times \dfrac{1}{2}$

$$
\begin{aligned}
&= \$5200 \times 0.06 \times \frac{1}{2} \\
&= \$312 \times \frac{1}{2} \\
&= \$156
\end{aligned}
$$

The interest earned is \$156. The amount in the account is the principal plus the interest: $\$5200 + \$156 = \$5356$.

**16.** $A = P \cdot \left(1 + \dfrac{r}{n}\right)^{n \cdot t}$

$$
\begin{aligned}
&= \$1000 \cdot \left(1 + \frac{5\frac{3}{8}\%}{1}\right)^{1 \cdot 2} \\
&= \$1000 \cdot \left(1 + \frac{0.05375}{1}\right)^{2} \\
&= \$1000(1.05375)^{2} \\
&\approx \$1110.39
\end{aligned}
$$

**17.** $A = P \cdot \left(1 + \dfrac{r}{n}\right)^{n \cdot t}$

$$
\begin{aligned}
&= \$10,000 \cdot \left(1 + \frac{4.9\%}{12}\right)^{12 \cdot 3} \\
&= \$10,000 \cdot \left(1 + \frac{0.049}{12}\right)^{36} \\
&\approx \$11,580.07
\end{aligned}
$$

**18. Plumber**: We add to find the number of jobs in 2016: $705,000 + 52,000 = 757,000$, so we project that there will be 757,000 jobs for plumbers in 2016. We solve an equation to find the percent of increase, $p$.

$$52,000 \ \text{ is } \ \underbrace{\text{what percent}} \ \text{ of } \ 705,000?$$

$$\downarrow \quad \downarrow \qquad\quad \downarrow \qquad\quad \downarrow \qquad \downarrow$$

$$52,000 \ = \qquad p \qquad \cdot \quad 705,000$$

Solve.

$$
\begin{aligned}
\frac{52,000}{705,000} &= \frac{p \cdot 705,000}{705,000} \\
0.074 &\approx p \\
7.4\% &\approx p
\end{aligned}
$$

The percent of increase is about 7.4%.

**Veterinary Assistant**: We subtract to find the change: $100,000 - 71,000 = 29,000$, so we project that the change will be 29,000. We solve an equation to find the percent of increase, $p$.

$$29,000 \ \text{ is } \ \underbrace{\text{what percent}} \ \text{ of } \ 71,000?$$

$$\downarrow \quad \downarrow \qquad\quad \downarrow \qquad\quad \downarrow \qquad \downarrow$$

$$29,000 \ = \qquad p \qquad \cdot \quad 71,000$$

Solve.

$$
\begin{aligned}
\frac{29,000}{71,000} &= \frac{p \cdot 71,000}{71,000} \\
0.408 &\approx p \\
40.8\% &\approx p
\end{aligned}
$$

The percent of increase is about 40.8%.

**Motorcycle Repair Technician**: We subtract to find the number of jobs in 2006; $24,000 - 3000 = 21,000$, so there were 21,000 jobs for motorcycle repair technicians in 2006. We solve an equation to find the percent of increase, $p$.

$$3000 \ \text{ is } \ \underbrace{\text{what percent}} \ \text{ of } \ 21,000?$$

$$\downarrow \quad \downarrow \qquad\quad \downarrow \qquad\quad \downarrow \qquad \downarrow$$

$$3000 \ = \qquad p \qquad \cdot \quad 21,000$$

Solve.

$$
\begin{aligned}
\frac{3000}{21,000} &= \frac{p \cdot 21,000}{21,000} \\
0.143 &\approx p \\
14.3\% &\approx p
\end{aligned}
$$

The percent of increase is about 14.3%.

**Fitness Professional**: Let $n =$ the number of jobs in 2006. If the number of jobs increases by 26.8%, then the number of jobs in 2016 represents 100% of $n$ plus 26.8% of $n$, or 126.8% of $n$, or $1.268n$. We solve an equation to find $n$.

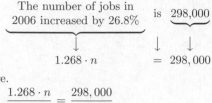

$$1.268 \cdot n = 298{,}000$$

Solve.

$$\frac{1.268 \cdot n}{1.268} = \frac{298{,}000}{1.268}$$

$$n \approx 235{,}000$$

There were about 235,000 jobs for fitness professionals in 2006.

We subtract to find the change; $298{,}000 - 235{,}000 = 63{,}000$, so we project that the change will be 63,000 jobs.

**19.** Discount = Original price − Sale price

$$D = 349.99 - 299.99$$

We subtract:    3 4 9 . 9 9
               − 2 9 9 . 9 9
               _____
                 5 0 . 0 0

The discount is $50.

Discount = Rate of discount × Original price

$$50 = R \times 349.99$$

To solve the equation we divide on both sides by 349.99.

$$\frac{50}{349.99} = \frac{R \times 349.99}{349.99}$$

$$0.143 \approx R$$

$$14.3\% \approx R$$

To check, note that a discount rate of 14.3% means that 85.7% of the original price is paid: $0.857 \times 349.99 \approx 299.94 \approx 299.99$. Since that is the sale price, the answer checks.

The rate of discount is about 14.3%.

**20.** We first use the formula $I = P \cdot r \cdot t$ to find the amount of interest paid.

$$I = P \cdot r \cdot t = \$2704.27 \cdot 0.163 \cdot \frac{1}{12} \approx \$36.73.$$

Then the amount by which the principal is reduced is

$$\$54 - \$36.73 = \$17.27.$$

Finally, we find that the principal after the payment is

$$\$2704.27 - \$17.27 = \$2687.$$

**21.** 0.75% of what number is 300?

*Translate:*   $300 = 0.75\% \times b$

*Solve:*   We divide on both sides by 0.75%:

$$\frac{300}{0.75\%} = \frac{0.75\% \times b}{0.75\%}$$

$$\frac{300}{0.0075} = b \quad (0.75\% = 0.0075)$$

$$40{,}000 = b$$

300 is 0.75% of 40,000. The correct answer is B.

**22.** *Familiarize.* Let $p =$ the price for which a realtor would have to sell the house in order for Juan and Marie to receive $180,000 from the sale. The realtor's commission would be

$7.5\% \cdot p$, or $0.075 \cdot p$, and Juan and Marie would receive 100% of $p - 7.5\%$ of $p$, or 92.5% of $p$, or $0.925 \cdot p$.

*Translate.*

Amount Juan and Marie receive is $180,000

$$0.925 \cdot p = 180{,}000$$

*Solve.*

$$\frac{0.925 \cdot p}{0.925} = \frac{180{,}000}{0.925}$$

$$p \approx 194{,}600 \quad \text{Rounding to the nearest hundred}$$

*Check.* 7.5% of $194{,}600 = 0.075 \cdot \$194{,}600 = \$14{,}595$ and $\$194{,}600 - \$14{,}595 = \$180{,}005 \approx \$180{,}000$. The answer checks.

*State.* A realtor would need to sell the house for about $194,600.

**23.** First we find the commission.

Commission = Commission rate × Sales

$$C = 16\% \times \$15{,}000$$

$$C = 0.16 \times \$15{,}000$$

$$C = \$2400$$

Now we find the amount in the account after 6 months.

$$A = P \cdot \left(1 + \frac{r}{n}\right)^{n \cdot t}$$

$$= \$2400 \cdot \left(1 + \frac{12\%}{4}\right)^{4 \cdot \frac{1}{2}}$$

$$= \$2400 \cdot \left(1 + \frac{0.12}{4}\right)^{2}$$

$$= \$2400 \cdot (1 + 0.03)^{2}$$

$$= \$2400 \cdot (1.03)^{2}$$

$$= \$2400(1.0609)$$

$$= \$2546.16$$

## Cumulative Review Chapters 1 - 7

**1.** 44%

a) Replace the percent symbol with $\times 0.01$.

$$44 \times 0.01$$

b) Move the decimal point two places to the left.

$$44\% = 0.44$$

30%

a) Replace the percent symbol with $\times 0.01$.

$$30 \times 0.01$$

b) Move the decimal point two places to the left.

$$30\% = 0.3$$

**2.** 0.269

a) Move the decimal point two places to the right.

$$26.9$$

b) Write a percent symbol.

$$0.269 = 26.9\%$$

**3.** First divide to find decimal notation.

$$
\begin{array}{r}
1.1\,2\,5 \\
8\,\overline{)\,9.0\,0\,0} \\
\underline{8} \\
10 \\
\underline{\phantom{0}8} \\
2\,0 \\
\underline{1\,6} \\
4\,0 \\
\underline{4\,0} \\
0
\end{array}
$$

$\dfrac{9}{8} = 1.125$

Convert to percent notation by moving the decimal point two places to the right and writing a percent symbol.

$\dfrac{9}{8} = 112.5\%$

**4.** First we consider $\dfrac{13}{6}$.

$\dfrac{13}{6} = 13 \div 6$

$$
\begin{array}{r}
2.1\,6\,6 \\
6\,\overline{)\,1\,3.0\,0\,0} \\
\underline{1\,2} \\
10 \\
\underline{\phantom{0}6} \\
4\,0 \\
\underline{3\,6} \\
4\,0 \\
\underline{3\,6} \\
4
\end{array}
$$

Since 4 keeps reappearing as a remainder, the digits repeat and $\dfrac{13}{6} = 2.166\ldots$, or $2.1\overline{6}$. Then $-\dfrac{13}{6} = -2.1\overline{6}$.

**5.** $\dfrac{5}{0.5} = \dfrac{5}{0.5} \cdot \dfrac{10}{10} = \dfrac{50}{5} = \dfrac{5 \cdot 10}{5 \cdot 1} = \dfrac{5}{5} \cdot \dfrac{10}{1} = \dfrac{10}{1}$

**6.** $\dfrac{350 \text{ km}}{15 \text{ hr}} = \dfrac{350}{15} \cdot \dfrac{\text{km}}{\text{hr}} = \dfrac{5 \cdot 70}{3 \cdot 5} \cdot \dfrac{\text{km}}{\text{hr}} = \dfrac{5}{5} \cdot \dfrac{70}{3} \cdot \dfrac{\text{km}}{\text{hr}} =$
$\dfrac{70}{3}$ km/h, or $23\dfrac{1}{3}$ km/h, or $23.\overline{3}$ km/h

**7.** The LCD is 56.

$\dfrac{5}{7} \cdot \dfrac{8}{8} = \dfrac{40}{56}$

$\dfrac{6}{8} \cdot \dfrac{7}{7} = \dfrac{42}{56}$

Since $40 < 42$, it follows that $\dfrac{40}{56} < \dfrac{42}{56}$, so $\dfrac{5}{7} < \dfrac{6}{8}$.

**8.** The LCD is 350.

$\dfrac{-15}{25} \cdot \dfrac{14}{14} = \dfrac{-210}{350}$

$\dfrac{-6}{14} \cdot \dfrac{25}{25} = \dfrac{-150}{350}$

Since $-210 < -150$, it follows that $\dfrac{-210}{350} < \dfrac{-150}{350}$, so $\dfrac{-15}{25} < \dfrac{-6}{14}$.

**9.** $263,961 + 32,090 + 127.89 \approx 264,000 + 32,100 + 100 = 296,200$

**10.** $73,510 - 23,450 \approx 73,500 - 23,500 = 50,000$

**11.**
$$
\begin{aligned}
&46 - [4(6 + 4 \div 2) + 2 \times 3 - 5] \\
&= 46 - [4(6 + 2) + 2 \times 3 - 5] \\
&= 46 - [4(8) + 2 \times 3 - 5] \\
&= 46 - [32 + 6 - 5] \\
&= 46 - [38 - 5] \\
&= 46 - 33 \\
&= 13
\end{aligned}
$$

**12.**
$$
\begin{aligned}
&[-0.8(1.5 - 9.8 \div 49) - (1 + 0.1)^2] \div 1.5 \\
&= [-0.8(1.5 - 0.2) - (1.1)^2] \div 1.5 \\
&= [-0.8(1.3) - 1.21] \div 1.5 \\
&= [-1.04 - 1.21] \div 1.5 \\
&= -2.25 \div 1.5 \\
&= -1.5
\end{aligned}
$$

**13.**
$$
\begin{aligned}
\dfrac{6}{5} + 1\dfrac{5}{6} &= \dfrac{6}{5} + \dfrac{11}{6} \\
&= \dfrac{6}{5} \cdot \dfrac{6}{6} + \dfrac{11}{6} \cdot \dfrac{5}{5} \\
&= \dfrac{36}{30} + \dfrac{55}{30} \\
&= \dfrac{91}{30}, \text{ or } 3\dfrac{1}{30}
\end{aligned}
$$

**14.** First we find the difference in the absolute values.

$$
\begin{array}{r}
{\scriptstyle 1\;8\;10} \\
4\,6.\cancel{9}\,\cancel{0} \\
-\phantom{00}2.\,8\,4 \\
\hline
4\,4.\,0\,6
\end{array}
$$

The negative number has the larger absolute value, so the answer is negative.

$-46.9 + 2.84 = -44.06$

**15.**
$$
\begin{array}{r}
{\scriptstyle 1\;\;1\;\;1\;\;1\;\;1} \\
4\,8\,7,0\,9\,4 \\
6,9\,3\,6 \\
+\phantom{0}2\,1,1\,2\,0 \\
\hline
5\,1\,5,1\,5\,0
\end{array}
$$

**16.**
$$
\begin{array}{r}
{\scriptstyle 4\;\;9\;10} \\
3\,5.\cancel{0}\,\cancel{0} \\
-\,3\,4.\,9\,8 \\
\hline
0.\,0\,2
\end{array}
$$

**17.** $3\dfrac{1}{3} - 2\dfrac{2}{3} = 2\dfrac{4}{3} - 2\dfrac{2}{3} = \dfrac{2}{3}$

**18.**
$$
\begin{aligned}
-\dfrac{8}{9} - \dfrac{6}{7} &= -\dfrac{8}{9} \cdot \dfrac{7}{7} - \dfrac{6}{7} \cdot \dfrac{9}{9} \\
&= -\dfrac{56}{63} - \dfrac{54}{63} \\
&= \dfrac{-110}{63}, \text{ or } -\dfrac{110}{63}
\end{aligned}
$$

**19.** $-\dfrac{7}{9} \cdot \left(-\dfrac{3}{14}\right) = \dfrac{7 \cdot 3}{9 \cdot 14} = \dfrac{7 \cdot 3 \cdot 1}{3 \cdot 3 \cdot 2 \cdot 7} = \dfrac{3 \cdot 7}{3 \cdot 7} \cdot \dfrac{1}{3 \cdot 2} = \dfrac{1}{3 \cdot 2} = \dfrac{1}{6}$

**20.**
$$
\begin{array}{r}
2\,3\,6,9\,8\,4 \\
\times\quad 3,6\,0\,0 \\
\hline
1\,4\,2\,1\,9\,0\,4\,0\,0 \\
7\,1\,0\,9\,5\,2\,0\,0\,0 \\
\hline
8\,5\,3,1\,4\,2,4\,0\,0
\end{array}
$$

**21.**
$$
\begin{array}{r}
4\,6.\,0\,1\,2 \\
\times\quad 0.\,0\,3 \\
\hline
1.\,3\,8\,0\,3\,6
\end{array}
$$

**22.** $-6\dfrac{3}{5} \div 4\dfrac{2}{5} = -\dfrac{33}{5} \div \dfrac{22}{5}$

$\qquad = -\dfrac{33}{5} \cdot \dfrac{5}{22}$

$\qquad = -\dfrac{33 \cdot 5}{5 \cdot 22}$

$\qquad = -\dfrac{3 \cdot 11 \cdot 5}{5 \cdot 2 \cdot 11}$

$\qquad = -\dfrac{3}{2} \cdot \dfrac{5 \cdot 11}{5 \cdot 11}$

$\qquad = -\dfrac{3}{2}, \text{ or } -1\dfrac{1}{2}$

**23.**
$$
\begin{array}{r}
\phantom{35.2\lceil}1\,2.2\,5 \\
35.2_{\wedge}\overline{\smash{)}\,4\,3\,1.2_{\wedge}0\,0} \\
\underline{3\,5\,2}\phantom{00000} \\
7\,9\,2\phantom{0000} \\
\underline{7\,0\,4}\phantom{0000} \\
8\,8\,0\phantom{000} \\
\underline{7\,0\,4}\phantom{000} \\
1\,7\,6\,0\phantom{0} \\
\underline{1\,7\,6\,0}\phantom{0} \\
0
\end{array}
$$

The answer is 12.25.

**24.**
$$
\begin{array}{r}
\phantom{15\lceil}1\,2\,3 \\
15\,\overline{\smash{)}\,1\,8\,5\,0} \\
\underline{1\,5}\phantom{000} \\
3\,5\phantom{00} \\
\underline{3\,0}\phantom{00} \\
5\,0\phantom{0} \\
\underline{4\,5}\phantom{0} \\
5
\end{array}
$$

The answer is 123 R 5.

**25.** $36 \cdot x = 3420$

$\qquad \dfrac{36 \cdot x}{36} = \dfrac{3420}{36}$

$\qquad\qquad x = 95$

The solution is 95.

**26.** $\qquad y + 142.87 = -151$

$y + 142.87 - 142.87 = -151 - 142.87$

$\qquad\qquad\qquad y = -293.87$

The solution is $-293.87$.

**27.** $-\dfrac{2}{15} \cdot t = \dfrac{6}{5}$

$\qquad t = \dfrac{6}{5} \div \left(-\dfrac{2}{15}\right)$

$\qquad t = \dfrac{6}{5} \cdot \left(-\dfrac{15}{2}\right)$

$\qquad = -\dfrac{6 \cdot 15}{5 \cdot 2} = -\dfrac{2 \cdot 3 \cdot 3 \cdot 5}{5 \cdot 2 \cdot 1}$

$\qquad = -\dfrac{3 \cdot 3}{1} \cdot \dfrac{2 \cdot 5}{2 \cdot 5} = -\dfrac{3 \cdot 3}{1}$

$\qquad = -9$

The solution is $-9$.

**28.** $\qquad \dfrac{3}{4} + x = \dfrac{5}{6}$

$\dfrac{3}{4} + x - \dfrac{3}{4} = \dfrac{5}{6} - \dfrac{3}{4}$

$\qquad\quad x = \dfrac{5}{6} \cdot \dfrac{2}{2} - \dfrac{3}{4} \cdot \dfrac{3}{3}$

$\qquad\quad x = \dfrac{10}{12} - \dfrac{9}{12}$

$\qquad\quad x = \dfrac{1}{12}$

The solution is $\dfrac{1}{12}$.

**29.** $\qquad \dfrac{y}{25} = \dfrac{24}{15}$

$\quad y \cdot 15 = 25 \cdot 24 \qquad$ Equating cross products

$\quad \dfrac{y \cdot 15}{15} = \dfrac{25 \cdot 24}{15}$

$\qquad\quad y = 40$

The solution is 40.

**30.** $\qquad \dfrac{16}{n} = \dfrac{21}{11}$

$\quad 16 \cdot 11 = n \cdot 21 \qquad$ Equating cross products

$\quad \dfrac{16 \cdot 11}{21} = \dfrac{n \cdot 21}{21}$

$\qquad \dfrac{176}{21} = n$

The solution is $\dfrac{176}{21}$.

**31.** *Familiarize.* Let $a$ = the total attendance.

*Translate.* We translate to an equation.

$$
\begin{array}{ccccc}
\text{What} & \text{is} & 30.31\% & \text{of} & 1,525,104? \\
\downarrow & \downarrow & \downarrow & \downarrow & \downarrow \\
a & = & 30.31\% & \times & 1,525,104
\end{array}
$$

*Solve.* We convert 30.31% to decimal notation and carry out the multiplication.

$$
\begin{array}{r}
1,5\,2\,5,1\,0\,4 \\
\times\quad 0.\,3\,0\,3\,1 \\
\hline
1\,5\,2\,5\,1\,0\,4 \\
4\,5\,7\,5\,3\,1\,2\,0 \\
4\,5\,7\,5\,3\,1\,2\,0\,0\,0 \\
\hline
4\,6\,2\,2\,5\,9.\,0\,2\,2\,4
\end{array}
$$

Thus, $a \approx 462,259$.

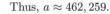

**Check**. We can repeat the calculation. We can also estimate $30.31\% \times 1,525,104 \approx 30\% \times 1,500,000 = 0.3 \times 1,500,000 = 450,000$. Since $450,000$ is close to $462,259$, the answer seems reasonable.

**State**. The total attendance at the museum was 462,259 visitors.

**32.** **Familiarize**. Let $n =$ the starting salary of a neurologist.

$$\underbrace{\text{Neurologist's salary}}_{\displaystyle n} \underbrace{\text{plus}}_{\displaystyle +} \underbrace{\text{How much more}}_{\displaystyle 172,500} \underbrace{\text{is}}_{\displaystyle =} \underbrace{\text{Radiologist's salary}}_{\displaystyle 350,000}$$

**Solve**. We subtract 172,500 on both sides of the equation.
$$n + 172,500 = 350,000$$
$$n + 172,500 - 172,500 = 350,000 - 172,500$$
$$n = 177,500$$

**Check**. Since $177,500 + 172,500 = 350,000$, the answer checks.

**State**. A neurologist's starting salary was $177,500 in 2007.

**33.** **Familiarize**. Let $g =$ the number of games that would be won in the entire season.

**Translate**. We translate to a proportion.
$$\begin{array}{l}\text{Wins} \rightarrow \\ \text{Games} \rightarrow \end{array} \frac{39}{49} = \frac{g}{82} \begin{array}{l}\leftarrow \text{Wins} \\ \leftarrow \text{Games}\end{array}$$

**Solve**.
$$39 \cdot 82 = 49 \cdot g \quad \text{Equating cross products}$$
$$\frac{39 \cdot 82}{49} = \frac{49 \cdot g}{49}$$
$$65 \approx g$$

**Check**. We substitute in the proportion and check cross products.
$$\frac{39}{49} = \frac{65}{82}; \ 39 \cdot 82 = 3198; \ 49 \cdot 65 = 3185$$
$3198 \approx 3185$, so the answer checks. (Remember that we rounded the value of $g$.)

**State**. At the given rate, the Cavaliers would win 65 games in the entire season.

**34.** **Familiarize**. Let $t =$ the price of each shirt.

**Translate**.
$$\underbrace{\text{Price per shirt}}_{\displaystyle t} \underbrace{\text{times}}_{\displaystyle \cdot} \underbrace{\text{Number of shirts}}_{\displaystyle 5} \underbrace{\text{is}}_{\displaystyle =} \underbrace{\text{Total price}}_{\displaystyle 424.75}$$

**Solve**.
$$t \cdot 5 = 424.75$$
$$\frac{t \cdot 5}{5} = \frac{424.75}{5}$$
$$t = 84.95$$

**Check**. Since $5 \cdot \$84.95 = \$424.75$, the answer checks.

**State**. Each shirt cost $84.95.

**35.** $\dfrac{\$14.99}{200 \text{ oz}} = \dfrac{1499\cancel{c}}{200 \text{ oz}} = 7.495\cancel{c}/\text{oz}$

**36.** **Familiarize**. Let $d =$ the total distance Patty walked, in miles.

**Translate**.
$$\underbrace{\text{Distance to school}}_{\displaystyle \frac{7}{10}} \underbrace{\text{plus}}_{\displaystyle +} \underbrace{\text{Distance to library}}_{\displaystyle \frac{8}{10}} \underbrace{\text{is}}_{\displaystyle =} \underbrace{\text{Total distance}}_{\displaystyle d}$$

**Solve**. We carry out the addition.
$$\frac{7}{10} + \frac{8}{10} = \frac{15}{10} = \frac{3 \cdot 5}{2 \cdot 5} = \frac{3}{2} \cdot \frac{5}{5} = \frac{3}{2} = 1\frac{1}{2}$$

**Check**. We repeat the calculation. The answer checks.

**State**. Patty walked $1\frac{1}{2}$ miles.

**37.** **Familiarize**. Let $d =$ the number of miles represented by $\frac{3}{4}$ in.

**Translate**. We translate to a proportion.
$$\begin{array}{l}\text{Map distance} \rightarrow \\ \text{Actual distance} \rightarrow \end{array} \frac{1}{80} = \frac{\frac{3}{4}}{d} \begin{array}{l}\leftarrow \text{Map distance} \\ \leftarrow \text{Actual distance}\end{array}$$

**Solve**.
$$1 \cdot d = 80 \cdot \frac{3}{4} \quad \text{Equating cross products}$$
$$d = \frac{80 \cdot 3}{4}$$
$$d = 60$$

**Check**. We substitute in the proportion and check cross products.
$$\frac{1}{80} = \frac{\frac{3}{4}}{60}; \ 1 \cdot 60 = 60; \ 80 \cdot \frac{3}{4} = 60$$
The cross products are the same, so the answer checks.

**State**. $\frac{3}{4}$ in. represents 60 mi.

**38.**
$$A = P \cdot \left(1 + \frac{r}{n}\right)^{n \cdot t}$$
$$= \$8500 \cdot \left(1 + \frac{8\%}{12}\right)^{12 \cdot 5}$$
$$= \$8500 \cdot \left(1 + \frac{0.08}{12}\right)^{60}$$
$$\approx \$12,663.69$$

**39.** **Familiarize**. Let $p =$ the number of pieces of ribbon that can be cut.

**Translate**.
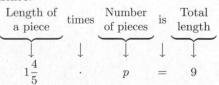
$$\underbrace{\text{Length of a piece}}_{\displaystyle 1\frac{4}{5}} \underbrace{\text{times}}_{\displaystyle \cdot} \underbrace{\text{Number of pieces}}_{\displaystyle p} \underbrace{\text{is}}_{\displaystyle =} \underbrace{\text{Total length}}_{\displaystyle 9}$$

*Solve.*

$$1\frac{4}{5} \cdot p = 9$$

$$\frac{9}{5} \cdot p = 9$$

$$p = 9 \div \frac{9}{5}$$

$$p = 9 \cdot \frac{5}{9} = \frac{9 \cdot 5}{9 \cdot 1}$$

$$p = \frac{9}{9} \cdot \frac{5}{1}$$

$$p = 5$$

**Check**. $1\frac{4}{5} \cdot 5 = \frac{9}{5} \cdot 5 = 9$, so the answer checks.

**State**. 5 pieces can be cut.

**40. Familiarize**. We will solve an equation to find the percent of increase. Then we will add to find the number of technicians in 2016.

**Translate**.

110,000  is  $\underbrace{\text{what percent}}$  of  773,000?

$\downarrow$   $\downarrow$   $\downarrow$   $\downarrow$   $\downarrow$

110,000  =  $p$  $\cdot$  773,000

**Solve**.

$$110,000 = p \cdot 773,000$$

$$\frac{110,000}{773,000} = \frac{p \cdot 773,000}{773,000}$$

$$\frac{110,000}{773,000} = p$$

$$0.142 \approx p$$

$$14.2\% \approx p$$

We add to find the number of technicians in 2016:

$$773,000 + 110,000 = 883,000.$$

**Check**. We can do the addition again. If the percent of increase is 14.2%, then the number of technicians in 2016 would be 114.2% of the number in 2006:
$114.2\% \times 773,000 = 1.142 \times 773,000 = 882,776 \approx 883,000.$
The answer checks.

**State**. The percent of increase is about 14.2%. There will be 883,000 auto repair technicians in 2016.

**41.** $-\dfrac{14}{25} + \dfrac{3}{20} = -\dfrac{14}{25} \cdot \dfrac{4}{4} + \dfrac{3}{20} \cdot \dfrac{5}{5} = -\dfrac{56}{100} + \dfrac{15}{100} = \dfrac{-41}{100}$, or

$-\dfrac{41}{100}$

Answer C is correct.

**42.** We find the amount of decrease.

$$\begin{array}{r} 4,4\,6\,8,9\,7\,6 \\ -\,4,2\,8\,7,7\,6\,8 \\ \hline 1\,8\,1,2\,0\,8 \end{array}$$

Now we find the percent of decrease, $p$.

181,208  is  $\underbrace{\text{what percent}}$  of  4,468,976?

$\downarrow$   $\downarrow$   $\downarrow$   $\downarrow$   $\downarrow$

181,208  =  $p$  $\cdot$  4,468,976

$$\frac{181,208}{4,468,976} = \frac{p \cdot 4,468,976}{4,468,976}$$

$$\frac{181,208}{4,468,976} = p$$

$$0.041 \approx p$$

$$4.1\% \approx p$$

Answer A is correct.

**43.** $\qquad a = 50\% \cdot b$

$$a = \frac{1}{2}b \qquad \left(50\% = \frac{1}{2}\right)$$

$$2 \cdot a = 2 \cdot \frac{1}{2}b$$

$$2 \cdot a = b$$

$$200\% \cdot a = b \qquad (2 = 200\%)$$

Thus, $b$ is 200% of $a$.

# Chapter 8

# Data, Graphs, and Statistics

## Exercise Set 8.1

**1.** To find the average, we first add the numbers. Then divide by the number of addends.

$$\frac{2 + 2 + 17 + 30 + 90 + 110 + 52 + 21 + 5}{9} = \frac{329}{9} = 36.\overline{5}$$

The average is $36.\overline{5}$.

To find the median, first list the numbers in order from smallest to largest. Then locate the middle number.

$$2, 2, 5, 17, 21, 30, 52, 90, 110$$
$$\uparrow$$
Middle number

The median is 21.

Find the mode:

The number that occurs most often is 2. The mode is 2.

**3.** To find the average, add the numbers. Then divide by the number of addends.

$$\frac{17 + 19 + 29 + 18 + 14 + 29}{6} = \frac{126}{6} = 21$$

The average is 21.

To find the median, first list the numbers in order from smallest to largest. Then locate the middle number.

$$14, 17, 18, 19, 29, 29$$
$$\uparrow$$
Middle number

The median is halfway between 18 and 19. It is the average of the two middle numbers:

$$\frac{18 + 19}{2} = \frac{37}{2} = 18.5$$

Find the mode:

The number that occurs most often is 29. The mode is 29.

**5.** To find the average, add the numbers. Then divide by the number of addends.

$$\frac{5 + 37 + 20 + 20 + 35 + 5 + 25}{7} = \frac{147}{7} = 21$$

The average is 21.

To find the median, first list the numbers in order from smallest to largest. Then locate the middle number.

$$5, 5, 20, 20, 25, 35, 37$$
$$\uparrow$$
Middle number

The median is 20.

Find the mode:

There are two numbers that occur most often, 5 and 20. Thus the modes are 5 and 20.

**7.** Find the average:

$$\frac{4.3 + 7.4 + 1.2 + 5.7 + 8.3}{5} = \frac{26.9}{5} = 5.38$$

The average is 5.38.

Find the median:

$$1.2, 4.3, 5.7, 7.4, 8.3$$
$$\uparrow$$
Middle number

The median is 5.7.

Find the mode:

All the numbers are equally represented. No mode exists.

**9.** Find the average:

$$\frac{234 + 228 + 234 + 229 + 234 + 278}{6} = \frac{1437}{6} = 239.5$$

The average is 239.5.

Find the median:

$$228, 229, 234, 234, 234, 278$$
$$\uparrow$$
Middle number

The median is halfway between 234 and 234. Although it seems clear that this is 234, we can compute it as follows:

$$\frac{234 + 234}{2} = \frac{468}{2} = 234$$

The median is 234.

Find the mode:

The number that occurs most often is 234. The mode is 234.

**11.** We divide the total number of miles, 253, by the number of gallons, 11.

$$\frac{253}{11} = 23$$

The average was 23 miles per gallon.

**13.** To find the GPA we first add the grade point values for each hour taken. This is done by first multiplying the grade point value by the number of hours in the course and then adding as follows:

B  $3.0 \cdot 4 = 12$
A  $4.0 \cdot 5 = 20$
D  $1.0 \cdot 3 = 3$
C  $2.0 \cdot 4 = 8$
$\overline{43}$ (Total)

The total number of hours taken is

4 + 5 + 3 + 4, or 16.

We divide 43 by 16 and round to the nearest tenth.

$$\frac{43}{16} = 2.6875 \approx 2.7$$

The student's grade point average is 2.7.

**15.** Find the average:

$$\frac{\$3.99 + \$4.49 + \$4.99 + \$3.99 + \$3.49}{5} = \frac{\$20.95}{5} = \$4.19$$

Find the median:

$$\$3.49, \$3.99, \$3.99, \$4.49, \$4.99$$
$$\uparrow$$
Middle number

The median is $3.99.

Find the mode:

The number that occurs most often is $3.99. The mode is $3.99.

**17.** We can find the total of the five scores needed as follows:

80 + 80 + 80 + 80 + 80 = 400.

The total of the scores on the first four tests is

80 + 74 + 81 + 75 = 310.

Thus Rich needs to get at least

400 − 310, or 90

to get a B. We can check this as follows:

$$\frac{80 + 74 + 81 + 75 + 90}{5} = \frac{400}{5} = 80.$$

**19.** We can find the total number of days needed as follows:

266 + 266 + 266 + 266 = 1064.

The total number of days for Marta's first three pregnancies is

270 + 259 + 272 = 801.

Thus, Marta's fourth pregnancy must last

1064 − 801 = 263 days

in order to equal the worldwide average.

We can check this as follows:

$$\frac{270 + 259 + 272 + 263}{4} = \frac{1064}{4} = 266.$$

**21.** Compare the averages of the two sets of data.

Bulb A: Average = (983 + 964 + 1214 + 1417 + 1211 + 1521 + 1084 + 1075 + 892 + 1423 + 949 + 1322)/12 = 1171.25

Bulb B: Average = (979 + 1083 + 1344 + 984 + 1445 + 975 + 1492 + 1325 + 1283 + 1325 + 1352 + 1432)/12 ≈ 1251.58

Since the average life of Bulb A is 1171.25 hr and of Bulb B is about 1251.58 hr, Bulb B is better.

**23.**
```
    1 2 . 8 6    (2 decimal places)
  ×   1 7 . 5    (1 decimal place)
    6 4 3 0
  9 0 0 2 0
1 2 8 6 0 0
2 2 5 . 0 5 0    (3 decimal places)
```

**25.**
$$\frac{4}{5} \cdot \frac{3}{28} = \frac{4 \cdot 3}{5 \cdot 28}$$
$$= \frac{4 \cdot 3}{5 \cdot 4 \cdot 7}$$
$$= \frac{4}{4} \cdot \frac{3}{5 \cdot 7}$$
$$= \frac{3}{35}$$

**27.** First we divide to find the decimal notation.
```
      1. 1 8 7 5
1 6 [ 1 9. 0 0 0 0
      1 6
        3 0
        1 6
        1 4 0
        1 2 8
          1 2 0
          1 1 2
            8 0
            8 0
             0
```

Then we move the decimal point two places to the right and write a percent symbol.

$$\frac{19}{16} = 1.1875 = 118.75\%$$

**29.** First we divide to find the decimal notation.
```
        0. 5 1 2
1 2 5 [ 6 4. 0 0 0
        6 2 5
        1 5 0
        1 2 5
          2 5 0
          2 5 0
             0
```

Then we move the decimal point two places to the right and write a percent symbol.

$$\frac{64}{125} = 0.512 = 51.2\%$$

**31.** Since $a$ is the middle number, it is the median, 30. Now find the average.

$$\frac{18 + 21 + 24 + 30 + 36 + 37 + b}{7} = 32$$
$$\frac{166 + b}{7} = 32$$
$$7\left(\frac{166 + b}{7}\right) = 7 \cdot 32$$
$$166 + b = 224$$
$$166 + b - 166 = 224 - 166$$
$$b = 58$$

Thus we have $a = 30$ and $b = 58$.

---

**33.** Amy's second offer: $\dfrac{\$3600 + \$3200}{2} = \$3400$

Jim's second offer: $\dfrac{\$3400 + \$3600}{2} = \$3500$

Amy's third offer: $\dfrac{\$3500 + \$3400}{2} = \$3450$

Jim's third offer: $\dfrac{\$3500 + \$3450}{2} = \$3475$

Amy will pay $3475 for the car.

## Exercise Set 8.2

**1.** Go down the Product column to Franklin Farms Portabella Fresh. Then go across to the column headed Calories and read the entry, 100. There are 100 calories in a Franklin Farms Portabella Fresh veggie burger.

**3.** Go down the column headed Calories. For each entry that is less than 110, go across to the Product column and read the entry. The veggie burgers with less than 110 calories are the Boca All American Flame Grilled Meatless, the Franklin Farms Portabella Fresh, and the Gardenburger Portabella. (Note that we also knew from Exercise 1 that the Franklin Farms Portabella Fresh veggie burger has less than 110 calories.)

**5.** Find the average fiber content:

$$\frac{3+3+4+3+5+3+3+4}{8} = \frac{28}{8} = 3.5$$

The average fiber content is 3.5 g.

Find the median:

$$3, 3, 3, 3, 3, 4, 4, 5$$
$$\uparrow$$
Middle number

The median is halfway between 3 and 3. It is the average of the two numbers.

$$\frac{3+3}{2} = \frac{6}{2} = 3$$

The median fiber content is 3 g.

Find the mode:

The number that occurs most often is 3. The mode is 3 g.

**7.** To find the most expensive veggie burger, find the largest number in the Cost column. It is 1.48. Go across to the Product column and read the entry. We see that the Lightlife Meatless Light veggie burger is the most expensive at $1.48 per patty.

To find the least expensive veggie burger, find the smallest number in the Cost column. It is 0.96. Go across to the Product column and read the entry. We see that the Boca All American Flame Grilled Meatless veggie burger is the least expensive at $0.96 per patty.

**9.** To find which veggie burger contains the most fat, find the largest number in the Fat column. It is 6.0. Go across to the Product column and read the entry. We see that the Veggie Patch Garlic Portabella veggie burger contains the most fat, 6.0 g per patty.

To find which veggie burger contains the least fat, find the smallest number in the Fat column. It is 1.5 and it occurs twice. Go across from each of these entries to the Product column and read the entries there. We see that the Franklin Farms Portabella Fresh and the Lightlife Meatless Light veggie burgers contain the least fat, 1.5 g per patty.

**11.** Go down the column headed Actual Temperature (°F) to 80°. Then go across to the Relative Humidity column headed 60%. The entry is 92, so the apparent temperature is 92°F.

**13.** Go down the column headed Actual Temperature (°F) to 85°. Then go across the Relative Humidity column headed 90%. The entry is 108, so the apparent temperature is 108°F.

**15.** The number 100 appears in the columns headed Apparent Temperature (°F) 3 times, so there are 3 temperature-humidity combinations that given an apparent temperature of 100°. They are 85° and 60% humidity, 90° and 40% humidity, and 100° and 10% humidity.

**17.** Go down the Relative Humidity column headed 50% and find all the entries greater than 100. The last 4 entries are greater than 100. Then go across to the column headed Actual Temperature (°F) and read the temperatures that correspond to these entries. At 50% humidity, the actual temperatures 90° and higher give an apparent temperature above 100°.

**19.** Go down the column headed Actual Temperature (°F) to 95°. Then read across to locate the entries greater than 100. All of the entries except the first two are greater than 100. Go up from each entry to find the corresponding relative humidity. At an actual temperature of 95°, relative humidities of 30% and higher give an apparent temperature above 100°.

**21.** Go down the column headed Actual Temperature (°F) to 85°, then across to 94, and up to find that the corresponding relative humidity is 40%. Similarly, go down to 85°, across to 108, and up to 90%. At an actual temperature of 85°, difference in humidities required to raise the apparent temperature from 94° to 108° is

$$90\% - 40\%, \text{ or } 50\%.$$

**23.** Go down the Planet column to Jupiter. Then go across to the column headed Average Distance from Sun (in miles) and read the entry, 483,612,200. The average distance from the sun to Jupiter is 483,612,200 miles.

**25.** Go down the column headed Time of Revolution in Earth Time (in years) to 164.78. Then go across the Planet column. The entry there is Neptune, so Neptune has a time of revolution of 164.78 days.

**27.** All of the entries in the column headed Average Distance from Sun (in miles) are greater than 1,000,000. Thus, all of the planets have an average distance from the sun that is greater than 1,000,000 mi.

**29.** Go down the Planet column to earth and then across to the Diameter (in miles) column to find that the diameter of Earth is 7926 mi. Similarly, find that the diameter of Jupiter is 88,846 mi. Then divide:

$$\frac{88,846}{7926} \approx 11$$

It would take about 11 Earth diameters to equal one Jupiter diameter.

**31.** Find the average of all the numbers in the column headed Diameter (in miles):

$(3031 + 7520 + 7926 + 4221 + 88,846 + 74,898 + 31,763 + 31,329)/8 = 31,191.75.$

The average of the diameters of the planets is 31,191.75 mi.

To find the median of the diameters of the planets we first list the diameters in order from smallest to largest:

3031, 4221, 7520, 7926, 31,329, 31,763, 74,898, 88,846.

The median is the average of the two middle numbers, 7926 and 31,329:

$$\frac{7926 + 31,329}{2} = 19,627.5.$$

The median of the diameters is 19,627.5 mi.

Since no number appears more than once in the Diameter (in miles) column, there is no mode.

**33.** The white rhino is represented by the greatest number of symbols, so this species has the greatest number of rhinos.

**35.** From the graph we see that there are about $12.5 \times 300$, or 3750, black rhinos and $8 \times 300$, or 2400, Indian rhinos. Thus there are about $3750 - 2400$, or 1350, more black rhinos than Indian rhinos.

**37.** From Exercise 35, we know that there are about 3750 black rhinos and 2400 Indian rhinos. Now we find the number of rhinos in the other three species.

White rhinos: $48.5 \times 300 = 14,550$

Javan rhino: $\frac{1}{6} \times 300 = 50$

Sumatran rhino: $\frac{1}{10} \times 300 = 30$

We find the average of the five numbers.

$$\frac{3750 + 2400 + 14,550 + 50 + 30}{5} = \frac{20,780}{5} = 4156$$

The average number of rhinos in these five species is about 4156 rhinos. (Answers may vary depending on how partial symbols are interpreted.)

**39.** Cabinets: 50% of $\$26,888 = 0.5(\$26,888) = \$13,444$

Countertops: 15% of $\$26,888 = 0.15(\$26,888) = \$4033.20$

Appliances: 8% of $\$26,888 = 0.08(\$26,888) = \$2151.04$

Fixtures: 3% of $\$26,888 = 0.03(\$26,888) = \$806.64$

## Chapter 8 Mid-Chapter Review

**1.** The given statement is true. See pages 465 and 469 in the text.

**2.** Consider the set of data 3, 3, 3. The average is

$$\frac{3 + 3 + 3}{3} = \frac{9}{3} = 3$$

The median is 3 and the mode is also 3. Thus, it is true that it is possible for the average, the median, and the mode of a set of data to be the same number.

**3.** The given statement is false. If there is an even number of items in a set of data, the median is the average of the two middle numbers.

**4.** $\dfrac{60 + 45 + 115 + 15 + 35}{5} = \dfrac{270}{5} = 54$

**5.** We first arrange the numbers from smallest to largest.

2.1, 4.8, 6.3, 8.7, 11.3, 14.5

The median is the average of the two middle numbers, 6.3 and 8.7.

$$\frac{6.3 + 8.7}{2} = \frac{15}{2} = 7.5.$$

The median is 7.5.

**6.** Find the average:

$$\frac{56 + 29 + 45 + 240 + 175 + 7 + 29}{7} = \frac{581}{7} = 83$$

To find the median, we first arrange the numbers from smallest to largest.

7, 29, 29, 45, 56, 175, 240

The median is the middle number, 45.

Find the mode:

The number that occurs most often is 29. The mode is 29.

**7.** Find the average:

$$\frac{2.12 + 18.42 + 9.37 + 43.89}{4} = \frac{73.8}{4} = 18.45$$

To find the median, we first arrange the numbers from smallest to largest.

2.12, 9.37, 18.42, 43.89

The median is the average of the two middle numbers, 9.37 and 18.42.

$$\frac{9.37 + 18.42}{2} = \frac{27.79}{2} = 13.895$$

Find the mode:

Each number occurs the same number of times, so no mode exists.

**8.** Find the average:

$$\frac{\frac{5}{10} + \frac{1}{10} + \frac{7}{10} + \frac{9}{10} + \frac{3}{10}}{5} = \frac{\frac{25}{10}}{5} = \frac{25}{10} \cdot \frac{1}{5} = \frac{25 \cdot 1}{10 \cdot 5} =$$

$$\frac{5 \cdot 5 \cdot 1}{2 \cdot 5 \cdot 5} = \frac{5 \cdot 5}{5 \cdot 5} \cdot \frac{1}{2} = \frac{1}{2}$$

To find the median, we first arrange the numbers from smallest to largest.

$$\frac{1}{10}, \frac{3}{10}, \frac{5}{10}, \frac{7}{10}, \frac{9}{10}$$

The median is the middle number, $\frac{5}{10}$. (We can simplify $\frac{5}{10}$, obtaining $\frac{1}{2}$.)

Find the mode:

Each number occurs the same number of times, so no mode exists.

**9.** Find the average:

$$\frac{160 + 102 + 102 + 116 + 160 + 116}{6} = \frac{756}{6} = 126$$

To find the median, we first arrange the numbers from smallest to largest.

$$102, 102, 116, 116, 160, 160$$

The median is the average of the two middle numbers, 116 and 116.

$$\frac{116 + 116}{2} = \frac{232}{2} - 116$$

Find the mode:

Each number occurs the same number of times, so no mode exists.

**10.** Find the average:

$$\frac{\$4.96 + \$5.24 + \$4.96 + \$10.05 + \$5.24}{5} = \frac{\$30.45}{5} = \$6.09$$

To find the median, we first arrange the numbers from smallest to largest.

$$\$4.96, \$4.96, \$5.24, \$5.24, \$10.05$$

The median is the middle number, $5.24.

Find the mode:

The numbers $4.96 and $5.24 each occur two times while $10.05 occurs only one time. Thus, the modes are $4.96 and $5.24.

**11.** Find the average:

$$\frac{\frac{1}{2} + \frac{3}{4} + \frac{7}{8} + \frac{5}{4}}{4} = \frac{\frac{4}{8} + \frac{6}{8} + \frac{7}{8} + \frac{10}{8}}{4} = \frac{\frac{27}{8}}{4} = \frac{27}{8} \cdot \frac{1}{4} = \frac{27}{32}$$

To find the median, we first arrange the numbers from smallest to largest. We will write the fractions with a common denominator as we did above.

$$\frac{4}{8}, \frac{6}{8}, \frac{7}{8}, \frac{10}{8}$$

The median is the average of the two middle numbers, $\frac{6}{8}$ and $\frac{7}{8}$.

$$\frac{\frac{6}{8} + \frac{7}{8}}{2} = \frac{\frac{13}{8}}{2} = \frac{13}{8} \cdot \frac{1}{2} = \frac{13}{16}$$

Find the mode:

Each number occurs the same number of times, so no mode exists.

**12.** Find the average:

$$\frac{2 + 5 + 7 + 7 + 8 + 5 + 5 + 7 + 8}{9} = \frac{54}{9} = 6$$

To find the median, we first arrange the numbers from smallest to largest.

$$2, 5, 5, 5, 7, 7, 7, 8, 8$$

The median is the middle number, 7.

Find the mode: There are two numbers that occur most often, 5 and 7. Thus the modes are 5 and 7.

**13.** Find the average:

$$\frac{38.2 + 38.2 + 38.2 + 38.2}{4} = \frac{152.8}{4} = 38.2$$

Find the median:

The numbers are arranged from smallest to largest. The median is the average of the two middle numbers, 38.2 and 38.2.

$$\frac{38.2 + 38.2}{2} = \frac{76.4}{2} = 38.2$$

Find the mode:

Each number occurs the same number of times, so no mode exists.

**14.** Find Breyer's ice cream in the Product column and go across to the Size column. We see that the old package contained 56 oz and the new package contains 48 oz. The difference is

$$56 \text{ oz} - 48 \text{ oz, or } 8 \text{ oz.}$$

**15.** Find Hellman's mayonnaise in the Product column and go across to the Percent Smaller column. We see that the new package is 6% smaller than the old package.

**16.** Find the largest number in the Percent Smaller column. It is 15. Go across to the Product column and read the entry. We see that Hershey's Special Dark chocolate bar has the greatest percent of decrease.

**17.** Find Tropicana orange juice in the Product column and go across to the Size column. We see that the old package contained 96 oz and the new package contains 89 oz. The difference is

$$96 \text{ oz} - 89 \text{ oz} = 7 \text{ oz.}$$

**18.** Find the smallest number in the Percent Smaller column. It is 5. Go across to the Product column and read the entry. We see that Nabisco Chips Ahoy cookies have the smallest percent of decrease.

**19.** The United States is represented by 18 symbols. Thus, the gun ownership rate is $18 \times 5$, or 90 guns per 100 citizens.

**20.** From Exercise 19 we know that the gun ownership rate in the United States is 90 guns per 100 citizens. Canada is represented by 6 symbols, so the gun ownership rate is $6 \times 5$, or 30 guns per 100 citizens. We find that the gun ownership in the United States is $90 - 30$, or 60 more guns per 100 citizens than in Canada.

**21.** Switzerland is represented by 9 symbols, so the gun ownership rate is $9 \times 5$, or 45 guns per 100 citizens.

Finland is represented by 11 symbols, so the gun ownership rate is $11 \times 5$, or 55 guns per 100 citizens.

**22.** The country with the smallest number of symbols is India. Of the countries represented, India has the lowest gun ownership rate.

**23.** From our work in the preceding exercises we know the following gun ownership rates:

Canada: 30 guns per 100 citizens

Finland: 55 guns per 100 citizens

Switzerland: 45 guns per 100 citizens

United States: 90 guns per 100 citizens

We find the rates for the other two countries.

India: $1 \times 5 = 5$ guns per 100 citizens

Yemen: $12 \times 5 = 60$ guns per 100 citizens

We find the average:
$$\frac{30 + 55 + 45 + 90 + 5 + 60}{6} = \frac{285}{6} = 47.5$$
The average gun ownership rate for the six countries is 47.5 guns per 100 citizens.

**24.** We list the rates from smallest to largest.

5, 30, 45, 55, 60, 90

The median is the average of the two middle numbers, 45 and 55.
$$\frac{45 + 55}{2} = \frac{100}{2} = 50 \text{ guns per 100 citizens}$$

**25.** Yes; if the trip took $1\frac{1}{2}$ hr then the average speed would be 20 mph. ($30 \text{ mi} \div 1\frac{1}{2}$ hr $= 30 \text{ mi} \div \frac{3}{2}$ hr $= 30 \text{ mi} \cdot \frac{2}{3}$ hr $= 20$ mph) But the driver could have driven at a speed of 75 mph for a brief period during that time.

**26.** Answers may vary. Some would ask for the average salary since it is a center point that places equal emphasis on all the salaries in the firm. Some would ask for the median salary since it is a center point that deemphasizes the extremely high and extremely low salaries. Some would ask for the mode of the salaries since it might indicate the salary you are most likely to earn.

## Exercise Set 8.3

**1.** Go across from 17 on the vertical scale to a bar whose shading begins at this level. Then go down to the horizontal scale and read the variety. We see that the miniature tall bearded iris has a minimum height of 17 in.

**3.** From the top of each blossom go across to the vertical scale and read the corresponding height. The maximum heights are 9 in., 26 in., 34 in., 27 in., 15 in., and 28 in. We find the average
$$\frac{9 + 26 + 34 + 27 + 15 + 28}{6} = \frac{139}{6} \approx 23.2$$
The average of the maximum heights is about 23.2 in.

**5.** Find the bar representing the border bearded iris. From the top and the bottom of the bar, go across to the vertical scale and read the corresponding heights. We see that this iris ranges in height from 16 in. to 26 in.

**7.** Find the shortest bar and blossom and read the corresponding variety from the horizontal scale. We see that the tall bearded iris has the smallest range in heights.

**9.** The tallest bar and blossom represent the tall bearded iris. The top of this bar corresponds to a height of 34 in. This is the height of the tallest iris with the largest maximum height.

The shortest bar and blossom represent the miniature dwarf bearded iris. The top of this bar corresponds to a height of 9 in.

We subtract to find the difference in heights:

35 in. $-$ 9 in. $=$ 25 in.

**25.** Yes; if the trip took $1\frac{1}{2}$ hr then the average speed would be 20 mph. ($30 \text{ mi} \div 1\frac{1}{2}$ hr $= 30 \text{ mi} \div \frac{3}{2}$ hr $= 30 \text{ mi} \cdot \frac{2}{3}$ hr $= 20$ mph) But the driver could have driven at a speed of 75 mph for a brief period during that time.

**11.** Move to the right along the bar representing 1 cup of hot cocoa with skim milk. We read that there are about 185 calories in the cup of cocoa.

**13.** The longest bar is for 1 slice of·chocolate cake with fudge frosting. Thus, it has the highest caloric content.

**15.** We locate 460 calories at the bottom of the graph and then go up until we reach a bar that ends at approximately 460 calories. Now go across to the left and read the dessert, 1 cup of premium chocolate ice cream.

**17.** From the graph we see that 1 cup of hot cocoa made with whole milk has about 310 calories and 1 cup of hot cocoa made with skim milk has about 190 calories. We subtract to find the difference:

$310 - 190 = 120$

The cocoa made with whole milk has about 120 more calories than the cocoa made with skim milk.

**19.** Find the instances for which the bar representing men's degrees is taller than the corresponding bar representing women's degrees and then read the years they represent on the horizontal scale. We see that more bachelor's degrees were conferred on men than on women in 1950 and 1970.

**21.** Find 2000 on the horizontal scale, go across to the vertical scale from the top of each bar, and read the corresponding numbers. We see that about 530,000 men and 705,000 women received bachelor's degrees in 2000. The difference is

$705,000 - 530,000$, or $175,000$.

Thus, about 175,000 more bachelor's degrees were conferred on women than on men in 2000.

**23.** First write the seven items in the City column of the table in equally-spaced intervals on the vertical scale and title the scale "City." Next, scale the horizontal axis. We see that the average costs range from $32 to $107. We will start the horizontal scale at 30, indicating the gap from 0 to 30 with a jagged line, and extend it by 10's through 110. Label this scale "Average daily cost of adult day care." Finally, draw horizontal bars to show the average costs.

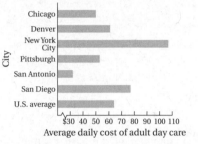

**25.** From the table we see that the average daily cost of adult day car is less than $75 in Chicago, Denver, Pittsburgh, and San Antonio. We could also read this from the bar graph by finding the cities associated with the bars whose lengths do not extend to 75 on the horizontal scale.

**27.** From the table we see that the average daily cost of adult day care in New York City is $107 and the national average cost is $64. The difference is $107 − $64, or $43. Thus, the average daily cost of adult day care in New York City is $43 higher than the national average cost.

**29.** From the table we see that New York has the greatest commuting time. We could also find this by observing that the longest bar in the graph drawn in Exercise 28 corresponds to New York.

**31.** We list the times from smallest to largest

21.6, 24.7, 25.9, 28.1, 33.1, 39.0

The median is the average of the two middle terms, 25.9 and 28.1.

$$\frac{25.9 + 28.1}{2} = \frac{54}{2} = 27$$

The median commuting time is 27 min.

**33.** Find 2004 on the horizontal scale, go up to the graph, and then go left to the vertical scale and read the value. We estimate that about 3.2 million international tourists visited Beijing in 2004.

Find 2007 on the horizontal scale, go up to the graph, and then go left to the vertical scale and read the value. We estimate that about 4.3 million international tourists visited Beijing in 2007.

**35.** We look for segments of the graph that slant down from left to right. The only such segment goes from the point associated with 2007 to the point associated with 2008. Thus, the number of international tourists to Beijing decreased between 2007 and 2008.

**37.** The highest point on the graph corresponds to the interval labeled 9-11 A.M. on the horizontal scale. This is the time interval in which most bank crimes occurred.

**39.** We see that 1900 bank crimes occurred between 9 A.M. and 11 A.M. while 1600 bank crimes occurred between 3 P.M. and 6 P.M. Thus, 1900 − 1600, or 300, more bank crimes occurred between 9 A.M. and 11 A.M. than between 3 P.M. and 6 P.M.

**41.** First indicate the years on the horizontal scale and label it "Year." The years range from 1980 to 2030 and increase by 10's. We could start the vertical scale at 0, but the graph will be more compact if we start at a higher number. The years lived beyond age 65 range from 14 to 17.5 so we choose to label the vertical scale from 13 to 18. We use a jagged line to indicate that we are not starting at 0. Label the vertical scale "Average number of years men are estimated to live beyond 65." Next, at the appropriate level above each year on the horizontal scale, mark the corresponding number of years. Finally, draw line segments connecting the points.

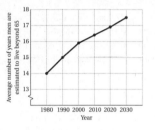

**43.** First we subtract to find the amount of the increase:

17.5 − 14 = 3.5

Let $p$ = the percent of increase. We write and solve an equation to find $p$.

$$3.5 = p \cdot 14$$
$$\frac{3.5}{14} = p$$
$$0.25 = p$$
$$25\% = p$$

Longevity is estimated to increase 25% between 1980 and 2030.

**45.** First we subtract to find the amount of increase:

17.5 − 15.9 = 1.6

Let $p$ = the percent of increase. We write and solve an equation to find $p$.

$$1.6 = p \cdot 15.9$$
$$\frac{1.6}{15.9} = p$$
$$0.101 \approx p$$
$$10.1\% \approx p$$

Longevity is estimated to increase about 10.1% between 2000 and 2030.

**47.** To find the average of a set of numbers, <u>add</u> the numbers and then <u>divide</u> by the number of items of data.

**49.** If an item has a regular price of $100 and is on sale for 25% off, $100 is called the <u>marked price</u>, 25% the <u>rate of discount</u>, 25% of $100, or $25, the <u>discount</u> and $100 − $25, or $75, the <u>sale price</u>.

**51.** The decimal $0.\overline{1518}$ is an example of a <u>repeating</u> decimal.

**53.** The statement $x + t = t + x$ illustrates the <u>commutative</u> law.

## Exercise Set 8.4

**1.** We see from the graph that 11% of foreign students were from South Korea.

**3.** We see from the graph that 15% of foreign students were from India. Find 15% of 625,000:

$$0.15 \times 625,000 = 93,750 \text{ students}$$

**5.** The section of the graph labeled 6% represents the foreign students from Japan.

**7.** From the graph we see that 18% of those who participate in outdoor activities look forward most to forgetting about work.

**9.** We see from the graph that 38% of the participants in the survey listed exercise as their main reason for looking forward to outdoor recreation. We find 38% of the 1027 participants:

$$0.38 \times 1027 \approx 390 \text{ people}$$

**11.** We first draw a line from the center of the circle to any tick mark. From that tick mark we count off 57 tick marks to graph 57% and label the wedge "College." We continue in this manner with the remaining items in the table. The graph is shown below.

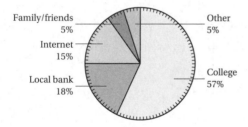

**13.** We first draw a line from the center of the circle to any tick mark. From that tick mark we count off 5 tick marks to graph 5% and label this wedge "12-17." We continue in this manner with the remaining items in the table. The graph is shown below.

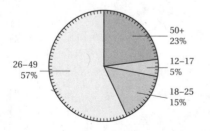

## Chapter 8 Concept Reinforcement

**1.** The statement is false. To find the average of a set of numbers, add the numbers and then *divide* by the number of items of data.

**2.** The statement is true. See page 469 in the text.

**3.** The statement is true. See page 469 in the text.

## Chapter 8 Important Concepts

**1.** Find the average:

$$\frac{8 + 13 + 1 + 4 + 8 + 7 + 15}{7} = \frac{56}{7} = 8$$

To find the median we first arrange the numbers from smallest to largest.

$$1, 4, 7, 8, 8, 13, 15$$

The median is the middle number, 8.

Find the mode:

The number 8 occurs most often. It is the mode.

**2.** Find the largest number in the Cost column, 10.54, and then read the corresponding entry in the Product column. We see that Quaker Organic Maple & Brown Sugar oatmeal has the highest cost per packet or serving.

**3.** Find Kashi oatmeal in the Product column and then read the corresponding number in the Sugar column, 12. We see that Kashi oatmeal has 12 g of sugar per packet or serving.

**4.** Read the names of the stadiums corresponding to the bars that do not extend to $100 million. We see that the building costs of Arrowhead Stadium and Candlestick Park were less than $100 million.

**5.** The bar representing Invesco Field extends to about $360 million while the bar representing Bank of America Stadium extends to about $250 million. Thus, Invesco Field cost about

$$\$360 \text{ million} - \$250 \text{ million, or } \$110 \text{ million,}$$

more to build than Bank of America Stadium.

**6.** The smallest wedge represents the 80 and older age group. This is the group with the fewest people.

**7.** We see from the graph that 27% of the population is under 20 years old.

## Chapter 8 Review Exercises

**1.** $\dfrac{26 + 34 + 43 + 51}{4} = \dfrac{154}{4} = 38.5$

**2.** $\dfrac{11 + 14 + 17 + 18 + 7}{5} = \dfrac{67}{5} = 13.4$

**3.** $\dfrac{0.2 + 1.7 + 1.9 + 2.4}{4} = \dfrac{6.2}{4} = 1.55$

**4.** $\dfrac{700 + 2700 + 3000 + 900 + 1900}{5} = \dfrac{9200}{5} = 1840$

**5.** $\dfrac{\$2 + \$14 + \$17 + \$17 + \$21 + \$29}{6} = \dfrac{\$100}{6} = \$16.\overline{6}$

**6.** $\dfrac{20 + 190 + 280 + 470 + 470 + 500}{6} = \dfrac{1930}{6} = 321.\overline{6}$

**7.** We can find the total of the four scores needed as follows:

$$90 + 90 + 90 + 90 = 360.$$

The total of the scores on the first three tests is

$$94 + 78 + 92 = 264.$$

Thus the student needs to get at least

$$360 - 264 = 96$$

to get an A. We can check this as follows:

$$\dfrac{90 + 78 + 92 + 96}{4} = \dfrac{360}{4} = 90.$$

**8.** We divide the number of miles, 532, by the number of gallons, 19.

$$\dfrac{532}{19} = 28$$

The average was 28 miles per gallon.

**9.** To find the GPA we first add the grade point values for each hour taken. This is done by first multiplying the grade point value by the number of hours in the course and then adding as follows:

$$
\begin{array}{lll}
\text{A} & 4.0 \cdot 5 = & 20 \\
\text{B} & 3.0 \cdot 3 = & 9 \\
\text{C} & 2.0 \cdot 4 = & 8 \\
\text{B} & 3.0 \cdot 3 = & 9 \\
\text{B} & 3.0 \cdot 1 = & 3 \\
\hline
& & 49 \ \text{(Total)}
\end{array}
$$

The total number of hours taken is

$$5 + 3 + 4 + 3 + 1, \text{ or } 16.$$

We divide 49 by 16 and round to the nearest tenth.

$$\dfrac{49}{16} = 3.0625 \approx 3.1$$

The student's grade point average is 3.1.

**10.** $26, 34, 43, 51$

$\uparrow$

Middle number

The median is halfway between 34 and 43. It is the average of the two middle numbers.

$$\dfrac{34 + 43}{2} = \dfrac{77}{2} = 38.5$$

The median is 38.5.

**11.** $7, 11, 14, 17, 18$

$\uparrow$

Middle number

The median is 14.

**12.** $0.2, 1.7, 1.9, 2.4$

$\uparrow$

Middle number

The median is halfway between 1.7 and 1.9. It is the average of the two middle numbers.

$$\dfrac{1.7 + 1.9}{2} = \dfrac{3.6}{2} = 1.8$$

The median is 1.8.

**13.** $700, 900, 1900, 2700, 3000$

$\uparrow$

Middle number

The median is 1900.

**14.** We arrange the numbers from smallest to largest.

$\$2, \$14, \$17, \$17, \$21, \$29$

$\uparrow$

Middle number

The median is halfway between $17 and $17. Although it seems clear that this is $17, we can compute it as follows:

$$\dfrac{\$17 + \$17}{2} = \dfrac{\$34}{2} = \$17$$

The median is $17.

**15.** We arrange the numbers from smallest to largest.

$20, 190, 280, 470, 470, 500$

$\uparrow$

Middle number

The median is halfway between 280 and 470. It is the average of the two middle numbers.

$$\dfrac{280 + 470}{2} = \dfrac{750}{2} = 375$$

The median is 375.

**16.** Find the average:

$$\dfrac{\$360 + \$192 + \$240 + \$216 + \$420 + \$132}{6} = \dfrac{\$1560}{6} = \$260$$

To find the median we first arrange the numbers from smallest to largest.

$\$132, \$192, \$216, \$240, \$360, \$420$

The median is the average of the two middle numbers, $216 and $240.

$$\dfrac{\$216 + \$240}{2} = \dfrac{\$456}{2} = \$228$$

**17.** The number that occurs most often is 26, so 26 is the mode.

**18.** The numbers that occur most often are 11 and 17. They are the modes.

**19.** The number that occurs most often is 0.2, so 0.2 is the mode.

**20.** The numbers that occur most often are 700 and 800. They are the modes.

**21.** The number that occurs most often is $17, so $17 is the mode.

**22.** The number that occurs most often is 20, so 20 is the mode.

**23.** Battery A:

$(38.9 + 39.3 + 40.4 + 53.1 + 41.7 + 38.0 + 36.8 + 47.7 +$

$48.1 + 38.2 + 46.9 + 47.4) \div 12 = \dfrac{516.5}{12} \approx 43.04$

Battery B:

$(39.3 + 38.6 + 38.8 + 37.4 + 47.6 + 37.9 + 46.9 + 37.8 +$

$38.1 + 47.9 + 50.1 + 38.2) \div 12 = \dfrac{498.6}{12} \approx 41.55$

Because the average time for Battery A is longer, it is the better battery.

**24.** Go down the UPS Package column to 5 and then go across to the Next Day Air Delivery column and read the cost, $30.37.

**25.** Go down the UPS Package column to 8 and then go across to the UPS Ground column and read the cost, $10.85.

**26.** First go down the UPS Package column to 5 and then go across to the Next Day Air Saver Delivery column and read the cost, $26.61. The amount saved is the difference in the costs, $30.37 − $26.61, or $3.76.

**27.** The 25-29 years age group is represented by 18 symbols, so there are 18 PGA champions in this age group.

**28.** The 30-34 years age group is represented by the most symbols, so this is the age group with the most PGA champions.

**29.** We see from the graph that the 30-34 years age group has 39 PGA champions while the 35-39 years age group has 20 champions. Thus, the 30-34 years age group has 39-20, or 19, more PGA champions than the 35-39 years age group.

**30.** We find the years associated with the bars that do not extend to 100. They are 2004, 2007, and 2008.

**31.** From the top of the bar for 2008, go to the vertical scale and read that the runoff was 60%.

**32.** Reading from the graph as described in Exercise 31, we see that the runoff in 2005 was about 105% and it was about 53% in 2007. Thus the difference in the runoffs was 105% − 53%, or 52%.

**33.** The only bar that extends beyond 170 on the vertical scale is associated with 2006. This is the year in which the runoff was over 170%.

**34.** The lowest point on the graph corresponds to 2001. This is the year in which Tiger Woods had his lowest score.

**35.** The highest point on the graph corresponds to 2007. This is the year in which Tiger Woods got his highest score.

**36.** The points on the graph that lie below 280 correspond to 2001, 2002, and 2005. These are the years in which Tiger Woods scored less than 280.

**37.** Tiger Woods' total score was 276 in the years that correspond to 276 on the graph. They are 2002 and 2005.

**38.** From the graph we see that the highest and lowest scores were 291 and 272, respectively. Thus, the highest score was 291 − 272, or 19, higher than the lowest score.

**39.** We read the scores from the graph and arrange them from smallest to largest.

$$272, \ 276, \ 276, \ 283, \ 284, \ 284, \ 290, \ 290, \ 291$$

The median is the middle score, 284.

**40.** We find the average of the scores in Exercise 39.
$$\frac{272 + 276 + 276 + 283 + 284 + 284 + 290 + 290 + 291}{9} =$$
$$\frac{2476}{9} \approx 283$$

**41.** From the graph we see that the Army is 36% of the Armed Forces.

**42.** From the graph we see that the Marines represent 13% of the Armed Forces.

**43.** From the graph we see that 25% of the Armed Forces were in the Air Force. We find 25% of 1,400,000:
$$0.25 \times 1,400,000 = 350,000$$

**44.** The Army and Navy represent 36% and 26% of the Armed Forces, respectively. Thus, the Army represents 36% − 26%, or 10% more of the Armed Forces than the Navy does.

**45.** 6, 9, 6, 8, 8, 5, 10, 5, 9, 10

Each number occurs the same number of times, so no mode exists. Answer D is correct.

**46.** $\dfrac{\frac{1}{2} + \frac{1}{3} + \frac{1}{4} + \frac{1}{5}}{4} = \dfrac{\frac{30}{60} + \frac{20}{60} + \frac{15}{60} + \frac{12}{60}}{4} = \dfrac{\frac{77}{60}}{4} = \dfrac{77}{60} \cdot \dfrac{1}{4} = \dfrac{77}{240}$

Answer A is correct.

**47.** On the horizontal scale in eight equally spaced intervals indicate the years. Label this scale "Year." Then label the vertical scale "Cost of first-class postage (in cents)." The smallest cost is 32¢ and the largest is 44¢, so we start the vertical scale at 0 and extend it to 50¢, labeling it by 10's. Finally, draw vertical bars above the years to show the cost of the postage.

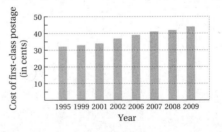

**48.** Prepare horizontal and vertical scales as described in Exercise 47. Then, at the appropriate level above each year, mark the corresponding postage. Finally, draw line segments connecting the points.

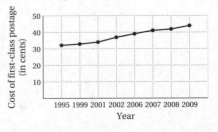

**49.** Using a circle with 100 equally-spaced tick marks, first draw a line from the center to any tick mark. Then count off 8 tick marks to graph 8% and label the wedge "20-24 years." Continue in this manner with the remaining age groups. The graph is shown below.

**PGA Champions by Age**

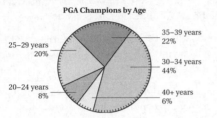

25–29 years 20%
35–39 years 22%
30–34 years 44%
20–24 years 8%
40+ years 6%

**50.** $a$ is the middle number and the median is 316, so $a = 316$.

The average is 326 so the data must add to
$326 + 326 + 326 + 326 + 326 + 326 + 326$, or 2282.

The sum of the known data items, including $a$, is
$298 + 301 + 305 + 316 + 323 + 390$, or 1933.

We subtract to find $b$:

$$b = 2282 - 1933 = 349$$

## Chapter 8 Discussion and Writing Exercises

**1.** The equation could represent a person's average income during a 4-yr period. Answers may vary.

**2.** Bar graphs that show change over time, such as the one in Exercise 47 in the Review Exercises, can be successfully converted to line graphs. Other bar graphs, such as the one in Example 1 in Section 8.3, cannot be successfully converted to line graphs.

**3.** We can use circle graphs to visualize how the number of items in various categories compare in size.

**4.** A bar graph is convenient for showing comparisons. A line graph is convenient for showing a change over time as well as to indicate patterns or trends. The choice of which to use to graph a particular set of data would probably depend on the type of data analysis desired.

**5.** The average, the median, and the mode are "center points" that characterize a set of data. You might use the average to find a center point that is midway between the extreme values of the data. The median is a center point that is in the middle of all the data. That is, there are as many values less than the median as there are values greater than the median. The mode is a center point that represents the value or values that occur most frequently.

**6.** Circle graphs are similar to bar graphs in that both allow us to tell at a glance how items in various categories compare in size. They differ in that circle graphs show percents whereas bar graphs show actual numbers of items in a given category.

## Chapter 8 Test

**1.** We add the numbers and then divide by the number of items of data.
$$\frac{45 + 49 + 52 + 52}{4} = \frac{198}{4} = 49.5$$

**2.** We add the numbers and then divide by the number of items of data.
$$\frac{1 + 1 + 3 + 5 + 3}{5} = \frac{13}{5} = 2.6$$

**3.** We add the numbers and then divide by the number of items of data.
$$\frac{3 + 17 + 17 + 18 + 18 + 20}{6} = \frac{93}{6} = 15.5$$

**4.** 45, 49, 52, 53

Find the median: There is an even number of numbers. The median is the average of the two middle numbers:
$$\frac{49 + 52}{2} = \frac{101}{2} = 50.5$$
Find the mode: Each number occurs only once, so there is no mode.

**5.** Find the median: First we rearrange the numbers from the smallest to largest.

$$1, 1, 3, 3, 5$$
$$\uparrow$$
Middle number

The median is 3.

Find the mode: There are two numbers that occur most often, 1 and 3. They are the modes.

**6.** 3, 17, 17, 18, 18, 20

Find the median: There is an even number of numbers. The median is the average of the two middle numbers:
$$\frac{17 + 18}{2} = \frac{35}{2} = 17.5$$
Find the mode: There are two numbers that occur most often, 17 and 18. They are the modes.

**7.** We divide the number of miles by the number of gallons.
$$\frac{462}{14} = 33 \text{ mpg}$$

**8.** The total of the four scores needed is
$$70 + 70 + 70 + 70 = 4 \cdot 70, \text{ or } 280.$$
The total of the scores on the first three tests is
$$68 + 71 + 65 = 204.$$
Thus the student needs to get at least
$$280 - 204, \text{ or } 76$$
on the fourth test.

9. To find the GPA we first add the grade point values for each class taken. This is done by first multiplying the grade point value by the number of hours in the course and then adding as follows:

$$\begin{array}{lll} B & 3.0 \cdot 3 = & 9 \\ A & 4.0 \cdot 3 = & 12 \\ C & 2.0 \cdot 4 = & 8 \\ B & 3.0 \cdot 3 = & 9 \\ B & 3.0 \cdot 2 = & 6 \\ \hline & & 44 \ (\text{Total}) \end{array}$$

The total number of hours taken is

$3 + 3 + 4 + 3 + 2$, or 15.

We divide 44 by 15 and round to the nearest tenth.

$$\frac{44}{15} = 2.9\overline{3} \approx 2.9$$

The grade point average is 2.9.

10. We find the average of each set of ratings.

Pecan:

$$\frac{9 + 10 + 8 + 10 + 9 + 7 + 6 + 9 + 10 + 7 + 8 + 8}{12} =$$

$$\frac{101}{12} \approx 8.417$$

Hazelnut:

$$\frac{10 + 6 + 8 + 9 + 10 + 10 + 8 + 7 + 6 + 9 + 10 + 8}{12} =$$

$$\frac{101}{12} \approx 8.417$$

Since the averages are equal, the chocolate bars are of equal quality.

11. Go down the column in the first table labeled "Height" to the entry "6 ft 1 in." Then go to the right and read the entry in the column headed "Medium Frame." We see that the desirable weight is 179 lb.

12. Locate the number 120 in the second table and observe that it is in the column headed "Medium Frame." Then go to the left and observe that the corresponding entry in the "Height" column is 5 ft 3 in. Thus a 5 ft 3 in. woman with a medium frame has a desirable weight of 120 lb.

13. From Exercise 12, we know that the desirable weight for a 5 ft 3 in. woman with a medium frame is 120 lb. To find the desirable weight for a 5 ft 3 in. woman with a small frame, first locate 5 ft 3 in. in the "Height" column. Then go to the right and read the entry in the "Small Frame" column. We see that the desirable weight is 111 lb. We subtract to find the difference: 120 lb − 111 lb = 9 lb. Thus, the desirable weight for a 5 ft 3 in. woman with a medium frame is 9 lb more than for a woman of the same height with a small frame.

14. To find the desirable weight for a 6 ft 3 in. man with a large frame, first locate 6 ft 3 in. in the "Height" column. Then go to the right and read the entry in the "Large Frame" column, 204 lb.

   To find the desirable weight for a 6 ft 3 in. man with a small frame, locate 6 ft 3 in. in the "Height" column and then go to the right and read the entry in the "Small Frame"

column, 172 lb. We subtract to find the difference: 204 lb− 172 lb = 32 lb. Thus, the desirable weight for a 6 ft 3 in. man with a large frame is 32 lb more than for a man of the same height with a small frame.

15. Since $1300 \div 100 = 13$, we look for a country represented by 13 symbols. We find that it is Spain.

16. Since $1500 \div 100 = 15$, we look for countries represented by more than 15 symbols. Those countries are Norway and the United States.

17. Canada is represented by 9 symbols, so $9 \cdot 100$, or 900 lb, of waste is generated per person per year in Canada.

18. The United States is represented by 17 symbols, so $17 \cdot 100$, or 1700 lb, of waste is generated per person per year in the United States. Mexico is represented by 7 symbols, so $7 \cdot 100$, or 700 lb, of waste is generated per person per year in Mexico. We subtract to find the difference: 1700 lb − 700 lb = 1000 lb. Thus, in the United States, 1000 lb more of waste is generated per person per year than in Mexico.

19. First indicate the names of the animals in eight equally spaced intervals on the horizontal scale. Title this scale "Animal." Now note that the lowest speed is 9 mph and the highest is 70 mph. We start the vertical scaling at 0 and label the marks on the scale by 10's from 0 to 80. Title this scale "Maximum speed (in miles per hour)." Finally, draw vertical bars above the names of the animals to show the speeds.

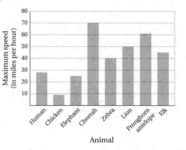

20. From the table or the bar graph, we see that the slowest speed is 9 mph and the fastest is 70 mph. Then the fastest speed exceeds the slowest by

   $70 − 9$, or 61 mph.

21. The fastest human's maximum speed is 28 mph and a zebra's maximum speed is 40 mph. Thus a human cannot outrun a zebra because a zebra can run $40 − 28$, or 12 mph, faster than a human.

22. We add the speeds and then divide by the number of speeds.

$$\frac{28 + 9 + 25 + 70 + 40 + 50 + 61 + 45}{8} = \frac{328}{8} = 41 \text{ mph}$$

23. First we write the numbers from smallest to largest.

   9, 25, 28, 40, 45, 50, 61, 70

   There is an even number of numbers. The median is the average of the two middle numbers.

$$\frac{40 + 45}{2} = \frac{85}{2} = 42.5 \text{ mph}$$

**24.** Find 2010 on the bottom scale and move up from there to the line. The line is labeled 53% at that point, so 53% of food dollars will be spent away from home in 2010.

**25.** Find 1985 halfway between 1980 and 1990 on the bottom scale and move up from that point to the line. Then go straight across to the left and find that about 41% of food dollars were spent away from home in 1985.

**26.** Locate 30% on the vertical scale. Then move to the right to the line. Look down to the bottom scale and observe that the year 1967 corresponds to this point.

**27.** Locate 50% on the vertical scale. Then move to the right to the line. Look down to the bottom scale and observe that the year 2006 corresponds to this point.

**28.** We will make a vertical bar graph. First indicate the days in seven equally-spaced intervals on the horizontal scale and title this scale "Day." Now note that the number of books ranges from 160 to 420. We start the vertical scaling with 150, using a jagged line to indicate that we are not showing the range of numbers from 0 to 150, and label the marks by 50's up to 450. Title this scale "Number of Books." Finally, draw vertical bars to show the number of books checked out each day.

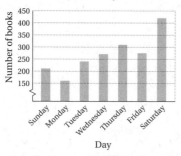

**29.** First indicate the days on the horizontal scale and title this scale "Day." We scale the vertical axis by 50's from 150 to 450, using a jagged line to indicate that we are not showing the range of numbers from 0 to 150. Next mark the number of books checked out above the days. Then draw line segments connecting adjacent points.

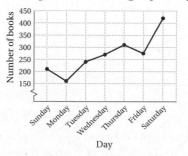

**30.** Using a circle with 100 equally spaced tick marks, we first draw a line from the center to any tick mark. From that tick mark, count off 23 tick marks and draw another line to graph 23%. Label this wedge "Meat, poultry, fish, and eggs" and "23%." Continue in this manner with the other types of food categories.

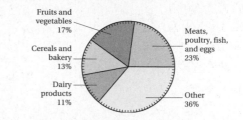

**31.** 13% of $664 = 0.13($664) = $86.32

Answer D is correct.

**32.** $a$ is the middle number in the ordered set of data, so $a$ is the median, 74.

Since the mean, or average, is 82, the total of the seven numbers is $7 \cdot 82$, or 574.

The total of the known numbers is

$$69 + 71 + 73 + 74 + 78 + 98, \text{ or } 463.$$

Then $b = 574 - 463 = 111$.

---

## Cumulative Review Chapters 1 - 8

**1.** 62.1 billion $= 62.1 \times 1$ billion
$$= 62.1 \times 1,000,000,000$$
$$= 62,100,000,000$$

**2.** Divide the number of miles, 312, by the number of gallons, 13.
$$\frac{312}{13} = 24$$
The mileage is 24 miles per gallon.

**3.** 402,513

The digit 5 means 5 hundreds.

**4.** $3 + 5^3 = 3 + 125 = 128$

**5.** Find as many two-factor factorizations as possible.

$60 = 1 \cdot 60 \qquad 60 = 4 \cdot 15$
$60 = 2 \cdot 30 \qquad 60 = 5 \cdot 12$
$60 = 3 \cdot 20 \qquad 60 = 6 \cdot 10$

The factors of 60 are 1, 2, 3, 4, 5, 6, 10, 12, 15, 20, 30, and 60.

**6.**

52.0 $\boxed{4}$5    Hundredths digit is 4 or lower.
↓                    Round down.
52.0

**7.** $3\frac{3}{10} = \frac{33}{10}$    $(3 \cdot 10 = 30 \text{ and } 30 + 3 = 33)$

**8.** Move 2 places to the left.

$2.10.¢

Change from ¢ sign at end to $ sign in front.

$210¢ = $2.10$

**9.** $\frac{7}{20} = \frac{7}{20} \cdot \frac{5}{5} = \frac{35}{100} = 35\%$

**10.** We can use cross products:

$$11 \cdot 12 = 132 \quad \underset{30 \quad 12}{\overset{11 \quad 4}{\bowtie}} \quad 30 \cdot 4 = 120$$

Since the cross products are not the same, $132 \neq 120$, we know that the numbers are not proportional.

**11.**
$$2\,\boxed{\frac{2}{5} \cdot \frac{2}{2}} = 2\,\frac{4}{10}$$
$$+\,4\,\frac{3}{10} = +\,4\,\frac{3}{10}$$
$$\rule{3cm}{0.4pt} \quad \rule{1.5cm}{0.4pt}$$
$$6\,\frac{7}{10}$$

**12.** $41.063 + (-43.5721)$

First we find the difference of the absolute values.

$$
\begin{array}{r}
\overset{6\ 12}{4\,3.5\,7\,\cancel{2}\,1} \\
-\,4\,1.0\,6\,3\,0 \\
\hline
2.5\,0\,9\,1
\end{array}
$$

The negative number has the larger absolute value, so the answer is negative.

$41.063 + (-43.5721) = -2.5091$

**13.** $-\frac{11}{15} - \frac{3}{5} = -\frac{11}{15} - \frac{3}{5} \cdot \frac{3}{3} = -\frac{11}{15} - \frac{9}{15} = -\frac{20}{15} =$

$-\frac{4 \cdot 5}{3 \cdot 5} = -\frac{4}{3} \cdot \frac{5}{5} = -\frac{4}{3}$

**14.**
$$
\begin{array}{r}
\overset{4\ 9\ 9\ 10}{3\,5\,0.0\,\cancel{0}} \\
-\quad 2\,4.5\,7 \\
\hline
3\,2\,5.4\,3
\end{array}
$$

**15.** $3\frac{3}{7} \cdot 4\frac{3}{8} = \frac{24}{7} \cdot \frac{35}{8} = \frac{24 \cdot 35}{7 \cdot 8} = \frac{3 \cdot 8 \cdot 5 \cdot 7}{7 \cdot 8 \cdot 1} =$

$\frac{7 \cdot 8}{7 \cdot 8} \cdot \frac{3 \cdot 5}{1} = \frac{3 \cdot 5}{1} = 15$

**16.**
$$
\begin{array}{r}
1\,2,4\,5\,6 \\
\times \quad\quad 2\,2\,0 \\
\hline
2\,4\,9\,1\,2\,0 \\
2\,4\,9\,1\,2\,0\,0 \\
\hline
2,7\,4\,0,3\,2\,0
\end{array}
$$

**17.** $-\frac{13}{15} \div \left(-\frac{26}{27}\right) = -\frac{13}{15} \cdot \left(-\frac{27}{26}\right) = \frac{13 \cdot 27}{15 \cdot 26} = \frac{13 \cdot 3 \cdot 9}{3 \cdot 5 \cdot 2 \cdot 13} =$

$\frac{3 \cdot 13}{3 \cdot 13} \cdot \frac{9}{5 \cdot 2} = \frac{9}{5 \cdot 2} = \frac{9}{10}$

**18.**
$$
\begin{array}{r}
4\,3\,6\,1 \\
2\,4\,\overline{)\,1\,0\,4,6\,7\,6} \\
\underline{9\,6\phantom{\,0,6\,7\,6}} \\
8\,6\phantom{\,6\,7\,6} \\
\underline{7\,2\phantom{\,6\,7\,6}} \\
1\,4\,7\phantom{\,6} \\
\underline{1\,4\,4\phantom{\,6}} \\
3\,6 \\
\underline{2\,4} \\
1\,2
\end{array}
$$

The answer is 4361 R 12, or $4361\frac{12}{24} = 4361\frac{1}{2}$, or $4361.5$.

**19.** $\frac{5}{8} = \frac{6}{x}$

$5 \cdot x = 8 \cdot 6$    Equating cross products

$\frac{5 \cdot x}{5} = \frac{8 \cdot 6}{5}$

$x = \frac{48}{5}$, or $9\frac{3}{5}$

**20.** $\frac{2}{5} \cdot y = \frac{3}{10}$

$y = \frac{3}{10} \div \frac{2}{5}$    Dividing by $\frac{2}{5}$

$y = \frac{3}{10} \cdot \frac{5}{2}$

$= \frac{3 \cdot 5}{10 \cdot 2} = \frac{3 \cdot 5}{2 \cdot 5 \cdot 2} = \frac{5}{5} \cdot \frac{3}{2 \cdot 2}$

$= \frac{3}{4}$

The solution is $\frac{3}{4}$.

**21.** $21.5 \cdot y = -146.2$

$\frac{21.5 \cdot y}{21.5} = \frac{-146.2}{21.5}$

$y = -6.8$

The solution is $-6.8$.

**22.** $x = 398,112 \div 26$

$x = 15,312$    Carrying out the division

The solution is $15,312$.

**23.** $\frac{\$2.99}{14.5 \text{ oz}} = \frac{299\cancel{c}}{14.5 \text{ oz}} \approx 20.6\cancel{c}/\text{oz}$

**24.** *Familiarize.* Let $c =$ the number of students who own a car.

*Translate.*

$$\underbrace{\text{What number}}_{c} \text{ is } 55.4\% \text{ of } 6000?$$
$$c = 55.4\% \cdot 6000$$

*Solve.* We convert 55.4% decimal notation and multiply.

$$
\begin{array}{r}
6\,0\,0\,0 \\
\times\,0.5\,5\,4 \\
\hline
2\,4\,0\,0\,0 \\
3\,0\,0\,0\,0\,0 \\
3\,0\,0\,0\,0\,0\,0 \\
\hline
3\,3\,2\,4.0\,0\,0
\end{array}
$$

Thus, $c = 3324$.

*Check.* We repeat the calculation. The answer checks.

*State.* 3324 students own a car.

**25.** *Familiarize.* Let $s =$ the length of each strip, in yards.

*Translate.* We translate to a division sentence.

$s = 1\frac{3}{4} \div 7$

*Solve.* We carry out the division.

$s = 1\frac{3}{4} \div 7 = \frac{7}{4} \div 7 = \frac{7}{4} \cdot \frac{1}{7} = \frac{7 \cdot 1}{4 \cdot 7} = \frac{7}{7} \cdot \frac{1}{4} = \frac{1}{4}$

**Check**. Since $7 \cdot \frac{1}{4} = \frac{7}{4} = 1\frac{3}{4}$, the answer checks.

**State**. Each strip is $\frac{1}{4}$ yd long.

26. **Familiarize**. Let $s =$ the number of cups of sugar that should be used for $\frac{1}{2}$ of the recipe.

    **Translate**. We translate to a multiplication sentence.
    $$s = \frac{1}{2} \cdot \frac{3}{4}$$

    **Solve**. We carry out the multiplication.
    $$s = \frac{1}{2} \cdot \frac{3}{4} = \frac{1 \cdot 3}{2 \cdot 4} = \frac{3}{8}$$

    **Check**. We repeat the calculation. The answer checks.

    **State**. $\frac{3}{8}$ cup of sugar should be used for $\frac{1}{2}$ of the recipe.

27. **Familiarize**. Let $p =$ the number of pounds of peanuts and products containing peanuts the average American eats in one year.

    **Translate**. We add the individual amounts to find $p$.
    $$p = 2.7 + 1.5 + 1.2 + 0.7 + 0.1$$

    **Solve**. We carry out the addition.

    $$\begin{array}{r} 2 \phantom{..} \\ 2.\,7 \\ 1.\,5 \\ 1.\,2 \\ 0.\,7 \\ + \,0.\,1 \\ \hline 6.\,2 \end{array}$$

    Thus, $p = 6.2$.

    **Check**. We repeat the calculation. The answer checks.

    **State**. The average American eats 6.2 lb of peanuts and products containing peanuts in one year.

28. **Familiarize**. Let $k =$ the number of kilowatt-hours, in billions, generated by American utility companies in the given year.

    **Translate**. We add the individual amounts to find $k$.
    $$k = 1464 + 455 + 273 + 250 + 118 + 12$$

    **Solve**. We carry out the addition.

    $$\begin{array}{r} 1\,2\,2 \phantom{..} \\ 1\,4\,6\,4 \\ 4\,5\,5 \\ 2\,7\,3 \\ 2\,5\,0 \\ 1\,1\,8 \\ + \phantom{1\,}1\,2 \\ \hline 2\,5\,7\,2 \end{array}$$

    Thus, $k = 2572$.

    **Check**. We repeat the calculation. The answer checks.

    **State**. American utility companies generated 2572 billion kilowatt-hours of electricity.

29. **Familiarize**. Let $c =$ the percent of people in the U. S. who have coronary heart disease, and let $h =$ the percent who die of heart attacks each year.

    **Translate**. We translate to two equations. We will express 509,000 as 0.509 million.

    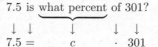

    7.5 is what percent of 301?
    $$7.5 = \quad c \quad \cdot \ 301$$

    0.509 is what percent of 301?
    $$0.509 = \quad h \quad \cdot \ 301$$

    **Solve**. We solve the first equation for $c$.
    $$7.5 = c \cdot 301$$
    $$\frac{7.5}{301} = \frac{c \cdot 301}{301}$$
    $$0.025 \approx c$$
    $$2.5\% \approx c$$

    Next we solve the second equation for $h$.
    $$0.509 = h \cdot 301$$
    $$\frac{0.509}{301} = \frac{h \cdot 301}{301}$$
    $$0.002 \approx h$$
    $$0.2\% \approx h$$

    **Check**. 2.5% of 301 is $0.025 \cdot 301 = 7.525 \approx 7.5$; 0.2% of 301 is $0.002 \cdot 301 = 0.602 \approx 0.509$. The answers check.

    **State**. About 2.5% of the people have coronary heart disease, and about 0.2% die of heart attacks each year.

30. **Familiarize**. First we subtract to find the amount of the increase, in billions of dollars.

    $$\begin{array}{r} {\scriptstyle 2\ 11} \phantom{.} \\ 3.\,\cancel{3}\,\cancel{1} \\ - \,3.\,1\,8 \\ \hline 0.\,1\,3 \end{array}$$

    Now let $p =$ the percent of increase.

    **Translate**.

    0.13 is what percent of 3.18?
    $$0.13 = \quad p \quad \cdot \ 3.18$$

    **Solve**.
    $$0.13 = p \cdot 3.18$$
    $$\frac{0.13}{3.18} = \frac{p \cdot 3.18}{3.18}$$
    $$0.041 \approx p$$
    $$4.1\% \approx p$$

    **Check**. 4.1% of $3.18 billion is $0.041 \cdot$ \$3.18 billion = \$0.13038 billion $\approx$ \$0.13 billion, so the answer checks.

    **State**. The percent of increase was about 4.1%.

**31. *Familiarize*.** Let $b =$ the fraction of the business owned by the fourth person.

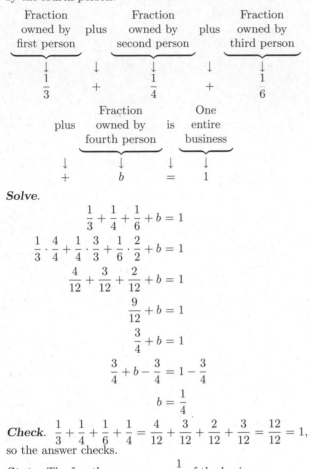

| Fraction owned by first person | plus | Fraction owned by second person | plus | Fraction owned by third person |
|---|---|---|---|---|
| ↓ | ↓ | ↓ | ↓ | ↓ |
| $\frac{1}{3}$ | $+$ | $\frac{1}{4}$ | $+$ | $\frac{1}{6}$ |

| plus | Fraction owned by fourth person | is | One entire business |
|---|---|---|---|
| ↓ | ↓ | ↓ | ↓ |
| $+$ | $b$ | $=$ | $1$ |

*Solve*.

$$\frac{1}{3} + \frac{1}{4} + \frac{1}{6} + b = 1$$

$$\frac{1}{3} \cdot \frac{4}{4} + \frac{1}{4} \cdot \frac{3}{3} + \frac{1}{6} \cdot \frac{2}{2} + b = 1$$

$$\frac{4}{12} + \frac{3}{12} + \frac{2}{12} + b = 1$$

$$\frac{9}{12} + b = 1$$

$$\frac{3}{4} + b = 1$$

$$\frac{3}{4} + b - \frac{3}{4} = 1 - \frac{3}{4}$$

$$b = \frac{1}{4}$$

**Check.** $\frac{1}{3} + \frac{1}{4} + \frac{1}{6} + \frac{1}{4} = \frac{4}{12} + \frac{3}{12} + \frac{2}{12} + \frac{3}{12} = \frac{12}{12} = 1$, so the answer checks.

**State.** The fourth person owns $\frac{1}{4}$ of the business.

**32. *Familiarize*.** Let $d =$ the number of defective valves that can be expected in a lot of 5049.

**Translate.** We translate to a proportion.

$$\begin{array}{c} \text{Defective} \rightarrow \\ \text{Total valves} \rightarrow \end{array} \frac{4}{18} = \frac{d}{5049} \begin{array}{c} \leftarrow \text{Defective} \\ \leftarrow \text{Total valves} \end{array}$$

**Solve.**

$$\frac{4}{18} = \frac{d}{5049}$$

$$4 \cdot 5049 = 18 \cdot d \quad \text{Equating cross products}$$

$$\frac{4 \cdot 5049}{18} = \frac{18 \cdot d}{18}$$

$$1122 = d$$

**Check.** We substitute 1122 for $d$ in the proportion and compare the cross products.

$$\frac{4}{18} = \frac{1122}{5049}; \ 4 \cdot 5049 = 20,196; \ 18 \cdot 1122 = 20,196$$

The cross products are the same so the answer checks.

**State.** In a lot of 5049 valves, 1122 could be expected to be defective.

**33. *Familiarize*.** Let $c =$ the cost of each tree.

| Cost of each tree | times | Number of trees | is | Total cost |
|---|---|---|---|---|
| ↓ | ↓ | ↓ | ↓ | ↓ |
| $c$ | $\cdot$ | $22$ | $=$ | $210$ |

**Solve.**

$$c \cdot 22 = 210$$

$$\frac{c \cdot 22}{22} = \frac{210}{22}$$

$$c \approx 9.55$$

**Check.** $22 \cdot \$9.55 = \$210.10 \approx \$210$, so the answer checks.

**State.** Each tree cost about $9.55.

**34.** Commission = Commission rate × Sales

$$182 \quad = \quad r \quad \times 2600$$

We solve the equation.

$$182 = r \times 2600$$

$$\frac{182}{2600} = \frac{r \times 2600}{2600}$$

$$0.07 = r$$

$$7\% = r$$

The commission rate is 7%.

**35.** Find the average:

$(\$38.08 + \$41.86 + \$46.11 + \$50.25 + \$54.69 + \$58.83 + \$62.97 + \$66.86 + \$71.00 + \$74.59) \div 10 = \dfrac{\$565.24}{10} \approx \$56.52$

Find the median:

The 10 numbers in the table are listed from smallest to largest. The median is the average of the two middle numbers, $54.69 and $58.83.

$$\frac{\$54.69 + \$58.83}{2} = \frac{\$113.52}{2} = \$56.76$$

**36.** First indicate the weights on the horizontal scale and label it "Weight (in pounds)." The costs range from $38.08 to $74.59. We label the marks on the vertical scale by 10's, ranging from 0 to $80, and label the scale Cost of FedEx Priority Overnight. Finally, draw vertical bars to show the costs associated with the weights.

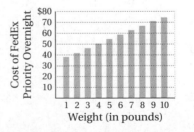

**37.** First indicate the weights on the horizontal scale and label it "Weight (in pounds)." The costs range from $38.08 to $74.59. We label the marks on the vertical scale by 10's, ranging from 0 to $80, and label the scale Cost of FedEx Priority Overnight. Next, at the appropriate level above each year, mark the corresponding cost. Finally, draw line segments connecting the points.

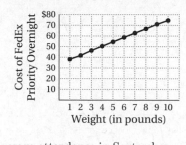

**38.** Average attendance in September:

$$\frac{28 + 23 + 26 + 23}{4} = \frac{100}{4} = 25$$

Average attendance in October:

$$\frac{26 + 20 + 14 + 28}{4} = 22$$

The average attendance decreased from September to October. We find the amount of the decrease: $25 - 22 = 3$. Now let $p =$ the percent of decrease.

*Translate.* 3 is what percent of 25?

$$
\begin{array}{ccccc}
\downarrow \downarrow & & \downarrow & & \downarrow \quad \downarrow \\
3 = & & p & \cdot & 25
\end{array}
$$

*Solve.*

$$3 = p \cdot 25$$

$$\frac{3}{25} = \frac{p \cdot 25}{25}$$

$$0.12 = p$$

$$12\% = p$$

Attendance decreased 12% from September to October.

# Chapter 9

# Measurement

**1.** 1 foot = 12 in.

This is the relation stated on page 512 of the text.

**3.** $1 \text{ in.} = 1 \text{ in.} \times \dfrac{1 \text{ ft}}{12 \text{ in.}}$    Multiplying by 1 using

$\dfrac{1 \text{ ft}}{12 \text{ in.}}$ to eliminate in.

$= \dfrac{1 \text{ in.}}{12 \text{ in.}} \times 1 \text{ ft}$

$= \dfrac{1}{12} \times \dfrac{\text{in.}}{\text{in.}} \times 1 \text{ ft}$

$= \dfrac{1}{12} \times 1 \text{ ft}$    The $\dfrac{\text{in.}}{\text{in.}}$ acts like 1, so we can omit it.

$= \dfrac{1}{12} \text{ ft}$

**5.** 1 mi = 5280 ft

This is the relation stated on page 512 of the text.

**7.** 3 yd = 3 × 1 yd

      = 3 × 36 in.    Substituting 36 in. for 1 yd

      = 108 in.    Multiplying

**9.** $84 \text{ in.} = \dfrac{84 \text{ in.}}{1} \times \dfrac{1 \text{ ft}}{12 \text{ in.}}$    Multiplying by 1 using

$\dfrac{1 \text{ ft}}{12 \text{ in.}}$

$= \dfrac{84 \text{ in.}}{12 \text{ in.}} \times 1 \text{ ft}$

$= \dfrac{84}{12} \times \dfrac{\text{in.}}{\text{in.}} \times 1 \text{ ft}$

$= 7 \times 1 \text{ ft}$    The $\dfrac{\text{in.}}{\text{in.}}$ acts like 1, so we can omit it.

$= 7 \text{ ft}$

**11.** $18 \text{ in.} = \dfrac{18 \text{ in.}}{1} \times \dfrac{1 \text{ ft}}{12 \text{ in.}}$    Multiplying by 1 using

$\dfrac{1 \text{ ft}}{12 \text{ in.}}$

$= \dfrac{18 \text{ in.}}{12 \text{ in.}} \times 1 \text{ ft}$

$= \dfrac{18}{12} \times \dfrac{\text{in.}}{\text{in.}} \times 1 \text{ ft}$

$= \dfrac{3}{2} \times 1 \text{ ft}$    The $\dfrac{\text{in.}}{\text{in.}}$ acts like 1, so we can omit it.

$= \dfrac{3}{2} \text{ ft, or } 1\dfrac{1}{2} \text{ ft, or 1.5 ft}$

**13.** 5 mi = 5 × 1 mi

     = 5 × 5280 ft    Substituting 5280 ft for 1 mi

     = 26,400 ft    Multiplying

**15.** $63 \text{ in.} = \dfrac{63 \text{ in.}}{1} \times \dfrac{1 \text{ ft}}{12 \text{ in.}}$    Multiplying by 1 using

$\dfrac{1 \text{ ft}}{12 \text{ in.}}$

$= \dfrac{63 \text{ in.}}{12 \text{ in.}} \times 1 \text{ ft}$

$= \dfrac{63}{12} \times \dfrac{\text{in.}}{\text{in.}} \times 1 \text{ ft}$

$= \dfrac{3 \cdot 21}{3 \cdot 4} \times 1 \text{ ft}$    The $\dfrac{\text{in.}}{\text{in.}}$ acts like 1, so we can omit it.

$= \dfrac{3}{3} \cdot \dfrac{21}{4} \text{ ft}$

$= \dfrac{21}{4} \text{ ft, or } 5\dfrac{1}{4} \text{ ft, or 5.25 ft}$

**17.** $10 \text{ ft} = 10 \text{ ft} \times \dfrac{1 \text{ yd}}{3 \text{ ft}}$    Multiplying by 1 using

$\dfrac{1 \text{ yd}}{3 \text{ ft}}$

$= \dfrac{10}{3} \times \dfrac{\text{ft}}{\text{ft}} \times 1 \text{ yd}$

$= \dfrac{10}{3} \times 1 \text{ yd}$

$= \dfrac{10}{3} \text{ yd, or } 3\dfrac{1}{3} \text{ yd}$

**19.** 7.1 mi = 7.1 × 1 mi

      = 7.1 × 5280 ft    Substituting 5280 ft for 1 mi

      = 37,488 ft    Multiplying

**21.** $4\dfrac{1}{2} \text{ ft} = 4\dfrac{1}{2} \text{ ft} \times \dfrac{1 \text{ yd}}{3 \text{ ft}}$

$= \dfrac{9}{2} \text{ ft} \times \dfrac{1 \text{ yd}}{3 \text{ ft}}$

$= \dfrac{9}{6} \times \dfrac{\text{ft}}{\text{ft}} \times 1 \text{ yd}$

$= \dfrac{3}{2} \times 1 \text{ yd}$

$= \dfrac{3}{2} \text{ yd, or } 1\dfrac{1}{2} \text{ yd, or 1.5 yd}$

**23.** $45 \text{ in.} = 45 \text{ in.} \times \dfrac{1 \text{ ft}}{12 \text{ in.}} \times \dfrac{1 \text{ yd}}{3 \text{ ft}}$

$\quad = \dfrac{45}{12 \cdot 3} \times \dfrac{\text{in.}}{\text{in.}} \times \dfrac{\text{ft}}{\text{ft}} \times 1 \text{ yd}$

$\quad = \dfrac{5 \cdot 3 \cdot 3}{4 \cdot 3 \cdot 3} \times 1 \text{ yd}$

$\quad = \dfrac{5}{4} \cdot \dfrac{3 \cdot 3}{3 \cdot 3} \times 1 \text{ yd}$

$\quad = \dfrac{5}{4}, \text{ or } 1\dfrac{1}{4} \text{ yd, or } 1.25 \text{ yd}$

**25.** $330 \text{ ft} = 330 \text{ ft} \times \dfrac{1 \text{ yd}}{3 \text{ ft}}$

$\quad = \dfrac{330}{3} \times \dfrac{\text{ft}}{\text{ft}} \times 1 \text{ yd}$

$\quad = 110 \times 1 \text{ yd}$
$\quad = 110 \text{ yd}$

**27.** $3520 \text{ yd} = 3520 \times 1 \text{ yd} \times \dfrac{3 \text{ ft}}{1 \text{ yd}} \times \dfrac{1 \text{ mi}}{5280 \text{ ft}}$

$\quad = \dfrac{3520 \cdot 3}{5280} \times \dfrac{\text{yd}}{\text{yd}} \times \dfrac{\text{ft}}{\text{ft}} \times 1 \text{ mi}$

$\quad = \dfrac{10,560}{5280} \times 1 \text{ mi}$

$\quad = 2 \times 1 \text{ mi}$
$\quad = 2 \text{ mi}$

**29.** $100 \text{ yd} = 100 \times 1 \text{ yd}$
$\quad = 100 \times 3 \text{ ft}$
$\quad = 300 \text{ ft}$

**31.** $360 \text{ in.} = 360 \text{ in.} \times \dfrac{1 \text{ ft}}{12 \text{ in.}}$

$\quad = \dfrac{360}{12} \times \dfrac{\text{in.}}{\text{in.}} \times 1 \text{ ft}$

$\quad = 30 \times 1 \text{ ft}$
$\quad = 30 \text{ ft}$

**33.** $1 \text{ in.} = 1 \text{ in.} \times \dfrac{1 \text{ ft}}{12 \text{ in.}} \times \dfrac{1 \text{ yd}}{3 \text{ ft}}$

$\quad = \dfrac{1}{12 \cdot 3} \times \dfrac{\text{in.}}{\text{in.}} \times \dfrac{\text{ft}}{\text{ft}} \times 1 \text{ yd}$

$\quad = \dfrac{1}{36} \times 1 \text{ yd}$

$\quad = \dfrac{1}{36} \text{ yd}$

**35.** $2 \text{ mi} = 2 \times 1 \text{ mi}$
$\quad = 2 \times 5280 \times 1 \text{ ft}$
$\quad = 2 \times 5280 \times 12 \text{ in.}$
$\quad = 126,720 \text{ in.}$

**37.** $83 \text{ yd} = 83 \times 1 \text{ yd}$
$\quad = 83 \times 36 \text{ in.}$
$\quad = 2988 \text{ in.}$

**39.** $9.25\% = \dfrac{9.25}{100}$

$\quad = \dfrac{9.25}{100} \cdot \dfrac{100}{100}$

$\quad = \dfrac{925}{10,000}$

$\quad = \dfrac{25 \cdot 37}{25 \cdot 400}$

$\quad = \dfrac{25}{25} \cdot \dfrac{37}{400}$

$\quad = \dfrac{37}{400}$

**41.** $27.5\% = \dfrac{27.5}{100} = \dfrac{27.5}{100} \cdot \dfrac{10}{10} = \dfrac{275}{1000} = \dfrac{25 \cdot 11}{25 \cdot 40} = \dfrac{25}{25} \cdot \dfrac{11}{40} = \dfrac{11}{40}$

**43.** a) First find decimal notation by division.

$$\begin{array}{r} 1.375 \\ 8\overline{\smash{)}11.000} \\ \underline{8\phantom{.000}} \\ 30\phantom{00} \\ \underline{24\phantom{00}} \\ 60\phantom{0} \\ \underline{56\phantom{0}} \\ 40 \\ \underline{40} \\ 0 \end{array}$$

$\dfrac{11}{8} = 1.375$

b) Convert the decimal notation to percent notation. Move the decimal point two places to the right and write a % symbol.

$1.37.5$

$\dfrac{11}{8} = 137.5\%, \text{ or } 137\dfrac{1}{2}\%$

**45.** $\dfrac{1}{4} = \dfrac{1}{4} \cdot \dfrac{25}{25} = \dfrac{25}{100} = 25\%$

**47.** The ratio of the number of trout to the number of catfish is $\dfrac{49.2}{368.7}$.

We can also express this without decimals as follows:

$\dfrac{49.2}{368.7} = \dfrac{49.2}{368.7} \cdot \dfrac{10}{10} = \dfrac{492}{3687}$.

The last fraction can also be simplified:

$\dfrac{492}{3687} = \dfrac{3 \cdot 164}{3 \cdot 1229} = \dfrac{3}{3} \cdot \dfrac{164}{1229} = \dfrac{164}{1229}$.

The ratio of the number of catfish to the number of trout is $\dfrac{368.7}{49.2}$.

Simplifying as we did above, we can also express this ratio as $\dfrac{3687}{492}$ or as $\dfrac{1229}{164}$.

**49.**  300 cubits = 300 × 1 cubit

$$\approx 300 \times 18 \text{ in.}$$

$$\approx 5400 \text{ in.}$$

50 cubits = 50 × 1 cubit

$$\approx 50 \times 18 \text{ in.}$$

$$\approx 900 \text{ in.}$$

30 cubits = 30 × 1 cubit

$$\approx 30 \times 18 \text{ in.}$$

$$\approx 540 \text{ in.}$$

In inches, the length of Noah's ark was about 5400 in., the breadth was about 900 in., and the height was about 540 in.

Now we convert these dimensions to feet.

$$5400 \text{ in.} = 5400 \text{ in.} \times \frac{1 \text{ ft}}{12 \text{ in.}}$$

$$= \frac{5400}{12} \times \frac{\text{in.}}{\text{in.}} \times 1 \text{ ft}$$

$$= 450 \times 1 \text{ ft}$$

$$= 450 \text{ ft}$$

$$900 \text{ in.} = 900 \text{ in.} \times \frac{1 \text{ ft}}{12 \text{ in.}}$$

$$= \frac{900}{12} \times \frac{\text{in.}}{\text{in.}} \times 1 \text{ ft}$$

$$= 75 \times 1 \text{ ft}$$

$$= 75 \text{ ft}$$

$$540 \text{ in.} = 540 \text{ in.} \times \frac{1 \text{ ft}}{12 \text{ in.}}$$

$$= \frac{540}{12} \times \frac{\text{in.}}{\text{in.}} \times 1 \text{ ft}$$

$$= 45 \times 1 \text{ ft}$$

$$= 45 \text{ ft}$$

In feet, the length of Noah's ark was 450 ft, the breadth was 75 ft, and the height was 45 ft.

---

## Exercise Set 9.2

**1.** a) 1 km = _____ m

Think: To go from km to m in the table is a move of 3 places to the right. Thus, we move the decimal point 3 places to the right.

1   1.000.
  └____↑

1 km = 1000 m

b) 1 m = _____ km

Think: To go from m to km in the table is a move of 3 places to the left. Thus, we move the decimal point 3 places to the left.

1   0.001.
   ↑____┘

1 m = 0.001 km

**3.** a) 1 dam = _____ m

Think: To go from dam to m in the table is a move of 1 place to the right. Thus, we move the decimal point 1 place to the right.

1   1.0.
  └↑

1 dam = 10 m

b) 1 m = _____ dam

Think: To go from m to dam in the table is a move of 1 place to the left. Thus, we move the decimal point 1 place to the left.

1   0.1.
  ↑┘

1 m = 0.1 dam

**5.** a) 1 cm = _____ m

Think: To go from cm to m in the table is a move of 2 places to the left. Thus, we move the decimal point 2 places to the left.

1   0.01.
  ↑__┘

1 cm = 0.01 m

b) 1 m = _____ cm

Think: To go from m to cm in the table is a move of 2 places to the right. Thus, we move the decimal point 2 places to the right.

1   1.00.
  └__↑

1 m = 100 cm

**7.** 6.7 km = _____ m

Think: To go from km to m in the table is a move of 3 places to the right. Thus, we move the decimal point 3 places to the right.

6.7   6.700.
    └___↑

6.7 km = 6700 m

**9.** 98 cm = _____ m

Think: To go from cm to m in the table is a move of 2 places to the left. Thus, we move the decimal point 2 places to the left.

98   0.98.
   ↑__┘

98 cm = 0.98 m

**11.** 8921 m = _____ km

Think: To go from m to km in the table is a move of 3 places to the left. Thus, we move the decimal point 3 places to the left.

8921   8.921.
     ↑___┘

8921 m = 8.921 km

**13.** 56.66 m = _____ km

Think: To go from m to km in the table is a move of
3 places to the left.  Thus, we move the decimal point 3
places to the left.

   56.66   0.056.66

56.66 m = 0.05666 km

**15.** 5666 m = _____ cm

Think: To go from m to cm in the table is a move of 2
places to the right.  Thus, we move the decimal point 2
places to the right.

   5666   5666.00.

5666 m = 566,600 cm

**17.** 477 cm = _____ m

Think: To go from cm to m in the table is a move of
2 places to the left.  Thus, we move the decimal point 2
places to the left.

   477   4.77.

477 cm = 4.77 m

**19.** 6.88 m = _____ cm

Think: To go from m to cm in the table is a move of 2
places to the right.  Thus, we move the decimal point 2
places to the right.

   6.88   6.88.

6.88 m = 688 cm

**21.** 1 mm = _____ cm

Think: To go from mm to cm in the table is a move of 1
place to the left. Thus, we move the decimal point 1 place
to the left.

   1   0.1.

1 mm = 0.1 cm

**23.** 1 km = _____ cm

Think: To go from km to cm in the table is a move of 5
places to the right.  Thus, we move the decimal point 5
places to the right.

   1   1.00000.

1 km = 100,000 cm

**25.** 14.2 cm = _____ mm

Think: To go from cm to mm in the table is a move of
1 place to the right.  Thus, we move the decimal point 1
place to the right.

   14.2   14.2.

14.2 cm = 142 mm

**27.** 8.2 mm = _____ cm

Think: To go from mm to cm in the table is a move of 1
place to the left. Thus, we move the decimal point 1 place
to the left.

   8.2   0.8.2

8.2 mm = 0.82 cm

**29.** 4500 mm = _____ cm

Think: To go from mm to cm in the table is a move of 1
place to the left. Thus, we move the decimal point 1 place
to the left.

   4500   450.0.

4500 mm = 450 cm

**31.** 0.024 mm = _____ m

Think: To go from mm to m in the table is a move of
3 places to the left.  Thus, we move the decimal point 3
places to the left.

   0.024   0.000.024

0.024 mm = 0.000024 m

**33.** 6.88 m = _____ dam

Think: To go from m to dam in the table is a move of 1
place to the left. Thus, we move the decimal point 1 place
to the left.

   6.88   0.6.88

6.88 m = 0.688 dam

**35.** 2.3 dam = _____ dm

Think: To go from dam to dm in the table is a move of
2 places to the right.  Thus, we move the decimal point 2
places to the right.

   2.3   2.30.

2.3 dam = 230 dm

**37.** 392 dam = _____ km

Think: To go from dam to km in the table is a move of
2 places to the left.  Thus, we move the decimal point 2
places to the left.

   392   3.92.

392 dam = 3.92 km

**39.** 18 cm = _____ mm

Think: To go from cm to mm in the table is a move of
1 place to the right.  Thus, we move the decimal point 1
place to the right.

18 cm = 180 mm

18 cm = _____ m

Think: To move from cm to m in the table is a move of
2 places to the left.  Thus, we move the decimal point
2 places to the left.

18 cm = 0.18 m

**41.** 0.278 m = _____ mm

Think: To go from m to mm in the table is a move of 3 places to the right. Thus, we move the decimal point 3 places to the right.

0.278 m = 278 mm

0.278 m = _____ cm

Think: to move from m to cm in the table is a move of 2 places to the right. Thus, we move the decimal point 2 places to the right.

0.278 m = 27.8 cm

**43.** 4844 cm = _____ mm

Think: To go from cm to mm in the table is a move of 1 place to the right. Thus, we move the decimal point 1 place to the right.

4844 cm = 48,440 mm

4844 cm = _____ m

Think: To move from cm to m in the table is a move of 2 places to the left. Thus, we move the decimal point 2 places to the left.

4844 cm = 48.44 m

**45.** 4 m = _____ mm

Think: To go from m to mm in the table is a move of 3 places to the right. Thus, we move the decimal point 3 places to the right.

4 m = 4000 mm

4 m = _____ cm

Think: to move from m to cm in the table is a move of 2 places to the right. Thus, we move the decimal point 2 places to the right.

4 m = 400 cm

**47.** 0.27 mm = _____ cm

Think: To move from mm to cm in the table is a move of 1 place to the left. Thus, we move the decimal point 1 place to the left.

0.27 mm = 0.027 cm

0.27 mm = _____ m

Think: To move from mm to m in the table is a move of 3 places to the left. Thus, we move the decimal point 3 places to the left.

0.27 mm = 0.00027 m

**49.** 442 m = _____ mm

Think: To go from m to mm in the table is a move of 3 places to the right. Thus, we move the decimal point 3 places to the right.

442 m = 442,000 mm

442 m = _____ cm

Think: to move from m to cm in the table is a move of 2 places to the right. Thus, we move the decimal point 2 places to the right.

442 m = 44,200 cm

**51.** To divide by 100, move the decimal point 2 places to the left.

23.4   0.23.4

$23.4 \div 100 = 0.234$

**53.**

$$
\begin{array}{r}
3\,.1\,4 \quad \text{(2 decimal places)} \\
\times\, 4\,.4\,1 \quad \text{(2 decimal places)} \\
\hline
3\,1\,4 \\
1\,2\,5\,6\,0 \\
1\,2\,5\,6\,0\,0 \\
\hline
1\,3\,.8\,4\,7\,4 \quad \text{(4 decimal places)}
\end{array}
$$

**55.** a) Find decimal notation using long division.

$$
\begin{array}{r}
0\,.3\,7\,5 \\
8\,\overline{)\,3\,.0\,0\,0} \\
2\,4 \\
\hline
6\,0 \\
5\,6 \\
\hline
4\,0 \\
4\,0 \\
\hline
0
\end{array}
$$

$\dfrac{3}{8} = 0.375$

b) Convert the decimal notation to percent notation. Move the decimal point two places to the right, and write a % symbol.

0.37.5

$\dfrac{3}{8} = 37.5\%$

**57.** a) Find decimal notation using long division.

$$
\begin{array}{r}
0\,.6\,6 \\
3\,\overline{)\,2\,.0\,0} \\
1\,8 \\
\hline
2\,0 \\
1\,8 \\
\hline
2
\end{array}
$$

$\dfrac{2}{3} = 0.\overline{6}$

b) Convert the decimal notation to percent notation. Move the decimal point two places to the right, and write a % symbol.

0.66.$\overline{6}$

$\dfrac{2}{3} = 66.\overline{6}\%$, or $66\dfrac{2}{3}\%$

**59.** 90%

a) Replace the percent symbol with ×0.01.

90 × 0.01

b) Move the decimal point 2 places to the left.

0.90.

Thus, 90% = 0.9.

**61.** $\dfrac{7}{15} + \dfrac{4}{25} = \dfrac{7}{15} \cdot \dfrac{5}{5} + \dfrac{4}{25} \cdot \dfrac{3}{3}$    LCD is 75

$= \dfrac{35}{75} + \dfrac{12}{75}$

$= \dfrac{47}{75}$

**63.** Since a meter is just over a yard we place the decimal point as follows: 1.0 m.

**65.** Since a centimeter is about 0.3937 inch, we place the decimal point as follows: 1.4 cm.

## Exercise Set 9.3

**1.** 330 ft $= 330 \times 1$ ft
$\approx 330 \times 0.305$ m
$= 100.65$ m

**3.** 1171.4 km $= 1171.4 \times 1$ km
$\approx 1171.4 \times 0.621$ mi
$= 727.4394$ mi

**5.** 65 mph $= 65\dfrac{\text{mi}}{\text{hr}} = 65 \times \dfrac{1\ \text{mi}}{\text{hr}} \approx 65 \times \dfrac{1.609\ \text{km}}{\text{hr}} = $ 104.585 km/h

**7.** 180 mi $= 180 \times 1$ mi
$\approx 180 \times 1.609$ km
$\approx 289.62$ km

**9.** 70 mph $= 70\dfrac{\text{mi}}{\text{hr}} = 70 \times \dfrac{1\ \text{mi}}{\text{hr}} \approx$
$70 \times \dfrac{1.609\ \text{km}}{\text{hr}} \approx 112.63$ km/h

**11.** 10 yd $= 10 \times 1$ yd
$\approx 10 \times 0.914$ m
$= 9.14$ m

**13.** 2.08 m $= 2.08 \times 1$ m
$\approx 2.08 \times 39.370$ in.
$= 81.8896$ in.

**15.** 381 m $= 381 \times 1$ m
$\approx 381 \times 3.281$ ft
$= 1250.061$ ft

**17.** 15.7 cm $\approx 15.7$ cm $\times \dfrac{1\ \text{in.}}{2.540\ \text{cm}}$
$= \dfrac{15.7}{2.540} \times \dfrac{\text{cm}}{\text{cm}} \times 1$ in.
$\approx 6.1811$ in.

**19.** 2216 km $= 2216 \times 1$ km
$\approx 2216 \times 0.621$ mi
$= 1376.136$ mi

**21.** 13 mm $= 13 \times 1$ mm
$= 13 \times 0.001$ m    Substituting 0.001 m
for 1 mm
$= 0.013$ m
$= 0.013 \times 1$ m
$\approx 0.013 \times 39.370$ in.
$= 0.51181$ in.

**23.** Convert from inches to yards:
$8\dfrac{1}{2}$ in. $= 8\dfrac{1}{2}$ in. $\times \dfrac{1\ \text{yd}}{36\ \text{in.}}$

$= \dfrac{\frac{17}{2}}{36} \times \dfrac{\text{in.}}{\text{in.}} \times 1$ yd

$= \dfrac{17}{2} \cdot \dfrac{1}{36} \times 1$ yd

$= \dfrac{17}{72} \times 1$ yd

$\approx 0.2361$ yd

Convert from inches to centimeters:
$8\dfrac{1}{2}$ in. $= 8\dfrac{1}{2} \times 1$ in.

$\approx 8.5 \times 2.540$ cm
$= 21.59$ cm

Convert from inches to meters: From the calculation immediately above we know that $8\dfrac{1}{2}$ in. $= 21.59$ cm. To convert this quantity to meters, move the decimal point two places to the left: $8\dfrac{1}{2}$ in. $= 21.59$ cm $= 0.2159$ m.

Convert from inches to millimeters: From one of the calculations above we know that $8\dfrac{1}{2}$ in. $= 21.59$ cm. To convert this quantity to millimeters, move the decimal point one place to the right: $8\dfrac{1}{2}$ in. $= 21.59$ cm $= 215.9$ mm.

**25.** Since we can easily convert from meters to yards and to inches, we first convert 23.8 cm to meters by moving the decimal point 2 places to the left: 4844 cm $= 48.44$ m.

Convert from centimeters to yards:
4844 cm $= 48.44$ m $= 48.44 \times 1$ m
$\approx 48.44 \times 1.094$ yd
$= 52.9934$ yd

Convert from centimeters to inches:
4844 cm $= 48.44$ m $= 48.44 \times 1$ m
$\approx 48.44 \times 39.370$ in.
$= 1907.0828$ in.

To convert from centimeters to millimeters, move the decimal point 1 place to the right: 4844 cm $= 48,440$ mm

**27.** First we convert yards to inches:
4 yd $= 4 \times 1$ yd
$= 4 \times 36$ in.
$= 144$ in.

Since we can easily convert yards to meters and then meters to centimeters and to millimeters, we next convert yards to meters.

4 yd $= 4 \times 1$ yd
$\approx 4 \times 0.914$ m
$= 3.656$ m

To convert meters to centimeters, move the decimal point 2 places to the right: 3.656 m $= 365.6$ cm.

To convert meters to millimeters, move the decimal point 3 places to the right: 3.656 m $= 3656$ mm.

**29.** Convert from meters to yards:
$$0.00027 \text{ m} = 0.00027 \times 1 \text{ m}$$
$$\approx 0.00027 \times 1.094 \text{ yd}$$
$$\approx 0.000295 \text{ yd}$$

To convert meters to centimeters, move the decimal point 2 places to the right: $0.00027$ m $= 0.027$ cm.

Convert meters to inches:
$$0.00027 \text{ m} = 0.00027 \times 1 \text{ m}$$
$$\approx 0.00027 \times 39.370 \text{ in.}$$
$$= 0.0106299 \text{ in.}$$

To convert from meters to millimeters, move the decimal point 3 places to the right: $0.00027$ m $= 0.27$ mm.

**31.** Convert from meters to yards:
$$442 \text{ m} = 442 \times 1 \text{ m}$$
$$\approx 442 \times 1.094 \text{ yd}$$
$$= 483.548 \text{ yd}$$

To convert meters to centimeters, move the decimal point 2 places to the right: $442$ m $= 44{,}200$ cm.

Convert meters to inches:
$$442 \text{ m} = 442 \times 1 \text{ m}$$
$$\approx 442 \times 39.370 \text{ in.}$$
$$= 17{,}401.54 \text{ in.}$$

To convert meters to millimeters, move the decimal point 3 places to the right: $442$ m $= 442{,}000$ mm.

**33.**
$$28.7 \text{ million} = 28.7 \times 1 \text{ million}$$
$$= 28.7 \times 1{,}000{,}000$$
$$= 28{,}700{,}000$$

**35.** $1$ in. $\approx 2.540$ cm $= 25.40$ mm

Thus, we have $1$ in. $\approx 25.4$ mm.

**37.** Since we know that $1$ km $\approx 0.621$ mi, we first convert $100$ m to kilometers by moving the decimal point 3 places to the left:
$$100 \text{ m} = 0.1 \text{ km}$$

Now we convert the speed to miles per hour.
$$\frac{0.1 \text{ km}}{10.49 \text{ sec}} = \frac{0.1 \text{ km}}{10.49 \text{ sec}} \times \frac{0.621 \text{ mi}}{1 \text{ km}} \times \frac{60 \text{ sec}}{1 \text{ min}} \times \frac{60 \text{ min}}{1 \text{ hr}}$$
$$= \frac{0.1 \times 0.621 \times 60 \times 60}{10.49} \times \frac{\text{mi}}{\text{hr}}$$
$$\approx 21.3 \text{ mph}$$

## Chapter 9 Mid-Chapter Review

**1.** Since $1$ mi $\approx 1.609$ km and $1$ m $= 0.001$ km, distances that are measured in miles in the American system would probably be measured in kilometers in the metric system. Thus, the given statement is false.

**2.** $1$ m $\approx 3.281$ ft and $1$ yd $= 3$ ft, so the statement is true.

**3.** $1$ km $\approx 0.621$ mi, so the statement is false.

**4.** When converting from meters to centimeters, we multiply by 100. This is equivalent to moving the decimal point two places to the right, so the given statement is true.

**5.** $1$ ft $\approx 0.305$ m $\approx 30.5$ cm $\approx 30$ cm, so the statement is true.

**6.** $16\frac{2}{3}$ yd $= 16\frac{2}{3} \times 1$ yd $= \frac{50}{3} \times 3$ ft $= 50$ ft

**7.** $13{,}200$ ft $= 13{,}200$ ft $\times \dfrac{1 \text{ mi}}{5280 \text{ ft}} = 2.5$ mi

**8.** $520$ mm $= 520$ mm $\times \dfrac{1 \text{ m}}{1000 \text{ mm}} = 0.52$ m $\times \dfrac{1 \text{ km}}{1000 \text{ m}} = 0.00052$ km

**9.** $10{,}200$ mm $= 10{,}200$ mm $\times \dfrac{1 \text{ m}}{1000 \text{ mm}} \approx 10.2$ m $\times \dfrac{3.281 \text{ ft}}{1 \text{ m}} = 33.4662$ ft

**10.**
$$5\frac{1}{2} \text{ mi} = 5\frac{1}{2} \times 1 \text{ mi}$$
$$= \frac{11}{2} \times 5280 \text{ ft} \times \frac{1 \text{ yd}}{3 \text{ ft}}$$
$$= \frac{11 \times 5280}{2 \times 3} \times \frac{\text{ft}}{\text{ft}} \times 1 \text{ yd}$$
$$= 9680 \text{ yd}$$

**11.**
$$840 \text{ in.} = 840 \text{ in.} \times \frac{1 \text{ ft}}{12 \text{ in.}}$$
$$= \frac{840 \text{ in.}}{12 \text{ in.}} \times 1 \text{ ft}$$
$$= \frac{840}{12} \times \frac{\text{in.}}{\text{in.}} \times 1 \text{ ft}$$
$$= 70 \text{ ft}$$

**12.** $24.05$ cm $= \underline{\hspace{1cm}}$ dm

Think: To go from cm to dm in the table is a move of 1 place to the left. Thus, we move the decimal point 1 place to the left.

$24.05$ cm $= 2.405$ dm

**13.** $0.15$ m $= \underline{\hspace{1cm}}$ km

Think: To go from m to km in the table is a move of 3 places to the left. Thus, we move the decimal point 3 places to the left.

$0.15$ m $= 0.00015$ km

**14.**
$$630 \text{ yd} = 630 \times 1 \text{ yd}$$
$$= 630 \times 36 \text{ in.}$$
$$= 22{,}680 \text{ in.}$$

**15.**
$$100 \text{ ft} = 100 \times 1 \text{ ft}$$
$$= 100 \times 12 \text{ in.}$$
$$= 1200 \text{ in.}$$

**16.** $6000$ dam $= \underline{\hspace{1cm}}$ m

Think: To go from dam to m in the table is a move of 1 place to the right. Thus, we move the decimal point 1 place to the right.

$6000$ dam $= 60{,}000$ m

**17.** 85,000 mm = _____ dm

Think: To go from mm to dm in the table is a move of 2 places to the left. Thus, we move the decimal point 2 places to the left.

85,000 mm = 850 dm

**18.**  $26,400 \text{ ft} = 26,400 \text{ ft} \times \dfrac{1 \text{ mi}}{5280 \text{ ft}}$

$= \dfrac{26,400 \text{ ft}}{5280 \text{ ft}} \times 1 \text{ mi}$

$= \dfrac{26,400}{5280} \times \dfrac{\text{ft}}{\text{ft}} \times 1 \text{ mi}$

$= 5 \text{ mi}$

**19.**  $3753 \text{ ft} = 3753 \text{ ft} \times \dfrac{1 \text{ yd}}{3 \text{ ft}}$

$= \dfrac{3753 \text{ ft}}{3 \text{ ft}} \times 1 \text{ yd}$

$= \dfrac{3753}{3} \times \dfrac{\text{ft}}{\text{ft}} \times 1 \text{ yd}$

$= 1251 \text{ yd}$

**20.**  $10 \text{ mi} = 10 \times 1 \text{ mi}$

$= 10 \times 5280 \text{ ft}$

$= 52,800 \text{ ft}$

**21.** 1800 m = _____ cm

Think: To go from m to cm in the table is a move of 2 places to the right. Thus, we move the decimal point 2 places to the right.

1800 m = 180,000 cm

**22.** 8.4 km = _____ dm

Think: To go from km to dm in the table is a move of 4 places to the right. Thus, we move the decimal point 4 places to the right.

8.4 km = 84,000 dm

**23.** 0.007 km = _____ cm

Think: To go from km to cm in the table is a move of 5 places to the right. Thus, we move the decimal point 5 places to the right.

0.007 km = 700 cm

**24.** 40 dm = _____ dam

Think: To go from dm to dam in the table is a move of 2 places to the left. Thus, we move the decimal point 2 places to the left.

40 dm = 0.4 dam

**25.** 80.09 cm = _____ m

Think: To go from cm to m in the table is a move of 2 places to the left. Thus, we move the decimal point 2 places to the left.

80.09 cm = 0.8009 m

**26.**  $360 \text{ in.} = 360 \text{ in.} \times \dfrac{1 \text{ yd}}{36 \text{ in.}}$

$= \dfrac{360 \text{ in.}}{36 \text{ in.}} \times 1 \text{ yd}$

$= \dfrac{360}{36} \times \dfrac{\text{in.}}{\text{in.}} \times 1 \text{ yd}$

$= 10 \text{ yd}$

**27.** 19.2 m = _____ mm

Think: To go from m to mm in the table is a move of 3 places to the right. Thus, we move the decimal point 3 places to the right.

19.2 m = 19,200 mm

**28.**  $1200 \text{ in.} = 1200 \text{ in.} \times \dfrac{1 \text{ ft}}{12 \text{ in.}}$

$= \dfrac{1200 \text{ in.}}{12 \text{ in.}} \times 1 \text{ ft}$

$= \dfrac{1200}{12} \times \dfrac{\text{in.}}{\text{in.}} \times 1 \text{ ft}$

$= 100 \text{ ft}$

**29.** 0.0001 mm = _____ hm

Think: To go from mm to hm in the table is a move of 5 places to the left. Thus, we move the decimal point 5 places to the left.

0.0001 mm = 0.000000001 hm

**30.** 4 km = _____ cm

Think: To go from km to cm in the table is a move of 5 places to the right. Thus, we move the decimal point 5 places to the right.

4 km = 400,000 cm

**31.**  $12 \text{ mi} = 12 \times 1 \text{ mi}$

$= 12 \times 5280 \text{ ft}$

$= 12 \times 5280 \times 1 \text{ ft}$

$= 12 \times 5280 \times 12 \text{ in.}$

$= 760,320 \text{ in.}$

**32.**  $36 \text{ m} = 36 \times 1 \text{ m}$

$\approx 36 \times 3.281 \text{ ft}$

$= 118.116 \text{ ft}$

**33.** 80 dm = _____ dam

Think: To go from dm to dam in the table is a move of 2 places to the left. Thus, we move the decimal point 2 places to the left.

80 dm = 0.8 dam

**34.**  $2.5 \text{ yd} = 2.5 \times 1 \text{ yd}$

$\approx 2.5 \times 0.914 \text{ m}$

$= 2.285 \text{ m}$

**35.** 6000 mm = _____ dm

Think: To go from mm to dm in the table is a move of 2 places to the left. Thus, we move the decimal point 2 places to the left.

6000 mm = 60 dm

**36.** In order to use the conversion from meters to inches, we will first convert 0.0635 mm to meters. To go from mm to m in the table is a move of 3 places to the left. Thus, we move the decimal point 3 places to the left: 0.0635 mm = 0.0000635 m. Now we make the conversion to inches.

$$0.0000635 \text{ m} = 0.0000635 \times 1 \text{ m}$$
$$\approx 0.0000635 \times 39.370 \text{ in.}$$
$$\approx 0.0025 \text{ in.}$$

**37.** $\frac{1}{4} \text{ yd} = \frac{1}{4} \times 1 \text{ y}$

$= \frac{1}{4} \times 36 \text{ in.}$

$= 9 \text{ in.}$

$144 \text{ in.} = 144 \text{ in.} \times \frac{1 \text{ ft}}{12 \text{ in.}}$

$= \frac{144 \text{ in.}}{12 \text{ in.}} \times 1 \text{ ft}$

$= \frac{144}{12} \times \frac{\text{in.}}{\text{in.}} \times 1 \text{ ft}$

$= 12 \text{ ft}$

To go from m to dm in the table is a move of 1 place to the right, so we move the decimal 1 place to the right to convert m to dm. Thus, 2400 m = 24,000 dm.

$0.75 \text{ mi} = 0.75 \times 1 \text{ mi}$

$= 0.75 \times 5280 \text{ ft} \times \frac{1 \text{ yd}}{3 \text{ ft}}$

$= \frac{0.75 \times 5280}{3} \times \frac{\text{ft}}{\text{ft}} \times 1 \text{ yd}$

$= 1320 \text{ yd}$

To go from m to km in the table is a move of 3 places to the left, so we move the decimal point 3 places to the left to convert m to km. Thus, 24 m = 0.024 km.

To go from cm to mm in the table is a move of 1 place to the right, so we move the decimal point 1 place to the right to convert cm to mm. Thus, 240 cm = 2400 mm.

**38.** We first express each measure using the same unit. We will use feet as the common unit. We have 100 in. = 8.$\overline{3}$ ft, 430 ft, $\frac{1}{100}$ mi = 52.8 ft, 3.5 ft, 6000 in. = 83.$\overline{3}$ ft, 2 yd = 6 ft. Now we arrange these measures from smallest to largest.

3.5 ft, 2 yd, 100 in., $\frac{1}{100}$ mi, 1000 in., 430 ft, 6000 ft

**39.** We first express each measure using the same unit. We will use meters as the common unit. We have 3240 cm = 32.4 m, 300 m, 250 dm = 25 m, 150 hm = 15,000 m, 33,000 mm = 33 m, 310 dam = 3100 m, 13 km = 13,000 m. Now we arrange these measures from largest to smallest.

150 hm, 13 km, 310 dam, 300 m, 33,000 mm, 3240 cm, 250 dm

**40.** We first express each measure using the same unit. We will use meters as the common unit. We have 2 yd ≈ 1.828 m, 1.5 mi ≈ 2.4135 km = 2413.5 m, 65 cm = 0.65 m,

$\frac{1}{2}$ ft ≈ 0.1525 m, 3 km = 3000 m, 2.5 m. Now we arrange these measures from smallest to largest.

$\frac{1}{2}$ ft, 65 cm, 2 yd, 2.5 m, 1.5 mi, 3 km

**41.** The student should have multiplied by $\frac{1}{12}$ (or divided by 12) to convert inches to feet. The correct procedure is as follows:

$$23 \text{ in.} = 23 \text{ in.} \times \frac{1 \text{ ft}}{12 \text{ in.}} = \frac{23 \text{ in.}}{12 \text{ in.}} \times 1 \text{ ft} =$$
$$\frac{23}{12} \times \frac{\text{in.}}{\text{in.}} \times 1 \text{ ft} = \frac{23}{12} \times 1 \text{ ft} = \frac{23}{12} \text{ ft}$$

**42.** Metric units are based on tens, so computations and conversions with metric units can be done by moving a decimal point. American units, which are not based on tens, require more complicated arithmetic in computations and conversions.

**43.** Note that a larger unit can be expressed as an equivalent number of smaller units. Then when converting from a larger unit to a smaller unit, we can express the quantity as a number times 1 unit and then substitute the equivalent number of smaller units for the larger unit. When converting from a smaller unit to a larger unit, we multiply by 1 using one larger unit in the numerator and the equivalent number of smaller units in the denominator. This allows us to "cancel" the smaller units, leaving the larger units.

**44.** Answers may vary.

---

## Exercise Set 9.4

**1.** 1 T = 2000 lb

This conversion relation is given in the text on page 531.

**3.** $6000 \text{ lb} = 6000 \text{ lb} \times \frac{1 \text{ T}}{2000 \text{ lb}}$    Multiplying by 1 using $\frac{1 \text{ T}}{2000 \text{ lb}}$

$= \frac{6000}{2000} \times \frac{\text{lb}}{\text{lb}} \times 1 \text{ T}$

$= 3 \times 1 \text{ T}$    The $\frac{\text{lb}}{\text{lb}}$ acts like 1, so we can omit it.

$= 3 \text{ T}$

**5.** $4 \text{ lb} = 4 \times 1 \text{ lb}$
$= 4 \times 16 \text{ oz}$    Substituting 16 oz for 1 lb
$= 64 \text{ oz}$

**7.** $6.32 \text{ T} = 6.32 \times 1 \text{ T}$
$= 6.32 \times 2000 \text{ lb}$    Substituting 2000 lb for 1 T
$= 12,640 \text{ lb}$

**9.** $3200 \text{ oz} = 3200 \text{ oz} \times \frac{1 \text{ lb}}{16 \text{ oz}} \times \frac{1 \text{ T}}{2000 \text{ lb}}$

$= \frac{3200}{16 \times 2000} \text{ T}$

$= \frac{1}{10} \text{ T, or } 0.1 \text{ T}$

**11.**  $80 \text{ oz} = 80 \text{ oz} \times \dfrac{1 \text{ lb}}{16 \text{ oz}}$

$\phantom{80 \text{ oz}} = \dfrac{80}{16} \text{ lb}$

$\phantom{80 \text{ oz}} = 5 \text{ lb}$

**13.**  $13,000,000 \text{ tons} = 13,000,000 \times 1 \text{ ton}$

$\phantom{13,000,000 \text{ tons}} = 13,000,000 \times 2000 \text{ lb}$

$\phantom{13,000,000 \text{ tons}} = 26,000,000,000 \text{ lb}$

**15.**  1 kg = _____ g

Think: To go from kg to g in the table is a move of 3 places to the right. Thus, we move the decimal point 3 places to the right.

1    1.000.

1 kg = 1000 g

**17.**  1 dag = _____ g

Think: To go from dag to g in the table is a move of 1 place to the right. Thus, we move the decimal point 1 place to the right.

1    1.0.

1 dag = 10 g

**19.**  1 cg = _____ g

Think: To go from cg to g in the table is a move of 2 places to the left. Thus, we move the decimal point 2 places to the left.

1    0.01.

1 cg = 0.01 g

**21.**  1 g = _____ mg

Think: To go from g to mg in the table is a move of 3 places to the right. Thus, we move the decimal point 3 places to the right.

1    1.000.

1 g = 1000 mg

**23.**  1 g = _____ dg

Think: To go from g to dg in the table is a move of 1 place to the right. Thus, we move the decimal point 1 place to the right.

1    1.0.

1 g = 10 dg

**25.**  Complete: 234 kg = _____ g

Think: To go from kg to g in the table is a move of 3 places to the right. Thus, we move the decimal point 3 places to the right.

234    234.000.

234 kg = 234,000 g

**27.**  Complete: 5200 g = _____ kg

Think: To go from g to kg in the table is a move of 3 places to the left. Thus, we move the decimal point 3 places to the left.

5200    5.200.

5200 g = 5.2 kg

**29.**  Complete: 67 hg = _____ kg

Think: To go from hg to kg in the table is a move of 1 place to the left. Thus, we move the decimal point 1 place to the left.

67    6.7.

67 hg = 6.7 kg

**31.**  Complete: 0.502 dg = _____ g

Think: To go from dg to g in the table is a move of 1 place to the left. Thus, we move the decimal point 1 place to the left.

0.502    0.0.502

0.502 dg = 0.0502 g

**33.**  Complete: 8492 g = _____ kg

Think: To go from g to kg in the table is a move of 3 places to the left. Thus, we move the decimal point 3 places to the left.

8492    8.492.

8492 g = 8.492 kg

**35.**  Complete: 585 mg = _____ cg

Think: To go from mg to cg in the table is a move of 1 place to the left. Thus, we move the decimal point 1 place to the left.

585    58.5.

585 mg = 58.5 cg

**37.**  Complete: 8 kg = _____ cg

Think: To go from kg to cg in the table is a move of 5 places to the right. Thus, we move the decimal point 5 places to the right.

8    8.00000.

8 kg = 800,000 cg

**39.**  1 t = 1000 kg

This conversion relation is given in the text on page 532.

**41.**  Complete: 3.4 cg = _____ dag

Think: To go from cg to dag in the table is a move of 3 places to the left. Thus, we move the decimal point 3 places to the left.

3.4    0.003.4

3.4 cg = 0.0034 dag

**43.** $60.3 \text{ kg} = 60.3 \text{ kg} \times \dfrac{1 \text{ t}}{1000 \text{ kg}}$

$\qquad = \dfrac{60.3 \text{ kg}}{1000 \text{ kg}} \times 1 \text{ t}$

$\qquad = \dfrac{60.3}{1000} \times \dfrac{\text{kg}}{\text{kg}} \times 1 \text{ t}$

$\qquad = 0.0603 \text{ t}$

**45.** $1 \text{ mg} = 0.001 \text{ g}$

$\qquad = 0.001 \times 1 \text{ g}$

$\qquad = 0.001 \times 1,000,000 \text{ mcg}$

$\qquad = 1000 \text{ mcg}$

**47.** $325 \text{ mcg} = 325 \times 1 \text{ mcg}$

$\qquad = 325 \times \dfrac{1}{1,000,000} \text{ g}$

$\qquad = 0.000325 \text{ g}$

$\qquad = 0.325 \text{ mg}$

**49.** $210.6 \text{ mg} = 210.6 \times 1 \text{ mg}$

$\qquad = 210.6 \times 0.001 \text{ g}$

$\qquad = 210.6 \times 0.001 \times 1 \text{ g}$

$\qquad = 210.6 \times 0.001 \times 1,000,000 \text{ mcg}$

$\qquad = 210,600 \text{ mcg}$

**51.** $4.9 \text{ mcg} = 4.9 \times 1 \text{ mcg}$

$\qquad = 4.9 \times \dfrac{1}{1,000,000} \text{ g}$

$\qquad = 0.0000049 \text{ g}$

$\qquad = 0.0049 \text{ mg}$

**53.** $0.125 \text{ mg} = 0.000125 \text{ g}$

$\qquad = 0.000125 \times 1 \text{ g}$

$\qquad = 0.000125 \times 1,000,000 \text{ mcg}$

$\qquad = 125 \text{ mcg}$

**55.** We multiply to find the number of milligrams that will be ingested.

$$\begin{array}{r} 0.125 \\ \times \quad\quad 7 \\ \hline 0.875 \end{array}$$

The patient will ingest 0.875 mg of Triazolam. Now convert 0.875 mg to micrograms.

$0.875 \text{ mg} = 0.000875 \text{ g}$

$\qquad = 0.000875 \times 1 \text{ g}$

$\qquad = 0.000875 \times 1,000,000 \text{ mcg}$

$\qquad = 875 \text{ mcg}$

**57.** First convert 500 mg to grams by moving the decimal point three places to the left: 500 mg = 0.5 g.

Then divide to determine the number of 500 mg tablets that would have to be taken.

$$0.5_\wedge \overline{\smash{)}2.0_\wedge} \phantom{0} \begin{array}{l} 4 \, . \\ \phantom{0} \\ \underline{2\,0} \\ \phantom{0}0 \end{array}$$

The patient would have to take 4 tablets per day.

**59.** We use a proportion. Let $a =$ the number of cubic centimeters of amoxicillin to be administered.

$$\dfrac{250}{400} = \dfrac{5}{a}$$

$$250 \cdot a = 400 \cdot 5$$

$$a = \dfrac{400 \cdot 5}{250}$$

$$a = \dfrac{50 \cdot 8 \cdot 5}{50 \cdot 5 \cdot 1} = \dfrac{50 \cdot 5}{50 \cdot 5} \cdot \dfrac{8}{1}$$

$$a = 8$$

The child's mother needs to administer 8 cc of amoxicillin.

**61.** $35\% = \dfrac{35}{100} = \dfrac{5 \cdot 7}{5 \cdot 20} = \dfrac{5}{5} \cdot \dfrac{7}{20} = \dfrac{7}{20}$

**63.** $85.5\% = \dfrac{85.5}{100} = \dfrac{85.5}{100} \cdot \dfrac{10}{10} = \dfrac{855}{1000} = \dfrac{5 \cdot 171}{5 \cdot 200} = \dfrac{171}{200}$

**65.** $37\frac{1}{2}\% = \dfrac{75}{2}\% = \dfrac{75}{2} \times \dfrac{1}{100} = \dfrac{75}{2 \cdot 100} = $

$\dfrac{25 \cdot 3}{2 \cdot 25 \cdot 4} = \dfrac{25}{25} \cdot \dfrac{3}{2 \cdot 4} = \dfrac{3}{8}$

**67.** $83.\overline{3}\% = 83\frac{1}{3}\% = \dfrac{250}{3}\% = \dfrac{250}{3} \times \dfrac{1}{100} = $

$\dfrac{250 \cdot 1}{3 \cdot 100} = \dfrac{5 \cdot 50 \cdot 1}{3 \cdot 2 \cdot 50} = \dfrac{50}{50} \cdot \dfrac{5 \cdot 1}{3 \cdot 2} = \dfrac{5}{6}$

**69.** *Familiarize*. This is a two-step problem. First we find the amount of the increase. Let $a =$ the amount by which the population increases.

*Translate*. We rephrase the question and translate.

$$\begin{array}{ccccc} \text{What} & \text{is} & 4\% & \text{of} & 180,000? \\ \downarrow & \downarrow & \downarrow & \downarrow & \downarrow \\ a & = & 4\% & \times & 180,000 \end{array}$$

*Solve*. Convert 4% to decimal notation and multiply.

$$a = 4\% \times 180,000 = 0.04 \times 180,000 = 7200$$

Now we add 7200 to the former population to find the new population.

$$180,000 + 7200 = 187,200$$

*Check*. We can do a partial check by estimating. The old population is approximately 200,000 and 4% of 200,000 is $0.04 \times 200,000$, or 8000. The new population would be about $180,000 + 8000$, or 188,000. Since 188,000 is close to 187,200, we have a partial check. We can also repeat the calculations. The answer checks.

*State*. The population will be 187,200.

**71.** *Familiarize*. Let $m =$ the meals tax.

*Translate*. We translate to a percent equation.

$$\begin{array}{ccccc} \text{What} & \text{is} & 4\frac{1}{2}\% & \text{of} & \$540? \\ \downarrow & \downarrow & \downarrow & \downarrow & \downarrow \\ m & = & 4\frac{1}{2}\% & \cdot & 540 \end{array}$$

*Solve*. Convert $4\frac{1}{2}\%$ to decimal notation and multiply.

$$m = 4\frac{1}{2}\% \cdot 540 = 0.045 \cdot 540 = 24.30$$

*Check*. We can repeat the calculation. The answer checks.

*State*. The meals tax is \$24.30.

**73.** ***Familiarize***. Let $s$ = the number of sheets in 15 reams of paper. Repeated addition works well here.

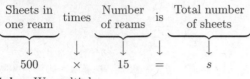

15 addends

***Translate***.

| Sheets in one ream | times | Number of reams | is | Total number of sheets |
|---|---|---|---|---|
| ↓ | ↓ | ↓ | ↓ | ↓ |
| 500 | × | 15 | = | $s$ |

***Solve***. We multiply.

$500 \times 15 = 7500$, so $7500 = s$, or $s = 7500$.

***Check***. We can repeat the calculation. The answer checks.

***State***. There are 7500 sheets in 15 reams of paper.

**75.** First convert $15\frac{3}{4}$ lb to ounces.

$$15\frac{3}{4} \text{ lb} = 15\frac{3}{4} \times 1 \text{ lb}$$
$$= \frac{63}{4} \times 16 \text{ oz}$$
$$= \frac{63 \times 16}{4} \text{ oz}$$
$$= 252 \text{ oz}$$

Now we divide to find the number of packages in the box.

$$252 \div 1\frac{3}{4} = 252 \div \frac{7}{4}$$
$$= 252 \cdot \frac{4}{7}$$
$$= \frac{252 \cdot 4}{7} = \frac{7 \cdot 36 \cdot 4}{7 \cdot 1}$$
$$= \frac{7}{7} \cdot \frac{36 \cdot 4}{1}$$
$$= 144$$

There are 144 packages in the box.

**77.** a) First we find how many milligrams the Golden Jubilee Diamond weighs.

$$545.67 \text{ carats} = 545.67 \times 1 \text{ carat}$$
$$= 545.67 \times 200 \text{ mg}$$
$$= 109,134 \text{ mg}$$

To go from mg to g in the table is a move of 3 places to the left. Thus, we move the decimal point 3 places to the left:

$$545.67 \text{ carats} = 109,134 \text{ mg} = 109.134 \text{ g}$$

b) First we find how many milligrams the Hope Diamond weighs.

$$45.52 \text{ carats} = 45.52 \times 1 \text{ carat}$$
$$= 45.52 \times 200 \text{ mg}$$
$$= 9104 \text{ mg}$$

To go from mg to g in the table is a move of 3 places to the left. Thus, we move the decimal point 3 places to the left:

$$45.52 \text{ carats} = 9104 \text{ mg} = 9.104 \text{ g}$$

c) Golden Jubilee Diamond:

$$109.134 \text{ g} = 109.134 \text{ g} \times \frac{1 \text{ lb}}{453.6 \text{ g}} \times \frac{16 \text{ oz}}{1 \text{ lb}}$$
$$= \frac{109.134 \times 16}{453.6} \times \frac{\text{g}}{\text{g}} \times \frac{\text{lb}}{\text{lb}} \times 1 \text{ oz}$$
$$\approx 3.85 \text{ oz}$$

Hope Diamond:

$$9.104 \text{ g} = 9.104 \text{ g} \times \frac{1 \text{ lb}}{453.6 \text{ g}} \times \frac{16 \text{ oz}}{1 \text{ lb}}$$
$$= \frac{9.104 \times 16}{453.6} \times \frac{\text{g}}{\text{g}} \times \frac{\text{lb}}{\text{lb}} \times 1 \text{ oz}$$
$$\approx 0.321 \text{ oz}$$

## Exercise Set 9.5

**1.** $1 \text{ L} = 1000 \text{ mL} = 1000 \text{ cm}^3$

These conversion relations appear in the text on page 540.

**3.** $\begin{aligned} 87 \text{ L} &= 87 \times (1 \text{ L}) \\ &= 87 \times (1000 \text{ mL}) \\ &= 87,000 \text{ mL} \end{aligned}$

**5.** $\begin{aligned} 49 \text{ mL} &= 49 \times (1 \text{ mL}) \\ &= 49 \times (0.001 \text{ L}) \\ &= 0.049 \text{ L} \end{aligned}$

**7.** $\begin{aligned} 0.401 \text{ mL} &= 0.401 \times (1 \text{ mL}) \\ &= 0.401 \times (0.001 \text{ L}) \\ &= 0.000401 \text{ L} \end{aligned}$

**9.** $\begin{aligned} 78.1 \text{ L} &= 78.1 \times (1 \text{ L}) \\ &= 78.1 \times (1000 \text{ cm}^3) \\ &= 78,100 \text{ cm}^3 \end{aligned}$

**11.** $\begin{aligned} 10 \text{ qt} &= 10 \times 1 \text{ qt} \\ &= 10 \times 2 \text{ pt} \\ &= 10 \times 2 \times 1 \text{ pt} \\ &= 10 \times 2 \times 16 \text{ oz} \\ &= 320 \text{ oz} \end{aligned}$

**13.** $20 \text{ cups} = 20 \text{ cups} \cdot \frac{1 \text{ pt}}{2 \text{ cups}} = \frac{20}{2} \cdot 1 \text{ pt} = 10 \text{ pt}$

**15.** $\begin{aligned} 8 \text{ gal} &= 8 \times 1 \text{ gal} \\ &= 8 \times 4 \text{ qt} \\ &= 32 \text{ qt} \end{aligned}$

**17.** $\begin{aligned} 5 \text{ gal} &= 5 \times 1 \text{ gal} \\ &= 5 \times 4 \text{ qt} \\ &= 20 \text{ qt} \end{aligned}$

**19.** $56 \text{ qt} = 56 \text{ qt} \times \frac{1 \text{ gal}}{4 \text{ qt}} = \frac{56}{4} \cdot 1 \text{ gal} = 14 \text{ gal}$

**21.** $\begin{aligned} 11 \text{ gal} &= 11 \cdot 1 \text{ gal} \\ &= 11 \cdot 4 \text{ qt} \\ &= 11 \cdot 4 \cdot 1 \text{ qt} \\ &= 11 \cdot 4 \cdot 2 \text{ pt} \\ &= 88 \text{ pt} \end{aligned}$

**23.** Convert to gallons:

$$144 \text{ oz} = 144 \text{ oz} \cdot \frac{1 \text{ pt}}{16 \text{ oz}} \cdot \frac{1 \text{ qt}}{2 \text{ pt}} \cdot \frac{1 \text{ gal}}{4 \text{ qt}} =$$

$$\frac{144}{16 \cdot 2 \cdot 4} \cdot 1 \text{ gal} = \frac{144}{128} \text{ gal} = 1.125 \text{ gal}$$

Convert gallons to quarts:

$$1.125 \text{ gal} = 1.125 \cdot 1 \text{ gal}$$
$$= 1.125 \cdot 4 \text{ qt}$$
$$= 4.5 \text{ qt}$$

Convert from quarts to pints:

$$4.5 \text{ qt} = 4.5 \cdot 1 \text{ qt}$$
$$= 4.5 \cdot 2 \text{ pt}$$
$$= 9 \text{ pt}$$

Convert from pints to cups:

$$9 \text{ pt} = 9 \cdot 1 \text{ pt}$$
$$= 9 \cdot 2 \text{ cups}$$
$$= 18 \text{ cups}$$

**25.** Convert from gallons to quarts:

$$16 \text{ gal} = 16 \cdot 1 \text{ gal}$$
$$= 16 \cdot 4 \text{ qt}$$
$$= 64 \text{ qt}$$

Convert from quarts to pints:

$$64 \text{ qt} = 64 \cdot 1 \text{ qt}$$
$$= 64 \cdot 2 \text{ pt}$$
$$= 128 \text{ pt}$$

Convert from pints to cups:

$$128 \text{ pt} = 128 \cdot 1 \text{ pt}$$
$$= 128 \cdot 2 \text{ cups}$$
$$= 256 \text{ cups}$$

Convert from cups to ounces:

$$256 \text{ cups} = 256 \cdot 1 \text{ cup}$$
$$= 256 \cdot 8 \text{ oz}$$
$$= 2048 \text{ oz}$$

**27.** First convert cups to gallons:

$$4 \text{ cups} = 4 \text{ cups} \cdot \frac{1 \text{ pt}}{2 \text{ cups}} \cdot \frac{1 \text{ qt}}{2 \text{ pt}} \cdot \frac{1 \text{ gal}}{4 \text{ qt}} =$$

$$\frac{4}{2 \cdot 2 \cdot 4} = \frac{1}{4} \text{ gal, or } 0.25 \text{ gal}$$

Convert gallons to quarts:

$$0.25 \text{ gal} = 0.25 \cdot 1 \text{ gal}$$
$$= 0.25 \cdot 4 \text{ qt}$$
$$= 1 \text{ qt}$$

From the list of conversions on page 539 of the text, we know that 1 qt = 2 pt.

Convert pints to ounces:

$$2 \text{ pt} = 2 \times 1 \text{ pt}$$
$$= 2 \times 16 \text{ oz}$$
$$= 32 \text{ oz}$$

**29.** First convert ounces to gallons:

$$51 \text{ oz} = 51 \text{ oz} \cdot \frac{1 \text{ pt}}{16 \text{ oz}} \cdot \frac{1 \text{ qt}}{2 \text{ pt}} \cdot \frac{1 \text{ gal}}{4 \text{ qt}} =$$

$$\frac{51}{16 \cdot 2 \cdot 4} = 0.3984375 \text{ gal}$$

Convert gallons to quarts:

$$0.3984375 \text{ gal} = 0.3984375 \times 1 \text{ gal}$$
$$= 0.3984375 \times 4 \text{ qt}$$
$$= 1.59375 \text{ qt}$$

Convert quarts to pints:

$$1.59375 \text{ qt} = 1.59375 \times 1 \text{ qt}$$
$$= 1.59375 \times 2 \text{ pt}$$
$$= 3.1875 \text{ pt}$$

Convert pints to cups:

$$3.1875 \text{ pt} = 3.1875 \times 1 \text{ pt}$$
$$= 3.1875 \times 2 \text{ cups}$$
$$= 6.375 \text{ cups}$$

**31.** To convert from L to mL, move the decimal point 3 places to the right. We also know that 1 mL= 1 cc = 1 cm$^3$. Thus, we have 2 L = 2000 mL = 2000 cc = 2000 cm$^3$.

**33.** To convert from L to mL, move the decimal point 3 places to the right. We also know that 1 mL = 1 cc = 1 cm$^3$. Thus, we have 64 L = 64,000 mL = 64,000 cc = 64,000 cm$^3$.

**35.** To convert from cc to L, move the decimal point 3 places to the left: 207 cc = 0.207 L. We know that 1 cc = 1 mL = 1 cm$^3$, so we also have 207 cc = 207 mL = 207 cm$^3$.

**37.** To convert L to mL, move the decimal point 3 places to the right: 2.0 L = 2000 mL.

**39.** To convert from mL to L, move the decimal point 3 places to the left: 320 mL = 0.32 L.

**41.** First we multiply to find the number of ounces ingested in a day: 0.5 oz $\times$ 4 = 2 oz.

Now we convert ounces to milliliters:

2 oz = 2 $\times$ 1 oz $\approx$ 2 $\times$ 29.57 mL = 59.14 mL

**43.** We convert 0.5 L to milliliters:
$$0.5 \text{ L} = 0.5 \times 1 \text{ L}$$
$$= 0.5 \times 1000 \text{ mL}$$
$$= 500 \text{ mL}$$

**45.** To convert L to mL, move the decimal point 3 places to the right: 3.0 L = 3000 mL. Now we divide to find the number of mL administered per hour:

$$\frac{3000 \text{ mL}}{24 \text{ hr}} = 125 \text{ mL/hr}$$

**47.** $45 \text{ mL} = 45 \text{ mL} \cdot \dfrac{1 \text{ tsp}}{5 \text{ mL}} = \dfrac{45}{5} \text{ tsp} = 9 \text{ tsp}$

**49.** $1 \text{ mL} = 1 \text{ mL} \cdot \dfrac{1 \text{ tsp}}{5 \text{ mL}} = \dfrac{1}{5} \text{ tsp}$

**51.** $2 \text{ T} = 2 \times 1 \text{ T}$
$$= 2 \times 3 \text{ tsp}$$
$$= 6 \text{ tsp}$$

**53.** $1 \text{ T} = 1 \cancel{\text{T}} \cdot \dfrac{3 \cancel{\text{tsp}}}{1 \cancel{\text{T}}} \cdot \dfrac{5 \text{ mL}}{1 \cancel{\text{tsp}}} = 15 \text{ mL}$

**55.**    $0.452$          $0.45.2$          Move the decimal point
                              $\underset{\llcorner\uparrow}{\phantom{x}}$          2 places to the right.

Write a % symbol: $45.2\%$

$0.452 = 45.2\%$

**57.** $\dfrac{1}{3} = 0.33\overline{3}$

$0.33.\overline{3}$          Move the decimal point
$\underset{\llcorner\uparrow}{\phantom{x}}$          2 places to the right.

Write a % symbol: $33.\overline{3}\%$

$\dfrac{1}{3} = 33.\overline{3}\%, \text{ or } 33\dfrac{1}{3}\%$

**59.** We multiply by 1 to get 100 in the denominator.

$\dfrac{11}{20} = \dfrac{11}{20} \cdot \dfrac{5}{5} = \dfrac{55}{100} = 55\%$

**61.** $\dfrac{22}{25} = \dfrac{22}{25} \cdot \dfrac{4}{4} = \dfrac{88}{100}, \text{ so } \dfrac{22}{25} = 88\%.$

**63.** The ratio of hardwood board feet to total board feet is $\dfrac{10.6}{71.3}$. We can express this without decimals as follows:

$\dfrac{10.6}{71.3} = \dfrac{10.6}{71.3} \cdot \dfrac{10}{10} = \dfrac{106}{713}.$

The ratio of softwood board feet to hardwood board feet is $\dfrac{60.7}{10.6}$. We can express this without decimals as follows:

$\dfrac{60.7}{10.6} = \dfrac{60.7}{10.6} \cdot \dfrac{10}{10} = \dfrac{607}{106}.$

**65.** We will express the weight in ounces, so we first convert 100 lb to ounces:

$100 \text{ lb} = 100 \times 1 \text{ lb} = 100 \times 16 \text{ oz} = 1600 \text{ oz}$

We divide to find the number of ounces of honey produced by each honey bee: $1600 \div 60,000 = 0.02\overline{6}$. Since each bee produces $\dfrac{1}{8}$ tsp of honey, we know that $\dfrac{1}{8}$ tsp of honey weighs $0.02\overline{6}$ oz. Now multiply to find the weight a tsp of honey: $0.02\overline{6} \text{ oz} \cdot 8 = 0.21\overline{3} \text{ oz}.$

A teaspoon of honey weighs $0.21\overline{3}$ oz.

**67.** $\$2.54/\text{gallon} \approx \dfrac{\$2.54}{1 \text{ gal}} \cdot \dfrac{1 \text{ gal}}{4 \text{ qt}} \cdot \dfrac{1.057 \text{ qt}}{1 \text{ L}} \approx \$0.671/\text{L}$

---

## Exercise Set 9.6

**1.** $1 \text{ day} = 24 \text{ hr}$

This conversion relation is given in the text on page 545.

**3.** $1 \text{ min} = 60 \text{ sec}$

This conversion relation is given in the text on page 545.

**5.** $1 \text{ yr} = 365\dfrac{1}{4} \text{ days}$

This conversion relation is given in the text on page 545.

**7.** $180 \text{ sec} = 180 \cancel{\text{ sec}} \cdot \dfrac{1 \text{ min}}{60 \cancel{\text{ sec}}} \cdot \dfrac{1 \text{ hr}}{60 \cancel{\text{ min}}}$

$\phantom{180 \text{ sec}} = \dfrac{180}{60 \cdot 60} \text{ hr}$

$\phantom{180 \text{ sec}} = 0.05 \text{ hr}$

**9.** $492 \text{ sec} = 492 \cancel{\text{ sec}} \times \dfrac{1 \text{ min}}{60 \cancel{\text{ sec}}}$

$\phantom{492 \text{ sec}} = \dfrac{492}{60} \text{ min}$

$\phantom{492 \text{ sec}} = 8.2 \text{ min}$

**11.** $156 \text{ hr} = 156 \cancel{\text{ hr}} \cdot \dfrac{1 \text{ day}}{24 \cancel{\text{ hr}}}$

$\phantom{156 \text{ hr}} = \dfrac{156}{24} \text{ days}$

$\phantom{156 \text{ hr}} = 6.5 \text{ days}$

**13.** $645 \text{ min} = 645 \cancel{\text{ min}} \cdot \dfrac{1 \text{ hr}}{60 \cancel{\text{ min}}}$

$\phantom{645 \text{ min}} = \dfrac{645}{60} \text{ hr}$

$\phantom{645 \text{ min}} = 10.75 \text{ hr}$

**15.** $2 \text{ wk} = 2 \times 1 \text{ wk}$

$\phantom{2 \text{ wk}} = 2 \times 7 \text{ days} \qquad \text{Substituting 7 days for 1 wk}$

$\phantom{2 \text{ wk}} = 14 \text{ days}$

$\phantom{2 \text{ wk}} = 14 \times 1 \text{ day}$

$\phantom{2 \text{ wk}} = 14 \times 24 \text{ hr} \qquad \text{Substituting 24 hr for 1 day}$

$\phantom{2 \text{ wk}} = 336 \text{ hr}$

**17.** $756 \text{ hr} = 756 \cancel{\text{ hr}} \cdot \dfrac{1 \cancel{\text{ day}}}{24 \cancel{\text{ hr}}} \cdot \dfrac{1 \text{ wk}}{7 \cancel{\text{days}}}$

$\phantom{756 \text{ hr}} = \dfrac{756}{24 \cdot 7} \text{ wk}$

$\phantom{756 \text{ hr}} = 4.5 \text{ wk}$

**19.** $2922 \text{ wk} = 2922 \cancel{\text{ wk}} \cdot \dfrac{7 \cancel{\text{ days}}}{1 \cancel{\text{ wk}}} \cdot \dfrac{1 \text{ yr}}{365\frac{1}{4} \cancel{\text{ days}}}$

$\phantom{2922 \text{ wk}} = \dfrac{2922 \cdot 7}{365\frac{1}{4}} \text{ yr}$

$\phantom{2922 \text{ wk}} = 56 \text{ yr}$

**21.** First find the number of seconds in 23 hours:

$23 \text{ hr} = 23 \times 1 \text{ hr}$

$\phantom{23 \text{ hr}} = 23 \times 60 \text{ min}$

$\phantom{23 \text{ hr}} = 1380 \text{ min}$

$\phantom{23 \text{ hr}} = 1380 \times 1 \text{ min}$

$\phantom{23 \text{ hr}} = 1380 \times 60 \text{ sec}$

$\phantom{23 \text{ hr}} = 82,800 \text{ sec}$

Next find the number of seconds in 56 minutes:

$56 \text{ min} = 56 \times 1 \text{ min}$

$\phantom{56 \text{ min}} = 56 \times 60 \text{ sec}$

$\phantom{56 \text{ min}} = 3360 \text{ sec}$

Finally, we add to find the number of seconds in a day:

$82,800 + 3360 + 4.2 = 86,164.2 \text{ sec}$

**23.** $F = \dfrac{9}{5} \cdot C + 32$

$\quad F = \dfrac{9}{5} \cdot 25 + 32$

$\quad\quad = 45 + 32$

$\quad\quad = 77$

Thus, $25°C = 77°F$.

**25.** $F = \dfrac{9}{5} \cdot C + 32$

$\quad F = \dfrac{9}{5} \cdot 40 + 32$

$\quad\quad = 72 + 32$

$\quad\quad = 104$

Thus, $40°C = 104°F$.

**27.** $F = 1.8C + 32$

$\quad F = 1.8 \cdot 86 + 32$

$\quad\quad = 154.8 + 32$

$\quad\quad = 186.8$

Thus, $86°C = 186.8°F$.

**29.** $F = 1.8C + 32$

$\quad F = 1.8 \cdot (-20) + 32$

$\quad\quad = -36 + 32$

$\quad\quad = -4$

Thus, $-20°C = -4°F$.

**31.** $F = 1.8C + 32$

$\quad F = 1.8 \cdot 2 + 32$

$\quad\quad = 3.6 + 32$

$\quad\quad = 35.6$

Thus, $2°C = 35.6°F$.

**33.** $F = 1.8C + 32$

$\quad F = 1.8 \cdot (-24) + 32$

$\quad\quad = -43.2 + 32$

$\quad\quad = -11.2$

Thus, $-24°C = -11.2°F$.

**35.** $F = \dfrac{9}{5} \cdot C + 32$

$\quad F = \dfrac{9}{5} \cdot 3000 + 32$

$\quad\quad = 5400 + 32$

$\quad\quad = 5432$

Thus, $3000°C = 5432°F$.

**37.** $C = \dfrac{5}{9} \cdot (F - 32)$

$\quad C = \dfrac{5}{9} \cdot (86 - 32)$

$\quad\quad = \dfrac{5}{9} \cdot 54$

$\quad\quad = 30$

Thus, $86°F = 30°C$.

**39.** $C = \dfrac{5}{9} \cdot (F - 32)$

$\quad C = \dfrac{5}{9} \cdot (-13 - 32)$

$\quad\quad = \dfrac{5}{9} \cdot (-45)$

$\quad\quad = -25$

Thus, $-13°F = -25°C$.

**41.** $C = \dfrac{F - 32}{1.8}$

$\quad C = \dfrac{178 - 32}{1.8}$

$\quad\quad = \dfrac{146}{1.8}$

$\quad\quad = 81.\overline{1}$

Thus, $178°F = 81.\overline{1}°C$.

**43.** $C = \dfrac{F - 32}{1.8}$

$\quad C = \dfrac{140 - 32}{1.8}$

$\quad\quad = \dfrac{108}{1.8}$

$\quad\quad = 60$

Thus, $140°F = 60°C$.

**45.** $C = \dfrac{F - 32}{1.8}$

$\quad C = \dfrac{68 - 32}{1.8}$

$\quad\quad = \dfrac{36}{1.8}$

$\quad\quad = 20$

Thus, $68°F = 20°C$.

**47.** $C = \dfrac{F - 32}{1.8}$

$\quad C = \dfrac{10 - 32}{1.8}$

$\quad\quad = \dfrac{-22}{1.8}$

$\quad\quad = -12.\overline{2}$

Thus, $10°F = -12.\overline{2}°C$.

**49.** $C = \dfrac{5}{9} \cdot (F - 32)$

$\quad C = \dfrac{5}{9} \cdot (98.6 - 32)$

$\quad\quad = \dfrac{5}{9} \cdot 66.6$

$\quad\quad = 37$

Thus, $98.6°F = 37°C$.

**51.** a) $C = \dfrac{F-32}{1.8}$

$\qquad C = \dfrac{136-32}{1.8}$

$\qquad\quad = \dfrac{104}{1.8}$

$\qquad\quad = 57.\overline{7}$

Thus, $136°F = 57.\overline{7}°C$.

$\qquad F = \dfrac{9}{5} \cdot C + 32$

$\qquad F = \dfrac{9}{5} \cdot 56\dfrac{2}{3} + 32$

$\qquad\quad = \dfrac{9}{5} \cdot \dfrac{170}{3} + 32$

$\qquad\quad = 102 + 32$

$\qquad\quad = 134$

Thus, $56\dfrac{2}{3}°C = 134°F$.

b) $136°F - 134°F = 2°F$

The world record is $2°F$ higher than the U. S. record.

**53.** $C = \dfrac{5}{9} \cdot (F-32)$

$\qquad C = \dfrac{5}{9} \cdot (12-32)$

$\qquad\quad = \dfrac{5}{9} \cdot (-20)$

$\qquad\quad \approx -11.1$

Thus, $12°F \approx -11.1°C$.

**55.** When interest is paid on interest, it is called <u>compound</u> interest.

**57.** The <u>median</u> of a set of data is the middle number if there is an odd number of data items.

**59.** In <u>similar</u> triangles, the lengths of their corresponding sides have the same ratio.

**61.** A natural number, other than 1, that is not prime is <u>composite</u>.

**63.** $\quad 1,000,000 \text{ sec}$

$= 1,000,000 \,\cancel{\text{sec}} \times \dfrac{1 \,\cancel{\text{min}}}{60 \,\cancel{\text{sec}}} \times \dfrac{1 \,\cancel{\text{hr}}}{60 \,\cancel{\text{min}}} \times \dfrac{1 \,\cancel{\text{day}}}{24 \,\cancel{\text{hr}}} \times \dfrac{1 \text{ yr}}{365\dfrac{1}{4} \,\cancel{\text{days}}}$

$\approx 0.03 \text{ yr}$

**65.** $\quad 1,000,000,000,000 \text{ sec}$

$= 1,000,000,000,000 \,\cancel{\text{sec}} \times \dfrac{1 \,\cancel{\text{min}}}{60 \,\cancel{\text{sec}}} \times \dfrac{1 \,\cancel{\text{hr}}}{60 \,\cancel{\text{min}}} \times \dfrac{1 \,\cancel{\text{day}}}{24 \,\cancel{\text{hr}}} \times$

$\quad \dfrac{1 \text{ yr}}{365\dfrac{1}{4} \,\cancel{\text{days}}}$

$\approx 31,688 \text{ yr}$

**67.** $0.9\dfrac{\text{L}}{\text{hr}} = 0.9\dfrac{\cancel{\text{L}}}{\cancel{\text{hr}}} \cdot \dfrac{1000 \text{ mL}}{1\cancel{\text{L}}} \cdot \dfrac{1 \,\cancel{\text{hr}}}{60 \,\cancel{\text{min}}} \cdot \dfrac{1 \,\cancel{\text{min}}}{60 \text{ sec}} = 0.25\dfrac{\text{mL}}{\text{sec}}$

## Exercise Set 9.7

**1.** $1 \text{ ft}^2 = 144 \text{ in}^2$

This conversion relation is given in the text on page 551.

**3.** $1 \text{ mi}^2 = 640 \text{ acres}$

This conversion relation is given in the text on page 551.

**5.** $\quad 1 \text{ in}^2 = 1 \text{ in}^2 \times \dfrac{1 \text{ ft}^2}{144 \text{ in}^2}\qquad$ Multiplying by 1

$\qquad\qquad\qquad\qquad\qquad\qquad\quad$ using $\dfrac{1 \text{ ft}^2}{144 \text{ in}^2}$

$\qquad\quad = \dfrac{1}{144} \times \dfrac{\text{in}^2}{\text{in}^2} \times 1 \text{ ft}^2$

$\qquad\quad = \dfrac{1}{144} \text{ ft}^2$

**7.** $\quad 22 \text{ yd}^2 = 22 \times 1 \text{ yd}^2$

$\qquad\qquad\quad = 22 \times 9 \text{ ft}^2\qquad$ Substituting $9 \text{ ft}^2$

$\qquad\qquad\qquad\qquad\qquad\qquad$ for $1 \text{ yd}^2$

$\qquad\qquad\quad = 198 \text{ ft}^2$

**9.** $\quad 44 \text{ yd}^2 = 44 \cdot 1 \text{ yd}^2$

$\qquad\qquad\quad = 44 \cdot 9 \text{ ft}^2\qquad$ Substituting $9 \text{ ft}^2$

$\qquad\qquad\qquad\qquad\qquad\qquad$ for $1 \text{ yd}^2$

$\qquad\qquad\quad = 396 \text{ ft}^2$

**11.** $20 \text{ mi}^2 = 20 \times 1 \text{ mi}^2$

$\qquad\qquad\quad = 20 \cdot 640 \text{ acres}\qquad$ Substituting $640$ acres

$\qquad\qquad\qquad\qquad\qquad\qquad\qquad$ for $1 \text{ mi}^2$

$\qquad\qquad\quad = 12,800 \text{ acres}$

**13.** $\quad 1 \text{ mi}^2 = 1 \cdot (1 \text{ mi})^2$

$\qquad\qquad = 1 \cdot (5280 \text{ ft})^2\qquad$ Substituting $5280$ ft

$\qquad\qquad\qquad\qquad\qquad\qquad\quad$ for $1 \text{ mi}$

$\qquad\qquad = 5280 \text{ ft} \cdot 5280 \text{ ft}$

$\qquad\qquad = 27,878,400 \text{ ft}^2$

**15.** $720 \text{ in}^2 = 720 \text{ in}^2 \times \dfrac{1 \text{ ft}^2}{144 \text{ in}^2}\qquad$ Multiplying by 1

$\qquad\qquad\qquad\qquad\qquad\qquad\qquad\quad$ using $\dfrac{1 \text{ ft}^2}{144 \text{ in}^2}$

$\qquad\qquad\quad = \dfrac{720}{144} \times \dfrac{\text{in}^2}{\text{in}^2} \times 1 \text{ ft}^2$

$\qquad\qquad\quad = 5 \text{ ft}^2$

**17.** $144 \text{ in}^2 = 1 \text{ ft}^2$

This conversion relation is given in the text on page 551.

**19.** $\quad 1 \text{ acre} = 1 \text{ acre} \cdot \dfrac{1 \text{ mi}^2}{640 \text{ acres}}$

$\qquad\qquad\quad = \dfrac{1}{640} \cdot \dfrac{\text{acres}}{\text{acres}} \cdot 1 \text{ mi}^2$

$\qquad\qquad\quad = \dfrac{1}{640} \text{ mi}^2, \text{ or } 0.0015625 \text{ mi}^2$

**21.** $40.3 \text{ mi}^2 = 40.3 \times 1 \text{ mi}^2$

$\qquad\qquad\quad = 40.3 \times 640 \text{ acres}$

$\qquad\qquad\quad = 25,792 \text{ acres}$

**23.** $5.21 \text{ km}^2 = \underline{\hspace{1cm}} \text{ m}^2$

Think: To go from km to m in the diagram is a move of 3 places to the right. So we move the decimal point $2 \cdot 3$, or 6 places to the right.

5.21    5.210000.

$5.21 \text{ km}^2 = 5{,}210{,}000 \text{ m}^2$

**25.** $0.014 \text{ m}^2 = \underline{\hspace{1cm}} \text{ cm}^2$

Think: To go from m to cm in the diagram is a move of 2 places to the right. So we move the decimal point $2 \cdot 2$, or 4 places to the right.

0.014    0.0140.

$0.014 \text{ m}^2 = 140 \text{ cm}^2$

**27.** $2345.6 \text{ mm}^2 = \underline{\hspace{1cm}} \text{ cm}^2$

Think: To go from mm to cm in the diagram is a move of 1 place to the left. So we move the decimal point $2 \cdot 1$, or 2 places to the left.

2345.6    23.45.6

$2345.6 \text{ mm}^2 = 23.456 \text{ cm}^2$

**29.** $852.14 \text{ cm}^2 = \underline{\hspace{1cm}} \text{ m}^2$

Think: To go from cm to m in the diagram is a move of 2 places to the left. So we move the decimal point $2 \cdot 2$, or 4 places to the left.

852.14    0.0852.14

$852.14 \text{ cm}^2 = 0.085214 \text{ m}^2$

**31.** $250{,}000 \text{ mm}^2 = \underline{\hspace{1cm}} \text{ cm}^2$

Think: To go from mm to cm in the diagram is a move of 1 place to the left. So we move the decimal point $2 \cdot 1$, or 2 places to the left.

250,000    2500.00.

$250{,}000 \text{ mm}^2 = 2500 \text{ cm}^2$

**33.** $472{,}800 \text{ m}^2 = \underline{\hspace{1cm}} \text{ km}^2$

Think: To go from m to km in the diagram is a move of 3 places to the left. So we move the decimal point $2 \cdot 3$, or 6 places to the left.

472,800    0.472800.

$472{,}800 \text{ m}^2 = 0.4728 \text{ km}^2$

**35.** $\text{Interest} = P \cdot r \cdot t$
$$= \$2000 \times 8\% \times 1.5$$
$$= \$2000 \times 0.08 \times 1.5$$
$$= \$240$$

The interest is $240.

**37.** a) $I = P \cdot r \cdot t$
$$= \$15{,}500 \times 9.5\% \times \frac{120}{365}$$
$$= \$15{,}500 \times 0.095 \times \frac{120}{365}$$
$$\approx \$484.11$$

The amount of simple interest due is $484.11.

b) The total amount that must be paid back is the amount borrowed plus the interest:
$$\$15{,}500 + \$484.11 = \$15{,}984.11$$

**39.** a) $I = P \cdot r \cdot t$
$$= \$6400 \times 8.4\% \times \frac{150}{365}$$
$$= \$6400 \times 0.084 \times \frac{150}{365}$$
$$\approx \$220.93$$

The amount of simple interest due is $220.93.

b) The total amount that must be paid back is the amount borrowed plus the interest:
$$\$6400 + \$220.93 = \$6620.93$$

**41.** $1 \text{ m}^2 = 1 \times 1 \text{ m} \times 1 \text{ m}$
$$\approx 1 \times 3.281 \text{ ft} \times 3.281 \text{ ft}$$
$$= 1 \times 3.281 \times 3.281 \times \text{ ft} \times \text{ ft}$$
$$\approx 10.76 \text{ ft}^2$$

**43.** $2 \text{ yd}^2 = 2 \times 1 \text{ yd} \times 1 \text{ yd}$
$$\approx 2 \times 3\,\cancel{\text{ft}} \times 3\,\cancel{\text{ft}} \times \frac{1 \text{ m}}{3.281\,\cancel{\text{ft}}} \times \frac{1 \text{ m}}{3.281\,\cancel{\text{ft}}}$$
$$= \frac{2 \times 3 \times 3}{3.281 \times 3.281} \times \text{ m} \times \text{ m}$$
$$\approx 1.67 \text{ m}^2$$

**45.** $153{,}000 \text{ ft}^2 = 153{,}000 \times 1 \text{ ft} \times 1 \text{ ft}$
$$\approx 153{,}000 \times 0.305 \text{ m} \times 0.305 \text{ m}$$
$$\approx 14{,}233 \text{ m}^2$$

## Chapter 9 Concept Reinforcement

**1.** Since a meter is just over a yard, or 3 feet, the given statement is true.

**2.** When converting from grams to milligrams, move the decimal point three places to the right. The given statement is false.

**3.** To convert mm to cm, we move the decimal point 1 place to the left. Then it follows that, to convert $\text{mm}^2$ to $\text{cm}^2$, we move the decimal point $2 \cdot 1$, or 2 places to the left. The given statement is true.

**4.** The statement is false. See Example 2 on page 551 in the text.

**5.** $40°\text{C} = 104°\text{F}$, so the given statement is false.

**6.** $10°\text{C} = 50°\text{F}$ and water does not freeze for temperatures above $32°\text{F}$, so the given statement is false.

## Chapter 9 Important Concepts

**1.**  $7 \text{ ft} = 7 \text{ ft} \times \dfrac{1 \text{ yd}}{3 \text{ ft}}$

$\quad = \dfrac{7 \text{ ft}}{3 \text{ ft}} \times 1 \text{ yd}$

$\quad = \dfrac{7}{3} \times \dfrac{\text{ft}}{\text{ft}} \times 1 \text{ yd}$

$\quad = \dfrac{7}{3} \text{ yd, or } 2\dfrac{1}{3} \text{ yd, or } 2.\overline{3} \text{ yd}$

**2.**  $2\dfrac{1}{2} \text{ mi} = 2\dfrac{1}{2} \times 1 \text{ mi}$

$\quad = \dfrac{5}{2} \times 5280 \text{ ft}$

$\quad = \dfrac{5 \times 5280}{2} \text{ ft}$

$\quad = 13,200 \text{ ft}$

**3.**  12 hm = _____ m

Think: To go from hm to m in the table is a move of 2 places to the right.  Thus, we move the decimal point 2 places to the right.

12 hm = 1200 m

**4.**  4.46 cm = _____ km

Think: To go from cm to km in the table is a move of 5 places to the left.  Thus, we move the decimal point 5 places to the left.

4.46 cm = 0.0000446 km

**5.**  $10 \text{ m} = 10 \times 1 \text{ m}$

$\quad \approx 10 \times 1.094 \text{ yd}$

$\quad = 10.94 \text{ yd}$

**6.**  $10,280 \text{ lb} = 10,280 \text{ lb} \times \dfrac{1 \text{ T}}{2000 \text{ lb}}$

$\quad = \dfrac{10,280 \text{ lb}}{2000 \text{ lb}} \times 1 \text{ T}$

$\quad = \dfrac{10,280}{2000} \times \dfrac{\text{lb}}{\text{lb}} \times 1 \text{ T}$

$\quad = 5.14 \text{ T}$

**7.**  9.78 mg = _____ g

Think: To go from mg to g in the table is a move of 3 places to the left.  Thus, we move the decimal point 3 places to the left.

9.78 mg = 0.00978 g

**8.**  $16 \text{ qt} = 16 \times 1 \text{ qt}$

$\quad = 16 \times 2 \text{ pt}$

$\quad = 16 \times 2 \times 1 \text{ pt}$

$\quad = 16 \times 2 \times 2 \text{ cups}$

$\quad = 64 \text{ cups}$

**9.**  42,670 mL = _____ L

Think: To go from mL to L in the table is a move of 3 places to the left.  Thus, we move the decimal point 3 places to the left.

42,670 mL = 42.67 L

**10.**  $3600 \text{ sec} = 3600 \text{ sec} \times \dfrac{1 \text{ min}}{60 \text{ sec}} \times \dfrac{1 \text{ hr}}{60 \text{ min}}$

$\quad = \dfrac{3600 \times 1 \times 1}{60 \times 60} \times \dfrac{\text{sec}}{\text{sec}} \times \dfrac{\text{min}}{\text{min}} \times 1 \text{ hr}$

$\quad = 1 \text{ hr}$

**11.**  $F = 1.8C + 32 = 1.8 \cdot 68 + 32$

$\quad\quad\quad\quad\quad\quad\quad = 122.4 + 32$

$\quad\quad\quad\quad\quad\quad\quad = 154.4$

Thus, $68°C = 154.4°F$.

**12.**  $C = \dfrac{5}{9}(F - 32) = \dfrac{5}{9}(104 - 32)$

$\quad\quad\quad\quad\quad\quad = \dfrac{5}{9}(72)$

$\quad\quad\quad\quad\quad\quad = 40$

Thus, $104°F = 40°C$.

**13.**  $81 \text{ ft}^2 = 81 \text{ ft}^2 \times \dfrac{1 \text{ yd}^2}{9 \text{ ft}^2}$

$\quad = \dfrac{81 \text{ ft}^2}{9 \text{ ft}^2} \times 1 \text{ yd}^2$

$\quad = \dfrac{81}{9} \times \dfrac{\text{ft}^2}{\text{ft}^2} \times 1 \text{ yd}^2$

$\quad = 9 \text{ yd}^2$

**14.**  52.4 cm² = _____ mm²

Think: To go from cm to mm in the table is a move of 1 place to the right. Thus, we move the decimal point $2 \cdot 1$, or 2 places to the right to convert cm² to mm².

52.4 cm² = 5240 mm²

## Chapter 9 Review Exercises

**1.**  $8 \text{ ft} = 8 \text{ ft} \times \dfrac{1 \text{ yd}}{3 \text{ ft}}$

$\quad = \dfrac{8}{3} \times 1 \text{ yd}$

$\quad = 2\dfrac{2}{3} \text{ yd}$

**2.**  $\dfrac{5}{6} \text{ yd} = \dfrac{5}{6} \times 1 \text{ yd}$

$\quad = \dfrac{5}{6} \times 36 \text{ in.}$

$\quad = \dfrac{5 \times 36}{6} \times 1 \text{ in.}$

$\quad = 30 \text{ in.}$

**3.** 0.3 mm = _____ cm

Think: To go from mm to cm in the table is a move of 1 place to the left. Thus, we move the decimal point 1 place to the left.

$$0.3 \quad 0.0.3$$
$$\overset{\curvearrowleft}{\sqcup}$$

0.3 mm = 0.03 cm

**4.** 4 m = _____ km

Think: To go from m to km in the table is a move of 3 places to the left. Thus, we move the decimal point 3 places to the left.

$$4 \quad 0.004.$$
$$\overset{\curvearrowleft}{\sqcup\!\!\sqcup}$$

4 m = 0.004 km

**5.**  2 yd = 2 × 1 yd
        = 2 × 36 in.
        = 72 in.

**6.** 4 km = _____ cm

Think: To go from km to cm in the table is a move of 5 places to the right. Thus, we move the decimal point 5 places to the right.

$$4 \quad 4.00000.$$
$$\underset{\sqcup\!\!\!\longrightarrow}{}$$

4 km = 400,000 cm

**7.**  14 in. = $14 \text{ in.} \times \dfrac{1 \text{ ft}}{12 \text{ in.}}$

        $= \dfrac{14}{12} \times 1 \text{ ft}$

        $= \dfrac{7}{6} \text{ ft, or } 1\dfrac{1}{6} \text{ ft}$

**8.** 15 cm = _____ m

Think: To go from cm to m in the table is a move of 2 places to the left. Thus, we move the decimal point 2 places to the left.

$$15 \quad 0.15.$$
$$\overset{\curvearrowleft}{\sqcup\!\!\sqcup}$$

15 cm = 0.15 m

**9.**  200 m = 200 × 1 m
        ≈ 200 × 1.094 yd
        = 218.8 yd

**10.**  20 mi = 20 × 1 mi
         ≈ 20 × 1.609 km
         = 32.18 km

**11.** 1 cm = _____ mm

Think: To go from cm to mm in the table is a move of 1 place to the right. Thus, we move the decimal point 1 place to the right.

1 cm = 10 mm

1 cm = _____ m

Think: To move from cm to m in the table is a move of 2 places to the left. Thus, we move the decimal point 2 places to the left.

1 cm = 0.01 m

**12.** 305 m = _____ mm

Think: To go from m to mm in the table is a move of 3 places to the right. Thus, we move the decimal point 3 places to the right.

305 m = 305,000 mm

305 m = _____ cm

Think: to move from m to cm in the table is a move of 2 places to the right. Thus, we move the decimal point 2 places to the right.

305 m = 30,500 cm

**13.**  7 lb = 7 × 1 lb
         = 7 × 16 oz
         = 112 oz

**14.** Complete: 4 g = _____ kg

Think: To go from g to kg in the table is a move of 3 places to the left. Thus, we move the decimal point 3 places to the left.

$$4 \quad 0.004.$$
$$\overset{\curvearrowleft}{\sqcup\!\!\sqcup}$$

4 g = 0.004 kg

**15.**  16 min = $16 \text{ min} \times \dfrac{1 \text{ hr}}{60 \text{ min}}$

         $= \dfrac{16}{60} \times 1 \text{ hr}$

         $= \dfrac{4}{15} \text{ hr, or } 0.2\overline{6} \text{ hr}$

**16.**  464 mL = 464 × 1 mL
         = 464 × 0.001 L
         = 0.464 L

**17.**  3 min = 3 × 1 min
         = 3 × 60 sec
         = 180 sec

**18.** Complete: 4.7 kg = _____ g

Think: To go from kg to g in the table is a move of 3 places to the right. Thus, we move the decimal point 3 places to the right.

$$4.7 \quad 4.700.$$
$$\underset{\sqcup\!\!\longrightarrow}{}$$

4.7 kg = 4700 g

**19.**  8.07 T = 8.07 × 1 T
         = 8.07 × 2000 lb
         = 16,140 lb

**20.**
$$0.83 \text{ L} = 0.83 \times 1 \text{ L}$$
$$= 0.83 \times 1000 \text{ mL}$$
$$= 830 \text{ mL}$$

**21.**
$$6 \text{ hr} = 6 \cancel{\text{hr}} \times \frac{1 \text{ day}}{24 \cancel{\text{hr}}}$$
$$= \frac{6}{24} \times 1 \text{ day}$$
$$= \frac{1}{4} \text{ day, or } 0.25 \text{ day}$$

**22.** $4 \text{ cg} = \underline{\hspace{1cm}} \text{ g}$

Think: To go from cg to g in the table is a move of 2 places to the left. Thus, we move the decimal point 2 places to the left.

    4   0.04.

$4 \text{ cg} = 0.04 \text{ g}$

**23.** $0.2 \text{ g} = \underline{\hspace{1cm}} \text{ mg}$

Think: To go from g to mg in the table is a move of 3 places to the right. Thus, we move the decimal point 3 places to the right.

    0.2   0.200.

$0.2 \text{ g} = 200 \text{ mg}$

**24.** Complete: $0.0003 \text{ kg} = \underline{\hspace{1cm}} \text{ cg}$

Think: To go from kg to cg in the table is a move of 5 places to the right. Thus, we move the decimal point 5 places to the right.

    0.0003   0.00030.

$0.0003 \text{ kg} = 30 \text{ cg}$

**25.**
$$0.7 \text{ mL} = 0.7 \times 1 \text{ mL}$$
$$= 0.7 \times 0.001 \text{ L}$$
$$= 0.0007 \text{ L}$$

**26.**
$$60 \text{ mL} = 60 \times 1 \text{ mL}$$
$$= 60 \times 0.001 \text{ L}$$
$$= 0.06 \text{ L}$$

**27.**
$$0.8 \text{ T} = 0.8 \times 1 \text{ T}$$
$$= 0.8 \times 2000 \text{ lb}$$
$$= 1600 \text{ lb}$$

**28.**
$$0.4 \text{ L} = 0.4 \times 1 \text{ L}$$
$$= 0.4 \times 1000 \text{ mL}$$
$$= 400 \text{ mL}$$

**29.**
$$20 \text{ oz} = 20 \cancel{\text{oz}} \times \frac{1 \text{ lb}}{16 \cancel{\text{oz}}}$$
$$= \frac{20}{16} \times 1 \text{ lb}$$
$$= \frac{5}{4} \text{ lb, or } 1.25 \text{ lb}$$

**30.**
$$\frac{5}{6} \text{ min} = \frac{5}{6} \cancel{\text{min}} \times \frac{60 \text{ sec}}{1 \cancel{\text{min}}}$$
$$= \frac{5 \times 60}{6} \times 1 \text{ sec}$$
$$= 50 \text{ sec}$$

**31.**
$$20 \text{ gal} = 20 \times 1 \text{ gal}$$
$$= 20 \times 4 \text{ qt}$$
$$= 20 \times 4 \times 1 \text{ qt}$$
$$= 20 \times 4 \times 2 \text{ pt}$$
$$= 160 \text{ pt}$$

**32.**
$$960 \text{ oz} = 960 \cancel{\text{oz}} \times \frac{1 \cancel{\text{pt}}}{16 \cancel{\text{oz}}} \times \frac{1 \cancel{\text{qt}}}{2 \cancel{\text{pt}}} \times \frac{1 \text{ gal}}{4 \cancel{\text{qt}}}$$
$$= \frac{960}{16 \times 2 \times 4} \times 1 \text{ gal}$$
$$= 7.5 \text{ gal}$$

**33.**
$$54 \text{ qt} = 54 \cancel{\text{qt}} \times \frac{1 \text{ gal}}{4 \cancel{\text{qt}}}$$
$$= \frac{54}{4} \times 1 \text{ gal}$$
$$= 13.5 \text{ gal}$$

**34.**
$$2.5 \text{ day} = 2.5 \times 1 \text{ day}$$
$$= 2.5 \times 24 \text{ hr}$$
$$= 60 \text{ hr}$$

**35.** Complete: $3020 \text{ cg} = \underline{\hspace{1cm}} \text{ kg}$

Think: To go from cg to kg in the table is a move of 5 places to the left. Thus, we move the decimal point 5 places to the left.

    3020   0.03020.

$3020 \text{ cg} = 0.0302 \text{ kg}$

**36.**
$$10,500 \text{ lb} = 10,500 \cancel{\text{lb}} \times \frac{1 \text{ T}}{2000 \cancel{\text{lb}}}$$
$$= \frac{10,500}{2000} \times 1 \text{ T}$$
$$= 5.25 \text{ T}$$

**37.** We use a proportion. Let $a =$ the number of mL of amoxicillin to be administered.
$$\frac{125}{150} = \frac{5}{a}$$
$$125 \cdot a = 150 \cdot 5$$
$$a = \frac{150 \cdot 5}{125}$$
$$a = \frac{25 \cdot 6 \cdot 5}{25 \cdot 5 \cdot 1} = \frac{25 \cdot 5}{25 \cdot 5} \cdot \frac{6}{1}$$
$$a = 6$$

The parent should administer 6 mL of amoxicillin.

**38.**
$$3 \text{ L} = 3 \times 1 \text{ L}$$
$$= 3 \times 1000 \text{ mL}$$
$$= 3000 \text{ mL}$$

**39.** 0.25 mg = 0.00025 g

$= 0.00025 \times 1$ g

$= 0.00025 \times 1,000,000$ mcg

$= 250$ mcg

**40.** $F = 1.8 \cdot C + 32$

$F = 1.8 \cdot (-6) + 32$

$= -10.8 + 32$

$= -21.2$

Thus, $-6°C = -21.2°F$.

**41.** $F = 1.8 \cdot C + 32$

$F = 1.8 \cdot 45 + 32$

$= 81 + 32$

$- 113$

Thus, $45°C = 113°F$.

**42.** $C = \dfrac{5}{9} \cdot (F - 32)$

$C = \dfrac{5}{9} \cdot (68 - 32)$

$= \dfrac{5}{9} \cdot 36$

$= 20$

Thus, $68°F = 20°C$.

**43.** $C = \dfrac{5}{9} \cdot (F - 32)$

$C = \dfrac{5}{9} \cdot (-20 - 32)$

$= \dfrac{5}{9} \cdot (-52)$

$= -28.\overline{8}$

Thus, $-20°F = -28.\overline{8}°C$.

**44.** $4 \text{ yd}^2 = 4 \times 1 \text{ yd}^2$

$= 4 \times 9 \text{ ft}^2$

$= 36 \text{ ft}^2$

**45.** $0.3 \text{ km}^2 = \underline{\hspace{1cm}} \text{ m}^2$

Think: To go from km to m in the diagram is a move of 3 places to the right. So we move the decimal point $2 \cdot 3$, or 6 places to the right.

0.3    0.300000.

$0.3 \text{ km}^2 = 300,000 \text{ m}^2$

**46.** $2070 \text{ in}^2 = 2070 \text{ in}^2 \times \dfrac{1 \text{ ft}^2}{144 \text{ in}^2}$

$= \dfrac{2070}{144} \times 1 \text{ ft}^2$

$= 14.375 \text{ ft}^2$

**47.** $600 \text{ cm}^2 = \underline{\hspace{1cm}} \text{ m}^2$

Think: To go from cm to m in the diagram is a move of 2 places to the left. So we move the decimal point $2 \cdot 2$, or 4 places to the left.

600    0.0600.

$600 \text{ cm}^2 = 0.06 \text{ m}^2$

**48.** $172.6 \text{ cm} = \underline{\hspace{1cm}} \text{ hm}$

Think: To go from cm to hm in the table is a move of 4 places to the left. Thus, we move the decimal point 4 places to the left.

$172.6 \text{ cm} = 0.01726 \text{ hm}$

Answer C is correct.

**49.** $0.16 \text{ gal} = 0.16 \times 1 \text{ gal}$

$= 0.16 \times 4 \text{ qt} = 0.16 \times 4 \times 1 \text{ qt}$

$= 0.16 \times 4 \times 2 \text{ pt} = 0.16 \times 4 \times 2 \times 1 \text{ pt}$

$= 0.16 \times 4 \times 2 \times 2 \text{ cups} = 2.56 \text{ cups}$

Answer B is correct.

**50.** Bolt's speed was $\dfrac{200 \text{ m}}{19.30 \text{ sec}}$ or approximately 10.3627 m/sec.

We convert 200 yd to meters:

$200 \text{ yd} = 200 \times 1 \text{ yd}$

$\approx 200 \times 0.914 \text{ m}$

$= 182.8 \text{ m}$

At a rate of 10.3627 m/sec, Bolt would run 200 yd, or 182.8 m, in a time of $\dfrac{182.8 \text{ m}}{10.3627 \text{ m/sec}}$, or about 17.64 sec.

**51.** 1 gal = 128 oz, so 1 oz of water (as capacity) weighs $\dfrac{8.3453}{128}$ lb, or about 0.0652 lb. An ounce of pennies weighs $\dfrac{1}{16}$ lb, or 0.0625 lb. Thus an ounce of water (as capacity) weighs more than an ounce of pennies.

## Chapter 9 Discussion and Writing Exercises

**1.** Grams are more easily converted to other units of mass than ounces. Since 1 gram is much smaller than 1 ounce, masses that might be expressed using fractional or decimal parts of ounces can often be expressed by whole numbers when grams are used.

**2.** A single container is all that is required for both types of measure.

**3.** Consider the diagram on page 552 in the text. Moving one place in the table corresponds to moving the decimal point one place. To convert units of length, we determine the corresponding number of moves in the diagram and then move the decimal point that number of places. Since area involves square units we multiply the number of moves by 2 and move the decimal point that number of places when converting units of area.

**4.** Since metric units are based on 10, they are more easily converted than American units.

**5.** a) $23°C = 73.4°F$, so you would want to play golf.

   b) $10°C = 50°F$, so you would not want to take a bath.

   c) Since $0°C = 32°F$, the freezing point of water, then the lake would certainly be frozen at the lower temperatures of $-10°$ C and it would be safe to go ice skating.

**6.** $1 \text{ m} \approx 3.281 \text{ ft}$, so 1 square meter $\approx 3.281 \text{ ft} \times 3.281 \text{ ft} = 10.764961$ square feet. Thus, one square meter is larger than 1 square foot.

## Chapter 9 Test

**1.**  $4 \text{ ft} = 4 \times 1 \text{ ft}$
   $= 4 \times 12 \text{ in.}$
   $= 48 \text{ in.}$

**2.**  $4 \text{ in.} = 4 \text{ in.} \times \dfrac{1 \text{ ft}}{12 \text{ in.}}$
   $= \dfrac{4 \text{ in.}}{12 \text{ in.}} \times 1 \text{ ft}$
   $= \dfrac{4}{12} \times \dfrac{\text{in.}}{\text{in.}} \times 1 \text{ ft}$
   $= \dfrac{1}{3} \times 1 \text{ ft}$
   $= \dfrac{1}{3} \text{ ft}$

**3.** a) 6 km = _____ m

Think: To go from km to m in the table is a move of 3 places to the right. Thus, we move the decimal point 3 places to the right.

   6   6.000.

6 km = 6000 m

**4.** 8.7 mm = _____ cm

Think: To go from mm to cm in the table is a move of 1 place to the left. Thus, we move the decimal point 1 place to the left.

   8.7   0.8.7

8.7 mm = 0.87 cm

**5.**  200 yd $= 200 \times 1 \text{ yd}$
   $\approx 200 \times 0.914 \text{ m}$
   $= 182.8 \text{ m}$

**6.**  2400 km $= 2400 \times 1 \text{ km}$
   $\approx 2400 \times 0.621 \text{ mi}$
   $= 1490.4 \text{ mi}$

**7.** 0.5 cm = _____ mm

Think: To go from cm to mm in the table is a move of 1 place to the right. Thus, we move the decimal point 1 place to the right.

0.5 cm = 5 mm

0.5 cm = _____ m

Think: To go from cm to m in the table is a move of 2 places to the left. Thus, we move the decimal point 2 places to the left.

0.5 cm = 0.005 m

**8.** 1.8542 m = _____ mm

Think: To go from m to mm in the table is a move of 3 places to the right. Thus, we move the decimal point 3 places to the right.

1.8542 m = 1854.2 mm

1.8542 m = _____ cm

Think: To go from m to cm in the table is a move of 2 places to the right. Thus, we move the decimal point 2 places to the right.

1.8542 m = 185.42 cm

**9.**  3080 mL $= 3080 \times 1 \text{ mL}$
   $= 3080 \times 0.001 \text{ L}$
   $= 3.08 \text{ L}$

**10.**  0.24 L $= 0.24 \times 1 \text{ L}$
   $= 0.24 \times 1000 \text{ mL}$
   $= 240 \text{ mL}$

**11.**  4 lb $= 4 \times 1 \text{ lb}$
   $= 4 \times 16 \text{ oz}$
   $= 64 \text{ oz}$

**12.**  4.11 T $= 4.11 \times 1 \text{ T}$
   $= 4.11 \times 2000 \text{ lb}$
   $= 8220 \text{ lb}$

**13.** 3.8 kg = _____ g

Think: To go from kg to g in the table is a move of 3 places to the right. Thus, we move the decimal point 3 places to the right.

3.8 kg = 3800 g

**14.** Complete: 4.325 mg = _____ cg

Think: To go from mg to cg in the table is a move of 1 place to the left. Thus, we move the decimal point 1 place to the left.

4.325 mg = 0.4325 cg

**15.** 2200 mg = _____ g

Think: To go from mg to g in the table is a move of 3 places to the left. Thus, we move the decimal point 3 places to the left.

2200 mg = 2.2 g

**16.** 5 hr = 5 × 1 hr

$\qquad$ = 5 × 60 min

$\qquad$ = 300 min

**17.** 15 days = 15 × 1 day

$\qquad$ = 15 × 24 hr

$\qquad$ = 360 hr

**18.** 64 pt = 64 ~~pt~~ × $\dfrac{1 \text{ qt}}{2 \text{ ~~pt~~}}$

$\qquad$ = $\dfrac{64}{2}$ × 1 qt

$\qquad$ = 32 qt

**19.** 10 gal = 10 × 1 gal = 10 × 4 qt

$\qquad$ = 10 × 4 × 1 qt = 10 × 4 × 2 pt

$\qquad$ = 10 × 4 × 2 × 1 pt = 10 × 4 × 2 × 16 oz

$\qquad$ = 1280 oz

**20.** 5 cups = 5 × 1 cup

$\qquad$ = 5 × 8 oz

$\qquad$ = 40 oz

**21.** 0.37 mg = 0.00037 g

$\qquad$ = 0.00037 × 1 g

$\qquad$ = 0.00037 × 1,000,000 mcg

$\qquad$ = 370 mcg

**22.** $C = \dfrac{F - 32}{1.8}$

$\qquad C = \dfrac{95 - 32}{1.8}$

$\qquad = \dfrac{63}{1.8}$

$\qquad = 35$

Thus, 95°F = 35°C.

**23.** $F = 1.8 \cdot C + 32$

$\qquad F = 1.8 \cdot 59 + 32$

$\qquad = 106.2 + 32$

$\qquad = 138.2$

Thus, 59°C = 138.2°F.

**24.** The table on page 525 of the text shows that 1 m ≈ 1.094 yd.

To convert m to cm we move the decimal point 2 places to the right.

$\qquad$ 1 m = 100 cm

The table on page 525 of the text shows that

$\qquad$ 1 m ≈ 39.370 in.

To convert m to mm we move the decimal point 3 places to the right.

$\qquad$ 1 m = 1000 mm

**25.** 398 yd = 398 × 1 yd ≈ 398 × 0.914 m

$\qquad$ = 363.772 m

$\qquad$ = 36,377.2 cm $\quad$ Moving the decimal point

$\qquad\qquad\qquad\qquad\qquad$ 2 places to the right

$\qquad$ 398 yd = 398 × 1 yd

$\qquad\qquad$ = 398 × 36 in.

$\qquad\qquad$ = 14,328 in.

From the first conversion above, we see that 398 yd = 363.772 m.

398 yd = 363.772 m

$\qquad$ = 363,772 mm $\quad$ Moving the decimal point

$\qquad\qquad\qquad\qquad\qquad$ 3 places to the right

**26.** To convert from L to mL, move the decimal point 3 places to the right.

$\qquad$ 2.5 L = 2500 mL

**27.** We multiply to find how many milligrams of the drug will be ingested each day:

$\qquad$ 0.5 · 3 = 1.5 mg

Now convert 1.5 mg to micrograms.

$\qquad$ 1.5 mg = 0.0015 g

$\qquad\qquad$ = 0.0015 × 1 g

$\qquad\qquad$ = 0.0015 × 1,000,000 mcg

$\qquad\qquad$ = 1500 mcg

**28.** 4 oz = 4 × 1 oz

$\qquad$ ≈ 4 × 29.57 mL

$\qquad$ = 118.28 mL

**29.** 12 ft$^2$ = 12 × 1 ft$^2$

$\qquad$ = 12 × 144 in$^2$

$\qquad$ = 1728 in$^2$

**30.** 3 cm$^2$ = _____ m$^2$

Think: To go from cm to m in the diagram is a move of 2 places to the left. So we move the decimal point 2 · 2, or 4 places to the left.

3 cm$^2$ = 0.0003 m$^2$

**31.** $F = 1.8 \cdot C + 32$

$\qquad$ = 1.8(45.5) + 32

$\qquad$ = 81.9 + 32

$\qquad$ = 113.9

Thus, 45.5°C = 113.9°F. Answer D is correct.

**32.** Johnson's speed was $\dfrac{400 \text{ m}}{43.18 \text{ sec}}$, or approximately 9.2635 m/sec.

We convert 400 yd to meters:

$\qquad$ 400 yd = 400 × 1 yd

$\qquad\qquad$ ≈ 400 × 0.914 m

$\qquad\qquad$ = 365.6 m

At a rate of 9.2635 m/sec, Johnson would run 400 yd, or 365.6 m, in a time of $\dfrac{365.6 \text{ m}}{9.2635 \text{ m/sec}}$, or about 39.47 sec.

## Cumulative Review Chapters 1 - 9

1. 37.3 million = 37.3 × 1 million
   $$= 37.3 \times 1,000,000$$
   $$= 37,300,000$$

   54.6 million = 54.6 × 1 million
   $$= 54.6 \times 1,000,000$$
   $$= 54,600,000$$

2. $\dfrac{561 \text{ mi}}{17 \text{ gal}} = 33$ mi/gal, or 33 mpg

3. *Familiarize*. First we find the amount of increase.
   $$\begin{array}{r} 2\,4,0\,0\,0 \\ -\,1\,9,0\,0\,0 \\ \hline 5\,0\,0\,0 \end{array}$$

   *Translate*. We translate to a percent equation.

   5000 is $\underbrace{\text{what percent}}$ of 19,000?

   $\downarrow \quad \downarrow \qquad \downarrow \qquad \downarrow \quad \downarrow$
   $5000 = \qquad p \quad \times \ 19,000$

   *Solve*.
   $$5000 = p \times 19,000$$
   $$\frac{5000}{19,000} = \frac{p \times 19,000}{19,000} \quad \text{Dividing by 19,000}$$
   $$0.263 \approx p$$
   $$26.3\% \approx p$$

   *Check*. With a 26.3% increase, the number of jobs in 2016 should be 100%+26.3%, or 126.3%, of the number in 2006.

   $126.3\% \times 19,000 = 1.263 \times 19,000 = 23,997 \approx 24,000$

   The answer checks.

   *State*. The percent of increase is about 26.3%.

4. *Familiarize*. First we find the amount of decrease.
   $$\begin{array}{r} 5\,2,0\,0\,0 \\ -\,5\,1,0\,0\,0 \\ \hline 1\,0\,0\,0 \end{array}$$

   *Translate*. We translate to a proportion.
   $$\frac{N}{100} = \frac{1000}{52,000}$$

   *Solve*.
   $$\frac{N}{100} = \frac{1000}{52,000}$$
   $$N \cdot 52,000 = 100 \cdot 1000 \quad \text{Equating cross products}$$
   $$\frac{N \cdot 52,000}{52,000} = \frac{100 \cdot 1000}{52,000} \quad \text{Dividing by 52,000}$$
   $$N \approx 1.9$$

   *Check*. With a 1.9% decrease, the number of jobs in 2016 should be 100% − 1.9%, or 98.1%, of the number in 2006.

   $98.1\% \times 52,000 = 0.981 \times 52,000 = 51,012 \approx 51,000$

   The answer checks.

   *State*. The percent of decrease is about 1.9%.

5.
$$\begin{array}{r} 4\,6,2\,3\,1 \\ \times \quad 1\,1\,0\,0 \\ \hline 4\,6\,2\,3\,1\,0\,0 \\ 4\,6\,2\,3\,1\,0\,0\,0 \\ \hline 5\,0,8\,5\,4,1\,0\,0 \end{array}$$

6. $\dfrac{1}{10} \cdot \left(-\dfrac{5}{6}\right) = -\dfrac{1 \cdot 5}{10 \cdot 6} = -\dfrac{1 \cdot 5}{2 \cdot 5 \cdot 6} = -\dfrac{1}{2 \cdot 6} \cdot \dfrac{5}{5} = -\dfrac{1}{12}$

7. $-14.5 + \dfrac{4}{5} - 0.1 = -14.5 + 0.8 - 0.1$
   $$= -13.7 - 0.1$$
   $$= -13.8$$

8. $-2\dfrac{3}{5} \div \left(-3\dfrac{9}{10}\right) = -\dfrac{13}{5} \div \left(-\dfrac{39}{10}\right) = -\dfrac{13}{5} \cdot \left(-\dfrac{10}{39}\right) =$
   $$\dfrac{13 \cdot 10}{5 \cdot 39} = \dfrac{13 \cdot 2 \cdot 5}{5 \cdot 3 \cdot 13} = \dfrac{13 \cdot 5}{13 \cdot 5} \cdot \dfrac{2}{3} = \dfrac{2}{3}$$

9. $0.1 \overline{)3.5\,6}$

   The divisor, 0.1, has 1 decimal place. To divide, we move the decimal point in the dividend 1 place to the right. The quotient is 35.6.

10.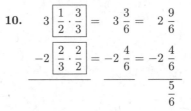

11. 1,298,032 is divisible by 8 because the number named by its last three digits, 32, is divisible by 8.

12. The sum of the digits, $5 + 0 + 2 + 4 + 1 + 2 + 0$, or 14, is not divisible by 3, so 5,024,120 is not divisible by 3.

13.
$$\begin{array}{r} 1\,1 \\ 3\,\overline{)\,3\,3} \\ 3\,\overline{)\,9\,9} \end{array}$$
   $99 = 3 \cdot 3 \cdot 11$

14. $35 = 5 \cdot 7$
    $49 = 7 \cdot 7$

    We use each factor the greatest number of times it occurs in either factorization.

    The LCM is $5 \cdot 7 \cdot 7$, or 245.

15. $35.\overline{7} = 35.777\ldots$

    The digit in the tenths place is 7. The next digit to the right, 7, is 5 or higher so we round up. We get 35.8.

16. One hundred three and sixty-four thousandths

17. Find the average:
    $$\frac{29 + 21 + 9 + 13 + 17 + 18}{6} = \frac{107}{6} = 17.8\overline{3}$$
    To find the median we first arrange the numbers from smallest to largest.

    9, 13, 17, 18, 21, 29

    The median is the average of the two middle numbers.
    $$\frac{17 + 18}{2} = \frac{35}{2} = 17.5$$

**18.** Move the decimal point two places to the right and write a % symbol.

$$0.08 = 8\%$$

**19.** $\dfrac{3}{5} = \dfrac{3}{5} \cdot \dfrac{20}{20} = \dfrac{60}{100} = 60\%$

**20.** $\begin{aligned} 2 \text{ yd} &= 2 \times 1 \text{ yd} \\ &= 2 \times 3 \text{ ft} \\ &= 6 \text{ ft} \end{aligned}$

**21.** $\begin{aligned} 6 \text{ oz} &= 6 \text{ oz} \times \dfrac{1 \text{ lb}}{16 \text{ oz}} \\ &= \dfrac{6}{16} \times \dfrac{\text{oz}}{\text{oz}} \times 1 \text{ lb} \\ &= \dfrac{3}{8} \text{ lb, or } 0.375 \text{ lb} \end{aligned}$

**22.** $\begin{aligned} \text{F} = 1.8^\circ\text{C} + 32 &= 1.8 \cdot 15 + 32 \\ &= 27 + 32 = 59 \end{aligned}$

Thus, $15^\circ\text{C} = 59^\circ\text{F}$.

**23.** $\begin{aligned} 0.087 \text{ L} &= 0.087 \times 1 \text{ L} \\ &= 0.087 \times 1000 \text{ mL} \\ &= 87 \text{ mL} \end{aligned}$

**24.** $\begin{aligned} 9 \text{ sec} &= 9 \text{ sec} \times \dfrac{1 \text{ min}}{60 \text{ sec}} \\ &= \dfrac{9}{60} \times \dfrac{\text{sec}}{\text{sec}} \times 1 \text{ min} \\ &= \dfrac{3}{20} \text{ min, or } 0.15 \text{ min} \end{aligned}$

**25.** $17 \text{ cm} = \underline{\phantom{xxxx}} \text{ m}$

Think: To go from cm to m in the table is a move of 2 places to the left. Thus, we move the decimal point 2 places to the left.

$$17 \text{ m} = 0.17 \text{ m}$$

**26.** $\begin{aligned} 2200 \text{ mi} &= 2200 \times 1 \text{ mi} \\ &\approx 2200 \times 1.609 \text{ km} \\ &= 3539.8 \text{ km} \end{aligned}$

**27.** $\begin{aligned} 2000 \text{ mL} &= 2000 \times 1 \text{ mL} \\ &= 2000 \times 0.001 \text{ L} \\ &= 2 \text{ L} \end{aligned}$

**28.** $\begin{aligned} 0.23 \text{ mg} &= 0.23 \times 1 \text{ mg} \\ &= 0.23 \times 0.001 \text{ g} \\ &= 0.23 \times 0.001 \times 1 \text{ g} \\ &= 0.23 \times 0.001 \times 1,000,000 \text{ mcg} \\ &= 230 \text{ mcg} \end{aligned}$

**29.** $\begin{aligned} 12 \text{ yd}^2 &= 12 \times 1 \text{ yd}^2 \\ &= 12 \times 9 \text{ ft}^2 \\ &= 108 \text{ ft}^2 \end{aligned}$

**30.** $\begin{aligned} 0.07 \cdot x &= -10.535 \\ \dfrac{0.07 \cdot x}{0.07} &= \dfrac{-10.535}{0.07} \\ x &= -150.5 \end{aligned}$

The solution is $-150.5$.

**31.** $\begin{aligned} x + 12{,}843 &= 32{,}091 \\ x + 12{,}843 - 12{,}843 &= 32{,}091 - 12{,}843 \\ x &= 19{,}248 \end{aligned}$

The solution is 19,248.

**32.** $\begin{aligned} \dfrac{2}{3} \cdot y &= 5 \\ \dfrac{\frac{2}{3} \cdot y}{\frac{2}{3}} &= \dfrac{5}{\frac{2}{3}} \\ y &= 5 \cdot \dfrac{3}{2} \\ y &= \dfrac{15}{2} \end{aligned}$

The solution is $\dfrac{15}{2}$.

**33.** $\begin{aligned} \dfrac{4}{5} + y &= -\dfrac{6}{7} \\ \dfrac{4}{5} + y - \dfrac{4}{5} &= -\dfrac{6}{7} - \dfrac{4}{5} \\ y &= -\dfrac{6}{7} \cdot \dfrac{5}{5} - \dfrac{4}{5} \cdot \dfrac{7}{7} \\ y &= -\dfrac{30}{35} - \dfrac{28}{35} \\ y &= -\dfrac{58}{35} \end{aligned}$

The solution is $-\dfrac{58}{35}$.

**34. *Familiarize*.** Let $h = $ the number of hours the mechanic worked on the car.

***Translate*.** We add the number of hours spent on each task to find the total time.

$$h = \dfrac{1}{3} + \dfrac{1}{2} + \dfrac{1}{10} + \dfrac{1}{4} + \dfrac{1}{15}$$

***Solve*.** We carry out the addition. The LCD is 60.

$$\dfrac{1}{3} + \dfrac{1}{2} + \dfrac{1}{10} + \dfrac{1}{4} + \dfrac{1}{15}$$
$$= \dfrac{1}{3} \cdot \dfrac{20}{20} + \dfrac{1}{2} \cdot \dfrac{30}{30} + \dfrac{1}{10} \cdot \dfrac{6}{6} + \dfrac{1}{4} \cdot \dfrac{15}{15} + \dfrac{1}{15} \cdot \dfrac{4}{4}$$
$$= \dfrac{20}{60} + \dfrac{30}{60} + \dfrac{6}{60} + \dfrac{15}{60} + \dfrac{4}{60}$$
$$= \dfrac{75}{60} = \dfrac{5 \cdot 15}{4 \cdot 15} = \dfrac{5}{4} \cdot \dfrac{15}{15}$$
$$= \dfrac{5}{4}, \text{ or } 1\dfrac{1}{4}$$

Thus, $h = \dfrac{5}{4}$, or $1\dfrac{1}{4}$.

***Check*.** We repeat the calculation. The answer checks.

***State*.** The mechanic spent $\dfrac{5}{4}$ hr, or $1\dfrac{1}{4}$ hr, working on the car.

**35. Familiarize.** Let $m$ = total milk production per year in America, in pounds. We express 9.112 million as 9,112,000.

**Translate.**

$$\underbrace{\text{Average production}}_{19,950} \underset{\text{times}}{\cdot} \underbrace{\text{Number}}_{9,112,000} \underset{\text{is}}{=} \underbrace{\text{Total}}_{m}$$

Average production per cow times Number of cows is Total production

$$19,950 \cdot 9,112,000 = m$$

**Solve.** We carry out the multiplication.

$$19,950 \cdot 9,112,000 = m$$
$$181,784,400,000 = m$$

**Check.** We can estimate the answer.

$$19,950 \cdot 9,112,000 \approx 20,000 \cdot 9,000,000 = 180,000,000,000$$

The estimate is close to our answer, so the answer seems reasonable.

**State.** 181,784,400,000 lb of milk are produced per year n America.

**36. Familiarize.** First we will find the number of miles $m$ that have been driven. Then we will find the gas mileage $g$.

**Translate and Solve.**

First odometer reading plus Number of miles driven is Second odometer reading

$$86,897.2 + m = 87,153.0$$

To solve the equation we subtract 86,897.2 on both sides.

$$86,897.2 + m - 86,897.2 = 87,153.0 - 86,897.2$$
$$m = 255.8$$

Next, we divide the number of miles driven by the number of gallons to find the gas mileage.

$$m = \frac{255.8}{16} = 15.9875$$

**Check.** First multiply the mileage by the number of gallons.

$$16 \times 15.9875 = 255.8$$

Now add 255.8 to the first odometer reading.

$$86,897.2 + 255.8 = 87,153.0$$

We get the second odometer reading, so the answer checks.

**State.** 255.8 mi had been driven. The gas mileage was 15.9875 mpg.

**37.** Commission = Commission rate $\times$ Sales

$$\begin{array}{llll} C & = & 7\frac{1}{2}\% & \times 215,000 \\ C & = & 0.075 & \times 215,000 \\ C & = & 16,125 \end{array}$$

The commission is $16,125.

**38.** $A = P \cdot \left(1 + \frac{r}{n}\right)^{n \cdot t}$

$$= \$2000 \cdot \left(1 + \frac{6\%}{2}\right)^{2 \cdot 3}$$

$$= \$2000 \cdot \left(1 + \frac{0.06}{2}\right)^{6}$$

$$= \$2000 \cdot (1 + 0.03)^{6}$$

$$= \$2000(1.03)^{6}$$

$$\approx \$2388.10$$

**39. Familiarize.** Let $p$ = the number of pounds that will be lost in 5 weeks.

**Translate.** We translate to a proportion.

$$\begin{array}{l} \text{Pounds} \rightarrow \\ \text{Weeks} \rightarrow \end{array} \frac{3\frac{1}{2}}{2} = \frac{p}{5} \begin{array}{l} \leftarrow \text{Pounds} \\ \leftarrow \text{Weeks} \end{array}$$

**Solve.**

$$3\frac{1}{2} \cdot 5 = 2 \cdot p \qquad \text{Equating cross products}$$
$$\frac{7}{2} \cdot 5 = 2 \cdot p$$
$$\frac{35}{2} = 2 \cdot p$$
$$\frac{\frac{35}{2}}{2} = \frac{2 \cdot p}{2}$$
$$\frac{35}{2} \cdot \frac{1}{2} = p$$
$$\frac{35}{4} = p, \text{ or}$$
$$8\frac{3}{4} = p$$

**Check.** We substitute in the proportion and check cross products.

$$\frac{3\frac{1}{2}}{2} = \frac{8\frac{3}{4}}{5}$$

$$3\frac{1}{2} \cdot 5 = \frac{7}{2} \cdot 5 = \frac{35}{2}; \ 2 \cdot 8\frac{3}{4} = 2 \cdot \frac{35}{4} = \frac{35}{2}$$

The cross products are the same, so the answer checks.

**State.** At the given rate, $8\frac{3}{4}$ lb would be lost in 5 weeks.

**40. Familiarize.** Let $f$ = the amount the family spends for food.

**Translate.**

$\frac{1}{4}$ of income is amount spent for food

$$\frac{1}{4} \cdot 52,800 = f$$

**Solve.** We carry out the multiplication.

$$\frac{1}{4} \cdot 52,800 = \frac{52,800}{4} = \frac{4 \cdot 13,200}{4 \cdot 1} = \frac{4}{4} \cdot \frac{13,200}{1} = 13,200$$

Thus, $13,200 = f$.

**Check.** We can repeat the calculation. The answer checks.

**State.** $13,200 is spent for food.

**41. Familiarize.** Let $p$ = the percent of the seeds that sprout.

**Translate.** We translate to a percent equation.

417 is what percent of 500?

$$417 = p \cdot 500$$

**Solve.** We divide by 500 on both sides.

$$417 = p \cdot 500$$

$$\frac{417}{500} = \frac{p \cdot 500}{500}$$

$$0.834 = p$$

$$83.4\% = p$$

**Check.** $83.4\%$ of $500 \approx 0.8 \cdot 500 = 400 \approx 417$. Since 400 is close to 417, the answer seems reasonable.

**State.** Since more than 80% of the seeds sprouted, the seeds pass government standards.

**42.**  $15 \text{ lb} = 15 \times 1 \text{ lb}$

$$= 15 \times 16 \text{ oz}$$

$$= 240 \text{ oz}$$

**43. Familiarize.** Let $d$ = the number of dips in a tub of ice cream. From Exercise 42 we know that a tub contains 240 oz of ice cream.

**Translate.**

Ounces in a dip times Number of dips in tub is Ounces in tub

$$4 \cdot d = 240$$

**Solve.**

$$4 \cdot d = 240$$

$$\frac{4 \cdot d}{4} = \frac{240}{4}$$

$$d = 60$$

**Check.** If we have 60 4-oz dips, then we have $60 \cdot 4$, or 240 oz, of ice cream. This is the amount of ice cream in a tub, so the answer checks.

**State.** There are 60 dips in a tub of ice cream.

**44. Familiarize.** First we will find the number $n$ of pounds of ice cream that was sold.

**Translate and Solve.** We can subtract to find $n$.

$$n = 15 - 8\frac{5}{8} = 14\frac{8}{8} - 8\frac{5}{8} = 6\frac{3}{8}$$

Now we convert $6\frac{3}{8}$ lb to ounces.

$$6\frac{3}{8} \text{ lb} = 6\frac{3}{8} \times 1 \text{ lb}$$

$$= \frac{51}{8} \times 16 \text{ oz}$$

$$= \frac{51 \times 16}{8} \text{ oz}$$

$$= 102 \text{ oz}$$

Finally, we divide to find the number $d$ of 4-oz dips in 102 oz.

$$d = 102 \div 4 = 25\frac{1}{2}$$

**Check.** We can repeat the calculations. The answer checks.

**State.** Since the result is $25\frac{1}{2}$ dips, we say that 25 or 26 dips were served.

**45. Familiarize.** Let $s$ = the amount taken in from the sale of a tub of ice cream that is used to make 1-dip cones. From Exercise 43 we know that there are 60 dips in a tub.

**Translate.**

Price per dip times Number of dips is Total sales

$$1.99 \cdot 60 = s$$

**Solve.** We carry out the multiplication.

$$\begin{array}{r} 1.99 \\ \times \quad 60 \\ \hline 119.40 \end{array}$$

Thus, $s = 119.40$.

**Check.** $1.99 \times 60 \approx 2 \times 60 = 120 \approx 119.40$. The answer seems reasonable.

**State.** $119.40 is taken in from the sale of a tub of ice cream that is used to make 1-dip cones.

**46. Familiarize.** Let $s$ = the amount taken in from the sale of a tub of ice cream that is used to make 2-dip cones. From Exercise 43 we know that there are 60 dips in a tub. Then $60 \div 2$, or 30, 2-dip cones can be made from a tub of ice cream.

**Translate.**

Price per dip times Number of dips is Total sales

$$2.99 \cdot 30 = s$$

**Solve.** We carry out the multiplication.

$$\begin{array}{r} 2.99 \\ \times \quad 30 \\ \hline 89.70 \end{array}$$

**Check.** $2.99 \times 30 \approx 3 \times 30 = 90 \approx 89.70$. The answer seems reasonable.

**State.** $89.70 is taken in from the sale of a tub of ice cream that is used to make 2-dip cones.

**47.**

$$r = \frac{2}{5} \cdot q$$

$$\frac{r}{\frac{2}{5}} = \frac{\frac{2}{5} \cdot q}{\frac{2}{5}}$$

$$r \cdot \frac{5}{2} = q$$

Thus $q$ is $\frac{5}{2}$ of $r$.

# Chapter 10

# Geometry

1. Perimeter $= 4$ mm $+ 6$ mm $+ 7$ mm
$= (4 + 6 + 7)$ mm
$= 17$ mm

3. Perimeter $= 3.5$ in. $+ 3.5$ in. $+ 4.25$ in. $+$
$0.5$ in. $+ 3.5$ in.
$= (3.5 + 3.5 + 4.25 + 0.5 + 3.5)$ in.
$= 15.25$ in.

5. $P = 2 \cdot (l + w)$       Perimeter of a rectangle
$P = 2 \cdot (5.6$ km $+ 3.4$ km$)$
$P = 2 \cdot (9$ km$)$
$P = 18$ km

7. $P = 2 \cdot (l + w)$       Perimeter of a rectangle
$P = 2 \cdot (5$ ft $+ 10$ ft$)$
$P = 2 \cdot (15$ ft$)$
$P = 30$ ft

9. $P = 2 \cdot (l + w)$       Perimeter of a rectangle
$P = 2 \cdot \left(3\frac{1}{2}\ \text{yd} + 4\frac{1}{2}\ \text{yd}\right)$
$P = 2 \cdot (8$ yd$)$
$P = 16$ yd

11. $P = 4 \cdot s$       Perimeter of a square
$P = 4 \cdot 22$ ft
$P = 88$ ft

13. $P = 4 \cdot s$       Perimeter of a square
$P = 4 \cdot 45.5$ mm
$P = 182$ mm

15. **Familiarize.** We let $P =$ the perimeter. We also make a drawing.

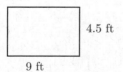

4.5 ft

9 ft

**Translate.** The perimeter of the billiard table is given by

$$P = 2 \cdot (l + w) = 2 \cdot (9\ \text{ft} + 4.5\ \text{ft}).$$

**Solve.** We calculate the perimeter.

$$P = 2 \cdot (9\ \text{ft} + 4.5\ \text{ft}) = 2 \cdot (13.5\ \text{ft}) = 27\ \text{ft}$$

**Check.** Repeat the calculations. The answer checks.

**State.** The perimeter of the table is 27 ft.

17. **Familiarize.** We make a drawing and let $P =$ the perimeter.

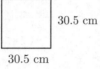

30.5 cm

30.5 cm

**Translate.** The perimeter of the square is given by

$$P = 4 \cdot s = 4 \cdot (30.5\ \text{cm}).$$

**Solve.** We do the calculation.

$$P = 4 \cdot (30.5\ \text{cm}) = 122\ \text{cm}.$$

**Check.** Repeat the calculation. The answer checks.

**State.** The perimeter of the tile is 122 cm.

19. **Familiarize.** We make a drawing and let $P =$ the perimeter. We will express 2 ft 8 in. as 32 in. and 4 ft 6 in. as 54 in.

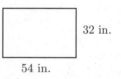

32 in.

54 in.

**Translate.** The perimeter of the backboard is given by

$$P = 2 \cdot (l + w) = 2 \cdot (54\ \text{in.} + 32\ \text{in.})$$

**Solve.** We calculate the perimeter.

$$P = 2 \cdot (54\ \text{in.} + 32\ \text{in.}) = 2 \cdot (86\ \text{in.}) = 172\ \text{in.}$$

**Check.** Repeat the calculation. The answer checks.

**State.** The perimeter of the backboard is 172 in., or 14 ft 4 in.

21. **Familiarize.** We label the missing lengths on the drawing and let $P =$ the perimeter.

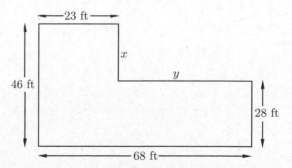

$\leftarrow$ 23 ft $\rightarrow$

$x$

$y$

46 ft

28 ft

$\leftarrow$ 68 ft $\rightarrow$

***Translate.*** First we find the missing lengths $x$ and $y$.

28 ft plus how many more ft is 46 ft

$$28 + x = 46$$

23 ft plus how many more ft is 68 ft

$$23 + y = 68$$

***Solve.*** We solve for $x$ and $y$.

$$28 + x = 46 \qquad 23 + y = 68$$
$$x = 46 - 28 \qquad y = 68 - 23$$
$$x = 18 \qquad y = 45$$

a) To find the perimeter we add the lengths of the sides of the house.

$$P = 23 \text{ ft} + 18 \text{ ft} + 45 \text{ ft} + 28 \text{ ft} + 68 \text{ ft} + 46 \text{ ft}$$
$$= (23 + 18 + 45 + 28 + 68 + 46) \text{ ft}$$
$$= 228 \text{ ft}$$

b) Next we find $t$, the total cost of the gutter.

Cost per foot times Number of feet is Total cost

$$4.59 \times 228 = t$$

We carry out the multiplication.

$$
\begin{array}{r}
2\ 2\ 8 \\
\times\ 4\ .5\ 9 \\
\hline
2\ 0\ 5\ 2 \\
1\ 1\ 4\ 0\ 0 \\
9\ 1\ 2\ 0\ 0 \\
\hline
1\ 0\ 4\ 6\ .5\ 2
\end{array}
$$

Thus, $t = 1046.52$.

***Check.*** We can repeat the calculations. The answer checks.

***State.*** (a) The perimeter of the house is 228 ft. (b) The total cost of the gutter is \$1046.52.

**23.** Interest $= P \cdot r \cdot t$

$$= \$600 \times 6.4\% \times \frac{1}{2}$$
$$= \frac{\$600 \times 0.064}{2}$$
$$= \$19.20$$

The interest is \$19.20.

**25.** $10^3 = 10 \cdot 10 \cdot 10 = 1000$

**27.** $15^2 = 15 \cdot 15 = 225$

**29.** $7^2 = 7 \cdot 7 = 49$

**31.** *Rephrase:* Sales tax is what percent of purchase price?

*Translate:* $878 = r \times 17{,}560$

To solve the equation we divide on both sides by 17,560.

$$\frac{878}{17{,}560} = \frac{r \times 17{,}560}{17{,}560}$$
$$0.05 = r$$
$$5\% = r$$

The sales tax rate is 5%.

**33.** Excluding the amount of ribbon required for the bow, the ribbon needed is equivalent to the perimeters of two rectangles, one measuring 8 in. by 4 in. and the other measuring 7 in. by 4 in.

For the 8 in. by 4 in. rectangle,

$$P = 2 \cdot (8 \text{ in.} + 4 \text{ in.}) = 2 \cdot (12 \text{ in.}) = 24 \text{ in.}$$

For the 7 in. by 4 in. rectangle,

$$P = 2 \cdot (7 \text{ in.} + 4 \text{ in.}) = 2 \cdot (11 \text{ in.}) = 22 \text{ in.}$$

Then, including the bow, the amount of ribbon required is

$$18 \text{ in.} + 24 \text{ in.} + 22 \text{ in.} = 64 \text{ in.}$$

## Exercise Set 10.2

**1.** $A = l \cdot w$ \qquad Area of a rectangular region
$A = (5 \text{ km}) \cdot (3 \text{ km})$
$A = 5 \cdot 3 \cdot \text{ km} \cdot \text{ km}$
$A = 15 \text{ km}^2$

**3.** $A = l \cdot w$ \qquad Area of a rectangular region
$A = (2 \text{ in.}) \cdot (0.7 \text{ in.})$
$A = 2 \cdot 0.7 \cdot \text{ in.} \cdot \text{ in.}$
$A = 1.4 \text{ in}^2$

**5.** $A = s \cdot s$ \qquad Area of a square
$A = \left(2\frac{1}{2} \text{ yd}\right) \cdot \left(2\frac{1}{2} \text{ yd}\right)$
$A = \left(\frac{5}{2} \text{ yd}\right) \cdot \left(\frac{5}{2} \text{ yd}\right)$
$A = \frac{5}{2} \cdot \frac{5}{2} \cdot \text{ yd} \cdot \text{ yd}$
$A = \frac{25}{4} \text{ yd}^2$, or $6\frac{1}{4} \text{ yd}^2$

**7.** $A = s \cdot s$ \qquad Area of a square
$A = (90 \text{ ft}) \cdot (90 \text{ ft})$
$A = 90 \cdot 90 \cdot \text{ ft} \cdot \text{ ft}$
$A = 8100 \text{ ft}^2$

**9.** $A = l \cdot w$ \qquad Area of a rectangular region
$A = (10 \text{ ft}) \cdot (5 \text{ ft})$
$A = 10 \cdot 5 \cdot \text{ ft} \cdot \text{ ft}$
$A = 50 \text{ ft}^2$

**11.** $A = l \cdot w$ \qquad Area of a rectangular region
$A = (34.67 \text{ cm}) \cdot (4.9 \text{ cm})$
$A = 34.67 \cdot 4.9 \cdot \text{ cm} \cdot \text{ cm}$
$A = 169.883 \text{ cm}^2$

**13.** $A = l \cdot w$    Area of a rectangular region

$A = \left(4\frac{2}{3} \text{ in.}\right) \cdot \left(8\frac{5}{6} \text{ in.}\right)$

$A = \left(\frac{14}{3} \text{ in.}\right) \cdot \left(\frac{53}{6} \text{ in.}\right)$

$A = \frac{14}{3} \cdot \frac{53}{6} \cdot \text{ in.} \cdot \text{ in.}$

$A = \frac{2 \cdot 7 \cdot 53}{3 \cdot 2 \cdot 3} \text{ in}^2$

$A = \frac{2}{2} \cdot \frac{7 \cdot 53}{3 \cdot 3} \text{ in}^2$

$A = \frac{371}{9} \text{ in}^2$, or $41\frac{2}{9} \text{ in}^2$

**15.** $A = s \cdot s$           Area of a square
$A = (22 \text{ ft}) \cdot (22 \text{ ft})$
$A = 22 \cdot 22 \cdot \text{ ft} \cdot \text{ ft}$
$A = 484 \text{ ft}^2$

**17.** $A = s \cdot s$           Area of a square
$A = (56.9 \text{ km}) \cdot (56.9 \text{ km})$
$A = 56.9 \cdot 56.9 \cdot \text{ km} \cdot \text{ km}$
$A = 3237.61 \text{ km}^2$

**19.** $A = s \cdot s$           Area of a square

$A = \left(5\frac{3}{8} \text{ yd}\right) \cdot \left(5\frac{3}{8} \text{ yd}\right)$

$A = \left(\frac{43}{8} \text{ yd}\right) \cdot \left(\frac{43}{8} \text{ yd}\right)$

$A = \frac{43}{8} \cdot \frac{43}{8} \cdot \text{ yd} \cdot \text{ yd}$

$A = \frac{1849}{64} \text{ yd}^2$, or $28\frac{57}{64} \text{ yd}^2$

**21.** $A = b \cdot h$      Area of a parallelogram
$A = 8 \text{ cm} \cdot 4 \text{ cm}$    Substituting 8 cm for $b$ and
                         4 cm for $h$
$A = 32 \text{ cm}^2$

**23.** $A = \frac{1}{2} \cdot b \cdot h$           Area of a triangle

$A = \frac{1}{2} \cdot 15 \text{ in.} \cdot 8 \text{ in.}$    Substituting 15 in. for $b$ and
                              8 in. for $h$
$A = 60 \text{ in}^2$

**25.** $A = \frac{1}{2} \cdot h \cdot (a + b)$       Area of a trapezoid

$A = \frac{1}{2} \cdot 8 \text{ ft} \cdot (6 + 20) \text{ ft}$    Substituting 8 ft for $h$, 6 ft
                                    for $a$, and 20 ft for $b$

$A = \frac{8 \cdot 26}{2} \text{ ft}^2$

$A = 104 \text{ ft}^2$

**27.** $A = \frac{1}{2} \cdot h \cdot (a + b)$       Area of a trapezoid

$A = \frac{1}{2} \cdot 7 \text{ in.} \cdot (4.5 + 8.5) \text{ in.}$    Substituting 7 in. for $h$,
                                      4.5 in. for $a$, and 8.5 in.
                                      for $b$

$A = \frac{7 \cdot 13}{2} \text{ in}^2$

$A = \frac{91}{2} \text{ in}^2$

$A = 45.5 \text{ in}^2$

**29.** $A = b \cdot h$              Area of a parallelogram
$A = 2.3 \text{ cm} \cdot 3.5 \text{ cm}$    Substituting 2.3 cm for $b$
                          and 3.5 cm for $h$

$A = 8.05 \text{ cm}^2$

**31.** $A = \frac{1}{2} \cdot h \cdot (a + b)$          Area of a trapezoid

$A = \frac{1}{2} \cdot 18 \text{ cm} \cdot (9 + 24) \text{ cm}$    Substituting 18 cm for
                                       $h$, 9 cm for $a$, and 24
                                       cm for $b$

$A = \frac{18 \cdot 33}{2} \text{ cm}^2$

$A = 297 \text{ cm}^2$

**33.** $A = \frac{1}{2} \cdot b \cdot h$          Area of a triangle

$A = \frac{1}{2} \cdot 4 \text{ m} \cdot 3.5 \text{ m}$    Substituting 4 m for $b$ and
                               3.5 m for $h$

$A = \frac{4 \cdot 3.5}{2} \text{ m}^2$

$A = 7 \text{ m}^2$

**35.** ***Familiarize.*** We draw a picture.

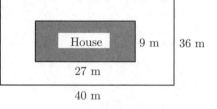

***Translate.*** We let $A$ = the area left over.

| Area left over | is | Area of lot | minus | Area of house |
|---|---|---|---|---|
| $\downarrow$ | $\downarrow$ | $\downarrow$ | $\downarrow$ | $\downarrow$ |
| $A$ | $=$ | $(40 \text{ m}) \cdot (36 \text{ m})$ | $-$ | $(27 \text{ m}) \cdot (9 \text{ m})$ |

***Solve.*** The area of the lot is

$(40 \text{ m}) \cdot (36 \text{ m}) = 40 \cdot 36 \cdot \text{ m} \cdot \text{ m} = 1440 \text{ m}^2.$

The area of the house is

$(27 \text{ m}) \cdot (9 \text{ m}) = 27 \cdot 9 \cdot \text{ m} \cdot \text{ m} = 243 \text{ m}^2.$

The area left over is

$A = 1440 \text{ m}^2 - 243 \text{ m}^2 = 1197 \text{ m}^2.$

***Check.*** Repeat the calculations. The answer checks.

***State.*** The area left over for the lawn is 1197 m$^2$.

**37.** a) First find the area of the entire yard, including the basketball court:

$$A = l \cdot w = \left(110\frac{2}{3} \text{ ft}\right) \cdot (80 \text{ ft})$$

$$= \left(\frac{332}{3} \text{ ft}\right) \cdot (80 \text{ ft})$$

$$= \frac{26,560}{3} \text{ ft}^2$$

$$= 8853\frac{1}{3} \text{ ft}^2$$

Now find the area of the basketball court:

$$A = s \cdot s = \left(19\frac{1}{2} \text{ ft}\right) \cdot \left(19\frac{1}{2} \text{ ft}\right) =$$

$$\frac{39}{2} \cdot \frac{39}{2} \text{ ft}^2 = \frac{1521}{4} \text{ ft}^2 = 380\frac{1}{4} \text{ ft}^2$$

Finally, subtract to find the area of the lawn:

$$8853\frac{1}{3} \text{ ft}^2 - 380\frac{1}{4} \text{ ft}^2 = 8853\frac{4}{12} \text{ ft}^2 - 380\frac{3}{12} \text{ ft}^2 =$$

$$8473\frac{1}{12} \text{ ft}^2 \approx 8473 \text{ ft}^2$$

b) Let $c$ = the cost of mowing the lawn. We translate to an equation.

| The cost of mowing | is | $0.012 | times | the area of the lawn. |
|---|---|---|---|---|
| ↓ | ↓ | ↓ | ↓ | ↓ |
| $c$ | = | 0.012 | · | 8473 |

We multiply to solve the equation.

$$c = 0.012 \cdot 8473 \approx \$102$$

The total cost of the mowing is about $102.

**39. Familiarize.** We use the drawing in the text.

**Translate.** We let $A$ = the area of the sidewalk, in square feet.

| Area of sidewalk | is | Total area | minus | Area of building |
|---|---|---|---|---|
| ↓ | ↓ | ↓ | ↓ | ↓ |
| $A$ | = | (113.4 ft) × (75.4 ft) | − | (110 ft) × (72 ft) |

**Solve.** The total area is

$$(113.4 \text{ ft}) \times (75.4 \text{ ft}) = 113.4 \times 75.4 \times \text{ ft} \times \text{ ft} = 8550.36 \text{ ft}^2.$$

The area of the building is

$$(110 \text{ ft}) \times (72 \text{ ft}) = 110 \times 72 \times \text{ ft} \times \text{ ft} = 7920 \text{ ft}^2.$$

The area of the sidewalk is

$$A = 8550.36 \text{ ft}^2 - 7920 \text{ ft}^2 = 630.36 \text{ ft}^2.$$

**Check.** Repeat the calculations. The answer checks.

**State.** The area of the sidewalk is 630.36 ft².

**41. Familiarize.** The dimensions are as follows:

Two walls are 15 ft by 8 ft.

Two walls are 20 ft by 8 ft.

The ceiling is 15 ft by 20 ft.

The total area of the walls and ceiling is the total area of the rectangles described above less the area of the windows and the door.

**Translate.** a) We let $A$ = the total area of the walls and ceiling. The total area of the two 15 ft by 8 ft walls is

$$2 \cdot (15 \text{ ft}) \cdot (8 \text{ ft}) = 2 \cdot 15 \cdot 8 \cdot \text{ ft} \cdot \text{ ft} = 240 \text{ ft}^2$$

The total area of the two 20 ft by 8 ft walls is

$$2 \cdot (20 \text{ ft}) \cdot (8 \text{ ft}) = 2 \cdot 20 \cdot 8 \cdot \text{ ft} \cdot \text{ ft} = 320 \text{ ft}^2$$

The area of the ceiling is

$$(15 \text{ ft}) \cdot (20 \text{ ft}) = 15 \cdot 20 \cdot \text{ ft} \cdot \text{ ft} = 300 \text{ ft}^2$$

The area of the two windows is

$$2 \cdot (3 \text{ ft}) \cdot (4 \text{ ft}) = 2 \cdot 3 \cdot 4 \cdot \text{ ft} \cdot \text{ ft} = 24 \text{ ft}^2$$

The area of the door is

$$\left(2\frac{1}{2} \text{ ft}\right) \cdot \left(6\frac{1}{2} \text{ ft}\right) = \left(\frac{5}{2} \text{ ft}\right) \cdot \left(\frac{13}{2} \text{ ft}\right)$$

$$= \frac{5}{2} \cdot \frac{13}{2} \cdot \text{ ft} \cdot \text{ ft}$$

$$= \frac{65}{4} \text{ ft}^2, \text{ or } 16\frac{1}{4} \text{ ft}^2$$

Thus

$$A = 240 \text{ ft}^2 + 320 \text{ ft}^2 + 300 \text{ ft}^2 - 24 \text{ ft}^2 - 16\frac{1}{4} \text{ ft}^2$$

$$= 819\frac{3}{4} \text{ ft}^2, \text{ or } 819.75 \text{ ft}^2$$

b) We divide to find how many gallons of paint are needed.

$$819.75 \div 360.625 \approx 2.27$$

It will be necessary to buy 3 gallons of paint in order to have the required 2.27 gallons.

c) We multiply to find the cost of the paint.

$$3 \times \$24.95 = \$74.85$$

**Check.** We repeat the calculations. The answer checks.

**State.** (a) The total area of the walls and ceiling is 819.75 ft². (b) 3 gallons of paint are needed. (c) It will cost $74.85 to paint the room.

**43.**

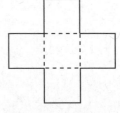

Each side is 4 cm.

The region is composed of 5 squares, each with sides of length 4 cm. The area is

$$A = 5 \cdot (s \cdot s) = 5 \cdot (4 \text{ cm} \cdot 4 \text{ cm}) = 5 \cdot 4 \cdot 4 \text{ cm} \cdot \text{ cm} = 80 \text{ cm}^2$$

**45. Familiarize.** We look for the kinds of figures whose areas we can calculate using area formulas that we already know.

**Translate.** The shaded region consists of a square region with a triangular region removed from it. The sides of the square are 30 cm, and the triangle has base 30 cm and height 15 cm. We find the area of the square using the formula $A = s \cdot s$, and the area of the triangle using $A = \frac{1}{2} \cdot b \cdot h$. Then we subtract.

**Solve.** Area of the square: $A = 30 \text{ cm} \cdot 30 \text{ cm} = 900 \text{ cm}^2$.

Area of the triangle: $A = \frac{1}{2} \cdot 30 \text{ cm} \cdot 15 \text{ cm} = 225 \text{ cm}^2$.

Area of the shaded region: $A = 900 \text{ cm}^2 - 225 \text{ cm}^2 = 675 \text{ cm}^2$.

**Check.** We repeat the calculations. The answer checks.

**State.** The area of the shaded region is 675 cm$^2$.

**47. Familiarize.** We have one large triangle with height and base each 6 cm. We also have 6 small triangles, each with height and base 1 cm.

**Translate.** We will find the area of each type of triangle using the formula $A = \frac{1}{2} \cdot b \cdot h$. Next we will multiply the area of the smaller triangle by 6. And, finally, we will add this product to the area of the larger triangle to find the total area.

**Solve.**

For the large triangle: $A = \frac{1}{2} \cdot 6 \text{ cm} \cdot 6 \text{ cm} = 18 \text{ cm}^2$

For one small triangle: $A = \frac{1}{2} \cdot 1 \text{ cm} \cdot 1 \text{ cm} = \frac{1}{2} \text{ cm}^2$

Find the area of the 6 small triangles: $6 \cdot \frac{1}{2} \text{ cm}^2 = 3 \text{ cm}^2$

Add to find the total area: $18 \text{ cm}^2 + 3 \text{ cm}^2 = 21 \text{ cm}^2$

**Check.** We repeat the calculations.

**State.** The area of the shaded region is 21 cm$^2$.

**49. Familiarize.** The sail consists of a triangle with base 9 ft and height 12 ft as well as three rectangles, one that measures $\frac{1}{2}$ ft by 9 ft, one that measures $\frac{1}{2}$ ft by 12 ft, and one that measures $\frac{1}{2}$ ft by 15 ft. The piece of sailcloth from which the sail is cut is a rectangle that measures 18 ft by 12 ft.

**Translate.** We will use the formula $A = \frac{1}{2} \cdot b \cdot h$ to find the area of the triangle and the formula $A = l \cdot w$ to find the area of the rectangles. Then we will add to find the area of the sail and, finally, subtract to find how much fabric is left over.

**Solve.**

Area of the triangle: $A = \frac{1}{2} \cdot 9 \text{ ft} \cdot 12 \text{ ft} = 54 \text{ ft}^2$

Area of the $\frac{1}{2}$ ft by 9 ft rectangle: $A = \frac{1}{2} \text{ ft} \cdot 9 \text{ ft} = \frac{9}{2} \text{ ft}^2$

Area of the $\frac{1}{2}$ ft by 12 ft rectangle:

$A = \frac{1}{2} \text{ ft} \cdot 12 \text{ ft} = 6 \text{ ft}^2$

Area of the $\frac{1}{2}$ ft by 15 ft rectangle:

$A = \frac{1}{2} \text{ ft} \cdot 15 \text{ ft} = \frac{15}{2} \text{ ft}^2$

Area of the 18 ft by 12 ft rectangle:
$A = 18 \text{ ft} \cdot 12 \text{ ft} = 216 \text{ ft}^2$

Total area of the sail:

$54 \text{ ft}^2 + \frac{9}{2} \text{ ft}^2 + 6 \text{ ft}^2 + \frac{15}{2} \text{ ft}^2 = 72 \text{ ft}^2$

Area of fabric left over: $216 \text{ ft}^2 - 72 \text{ ft}^2 = 144 \text{ ft}^2$

**Check.** We repeat the calculations. The answer checks.

**State.** The area of the fabric left over is 144 ft$^2$.

**51.** 23.4 cm = _____ mm

Think: To go from cm to mm in the diagram is a move of 1 place to the right. Thus, we move the decimal point 1 place to the right.

23.4    23.4.
       ⌐↑

23.4 cm = 234 mm

**53.** 28 ft = 28 × 1 ft = 28 × 12 in. = 336 in.

**55.** 72.4 cm = _____ m

Think: To go from cm to m in the diagram is a move of 2 places to the left. Thus, we move the decimal point 2 places to the left.

72.4    0.72. 4
      ↑⌐⌐

72.4 cm = 0.724 m

**57.**  70 yd = 70 × 1 yd
           = 70 × 3 ft = 70 × 3 × 1 ft
           = 70 × 3 × 12 in.
           = 2520 in.

**59.**  $84 \text{ ft} = 84 \text{ ft} \times \frac{1 \text{ yd}}{3 \text{ ft}}$

           $= \frac{84}{3} \text{ yd} = 28 \text{ yd}$

**61.**  $144 \text{ in.} = 144 \text{ in.} \times \frac{1 \text{ ft}}{12 \text{ in.}}$

           $= \frac{144}{12} \text{ ft} = 12 \text{ ft}$

**63.**  $A = P \cdot \left(1 + \frac{r}{n}\right)^{n \cdot t}$

      $= \$25{,}000 \cdot \left(1 + \frac{4\%}{2}\right)^{2 \cdot 5}$

      $= \$25{,}000 \cdot \left(1 + \frac{0.04}{2}\right)^{10}$

      $= \$25{,}000(1.02)^{10}$

      $\approx \$30{,}474.86$

**65.** $A = P \cdot \left(1 + \dfrac{r}{n}\right)^{n \cdot t}$

$= \$150,000 \cdot \left(1 + \dfrac{7.4\%}{2}\right)^{2 \cdot 20}$

$= \$150,000 \cdot \left(1 + \dfrac{0.074}{2}\right)^{40}$

$= \$150,000(1.037)^{40}$

$\approx \$641,566.26$

**67.**

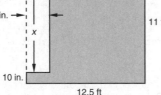

$2 \text{ ft} = 2 \times 1 \text{ ft} = 2 \times 12 \text{ in.} = 24 \text{ in.}$, so $2 \text{ ft, } 2 \text{ in.} = 2 \text{ ft} + 2 \text{ in.} = 24 \text{ in.} + 2 \text{ in.} = 26 \text{ in.}$

$11 \text{ ft} = 11 \times 1 \text{ ft} = 11 \times 12 \text{ in.} = 132 \text{ in.}$

$12.5 \text{ ft} = 12.5 \times 1 \text{ ft} = 12.5 \times 12 \text{ in.} = 150 \text{ in.}$

We solve an equation to find $x$, in inches:

$11 + x + 10 = 132$

$21 + x = 132$

$21 + x - 21 = 132 - 21$

$x = 111$

Then the area of the shaded region is the area of a 150 in. by 132 in. rectangle less the area of a 111 in. by 26 in. rectangle.

$A = (150 \text{ in.}) \cdot (132 \text{ in.}) - (111 \text{ in.}) \cdot (26 \text{ in.})$

$A = 19,800 \text{ in}^2 - 2886 \text{ in}^2$

$A = 16,914 \text{ in}^2$

## Exercise Set 10.3

**1.** $d = 2 \cdot r$

$d = 2 \cdot 7 \text{ cm} = 14 \text{ cm}$

$C = 2 \cdot \pi \cdot r$

$C \approx 2 \cdot \dfrac{22}{7} \cdot 7 \text{ cm} = \dfrac{2 \cdot 22 \cdot 7}{7} \text{ cm} = 44 \text{ cm}$

$A = \pi \cdot r \cdot r$

$A \approx \dfrac{22}{7} \cdot 7 \text{ cm} \cdot 7 \text{ cm} = \dfrac{22}{7} \cdot 49 \text{ cm}^2 = 154 \text{ cm}^2$

**3.** $d = 2 \cdot r$

$d = 2 \cdot \dfrac{3}{4} \text{ in.} = \dfrac{6}{4} \text{ in.} = \dfrac{3}{2} \text{ in.}$, or $1\dfrac{1}{2} \text{ in.}$

$C = 2 \cdot \pi \cdot r$

$C \approx 2 \cdot \dfrac{22}{7} \cdot \dfrac{3}{4} \text{ in.} = \dfrac{2 \cdot 22 \cdot 3}{7 \cdot 4} \text{ in.} = \dfrac{132}{28} \text{ in.} = \dfrac{33}{7} \text{ in.}$, or $4\dfrac{5}{7} \text{ in.}$

$A = \pi \cdot r \cdot r$

$A \approx \dfrac{22}{7} \cdot \dfrac{3}{4} \text{ in.} \cdot \dfrac{3}{4} \text{ in.} = \dfrac{22 \cdot 3 \cdot 3}{7 \cdot 4 \cdot 4} \text{ in}^2 = \dfrac{99}{56} \text{ in}^2$, or $1\dfrac{43}{56} \text{ in}^2$

**5.** $r = \dfrac{d}{2}$

$r = \dfrac{32 \text{ ft}}{2} = 16 \text{ ft}$

$C = \pi \cdot d$

$C \approx 3.14 \cdot 32 \text{ ft} = 100.48 \text{ ft}$

$A = \pi \cdot r \cdot r$

$A \approx 3.14 \cdot 16 \text{ ft} \cdot 16 \text{ ft} \qquad \left(r = \dfrac{d}{2}; r = \dfrac{32 \text{ ft}}{2} = 16 \text{ ft}\right)$

$A = 3.14 \cdot 256 \text{ ft}^2$

$A = 803.84 \text{ ft}^2$

**7.** $r = \dfrac{d}{2}$

$r = \dfrac{1.4 \text{ cm}}{2} = 0.7 \text{ cm}$

$C = \pi \cdot d$

$C \approx 3.14 \cdot 1.4 \text{ cm} = 4.396 \text{ cm}$

$A = \pi \cdot r \cdot r$

$A \approx 3.14 \cdot 0.7 \text{ cm} \cdot 0.7 \text{ cm}$

$\qquad \left(r = \dfrac{d}{2}; r = \dfrac{1.4 \text{ cm}}{2} = 0.7 \text{ cm}\right)$

$A = 3.14 \cdot 0.49 \text{ cm}^2 = 1.5386 \text{ cm}^2$

**9.** $d = 2 \cdot r$

$d = 2 \cdot 6.37 \text{ ft} = 12.74 \text{ ft}$

$C = 2 \cdot \pi \cdot r$

$C \approx 2 \cdot 3.14 \cdot 6.37 \text{ ft} \approx 40 \text{ ft}$

$A = \pi \cdot r \cdot r$

$A \approx 3.14 \cdot 6.37 \text{ ft} \cdot 6.37 \text{ ft} \approx 127.41 \text{ ft}^2$

The areas of the nets described in Example 10 in Section 10.2 are about 80.74 ft$^2$ and 107.65 ft$^2$. The differences in the areas are

$127.41 \text{ ft}^2 - 80.74 \text{ ft}^2$, or $46.67 \text{ ft}^2$

and

$127.41 \text{ ft}^2 - 107.65 \text{ ft}^2$, or $19.76 \text{ ft}^2$.

Thus, the Adventure II net is 46.67 ft$^2$ larger than the medium net and 19.76 ft$^2$ larger than the large net.

**11.** Area of circular pizza $\left(r = \dfrac{12 \text{ in.}}{2} = 6 \text{ in.}\right)$:

$\pi \cdot r \cdot r \approx 3.14 \cdot 6 \text{ in.} \cdot 6 \text{ in.} = 113.04 \text{ in}^2$

Area of square pizza:

$s \cdot s = 12 \text{ in.} \cdot 12 \text{ in.} = 144 \text{ in}^2$

Difference in areas:

$144 \text{ in}^2 - 113.04 \text{ in}^2 = 30.96 \text{ in}^2$

The square pizza is 30.96 in$^2$ larger.

**13.** $C = \pi \cdot d$

$C \approx 3.14 \cdot 7926.41 \approx 24,889 \text{ mi}$

15. Maximum circumference of the barrel:

$$C = \pi \cdot d$$

$$C \approx \frac{22}{7} \cdot 2\frac{3}{4} = \frac{22}{7} \cdot \frac{11}{4} = \frac{\cancel{2} \cdot 11 \cdot 11}{7 \cdot \cancel{2} \cdot 2} = \frac{121}{14}, \text{ or } 8\frac{9}{14} \text{ in.}$$

Minimum circumference of the handle:

$$C = \pi \cdot d$$

$$C \approx \frac{22}{7} \cdot \frac{16}{19} = \frac{352}{133}, \text{ or } 2\frac{86}{133} \text{ in.}$$

17. Find the area of the larger circle (pool plus walk). Its diameter is 1 yd + 20 yd + 1 yd, or 22 yd. Thus its radius is $\frac{22}{2}$ yd, or 11 yd.

$$A = \pi \cdot r \cdot r$$
$$A \approx 3.14 \cdot 11 \text{ yd} \cdot 11 \text{ yd} = 379.94 \text{ yd}^2$$

Find the area of the pool. Its diameter is 20 yd. Thus its radius is $\frac{20}{2}$ yd, or 10 yd.

$$A = \pi \cdot r \cdot r$$
$$A \approx 3.14 \cdot 10 \text{ yd} \cdot 10 \text{ yd} = 314 \text{ yd}^2$$

We subtract to find the area of the walk:

$$A = 379.94 \text{ yd}^2 - 314 \text{ yd}^2$$
$$A = 65.94 \text{ yd}^2$$

The area of the walk is 65.94 yd$^2$.

19. The perimeter consists of the circumferences of three semicircles, each with diameter 8 ft, and one side of a square of length 8 ft. We first find the circumference of one semicircle. This is one-half the circumference of a circle with diameter 8 ft:

$$\frac{1}{2} \cdot \pi \cdot d \approx \frac{1}{2} \cdot 3.14 \cdot 8 \text{ ft} = 12.56 \text{ ft}$$

Then we multiply by 3:

$$3 \cdot (12.56 \text{ ft}) = 37.68 \text{ ft}$$

Finally we add the circumferences of the semicircles and the length of the side of the square:

$$37.68 \text{ ft} + 8 \text{ ft} = 45.68 \text{ ft}$$

The perimeter is 45.68 ft.

21. The perimeter consists of three-fourths of the circumference of a circle with radius 4 yd and two sides of a square with sides of length 4 yd. We first find three-fourths of the circumference of the circle:

$$\frac{3}{4} \cdot 2 \cdot \pi \cdot r \approx 0.75 \cdot 2 \cdot 3.14 \cdot 4 \text{ yd} = 18.84 \text{ yd}$$

Then we add this length to the lengths of two sides of the square:

$$18.84 \text{ yd} + 4 \text{ yd} + 4 \text{ yd} = 26.84 \text{ yd}$$

The perimeter is 26.84 yd.

23. The perimeter consists of three-fourths of the perimeter of a square with side of length 10 yd and the circumference of a semicircle with diameter 10 yd. First we find three-fourths of the perimeter of the square:

$$\frac{3}{4} \cdot 4 \cdot s = \frac{3}{4} \cdot 4 \cdot 10 \text{ yd} = 30 \text{ yd}$$

Then we find one-half of the circumference of a circle with diameter 10 yd:

$$\frac{1}{2} \cdot \pi \cdot d \approx \frac{1}{2} \cdot 3.14 \cdot 10 \text{ yd} = 15.7 \text{ yd}$$

Then we add:

$$30 \text{ yd} + 15.7 \text{ yd} = 45.7 \text{ yd}$$

The perimeter is 45.7 yd.

25. The shaded region consists of a circle of radius 8 m, with two circles each of diameter 8 m, removed. First we find the area of the large circle:

$$A = \pi \cdot r \cdot r \approx 3.14 \cdot 8 \text{ m} \cdot 8 \text{ m} = 200.96 \text{ m}^2$$

Then we find the area of one of the small circles:

The radius is $\frac{8 \text{ m}}{2} = 4$ m.

$$A = \pi \cdot r \cdot r \approx 3.14 \cdot 4 \text{ m} \cdot 4 \text{ m} = 50.24 \text{ m}^2$$

We multiply this area by 2 to find the area of the two small circles:

$$2 \cdot 50.24 \text{ m}^2 = 100.48 \text{ m}^2$$

Finally we subtract to find the area of the shaded region:

$$200.96 \text{ m}^2 - 100.48 \text{ m}^2 = 100.48 \text{ m}^2$$

The area of the shaded region is 100.48 m$^2$.

27. The shaded region consists of one-half of a circle with diameter 2.8 cm and a triangle with base 2.8 cm and height 2.8 cm. First we find the area of the semicircle. The radius is $\frac{2.8 \text{ cm}}{2} = 1.4$ cm.

$$A = \frac{1}{2} \cdot \pi \cdot r \cdot r \approx \frac{1}{2} \cdot 3.14 \cdot 1.4 \text{ cm} \cdot 1.4 \text{ cm} = 3.0772 \text{ cm}^2$$

Then we find the area of the triangle:

$$A = \frac{1}{2} \cdot b \cdot h = \frac{1}{2} \cdot 2.8 \text{ cm} \cdot 2.8 \text{ cm} = 3.92 \text{ cm}^2$$

Finally we add to find the area of the shaded region:

$$3.0772 \text{ cm}^2 + 3.92 \text{ cm}^2 = 6.9972 \text{ cm}^2$$

The area of the shaded region is 6.9972 cm$^2$.

29. The shaded area consists of a rectangle of dimensions 11.4 in. by 14.6 in., with the area of two semicircles, each of diameter 11.4 in., removed. This is equivalent to removing

one circle with diameter 11.4 in. from the rectangle. First we find the area of the rectangle:

$$l \cdot w = (11.4 \text{ in.}) \cdot (14.6 \text{ in.}) = 166.44 \text{ in}^2$$

Then we find the area of the circle. The radius is $\frac{11.4 \text{ in.}}{2} = 5.7$ in.

$$\pi \cdot r \cdot r \approx 3.14 \cdot 5.7 \text{ in.} \cdot 5.7 \text{ in.} = 102.0186 \text{ in}^2$$

Finally we subtract to find the area of the shaded region:

$$166.44 \text{ in}^2 - 102.0186 \text{ in}^2 = 64.4214 \text{ in}^2$$

**31.** 5.43 m = _____ cm

Think: To go from m to cm in the diagram is a move of 2 places to the right. Thus, we move the decimal point 2 places to the right.

5.43    5.43.

5.43 m = 543 cm

**33. *Familiarize*.** Let $w$ = the weight of the brain, in pounds, for a 200-lb person.

***Translate***.

What is 2.5% of 200 lb?

$w$    = 2.5% ×    200

***Solve***. We convert 2.5% to decimal notation and multiply.

$$w = 0.025 \times 200 = 5$$

***Check***. We repeat the calculation. The answer checks.

***State***. For a 200-lb person, the brain weighs 5 lb.

**35.** $A = l \cdot w = 580.8 \text{ ft} \cdot 75 \text{ ft} = 43,560 \text{ ft}^2$

$P = 2 \cdot (l + w) = 2 \cdot (580.8 \text{ ft} + 75 \text{ ft}) = 2 \cdot (655.8 \text{ ft}) = 1311.6 \text{ ft}$

Number of rolls of fencing needed:

1311.6 ft ÷ 330 ft ≈ 3.97

Thus, 4 rolls of fencing must be purchased.

Cost of fencing: 4 · $149.99 = $599.96

**37.** $A = \pi \cdot r \cdot r \approx 3.14 \cdot 117.83 \text{ ft} \cdot 117.83 \text{ ft} = 43,595.47395 \text{ ft}^2$

$C = 2 \cdot \pi \cdot r \approx 2 \cdot 3.14 \cdot 117.83 \text{ ft} = 739.9724 \text{ ft}$

Number of rolls of fencing needed:

739.9724 ft ÷ 330 ft ≈ 2.24

Thus, 3 rolls of fencing must be purchased.

Cost of fencing: 3 · $149.99 = $449.97

**39.** $A = l \cdot w = 242 \text{ ft} \cdot 180 \text{ ft} = 43,560 \text{ ft}^2$

$P = 2 \cdot (l + w) = 2 \cdot (242 \text{ ft} + 180 \text{ ft}) = 2 \cdot (422 \text{ ft}) = 844 \text{ ft}$

Number of rolls of fencing needed:

844 ft ÷ 330 ft ≈ 2.58

Thus, 3 rolls of fencing must be purchased.

Cost of fencing: 3 · $149.99 = $449.97

## Chapter 10 Mid-Chapter Review

**1.** The area of a square is the square of the length of a side. The given statement is false.

**2.** The statement is true. Each area is 8 cm·5 cm, or 40 cm².

**3.** We find the area of the square by squaring 4 in. We find the area of the circle by multiplying the square of 4 by $\pi$, so the area of the square is less than the area of the circle. The given statement is true.

**4.** The perimeter of the rectangle is $2 \cdot (6 \text{ ft} + 3 \text{ ft}) = 2 \cdot (9 \text{ ft}) = 18$ ft. The circumference of the circle is $2 \cdot \pi \cdot 3 \text{ ft} \approx 2 \cdot 3.14 \cdot 3 \text{ ft} = 18.84 \text{ ft}$. The perimeter of the rectangle is less than the circumference of the circle, so the given statement is false.

**5.** $C = \pi \cdot d$, so $C/d = \pi$. The given statement is true.

**6.** $P = 2 \cdot (l + w)$

$P = 2 \cdot (10 \text{ ft} + 3 \text{ ft})$

$P = 2 \cdot (13 \text{ ft})$

$P = 26 \text{ ft}$

$A = l \cdot w$

$A = 10 \text{ ft} \cdot 3 \text{ ft}$

$A = 10 \cdot 3 \cdot \text{ft} \cdot \text{ft}$

$A = 30 \text{ ft}^2$

**7.** $A = \frac{1}{2} \cdot b \cdot h$

$A = \frac{1}{2} \cdot 12 \text{ cm} \cdot 8 \text{ cm}$

$A = \frac{12 \cdot 8}{2} \text{ cm}^2$

$A = \frac{96}{2} \text{ cm}^2 = 48 \text{ cm}^2$

**8.** $C = \pi \cdot d$

$C \approx 3.14 \cdot 10.2 \text{ in.}$

$C = 32.028 \text{ in.}$

$r = \frac{d}{2} = \frac{10.2 \text{ in.}}{2} = 5.1$

$A = \pi \cdot r \cdot r$

$A \approx 3.14 \cdot 5.1 \text{ in.} \cdot 5.1 \text{ in.}$

$A = 81.6714 \text{ in}^2$

**9.** Perimeter = 23 mm + 8 mm + 10 mm + 7 mm + 13 mm + 15 mm

= (23 + 8 + 10 + 7 + 13 + 15) mm

= 76 mm

**10.** $P = 4 \cdot s$

$= 4 \cdot 12\frac{2}{3} \text{ ft} = 4 \cdot \frac{38}{3} \text{ ft}$

$= \frac{152}{3} \text{ ft, or } 50\frac{2}{3} \text{ ft}$

$A = s \cdot s$

$= 12\frac{2}{3}$ ft $\cdot 12\frac{2}{3}$ ft

$= \frac{38}{3}$ ft $\cdot \frac{38}{3}$ ft $= \frac{38}{3} \cdot \frac{38}{3} \cdot$ ft $\cdot$ ft

$= \frac{1444}{9}$ ft$^2$, or $160\frac{4}{9}$ ft$^2$

**11.** $A = b \cdot h$

$A = 40$ in. $\cdot 20$ in.

$A = 800$ in$^2$

**12.** $A = \frac{1}{2} \cdot b \cdot h$

$A = \frac{1}{2} \cdot \frac{3}{4}$ yd $\cdot 1\frac{1}{2}$ yd $= \frac{1}{2} \cdot \frac{3}{4}$ yd $\cdot \frac{3}{2}$ yd

$A = \frac{3 \cdot 3}{2 \cdot 4 \cdot 2}$ yd$^2 = \frac{9}{16}$ yd$^2$

**13.** $A = \frac{1}{2} \cdot h \cdot (a + b)$

$A = \frac{1}{2} \cdot 6$ km $\cdot (13$ km $+ 9$ km$)$

$= \frac{6 \cdot 22}{2}$ km$^2$

$= 66$ km$^2$

**14.** $C = 2 \cdot \pi \cdot r$

$C \approx 2 \cdot 3.14 \cdot 7$ in. $= 43.96$ in.

$A = \pi \cdot r \cdot r$

$A \approx 3.14 \cdot 7$ in. $\cdot 7$ in. $= 153.86$ in$^2$

**15.** $C = \pi \cdot d$

$C \approx 3.14 \cdot 8.6$ cm $= 27.004$ cm

$r = \frac{d}{2} = \frac{8.6 \text{ cm}}{2} = 4.3$ cm

$A = \pi \cdot r \cdot r$

$A \approx 3.14 \cdot 4.3$ cm $\cdot 4.3$ cm $= 58.0586$ cm$^2$

**16.** Area of a circle with radius 4 ft:

$A = \pi \cdot 4$ ft $\cdot 4$ ft $= 16 \cdot \pi$ ft$^2$

Area of a square with side 4 ft:

$A = 4$ ft $\cdot 4$ ft $= 16$ ft$^2$

Circumference of a circle with radius 4 ft:

$C = 2 \cdot \pi \cdot 4$ ft $= 8 \cdot \pi$ ft

Area of a rectangle with length 8 ft and width 4 ft:

$A = 8$ ft $\cdot 4$ ft $= 32$ ft$^2$

Area of a triangle with base 4 ft and height 8 ft:

$A = \frac{1}{2} \cdot 4$ ft $\cdot 8$ ft $= 16$ ft$^2$

Perimeter of a square with side 4 ft:

$P = 4 \cdot 4$ ft $= 16$ ft

Perimeter of a rectangle with length 8 ft and width 4 ft:

$P = 2 \cdot (8$ ft $+ 4$ ft$) = 24$ ft

**17.** The area of a 16-in.-diameter pizza is approximately $3.14 \cdot 8$ in. $\cdot 8$in., or $200.96$ in$^2$. At \$16.25, its unit price is $\frac{\$16.25}{200.96 \text{ in}^2}$, or about \$0.08/in$^2$. The area of a 10-in.-diameter pizza is approximately $3.14 \cdot 5$ in. $\cdot 5$ in., or $78.5$ in$^2$. At \$7.85, its unit price is $\frac{\$7.85}{78.5 \text{ in}^2}$, or \$0.10/in$^2$. Since the 16-in.-diameter pizza has the lower unit price, it is a better buy.

**18.** No; let $l$ and $w$ represent the length and width of the smaller rectangle. Then $3 \cdot l$ and $3 \cdot w$ represent the length and width of the larger rectangle. The area of the first rectangle is $l \cdot w$, but the area of the second is $3 \cdot l \cdot 3 \cdot w = 3 \cdot 3 \cdot l \cdot w = 9 \cdot l \cdot w$, or 9 times the area of the smaller rectangle.

**19.** Yes; let $s =$ the length of a side of the larger square. Then $\frac{1}{2} \cdot s =$ the length of a side of the smaller square. The perimeter of the larger square is $4 \cdot s$, and the perimeter of the smaller square is $4 \cdot \frac{1}{2} \cdot s$, or $\frac{1}{2} \cdot (4 \cdot s)$, or $\frac{1}{2}$ the perimeter of the larger square.

**20.** For a rectangle with length $l$ and width $w$,

$P = l + w + l + w$

$P = (l + w) + (l + w)$

$P = 2 \cdot (l + w).$

We also have

$P = l + w + l + w$

$P = (l + l) + (w + w)$

$P = 2 \cdot l + 2 \cdot w.$

**21.** See page 573 of the text.

**22.** No; let $r =$ the radius of the smaller circle. Then its area is $\pi \cdot r \cdot r$, or $\pi r^2$. The radius of the larger circle is $2r$, and its area is $\pi \cdot 2r \cdot 2r$, or $4\pi r^2$, or $4 \cdot \pi r^2$. Thus, the area of the larger circle is 4 times the area of the smaller circle.

## Exercise Set 10.4

**1.** $V = l \cdot w \cdot h$

$V = 12$ cm $\cdot 8$ cm $\cdot 8$ cm

$V = 12 \cdot 64$ cm$^3$

$V = 768$ cm$^3$

**3.** $V = l \cdot w \cdot h$

$V = 7.5$ in. $\cdot 2$ in. $\cdot 3$ in.

$V = 7.5 \cdot 6$ in$^3$

$V = 45$ in$^3$

**5.** $V = l \cdot w \cdot h$

$V = 10$ m $\cdot 5$ m $\cdot 1.5$ m

$V = 10 \cdot 7.5$ m$^3$

$V = 75$ m$^3$

7. $V = l \cdot w \cdot h$

$V = 6\frac{1}{2} \text{ yd} \cdot 5\frac{1}{2} \text{ yd} \cdot 10 \text{ yd}$

$V = \frac{13}{2} \cdot \frac{11}{2} \cdot 10 \text{ yd}^3$

$V = \frac{715}{2} \text{ yd}^3$

$V = 357\frac{1}{2} \text{ yd}^3$

9. $V = Bh = \pi \cdot r^2 \cdot h$

$\approx 3.14 \times 8 \text{ in.} \times 8 \text{ in.} \times 4 \text{ in.}$

$= 803.84 \text{ in}^3$

11. $V = Bh = \pi \cdot r^2 \cdot h$

$\approx 3.14 \times 5 \text{ cm} \times 5 \text{ cm} \times 4.5 \text{ cm}$

$= 353.25 \text{ cm}^3$

13. $V = Bh = \pi \cdot r^2 \cdot h$

$\approx \frac{22}{7} \times 210 \text{ yd} \times 210 \text{ yd} \times 300 \text{ yd}$

$= 41,580,000 \text{ yd}^3$

15. $V = \frac{4}{3} \cdot \pi \cdot r^3$

$\approx \frac{4}{3} \times 3.14 \times (100 \text{ in.})^3$

$= \frac{4 \times 3.14 \times 1,000,000 \text{ in}^3}{3}$

$= 4,186,666\frac{2}{3} \text{ in}^3$

17. $V = \frac{4}{3} \cdot \pi \cdot r^3$

$\approx \frac{4}{3} \times 3.14 \times (3.1 \text{ m})^3$

$= \frac{4 \times 3.14 \times 29.791 \text{ m}^3}{3}$

$\approx 124.72 \text{ m}^3$

19. $V = \frac{4}{3} \cdot \pi \cdot r^3$

$\approx \frac{4}{3} \times \frac{22}{7} \times \left(7\frac{3}{4} \text{ ft}\right)^3$

$= \frac{4}{3} \times \frac{22}{7} \times \left(\frac{31}{4} \text{ ft}\right)^3$

$= \frac{4 \times 22 \times 29,791 \text{ ft}^3}{3 \times 7 \times 64}$

$\approx 1950\frac{101}{168} \text{ ft}^3$

21. $V = \frac{1}{3} \cdot \pi \cdot r^2 \cdot h$

$\approx \frac{1}{3} \times 3.14 \times 33 \text{ ft} \times 33 \text{ ft} \times 100 \text{ ft}$

$\approx 113,982 \text{ ft}^3$

23. $V = \frac{1}{3} \cdot \pi \cdot r^2 \cdot h$

$\approx \frac{1}{3} \times \frac{22}{7} \times 1.4 \text{ cm} \times 1.4 \text{ cm} \times 12 \text{ cm}$

$\approx 24.64 \text{ cm}^3$

25. $V = \frac{1}{3} \cdot \pi \cdot r^2 \cdot h$

$\approx \frac{1}{3} \times \frac{22}{7} \times \frac{3}{4} \text{ yd} \times \frac{3}{4} \text{ yd} \times \frac{7}{5} \text{ yd}$

$= \frac{33}{40} \text{ yd}^3$

27. We must find the radius of the base in order to use the formula for the volume of a circular cylinder.

$$r = \frac{d}{2} = \frac{12 \text{ cm}}{2} = 6 \text{ cm}$$
$$V = Bh = \pi \cdot r^2 \cdot h$$
$$\approx 3.14 \times 6 \text{ cm} \times 6 \text{ cm} \times 42 \text{ cm}$$
$$\approx 4747.68 \text{ cm}^3$$

29. We first find the radius of the ball in order to use the formula for the volume of a sphere.

$$r = \frac{d}{2} = \frac{12 \text{ ft}}{2} = 6 \text{ ft}$$
$$V = \frac{4}{3} \cdot \pi \cdot r^3$$
$$\approx \frac{4}{3} \times 3.14 \times (6 \text{ ft})^3$$
$$\approx 904 \text{ ft}^3$$

31. First we find the radius of the ball:

$$r = \frac{d}{2} = \frac{6.5 \text{ cm}}{2} = 3.25 \text{ cm}$$

Then we find the volume, using the formula for the volume of a sphere.

$$V = \frac{4}{3} \cdot \pi \cdot r^3$$
$$\approx \frac{4}{3} \cdot 3.14 \cdot (3.25 \text{ cm})^3$$
$$\approx 143.72 \text{ cm}^3$$

33. First we find the radius of the earth:

$$\frac{3980 \text{ mi}}{2} = 1990 \text{ mi}$$

Then we find the volume, using the formula for the volume of a sphere.

$$V = \frac{4}{3} \cdot \pi \cdot r^3$$
$$\approx \frac{4}{3} \cdot 3.14 \cdot (1990 \text{ mi})^3$$
$$\approx 32,993,440,000 \text{ mi}^3$$

35. First we find the radius of the base.

$$r = \frac{d}{2} = \frac{4.875 \text{ in.}}{2} = 2.4375 \text{ in.}$$
$$V = \frac{1}{3} \cdot \pi \cdot r^2 \cdot h$$
$$\approx \frac{1}{3} \times 3.14 \times 2.4375 \text{ in.} \times 2.4375 \text{ in.} \times 12.5 \text{ in.}$$
$$\approx 77.7 \text{ in}^3$$

**37.** $V = Bh = \pi \cdot r^2 \cdot h$

$\approx \dfrac{22}{7} \cdot 14 \text{ m} \cdot 14 \text{ m} \cdot 100 \text{ m}$

$= 61{,}600 \text{ m}^3$

**39.** A cube is a rectangular solid.

$V = l \cdot w \cdot h$

$= 18 \text{ yd} \cdot 18 \text{ yd} \cdot 18 \text{ yd}$

$= 5832 \text{ yd}^3$

**41.** First we find the radius of the can.

$r = \dfrac{d}{2} = \dfrac{6.5 \text{ cm}}{2} = 3.25 \text{ cm}$

The height of the can is the length of the diameters of 3 tennis balls.

$h = 3(6.5 \text{ cm}) = 19.5 \text{ cm}$

Now we find the volume.

$V = Bh = \pi \cdot r^2 \cdot h$

$\approx 3.14 \times 3.25 \text{ cm} \times 3.25 \text{ cm} \times 19.5 \text{ cm}$

$\approx 646.74 \text{ cm}^3$

**43.** $11 \text{ yd} = 11 \times 1 \text{ yd}$

$= 11 \times 3 \text{ ft}$

$= 11 \times 3 \times 1 \text{ ft}$

$= 11 \times 3 \times 12 \text{ in.}$

$= 396 \text{ in.}$

**45.** $42 \text{ ft} = 42 \text{ ft} \times \dfrac{1 \text{ yd}}{3 \text{ ft}}$

$= \dfrac{42 \text{ ft}}{3 \text{ ft}} \times 1 \text{ yd}$

$= \dfrac{42}{3} \times \dfrac{\text{ft}}{\text{ft}} \times 1 \text{ yd}$

$= 14 \times 1 \text{ yd}$

$= 14 \text{ yd}$

**47.** $144 \text{ in.} = 144 \text{ in.} \times \dfrac{1 \text{ ft}}{12 \text{ in.}}$

$= \dfrac{144 \text{ in.}}{12 \text{ in.}} \times 1 \text{ ft}$

$= \dfrac{144}{12} \times \dfrac{\text{in.}}{\text{in.}} \times 1 \text{ ft}$

$= 12 \times 1 \text{ ft}$

$= 12 \text{ ft}$

**49.** $6 \text{ gal} = 6 \times 1 \text{ gal}$

$= 6 \times 4 \text{ qt}$

$= 24 \text{ qt}$

**51.** We move the decimal point three places to the left: $566 \text{ mL} = 0.566 \text{ L}$.

**53.** *Familiarize.* Let $m = $ the number of cubic miles by which the volume of water in Lake Michigan exceeds the volume of water in Lake Erie.

*Translate.*

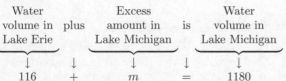

| Water volume in Lake Erie | plus | Excess amount in Lake Michigan | is | Water volume in Lake Michigan |
|---|---|---|---|---|
| ↓ | ↓ | ↓ | ↓ | ↓ |
| 116 | + | $m$ | = | 1180 |

*Solve.*

$116 + m = 1180$

$116 + m - 116 = 1180 - 116$

$m = 1064$

*Check.* Since $116 + 1064 = 1180$, the answer checks.

*State.* The volume of water in Lake Michigan is 1064 mi³ greater than the volume of water in Lake Erie.

**55.** $\dfrac{483 + 279 + 195 + 62 + 283}{5} = \dfrac{1302}{5} = 260.4 \text{ ft}$

**57.** $850 \text{ mi}^3 = 850 \times 1 \text{ mi} \times 1 \text{ mi} \times 1 \text{ mi}$

$\approx 850 \times 1.609 \text{ km} \times 1.609 \text{ km} \times 1.609 \text{ km}$

$\approx 3540.68 \text{ km}^3$

**59.** The length of a side of the cube is the length of a diameter of the sphere, 1 m.

$V = l \cdot w \cdot h$

$= 1 \text{ m} \cdot 1 \text{ m} \cdot 1 \text{ m} = 1 \text{ m}^3$

The radius of the sphere is $\dfrac{1 \text{ m}}{2}$, or 0.5 m.

$V = \dfrac{4}{3} \cdot \pi \cdot r^3$

$\approx \dfrac{4}{3} \cdot 3.14 \cdot (0.5 \text{ m})^3 \approx 0.523 \text{ m}^3$

$1 \text{ m}^3 - 0.523 \text{ m}^3 = 0.477 \text{ m}^3$, so there is 0.477 m³ more volume in the cube.

## Exercise Set 10.5

**1.** The angle can be named in five different ways:

angle $GHI$, angle $IHG$, $\angle GHI$, $\angle IHG$, or $\angle H$.

**3.** Place the $\triangle$ of the protractor at the vertex of the angle, and line up one of the sides at 0°. We choose the horizontal side. Since 0° is on the inside scale, we check where the other side of the angle crosses the inside scale. It crosses at 10°. Thus, the measure of the angle is 10°.

**5.** Place the $\triangle$ of the protractor at the vertex of the angle, point $B$. Line up one of the sides at 0°. We choose the side that contains point $A$. Since 0° is on the outside scale, we check where the other side crosses the outside scale. It crosses at 180°. Thus, the measure of the angle is 180°.

**7.** Place the $\triangle$ of the protractor at the vertex of the angle, and line up one of the sides at 0°. We choose the horizontal side. Since 0° is on the inside scale, we check where the other side crosses the inside scale. It crosses at 130°. Thus, the measure of the angle is 130°.

**9.** Using a protractor, we find that the measure of the angle in Exercise 1 is 148°. Since its measure is greater than 90° and less than 180°, it is an obtuse angle.

**11.** The measure of the angle in Exercise 3 is 10°. Since its measure is greater than 0° and less than 90°, it is an acute angle.

**13.** The measure of the angle in Exercise 5 is 180°. It is a straight angle.

15. The measure of the angle in Exercise 7 is 130°. Since its measure is greater than 90°and less than 180°, it is an obtuse angle.

17. The measure of the angle in Margin Exercise 1 is 30°. Since its measure is greater than 0°and less than 90°, it is an acute angle.

19. The measure of the angle in Margin Exercise 3 is 126°. Since its measure is greater than 90°and less than 180°, it is an obtuse angle.

21. Two angles are complementary if the sum of their measures is 90°.
$$90° - 11° = 79°.$$
The measure of a complement is 79°.

23. Two angles are complementary if the sum of their measures is 90°.
$$90° - 67° = 23°.$$
The measure of a complement is 23°.

25. Two angles are complementary if the sum of their measures is 90°.
$$90° - 58° = 32°.$$
The measure of a complement is 32°.

27. Two angles are complementary if the sum of their measures is 90°.
$$90° - 29° = 61°.$$
The measure of a complement is 61°.

29. Two angles are supplementary if the sum of their measures is 180°.
$$180° - 3° = 177°.$$
The measure of a supplement is 177°.

31. Two angles are supplementary if the sum of their measures is 180°.
$$180° - 139° = 41°.$$
The measure of a supplement is 41°.

33. Two angles are supplementary if the sum of their measures is 180°.
$$180° - 85° = 95°.$$
The measure of a supplement is 95°.

35. Two angles are supplementary if the sum of their measures is 180°.
$$180° - 102° = 78°.$$
The measure of a supplement is 78°.

37. All the sides are of different lengths. The triangle is a scalene triangle.

One angle is an obtuse angle. The triangle is an obtuse triangle.

39. All the sides are of different lengths. The triangle is a scalene triangle.

One angle is a right angle. The triangle is a right triangle.

41. All the sides are the same length. The triangle is an equilateral triangle.

All three angles are acute. The triangle is an acute triangle.

43. All the sides are of different lengths. The triangle is a scalene triangle.

One angle is an obtuse angle. The triangle is an obtuse triangle.

45. $m(\angle A) + m(\angle B) + m(\angle C) = 180°$
$$42° + 92° + x = 180°$$
$$134° + x = 180°$$
$$x = 180° - 134°$$
$$x = 46°$$

47. $31° + 29° + x = 180°$
$$60° + x = 180°$$
$$x = 180° - 60°$$
$$x = 120°$$

49. $37° + 85° + x = 180°$
$$122° + x = 180°$$
$$x = 180° - 122°$$
$$x = 58°$$

51. A number is divisible by 8 if the number named by the last three digits is divisible by 8.

53. In the metric system, the gram is the basic unit of mass.

55. An angle is a set of points consisting of two rays.

57. The perimeter of a polygon is the sum of the lengths of its sides.

59. We find $m \angle 2$:
$$m \angle 6 + m \angle 1 + m \angle 2 = 180°$$
$$33.07° + 79.8° + m \angle 2 = 180°$$
$$112.87° + m \angle 2 = 180°$$
$$m \angle 2 = 180° - 112.87°$$
$$m \angle 2 = 67.13°$$
The measure of angle 2 is 67.13°.

We find $m \angle 3$:
$$m \angle 1 + m \angle 2 + m \angle 3 = 180°$$
$$79.8° + 67.13° + m \angle 3 = 180°$$
$$146.93° + m \angle 3 = 180°$$
$$m \angle 3 = 180° - 146.93°$$
$$m \angle 3 = 33.07°$$
The measure of angle 3 is 33.07°.

We find $m \angle 4$:
$$m \angle 2 + m \angle 3 + m \angle 4 = 180°$$
$$67.13° + 33.07° + m \angle 4 = 180°$$
$$100.2° + m \angle 4 = 180°$$
$$m \angle 4 = 180° - 100.2°$$
$$m \angle 4 = 79.8°$$
The measure of angle 4 is 79.8°.

To find $m \angle 5$, note that $m \angle 6 + m \angle 1 + m \angle 5 = 180°$. Then to find $m \angle 5$ we follow the same procedure we used to find $m \angle 2$. Thus, the measure of angle 5 is 67.13°.

**61.** $\angle ACB$ and $\angle ACD$ are complementary angles. Since $m\angle ACD = 40°$ and $90° - 40° = 50°$, we have $m\angle ACB = 50°$.

Now consider triangle $ABC$. We know that the sum of the measures of the angles is $180°$. Then

$$m\angle ABC + m\angle BCA + m\angle CAB = 180°$$
$$50° + 90° + m\angle CAB = 180°$$
$$140° + m\angle CAB = 180°$$
$$m\angle CAB = 180° - 140°$$
$$m\angle CAB = 40°,$$

so $m\angle CAB = 40°$.

To find $m\angle EBC$ we first find $m\angle CEB$. We note that $\angle DEC$ and $\angle CEB$ are supplementary angles. Since $m\angle DEC = 100°$ and $180° - 100° = 80°$, we have $m\angle CEB = 80°$. Now consider triangle $BCE$. We know that the sum of the measures of the angles is $180°$. Note that $\angle ACB$ can also be named $\angle BCE$. Then

$$m\angle BCE + m\angle CEB + m\angle EBC = 180°$$
$$50° + 80° + m\angle EBC = 180°$$
$$130° + m\angle EBC = 180°$$
$$m\angle EBC = 180° - 130°$$
$$m\angle EBC = 50°,$$

so $m\angle EBC = 50°$.

$\angle EBA$ and $\angle EBC$ are complementary angles. Since $m\angle EBC = 50°$ and $90° - 50° = 40°$, we have $m\angle EBA = 40°$.

Now consider triangle $ABE$. We know that the sum of the measures of the angles is $180°$. Then

$$m\angle CAB + m\angle EBA + m\angle AEB = 180°$$
$$40° + 40° + m\angle AEB = 180°$$
$$80° + m\angle AEB = 180°$$
$$m\angle AEB = 180° - 80°$$
$$m\angle AEB = 100°,$$

so $m\angle AEB = 100°$.

To find $m\angle ADB$ we first find $m\angle EDC$. Consider triangle $CDE$. We know that the sum of the measures of the angles is $180°$. Then

$$m\angle DEC + m\angle ECD + m\angle EDC = 180°$$
$$100° + 40° + m\angle EDC = 180°$$
$$140° + m\angle EDC = 180°$$
$$m\angle EDC = 180° - 140°$$
$$m\angle EDC = 40°,$$

so $m\angle EDC = 40°$. We now note that $\angle ADB$ and $\angle EDC$ are complementary angles. Since $m\angle EDC = 40°$ and $90° - 40° = 50°$, we have $m\angle ADB = 50°$.

## Exercise Set 10.6

**1.** $\sqrt{100} = 10$

The square root of 100 is 10 because $10^2 = 100$.

**3.** $\sqrt{441} = 21$

The square root of 441 is 21 because $21^2 = 441$.

**5.** $\sqrt{625} = 25$

The square root of 625 is 25 because $25^2 = 625$.

**7.** $\sqrt{361} = 19$

The square root of 361 is 19 because $19^2 = 361$.

**9.** $\sqrt{529} = 23$

The square root of 529 is 23 because $23^2 = 529$.

**11.** $\sqrt{10,000} = 100$

The square root of 10,000 is 100 because $100^2 = 10,000$.

**13.** $\sqrt{48} \approx 6.928$

**15.** $\sqrt{8} \approx 2.828$

**17.** $\sqrt{18} \approx 4.243$

**19.** $\sqrt{6} \approx 2.449$

**21.** $\sqrt{10} \approx 3.162$

**23.** $\sqrt{75} \approx 8.660$

**25.** $\sqrt{196} = 14$

**27.** $\sqrt{183} \approx 13.528$

**29.**
$$\begin{aligned}
a^2 + b^2 &= c^2 && \text{Pythagorean equation} \\
3^2 + 5^2 &= c^2 && \text{Substituting} \\
9 + 25 &= c^2 \\
34 &= c^2 \\
\sqrt{34} &= c && \text{Exact answer} \\
5.831 &\approx c && \text{Approximation}
\end{aligned}$$

**31.**
$$\begin{aligned}
a^2 + b^2 &= c^2 && \text{Pythagorean equation} \\
7^2 + 7^2 &= c^2 && \text{Substituting} \\
49 + 49 &= c^2 \\
98 &= c^2 \\
\sqrt{98} &= c && \text{Exact answer} \\
9.899 &\approx c && \text{Approximation}
\end{aligned}$$

**33.**
$$\begin{aligned}
a^2 + b^2 &= c^2 \\
a^2 + 12^2 &= 13^2 \\
a^2 + 144 &= 169 \\
a^2 &= 169 - 144 = 25 \\
a &= 5
\end{aligned}$$

**35.**
$$\begin{aligned}
a^2 + b^2 &= c^2 \\
6^2 + b^2 &= 10^2 \\
36 + b^2 &= 100 \\
b^2 &= 100 - 36 = 64 \\
b &= 8
\end{aligned}$$

**37.**
$$\begin{aligned}
a^2 + b^2 &= c^2 \\
10^2 + 24^2 &= c^2 \\
100 + 576 &= c^2 \\
676 &= c^2 \\
26 &= c
\end{aligned}$$

**39.**
$$\begin{aligned}
a^2 + b^2 &= c^2 \\
9^2 + b^2 &= 15^2 \\
81 + b^2 &= 225 \\
81 + b^2 - 81 &= 225 - 81 \\
b^2 &= 225 - 81 \\
b^2 &= 144 \\
b &= 12
\end{aligned}$$

**41.**    $a^2 + b^2 = c^2$
$1^2 + b^2 = 32^2$
$1 + b^2 = 1024$
$1 + b^2 - 1 = 1024 - 1$
$b^2 = 1024 - 1$
$b^2 = 1023$
$b = \sqrt{1023}$      Exact answer
$b \approx 31.984$     Approximation

**43.**  $a^2 + b^2 = c^2$
$4^2 + 3^2 = c^2$
$16 + 9 = c^2$
$25 = c^2$
$5 = c$

**45. Familiarize.** We first make a drawing. In it we see a right triangle. We let $w =$ the length of the wire, in meters.

**Translate.** We substitute 9 for $a$, 13 for $b$, and $w$ for $c$ in the Pythagorean equation.

$$a^2 + b^2 = c^2$$
$$9^2 + 13^2 = w^2$$

**Solve.** We solve the equation for $w$.

$81 + 169 = w^2$
$250 = w^2$
$\sqrt{250} = w$      Exact answer
$15.8 \approx w$      Approximation

**Check.** $9^2 + 13^2 = 81 + 169 = 250 = (\sqrt{250})^2$
**State.** The length of the wire is $\sqrt{250}$ m, or about 15.8 m.

**47. Familiarize.** We refer to the drawing in the text. We let $d =$ the distance from home to second base, in feet.

**Translate.** We substitute 65 for $a$, 65 for $b$, and $d$ for $c$ in the Pythagorean equation.

$$a^2 + b^2 = c^2$$
$$65^2 + 65^2 = d^2$$

**Solve.** We solve the equation for $d$.

$4225 + 4225 = d^2$
$8450 = d^2$
$\sqrt{8450} = d$
$91.9 \approx d$

**Check.** $65^2 + 65^2 = 4225 + 4225 = 8450 = (\sqrt{8450})^2$
**State.** The distance from home to second base is $\sqrt{8450}$ ft, or about 91.9 ft.

**49. Familiarize.** We refer to the drawing in the text.
**Translate.** We substitute in the Pythagorean equation.

$$a^2 + b^2 = c^2$$
$$20^2 + h^2 = 30^2$$

**Solve.** We solve the equation for $h$.

$400 + h^2 = 900$
$h^2 = 900 - 400$
$h^2 = 500$
$h = \sqrt{500}$
$h \approx 22.4$

**Check.** $20^2 + (\sqrt{500})^2 = 400 + 500 = 900 = 30^2$
**State.** The height of the tree is $\sqrt{500}$ ft, or about 22.4 ft.

**51. Familiarize.** We refer to the drawing in the text. We let $h =$ the plane's horizontal distance from the airport.
**Translate.** We substitute 4100 for $a$, $h$ for $b$, and 15,100 for $c$ in the Pythagorean equation.

$$a^2 + b^2 = c^2$$
$$4100^2 + h^2 = 15,100^2$$

**Solve.** We solve the equation for $h$.

$16,810,000 + h^2 = 228,010,000$
$h^2 = 228,010,000 - 16,810,000$
$h^2 = 211,200,000$
$h = \sqrt{211,200,000}$
$h \approx 14,532.7$

**Check.** $4100^2 + (\sqrt{211,200,000})^2 = 16,810,000 + 211,200,000 = 228,010,000 = 15,100^2$
**State.** The plane's horizontal distance from the airport is $\sqrt{211,200,000}$ ft, or about 14,532.7 ft.

**53. Familiarize.** We first make a drawing. In it we see a right triangle. Let $l =$ the length of the string of lights, in feet.

**Translate.** Substitute 16 for $a$, 24 for $b$, and $l$ for $c$ in the Pythagorean equation.
$$a^2 + b^2 = c^2$$
$$16^2 + 24^2 = l^2$$
**Solve.** We solve the equation for $l$.
$256 + 576 = l^2$
$832 = l^2$
$\sqrt{832} = l$
$28.8 \approx l$

**Check.** $16^2 + 24^2 = 256 + 576 = 832 = (\sqrt{832})^2$
**State.** The string of lights is $\sqrt{832}$ ft, or about 28.8 ft long.

**55.** $10^3 = 10 \times 10 \times 10 = 1000$

**57.** $10^5 = 10 \times 10 \times 10 \times 10 \times 10 = 100,000$

**59.** $A = P \cdot \left(1 + \dfrac{r}{n}\right)^{n \cdot t}$

$= \$200,000\left(1 + \dfrac{6\frac{5}{8}\%}{1}\right)^{1 \cdot 25}$

$= \$200,000(1 + 0.06625)^{25}$

$= \$200,000(1.06625)^{25}$

$\approx \$994,274.04$

**61.** $A = P \cdot \left(1 + \dfrac{r}{n}\right)^{n \cdot t}$

$= \$45,000\left(1 + \dfrac{4.5\%}{1}\right)^{1 \cdot 15}$

$= \$45,000(1 + 0.045)^{15}$

$= \$45,000(1.045)^{15}$

$\approx \$87,087.71$

**63.** To find the areas we must first use the Pythagorean equation to find the height of each triangle and then use the formula for the area of a triangle.

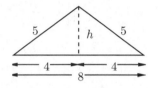

$a^2 + b^2 = c^2$      Pythagorean equation
$4^2 + h^2 = 5^2$      Substituting
$16 + h^2 = 25$
$h^2 = 25 - 16 = 9$
$h = 3$

$A = \dfrac{1}{2} \cdot b \cdot h$      Area of a triangle

$A = \dfrac{1}{2} \cdot 8 \cdot 3$      Substituting

$A = 12$

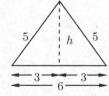

$a^2 + b^2 = c^2$      Pythagorean equation
$3^2 + h^2 = 5^2$      Substituting
$9 + h^2 = 25$
$h^2 = 25 - 9 = 16$
$h = 4$

$A = \dfrac{1}{2} \cdot b \cdot h$      Area of a triangle

$A = \dfrac{1}{2} \cdot 6 \cdot 4$      Substituting

$A = 12$

The areas of the triangles are the same (12 square units).

---

## Chapter 10 Concept Reinforcement

**1.** True; the sum of the measures of the acute angles of a right triangle is $180° - 90°$, or $90°$.

**2.** False; two angles are supplementary if the sum of their measures is $180°$.

**3.** Using a calculator, we find that $\pi \approx 3.14159$ and $\dfrac{22}{7} \approx 3.14288$. Thus, $\pi$ is greater than 3.14 but less than $\dfrac{22}{7}$. The given statement is false.

**4.** True; the volume of a sphere with diameter 6 ft, or radius 3 ft, is $\dfrac{4}{3} \cdot \pi \cdot (3 \text{ ft})^3 \approx \dfrac{4}{3} \times 3.14 \times (3 \text{ ft})^3 = 113.04 \text{ ft}^3$.

The volume of a rectangular solid that measures 6 ft by 6 ft by 6 ft is 6 ft $\times$ 6 ft $\times$ 6 ft $= 216 \text{ ft}^3$.

**5.** True; the measure of an obtuse angle is greater than $90°$ while the measure of an acute angle is less than $90°$.

**6.** The statement is true. The hypotenuse of a right triangle is the longest side of the triangle.

---

## Chapter 10 Important Concepts

**1.** $P = 2 \cdot (l + w)$

$= 2 \cdot (8.2 \text{ ft} + 5.7 \text{ ft})$

$= 2 \cdot (13.9 \text{ ft}) = 27.8 \text{ ft}$

$A = l \cdot w$

$= 8.2 \text{ ft} \cdot 5.7 \text{ ft}$

$= 8.2 \cdot 5.7 \cdot \text{ ft} \cdot \text{ ft} = 46.74 \text{ ft}^2$

**2.** $A = b \cdot h$

$= 6.2 \text{ m} \cdot 2.5 \text{ m}$

$= 6.2 \cdot 2.5 \cdot \text{ m} \cdot \text{ m} = 15.5 \text{ m}^2$

**3.** $A = \dfrac{1}{2} \cdot b \cdot h$

$= \dfrac{1}{2} \times 3.5 \text{ ft} \cdot 5 \text{ ft}$

$= \dfrac{3.5 \cdot 5}{2} \text{ ft}^2 = 8.75 \text{ ft}^2$

**4.** $A = \frac{1}{2} \cdot h \cdot (a+b)$

$= \frac{1}{2} \times 8 \text{ m} \times (5 \text{ m} + 15 \text{ m})$

$= \frac{1}{2} \times 8 \text{ m} \times (20 \text{ m})$

$= \frac{8 \times 20}{2} \text{ m}^2 = 80 \text{ m}^2$

**5.** $C = 2 \cdot \pi \cdot r$

$\approx 2 \cdot 3.14 \cdot 6 \text{ in.} = 37.68 \text{ in.}$

**6.** $A = \pi \cdot r \cdot r$

$\approx \frac{22}{7} \cdot 14 \text{ cm} \cdot 14 \text{ cm}$

$= \frac{22 \cdot 14 \cdot 14}{7} \text{ cm}^2$

$= 616 \text{ cm}^2$

**7.** $V = l \cdot w \cdot h$

$= 18.1 \text{ m} \times 15 \text{ m} \times 6.2 \text{ m}$

$= 1683.3 \text{ m}^3$

**8.** $V = \pi \cdot r^2 \cdot h$

$\approx \frac{22}{7} \times 1\frac{1}{3} \text{ ft} \times 1\frac{1}{3} \text{ ft} \times 5\frac{2}{5} \text{ ft}$

$= \frac{22}{7} \times \frac{4}{3} \text{ ft} \times \frac{4}{3} \text{ ft} \times \frac{27}{5} \text{ ft}$

$= \frac{22 \times 4 \times 4 \times 27}{7 \times 3 \times 3 \times 5} \text{ ft}^3$

$= \frac{1056}{35} \text{ ft}^3, \text{ or } 30\frac{6}{35} \text{ ft}^3$

**9.** $V = \frac{4}{3} \cdot \pi \cdot r^3$

$\approx \frac{4}{3} \times 3.14 \times (7.4 \text{ cm})^3$

$= 1696.537813 \text{ cm}^3$

**10.** $V = \frac{1}{3} \cdot \pi \cdot r^2 \cdot h$

$\approx \frac{1}{3} \times 3.14 \times 2.25 \text{ ft} \times 2.25 \text{ ft} \times 5 \text{ ft}$

$= 26.49375 \text{ ft}^3$

**11.** Measure of the complement: $90° - 38° = 52°$

Measure of the supplement: $180° - 38° = 142°$

**12.** $x + 21° + 72° = 180°$

$x + 93° = 180°$

$x = 180° - 93°$

$x = 87°$

**13.** $a^2 + b^2 = c^2$

$9^2 + b^2 = 17^2$

$81 + b^2 = 289$

$b^2 = 289 - 81$

$b^2 = 208$

$b = \sqrt{208}$

$b \approx 14.422$

## Chapter 10 Review Exercises

**1.** Perimeter $= 5 \text{ ft} + 7 \text{ ft} + 4 \text{ ft} + 4 \text{ ft} + 3 \text{ ft}$

$= (5 + 7 + 4 + 4 + 3) \text{ ft}$

$= 23 \text{ ft}$

**2.** Perimeter $= 0.5 \text{ m} + 1.9 \text{ m} + 1.2 \text{ m} + 0.8 \text{ m}$

$= (0.5 + 1.9 + 1.2 + 0.8) \text{ m}$

$= 4.4 \text{ m}$

**3.** $P = 2 \cdot l + 2 \cdot w$

$P = 2 \cdot 78 \text{ ft} + 2 \cdot 36 \text{ ft}$

$P = 156 \text{ ft} + 72 \text{ ft}$

$P = 228 \text{ ft}$

$A = l \cdot w$

$A = 78 \text{ ft} \cdot 36 \text{ ft}$

$A = 2808 \text{ ft}^2$

**4.** $P = 4 \cdot s$

$P = 4 \cdot 9 \text{ ft}$

$P = 36 \text{ ft}$

$A = s \cdot s$

$A = 9 \text{ ft} \cdot 9 \text{ ft}$

$A = 81 \text{ ft}^2$

**5.** $P = 2 \cdot (l + w)$

$P = 2 \cdot (7 \text{ cm} + 1.8 \text{ cm})$

$P = 2 \cdot (8.8 \text{ cm})$

$P = 17.6 \text{ cm}$

$A = l \cdot w$

$A = 7 \text{ cm} \cdot 1.8 \text{ cm}$

$A = 12.6 \text{ cm}^2$

**6.** $A = \frac{1}{2} \cdot b \cdot h$

$A = \frac{1}{2} \cdot 15 \text{ m} \cdot 3 \text{ m}$

$A = \frac{15 \cdot 3}{2} \text{ m}^2$

$A = 22.5 \text{ m}^2$

**7.** $A = \frac{1}{2} \cdot b \cdot h$

$A = \frac{1}{2} \cdot 11.4 \text{ yd} \cdot 5.2 \text{ yd}$

$A = \frac{11.4 \cdot 5.2}{2} \text{ yd}^2$

$A = 29.64 \text{ yd}^2$

**8.**   $A = \frac{1}{2} \cdot h \cdot (a+b)$

$A = \frac{1}{2} \cdot 8 \text{ m} \cdot (5+17) \text{ m}$

$A = \frac{8 \cdot 22}{2} \text{ m}^2$

$A = 88 \text{ m}^2$

**9.**   $A = b \cdot h$

$A = 21\frac{5}{6} \text{ in.} \cdot 6\frac{2}{3} \text{ in.}$

$A = \frac{131}{6} \cdot \frac{20}{3} \text{ in}^2$

$A = \frac{131 \cdot 20}{6 \cdot 3} \text{ in}^2$

$A = \frac{1310}{9} \text{ in}^2$

$A = 145\frac{5}{9} \text{ in}^2$

**10.**  *Familiarize*. The seeded area is the total area of the house and the seeded area less the area of the house. From the drawing in the text we see that the total area is the area of a rectangle with length 70 ft and width 25 ft + 7 ft, or 32 ft. The length of the rectangular house is 70 ft − 7 ft − 7 ft, or 56 ft, and its width is 25 ft. We let $A$ = the seeded area.

*Translate*.

$$\underbrace{\text{Seeded area}}_{\downarrow} \text{ is } \underbrace{\text{Total area}}_{\downarrow} \text{ minus } \underbrace{\text{Area of house}}_{\downarrow}$$

$$A \quad = \quad 70 \text{ ft} \cdot 32 \text{ ft} \quad - \quad 56 \text{ ft} \cdot 25 \text{ ft}$$

*Solve*.

$A = 70 \text{ ft} \cdot 32 \text{ ft} - 56 \text{ ft} \cdot 25 \text{ ft}$

$A = 2240 \text{ ft}^2 - 1400 \text{ ft}^2$

$A = 840 \text{ ft}^2$

*Check*. We can repeat the calculations. The answer checks.

*State*. The seeded area is 840 ft².

**11.**  $r = \frac{d}{2} = \frac{16 \text{ m}}{2} = 8 \text{ m}$

**12.**  $r = \frac{d}{2} = \frac{\frac{28}{11} \text{ in.}}{2} = \frac{28}{11} \text{ in.} \cdot \frac{1}{2}$

$= \frac{28}{11 \cdot 2} \text{ in.} = \frac{\cancel{2} \cdot 14}{11 \cdot \cancel{2}} \text{ in.}$

$= \frac{14}{11} \text{ in., or } 1\frac{3}{11} \text{ in.}$

**13.**  $d = 2 \cdot r = 2 \cdot 7 \text{ ft} = 14 \text{ ft}$

**14.**  $d = 2 \cdot r = 2 \cdot 10 \text{ cm} = 20 \text{ cm}$

**15.**  $C = \pi \cdot d$

$C \approx 3.14 \cdot 16 \text{ m}$

$= 50.24 \text{ m}$

**16.**  $C = \pi \cdot d$

$C \approx \frac{22}{7} \cdot \frac{28}{11} \text{ in.}$

$= \frac{22 \cdot 28}{7 \cdot 11} \text{ in.} = \frac{2 \cdot \cancel{11} \cdot 4 \cdot \cancel{7}}{\cancel{7} \cdot \cancel{11} \cdot 1} \text{ in.}$

$= 8 \text{ in.}$

**17.**  In Exercise 13 we found that the radius of the circle is 8 m.

$A = \pi \cdot r \cdot r$

$A \approx 3.14 \cdot 8 \text{ m} \cdot 8 \text{ m}$

$A = 200.96 \text{ m}^2$

**18.**  In Exercise 14 we found that the radius of the circle is $\frac{14}{11}$ in.

$A = \pi \cdot r \cdot r$

$A \approx \frac{22}{7} \cdot \frac{14}{11} \text{ in.} \cdot \frac{14}{11} \text{ in.}$

$A = \frac{22 \cdot 14 \cdot 14}{7 \cdot 11 \cdot 11} \text{ in}^2 = \frac{2 \cdot \cancel{11} \cdot 2 \cdot \cancel{7} \cdot 14}{\cancel{7} \cdot \cancel{11} \cdot 11}$

$A = \frac{56}{11} \text{ in}^2, \text{ or } 5\frac{1}{11} \text{ in}^2$

**19.**  The shaded area is the area of a circle with radius of 21 ft less the area of a circle with a diameter of 21 ft. The radius of the smaller circle is $\frac{21 \text{ ft}}{2}$, or 10.5 ft.

$A = \pi \cdot 21 \text{ ft} \cdot 21 \text{ ft} - \pi \cdot 10.5 \text{ ft} \cdot 10.5 \text{ ft}$

$A \approx 3.14 \cdot 21 \text{ ft} \cdot 21 \text{ ft} - 3.14 \cdot 10.5 \text{ ft} \cdot 10.5 \text{ ft}$

$A = 1384.74 \text{ ft}^2 - 346.185 \text{ ft}^2$

$A = 1038.555 \text{ ft}^2$

**20.**  $V = l \cdot w \cdot h$

$V = 12 \text{ yd} \cdot 3 \text{ yd} \cdot 2.6 \text{ yd}$

$V = 36 \cdot 2.6 \text{ yd}^3$

$V = 93.6 \text{ yd}^3$

**21.**  $V = l \cdot w \cdot h$

$V = 4.6 \text{ cm} \cdot 3 \text{ cm} \cdot 14 \text{ cm}$

$V = 13.8 \cdot 14 \text{ cm}^3$

$V = 193.2 \text{ cm}^3$

**22.**  $r = \frac{20 \text{ ft}}{2} = 10 \text{ ft}$

$V = B \cdot h = \pi \cdot r^2 \cdot h$

$\approx 3.14 \times 10 \text{ ft} \times 10 \text{ ft} \times 100 \text{ ft}$

$= 31,400 \text{ ft}^3$

**23.**  $V = \frac{4}{3} \cdot \pi \cdot r^3$

$\approx \frac{4}{3} \times 3.14 \times (2 \text{ cm})^3$

$= \frac{4 \times 3.14 \times 8 \text{ cm}^3}{3}$

$= 33.49\overline{3} \text{ cm}^3$

**24.** $V = \frac{1}{3} \cdot \pi \cdot r^2 \cdot h$

$\approx \frac{1}{3} \times 3.14 \times 1 \text{ in.} \times 1 \text{ in.} \times 4.5 \text{ in.}$

$= 4.71 \text{ in}^3$

**25.** $V = B \cdot h = \pi \cdot r^2 \cdot h$

$\approx 3.14 \times 5 \text{ cm} \times 5 \text{ cm} \times 12 \text{ cm}$

$= 942 \text{ cm}^3$

**26.** The window is composed of half of a circle with radius 2 ft and of a rectangle with length 5 ft and width twice the radius of the half circle, or $2 \cdot 2$ ft, or 4 ft. To find the area of the window we add one-half the area of a circle with radius 2 ft and the area of a rectangle with length 5 ft and width 4 ft.

$A = \frac{1}{2} \cdot \pi \cdot 2 \text{ ft} \cdot 2 \text{ ft} + 5 \text{ ft} \cdot 4 \text{ ft}$

$\approx \frac{1}{2} \cdot 3.14 \cdot 2 \text{ ft} \cdot 2 \text{ ft} + 5 \text{ ft} \cdot 4 \text{ ft}$

$= \frac{3.14 \cdot 2 \cdot 2}{2} \text{ ft}^2 + 20 \text{ ft}^2$

$= 6.28 \text{ ft}^2 + 20 \text{ ft}^2$

$= 26.28 \text{ ft}^2$

The perimeter is composed of one-half the circumference of a circle with radius 2 ft along with the lengths of three sides of the rectangle.

$P = \frac{1}{2} \cdot 2 \cdot \pi \cdot 2 \text{ ft} + 5 \text{ ft} + 4 \text{ ft} + 5 \text{ ft}$

$\approx \frac{1}{2} \cdot 2 \cdot 3.14 \cdot 2 \text{ ft} + 5 \text{ ft} + 4 \text{ ft} + 5 \text{ ft}$

$= \frac{2 \cdot 3.14 \cdot 2 \text{ ft}}{2} + 5 \text{ ft} + 4 \text{ ft} + 5 \text{ ft}$

$= 6.28 \text{ ft} + 5 \text{ ft} + 4 \text{ ft} + 5 \text{ ft}$

$= 20.28 \text{ ft}$

**27.** Place the $\triangle$ of the protractor at the vertex of the angle, and line up one of the sides at $0°$. We choose the nearly horizontal side. Since $0°$ is on the inside scale, we check where the other side of the angle crosses the inside scale. It crosses at $54°$. Thus, the measure of the angle is $54°$.

**28.** Place the $\triangle$ of the protractor at the vertex of the angle, point $B$. Line up one of the sides at $0°$. We choose the side that contains point $P$. Since $0°$ is on the outside scale, we check where the other side crosses the outside scale. It crosses at $180°$. Thus, the measure of the angle is $180°$.

**29.** Place the $\triangle$ of the protractor at the vertex of the angle, and line up one of the sides at $0°$. We choose the horizontal side. Since $0°$ is on the inside scale, we check where the other side crosses the inside scale. It crosses at $140°$. Thus, the measure of the angle is $140°$.

**30.** Place the $\triangle$ of the protractor at the vertex of the angle, and line up one of the sides at $0°$. We choose the horizontal side. Since $0°$ is on the inside scale, we check where the other side crosses the inside scale. It crosses at $90°$. Thus, the measure of the angle is $90°$.

**31.** The measure of the angle in Exercise 27 is $54°$. Since its measure is greater than $0°$ and less than $90°$, it is an acute angle.

**32.** The measure of the angle in Exercise 28 is $180°$. It is a straight angle.

**33.** The measure of the angle in Exercise 29 is $140°$. Since its measure is greater than $90°$ and less than $180°$, it is an obtuse angle.

**34.** The measure of the angle in Exercise 30 is $90°$. It is a right angle.

**35.** Two angles are complementary if the sum of their measures is $90°$.

$90° - 41° = 49°.$

The measure of a complement of $\angle BAC$ is $49°$.

**36.** Two angles are supplementary if the sum of their measures is $180°$.

$180° - 44° = 136°.$

The measure of a supplement is $136°$.

**37.** $30° + 90° + x = 180°$

$120° + x = 180°$

$x = 180° - 120°$

$x = 60°$

**38.** All the sides are of different lengths. The triangle is a scalene triangle.

**39.** One angle is a right angle. The triangle is a right triangle.

**40.** $\sqrt{64} = 8$ because $8 \cdot 8 = 64$.

**41.** $\sqrt{83} \approx 9.110$

**42.** $a^2 + b^2 = c^2$

$15^2 + 25^2 = c^2$

$225 + 625 = c^2$

$850 = c^2$

$\sqrt{850} = c$     Exact answer

$29.155 \approx c$     Approximation

**43.** $a^2 + b^2 = c^2$

$7^2 + b^2 = 10^2$

$49 + b^2 = 100$

$b^2 = 100 - 49$

$b^2 = 51$

$b = \sqrt{51}$     Exact answer

$b \approx 7.141$     Approximation

**44.** $a^2 + b^2 = c^2$

$5^2 + 8^2 = c^2$

$25 + 64 = c^2$

$89 = c^2$

$\sqrt{89} = c$

$9.434 \approx c$

$c = \sqrt{89}$ ft, or approximately 9.434 ft.

**45.** $a^2 + b^2 = c^2$

$a^2 + 18^2 = 20^2$

$a^2 + 324 = 400$

$a^2 = 400 - 324$

$a^2 = 76$

$a = \sqrt{76} \approx 8.718$

$a = \sqrt{76}$ cm, or approximately 8.718 cm

**46. *Familiarize*.** We first make a drawing. In it we see a right triangle. Let $l$ = the length of the wire, in feet.

*Translate*. Substitute 16 for $a$, 24 for $b$, and $l$ for $c$ in the Pythagorean equation.

$a^2 + b^2 = c^2$

$16^2 + 24^2 = l^2$

*Solve*. We solve the equation for $l$.

$256 + 576 = l^2$

$832 = l^2$

$\sqrt{832} = l$

$28.8 \approx l$

*Check*. $16^2 + 24^2 = 256 + 576 = 832 = (\sqrt{832})^2$

*State*. The wire is about 28.8 ft long.

**47. *Familiarize*.** Referring to the drawing in the text, we see that we have a right triangle and that $h$ = the height of the tree.

*Translate*. Substitute 24 for $a$, $h$ for $b$, and 83 for $c$ in the Pythagorean equation.

$a^2 + b^2 = c^2$

$24^2 + h^2 = 83^2$

*Solve*. We solve the equation for $h$.

$576 + h^2 = 6889$

$h^2 = 6313$

$h = \sqrt{6313} \approx 79$

*Check*. $24^2 + (\sqrt{6313})^2 = 576 + 6313 = 6899 = 83^2$

*State*. The tree is approximately 79 ft tall.

**48. *Familiarize*.** From the drawing in Exercise 3, we see that the diagonal is the hypotenuse of a right triangle with legs of length 36 ft and 78 ft. Let $d$ = the length of the diagonal, in feet.

*Translate*. We substitute 36 for $a$, 78 for $b$, and $d$ for $c$ in the Pythagorean equation.

$a^2 + b^2 = c^2$

$36^2 + 78^2 = d^2$

*Solve*. We solve the equation for $d$.

$1296 + 6084 = d^2$

$7380 = d^2$

$\sqrt{7380} = d$

$85.9 \approx d$

*Check*. $36^2 + 78^2 = 1296 + 6084 = 7380 = (\sqrt{7380})^2$

*State*. The length of a diagonal of the tennis court is about 85.9 ft.

**49.** $180° - 20\frac{3}{4}° = 179\frac{4}{4}° - 20\frac{3}{4}° = 159\frac{1}{4}°$

Answer B is correct.

**50.** $r = \dfrac{d}{2} = \dfrac{\frac{7}{9} \text{ in.}}{2} = \dfrac{7}{9} \text{ in.} \times \dfrac{1}{2} = \dfrac{7}{18} \text{ in.}$

$A = \pi \cdot r \cdot r$

$\approx \dfrac{22}{7} \cdot \dfrac{7}{18} \text{ in.} \cdot \dfrac{7}{18} \text{ in.}$

$= \dfrac{22 \cdot 7 \cdot 7}{7 \cdot 18 \cdot 18} \cdot \text{ in.} \cdot \text{ in.}$

$= \dfrac{77}{162} \text{ in}^2$

Answer B is correct.

**51. *Familiarize*.** Let $s$ = the length of a side of the square, in feet. When the square is cut in half the resulting rectangle has length $s$ and width $s/2$.

*Translate*.

$$\underbrace{\text{Perimeter of rectangle}}_{2 \cdot s + 2 \cdot \frac{s}{2}} \text{ is } \underbrace{30 \text{ ft.}}_{30}$$

*Solve*.

$2 \cdot s + 2 \cdot \dfrac{s}{2} = 30$

$2 \cdot s + s = 30$

$3 \cdot s = 30$

$s = 10$

If $s = 10$, then the area of the square is 10 ft · 10 ft, or 100 ft$^2$.

*Check*. If $s = 10$, then $s/2 = 10/2 = 5$ and the perimeter of a rectangle with length 10 ft and width 5 ft is $2 \cdot 10$ ft + $2 \cdot 5$ ft = 20 ft + 10 ft = 30 ft. We can also recheck the calculation for the area of the square. The answer checks.

*State*. The area of the square is 100 ft$^2$.

**52.** The area $A$ of the shaded region is the area of a square with sides 2.8 m less the areas of the four small squares cut out at each corner. Each of the small squares has sides of 1.8 mm, or 0.0018 m.

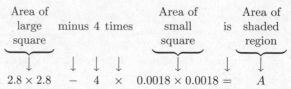

We carry out the calculations.

$7.84 - 4 \times 0.00000324 = 7.84 - 0.00001296 = 7.83998704$

The area of the shaded region is $7.83998704$ m$^2$.

**53.** The shaded region consists of one large triangle with base 84 mm, or 8.4 cm, and height 100 mm, or 10 cm, and 8 small triangles, each with height 1.25 cm and base 1.05 cm. Let $A =$ the area of the shaded region.

| Area of shaded region is | | Area of large triangle | plus | 8 | times | Area of small triangle |
|---|---|---|---|---|---|---|
| $A$ | $=$ | $\frac{1}{2} \cdot 8.4 \cdot 10$ | $+$ | $8$ | $\cdot$ | $\frac{1}{2} \cdot 1.25 \cdot 1.05$ |

We carry out the computations.

$$A = \frac{1}{2} \cdot 8.4 \cdot 10 + 8 \cdot \frac{1}{2} \cdot 1.25 \cdot 1.05$$

$$A = 42 + 5.25$$

$$A = 47.25$$

The area of the shaded region is $47.25$ cm$^2$.

## Chapter 10 Discussion and Writing Exercises

**1.** Add $90°$ to the measure of the angle's complement.

**2.** This could be done using the technique in Example 7 in Section 10.4. We could also approximate the volume with the volume of a similarly-shaped rectangular solid. Another method is to break the egg and measure the capacity of its contents.

**3.** Show that the sum of the squares of the lengths of the legs is the same as the square of the length of the hypotenuse.

**4.** Divide the figure into 3 triangles.

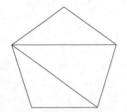

The sum of the measures of the angles of each triangle is $180°$, so the sum of the measures of the angles of the figure is $3 \cdot 180°$, or $540°$.

**5.** The volume of the cone is half the volume of the dome. It can be argued that a cone-cap is more energy-efficient since there is less air under it to be heated and cooled.

**6.** Volume of two spheres, each with radius $r$:

$2\left(\frac{4}{3}\pi r^3\right) = \frac{8}{3}\pi r^3$; volume of one sphere with radius $2r$:

$\frac{4}{3}\pi(2r)^3 = \frac{32}{3}\pi r^3$. The volume of the sphere with radius $2r$ is four times the volume of the two spheres, each with radius $r$: $\frac{32}{3}\pi r^3 = 4 \cdot \frac{8}{3}\pi r^3$.

## Chapter 10 Test

**1.** $P = 2 \cdot (l + w)$

$= 2 \cdot (9.4 \text{ cm} + 7.01 \text{ cm})$

$= 2 \cdot (16.41 \text{ cm})$

$= 32.82$ cm

$A = l \cdot w$

$= (9.4 \text{ cm}) \cdot (7.01 \text{ cm})$

$= 9.4 \cdot 7.01 \cdot \text{cm} \cdot \text{cm}$

$= 65.894$ cm$^2$

**2.** $P = 4 \cdot s$

$= 4 \cdot 4\frac{7}{8}$ in.

$= 4 \cdot \frac{39}{8}$ in.

$= \frac{4 \cdot 39}{8}$ in.

$= \frac{\cancel{4} \cdot 39}{2 \cdot \cancel{4}}$ in.

$= \frac{39}{2}$ in., or $19\frac{1}{2}$ in.

$A = s \cdot s$

$= \left(4\frac{7}{8} \text{ in.}\right) \cdot \left(4\frac{7}{8} \text{ in.}\right)$

$= 4\frac{7}{8} \cdot 4\frac{7}{8} \cdot \text{in.} \cdot \text{in.}$

$= \frac{39}{8} \cdot \frac{39}{8} \cdot \text{in}^2$

$= \frac{1521}{64}$ in$^2$, or $23\frac{49}{64}$ in$^2$

**3.** $A = b \cdot h$

$= 10 \text{ cm} \cdot 2.5 \text{ cm}$

$= 25$ cm$^2$

**4.** $A = \frac{1}{2} \cdot b \cdot h$

$= \frac{1}{2} \cdot 8 \text{ m} \cdot 3 \text{ m}$

$= \frac{8 \cdot 3}{2}$ m$^2$

$= 12$ m$^2$

**5.** $A = \frac{1}{2} \cdot h \cdot (a + b)$

$= \frac{1}{2} \cdot 3 \text{ ft} \cdot (8 \text{ ft} + 4 \text{ ft})$

$= \frac{1}{2} \cdot 3 \text{ ft} \cdot 12 \text{ ft}$

$= \frac{3 \cdot 12}{2}$ ft$^2$

$= 18$ ft$^2$

6. $d = 2 \cdot r = 2 \cdot \dfrac{1}{8}$ in. $= \dfrac{1}{4}$ in.

7. $r = \dfrac{d}{2} = \dfrac{18 \text{ cm}}{2} = 9$ cm

8. $C = 2 \cdot \pi \cdot r$

$\approx 2 \cdot \dfrac{22}{7} \cdot \dfrac{1}{8}$ in.

$= \dfrac{2 \cdot 22 \cdot 1}{7 \cdot 8}$ in.

$= \dfrac{\cancel{2} \cdot \cancel{2} \cdot 11 \cdot 1}{7 \cdot \cancel{2} \cdot \cancel{2} \cdot 2}$ in.

$= \dfrac{11}{14}$ in.

9. In Exercise 7 we found that the radius of the circle is 9 cm.

$A = \pi \cdot r \cdot r$

$\approx 3.14 \cdot 9$ cm $\cdot 9$ cm

$= 3.14 \cdot 81$ cm$^2$

$= 254.34$ cm$^2$

10. The perimeter of the shaded region consists of 2 sides of length 18.6 km and the circumferences of two semicircles with diameter 9.0 km. Note that the sum of the circumferences of the two semicircles is the same as the circumference of one circle with diameter 9.0 km.

The total length of the 2 sides of length 18.6 km is

$2 \cdot 18.6$ km $= 37.2$ km.

Next we find the perimeter, or circumference, of the circle.

$C = \pi \cdot d$

$\approx 3.14 \cdot 9.0$ km

$= 28.26$ km

Finally we add to find the perimeter of the shaded region.

$37.2$ km $+ 28.26$ km $= 65.46$ km

The shaded region is the area of a rectangle that is 18.6 km by 9.0 km less the area of two semicircles, each with diameter 9.0 km. Note that the two semicircles have the same area as one circle with diameter 9.0 km.

First we find the area of the rectangle.

$A = l \cdot w$

$= 18.6$ km $\cdot 9.0$ km

$= 167.4$ km$^2$

Now find the area of the circle. The radius is $\dfrac{9.0 \text{ km}}{2}$, or 4.5 km.

$A = \pi \cdot r \cdot r$

$\approx 3.14 \cdot 4.5$ km $\cdot 4.5$ km

$= 3.14 \cdot 20.25$ km$^2$

$= 63.585$ km$^2$

Finally, we subtract to find the area of the shaded region.

$167.4$ km$^2 - 63.585$ km$^2 = 103.815$ km$^2$

11. $V = l \cdot w \cdot h$

$= 4$ cm $\cdot 2$ cm $\cdot 10.5$ cm

$= 8 \cdot 10.5$ cm$^3$

$= 84$ cm$^3$

12. $V = l \cdot w \cdot h$

$= 10\dfrac{1}{2}$ in. $\cdot 8$ in. $\cdot 5$ in.

$= \dfrac{21}{2}$ in. $\cdot 8$ in. $\cdot 5$ in.

$= \dfrac{21 \cdot 8 \cdot 5}{2}$ in$^3$

$= \dfrac{21 \cdot \cancel{2} \cdot 4 \cdot 5}{\cancel{2} \cdot 1}$ in$^3$

$= 420$ in$^3$

13. $V = \pi \cdot r^2 \cdot h$

$\approx 3.14 \times 5$ ft $\times 5$ ft $\times 15$ ft

$= 1177.5$ ft$^3$

14. $r = \dfrac{d}{2} = \dfrac{20 \text{ yd}}{2} = 10$ yd

$V = \dfrac{4}{3} \cdot \pi \cdot r^3$

$\approx \dfrac{4}{3} \times 3.14 \times (10 \text{ yd})^3$

$= 4186.\overline{6}$ yd$^3$

15. $V = \dfrac{1}{3}\pi \cdot r^2 \cdot h$

$\approx \dfrac{1}{3} \times 3.14 \times 3$ cm $\times 3$ cm $\times 12$ cm

$= 113.04$ cm$^3$

16. Using a protractor, we find that the measure of the angle is 90°.

17. Using a protractor, we find that the measure of the angle is 35°.

18. Using a protractor, we find that the measure of the angle is 180°.

19. Using a protractor, we find that the measure of the angle is 113°.

20. The measure of the angle in Exercise 16 is 90°. It is a right angle.

21. The measure of the angle in Exercise 17 is 35°. Since its measure is greater than 0° and less than 90°, it is an acute angle.

22. The measure of the angle in Exercise 18 is 180°. It is a straight angle.

23. The measure of the angle in Exercise 19 is 113°. Since its measure is greater than 90° and less than 180°, it is an obtuse angle.

24. $m(\angle A) + m(\angle H) + m(\angle F) = 180°$

$35° + 110° + x = 180°$

$145° + x = 180°$

$x = 180° - 145°$

$x = 35°$

**25.** From the labels on the triangle, we see that two sides are the same length. By measuring we find that the third side is a different length. Thus, this is an isosceles triangle.

**26.** One angle is an obtuse angle, so this is an obtuse triangle.

**27.** $\angle CAD = 65°$

$90° - 65° = 25°$, so the measure of a complement is $25°$.

$180° - 65° = 115°$, so the measure of a supplement is $115°$.

**28.** $\sqrt{225} = 15$

The square root of 225 is 15 because $15^2 = 225$.

**29.** $\sqrt{87} \approx 9.327$

**30.**
$$a^2 + b^2 = c^2$$
$$24^2 + 32^2 = c^2$$
$$576 + 1024 = c^2$$
$$1600 = c^2$$
$$40 = c$$

**31.**
$$a^2 + b^2 = c^2$$
$$2^2 + b^2 = 8^2$$
$$4 + b^2 = 64$$
$$4 + b^2 - 4 = 64 - 4$$
$$b^2 = 60$$
$$b = \sqrt{60} \qquad \text{Exact answer}$$
$$b \approx 7.746 \qquad \text{Approximation}$$

**32.**
$$a^2 + b^2 = c^2$$
$$1^2 + 1^2 = c^2$$
$$1 + 1 = c^2$$
$$2 = c^2$$
$$\sqrt{2} = c \qquad \text{Exact answer}$$
$$1.414 \approx c \qquad \text{Approximation}$$

**33.**
$$a^2 + b^2 = c^2$$
$$7^2 + b^2 = 10^2$$
$$49 + b^2 = 100$$
$$49 + b^2 - 49 = 100 - 49$$
$$b^2 = 51$$
$$b = \sqrt{51} \qquad \text{Exact answer}$$
$$b \approx 7.141 \qquad \text{Approximation}$$

**34.** *Familiarize.* We first make a drawing. In it we see a right triangle. We let $w =$ the length of the wire, in meters.

*Translate.* We substitute 9 for $a$, 13 for $b$, and $w$ for $c$ in the Pythagorean equation.

$$a^2 + b^2 = c^2$$
$$9^2 + 13^2 = w^2$$

*Solve.* We solve the equation for $w$.

$$81 + 169 = w^2$$
$$250 = w^2$$
$$\sqrt{250} = w \qquad \text{Exact answer}$$
$$15.8 \approx w \qquad \text{Approximation}$$

*Check.* $9^2 + 13^2 = 81 + 169 = 250 = (\sqrt{250})^2$

*State.* The length of the wire is $\sqrt{250}$ m, or about 15.8 m.

**35.** $r = \dfrac{d}{2} = \dfrac{42 \text{ cm}}{2} = 21 \text{ cm}$

$$V = \frac{4}{3} \cdot \pi \cdot r^3$$
$$\approx \frac{4}{3} \cdot \frac{22}{7} \cdot (21 \text{ cm})^3$$
$$= 38,808 \text{ cm}^3$$

Answer D is correct.

**36.** First we convert 3 in. to feet.

$$3 \text{ in.} = 3 \text{ in.} \cdot \frac{1 \text{ ft}}{12 \text{ in.}}$$
$$= \frac{3}{12} \cdot \frac{\text{in.}}{\text{in.}} \cdot 1 \text{ ft}$$
$$= \frac{1}{4} \text{ ft}$$

Now we find the area of the rectangle.

$$A = l \cdot w$$
$$= 8 \text{ ft} \cdot \frac{1}{4} \text{ ft}$$
$$= \frac{8}{4} \text{ ft}^2$$
$$= 2 \text{ ft}^2$$

**37.** We convert both units of measure to feet. From Exercise 35 we know that 3 in. $= \dfrac{1}{4}$ ft. We also have

$$5 \text{ yd} = 5 \times 1 \text{ yd}$$
$$= 5 \times 3 \text{ ft}$$
$$= 15 \text{ ft.}$$

Now we find the area.

$$A = \frac{1}{2} \cdot b \cdot h$$
$$= \frac{1}{2} \cdot 15 \text{ ft} \cdot \frac{1}{4} \text{ ft}$$
$$= \frac{15}{2 \cdot 4} \text{ ft}^2$$
$$= \frac{15}{8} \text{ ft}^2, \text{ or } 1.875 \text{ ft}^2$$

**38.** First we convert 2.6 in. and 3 in. to feet.

$$2.6 \text{ in.} = 2.6 \text{ in.} \cdot \frac{1 \text{ ft}}{12 \text{ in.}}$$
$$= \frac{2.6}{12} \cdot \frac{\text{in.}}{\text{in.}} \cdot 1 \text{ ft}$$
$$= \frac{2.6}{12} \text{ ft}$$

From Exercise 35 we know that 3 in. = $\frac{1}{4}$ ft. Now we find the volume.

$$V = l \cdot w \cdot h$$
$$= 12 \text{ ft} \cdot \frac{1}{4} \text{ ft} \cdot \frac{2.6}{12} \text{ ft}$$
$$= \frac{12 \cdot 2.6}{4 \cdot 12} \text{ ft}^3$$
$$= 0.65 \text{ ft}^3$$

**39.** First we convert 1 in. to feet.

$$1 \text{ in.} = 1 \text{ in.} \cdot \frac{1 \text{ ft}}{12 \text{ in.}}$$
$$= \frac{1}{12} \cdot \frac{\text{in.}}{\text{in.}} \cdot 1 \text{ ft}$$
$$= \frac{1}{12} \text{ ft}$$

Now we find the volume.

$$V = \frac{1}{3}\pi \cdot r^2 \cdot h$$
$$\approx \frac{1}{3} \cdot 3.14 \cdot \frac{1}{12} \text{ ft} \cdot \frac{1}{12} \text{ ft} \cdot 4.5 \text{ ft}$$
$$= \frac{3.14 \cdot 4.5}{3 \cdot 12 \cdot 12} \text{ ft}^3$$
$$\approx 0.033 \text{ ft}^3$$

**40.** First we find the radius of the cylinder.

$$r = \frac{d}{2} = \frac{\frac{3}{4} \text{ in.}}{2} = \frac{3}{4} \text{ in.} \cdot \frac{1}{2} = \frac{3}{8} \text{ in.}$$

Now we convert $\frac{3}{8}$ in. to feet.

$$\frac{3}{8} \text{ in.} = \frac{3}{8} \text{ in.} \cdot \frac{1 \text{ ft}}{12 \text{ in.}}$$
$$= \frac{3}{8 \cdot 12} \cdot \frac{\text{in.}}{\text{in.}} \cdot 1 \text{ ft}$$
$$= \frac{1}{32} \text{ ft}$$

Finally, we find the volume.

$$V = B \cdot h = \pi \cdot r^2 \cdot h$$
$$\approx 3.14 \cdot \frac{1}{32} \text{ ft} \cdot \frac{1}{32} \text{ ft} \cdot 18 \text{ ft}$$
$$= \frac{3.14 \cdot 18}{32 \cdot 32} \text{ ft}^3$$
$$\approx 0.055 \text{ ft}^3$$

## Cumulative Review Chapters 1 - 10

**1.**  79.2 million $= 79.2 \times 1$ million
$$= 79.2 \times 1,000,000$$
$$= 79,200,000$$

**2. Familiarize.** First we will find the volume $V$ of the water. Then we will find its weight, $w$.

**Translate and Solve.**

$$V = l \cdot w \cdot h$$
$$V = 60 \text{ ft} \cdot 25 \text{ ft} \cdot 1 \text{ ft}$$
$$V = 1500 \text{ ft}^3$$

| Weight of water | is | Volume of water | times | Weight per cubic foot |
|---|---|---|---|---|
| $\downarrow$ | $\downarrow$ | $\downarrow$ | $\downarrow$ | $\downarrow$ |
| $w$ | $=$ | $1500$ | $\times$ | $62\frac{1}{2}$ |

We carry out the multiplication.

$$w = 1500 \times 62\frac{1}{2} = 1500 \times \frac{125}{2}$$
$$= \frac{1500 \times 125}{2} = \frac{2 \times 750 \times 125}{2 \times 1}$$
$$= 93,750$$

**Check.** We repeat the calculations. The answer checks.

**State.** The water weighs 93,750 lb.

**3.** The LCD is 6.

$$\begin{array}{r} 1\,\boxed{\frac{1}{2} \cdot \frac{3}{3}} = 1\frac{3}{6} \\ +2\,\boxed{\frac{2}{3} \cdot \frac{2}{2}} = +2\frac{4}{6} \\ \hline \end{array}$$

$$3\frac{7}{6} = 3 + \frac{7}{6}$$
$$= 3 + 1\frac{1}{6}$$
$$= 4\frac{1}{6}$$

**4.**
$$\begin{array}{r} {}^{11}\;{}^{14} \\ \cancel{1}\;\cancel{9}\;\cancel{4}\;{}^{10} \\ \cancel{1}\,\cancel{2}\,\cancel{0}.\cancel{5}\;\cancel{0} \\ -\quad 3\,2\,.\,9\,8 \\ \hline 8\,7\,.\,5\,2 \end{array}$$

**5.**
$$\begin{array}{r} 1\,2\,3\,4 \\ 2\,2\,\overline{)\,2\,7{,}1\,4\,8} \\ \underline{2\,2} \\ 5\,1 \\ \underline{4\,4} \\ 7\,4 \\ \underline{6\,6} \\ 8\,8 \\ \underline{8\,8} \\ 0 \end{array}$$

**6.**
$$8^3 - 45 \cdot 24 - 9^2 \div 3$$
$$= 512 - 45 \cdot 24 - 81 \div 3$$
$$= 512 - 1080 - 81 \div 3$$
$$= 512 - 1080 - 27$$
$$= -568 - 27$$
$$= -595$$

**7.**
$$\left(\frac{1}{4}\right)^2 \div \left(\frac{1}{2}\right)^3 \times 2^4 - (10.3)(4)$$
$$= \frac{1}{16} \div \frac{1}{8} \times 16 - (10.3)(4)$$
$$= \frac{1}{16} \cdot \frac{8}{1} \times 16 - (10.3)(4)$$
$$= \frac{8}{16} \times 16 - (10.3)(4)$$
$$= \frac{8 \times 16}{16} - (10.3)(4)$$
$$= \frac{8 \times \cancel{16}}{\cancel{16} \times 1} - (10.3)(4)$$
$$= 8 - 41.2$$
$$= -33.2$$

**8.**
$$14 \div [33 \div 11 + 8 \times 2 - (15 - 3)]$$
$$= 14 \div [33 \div 11 + 8 \times 2 - 12]$$
$$= 14 \div [3 + 8 \times 2 - 12]$$
$$= 14 \div [3 + 16 - 12]$$
$$= 14 \div [19 - 12]$$
$$= 14 \div 7$$
$$= 2$$

**9.**    1.$\underline{209}$       1.209.       $\dfrac{1209}{1000}$

   3 places   Move 3 places.   3 zeros

$$1.209 = \frac{1209}{1000}$$

**10.** We use the definition of percent.
$$17\% = \frac{17}{100}$$

**11.**
$$\frac{5}{6} = \frac{5}{6} \cdot \frac{4}{4} = \frac{20}{24}$$
$$\frac{7}{8} = \frac{7}{8} \cdot \frac{3}{3} = \frac{21}{24}$$
Since $20 < 21$, $\dfrac{20}{24} < \dfrac{21}{24}$ and thus $\dfrac{5}{6} < \dfrac{7}{8}$.

**12.** The LCD is 36.
$$\frac{-15}{18} \cdot \frac{2}{2} = \frac{-30}{36}$$
$$\frac{-10}{12} \cdot \frac{3}{3} = \frac{-30}{36}$$
Since $\dfrac{-30}{36} = \dfrac{-30}{36}$, we have $\dfrac{-15}{18} = \dfrac{-10}{12}$.

**13.**  $9 \text{ L} = 9 \times 1 \text{ L}$
$$= 9 \times 1000 \text{ mL}$$
$$= 9000 \text{ mL}$$
$$\approx 9000 \text{ m\cancel{L}} \times \frac{1 \text{ \cancel{oz}}}{29.57 \text{ m\cancel{L}}} \times \frac{1 \text{ \cancel{pt}}}{16 \text{ \cancel{oz}}} \times \frac{1 \text{ qt}}{2 \text{ \cancel{pt}}}$$
$$\approx 9.51 \text{ qt}$$

**14.**  $9 \text{ sec} = 9 \text{ \cancel{sec}} \times \dfrac{1 \text{ min}}{60 \text{ \cancel{sec}}}$
$$= \frac{9}{60} \times 1 \text{ min}$$
$$= \frac{3}{20} \text{ min}$$

**15.**  $F = 1.8 \cdot C + 32 = 1.8 \cdot 15 + 32 = 27 + 32 = 59$
Thus, $15°C = 59°F$.

**16.**  $0.087 \text{ L} = 0.087 \times 1 \text{ L}$
$$= 0.087 \times 1000 \text{ mL}$$
$$= 87 \text{ mL}$$

**17.**  $3 \text{ yd}^2 = 3 \times 1 \text{ yd}^2$
$$= 3 \times 9 \text{ ft}^2$$
$$= 27 \text{ ft}^2$$

**18.** We move the decimal point 2 places to the left.
$$17 \text{ cm} = 0.17 \text{ m}$$

**19.**
$$x + \frac{3}{4} = -\frac{7}{8}$$
$$x + \frac{3}{4} - \frac{3}{4} = -\frac{7}{8} - \frac{3}{4}$$
$$x = -\frac{7}{8} - \frac{3}{4} \cdot \frac{2}{2}$$
$$x = -\frac{7}{8} - \frac{6}{8}$$
$$x = -\frac{13}{8}$$
The solution is $-\dfrac{13}{8}$.

**20.**
$$\frac{3}{x} = \frac{7}{10}$$
$$3 \cdot 10 = x \cdot 7 \qquad \text{Equating cross products}$$
$$\frac{3 \cdot 10}{7} = \frac{x \cdot 7}{7}$$
$$\frac{30}{7} = x, \text{ or}$$
$$4\frac{2}{7} = x$$
The solution is $\dfrac{30}{7}$, or $4\dfrac{2}{7}$.

**21.**  $25 \cdot x = 2835$
$$\frac{25 \cdot x}{25} = \frac{2835}{25}$$
$$x = 113.4$$
The solution is 113.4.

**22.**

$$\frac{12}{15} = \frac{x}{18}$$

$12 \cdot 18 = 15 \cdot x$    Equating cross products

$$\frac{12 \cdot 18}{15} = \frac{15 \cdot x}{15}$$

$$\frac{\cancel{3} \cdot 4 \cdot 18}{\cancel{3} \cdot 5} = x$$

$$\frac{72}{5} = x, \text{ or}$$

$$14\frac{2}{5} = x$$

The solution is $\frac{72}{5}$, or $14\frac{2}{5}$.

**23.** $P = 80 \text{ cm} + 110 \text{ cm} + 80 \text{ cm} + 110 \text{ cm} = 380 \text{ cm}$

$A = b \cdot h$

$A = 110 \text{ cm} \times 50 \text{ cm}$

$A = 5500 \text{ cm}^2$

**24.**   $d = 2 \cdot r = 2 \cdot 35 \text{ in.} = 70 \text{ in.}$

$C = 2 \cdot \pi \cdot r$

$C \approx 2 \times \frac{22}{7} \times 35 \text{ in.}$

$C = \frac{2 \times 22 \times 35}{7} \text{ in.}$

$C = 220 \text{ in.}$

$A = \pi \cdot r \cdot r$

$A \approx \frac{22}{7} \times 35 \text{ in.} \times 35 \text{ in.}$

$A = \frac{22 \times 35 \times 35}{7} \text{ in}^2$

$A = 3850 \text{ in}^2$

**25.**   $V = \frac{4}{3} \cdot \pi \cdot r^3$

$V \approx \frac{4}{3} \times \frac{22}{7} \times (35 \text{ in.})^3$

$V = \frac{4 \times 22 \times 42,875 \text{ in}^3}{3 \times 7}$

$V = \frac{3,773,000}{21} \text{ in}^3$

$V = 179,666\frac{2}{3} \text{ in}^3$

**26.** The total score needed is

$$90 + 90 + 90 + 90 + 90 = 5 \cdot 90 = 450.$$

The total of the scores on the first four tests is

$$85 + 92 + 79 + 95 = 351.$$

Thus, the lowest score the student can get on the next test is

$$450 - 351 = 99$$

in order to get an A.

**27.**   $I = P \cdot r \cdot t$

$= \$8000 \cdot 4.2\% \cdot \frac{1}{4}$

$= \$8000 \cdot 0.042 \cdot \frac{1}{4}$

$= \$84$

**28.**   $A = P \cdot \left(1 + \frac{r}{n}\right)^{n \cdot t}$

$= \$8000 \cdot \left(1 + \frac{4.2\%}{1}\right)^{1 \cdot 25}$

$= \$8000 \cdot (1 + 0.042)^{25}$

$= \$8000 \cdot (1.042)^{25}$

$\approx \$22,376.03$

**29.** **Familiarize.** We first make a drawing. In it we see a right triangle. We let $r$ = the length of the rope, in meters.

**Translate.** We substitute 15 for $a$, 8 for $b$, and $r$ for $c$ in the Pythagorean equation.

$$a^2 + b^2 = c^2$$
$$15^2 + 8^2 = r^2$$

**Solve.** We solve the equation for $w$.

$$225 + 64 = r^2$$
$$289 = r^2$$
$$\sqrt{289} = r$$
$$17 = r$$

**Check.** $15^2 + 8^2 = 225 + 64 = 289 = (17)^2$

**State.** The length of the rope is 17 m.

**30.**   $\underbrace{\text{Sales tax}}$ is $\underbrace{\text{what percent}}$ of $\underbrace{\text{purchase price?}}$

     $\downarrow$    $\downarrow$      $\downarrow$      $\downarrow$      $\downarrow$

   $\$0.33$   $=$      $r$      $\times$     $\$5.50$

To solve the equation we divide by 5.50 on both sides.

$$\frac{0.33}{5.50} = \frac{r \times 5.50}{5.50}$$

$$0.06 = r$$

$$6\% = r$$

The sales tax rate is 6%.

**31.** **Familiarize.** Let $f$ = the number of yards of fabric remaining on the bolt.

**Translate.**

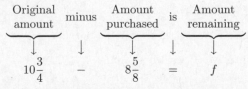

*Solve*. We carry out the subtraction.

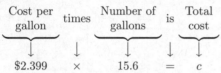

Thus, $f = 2\frac{1}{8}$.

*Check*. $8\frac{5}{8} + 2\frac{1}{8} = 10\frac{6}{8} = 10\frac{3}{4}$, so the answer checks.

*State*. $2\frac{1}{8}$ yd of fabric remains on the bolt.

**32.** *Familiarize*. Let $c$ = the cost of the gasoline. We express $239.9¢$ as $\$2.399$.

*Translate*.

| $\underbrace{\text{Cost per gallon}}$ | times | $\underbrace{\text{Number of gallons}}$ | is | $\underbrace{\text{Total cost}}$ |
|---|---|---|---|---|
| ↓ | ↓ | ↓ | ↓ | ↓ |
| $\$2.399$ | $\times$ | $15.6$ | $=$ | $c$ |

*Solve*. We carry out the multiplication.

$$
\begin{array}{r}
\$2.3\,9\,9 \\
\times\quad 1\,5.\,6 \\
\hline
1\,4\,3\,9\,4 \\
1\,1\,9\,9\,5\,0 \\
2\,3\,9\,9\,0\,0 \\
\hline
\$3\,7.\,4\,2\,4\,4
\end{array}
$$

Thus, $c \approx \$37.42$.

*Check*. We repeat the calculation. The answer checks.

*State*. The gasoline cost $\$37.42$.

**33.** $\dfrac{\$4.99}{20 \text{ qt}} = \dfrac{499¢}{20 \text{ qt}} = 24.95¢/\text{qt}$

$\dfrac{\$1.99}{8 \text{ qt}} = \dfrac{199¢}{8 \text{ qt}} = 24.875¢/\text{qt}$

The 8-qt box has the lower unit price.

**34.** *Familiarize*. Let $d$ = the distance Maria walked, in km. She walks $\frac{1}{4}$ of the distance from the dormitory to the library and then turns and walks the same distance back to the dormitory, so she walks a total of $\frac{1}{4} + \frac{1}{4}$, or $\frac{1}{2}$, of the distance from the dormitory to the library.

*Translate*.

| $\underbrace{\text{Distance walked}}$ | is | $\frac{1}{2}$ | of | $\frac{7}{10}$ km |
|---|---|---|---|---|
| ↓ | ↓ | ↓ | ↓ | ↓ |
| $d$ | $=$ | $\frac{1}{2}$ | $\cdot$ | $\frac{7}{10}$ |

*Solve*. We carry out the multiplication.

$$d = \frac{1}{2} \cdot \frac{7}{10} = \frac{7}{20}$$

*Check*. We repeat the calculation. The answer checks.

*State*. Maria walked $\frac{7}{20}$ km.

**35.**
$$130° + 20° + x = 180°$$
$$150° + x = 180°$$
$$x = 180° - 150°$$
$$x = 30°$$

**36.** Two sides are the same length. The triangle is an isosceles triangle.

**37.** One angle is an obtuse angle. The triangle is an obtuse triangle.

**38.** $100 \text{ yd} = 100 \times 1 \text{ yd} = 100 \times 3 \text{ ft} = 300 \text{ ft}$

$V = B \cdot h = \pi \cdot r^2 \cdot h$

$\quad \approx 3.14 \times 10 \text{ ft} \times 10 \text{ ft} \times 300 \text{ ft}$

$\quad = 94,200 \text{ ft}^3$

**39.** $3 \text{ in.} = 3 \text{ in.} \times \dfrac{1 \text{ ft}}{12 \text{ in.}} = \dfrac{3}{12} \times 1 \text{ ft} = \dfrac{1}{4} \text{ ft}$

$4.6 \text{ in.} = 4.6 \text{ in.} \times \dfrac{1 \text{ ft}}{12 \text{ in.}} = \dfrac{4.6}{12} \times 1 \text{ ft} = \dfrac{4.6}{12} \text{ ft}$

$V = l \cdot w \cdot h$

$\quad = \dfrac{4.6}{12} \text{ ft} \times \dfrac{1}{4} \text{ ft} \times 14 \text{ ft}$

$\quad = \dfrac{4.6 \times 14}{12 \times 4} \text{ ft}^3$

$\quad \approx 1.342 \text{ ft}^3$

# Chapter 11

# Algebra: Solving Equations and Problems

1. $6x = 6 \cdot 7 = 42$

3. $\dfrac{x}{y} = \dfrac{9}{3} = 3$

5. $\dfrac{3p}{q} = \dfrac{3(-2)}{6} = \dfrac{-6}{6} = -1$

7. $\dfrac{x+y}{5} = \dfrac{10+20}{5} = \dfrac{30}{5} = 6$

9. $ab = -5 \cdot 4 = -20$

11. $10(x+y) = 10(20+4) = 10 \cdot 24 = 240$
    $10x + 10y = 10 \cdot 20 + 10 \cdot 4 = 200 + 40 = 240$

13. $10(x-y) = 10(20-4) = 10 \cdot 16 = 160$
    $10x - 10y = 10 \cdot 20 - 10 \cdot 4 = 200 - 40 = 160$

15. $2(b+5) = 2 \cdot b + 2 \cdot 5 = 2b + 10$

17. $7(1-t) = 7 \cdot 1 - 7 \cdot t = 7 - 7t$

19. $6(5x+2) = 6 \cdot 5x + 6 \cdot 2 = 30x + 12$

21. $7(x+4+6y) = 7 \cdot x + 7 \cdot 4 + 7 \cdot 6y = 7x + 28 + 42y$

23. $-7(y-2) = -7 \cdot y - (-7) \cdot 2 = -7y - (-14) = -7y + 14$

25. $-9(-5x - 6y + 8) = -9(-5x) - (-9)6y + (-9)8 =$
    $45x - (-54y) + (-72) = 45x + 54y - 72$

27. $\dfrac{3}{4}(x - 3y - 2z) = \dfrac{3}{4} \cdot x - \dfrac{3}{4} \cdot 3y - \dfrac{3}{4} \cdot 2z =$
    $\dfrac{3}{4}x - \dfrac{9}{4}y - \dfrac{6}{4}z = \dfrac{3}{4}x - \dfrac{9}{4}y - \dfrac{3}{2}z$

29. $3.1(-1.2x + 3.2y - 1.1) = 3.1(-1.2x) + (3.1)3.2y - 3.1(1.1) =$
    $-3.72x + 9.92y - 3.41$

31. $2x + 4 = 2 \cdot x + 2 \cdot 2 = 2(x+2)$

33. $30 + 5y = 5 \cdot 6 + 5 \cdot y = 5(6+y)$

35. $14x + 21y = 7 \cdot 2x + 7 \cdot 3y = 7(2x + 3y)$

37. $5x + 10 + 15y = 5 \cdot x + 5 \cdot 2 + 5 \cdot 3y = 5(x + 2 + 3y)$

39. $8x - 24 = 8 \cdot x - 8 \cdot 3 = 8(x - 3)$

41. $32 - 4y = 4 \cdot 8 - 4 \cdot y = 4(8 - y)$

43. $8x + 10y - 22 = 2 \cdot 4x + 2 \cdot 5y - 2 \cdot 11 = 2(4x + 5y - 11)$

45. $-18x - 12y + 6 = -6 \cdot 3x - 6 \cdot 2y - 6 \cdot (-1) = -6(3x + 2y - 1),$
    or $-18x - 12y + 6 = 6 \cdot (-3x) + 6 \cdot (-2y) + 6 \cdot 1 =$
    $6(-3x - 2y + 1)$

47. $9a + 10a = (9 + 10)a = 19a$

49. $10a - a = 10a - 1 \cdot a = (10 - 1)a = 9a$

51. $2x + 9z + 6x = 2x + 6x + 9z$
    $\qquad = (2 + 6)x + 9z$
    $\qquad = 8x + 9z$

53. $41a + 90 - 60a - 2 = 41a - 60a + 90 - 2$
    $\qquad = (41 - 60)a + (90 - 2)$
    $\qquad = -19a + 88$

55. $23 + 5t + 7y - t - y - 27$
    $\qquad = 23 - 27 + 5t - 1 \cdot t + 7y - 1 \cdot y$
    $\qquad = (23 - 27) + (5 - 1)t + (7 - 1)y$
    $\qquad = -4 + 4t + 6y, \text{ or } 4t + 6y - 4$

57. $11x - 3x = (11 - 3)x = 8x$

59. $6n - n = (6 - 1)n = 5n$

61. $y - 17y = (1 - 17)y = -16y$

63. $-8 + 11a - 5b + 6a - 7b + 7$
    $\qquad = 11a + 6a - 5b - 7b - 8 + 7$
    $\qquad = (11 + 6)a + (-5 - 7)b + (-8 + 7)$
    $\qquad = 17a - 12b - 1$

65. $9x + 2y - 5x = (9 - 5)x + 2y = 4x + 2y$

67. $\qquad \dfrac{11}{4}x + \dfrac{2}{3}y - \dfrac{4}{5}x - \dfrac{1}{6}y + 12$
    $= \left(\dfrac{11}{4} - \dfrac{4}{5}\right)x + \left(\dfrac{2}{3} - \dfrac{1}{6}\right)y + 12$
    $= \left(\dfrac{55}{20} - \dfrac{16}{20}\right)x + \left(\dfrac{4}{6} - \dfrac{1}{6}\right)y + 12$
    $= \dfrac{39}{20}x + \dfrac{3}{6}y + 12$
    $= \dfrac{39}{20}x + \dfrac{1}{2}y + 12$

69. $2.7x + 2.3y - 1.9x - 1.8y = (2.7 - 1.9)x + (2.3 - 1.8)y =$
    $0.8x + 0.5y$

71. $d = 2 \cdot r = 2 \cdot 15 \text{ yd} = 30 \text{ yd}$
    $C = 2 \cdot \pi \cdot r \approx 2 \cdot 3.14 \cdot 15 \text{ yd} \approx 94.2 \text{ yd}$
    $A = \pi \cdot r \cdot r \approx 3.14 \cdot 15 \text{ yd} \cdot 15 \text{ yd} \approx 706.5 \text{ yd}^2$

**73.**  $d = 2 \cdot r = 2 \cdot 9\frac{1}{2}$ mi $= 2 \cdot \frac{19}{2}$ mi $= 19$ mi

$C = 2 \cdot \pi \cdot r \approx 2 \cdot 3.14 \cdot 9\frac{1}{2}$ mi $\approx 2 \cdot 3.14 \cdot \frac{19}{2}$ mi $\approx$
59.66 mi

$A = \pi \cdot r \cdot r \approx 3.14 \cdot 9\frac{1}{2}$ mi $\cdot 9\frac{1}{2}$ mi $\approx$

$3.14 \cdot \frac{19}{2}$ mi $\cdot \frac{19}{2}$ mi $\approx 283.385$ mi$^2$

**75.**  $r = \dfrac{d}{2} = \dfrac{20 \text{ mm}}{2} = 10$ mm

$C = \pi \cdot d \approx 3.14 \cdot 20$ mm $\approx 62.8$ mm

$A = \pi \cdot r \cdot r \approx 3.14 \cdot 10$ mm $\cdot 10$ mm $\approx 314$ mm$^2$

**77.**  $r = \dfrac{d}{2} = \dfrac{4.6 \text{ ft}}{2} = 2.3$ ft

$C = \pi \cdot d \approx 3.14 \cdot 4.6$ ft $\approx 14.444$ ft

$A = \pi \cdot r \cdot r \approx 3.14 \cdot 2.3$ ft $\cdot 2.3$ ft $\approx 16.6106$ ft$^2$

**79.**  $q + qr + qrs + qrst$
$= q \cdot 1 + q \cdot r + q \cdot rs + q \cdot rst$
$= q(1 + r + rs + rst)$

## Exercise Set 11.2

**1.**  $x + 2 = 6$
$x + 2 - 2 = 6 - 2$     Subtracting 2 on both sides
$x + 0 = 4$                  Simplifying
$x = 4$                       Identity property of zero

Check:   $\dfrac{x + 2 = 6}{}$
$\quad\quad 4 + 2 \ ? \ 6$
$\quad\quad\quad 6 \ |$         TRUE

The solution is 4.

**3.**  $x + 15 = -5$
$x + 15 - 15 = -5 - 15$     Subtracting 15 on both sides
$x + 0 = -20$                  Simplifying
$x = -20$                       Identity property of zero

Check:   $\dfrac{x + 15 = -5}{}$
$\quad -20 + 15 \ ? \ -5$
$\quad\quad\quad -5 \ |$         TRUE

The solution is $-20$.

**5.**  $x + 6 = 8$
$x + 6 - 6 = 8 - 6$     Subtracting 6 on both sides
$x + 0 = 2$                  Simplifying
$x = 2$                       Identity property of zero

Check:   $\dfrac{x + 6 = 8}{}$
$\quad\quad 2 + 6 \ ? \ 8$
$\quad\quad\quad 8 \ |$         TRUE

The solution is 2.

**7.**  $x + 5 = 12$
$x + 5 - 5 = 12 - 5$     Subtracting 5 on both sides
$x + 0 = 7$                  Simplifying
$x = 7$                       Identity property of zero

Check:   $\dfrac{x + 5 = 12}{}$
$\quad\quad 7 + 5 \ ? \ 12$
$\quad\quad\quad 12 \ |$         TRUE

The solution is 7.

**9.**  $11 = y + 7$
$11 - 7 = y + 7 - 7$     Subtracting 7 on both sides
$4 = y$

Check:   $\dfrac{11 = y + 7}{}$
$\quad\quad 11 \ ? \ 4 + 7$
$\quad\quad\quad | \ 11$         TRUE

The solution is 4.

**11.**  $-22 = t + 4$
$-22 - 4 = t + 4 - 4$     Subtracting 4 on both sides
$-26 = t$

Check:   $\dfrac{-22 = t + 4}{}$
$\quad -22 \ ? \ -26 + 4$
$\quad\quad\quad | \ -22$         TRUE

The solution is $-26$.

**13.**  $x + 16 = -2$
$x + 16 - 16 = -2 - 16$
$x = -18$

Check:   $\dfrac{x + 16 = -2}{}$
$\quad -18 + 16 \ ? \ -2$
$\quad\quad\quad -2 \ |$         TRUE

The solution is $-18$.

**15.**  $y + 9 = -9$
$y + 9 - 9 = -9 - 9$
$y = -18$

Check:   $\dfrac{y + 9 = -9}{}$
$\quad -18 + 9 \ ? \ -9$
$\quad\quad\quad -9 \ |$         TRUE

The solution is $-18$.

**17.**  $x - 9 = 6$
$x - 9 + 9 = 6 + 9$
$x = 15$

Check:   $\dfrac{x - 9 = 6}{}$
$\quad\quad 15 - 9 \ ? \ 6$
$\quad\quad\quad 6 \ |$         TRUE

The solution is 15.

**19.**  $t - 3 = 16$
$t - 3 + 3 = 16 + 3$
$t = 19$

Check:   $\dfrac{t - 3 = 16}{}$
$\quad\quad 19 - 3 \ ? \ 16$
$\quad\quad\quad 16 \ |$         TRUE

The solution is 19.

**21.**
$$y - 8 = -9$$
$$y - 8 + 8 = -9 + 8$$
$$y = -1$$

Check: 
$$\begin{array}{c} y - 8 = -9 \\ \hline -1 - 8 \;?\; -9 \\ -9 \;\big| \end{array}$$ TRUE

The solution is $-1$.

**23.**
$$x - 7 = -21$$
$$x - 7 + 7 = -21 + 7$$
$$x = -14$$

Check:
$$\begin{array}{c} x - 7 = -21 \\ \hline -14 - 7 \;?\; -21 \\ -21 \;\big| \end{array}$$ TRUE

The solution is $-14$.

**25.**
$$5 + t = 7$$
$$-5 + 5 + t = -5 + 7$$
$$t = 2$$

Check:
$$\begin{array}{c} 5 + t = 7 \\ \hline 5 + 2 \;?\; 7 \\ 7 \;\big| \end{array}$$ TRUE

The solution is $2$.

**27.**
$$-7 + y = 13$$
$$7 + (-7) + y = 7 + 13$$
$$y = 20$$

Check:
$$\begin{array}{c} -7 + y = 13 \\ \hline -7 + 20 \;?\; 13 \\ 13 \;\big| \end{array}$$ TRUE

The solution is $20$.

**29.**
$$-3 + t = -9$$
$$3 + (-3) + t = 3 + (-9)$$
$$t = -6$$

Check:
$$\begin{array}{c} -3 + t = -9 \\ \hline -3 + (-6) \;?\; -9 \\ -9 \;\big| \end{array}$$ TRUE

The solution is $-6$.

**31.**
$$r + \frac{1}{3} = \frac{8}{3}$$
$$r + \frac{1}{3} - \frac{1}{3} = \frac{8}{3} - \frac{1}{3}$$
$$r = \frac{7}{3}$$

Check:
$$\begin{array}{c} r + \dfrac{1}{3} = \dfrac{8}{3} \\ \hline \dfrac{7}{3} + \dfrac{1}{3} \;?\; \dfrac{8}{3} \\ \dfrac{8}{3} \;\Big| \end{array}$$ TRUE

The solution is $\dfrac{7}{3}$.

**33.**
$$m + \frac{5}{6} = -\frac{11}{12}$$
$$m + \frac{5}{6} - \frac{5}{6} = -\frac{11}{12} - \frac{5}{6}$$
$$m = -\frac{11}{12} - \frac{5}{6} \cdot \frac{2}{2}$$
$$m = -\frac{11}{12} - \frac{10}{12}$$
$$m = -\frac{21}{12} = -\frac{\cancel{3} \cdot 7}{\cancel{3} \cdot 4}$$
$$m = -\frac{7}{4}$$

Check:
$$\begin{array}{c} m + \dfrac{5}{6} = -\dfrac{11}{12} \\ \hline -\dfrac{7}{4} + \dfrac{5}{6} \;?\; -\dfrac{11}{12} \\ -\dfrac{21}{12} + \dfrac{10}{12} \;\Big| \\ -\dfrac{11}{12} \;\Big| \end{array}$$ TRUE

The solution is $-\dfrac{7}{4}$.

**35.**
$$x - \frac{5}{6} = \frac{7}{8}$$
$$x - \frac{5}{6} + \frac{5}{6} = \frac{7}{8} + \frac{5}{6}$$
$$x = \frac{7}{8} \cdot \frac{3}{3} + \frac{5}{6} \cdot \frac{4}{4}$$
$$x = \frac{21}{24} + \frac{20}{24}$$
$$x = \frac{41}{24}$$

Check:
$$\begin{array}{c} x - \dfrac{5}{6} = \dfrac{7}{8} \\ \hline \dfrac{41}{24} - \dfrac{5}{6} \;?\; \dfrac{7}{8} \\ \dfrac{41}{24} - \dfrac{20}{24} \;\Big|\; \dfrac{21}{24} \\ \dfrac{21}{24} \;\Big| \end{array}$$ TRUE

The solution is $\dfrac{41}{24}$.

**37.**
$$-\frac{1}{5} + z = -\frac{1}{4}$$
$$\frac{1}{5} - \frac{1}{5} + z = \frac{1}{5} - \frac{1}{4}$$
$$z = \frac{1}{5} \cdot \frac{4}{4} - \frac{1}{4} \cdot \frac{5}{5}$$
$$z = \frac{4}{20} - \frac{5}{20}$$
$$z = -\frac{1}{20}$$

Check:
$$-\frac{1}{5} + z = -\frac{1}{4}$$

$$-\frac{1}{5} + \left(-\frac{1}{20}\right) \ ? \ -\frac{1}{4}$$

$$-\frac{4}{20} + \left(-\frac{1}{20}\right) \ \bigg| \ -\frac{5}{20}$$

$$-\frac{5}{20} \ \bigg| \qquad \text{TRUE}$$

The solution is $-\frac{1}{20}$.

**39.**
$$x + 2.3 = 7.4$$
$$x + 2.3 - 2.3 = 7.4 - 2.3$$
$$x = 5.1$$

Check:
$$x + 2.3 = 7.4$$
$$5.1 + 2.3 \ ? \ 7.4$$
$$7.4 \ \bigg| \qquad \text{TRUE}$$

The solution is 5.1.

**41.**
$$7.6 = x - 4.8$$
$$7.6 + 4.8 = x - 4.8 + 4.8$$
$$12.4 = x$$

Check:
$$7.6 = x - 4.8$$
$$7.6 \ ? \ 12.4 - 4.8$$
$$\bigg| \ 7.6 \qquad \text{TRUE}$$

The solution is 12.4.

**43.**
$$-9.7 = -4.7 + y$$
$$4.7 + (-9.7) = 4.7 + (-4.7) + y$$
$$-5 = y$$

Check:
$$-9.7 = -4.7 + y$$
$$-9.7 \ ? \ -4.7 + (-5)$$
$$\bigg| \ -9.7 \qquad \text{TRUE}$$

The solution is $-5$.

**45.**
$$5\frac{1}{6} + x = 7$$
$$-5\frac{1}{6} + 5\frac{1}{6} + x = -5\frac{1}{6} + 7$$
$$x = -5\frac{1}{6} + 6\frac{6}{6}$$
$$x = 1\frac{5}{6}$$

Check:
$$5\frac{1}{6} + x = 7$$
$$5\frac{1}{6} + 1\frac{5}{6} \ ? \ 7$$
$$7 \ \bigg| \qquad \text{TRUE}$$

The solution is $1\frac{5}{6}$.

**47.**
$$q + \frac{1}{3} = -\frac{1}{7}$$
$$q + \frac{1}{3} - \frac{1}{3} = -\frac{1}{7} - \frac{1}{3}$$
$$q = -\frac{1}{7} \cdot \frac{3}{3} - \frac{1}{3} \cdot \frac{7}{7}$$
$$q = -\frac{3}{21} - \frac{7}{21}$$
$$q = -\frac{10}{21}$$

Check:
$$q + \frac{1}{3} = -\frac{1}{7}$$
$$-\frac{10}{21} + \frac{1}{3} \ ? \ -\frac{1}{7}$$
$$-\frac{10}{21} + \frac{7}{21} \ \bigg| \ -\frac{3}{21}$$
$$-\frac{3}{21} \ \bigg| \qquad \text{TRUE}$$

The solution is $-\frac{10}{21}$.

**49.** $-3 + (-8)$  Two negative numbers. We add the absolute values, getting 11, and make the answer negative.
$$-3 + (-8) = -11$$

**51.** $-14.3 + (-19.8)$ Two negative numbers. We add the absolute values, getting 34.1, and make the answer negative.
$$-14.3 + (-19.8) = -34.1$$

**53.** $-3 - (-8) = -3 + 8$   Adding the opposite of $-8$
$$= 5$$

**55.**
$$-14.3 - (-19.8)$$
$$= -14.3 + 19.8 \qquad \text{Adding the opposite of } -19.8$$
$$= 5.5$$

**57.** The product of two negative numbers is positive.
$$-3(-8) = 24$$

**59.** The product of two negative numbers is positive.
$$-14.3 \times (-19.8) = 283.14$$

**61.** The quotient of two negative numbers is positive.
$$\frac{-24}{-3} = 8$$

**63.** The numbers have different signs. The quotient is negative.
$$\frac{283.14}{-19.8} = -14.3$$

**65.**
$$-356.788 = -699.034 + t$$
$$699.034 + (-356.788) = 699.034 + (-699.034) + t$$
$$342.246 = t$$

The solution is 342.246.

**67.**
$$x + \frac{4}{5} = -\frac{2}{3} - \frac{4}{15}$$

$$x + \frac{4}{5} - \frac{4}{5} = -\frac{2}{3} - \frac{4}{15} - \frac{4}{5}$$

$$x = -\frac{2}{3} \cdot \frac{5}{5} - \frac{4}{15} - \frac{4}{5} \cdot \frac{3}{3}$$

$$x = -\frac{10}{15} - \frac{4}{15} - \frac{12}{15}$$

$$x = -\frac{26}{15}$$

The solution is $-\frac{26}{15}$.

**69.**
$$16 + x - 22 = -16$$

$$x - 6 = -16 \qquad \text{Adding on the left side}$$

$$x - 6 + 6 = -16 + 6$$

$$x = -10$$

The solution is $-10$.

**71.**
$$-\frac{3}{2} + x = -\frac{5}{17} - \frac{3}{2}$$

$$\frac{3}{2} - \frac{3}{2} + x = \frac{3}{2} - \frac{5}{17} - \frac{3}{2}$$

$$x = \left(\frac{3}{2} - \frac{3}{2}\right) - \frac{5}{17}$$

$$x = -\frac{5}{17}$$

The solution is $-\frac{5}{17}$.

---

## Exercise Set 11.3

**1.**
$$6x = 36$$

$$\frac{6x}{6} = \frac{36}{6} \qquad \text{Dividing by 6 on both sides}$$

$$1 \cdot x = 6 \qquad \text{Simplifying}$$

$$x = 6 \qquad \text{Identity property of 1}$$

Check: $\quad \dfrac{6x = 36}{}$

$$6 \cdot 6 \; ? \; 36$$

$$36 \; | \qquad \text{TRUE}$$

The solution is 6.

**3.**
$$5x = 45$$

$$\frac{5x}{5} = \frac{45}{5} \qquad \text{Dividing by 5 on both sides}$$

$$1 \cdot x = 9 \qquad \text{Simplifying}$$

$$x = 9 \qquad \text{Identity property of 1}$$

Check: $\quad \dfrac{5x = 45}{}$

$$5 \cdot 9 \; ? \; 45$$

$$45 \; | \qquad \text{TRUE}$$

The solution is 9.

**5.**
$$84 = 7x$$

$$\frac{84}{7} = \frac{7x}{7}$$

$$12 = 1 \cdot x$$

$$12 = x$$

Check: $\quad \dfrac{84 = 7x}{}$

$$84 \; ? \; 7 \cdot 12$$

$$| \; 84 \qquad \text{TRUE}$$

The solution is 12.

**7.**
$$-x = 40$$

$$-1 \cdot x = 40$$

$$-1 \cdot (-1 \cdot x) = -1 \cdot 40$$

$$1 \cdot x = -40$$

$$x = -40$$

Check: $\quad \dfrac{-x = 40}{}$

$$-(-40) \; ? \; 40$$

$$40 \; | \qquad \text{TRUE}$$

The solution is $-40$.

**9.**
$$6x = -42$$

$$\frac{6x}{6} = \frac{-42}{6}$$

$$1 \cdot x = -7$$

$$x = -7$$

Check: $\quad \dfrac{6x = -42}{}$

$$6(-7) \; ? \; -42$$

$$-42 \; | \qquad \text{TRUE}$$

The solution is $-7$.

**11.**
$$7x = -49$$

$$\frac{7x}{7} = \frac{-49}{7}$$

$$1 \cdot x = -7$$

$$x = -7$$

Check: $\quad \dfrac{7x = -49}{}$

$$7(-7) \; ? \; -49$$

$$-49 \; | \qquad \text{TRUE}$$

The solution is $-7$.

**13.**
$$-12x = 72$$

$$\frac{-12x}{-12} = \frac{72}{-12}$$

$$1 \cdot x = -6$$

$$x = -6$$

Check: $\quad \dfrac{-12x = 72}{}$

$$-12(-6) \; ? \; 72$$

$$72 \; | \qquad \text{TRUE}$$

The solution is $-6$.

**15.**
$$-9x = 45$$

$$\frac{-9x}{-9} = \frac{45}{-9}$$

$$1 \cdot x = -5$$

$$x = -5$$

Check: $\quad \dfrac{-9x = 45}{}$

$$-9(-5) \; ? \; 45$$

$$45 \; | \qquad \text{TRUE}$$

The solution is $-5$.

**17.**   $-21x = -126$

$$\frac{-21x}{-21} = \frac{-126}{-21}$$

$$1 \cdot x = 6$$

$$x = 6$$

Check:   $-21x = -126$

$$\frac{}{-21 \cdot 6 \ ? \ -126}$$

$$-126 \ |$$        TRUE

The solution is 6.

**19.**   $-2x = -10$

$$\frac{-2x}{-2} = \frac{-10}{-2}$$

$$1 \cdot x = 5$$

$$x = 5$$

Check:   $-2x = -10$

$$\frac{}{-2 \cdot 5 \ ? \ -10}$$

$$-10 \ |$$        TRUE

The solution is 5.

**21.**   $\frac{1}{7}t = -9$

$$7 \cdot \frac{1}{7}t = 7 \cdot (-9)$$

$$1 \cdot t = -63$$

$$t = -63$$

Check:   $\frac{1}{7}t = -9$

$$\frac{}{\frac{1}{7} \cdot (-63) \ ? \ -9}$$

$$-9 \ |$$        TRUE

The solution is $-63$.

**23.**   $\frac{3}{4}x = 27$

$$\frac{4}{3} \cdot \frac{3}{4}x = \frac{4}{3} \cdot 27$$

$$1 \cdot x = \frac{4 \cdot \cancel{3} \cdot 3 \cdot 3}{\cancel{3} \cdot 1}$$

$$x = 36$$

Check:   $\frac{3}{4}x = 27$

$$\frac{}{\frac{3}{4} \cdot 36 \ ? \ 27}$$

$$27 \ |$$        TRUE

The solution is 36.

**25.**   $-\frac{1}{3}t = 7$

$$-3 \cdot \left(-\frac{1}{3}\right) \cdot t = -3 \cdot 7$$

$$1 \cdot t = -21$$

$$t = -21$$

Check:   $-\frac{1}{3}t = 7$

$$\frac{}{-\frac{1}{3} \cdot (-21) \ ? \ 7}$$

$$7 \ |$$        TRUE

The solution is $-21$.

**27.**   $-\frac{1}{3}m = \frac{1}{5}$

$$-3 \cdot \left(-\frac{1}{3}m\right) = -3 \cdot \frac{1}{5}$$

$$1 \cdot m = -\frac{3}{5}$$

$$m = -\frac{3}{5}$$

Check:   $-\frac{1}{3}m = \frac{1}{5}$

$$\frac{}{-\frac{1}{3} \cdot \left(-\frac{3}{5}\right) \ ? \ \frac{1}{5}}$$

$$\frac{1}{5} \ |$$        TRUE

The solution is $-\frac{3}{5}$.

**29.**   $-\frac{3}{5}r = \frac{9}{10}$

$$-\frac{5}{3} \cdot \left(-\frac{3}{5}r\right) = -\frac{5}{3} \cdot \frac{9}{10}$$

$$1 \cdot r = -\frac{\cancel{5} \cdot \cancel{3} \cdot 3}{\cancel{3} \cdot \cancel{5} \cdot 2}$$

$$r = -\frac{3}{2}$$

Check:   $-\frac{3}{5}r = \frac{9}{10}$

$$\frac{}{-\frac{3}{5} \cdot \left(-\frac{3}{2}\right) \ ? \ \frac{9}{10}}$$

$$\frac{9}{10} \ |$$        TRUE

The solution is $-\frac{3}{2}$.

**31.**   $-\frac{3}{2}r = -\frac{27}{4}$

$$-\frac{2}{3} \cdot \left(-\frac{3}{2}r\right) = -\frac{2}{3} \cdot \left(-\frac{27}{4}\right)$$

$$1 \cdot r = \frac{\cancel{2} \cdot \cancel{3} \cdot 3 \cdot 3}{\cancel{3} \cdot \cancel{2} \cdot 2}$$

$$r = \frac{9}{2}$$

Check:   $-\frac{3}{2}r = -\frac{27}{4}$

$$\frac{}{-\frac{3}{2} \cdot \frac{9}{2} \ ? \ -\frac{27}{4}}$$

$$-\frac{27}{4} \ |$$        TRUE

The solution is $\frac{9}{2}$.

**33.** $6.3x = 44.1$

$\dfrac{6.3x}{6.3} = \dfrac{44.1}{6.3}$

$1 \cdot x = 7$

$x = 7$

Check: $\dfrac{6.3x = 44.1}{6.3 \cdot 7 \; ? \; 44.1}$

$\qquad\quad 44.1 \;\big|$      TRUE

The solution is 7.

**35.** $-3.1y = 21.7$

$\dfrac{-3.1y}{-3.1} = \dfrac{21.7}{-3.1}$

$1 \cdot y = -7$

$y = -7$

Check: $\dfrac{3.1y = 21.7}{-3.1(-7) \; ? \; 21.7}$

$\qquad\qquad 21.7 \;\big|$     TRUE

The solution is $-7$.

**37.** $-38.7m = 309.6$

$\dfrac{-38.7m}{-38.7} = \dfrac{309.6}{-38.7}$

$1 \cdot m = -8$

$m = -8$

Check: $\dfrac{-38.7m = 309.6}{-38.7(-8) \; ? \; 309.6}$

$\qquad\qquad 309.6 \;\big|$    TRUE

The solution is $-8$.

**39.** $-\dfrac{2}{3}y = -10.6$

$-\dfrac{3}{2} \cdot \left(-\dfrac{2}{3}y\right) = -\dfrac{3}{2} \cdot (-10.6)$

$1 \cdot y = \dfrac{31.8}{2}$

$y = 15.9$

Check: $-\dfrac{2}{3}y = -10.6$

$\dfrac{}{-\dfrac{2}{3} \cdot (15.9) \; ? \; -10.6}$

$\qquad -\dfrac{31.8}{3} \;\Big|$

$\qquad\quad -10.6 \;\big|$    TRUE

The solution is 15.9.

**41.** $\dfrac{-x}{5} = 10$

$5 \cdot \dfrac{-x}{5} = 5 \cdot 10$

$-x = 50$

$-1 \cdot (-x) = -1 \cdot 50$

$x = -50$

Check: $\dfrac{-x}{5} = 10$

$\dfrac{}{\dfrac{-(-50)}{5} \; ? \; 10}$

$\qquad \dfrac{50}{5} \;\Big|$

$\qquad 10 \;\big|$      TRUE

The solution is $-50$.

**43.** $\dfrac{t}{-2} = 7$

$-2 \cdot \dfrac{t}{-2} = -2 \cdot 7$

$t = -14$

Check: $\dfrac{t}{-2} = 7$

$\dfrac{}{\dfrac{-14}{-2} \; ? \; 7}$

$\qquad 7 \;\big|$      TRUE

The solution is $-14$.

**45.** $C = 2 \cdot \pi \cdot r$

$C \approx 2 \cdot 3.14 \cdot 10 \text{ ft} = 62.8 \text{ ft}$

$d = 2 \cdot r$

$d = 2 \cdot 10 \text{ ft} = 20 \text{ ft}$

$A = \pi \cdot r \cdot r$

$A \approx 3.14 \cdot 10 \text{ ft} \cdot 10 \text{ ft} = 314 \text{ ft}^2$

**47.** $V = \ell \cdot w \cdot h$

$V = 25 \text{ ft} \cdot 10 \text{ ft} \cdot 32 \text{ ft} = 8000 \text{ ft}^3$

**49.** $A = b \cdot h$

$= (6.3 \text{ cm}) \cdot (8.5 \text{ cm})$

$= (6.3) \cdot (8.5) \cdot \text{ cm} \cdot \text{ cm}$

$= 53.55 \text{ cm}^2$

**51.** $A = \dfrac{1}{2} \cdot h \cdot (a + b)$

$= \dfrac{1}{2} \cdot 8 \text{ in.} \cdot (6.5 \text{ in.} + 10.5 \text{ in.})$

$= \dfrac{1}{2} \cdot 8 \text{ in.} \cdot 17 \text{ in.}$

$= \dfrac{8 \cdot 17}{2} \cdot \text{ in.} \cdot \text{ in.}$

$= 68 \text{ in}^2$

**53.** $-0.2344m = 2028.732$

$\dfrac{-0.2344m}{-0.2344} = \dfrac{2028.732}{-0.2344}$

$1 \cdot m = -8655$

$m = -8655$

**55.** For all $x$, $0 \cdot x = 0$. There is no solution to $0 \cdot x = 9$.

**57.**     $2|x| = -12$

$$\frac{2|x|}{2} = \frac{-12}{2}$$

$$1 \cdot |x| = -6$$

$$|x| = -6$$

Absolute value cannot be negative. The equation has no solution.

## Chapter 11 Mid-Chapter Review

**1.** False; $2(x+3) = 2 \cdot x + 2 \cdot 3$, or $2x + 6 \neq 2 \cdot x + 3$.

**2.** True; see page 639 in the text.

**3.** True; see page 640 in the text.

**4.** False; $3 - x = 4x$ is equivalent to $3 - x + x = 4x + x$, or $3 = 5x$, or $x = \dfrac{3}{5}$; $5x = -3$ is equivalent to $x = -\dfrac{3}{5}$.

**5.** $6x - 3y + 18 = 3 \cdot 2x - 3 \cdot y + 3 \cdot 6 = 3(2x - y + 6)$

**6.**     $x + 5 = -3$

$$x + 5 - 5 = -3 - 5$$

$$x + 0 = -8$$

$$x = -8$$

**7.**   $-6x = 42$

$$\frac{-6x}{-6} = \frac{42}{-6}$$

$$1 \cdot x = -7$$

$$x = -7$$

**8.** $4x = 4(-7) = -28$

**9.** $\dfrac{a}{b} = \dfrac{56}{8} = 7$

**10.** $\dfrac{m-n}{3} = \dfrac{17-2}{3} = \dfrac{15}{3} = 5$

**11.** $3(x + 5) = 3 \cdot x + 3 \cdot 5 = 3x + 15$

**12.** $4(2y - 7) = 4 \cdot 2y - 4 \cdot 7 = 8y - 28$

**13.** $6(3x + 2y - 1) = 6 \cdot 3x + 6 \cdot 2y - 6 \cdot 1 = 18x + 2y - 6$

**14.** $-2(-3x - y + 8) = -2(-3x) - 2(-y) - 2 \cdot 8 = 6x + 2y - 16$

**15.** $3y + 21 = 3 \cdot y + 3 \cdot 7 = 3(y + 7)$

**16.** $5z + 45 = 5 \cdot z + 5 \cdot 9 = 5(z + 9)$

**17.** $9x - 36 = 9 \cdot x - 9 \cdot 4 = 9(x - 4)$

**18.** $24a - 8 = 8 \cdot 3a - 8 \cdot 1 = 8(3a - 1)$

**19.** $4x + 6y - 2 = 2 \cdot 2x + 2 \cdot 3y - 2 \cdot 1 = 2(2x + 3y - 1)$

**20.** $12x - 9y + 3 = 3 \cdot 4x - 3 \cdot 3y + 3 \cdot 1 = 3(4x - 3y + 1)$

**21.** $4a - 12b + 32 = 4 \cdot a - 4 \cdot 3b + 4 \cdot 8 = 4(a - 3b + 8)$

**22.** $30a - 18b - 24 = 6 \cdot 5a - 6 \cdot 3b - 6 \cdot 4 = 6(5a - 3b - 4)$

**23.** $7x + 8x = (7 + 8)x = 15x$

**24.** $3y - y = 3y - 1 \cdot y = (3 - 1)y = 2y$

**25.** $5x - 2y + 6 - 3x + y - 9 = 5x - 3x - 2y + y + 6 - 9$

$$= (5 - 3)x + (-2 + 1)y + (6 - 9)$$

$$= 2x - y - 3$$

**26.**     $x + 5 = 11$

$$x + 5 - 5 = 11 - 5$$

$$x = 6$$

The solution is 6.

**27.**     $x + 9 = -3$

$$x + 9 - 9 = -3 - 9$$

$$x = -12$$

The solution is $-12$.

**28.**     $8 = t + 1$

$$8 - 1 = t + 1 - 1$$

$$7 = t$$

The solution is 7.

**29.**     $-7 = y + 3$

$$-7 - 3 = y + 3 - 3$$

$$-10 = y$$

The solution is $-10$.

**30.**     $x - 6 = 14$

$$x - 6 + 6 = 14 + 6$$

$$x = 20$$

The solution is 20.

**31.**     $y - 7 = -2$

$$y - 7 + 7 = -2 + 7$$

$$y = 5$$

The solution is 5.

**32.**     $3 + t = 10$

$$3 + t - 3 = 10 - 3$$

$$t = 7$$

The solution is 7.

**33.**     $-5 + x = 5$

$$-5 + x + 5 = 5 + 5$$

$$x = 10$$

The solution is 10.

**34.**     $y + \dfrac{1}{3} = -\dfrac{1}{2}$

$$y + \frac{1}{3} - \frac{1}{3} = -\frac{1}{2} - \frac{1}{3}$$

$$y = -\frac{3}{6} - \frac{2}{6}$$

$$y = -\frac{5}{6}$$

The solution is $-\dfrac{5}{6}$.

**35.**
$$-\frac{3}{2} + x = -\frac{3}{4}$$
$$-\frac{3}{2} + x + \frac{3}{2} = -\frac{3}{4} + \frac{3}{2}$$
$$x = -\frac{3}{4} + \frac{6}{4}$$
$$x = \frac{3}{4}$$
THe solution is $\frac{3}{4}$.

**36.**
$$4.6 = x + 3.9$$
$$4.6 - 3.9 = x + 3.9 - 3.9$$
$$0.7 = x$$
The solution is 0.7.

**37.**
$$-3.3 = -1.9 + t$$
$$-3.3 + 1.9 = -1.9 + t + 1.9$$
$$-1.4 = t$$
The solution is $-1.4$.

**38.** $7x = 42$
$$\frac{7x}{7} = \frac{42}{7}$$
$$x = 6$$
The solution is 6.

**39.** $144 = 12y$
$$\frac{144}{12} = \frac{12y}{12}$$
$$12 = y$$
The solution is 12.

**40.**
$$17 = -t$$
$$-1 \cdot 17 = -1(-t)$$
$$-17 = t$$
The solution is $-17$.

**41.** $6x = -54$
$$\frac{6x}{6} = \frac{-54}{6}$$
$$x = -9$$
The solution is $-9$.

**42.** $-5y = -85$
$$\frac{-5y}{-5} = \frac{-85}{-5}$$
$$y = 17$$
The solution is 17.

**43.** $-8x = 48$
$$\frac{-8x}{-8} = \frac{48}{-8}$$
$$x = -6$$
The solution is $-6$.

**44.**
$$\frac{2}{3}x = 12$$
$$\frac{3}{2} \cdot \frac{2}{3}x = \frac{3}{2} \cdot 12$$
$$x = \frac{36}{2}$$
$$x = 18$$
The solution is 18.

**45.**
$$-\frac{1}{5}t = 3$$
$$-\frac{5}{1}\left(-\frac{1}{5}t\right) = -\frac{5}{1} \cdot 3$$
$$t = -15$$
The solution is $-15$.

**46.**
$$\frac{3}{4}x = -\frac{9}{8}$$
$$\frac{4}{3} \cdot \frac{3}{4}x = \frac{4}{3}\left(-\frac{9}{8}\right)$$
$$x = -\frac{36}{24}$$
$$x = -\frac{3}{2}$$
The solution is $-\frac{3}{2}$.

**47.**
$$-\frac{5}{6}t = -\frac{25}{18}$$
$$-\frac{6}{5}\left(-\frac{5}{6}t\right) = -\frac{6}{5}\left(-\frac{25}{18}\right)$$
$$t = \frac{6 \cdot 25}{5 \cdot 18} = \frac{\cancel{6} \cdot \cancel{5} \cdot 5}{\cancel{5} \cdot 3 \cdot \cancel{6}}$$
$$t = \frac{5}{3}$$
The solution is $\frac{5}{3}$.

**48.** $1.8y = -5.4$
$$\frac{1.8y}{1.8} = \frac{-5.4}{1.8}$$
$$y = -3$$
The solution is $-3$.

**49.**
$$\frac{-y}{7} = 5$$
$$7\left(\frac{-y}{7}\right) = 7 \cdot 5$$
$$-y = 35$$
$$-1(-y) = -1 \cdot 35$$
$$y = -35$$
The solution is $-35$.

**50.** They are not equivalent. For example, let $a = 2$ and $b = 3$. Then $(a+b)^2 = (2+3)^2 = 5^2 = 25$, but $a^2 + b^2 = 2^2 + 3^2 = 4 + 9 = 13$.

**51.** We use the distributive law when we collect like terms even though we might not always write this step.

**52.** The student probably added $\frac{1}{3}$ on both sides of the equation rather than adding $-\frac{1}{3}$ (or subtracting $\frac{1}{3}$) on both sides. The correct solution is $-2$.

**53.** The student apparently multiplied by $-\frac{2}{3}$ on both sides rather than dividing by $\frac{2}{3}$ on both sides. The correct solution is $-\frac{5}{2}$.

## Exercise Set 11.4

**1.**
$$5x + 6 = 31$$
$$5x + 6 - 6 = 31 - 6 \qquad \text{Subtracting 6 on both sides}$$
$$5x = 25 \qquad \text{Simplifying}$$
$$\frac{5x}{5} = \frac{25}{5} \qquad \text{Dividing by 5 on both sides}$$
$$x = 5 \qquad \text{Simplifying}$$

Check: $\quad \dfrac{5x + 6 = 31}{}$
$$5 \cdot 5 + 6 \ ? \ 31$$
$$25 + 6 \ \Big|$$
$$31 \ \Big| \qquad \text{TRUE}$$

The solution is 5.

**3.**
$$8x + 4 = 68$$
$$8x + 4 - 4 = 68 - 4 \qquad \text{Subtracting 4 on both sides}$$
$$8x = 64 \qquad \text{Simplifying}$$
$$\frac{8x}{8} = \frac{64}{8} \qquad \text{Dividing by 8 on both sides}$$
$$x = 8 \qquad \text{Simplifying}$$

Check: $\quad \dfrac{8x + 4 = 68}{}$
$$8 \cdot 8 + 4 \ ? \ 68$$
$$64 + 4 \ \Big|$$
$$68 \ \Big| \qquad \text{TRUE}$$

The solution is 8.

**5.**
$$4x - 6 = 34$$
$$4x - 6 + 6 = 34 + 6 \qquad \text{Adding 6 on both sides}$$
$$4x = 40$$
$$\frac{4x}{4} = \frac{40}{4} \qquad \text{Dividing by 4 on both sides}$$
$$x = 10$$

Check: $\quad \dfrac{4x - 6 = 34}{}$
$$4 \cdot 10 - 6 \ ? \ 34$$
$$40 - 6 \ \Big|$$
$$34 \ \Big| \qquad \text{TRUE}$$

The solution is 10.

**7.**
$$3x - 9 = 33$$
$$3x - 9 + 9 = 33 + 9$$
$$3x = 42$$
$$\frac{3x}{3} = \frac{42}{3}$$
$$x = 14$$

Check: $\quad \dfrac{3x - 9 = 33}{}$
$$3 \cdot 14 - 9 \ ? \ 33$$
$$42 - 9 \ \Big|$$
$$33 \ \Big| \qquad \text{TRUE}$$

The solution is 14.

**9.**
$$7x + 2 = -54$$
$$7x + 2 - 2 = -54 - 2$$
$$7x = -56$$
$$\frac{7x}{7} = \frac{-56}{7}$$
$$x = -8$$

Check: $\quad \dfrac{7x + 2 = -54}{}$
$$7(-8) + 2 \ ? \ -54$$
$$-56 + 2 \ \Big|$$
$$-54 \ \Big| \qquad \text{TRUE}$$

The solution is $-8$.

**11.**
$$-45 = 6y + 3$$
$$-45 - 3 = 6y + 3 - 3$$
$$-48 = 6y$$
$$\frac{-48}{6} = \frac{6y}{6}$$
$$-8 = y$$

Check: $\quad \dfrac{-45 = 6y + 3}{}$
$$-45 \ ? \ 6(-8) + 3$$
$$\Big| \ -48 + 3$$
$$\Big| \ -45 \qquad \text{TRUE}$$

The solution is $-8$.

**13.**
$$-4x + 7 = 35$$
$$-4x + 7 - 7 = 35 - 7$$
$$-4x = 28$$
$$\frac{-4x}{-4} = \frac{28}{-4}$$
$$x = -7$$

Check: $\quad \dfrac{-4x + 7 = 35}{}$
$$-4(-7) + 7 \ ? \ 35$$
$$28 + 7 \ \Big|$$
$$35 \ \Big| \qquad \text{TRUE}$$

The solution is $-7$.

**15.**
$$-7x - 24 = -129$$
$$-7x - 24 + 24 = -129 + 24$$
$$-7x = -105$$
$$\frac{-7x}{-7} = \frac{-105}{-7}$$
$$x = 15$$

Check: $\quad \dfrac{-7x - 24 = -129}{}$
$$-7 \cdot 15 - 24 \ ? \ -129$$
$$-105 - 24 \ \Big|$$
$$-129 \ \Big| \qquad \text{TRUE}$$

The solution is 15.

**17.** $5x + 7x = 72$
$\qquad 12x = 72 \qquad$ Collecting like terms

$\qquad \dfrac{12x}{12} = \dfrac{72}{12} \qquad$ Dividing by 12 on both sides

$\qquad x = 6$

Check: $\quad \dfrac{5x + 7x = 72}{}$

$\qquad 5 \cdot 6 + 7 \cdot 6 \ ? \ 72$
$\qquad 30 + 42 \ |$
$\qquad 72 \ | \qquad$ TRUE

The solution is 6.

**19.** $8x + 7x = 60$
$\qquad 15x = 60 \qquad$ Collecting like terms

$\qquad \dfrac{15x}{15} = \dfrac{60}{15} \qquad$ Dividing by 15 on both sides

$\qquad x = 4$

Check: $\quad \dfrac{8x + 7x = 60}{}$

$\qquad 8 \cdot 4 + 7 \cdot 4 \ ? \ 60$
$\qquad 32 + 28 \ |$
$\qquad 60 \ | \qquad$ TRUE

The solution is 4.

**21.** $4x + 3x = 42$
$\qquad 7x = 42$

$\qquad \dfrac{7x}{7} = \dfrac{42}{7}$

$\qquad x = 6$

Check: $\quad \dfrac{4x + 3x = 42}{}$

$\qquad 4 \cdot 6 + 3 \cdot 6 \ ? \ 42$
$\qquad 24 + 18 \ |$
$\qquad 42 \ | \qquad$ TRUE

The solution is 6.

**23.** $-6y - 3y = 27$
$\qquad -9y = 27$

$\qquad \dfrac{-9y}{-9} = \dfrac{27}{-9}$

$\qquad y = -3$

Check: $\quad \dfrac{-6y - 3y = 27}{}$

$\qquad -6(-3) - 3(-3) \ ? \ 27$
$\qquad 18 + 9 \ |$
$\qquad 27 \ | \qquad$ TRUE

The solution is $-3$.

**25.** $-7y - 8y = -15$
$\qquad -15y = -15$

$\qquad \dfrac{-15y}{-15} = \dfrac{-15}{-15}$

$\qquad y = 1$

Check: $\quad \dfrac{-7y - 8y = -15}{}$

$\qquad -7 \cdot 1 - 8 \cdot 1 \ ? \ -15$
$\qquad -7 - 8 \ |$
$\qquad -15 \ | \qquad$ TRUE

The solution is 1.

**27.** $10.2y - 7.3y = -58$
$\qquad 2.9y = -58$

$\qquad \dfrac{2.9y}{2.9} = \dfrac{-58}{2.9}$

$\qquad y = -20$

Check: $\quad \dfrac{10.2y - 7.3y = -58}{}$

$\qquad 10.2(-20) - 7.3(-20) \ ? \ -58$
$\qquad -204 + 146 \ |$
$\qquad -58 \ | \qquad$ TRUE

The solution is $-20$.

**29.** $x + \dfrac{1}{3}x = 8$

$\qquad \left(1 + \dfrac{1}{3}\right)x = 8$

$\qquad \dfrac{4}{3}x = 8$

$\qquad \dfrac{3}{4} \cdot \dfrac{4}{3}x = \dfrac{3}{4} \cdot 8$

$\qquad x = 6$

Check: $\quad \dfrac{x + \dfrac{1}{3}x = 8}{}$

$\qquad 6 + \dfrac{1}{3} \cdot 6 \ ? \ 8$
$\qquad 6 + 2 \ |$
$\qquad 8 \ | \qquad$ TRUE

The solution is 6.

**31.** $8y - 35 = 3y$
$\qquad 8y = 3y + 35 \qquad$ Adding 35 and simplifying
$\qquad 8y - 3y = 35 \qquad$ Subtracting $3y$ and simplifying
$\qquad 5y = 35 \qquad$ Collecting like terms

$\qquad \dfrac{5y}{5} = \dfrac{35}{5} \qquad$ Dividing by 5

$\qquad y = 7$

Check: $\quad \dfrac{8y - 35 = 3y}{}$

$\qquad 8 \cdot 7 - 35 \ ? \ 3 \cdot 7$
$\qquad 56 - 35 \ | \ 21$
$\qquad 21 \ | \qquad$ TRUE

The solution is 7.

**33.** $8x - 1 = 23 - 4x$
$\qquad 8x + 4x = 23 + 1 \qquad$ Adding 1 and $4x$
$\qquad 12x = 24 \qquad$ Collecting like terms

$\qquad \dfrac{12x}{12} = \dfrac{24}{12} \qquad$ Dividing by 12

$\qquad x = 2$

Check: $\quad \dfrac{8x - 1 = 23 - 4x}{}$

$\qquad 8 \cdot 2 - 1 \ ? \ 23 - 4 \cdot 2$
$\qquad 16 - 1 \ | \ 23 - 8$
$\qquad 15 \ | \ 15 \qquad$ TRUE

The solution is 2.

**35.** $2x - 1 = 4 + x$
$\qquad 2x - x = 4 + 1 \qquad$ Adding 1 and subtracting $x$
$\qquad x = 5 \qquad$ Collecting like terms

Check: $\dfrac{2x-1=4+x}{}$

$2 \cdot 5 - 1 \; ? \; 4 + 5$

$\begin{array}{c|c} 10 - 1 & 9 \\ 9 & \end{array}$     TRUE

The solution is 5.

**37.**    $6x + 3 = 2x + 11$

$6x - 2x = 11 - 3$

$4x = 8$

$\dfrac{4x}{4} = \dfrac{8}{4}$

$x = 2$

Check: $\dfrac{6x+3=2x+11}{}$

$6 \cdot 2 + 3 \; ? \; 2 \cdot 2 + 11$

$\begin{array}{c|c} 12 + 3 & 4 + 11 \\ 15 & 15 \end{array}$     TRUE

The solution is 2.

**39.**    $5 - 2x = 3x - 7x + 25$

$5 - 2x = -4x + 25$     Collecting like terms

$4x - 2x = 25 - 5$

$2x = 20$

$\dfrac{2x}{2} = \dfrac{20}{2}$

$x = 10$

Check: $\dfrac{5 - 2x = 3x - 7x + 25}{}$

$5 - 2 \cdot 10 \; ? \; 3 \cdot 10 - 7 \cdot 10 + 25$

$\begin{array}{c|c} 5 - 20 & 30 - 70 + 25 \\ -15 & -40 + 25 \\ & -15 \end{array}$     TRUE

The solution is 10.

**41.**    $4 + 3x - 6 = 3x + 2 - x$

$3x - 2 = 2x + 2$     Collecting like terms

$3x - 2x = 2 + 2$

$x = 4$

Check: $\dfrac{4 + 3x - 6 = 3x + 2 - x}{}$

$4 + 3 \cdot 4 - 6 \; ? \; 3 \cdot 4 + 2 - 4$

$\begin{array}{c|c} 4 + 12 - 6 & 12 + 2 - 4 \\ 16 - 6 & 14 - 4 \\ 10 & 10 \end{array}$     TRUE

The solution is 4.

**43.**    $4y - 4 + y + 24 = 6y + 20 - 4y$

$5y + 20 = 2y + 20$     Collecting like terms

$5y - 2y = 20 - 20$

$3y = 0$

$\dfrac{3y}{3} = \dfrac{0}{3}$

$y = 0$

Check: $\dfrac{4y - 4 + y + 24 = 6y + 20 - 4y}{}$

$4 \cdot 0 - 4 + 0 + 24 \; ? \; 6 \cdot 0 + 20 - 4 \cdot 0$

$\begin{array}{c|c} 0 - 4 + 0 + 24 & 0 + 20 - 0 \\ 20 & 20 \end{array}$     TRUE

The solution is 0.

**45.**    $\dfrac{7}{2}x + \dfrac{1}{2}x = 3x + \dfrac{3}{2} + \dfrac{5}{2}x$

The least common multiple of all the denominators is 2.
We multiply by 2 on both sides.

$$2\left(\dfrac{7}{2}x + \dfrac{1}{2}x\right) = 2\left(3x + \dfrac{3}{2} + \dfrac{5}{2}x\right)$$

$$2 \cdot \dfrac{7}{2}x + 2 \cdot \dfrac{1}{2}x = 2 \cdot 3x + 2 \cdot \dfrac{3}{2} + 2 \cdot \dfrac{5}{2}x$$

$$7x + x = 6x + 3 + 5x$$

$$8x = 11x + 3$$

$$8x - 11x = 3$$

$$-3x = 3$$

$$\dfrac{-3x}{-3} = \dfrac{3}{-3}$$

$$x = -1$$

Check: $\dfrac{\dfrac{7}{2}x + \dfrac{1}{2}x = 3x + \dfrac{3}{2} + \dfrac{5}{2}x}{}$

$\dfrac{7}{2}(-1) + \dfrac{1}{2}(-1) \; ? \; 3(-1) + \dfrac{3}{2} + \dfrac{5}{2}(-1)$

$\begin{array}{c|c} -\dfrac{7}{2} - \dfrac{1}{2} & -3 + \dfrac{3}{2} - \dfrac{5}{2} \\[2mm] -\dfrac{8}{2} & -\dfrac{3}{2} - \dfrac{5}{2} \\[2mm] -4 & -\dfrac{8}{2} \\[2mm] & -4 \end{array}$     TRUE

The solution is $-1$.

**47.**    $\dfrac{2}{3} + \dfrac{1}{4}t = \dfrac{1}{3}$

The least common multiple of all the denominators is 12.
We multiply by 12 on both sides.

$$12\left(\dfrac{2}{3} + \dfrac{1}{4}t\right) = 12 \cdot \dfrac{1}{3}$$

$$12 \cdot \dfrac{2}{3} + 12 \cdot \dfrac{1}{4}t = 12 \cdot \dfrac{1}{3}$$

$$8 + 3t = 4$$

$$3t = 4 - 8$$

$$3t = -4$$

$$\dfrac{3t}{3} = \dfrac{-4}{3}$$

$$t = -\dfrac{4}{3}$$

Check: $\dfrac{\dfrac{2}{3} + \dfrac{1}{4}t = \dfrac{1}{3}}{}$

$\dfrac{2}{3} + \dfrac{1}{4}\left(-\dfrac{4}{3}\right) \; ? \; \dfrac{1}{3}$

$\begin{array}{c|c} \dfrac{2}{3} - \dfrac{1}{3} & \\[2mm] \dfrac{1}{3} & \dfrac{1}{3} \end{array}$     TRUE

The solution is $-\dfrac{4}{3}$.

**49.**
$$\frac{2}{3} + 3y = 5y - \frac{2}{15}, \quad \text{LCM is 15}$$

$$15\left(\frac{2}{3} + 3y\right) = 15\left(5y - \frac{2}{15}\right)$$

$$15 \cdot \frac{2}{3} + 15 \cdot 3y = 15 \cdot 5y - 15 \cdot \frac{2}{15}$$

$$10 + 45y = 75y - 2$$
$$10 + 2 = 75y - 45y$$
$$12 = 30y$$

$$\frac{12}{30} = \frac{30y}{30}$$

$$\frac{2}{5} = y$$

Check:
$$\frac{2}{3} + 3y = 5y - \frac{2}{15}$$

$$\frac{2}{3} + 3 \cdot \frac{2}{5} \; ? \; 5 \cdot \frac{2}{5} - \frac{2}{15}$$

$$\frac{2}{3} + \frac{6}{5} \;\Big|\; 2 - \frac{2}{15}$$

$$\frac{10}{15} + \frac{18}{15} \;\Big|\; \frac{30}{15} - \frac{2}{15}$$

$$\frac{28}{15} \;\Big|\; \frac{28}{15} \quad \text{TRUE}$$

The solution is $\frac{2}{5}$.

**51.**
$$\frac{5}{3} + \frac{2}{3}x = \frac{25}{12} + \frac{5}{4}x + \frac{3}{4}, \quad \text{LCM is 12}$$

$$12\left(\frac{5}{3} + \frac{2}{3}x\right) = 12\left(\frac{25}{12} + \frac{5}{4}x + \frac{3}{4}\right)$$

$$12 \cdot \frac{5}{3} + 12 \cdot \frac{2}{3}x = 12 \cdot \frac{25}{12} + 12 \cdot \frac{5}{4}x + 12 \cdot \frac{3}{4}$$

$$20 + 8x = 25 + 15x + 9$$
$$20 + 8x = 15x + 34$$
$$20 - 34 = 15x - 8x$$
$$-14x = 7x$$

$$\frac{-14}{7} = \frac{7x}{7}$$
$$-2 = x$$

Check:
$$\frac{5}{3} + \frac{2}{3}x = \frac{25}{12} + \frac{5}{4}x + \frac{3}{4},$$

$$\frac{5}{3} + \frac{2}{3}(-2) \; ? \; \frac{25}{12} + \frac{5}{4}(-2) + \frac{3}{4}$$

$$\frac{5}{3} - \frac{4}{3} \;\Big|\; \frac{25}{12} - \frac{5}{2} + \frac{3}{4}$$

$$\frac{1}{3} \;\Big|\; \frac{25}{12} - \frac{30}{12} + \frac{9}{12}$$

$$\;\Big|\; \frac{4}{12}$$

$$\;\Big|\; \frac{1}{3} \quad \text{TRUE}$$

The solution is $-2$.

**53.**
$$2.1x + 45.2 = 3.2 - 8.4x$$
Greatest number of decimal places is 1
$$10(2.1x + 45.2) = 10(3.2 - 8.4x)$$
Multiplying by 10 to clear decimals
$$10(2.1x) + 10(45.2) = 10(3.2) - 10(8.4x)$$
$$21x + 452 = 32 - 84x$$
$$21x + 84x = 32 - 452$$
$$105x = -420$$

$$\frac{105x}{105} = \frac{-420}{105}$$
$$x = -4$$

Check:
$$2.1x + 45.2 = 3.2 - 8.4x$$

$$2.1(-4) + 45.2 \; ? \; 3.2 - 8.4(-4)$$

$$-8.4 + 45.2 \;\Big|\; 3.2 + 33.6$$

$$36.8 \;\Big|\; 36.8 \quad \text{TRUE}$$

The solution is $-4$.

**55.**
$$1.03 - 0.62x = 0.71 - 0.22x$$
Greatest number of decimal places is 2
$$100(1.03 - 0.62x) = 100(0.71 - 0.22x)$$
Multiplying by 100 to clear decimals
$$100(1.03) - 100(0.62x) = 100(0.71) - 100(0.22x)$$
$$103 - 62x = 71 - 22x$$
$$32 = 40x$$

$$\frac{32}{40} = \frac{40x}{40}$$

$$\frac{4}{5} = x, \text{ or}$$

$$0.8 = x$$

Check:
$$1.03 - 0.62x = 0.71 - 0.22x$$

$$1.03 - 0.62(0.8) \; ? \; 0.71 - 0.22(0.8)$$

$$1.03 - 0.496 \;\Big|\; 0.71 - 0.176$$

$$0.534 \;\Big|\; 0.534 \quad \text{TRUE}$$

The solution is $\frac{4}{5}$, or 0.8.

**57.**
$$\frac{2}{7}x - \frac{1}{2}x = \frac{3}{4}x + 1, \text{ LCM is 28}$$

$$28\left(\frac{2}{7}x - \frac{1}{2}x\right) = 28\left(\frac{3}{4}x + 1\right)$$

$$28 \cdot \frac{2}{7}x - 28 \cdot \frac{1}{2}x = 28 \cdot \frac{3}{4}x + 28 \cdot 1$$

$$8x - 14x = 21x + 28$$
$$-6x = 21x + 28$$
$$-6x - 21x = 28$$
$$-27x = 28$$

$$x = -\frac{28}{27}$$

Check:
$$\frac{2}{7}x - \frac{1}{2}x = \frac{3}{4}x + 1$$

$$\frac{2}{7}\left(-\frac{28}{27}\right) - \frac{1}{2}\left(-\frac{28}{27}\right) \; ? \; \frac{3}{4}\left(-\frac{28}{27}\right) + 1$$

$$-\frac{8}{27} + \frac{14}{27} \;\Big|\; -\frac{21}{27} + 1$$

$$\frac{6}{27} \;\Big|\; \frac{6}{27} \quad \text{TRUE}$$

The solution is $-\frac{28}{27}$.

**59.**  $3(2y - 3) = 27$
  $6y - 9 = 27$            Using a distributive law
  $6y = 27 + 9$         Adding 9
  $6y = 36$
  $y = 6$                   Dividing by 6

Check:    $3(2y - 3) = 27$
  $3(2 \cdot 6 - 3)$ ? $27$
  $3(12 - 3)$
  $3 \cdot 9$
  $27$                 TRUE

The solution is 6.

**61.**        $40 = 5(3x + 2)$
        $40 = 15x + 10$      Using a distributive law
  $40 - 10 = 15x$
  $30 = 15x$
  $2 = x$

Check:    $40 = 5(3x + 2)$
  $40$ ? $5(3 \cdot 2 + 2)$
  $5(6 + 2)$
  $5 \cdot 8$
  $40$                 TRUE

The solution is 2.

**63.**  $2(3 + 4m) - 9 = 45$
  $6 + 8m - 9 = 45$        Collecting like terms
  $8m - 3 = 45$
  $8m = 45 + 3$
  $8m = 48$
  $m = 6$

Check:    $2(3 + 4m) - 9 = 45$
  $2(3 + 4 \cdot 6) - 9$ ? $45$
  $2(3 + 24) - 9$
  $2 \cdot 27 - 9$
  $54 - 9$
  $45$                 TRUE

The solution is 6.

**65.**  $5r - (2r + 8) = 16$
  $5r - 2r - 8 = 16$
  $3r - 8 = 16$           Collecting like terms
  $3r = 16 + 8$
  $3r = 24$
  $r = 8$

Check:    $5r - (2r + 8) = 16$
  $5 \cdot 8 - (2 \cdot 8 + 8)$ ? $16$
  $40 - (16 + 8)$
  $40 - 24$
  $16$                 TRUE

The solution is 8.

**67.**  $6 - 2(3x - 1) = 2$
  $6 - 6x + 2 = 2$
  $8 - 6x = 2$
  $8 - 2 = 6x$          Adding $6x$ and subtract-
                              ing 2
  $6 = 6x$
  $1 = x$

Check:    $6 - 2(3x - 1) = 2$
  $6 - 2(3 \cdot 1 - 1)$ ? $2$
  $6 - 2(3 - 1)$
  $6 - 2 \cdot 2$
  $6 - 4$
  $2$                 TRUE

The solution is 1.

**69.**  $5(d + 4) = 7(d - 2)$
  $5d + 20 = 7d - 14$
  $20 + 14 = 7d - 5d$
  $34 = 2d$
  $17 = d$

Check:    $5(d + 4) = 7(d - 2)$
  $5(17 + 4)$ ? $7(17 - 2)$
  $5 \cdot 21$ | $7 \cdot 15$
  $105$ | $105$        TRUE

The solution is 17.

**71.**  $8(2t + 1) = 4(7t + 7)$
  $16t + 8 = 28t + 28$
  $16t - 28t = 28 - 8$
  $-12t = 20$
  $t = -\dfrac{20}{12}$
  $t = -\dfrac{5}{3}$

Check:    $8(2t + 1) = 4(7t + 7)$
  $8\left(2\left(-\dfrac{5}{3}\right) + 1\right)$ ? $4\left(7\left(-\dfrac{5}{3}\right) + 7\right)$
  $8\left(-\dfrac{10}{3} + 1\right)$ | $4\left(-\dfrac{35}{3} + 7\right)$
  $8\left(-\dfrac{7}{3}\right)$ | $4\left(-\dfrac{14}{3}\right)$
  $-\dfrac{56}{3}$ | $-\dfrac{56}{3}$   TRUE

The solution is $-\dfrac{5}{3}$.

**73.**  $3(r - 6) + 2 = 4(r + 2) - 21$
  $3r - 18 + 2 = 4r + 8 - 21$
  $3r - 16 = 4r - 13$
  $13 - 16 = 4r - 3r$
  $-3 = r$

Check:    $3(r - 6) + 2 = 4(r + 2) - 21$
  $3(-3 - 6) + 2$ ? $4(-3 + 2) - 21$
  $3(-9) + 2$ | $4(-1) - 21$
  $-27 + 2$ | $-4 - 21$
  $-25$ | $-25$      TRUE

The solution is $-3$.

**75.**  $19 - (2x + 3) = 2(x + 3) + x$
  $19 - 2x - 3 = 2x + 6 + x$
  $16 - 2x = 3x + 6$
  $16 - 6 = 3x + 2x$
  $10 = 5x$
  $2 = x$

Check:  $\dfrac{19-(2x+3)=2(x+3)+x}{}$

$$\begin{array}{c|c} 19-(2\cdot 2+3) \ ? \ 2(2+3)+2 & \\ 19-(4+3) & 2\cdot 5+2 \\ 19-7 & 10+2 \\ 12 & 12 \quad \text{TRUE} \end{array}$$

The solution is 2.

**77.**  $0.7(3x+6)=1.1-(x+2)$
$2.1x+4.2=1.1-x-2$
$10(2.1x+4.2)=10(1.1-x-2)$
                    Clearing decimals
$21x+42=11-10x-20$
$21x+42=-10x-9$
$21x+10x=-9-42$
$31x=-51$
$x=-\dfrac{51}{31}$

The check is left to the student.

The solution is $-\dfrac{51}{31}$.

**79.**  $a+(a-3)=(a+2)-(a+1)$
$a+a-3=a+2-a-1$
$2a-3=1$
$2a=1+3$
$2a=4$
$a=2$

Check:  $\dfrac{a+(a-3)-(a+2)-(a+1)}{}$

$$\begin{array}{c|c} 2+(2-3) \ ? \ (2+2)-(2+1) & \\ 2-1 & 4-3 \\ 1 & 1 \quad \text{TRUE} \end{array}$$

The solution is 2.

**81.** The <u>rational</u> numbers consist of all numbers that can be named in the form $\dfrac{a}{b}$, where $a$ and $b$ are integers.

**83.** Two numbers whose product is <u>one</u> are called reciprocals of each other.

**85.** An <u>obtuse</u> angle is an angle whose measure is greater than $90°$ and less than $180°$.

**87.** The basic unit of length in the metric system is the <u>meter</u>.

**89.**  $\dfrac{y-2}{3}=\dfrac{2-y}{5}$, LCM is 15

$15\left(\dfrac{y-2}{3}\right)=15\left(\dfrac{2-y}{5}\right)$
$5(y-2)=3(2-y)$
$5y-10=6-3y$
$5y+3y=6+10$
$8y=16$
$y=2$

The solution is 2.

**91.**  $\dfrac{5+2y}{3}=\dfrac{25}{12}+\dfrac{5y+3}{4}$, LCM is 12

$12\left(\dfrac{5+2y}{3}\right)=12\left(\dfrac{25}{12}+\dfrac{5y+3}{4}\right)$
$4(5+2y)=25+3(5y+3)$
$20+8y=25+15y+9$
$20+8y=34+15y$
$-7y=14$
$y=-2$

The solution is $-2$.

**93.**  $\dfrac{2}{3}(2x-1)=10$

$3\cdot\dfrac{2}{3}(2x-1)=3\cdot 10$
$2(2x-1)=30$
$4x-2=30$
$4x=32$
$x=8$

The solution is 8.

**95. Familiarize.** The perimeter $P$ is the sum of the lengths of the sides, so we have $P=\dfrac{5}{4}x+x+\dfrac{5}{2}+6+2.$

**Translate.** We substitute 15 for $P$.

$\dfrac{5}{4}x+x+\dfrac{5}{2}+6+2=15$

**Solve.** We solve the equation. We begin by collecting like terms on the left side.

$\dfrac{5}{4}x+x+\dfrac{5}{2}+6+2=15$

$\left(\dfrac{5}{4}+1\right)x+\dfrac{5}{2}+6\cdot\dfrac{2}{2}+2\cdot\dfrac{2}{2}=15$

$\left(\dfrac{5}{4}+\dfrac{4}{4}\right)x+\dfrac{5}{2}+\dfrac{12}{2}+\dfrac{4}{2}=15$

$\dfrac{9}{4}x+\dfrac{21}{2}=15$

$\dfrac{9}{4}x+\dfrac{21}{2}-\dfrac{21}{2}=15-\dfrac{21}{2}$

$\dfrac{9}{4}x+0=15\cdot\dfrac{2}{2}-\dfrac{21}{2}$

$\dfrac{9}{4}x=\dfrac{30}{2}-\dfrac{21}{2}$

$\dfrac{9}{4}x=\dfrac{9}{2}$

$\dfrac{4}{9}\cdot\dfrac{9}{4}x=\dfrac{4}{9}\cdot\dfrac{9}{2}$

$1x=\dfrac{2\cdot 2\cdot 9}{9\cdot 2}$

$x=2$

**Check.** $\dfrac{5}{4}\cdot 2+2+\dfrac{5}{2}+6+2=\dfrac{5}{2}+2+\dfrac{5}{2}+6+2=15$, so the result checks.

**State.** $x$ is 2 cm.

## Exercise Set 11.5

**1.** Let $x =$ the number; $2x - 3$.

**3.** Let $y =$ the number; $97\%y$, or $0.97y$

**5.** Let $x =$ the number; $5x + 4$, or $4 + 5x$

**7.** The shorter piece is one-third the length of the longer piece, so we have $\frac{1}{3}x$. Since the sum of the lengths is 240 in., we can also express the length of the shorter piece as $240 - x$.

**9.** *Familiarize*. Let $x =$ the number. Then "what number added to 85" translates to $x + 85$.

*Translate*.

$$\underbrace{\text{What number}}_{x} \ \underbrace{\text{added to}}_{+} \ \underset{85}{85} \ \underset{=}{\text{is}} \ \underset{117}{117?}$$

*Solve*. We solve the equation.

$$x + 85 = 117$$
$$x + 85 - 85 = 117 - 85 \qquad \text{Subtracting 85}$$
$$x = 32$$

*Check*. 32 added to 85, or $32 + 85$, is 117. The answer checks.

*State*. The number is 32.

**11.** *Familiarize*. Let $m =$ Florida's manatee population in 2006.

*Translate*.

$$\underbrace{\text{2007 population}}_{2817} \ \underset{=}{\text{is}} \ \underbrace{\text{2006 population}}_{m} \ \underset{-}{\text{minus 296}} \ \underset{296}{}$$

*Solve*. We solve the equation.

$$2817 = m - 296$$
$$2817 + 296 = m - 296 + 296 \quad \text{Adding 296}$$
$$3113 = m$$

*Check*. 296 fewer manatees than 3113 is $3113 - 296$, or 2817, the 2007 population. The answer checks.

*State*. Florida's manatee population was 3113 in 2006.

**13.** *Familiarize*. Let $h =$ the height of the Statue of Liberty.

*Translate*.

$$\underbrace{\substack{\text{Height of} \\ \text{Statue of Liberty}}}_{h} \ \underset{+}{\text{plus}} \ \underbrace{\substack{\text{Additional} \\ \text{height}}}_{669} \ \underset{=}{\text{is}} \ \underbrace{\substack{\text{Height of} \\ \text{Eiffel Tower}}}_{974}$$

*Solve*. We solve the equation.

$$h + 669 = 974$$
$$h + 669 - 669 = 974 - 669 \quad \text{Subtracting 669}$$
$$h = 305$$

*Check*. If we add 669 ft to 305 ft, we get 974 ft. The answer checks.

*State*. The height of the Statue of Liberty is 305 ft.

**15.** *Familiarize*. Let $x =$ the number. Then "four times the number" translates to $4x$, and "17 subtracted from four times the number" translates to $4x - 17$.

*Translate*. We reword the problem.

$$\text{Four times} \ \underbrace{\text{a number}}_{} \ \text{less 17 is} \ \ 211$$
$$\downarrow \qquad \downarrow \qquad \downarrow \qquad \downarrow \ \downarrow \ \downarrow \ \downarrow$$
$$4 \quad \cdot \quad x \quad - \ 17 = 211$$

*Solve*. We solve the equation.

$$4x - 17 = 211$$
$$4x = 228 \qquad \text{Adding 17}$$
$$\frac{4x}{4} = \frac{228}{4} \qquad \text{Dividing by 4}$$
$$x = 57$$

*Check*. Four times 57 is 228. Subtracting 17 from 228 we get 211. The answer checks.

*State*. The number is 57.

**17.** *Familiarize*. Let $y =$ the number.

*Translate*. We reword the problem.

$$\underbrace{\substack{\text{Two times} \\ \text{a number}}}_{2 \cdot y} \ \underset{+}{\text{plus}} \ \underset{16}{16} \ \underset{=}{\text{is}} \ \underbrace{\substack{\frac{2}{3} \text{ of the} \\ \text{number}}}_{\frac{2}{3} \cdot y}$$

*Solve*. We solve the equation.

$$2y + 16 = \frac{2}{3}y$$
$$3(2y + 16) = 3 \cdot \frac{2}{3}y \qquad \text{Clearing the fraction}$$
$$6y + 48 = 2y$$
$$48 = -4y \qquad \text{Subtracting 6y}$$
$$-12 = y \qquad \text{Dividing by } -4$$

*Check*. We double $-12$ and get $-24$. Adding 16, we get $-8$. Also, $\frac{2}{3}(-12) = -8$. The answer checks.

*State*. The number is $-12$.

**19.** *Familiarize*. Let $d =$ the musher's distance from Nome, in miles. Then $2d =$ the distance from Anchorage, in miles. This is the number of miles the musher has completed. The sum of the two distances is the length of the race, 1049 miles.

*Translate*.

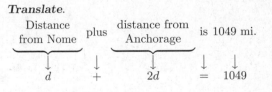

$$\underbrace{\substack{\text{Distance} \\ \text{from Nome}}}_{d} \ \underset{+}{\text{plus}} \ \underbrace{\substack{\text{distance from} \\ \text{Anchorage}}}_{2d} \ \underset{=}{\text{is 1049 mi.}} \ \underset{1049}{}$$

**Carry out**. We solve the equation.

$$d + 2d = 1049$$
$$3d = 1049 \qquad \text{Combining like terms}$$
$$d = \frac{1049}{3}$$

If $d = \frac{1049}{3}$, then $2d = 2 \cdot \frac{1049}{3} = \frac{2098}{3} = 699\frac{1}{3}$.

**Check**. $\frac{2098}{3}$ is twice $\frac{1049}{3}$, and $\frac{1049}{3} + \frac{2098}{3} =$

$\frac{3147}{3} = 1049$. The result checks.

**State**. The musher has traveled $699\frac{1}{3}$ miles.

21. **Familiarize**. Let $x =$ the length of the first piece, in meters. Then $3x =$ the length of the second piece, and $4 \cdot 3x$, or $12x =$ the length of the third piece. The sum of the lengths is 480 m.

**Translate**.

| Length of 1st piece | plus | Length of 2nd piece | plus | Length of 3rd piece | is 480 |
|---|---|---|---|---|---|
| ↓ | ↓ | ↓ | ↓ | ↓ | ↓ ↓ |
| $x$ | $+$ | $3x$ | $+$ | $12x$ | $= 480$ |

**Solve**. We solve the equation.

$$x + 3x + 12x = 480$$
$$16x = 480$$
$$x = 30$$

If $x = 30$, then $3x = 3 \cdot 30$, or 90 and $12x = 12 \cdot 30$, 360.

**Check**. 90 is three times 30 and 360 is four times 90. Also, $30 + 90 + 360 = 480$. The answer checks.

**State**. The first piece of pipe is 30 m long, the second piece is 90 m, and the third piece is 360 m.

23. **Familiarize**. We draw a picture. Let $w =$ the width of the court. Then $w + 44 =$ the length.

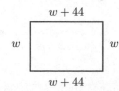

The perimeter $P$ of a rectangle is given by the formula $2l + 2w = P$, where $l =$ the length and $w =$ the width.

**Translate**. We substitute $w + 44$ for $l$ and 288 for $P$ in the formula for perimeter.

$$2l + 2w = P$$
$$2(w + 44) + 2w = 288$$

**Solve**. We solve the equation.

$$2(w + 44) + 2w = 288$$
$$2w + 88 + 2w = 288$$
$$4w + 88 = 288$$
$$4w = 200$$
$$w = 50$$

Possible dimensions are $w = 50$ ft and $w + 44 = 94$ ft.

**Check**. The length is 44 ft more than the width. The perimeter is $2 \cdot 94$ ft $+ 2 \cdot 50$ ft, or 288 ft. The result checks.

**State**. The width of the rectangle is 50 ft and the length is 94 ft.

25. **Familiarize**. We draw a picture. Let $w =$ the width of the rectangle in feet. Then $w + 100 =$ the length.

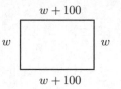

The perimeter of a rectangle is the sum of the lengths of the sides. The area is the product of the length and the width.

**Translate**. We use the definition of perimeter to write an equation that will allow us to find the width and length.

| Width | + | Width | + | Length | + | Length | = Perimeter. |
|---|---|---|---|---|---|---|---|
| ↓ | ↓ | ↓ | ↓ | ↓ | ↓ | ↓ | ↓ ↓ |
| $w$ | $+$ | $w$ | $+ (w + 100) +$ | | | $(w + 100) =$ | 860 |

**Carry out**. We solve the equation.

$$w + w + (w + 100) + (w + 100) = 860$$
$$4w + 200 = 860$$
$$4w = 660$$
$$w = 165$$

If $w = 165$, then $w + 100 = 165 + 100 = 265$, and the area is $265(165) = 43,725$.

**Check**. The length is 100 ft more than the width. The perimeter is $165 + 165 + 265 + 265 = 860$ ft. This checks. To check the area we recheck the computation. This also checks.

**State**. The width of the rectangle is 165 ft, the length is 265 ft, and the area is 43,725 ft$^2$.

27. **Familiarize**. The total cost is the daily charge plus the mileage charge. The mileage charge is the cost per mile times the number of miles driven. Let $m =$ the number of miles that can be driven for $200. We will express 40 cents as $0.40.

**Translate**. We reword the problem.

| Daily rate | plus | Cost per mile | times | Number of miles driven | is | Amount |
|---|---|---|---|---|---|---|
| ↓ | ↓ | ↓ | ↓ | ↓ | ↓ | ↓ |
| 69.95 | $+$ | 0.40 | $\cdot$ | $m$ | $=$ | 200 |

**Solve**. We solve the equation.

$$69.95 + 0.40m = 200$$
$$100(69.95 + 0.40m) = 100(200) \quad \text{Clearing decimals}$$
$$6995 + 40m = 20,000$$
$$40m = 13,005$$
$$m = 325.125$$

**Check**. The mileage cost is found by multiplying 325.125 by $0.40 obtaining $130.05. Then we add $130.05 to $69.95, the daily rate, and get $200. The answer checks.

**State.**  Rick can drive about 325 mi on the car-rental allotment.

**29. Familiarize.** We draw a picture. We let $x =$ the measure of the first angle. Then $4x =$ the measure of the second angle, and $(x + 4x) - 45$, or $5x - 45 =$ the measure of the third angle.

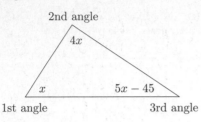

2nd angle

$4x$

$x$          $5x - 45$

1st angle          3rd angle

Recall that the measures of the angles of any triangle add up to $180°$.

**Translate.**

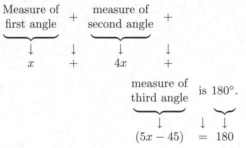

$$\underbrace{\text{Measure of first angle}} + \underbrace{\text{measure of second angle}} +$$

$$\downarrow \qquad \downarrow \qquad \downarrow \qquad \downarrow$$
$$x \qquad + \qquad 4x \qquad +$$

$$\underbrace{\text{measure of third angle}} \text{ is } 180°.$$

$$\downarrow \qquad \downarrow \quad \downarrow$$
$$(5x - 45) \quad = \quad 180$$

**Carry out.** We solve the equation.

$$x + 4x + (5x - 45) = 180$$
$$10x - 45 = 180$$
$$10x = 225$$
$$x = 22.5$$

Possible answers for the angle measures are as follows:

First angle:          $x = 22.5°$
Second angle:          $4x = 4(22.5) = 90°$
Third angle:     $5x - 45 = 5(22.5) - 45$
$$= 112.5 - 45 = 67.5°$$

**Check.**  Consider $22.5°$, $90°$, and $67.5°$. The second is four times the first, and the third is $45°$ less than five times the first. The sum is $180°$. These numbers check.

**State.**  The measure of the first angle is $22.5°$, the measure of the second angle is $90°$, and the measure of the third angle is $67.5°$.

**31. Familiarize.** Let $a =$ the amount Sarah invested. The investment grew by 28% of $a$, or $0.28a$.

**Translate.**

$$\underbrace{\text{Amount invested}} \text{ plus } \underbrace{\text{amount of growth}} \text{ is } \$448.$$

$$\downarrow \qquad \downarrow \qquad \downarrow \qquad \downarrow \quad \downarrow$$
$$a \qquad + \qquad 0.28a \qquad = \quad 448$$

**Carry out.** We solve the equation.

$$a + 0.28a = 448$$
$$1.28a = 448$$
$$a = 350$$

**Check.**  28% of $350 is $0.28(\$350)$, or $98, and $350 + $98 = $448. The answer checks.

**State.**  Sarah invested $350.

**33. Familiarize.** Let $b =$ the balance in the account at the beginning of the month. The balance grew by 2% of $b$, or $0.02b$.

**Translate.**

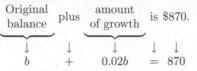

$$\underbrace{\text{Original balance}} \text{ plus } \underbrace{\text{amount of growth}} \text{ is } \$870.$$

$$\downarrow \qquad \downarrow \qquad \downarrow \qquad \downarrow \quad \downarrow$$
$$b \qquad + \qquad 0.02b \qquad = \quad 870$$

**Carry out.** We solve the equation.

$$b + 0.02b = 870$$
$$1.02b = 870$$
$$b \approx \$852.94$$

**Check.**  2% of $852.94 is $0.02(\$852.94)$, or $17.06, and $852.94 + $17.06 = $870. The answer checks.

**State.**  The balance at the beginning of the month was $852.94.

**35. Familiarize.** The total cost is the initial charge plus the mileage charge. Let $d =$ the distance, in miles, that Courtney can travel for $12. The mileage charge is the cost per mile times the number of miles traveled or $0.75d$.

**Translate.**

$$\underbrace{\text{Initial charge}} \text{ plus } \underbrace{\text{mileage charge}} \text{ is } \$12.$$

$$\downarrow \qquad \downarrow \qquad \downarrow \qquad \downarrow \quad \downarrow$$
$$3 \qquad + \qquad 0.75d \qquad = \quad 12$$

**Carry out.** We solve the equation.

$$3 + 0.75d = 12$$
$$0.75d = 9$$
$$d = 12$$

**Check.**  A 12-mi taxi ride from the airport would cost $3 + 12(\$0.75)$, or $3 + $9, or $12. The answer checks.

**State.**  Courtney can travel 12 mi from the airport for $12.

**37. Familiarize.** Let $c =$ the cost of the meal before the tip. Then the amount of the tip was 15% of $c$, or $0.15c$.

**Translate.**

$$\underbrace{\text{Cost of meal before tip}} \text{ plus Tip was } \$41.40$$

$$\downarrow \qquad \downarrow \quad \downarrow \quad \downarrow \quad \downarrow$$
$$c \qquad + \quad 0.15c \quad = \quad 41.40$$

**Solve.** We solve the equation.

$$c + 0.15c = 41.40$$
$$1.15c = 41.40$$
$$c = \frac{41.40}{1.15}$$
$$c = 36$$

**Check.**  15% of $36, or $0.15(\$36)$, is $5.40 and $36 + $5.40 = $41.40, so the answer checks.

**State.**  The cost of the meal before the tip was added was $36.

**39. Familiarize.** Using the labels on the drawing in the text, we let $w =$ the width and $3w + 6 =$ the length. The perimeter $P$ of a rectangle is given by the formula $2l + 2w = P$, where $l =$ the length and $w =$ the width.

**Translate.** Substitute $3w + 6$ for $l$ and 124 for $P$:

$$2l + 2w = P$$
$$2(3w + 6) + 2w = 124$$

**Solve.** We solve the equation.

$$2(3w + 6) + 2w = 124$$
$$6w + 12 + 2w = 124$$
$$8w + 12 = 124$$
$$8w + 12 - 12 = 124 - 12$$
$$8w = 112$$
$$\frac{8w}{8} = \frac{112}{8}$$
$$w = 14$$

The possible dimensions are $w = 14$ ft and $l = 3w + 6 = 3(14) + 6$, or 48 ft.

**Check.** The length, 48 ft, is 6 ft more than three times the width, 14 ft. The perimeter is $2(48 \text{ ft}) + 2(14 \text{ ft}) = 96 \text{ ft} + 28 \text{ ft} = 124 \text{ ft}$. The answer checks.

**State.** The width is 14 ft, and the length is 48 ft.

**41.** $-\dfrac{4}{5} - \left(\dfrac{3}{8}\right) = -\dfrac{4}{5} + \left(-\dfrac{3}{8}\right) = -\dfrac{32}{40} + \left(-\dfrac{15}{40}\right) = -\dfrac{47}{40}$

**43.** $-\dfrac{4}{5} \cdot \dfrac{3}{8} = -\dfrac{4 \cdot 3}{5 \cdot 8} = -\dfrac{4 \cdot 3}{5 \cdot 2 \cdot 4} = -\dfrac{\cancel{4} \cdot 3}{5 \cdot 2 \cdot \cancel{4}} = -\dfrac{3}{10}$

**45.** Do the long division. The answer is positive.

$$
\begin{array}{r}
1.\,6 \\
16\,\overline{)\,2\,5.\,6} \\
\underline{1\,6} \\
9\,6 \\
\underline{9\,6} \\
0
\end{array}
$$

$-25.6 \div (-16) = 1.6$

**47.** $-25.6 - (-16) = -25.6 + 16 = -9.6$

**49. Familiarize.** Let $s =$ one score. Then four score $= 4s$ and four score and seven $= 4s + 7$.

**Translate.** We reword.

| 1776 | plus | four score and seven | is | 1863 |
|------|------|----------------------|-----|------|
| ↓ | ↓ | ↓ | ↓ | ↓ |
| 1776 | + | $(4s + 7)$ | = | 1863 |

We could also translate this as $1863 - 1776 = 4s + 7$.

**51. Familiarize.** Let $l =$ the length of the rectangle. Then the width is $\frac{3}{4}l$. When the length and width are each increased by 2 cm they become $l + 2$ and $\frac{3}{4}l + 2$, respectively. We will use the formula for perimeter, $2l + 2w = P$.

**Translate.** We substitute $l + 2$ for $l$, $\frac{3}{4}l + 2$ for $w$, and 50 for $P$ in the formula.

$$2(l + 2) + 2\left(\frac{3}{4}l + 2\right) = 50$$

**Solve.** We solve the equation.

$$2(l + 2) + 2\left(\frac{3}{4}l + 2\right) = 50$$
$$2l + 4 + \frac{3}{2}l + 4 = 50$$
$$2\left(2l + 4 + \frac{3}{2}l + 4\right) = 2(50)$$

Clearing the fraction

$$4l + 8 + 3l + 8 = 100$$
$$7l + 16 = 100$$
$$7l = 84$$
$$l = 12$$

Possible dimensions are $l = 12$ cm and $w = \frac{3}{4} \cdot 12$ cm $= 9$ cm.

**Check.** If the length is 12 cm and the width is 9 cm, then they become $12 + 2$, or 14 cm, and $9 + 2$, or 11 cm, respectively, when they are each increased by 2 cm. The perimeter becomes $2 \cdot 14 + 2 \cdot 11$, or $28 + 22$, or 50 cm. This checks.

**State.** The length is 12 cm, and the width is 9 cm.

**53. Familiarize.** Let $x =$ the number of dimes. Then $2x =$ the number of quarters and $x + 10 =$ the number of nickels.

The value of $x$ dimes is $0.10x$.

The value of $2x$ quarters is $0.25(2x)$, or $0.50x$.

The value of $x + 10$ nickels is $0.05(x + 10)$, or $0.05x + 0.50$.

**Translate.** The total value of the coins is $20.

$$0.10x + 0.50x + 0.05x + 0.50 = 20$$

**Solve.**

$$0.10x + 0.50x + 0.05x + 0.50 = 20$$
$$0.65x + 0.50 = 20$$
$$100(0.65x + 0.50) = 100(20)$$

Clearing decimals

$$65x + 50 = 2000$$
$$65x = 1950$$
$$x = 30$$

Possible answers for the numbers of each type of coin:

Dimes $= x = 30$

Quarters $= 2x = 2 \cdot 30 = 60$

Nickels $= x + 10 = 30 + 10 = 40$

**Check.** The value of

30 dimes $= \$0.10(30) = \$3$

60 quarters $= \$0.25(60) = \$15$

40 nickels $= \$0.05(40) = \$2$

The total value of the coins is $\$3 + \$15 + \$2$, or $20. The answer checks.

**State.** Susanne got 60 quarters, 30 dimes, and 40 nickels.

## Chapter 11 Concept Reinforcement

1. True; for instance, when $x = 1$, we have $x - 7 = 1 - 7 = -6$ but $7 - x = 7 - 1 = 6$. The expressions are not equivalent.

2. False; the variable is not raised to the same power in both terms, so they are not like terms.

3.
$$x + 5 = 2$$
$$x + 5 - 5 = 2 - 5$$
$$x = -3$$

Since $x = -3$ and $x = 3$ are not equivalent, we know that $x + 5 = 2$ and $x = 3$ are not equivalent. The given statement is false.

4. This is true because division is the same as multiplying by a reciprocal.

## Chapter 11 Important Concepts

1. $4(x + 5y - 7) = 4 \cdot x + 4 \cdot 5y - 4 \cdot 7 = 4x + 20y - 28$

2. $24a - 8b + 16 = 8 \cdot 3a - 8 \cdot b + 8 \cdot 2 = 8(3a - b + 2)$

3. $7x + 3y - x - 6y = 7x - x + 3y - 6y$
$$= 7x - 1 \cdot x + 3y - 6y$$
$$= (7 - 1)x + (3 - 6)y$$
$$= 6x - 3y$$

4.
$$y - 4 = -2$$
$$y - 4 + 4 = -2 + 4$$
$$y + 0 = 2$$
$$y = 2$$

The solution is 2.

5.
$$9x = -72$$
$$\frac{9x}{9} = \frac{-72}{9}$$
$$1 \cdot x = -8$$
$$x = -8$$

The solution is $-8$.

6.
$$5y + 1 = 6$$
$$5y + 1 - 1 = 6 - 1$$
$$5y = 5$$
$$\frac{5y}{5} = \frac{5}{5}$$
$$y = 1$$

The solution is 1.

7.
$$6x - 4 - x = 2x - 10$$
$$5x - 4 = 2x - 10$$
$$5x - 4 - 2x = 2x - 10 - 2x$$
$$3x - 4 = -10$$
$$3x - 4 + 4 = -10 + 4$$
$$3x = -6$$
$$\frac{3x}{3} = \frac{-6}{3}$$
$$x = -2$$

The solution is $-2$.

8.
$$2(y - 1) = 5(y - 4)$$
$$2y - 2 = 5y - 20$$
$$2y - 2 - 5y = 5y - 20 - 5y$$
$$-3y - 2 = -20$$
$$-3y - 2 + 2 = -20 + 2$$
$$-3y = -18$$
$$\frac{-3y}{-3} = \frac{-18}{-3}$$
$$y = 6$$

The solution is 6.

9. Let $n =$ the number. We have $n + 5$, or $5 + n$.

## Chapter 11 Review Exercises

1. $\dfrac{x - y}{3} = \dfrac{17 - 5}{3} = \dfrac{12}{3} = 4$

2. $5(3x - 7) = 5 \cdot 3x - 5 \cdot 7 = 15x - 35$

3. $-2(4x - 5) = -2 \cdot 4x - (-2) \cdot 5 = -8x - (-10) = -8x + 10$

4. $10(0.4x + 1.5) = 10 \cdot 0.4x + 10 \cdot 1.5 = 4x + 15$

5. $-8(3 - 6x + 2y) = -8 \cdot 3 - 8(-6x) - 8(2y) = -24 + 48x - 16y$

6. $2x - 14 = 2 \cdot x - 2 \cdot 7 = 2(x - 7)$

7. $6x - 6 = 6 \cdot x - 6 \cdot 1 = 6(x - 1)$

8. $5x + 10 = 5 \cdot x + 5 \cdot 2 = 5(x + 2)$

9. $12 - 3x + 6z = 3 \cdot 4 - 3 \cdot x + 3 \cdot 2z = 3(4 - x + 2z)$

10. $11a + 2b - 4a - 5b = 11a - 4a + 2b - 5b$
$$= (11 - 4)a + (2 - 5)b$$
$$= 7a - 3b$$

11. $7x - 3y - 9x + 8y = 7x - 9x - 3y + 8y$
$$= (7 - 9)x + (-3 + 8)y$$
$$= -2x + 5y$$

12. $6x + 3y - x - 4y = 6x - x + 3y - 4y$
$$= (6 - 1)x + (3 - 4)y$$
$$= 5x - y$$

**13.** $-3a + 9b + 2a - b = -3a + 2a + 9b - b$
$$= (-3 + 2)a + (9 - 1)b$$
$$= -a + 8b$$

**14.** $\quad x + 5 = -17$
$$x + 5 - 5 = -17 - 5$$
$$x = -22$$
The number $-22$ checks. It is the solution.

**15.** $\quad -8x = -56$
$$\frac{-8x}{-8} = \frac{-56}{-8}$$
$$x = 7$$
The number 7 checks. It is the solution.

**16.** $\quad -\frac{x}{4} = 48$
$$-\frac{1}{4} \cdot x = 48$$
$$-4\left(-\frac{1}{4} \cdot x\right) = -4 \cdot 48$$
$$x = -192$$
The number $-192$ checks. It is the solution.

**17.** $\quad n - 7 = -6$
$$n - 7 + 7 = -6 + 7$$
$$n = 1$$
The number 1 checks. It is the solution.

**18.** $\quad 15x = -35$
$$\frac{15x}{15} = \frac{-35}{15}$$
$$x = -\frac{35}{15} = -\frac{5 \cdot 7}{3 \cdot 5} = -\frac{7}{3} \cdot \frac{5}{5}$$
$$x = -\frac{7}{3}$$
The number $-\frac{7}{3}$ checks. It is the solution.

**19.** $\quad x - 11 = 14$
$$x - 11 + 11 = 14 + 11$$
$$x = 25$$
The number 25 checks. It is the solution.

**20.** $\quad -\frac{2}{3} + x = -\frac{1}{6}$
$$-\frac{2}{3} + x + \frac{2}{3} = -\frac{1}{6} + \frac{2}{3}$$
$$x = -\frac{1}{6} + \frac{4}{6}$$
$$x = \frac{3}{6} = \frac{1}{2}$$
The number $\frac{1}{2}$ checks. It is the solution.

**21.** $\quad \frac{4}{5}y = -\frac{3}{16}$
$$\frac{5}{4} \cdot \frac{4}{5}y = \frac{5}{4} \cdot \left(-\frac{3}{16}\right)$$
$$y = -\frac{5 \cdot 3}{4 \cdot 16} = -\frac{15}{64}$$
The number $-\frac{15}{64}$ checks. It is the solution.

**22.** $\quad y - 0.9 = 9.09$
$$y - 0.9 + 0.9 = 9.09 + 0.9$$
$$y = 9.99$$
The number 9.99 checks. It is the solution.

**23.** $\quad 5 - x = 13$
$$5 - x - 5 = 13 - 5$$
$$-x = 8$$
$$-1 \cdot x = 8$$
$$-1 \cdot (-1 \cdot x) = -1 \cdot 8$$
$$x = -8$$
The number $-8$ checks. It is the solution.

**24.** $\quad 5t + 9 = 3t - 1$
$$5t + 9 - 3t = 3t - 1 - 3t$$
$$2t + 9 = -1$$
$$2t + 9 - 9 = -1 - 9$$
$$2t = -10$$
$$\frac{2t}{2} = \frac{-10}{2}$$
$$t = -5$$
The number $-5$ checks. It is the solution.

**25.** $\quad 7x - 6 = 25x$
$$7x - 6 - 7x = 25x - 7x$$
$$-6 = 18x$$
$$\frac{-6}{18} = \frac{18x}{18}$$
$$-\frac{1}{3} = x$$
The number $-\frac{1}{3}$ checks. It is the solution.

**26.** $\quad \frac{1}{4}x - \frac{5}{8} = \frac{3}{8}$
$$\frac{1}{4}x - \frac{5}{8} + \frac{5}{8} = \frac{3}{8} + \frac{5}{8}$$
$$\frac{1}{4}x = \frac{8}{8}$$
$$\frac{1}{4}x = 1$$
$$4 \cdot \frac{1}{4}x = 4 \cdot 1$$
$$x = 4$$
The number 4 checks. It is the solution.

**27.**
$$14y = 23y - 17 - 10$$
$$14y = 23y - 27$$
$$14y - 23y = 23y - 27 - 23y$$
$$-9y = -27$$
$$\frac{-9y}{-9} = \frac{-27}{-9}$$
$$y = 3$$
The number 3 checks. It is the solution.

**28.**
$$0.22y - 0.6 = 0.12y + 3 - 0.8y$$
$$0.22y - 0.6 = -0.68y + 3$$
$$0.22y - 0.6 + 0.68y = -0.68y + 3 + 0.68y$$
$$0.9y - 0.6 = 3$$
$$0.9y - 0.6 + 0.6 = 3 + 0.6$$
$$0.9y = 3.6$$
$$\frac{0.9y}{0.9} = \frac{3.6}{0.9}$$
$$y = 4$$
The number 4 checks. It is the solution.

**29.**
$$\frac{1}{4}x - \frac{1}{8}x = 3 - \frac{1}{16}x$$
$$\frac{2}{8}x - \frac{1}{8}x = 3 - \frac{1}{16}x$$
$$\frac{1}{8}x = 3 - \frac{1}{16}x$$
$$\frac{1}{8}x + \frac{1}{16}x = 3 - \frac{1}{16}x + \frac{1}{16}x$$
$$\frac{2}{16}x + \frac{1}{16}x = 3$$
$$\frac{3}{16}x = 3$$
$$\frac{16}{3} \cdot \frac{3}{16}x = \frac{16}{3} \cdot 3$$
$$x = \frac{16 \cdot 3}{3 \cdot 1} = \frac{3}{3} \cdot \frac{16}{1}$$
$$x = 16$$
The number 16 checks. It is the solution.

**30.**
$$4(x + 3) = 36$$
$$4x + 12 = 36$$
$$4x + 12 - 12 = 36 - 12$$
$$4x = 24$$
$$\frac{4x}{4} = \frac{24}{4}$$
$$x = 6$$
The number 6 checks. It is the solution.

**31.**
$$3(5x - 7) = -66$$
$$15x - 21 = -66$$
$$15x - 21 + 21 = -66 + 21$$
$$15x = -45$$
$$\frac{15x}{15} = \frac{-45}{15}$$
$$x = -3$$
The number $-3$ checks. It is the solution.

**32.**
$$8(x - 2) - 5(x + 4) = 20x + x$$
$$8x - 16 - 5x - 20 = 21x$$
$$3x - 36 = 21x$$
$$3x - 36 - 3x = 21x - 3x$$
$$-36 = 18x$$
$$\frac{-36}{18} = \frac{18x}{18}$$
$$-2 = x$$
The number $-2$ checks. It is the solution.

**33.**
$$-5x + 3(x + 8) = 16$$
$$-5x + 3x + 24 = 16$$
$$-2x + 24 = 16$$
$$-2x + 24 - 24 = 16 - 24$$
$$-2x = -8$$
$$\frac{-2x}{-2} = \frac{-8}{-2}$$
$$x = 4$$
The number 4 checks. It is the solution.

**34.** Let $x =$ the number; $19\%x$, or $0.19x$

**35.** *Familiarize.* Let $w =$ the width. Then $w + 90 =$ the length.

*Translate.* We use the formula for the perimeter of a rectangle, $P = 2 \cdot l + 2 \cdot w$.
$$1280 = 2 \cdot (w + 90) + 2 \cdot w$$

*Solve.*
$$1280 = 2 \cdot (w + 90) + 2 \cdot w$$
$$1280 = 2w + 180 + 2w$$
$$1280 = 4w + 180$$
$$1100 = 4w$$
$$275 = w$$
If $w = 275$, then $w + 90 = 275 + 90 = 365$.

*Check.* The length is 90 mi more than the width. The perimeter is $2 \cdot 365$ mi $+ 2 \cdot 275$ mi $= 730$ mi $+ 550$ mi $= 1280$ mi. The answer checks.

*State.* The length is 365 mi, and the width is 275 mi.

**36.** *Familiarize.* Let $f =$ the cost of the entertainment center in February.

*Translate.*

| Cost in June | is | $332 | more than | Cost in February |
|:---:|:---:|:---:|:---:|:---:|
| ↓ | ↓ | ↓ | ↓ | ↓ |
| 2449 | = | 332 | + | $f$ |

*Solve.*

$$2449 = 332 + f$$
$$2117 = f$$

*Check.* $332 more than $2117 is $2117 + $332, or $2449. The answer checks.

*State.* The entertainment center cost $2117 in February.

**37.** *Familiarize.* Let $a =$ the number of appliances Ty sold.

*Translate.*

| Commission | is | Commission for each appliance | times | Number of appliances sold |
|---|---|---|---|---|
| ↓ | ↓ | ↓ | ↓ | ↓ |
| 216 | = | 8 | · | $a$ |

*Solve.*

$$216 = 8a$$
$$27 = a$$

*Check.* $27 \cdot \$8 = \$216$, so the answer checks.

*State.* Ty sold 27 appliances.

**38.** *Familiarize.* Let $x =$ the measure of the first angle. Then $x + 50 =$ the measure of the second angle and $2x - 10 =$ the measure of the third angle.

*Translate.* The sum of the measures of the angles of a triangle is 180°, so we have

$$x + (x + 50) + (2x - 10) = 180.$$

*Solve.*

$$x + (x + 50) + (2x - 10) = 180$$
$$4x + 40 = 180$$
$$4x = 140$$
$$x = 35$$

If $x = 35$, then $x + 50 = 35 + 50 = 85$ and $2x - 10 = 2 \cdot 35 - 10 = 70 - 10 = 60$.

*Check.* The second angle, 85°, is 50° more than the first angle, 35°, and the third angle, 60°, is 10° less than twice the first angle. The sum of the measures is $35° + 85° + 60°$, or 180°. The answer checks.

*State.* The measure of the first angle is 35°, the measure of the second angle is 85°, and the measure of the third angle is 60°.

**39.** *Familiarize.* Let $p =$ the marked price of the bread maker.

*Translate.*

| Marked price | minus | 30% | of | Marked price | is | Sale price |
|---|---|---|---|---|---|---|
| ↓ | ↓ | ↓ | ↓ | ↓ | ↓ | ↓ |
| $p$ | − | 0.3 | · | $p$ | = | 154 |

*Solve.*

$$p - 0.3p = 154$$
$$0.7p = 154$$
$$p = 220$$

*Check.* 30% of $220 $= 0.3 \cdot \$220 = \$66$ and $220 − $66 = $154$. The answer checks.

*State.* The marked price of the bread maker was $220.

**40.** *Familiarize.* Let $a =$ the amount the organization actually owes. This is the cost of the office supplies without sales tax added.

*Translate.*

| Amount owed | is | Amount of bill | minus | 5% | of | Amount owed |
|---|---|---|---|---|---|---|
| ↓ | ↓ | ↓ | ↓ | ↓ | ↓ | ↓ |
| $a$ | = | 145.90 | − | 0.05 | · | $a$ |

*Solve.*

$$a = 145.90 - 0.05a$$
$$1.05a = 145.90$$
$$a \approx 138.95$$

*Check.* 5% of $138.95 $= 0.05 \cdot \$138.95 \approx \$6.95$ and $138.95 + $6.95 = $145.90. The answer checks.

*State.* The organization actually owes $138.95.

**41.** *Familiarize.* Let $s =$ the previous salary.

*Translate.*

| Previous salary | plus | 20% | of | Previous salary | is | New salary |
|---|---|---|---|---|---|---|
| ↓ | ↓ | ↓ | ↓ | ↓ | ↓ | ↓ |
| $s$ | + | 0.2 | · | $s$ | = | 90,000 |

*Solve.*

$$s + 0.2s = 90,000$$
$$1.2s = 90,000$$
$$s = 75,000$$

*Check.* 20% of $75,000 $= 0.2 \cdot \$75,000 = \$15,000$ and $75,000 + $15,000 = $90,000. The answer checks.

*State.* The previous salary was $75,000.

**42.** *Familiarize.* Let $c =$ the cost of the television in January.

*Translate.*

| Cost in May | is | Cost in January | less | $38 |
|---|---|---|---|---|
| ↓ | ↓ | ↓ | ↓ | ↓ |
| 829 | = | $c$ | − | 38 |

*Solve.*

$$829 = c - 38$$
$$829 + 38 = c - 38 + 38$$
$$867 = c$$

*Check.* $38 less than $867 is $867 − $38, or $829. This is the cost of the television in May, so the answer checks.

*State.* The television cost $867 in January.

**43.** *Familiarize.* Let $l =$ the length. Then $l - 6 =$ the width.

*Translate.* We use the formula for the perimeter of a rectangle, $P = 2 \cdot l + 2 \cdot w$.

$$56 = 2 \cdot l + 2 \cdot (l - 6)$$

*Solve*.

$$56 = 2l + 2(l - 6)$$
$$56 = 2l + 2l - 12$$
$$56 = 4l - 12$$
$$68 = 4l$$
$$17 = l$$

If $l = 17$, then $l - 6 = 17 - 6 = 11$.

**Check**. 11 cm is 6 cm less than 17 cm. The perimeter is $2 \cdot 17$ cm $+ 2 \cdot 11$ cm $= 34$ cm $+ 22$ cm $= 56$ cm. The answer checks.

**State**. The length is 17 cm, and the width is 11 cm.

**44. Familiarize**. The Nile River is 234 km longer than the Amazon River, so we let $l =$ the length of the Amazon River and $l + 234 =$ the length of the Nile River.

**Translate**.

| Length of Nile River | plus | Length of Amazon River | is | Total length |
|---|---|---|---|---|
| $\downarrow$ | $\downarrow$ | $\downarrow$ | $\downarrow$ | $\downarrow$ |
| $(l + 234)$ | $+$ | $l$ | $=$ | $13,108$ |

**Solve**.

$$(l + 234) + l = 13,108$$
$$2l + 234 = 13,108$$
$$2l = 12,874$$
$$l = 6437$$

If $l = 6437$, then $l + 234 = 6437 + 234 = 6671$.

**Check**. 6671 km is 234 km more than 6437 km, and 6671 km $+ 6437$ km $= 13,108$ km. The answer checks.

**State**. The length of the Amazon River is 6437 km, and the length of the Nile River is 6671 km.

**45.** $6a - 30b + 3 = 3 \cdot 2a - 3 \cdot 10b + 3 \cdot 1 = 3(2a - 10b + 1)$

Answer C is correct.

**46.** $3x - 2y + x - 5y = 3x + x - 2y - 5y$
$$= 3x + 1 \cdot x - 2y - 5y$$
$$= (3 + 1)x + (-2 - 5)y$$
$$= 4x - 7y$$

Answer A is correct.

**47.** $2|n| + 4 = 50$
$$2|n| = 46$$
$$|n| = 23$$

The solutions are the numbers whose distance from 0 is 23. Thus, $n = -23$ or $n = 23$. These are the solutions.

**48.** $|3n| = 60$

$3n$ is 60 units from 0, so we have:

$$3n = -60 \ \ or \ \ 3n = 60$$
$$n = -20 \ \ or \ \ n = 20$$

The solutions are $-20$ and $20$.

## Chapter 11 Discussion and Writing Exercises

**1.** The distributive laws are used to multiply, factor, and collect like terms in this chapter.

**2.** For an equation $x + a = b$, we add the opposite of $a$ on both sides of the equation to get $x$ alone.

**3.** For an equation $ax = b$, we multiply by the reciprocal of $a$ on both sides of the equation to get $x$ alone.

**4.** Add $-b$ (or subtract $b$) on both sides and simplify. Then multiply by the reciprocal of $c$ (or divide by $c$) on both sides and simplify.

## Chapter 11 Test

**1.** $\dfrac{3x}{y} = \dfrac{3 \cdot 10}{5} = \dfrac{30}{5} = 6$

**2.** $3(6 - x) = 3 \cdot 6 - 3 \cdot x = 18 - 3x$

**3.** $-5(y - 1) = -5 \cdot y - (-5)(1) = -5y - (-5) = -5y + 5$

**4.** $12 - 22x = 2 \cdot 6 - 2 \cdot 11x = 2(6 - 11x)$

**5.** $7x + 21 + 14y = 7 \cdot x + 7 \cdot 3 + 7 \cdot 2y = 7(x + 3 + 2y)$

**6.** $9x - 2y - 14x + y = 9x - 14x - 2y + y$
$$= 9x - 14x - 2y + 1 \cdot y$$
$$= (9 - 14)x + (-2 + 1)y$$
$$= -5x + (-y)$$
$$= -5x - y$$

**7.** $-a + 6b + 5a - b = -a + 5a + 6b - b$
$$= -1 \cdot a + 5a + 6b - 1 \cdot b$$
$$= (-1 + 5)a + (6 - 1)b$$
$$= 4a + 5b$$

**8.** $x + 7 = 15$

$x + 7 - 7 = 15 - 7$  Subtracting 7 on both sides

$x + 0 = 8$  Simplifying

$x = 8$  Identity property of 0

Check:  $\dfrac{x + 7 = 15}{8 + 7 \ ? \ 15}$

$15$ | TRUE

The solution is 8.

**9.** $t - 9 = 17$

$t - 9 + 9 = 17 + 9$  Adding 9 on both sides

$t = 26$

Check:  $\dfrac{t - 9 = 17}{26 - 9 \ ? \ 17}$

$17$ | TRUE

The solution is 26.

**10.** $3x = -18$

$\dfrac{3x}{3} = \dfrac{-18}{3}$ Dividing by 3 on both sides

$1 \cdot x = -6$ Simplifying

$x = -6$ Identity property of 1

The answer checks. The solution is $-6$.

**11.** $-\dfrac{4}{7}x = -28$

$-\dfrac{7}{4} \cdot \left(-\dfrac{4}{7}x\right) = -\dfrac{7}{4} \cdot (-28)$ Multiplying by the reciprocal of $-\dfrac{4}{7}$ to eliminate $-\dfrac{4}{7}$ on the left

$1 \cdot x = \dfrac{7 \cdot 28}{4}$

$x = 49$

The answer checks. The solution is 49.

**12.** $3t + 7 = 2t - 5$

$3t + 7 - 2t = 2t - 5 - 2t$

$t + 7 = -5$

$t + 7 - 7 = -5 - 7$

$t = -12$

The answer checks. The solution is $-12$.

**13.** $\dfrac{1}{2}x - \dfrac{3}{5} = \dfrac{2}{5}$

$\dfrac{1}{2}x - \dfrac{3}{5} + \dfrac{3}{5} = \dfrac{2}{5} + \dfrac{3}{5}$

$\dfrac{1}{2}x = 1$

$2 \cdot \dfrac{1}{2}x = 2 \cdot 1$

$x = 2$

The answer checks. The solution is 2.

**14.** $8 - y = 16$

$8 - y - 8 = 16 - 8$

$-y = 8$

$-1(-y) = -1 \cdot 8$

$y = -8$

The answer checks. The solution is $-8$.

**15.** $-\dfrac{2}{5} + x = -\dfrac{3}{4}$

$-\dfrac{2}{5} + x + \dfrac{2}{5} = -\dfrac{3}{4} + \dfrac{2}{5}$

$x = -\dfrac{3}{4} \cdot \dfrac{5}{5} + \dfrac{2}{5} \cdot \dfrac{4}{4}$

$x = -\dfrac{15}{20} + \dfrac{8}{20}$

$x = -\dfrac{7}{20}$

The answer checks. The solution is $-\dfrac{7}{20}$.

**16.** $0.4p + 0.2 = 4.2p - 7.8 - 0.6p$

$0.4p + 0.2 = 3.6p - 7.8$ Collecting like terms on the right

$0.4p + 0.2 - 0.4p = 3.6p - 7.8 - 0.4p$

$0.2 = 3.2p - 7.8$

$0.2 + 7.8 = 3.2p - 7.8 + 7.8$

$8 = 3.2p$

$\dfrac{8}{3.2} = \dfrac{3.2p}{3.2}$

$2.5 = p$

The answer checks. The solution is 2.5.

**17.** $3(x + 2) = 27$

$3x + 6 = 27$ Multiplying to remove parentheses

$3x + 6 - 6 = 27 - 6$

$3x = 21$

$\dfrac{3x}{3} = \dfrac{21}{3}$

$x = 7$

The answer checks. The solution is 7.

**18.** $-3x - 6(x - 4) = 9$

$-3x - 6x + 24 = 9$

$-9x + 24 = 9$

$-9x + 24 - 24 = 9 - 24$

$-9x = -15$

$\dfrac{-9x}{-9} = \dfrac{-15}{-9}$

$x = \dfrac{5}{3}$

The answer checks. The solution is $\dfrac{5}{3}$.

**19.** Let $x =$ the number; $x - 9$.

**20. *Familiarize*.** We draw a picture. Let $w =$ the width of the photograph, in cm. Then $w + 4 =$ the length.

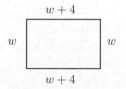

The perimeter $P$ of a rectangle is given by the formula $2l + 2w = P$, where $l =$ the length and $w =$ the width.

***Translate*.** We substitute $w + 4$ for $l$ and 36 for $P$ in the formula for perimeter.

$2l + 2w = P$

$2(w + 4) + 2w = 36$

***Solve*.** We solve the equation.

$2(w + 4) + 2w = 36$

$2w + 8 + 2w = 36$

$4w + 8 = 36$

$4w = 28$

$w = 7$

Possible dimensions are $w - 7$ cm and $w + 4 = 11$ cm.

**Check**.  The length is 4 cm more than the width.  The perimeter is $2 \cdot 11$ cm $+ 2 \cdot 7$ cm, or 36 cm.  The result checks.

**State**.  The width of the photograph is 7 cm and the length is 11 cm.

**21. Familiarize**.  Let $c =$ the amount of contributions to all charities, in billions of dollars.

**Translate**.

$$\underbrace{\text{Contribution to}\atop\text{religious charities}}\ \ \text{is 33\% of}\ \ \underbrace{\text{contributions to}\atop\text{all charities}}$$

$$102.3 \qquad = 33\% \cdot \qquad c$$

**Solve**.  We write 33% in decimal notation and solve the equation.

$$102.3 = 0.33 \cdot c$$
$$\frac{102.3}{0.33} = \frac{0.33 \cdot c}{0.33}$$
$$310 = c$$

**Check**.  We find 33% of $310 billion.

$$0.33(\$310 \text{ billion}) = \$102.3 \text{ billion}$$

The answer checks.

**State**.  In 2007, $310 billion was given to charities.

**22. Familiarize**.  Using the labels on the drawing in the text, we let $x$ and $x + 2$ represent the lengths of the pieces, in meters.

**Translate**.

$$\underbrace{\text{Length of}\atop\text{shorter piece}}\ \ \text{plus}\ \ \underbrace{\text{Length of}\atop\text{longer piece}}\ \ \text{is}\ \ \underbrace{\text{Length of}\atop\text{the board}}$$

$$x \qquad + \qquad x + 2 \qquad = \qquad 8$$

**Solve**.

$$x + x + 2 = 8$$
$$2x + 2 = 8$$
$$2x = 6 \quad \text{Subtracting 2}$$
$$x = 3 \quad \text{Dividing by 2}$$

If the length of the shorter piece is 3 m, then the length of the longer piece is $3 + 2$, or 5 m.

**Check**.  The 5-m piece is 2 m longer than the 3-m piece, and the sum of the lengths is $3 + 5$, or 8 m.  The answer checks.

**State**.  The pieces are 3 m and 5 m long.

**23. Familiarize**.  Let $a =$ the amount that was originally invested.  Then the interest earned is $5\%a$, or $0.05a$.

**Translate**.

$$\underbrace{\text{Amount}\atop\text{invested}}\ \ \text{plus}\ \ \underbrace{\text{Amount}\atop\text{of growth}}\ \ \text{is \$924}$$

$$a \qquad + \qquad 0.05a \qquad = 924$$

**Solve**.  First we collect terms on the left side.

$$a + 0.05a = 924$$
$$1.05a = 924$$
$$a = 880 \quad \text{Dividing by 1.05}$$

**Check**.  5% of $880 is $44 and $880 + $44 = $924, so the answer checks.

**State**.  $880 was originally invested.

**24. Familiarize**.  Let $n =$ the original number.

**Translate**.

$$\text{Three times}\ \underbrace{\text{a number}}\ \text{minus}\ 14\ \text{is}\ \tfrac{2}{3}\ \text{of}\ \underbrace{\text{the number}}$$

$$3 \qquad \cdot \qquad n \qquad - \quad 14 = \tfrac{2}{3} \cdot \qquad n$$

**Solve**.

$$3n - 14 = \frac{2}{3}n$$
$$-14 = -\frac{7}{3}n \qquad \text{Subtracting } 3n$$
$$-\frac{3}{7}(-14) = -\frac{3}{7}\left(-\frac{7}{3}n\right)$$
$$6 = n$$

**Check**.  $3 \cdot 6 - 14 = 18 - 14 = 4$ and $\frac{2}{3} \cdot 6 = 4$, so the answer checks.

**State**.  The original number is 6.

**25. Familiarize**.  We draw a picture.  We let $x =$ the measure of the first angle.  Then $3x =$ the measure of the second angle, and $(x + 3x) - 25$, or $4x - 25 =$ the measure of the third angle.

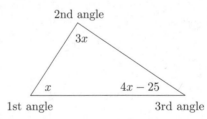

Recall that the measures of the angles of any triangle add up to 180°.

**Translate**.

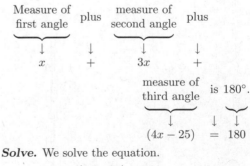

$$\underbrace{\text{Measure of}\atop\text{first angle}}\ \ \text{plus}\ \ \underbrace{\text{measure of}\atop\text{second angle}}\ \ \text{plus}$$

$$x \qquad + \qquad 3x \qquad +$$

$$\underbrace{\text{measure of}\atop\text{third angle}}\ \ \text{is 180°.}$$

$$(4x - 25) \qquad = \quad 180$$

**Solve**.  We solve the equation.

$$x + 3x + (4x - 25) = 180$$
$$8x - 25 = 180$$
$$8x = 205$$
$$x = 25.625$$

Although we are asked to find only the measure of the first angle, we find the measures of the other two angles as well so that we can check the answer.

Possible answers for the angle measures are as follows:

First angle:       $x = 25.625°$
Second angle:    $3x = 3(25.625) = 76.875°$
Third angle:    $4x - 25 = 4(25.625) - 25$
                          $= 102.5 - 25 = 77.5°$

**Check.** Consider $25.625°$, $76.875°$, and $77.5°$. The second is three times the first, and the third is $25°$ less than four times the first. The sum is $180°$. These numbers check.

**State.** The measure of the first angle is $25.625°$.

**26.**    $5y - 1 = 3y + 7$
$5y - 1 - 3y = 3y + 7 - 3y$
$2y - 1 = 7$
$2y - 1 + 1 = 7 + 1$
$2y = 8$
$\dfrac{2y}{2} = \dfrac{8}{2}$
$y = 4$

The answer checks. The solution is 4. Answer D is correct.

**27.**    $3|w| - 8 = 37$
$3|w| = 45$    Adding 8
$|w| = 15$    Dividing by 3

Since $|w| = 15$, the distance of $w$ from 0 on the number line is 15. Thus, $w = 15$ or $w = -15$.

**28. Familiarize.** Let $t$ = the number of tickets given away. Then the first person got $\dfrac{1}{3}t$ tickets, the second person got $\dfrac{1}{4}t$, the third person got $\dfrac{1}{5}t$, the fourth person got 8 tickets, and the fifth person got 5.

**Translate.** There were $t$ tickets given away, so we have
$$\dfrac{1}{3}t + \dfrac{1}{4}t + \dfrac{1}{5}t + 8 + 5 = t.$$

**Solve.** First we collect like terms on the left.
$$\dfrac{1}{3}t + \dfrac{1}{4}t + \dfrac{1}{5}t + 8 + 5 = t$$
$$\dfrac{20}{60}t + \dfrac{15}{60}t + \dfrac{12}{60}t + 13 = t$$
$$\dfrac{47}{60}t + 13 = t$$
$$13 = \dfrac{13}{60}t \qquad \text{Subtracting } \dfrac{47}{60}t$$
$$\dfrac{60}{13} \cdot 13 = \dfrac{60}{13} \cdot \dfrac{13}{60}t$$
$$60 = t$$

**Check.** $\dfrac{1}{3} \cdot 60 = 20$, $\dfrac{1}{4} \cdot 60 = 15$, and $\dfrac{1}{5} \cdot 60 = 12$. Since $20 + 15 + 12 + 8 + 5 = 60$, the answer checks.

**State.** 60 tickets were given away.

## Cumulative Review Chapters 1 - 11

**1.** 47,201

The digit 7 tells the number of thousands.

**2.** 7405 = 7 thousands + 4 hundreds + 0 tens + 5 ones, or 7 thousands + 4 hundreds + 5 ones

**3.** 7.463

a) Write a word name for the whole number.    $\boxed{\text{Seven}}$

b) Write "and" for the decimal point.    Seven
   $\boxed{\text{and}}$

c) Write a word name for the number to the right    Seven
   of the decimal point,    and
   followed by the place    $\boxed{\begin{array}{c}\text{four hundred}\\\text{sixty-three}\\\text{thousandths}\end{array}}$
   value of the last digit.

A word name for 7.463 is seven and four hundred sixty-three thousandths.

**4.**
$$\begin{array}{r} {}^{1}\phantom{0}\phantom{0} \\ 7\,4\,1 \\ +\ 2\,7\,1 \\ \hline 1\,0\,1\,2 \end{array}$$

**5.**
$$\begin{array}{r} {}^{2}\,{}^{1}\,{}^{1}\phantom{0} \\ 4\,9\,0\,3 \\ 5\,2\,7\,8 \\ 6\,3\,9\,1 \\ +\ 4\,5\,1\,3 \\ \hline 2\,1,0\,8\,5 \end{array}$$

**6.** $-\dfrac{2}{13} + \dfrac{1}{26} = -\dfrac{2}{13} \cdot \dfrac{2}{2} + \dfrac{1}{26}$
$\qquad\qquad = -\dfrac{4}{26} + \dfrac{1}{26}$
$\qquad\qquad = -\dfrac{3}{26}$

**7.**    $2\dfrac{4}{9} \qquad = \quad 2\dfrac{4}{9}$

$+3\boxed{\dfrac{1}{3} \cdot \dfrac{3}{3}} = +3\dfrac{3}{9}$
$\qquad\qquad\qquad\qquad \overline{\qquad 5\dfrac{7}{9}}$

**8.**
$$\begin{array}{r} {}^{1}\phantom{0}\ {}^{2}\phantom{0} \\ 2.0\,4\,8 \\ 6\,3.9\,1\,4 \\ +\,4\,2\,8.0\,0\,9 \\ \hline 4\,9\,3.9\,7\,1 \end{array}$$

**9.**
$$\begin{array}{r} {}^{1}\,{}^{1}\,{}^{1}\,{}^{1}\phantom{0} \\ 3\,4.5\,6\,0 \\ 2.7\,8\,3 \\ 0.4\,3\,3 \\ +\,7\,6\,5.1\,0\,0 \\ \hline 8\,0\,2.8\,7\,6 \end{array}$$

**10.**
$$\begin{array}{r} 6\ 7\ 4 \\ -\ 5\ 2\ 2 \\ \hline 1\ 5\ 2 \end{array}$$

**11.**
$$\begin{array}{r} {}^{8}\ {}^{13}\ {}^{3}\ {}^{16} \\ \not{9}\ \not{4}\ \not{6}\ 5 \\ -\ 8\ 7\ 9\ 1 \\ \hline 6\ 7\ 4 \end{array}$$

**12.** $\dfrac{7}{8} - \dfrac{2}{3} = \dfrac{7}{8} \cdot \dfrac{3}{3} - \dfrac{2}{3} \cdot \dfrac{8}{8}$

$\qquad = \dfrac{21}{24} - \dfrac{16}{24}$

$\qquad = \dfrac{5}{24}$

**13.** $4 \boxed{\dfrac{1}{3} \cdot \dfrac{8}{8}} = 4\dfrac{8}{24} = 3\dfrac{32}{24}$

$-1 \boxed{\dfrac{5}{8} \cdot \dfrac{3}{3}} = -1\dfrac{15}{24} = -1\dfrac{15}{24}$

$\qquad\qquad\qquad\qquad\qquad 2\dfrac{17}{24}$

**14.**
$$\begin{array}{r} {}^{1}\ {}^{9}\ {}^{9}\ {}^{9}\ {}^{9}\ {}^{10} \\ 2\ 0.\ 0\ 0\ 0\ \not{0} \\ -\ \ \ 0.\ 0\ 0\ 2\ 7 \\ \hline 1\ 9.\ 9\ 9\ 7\ 3 \end{array}$$

**15.** $40.03 - (-5.789) = 40.03 + 5.789$

We add.
$$\begin{array}{r} {}^{1} \\ 4\ 0.0\ 3\ 0 \\ +\ \ 5.7\ 8\ 9 \\ \hline 4\ 5.8\ 1\ 9 \end{array}$$

**16.** $\dfrac{21}{30} = \dfrac{3 \cdot 7}{3 \cdot 10} = \dfrac{3}{3} \cdot \dfrac{7}{10} = 1 \cdot \dfrac{7}{10} = \dfrac{7}{10}$

**17.** $\dfrac{275}{5} = \dfrac{5 \cdot 55}{5 \cdot 1} = \dfrac{5}{5} \cdot \dfrac{55}{1} = 1 \cdot \dfrac{55}{1} = 55$

**18.**
$$\begin{array}{r} 2\ 9\ 7 \\ \times\ \ \ 1\ 6 \\ \hline 1\ 7\ 8\ 2 \\ 2\ 9\ 7\ 0 \\ \hline 4\ 7\ 5\ 2 \end{array}$$

**19.**
$$\begin{array}{r} 3\ 4\ 9 \\ \times\ \ \ 7\ 6\ 3 \\ \hline 1\ 0\ 4\ 7 \\ 2\ 0\ 9\ 4\ 0 \\ 2\ 4\ 4\ 3\ 0\ 0 \\ \hline 2\ 6\ 6,2\ 8\ 7 \end{array}$$

**20.** $1\dfrac{3}{4} \cdot 2\dfrac{1}{3} = \dfrac{7}{4} \cdot \dfrac{7}{3} = \dfrac{7 \cdot 7}{4 \cdot 3} = \dfrac{49}{12} = 4\dfrac{1}{12}$

**21.** $\dfrac{9}{7} \cdot \dfrac{14}{15} = \dfrac{9 \cdot 14}{7 \cdot 15} = \dfrac{3 \cdot 3 \cdot 2 \cdot 7}{7 \cdot 3 \cdot 5} = \dfrac{3 \cdot 7}{3 \cdot 7} \cdot \dfrac{3 \cdot 2}{5} =$

$\dfrac{3 \cdot 2}{5} = \dfrac{6}{5}$

**22.** $12 \cdot \dfrac{5}{6} = \dfrac{12 \cdot 5}{6} = \dfrac{2 \cdot 6 \cdot 5}{6 \cdot 1} = \dfrac{6}{6} \cdot \dfrac{2 \cdot 5}{1} = \dfrac{2 \cdot 5}{1} = 10$

**23.**
$$\begin{array}{r} 3\ 4.\ 0\ 9 \quad \text{(2 decimal places)} \\ \times\ \ \ \ \ 7.\ 6 \quad \text{(1 decimal place)} \\ \hline 2\ 0\ 4\ 5\ 4 \\ 2\ 3\ 8\ 6\ 3\ 0 \\ \hline 2\ 5\ 9.\ 0\ 8\ 4 \quad \text{(3 decimal places)} \end{array}$$

**24.** To convert $\dfrac{18}{5}$ to a mixed numeral, we divide.

$$\begin{array}{r} 3 \\ 5\overline{\smash{)}1\ 8} \\ 1\ 5 \\ \hline 3 \end{array}$$

$\dfrac{18}{5} = 3\dfrac{3}{5}$

**25.**
$$\begin{array}{r} 5\ 7\ 3 \\ 6\overline{\smash{)}3\ 4\ 3\ 8} \\ 3\ 0 \\ \hline 4\ 3 \\ 4\ 2 \\ \hline 1\ 8 \\ 1\ 8 \\ \hline 0 \end{array}$$

The answer is 573.

**26.**
$$\begin{array}{r} 5\ 6 \\ 3\ 4\overline{\smash{)}1\ 9\ 1\ 4} \\ 1\ 7\ 0 \\ \hline 2\ 1\ 4 \\ 2\ 0\ 4 \\ \hline 1\ 0 \end{array}$$

The answer is 56 R 10.

**27.** A mixed numeral for the quotient in Exercise 26 is:

$56\dfrac{10}{34} = 56\dfrac{5}{17}$.

**28.** $\dfrac{4}{5} \div \dfrac{8}{15} = \dfrac{4}{5} \cdot \dfrac{15}{8} = \dfrac{4 \cdot 15}{5 \cdot 8} = \dfrac{4 \cdot 3 \cdot 5}{5 \cdot 2 \cdot 4} = \dfrac{4 \cdot 5}{4 \cdot 5} \cdot \dfrac{3}{2} = \dfrac{3}{2}$

**29.** $2\dfrac{1}{3} \div (-30) = \dfrac{7}{3} \div (-30) = \dfrac{7}{3} \cdot \left(-\dfrac{1}{30}\right) = -\dfrac{7}{90}$

**30.**
$$\begin{array}{r} 3\ 9. \\ 2.\ 7_\wedge\overline{\smash{)}1\ 0\ 5.\ 3_\wedge} \\ 8\ 1 \\ \hline 2\ 4\ 3 \\ 2\ 4\ 3 \\ \hline 0 \end{array}$$

The answer is 39.

**31.** $6\ 8,\ \boxed{4}\ 8\ 9$
$\qquad\quad \uparrow$

The digit 8 is in the thousands place. Consider the next digit to the right. Since the digit, 4, is 4 or lower round down, meaning that 8 thousands stay as 8 thousands. Then change all digits to the right of the thousands digit to zeros.

The answer is 68,000.

**32.**

$$0.427\boxed{5}$$  Ten-thousandths digit is 5 or higher.
$$\downarrow$$  Round up.
$$0.428$$

**33.** Round

21.8 3 $\boxed{8}$ 3 ... to the nearest hundredth.
$$\downarrow \quad \uparrow$$  Thousandths digit is 5 or higher.
2 1. 8 4  Round up.

**34.** A number is divisible by 6 if it is even and the sum of its digits is divisible by 3. The number 1368 is even. The sum of its digits, $1 + 3 + 6 + 8$, or 18, is divisible by 3, so 1368 is divisible by 6.

**35.** We find as many two-factor factorizations as we can.

$$15 = 1 \cdot 15$$
$$15 = 3 \cdot 5$$

The factors of 15 are 1, 3, 5, and 15.

**36.** $16 = 2 \cdot 2 \cdot 2 \cdot 2$
$25 = 5 \cdot 5$
$32 = 2 \cdot 2 \cdot 2 \cdot 2 \cdot 2$

The LCM is $2 \cdot 2 \cdot 2 \cdot 2 \cdot 2 \cdot 5 \cdot 5$, or 800.

**37.** We multiply these   We multiply these
two numbers:   two numbers:

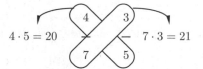

$4 \cdot 5 = 20$   $7 \cdot 3 = 21$

Since $20 \neq 21$, $\frac{4}{7} \neq \frac{3}{5}$.

**38.** $\frac{4}{7} = \frac{4}{7} \cdot \frac{5}{5} = \frac{20}{35}$
$\frac{3}{5} = \frac{3}{5} \cdot \frac{7}{7} = \frac{21}{35}$

Since $20 < 21$, it follows that $\frac{20}{35} < \frac{21}{35}$, so $\frac{4}{7} < \frac{3}{5}$.

**39.** To compare two negative numbers in decimal notation, start at the left and compare corresponding digits moving from left to right. When two digits differ, the number with the smaller digit is the larger of the two numbers.

$-1.001$
$\uparrow$   Different; 0 is smaller than 1.
$-0.9976$

Thus, $-0.9976$ is larger.

**40.** $\dfrac{\$0.95}{8\frac{1}{2}\text{ oz}} = \dfrac{95\text{¢}}{8.5\text{ oz}} \approx 11.176\text{¢/ oz}$

$\dfrac{\$1.66}{15\text{ oz}} = \dfrac{166\text{¢}}{15\text{ oz}} \approx 11.067\text{¢/ oz}$

$\dfrac{\$1.86}{15\frac{1}{4}\text{ oz}} = \dfrac{186\text{¢}}{15.25\text{ oz}} \approx 12.197\text{¢/ oz}$

$\dfrac{\$2.54}{24\text{ oz}} = \dfrac{254\text{¢}}{24\text{ oz}} \approx 10.583\text{¢/ oz}$

$\dfrac{\$3.07}{29\text{ oz}} = \dfrac{307\text{¢}}{29\text{ oz}} \approx 10.586\text{¢/ oz}$

Brand D has the lowest unit price.

**41.** a) $C = \pi \cdot d$

$C \approx \dfrac{22}{7} \cdot 1400 \text{ mi} = 4400 \text{ mi}$

b) First we find the radius.

$r = \dfrac{d}{2} = \dfrac{1400 \text{ mi}}{2} = 700 \text{ mi}$

Now we find the volume.

$V = \dfrac{4}{3} \cdot \pi \cdot r^3$

$\approx \dfrac{4}{3} \times \dfrac{22}{7} \times (700 \text{ mi})^3$

$= \dfrac{4 \times 22 \times 343,000,000 \text{ mi}^3}{3 \times 7}$

$\approx 1,437,333,333 \text{ mi}^3$

**42.** Let $c =$ the cost of the cabinets.

*Translate.*

$\underbrace{\text{What number}}$ is 40% of \$26,888?
$\downarrow \qquad \downarrow \ \downarrow \ \downarrow \qquad \downarrow$
$c \qquad = 40\% \ \cdot \ 26,888$

*Solve.* We convert 40% to decimal notation and multiply.

$$\begin{array}{r} 2\,6,8\,8\,8 \\ \times \qquad 0.\,4 \\ \hline 1\,0,7\,5\,5.\,2 \end{array}$$

The cabinets cost \$10,755.20.

**43.** Let $p =$ the percent of the cost represented by the countertops.

*Translate.*

\$4033.20 is $\underbrace{\text{what percent}}$ of \$26,888?
$4033.20 = \qquad p \qquad \cdot \quad 26,888$

*Solve.*

$4033.20 = p \cdot 26,888$

$\dfrac{4033.20}{26,888} = \dfrac{p \cdot 26,888}{26,888}$

$0.15 = p$

$15\% = p$

The countertops account for 15% of the total cost.

**44.** Let $a =$ the cost of the appliances.

*Translate.*

$\underbrace{\text{What number}}$ is 13% of \$26,888?
$\downarrow \qquad \downarrow \ \downarrow \ \downarrow \qquad \downarrow$
$a \qquad = 13\% \ \cdot \ 26,888$

*Solve.* Convert 13% to decimal notation and multiply.

$$\begin{array}{r} 2\,6,8\,8\,8 \\ \times \qquad 0.\,1\,3 \\ \hline 8\,0\,6\,6\,4 \\ 2\,6\,8\,8\,8\,0 \\ \hline 3\,4\,9\,5.\,4\,4 \end{array}$$

The appliances cost \$3495.44.

**45.** Let $p$ = the percent of the cost represented by the fixtures.

*Translate.*

$8066.40 is $\underbrace{\text{what percent}}$ of $26,888?

$$8066.40 = \underset{p}{\phantom{xx}} \cdot 26{,}888$$

*Solve.*

$$8066.40 = p \cdot 26{,}888$$

$$\frac{8066.40}{26{,}888} = \frac{p \cdot 26{,}888}{26{,}888}$$

$$0.3 = p$$

$$30\% = p$$

The fixtures account for 30% of the total cost.

**46.** Let $f$ = the cost of the flooring.

*Translate.*

$\underbrace{\text{What number}}$ is 2% of $26,888?

$$\underset{f}{\downarrow} \quad \underset{=}{\downarrow} \underset{2\%}{\downarrow} \underset{\cdot}{\downarrow} \quad \underset{26{,}888}{\downarrow}$$

*Solve.* Convert 2% to decimal notation and multiply.

$$\begin{array}{r} 2\,6,8\,8\,8 \\ \times \quad 0.\,0\,2 \\ \hline 5\,3\,7.\,7\,6 \end{array}$$

The flooring cost $537.76.

**47.** Since 987 is to the right of 879 on the number line, we have $987 > 879$.

**48.** The rectangle is divided into 5 equal parts. The unit is $\frac{1}{5}$. The denominator is 5. We have 3 parts shaded. This tells us that the numerator is 3. Thus, $\frac{3}{5}$ is shaded.

**49.**    $\frac{37}{1000}$         0.037.

3 zeros    Move 3 places.

$$\frac{37}{1000} = 0.037$$

**50.** $-\frac{13}{25} = -\frac{13}{25} \cdot \frac{4}{4} = -\frac{52}{100} = -0.52$

**51.** $\frac{8}{9} = 8 \div 9$

$$\begin{array}{r} 0.\,8\,8 \\ 9\overline{\smash{)}8.\,0\,0} \\ \underline{7\,2} \\ 8\,0 \\ \underline{7\,2} \\ 8 \end{array}$$

Since 8 keeps reappearing as a remainder, the digits repeat and $\frac{8}{9} = 0.888\ldots$, or $0.\overline{8}$.

**52.** 7%

a) Replace the percent symbol with $\times 0.01$.

$7 \times 0.01$

b) Move the decimal point two places to the left.

0.07.

Thus, $7\% = 0.07$.

**53.**    $4.\underline{63}$         4.63.         $\frac{463}{100}$

2 places   Move 2 places.   2 zeros

$$4.63 = \frac{463}{100}$$

**54.** First we consider $7\frac{1}{4}$.

$$7\frac{1}{4} = \frac{29}{4} \qquad (7 \cdot 4 = 28 \text{ and } 28 + 1 = 29)$$

Then $-7\frac{1}{4} = -\frac{29}{4}$.

**55.** $40\% = \frac{40}{100}$    Definition of percent

$$= \frac{2 \cdot 20}{5 \cdot 20}$$

$$= \frac{2}{5} \cdot \frac{20}{20}$$

$$= \frac{2}{5}$$

**56.** $\frac{17}{20} = \frac{17}{20} \cdot \frac{5}{5} = \frac{85}{100} = 85\%$

**57.** 1.5

a) Move the decimal point two places to the right.

1.50.

b) Write a percent symbol: 150%

Thus, $1.5 = 150\%$.

**58.**        $234 + y = 789$

$$234 + y - 234 = 789 - 234$$

$$y = 555$$

The number 555 checks. It is the solution.

**59.**    $3.9 \times y = 249.6$

$$\frac{3.9 \times y}{3.9} = \frac{249.6}{3.9}$$

$$y = 64$$

The number 64 checks. It is the solution.

**60.** $\frac{2}{3} \cdot t = \frac{5}{6}$

$$t = \frac{5}{6} \div \frac{2}{3} \qquad \text{Dividing both sides by } \frac{2}{3}$$

$$t = \frac{5}{6} \cdot \frac{3}{2} = \frac{5 \cdot 3}{6 \cdot 2}$$

$$= \frac{5 \cdot 3}{2 \cdot 3 \cdot 2} = \frac{3}{3} \cdot \frac{5}{2 \cdot 2}$$

$$= \frac{5}{4}$$

The number $\frac{5}{4}$ checks. It is the solution.

**61.**

$$\frac{8}{17} = \frac{36}{x}$$

$$8 \cdot x = 17 \cdot 36 \qquad \text{Equating cross products}$$

$$\frac{8 \cdot x}{8} = \frac{17 \cdot 36}{8}$$

$$x = \frac{17 \cdot 4 \cdot 9}{2 \cdot 4} = \frac{4}{4} \cdot \frac{17 \cdot 9}{2}$$

$$x = \frac{153}{2}, \text{ or } 76.5, \text{ or } 76\frac{1}{2}$$

**62.** Using a circle with 100 equally-spaced tick marks we first draw a line from the center to any tick mark. From that tick mark we count off 75 tick marks to graph 75% and label the wedge "Individuals." We continue in this manner with the other sources of donations. Finally we title the graph "Sources of Charitable Donations."

**Sources of Charitable Donations**

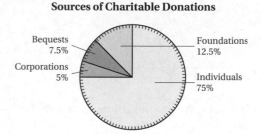

**63.** We will make a vertical bar graph. On the horizontal scale, in four equally-spaced intervals, indicate the sources of the donations. Then make ten equally-spaced tick marks on the vertical scale and label them by 10's. Label this scale "Percent." Draw vertical bars above the sources of the donations to show the percents. Finally, title the graph "Sources of Charitable Donations."

**Sources of Charitable Donations**

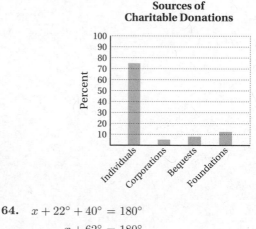

**64.**

$$x + 22° + 40° = 180°$$

$$x + 62° = 180°$$

$$x = 180° - 62°$$

$$x = 118°$$

**65.** From Exercise 64 we know that $m(\angle A) = 118°$, so $\angle A$ is an obtuse angle. Thus, the triangle is an obtuse triangle.

**66. *Familiarize.*** Let $d =$ the total donation.

***Translate.***

| First donation | plus | Second donation | is | Total donation |
|---|---|---|---|---|
| ↓ | ↓ | ↓ | ↓ | ↓ |
| 627 | + | 48 | = | d |

***Solve.*** We carry out the addition.

$$627 + 48 = d$$

$$675 = d$$

***Check.*** We can repeat the calculation. The answer checks.

***State.*** The total donation was $675.

**67. *Familiarize.*** Let $m =$ the number of minutes it takes to wrap 8710 candy bars.

***Translate.***

| Number of bars per minute | times | Number of minutes | is | Number of bars wrapped |
|---|---|---|---|---|
| ↓ | ↓ | ↓ | ↓ | ↓ |
| 134 | × | m | = | 8710 |

***Solve.***

$$134 \times m = 8710$$

$$\frac{134 \times m}{134} = \frac{8710}{134}$$

$$m = 65$$

***Check.*** $134 \cdot 65 = 8710$, so the answer checks.

***State.*** It takes 65 min to wrap 8710 candy bars.

**68. *Familiarize.*** Let $p =$ the price of the stock when it was resold.

***Translate.***

| Original price | minus | Drop in price | is | Price before resale |
|---|---|---|---|---|
| ↓ | ↓ | ↓ | ↓ | ↓ |
| 29.63 | − | 3.88 | = | p |

***Solve.*** We carry out the subtraction.

$$29.63 - 3.88 = p$$

$$25.75 = p$$

***Check.*** we can repeat the calculation. The answer checks.

***State.*** The price of the stock before it was resold was $25.75.

**69. *Familiarize.*** Let $t =$ the length of the trip, in miles.

***Translate.***

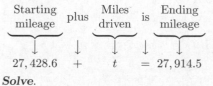

| Starting mileage | plus | Miles driven | is | Ending mileage |
|---|---|---|---|---|
| ↓ | ↓ | ↓ | ↓ | ↓ |
| 27,428.6 | + | t | = | 27,914.5 |

***Solve.***

$$27,428.6 + t = 27,914.5$$

$$27,428.6 + t - 27,428.6 = 27,914.5 - 27,428.6$$

$$t = 485.9$$

**Check**. $27,428.6 + 485.9 = 27,914.5$, so the answer checks.

**State**. The trip was 485.9 mi long.

**70. Familiarize**. Let $a =$ the amount that remains after the taxes are paid.

**Translate**.

| Income | minus | Federal taxes | minus | State taxes | is | Amount remaining |
|--------|-------|---------------|-------|-------------|----|------------------|
| ↓ | ↓ | ↓ | ↓ | ↓ | ↓ | ↓ |
| 12,000 | − | 2300 | − | 1600 | = | $t$ |

**Solve**. We carry out the calculations on the left side of the equation.

$$12,000 - 2300 - 1600 = t$$
$$9700 - 1600 = t$$
$$8100 = t$$

**Check**. The total taxes paid were $2300 + $1600$, or $3900, and $12,000 - $3900 = $8100$ so the answer checks.

**State**. $8100 remains after the taxes are paid.

**71. Familiarize**. Let $p =$ the amount the teacher was paid.

**Translate**.

| Daily pay | times | Number of days | is | Amount paid |
|-----------|-------|----------------|----|-----------| 
| ↓ | ↓ | ↓ | ↓ | ↓ |
| 87 | × | 9 | = | $p$ |

**Solve**. We carry out the multiplication.

$$87 \times 9 = p$$
$$783 = p$$

**Check**. We can repeat the calculation. The answer checks.

**State**. The teacher was paid $783.

**72. Familiarize**. Let $d =$ the distance Celeste would walk in $\frac{1}{2}$ hr, in kilometers.

**Translate**.

| Speed | times | Time | is | Distance |
|-------|-------|------|----|----------| 
| ↓ | ↓ | ↓ | ↓ | ↓ |
| $\frac{3}{5}$ | × | $\frac{1}{2}$ | = | $d$ |

**Solve**. We carry out the multiplication.

$$\frac{3}{5} \times \frac{1}{2} = d$$
$$\frac{3}{10} = d$$

**Check**. We can repeat the calculation. The answer checks.

**State**. Celeste would walk $\frac{3}{10}$ km in $\frac{1}{2}$ hr.

**73. Familiarize**. Let $s =$ the cost of each sweater.

**Translate**.

| Cost of each sweater | times | Number of sweaters | is | Total cost |
|----------------------|-------|--------------------|----|-----------| 
| ↓ | ↓ | ↓ | | ↓ |
| $s$ | × | 8 | = | 679.68 |

**Solve**.

$$s \times 8 = 679.68$$
$$\frac{s \times 8}{8} = \frac{679.68}{8}$$
$$s = 84.96$$

**Check**. $8 \cdot \$84.96 = \$679.68$, so the answer checks.

**State**. Each sweater cost $84.96.

**74. Familiarize**. Let $p =$ the number of gallons of paint needed to cover 650 ft². 

**Translate**. We translate to a proportion.

$$\text{Gallons} \rightarrow \frac{8}{400} = \frac{p}{650} \leftarrow \text{Gallons}$$
$$\text{Area covered} \rightarrow \quad\quad\quad \leftarrow \text{Area covered}$$

**Solve**. We equate cross products.

$$\frac{8}{400} = \frac{p}{650}$$
$$8 \cdot 650 = 400 \cdot p$$
$$\frac{8 \cdot 650}{400} = \frac{400 \cdot p}{400}$$
$$13 = p$$

**Check**. We can substitute in the proportion and check the cross products.

$$\frac{8}{400} = \frac{13}{650}; \quad 8 \cdot 650 = 5200; \quad 400 \cdot 13 = 5200$$

The cross products are the same so the answer checks.

**State**. 13 gal of paint is needed to cover 650 ft².

**75.** $I = P \cdot r \cdot t$

$$= \$4000 \times 5\% \times \frac{3}{4}$$
$$= \$4000 \times 0.05 \times \frac{3}{4}$$
$$= \$150$$

**76.** Commission = Commission rate × Sales

    5800      =      $r$      × 84,000

We divide both sides of the equation by 84,000 to find $r$.

$$\frac{5880}{84,000} = \frac{r \times 84,000}{84,000}$$
$$0.07 = r$$
$$7\% = r$$

The commission rate is 7%.

**77. Familiarize**. Let $p =$ the population after a year.

**Translate**.

| Current population | plus 4% of | Current population | is | Population after a year |
|--------------------|-----------|--------------------|----|------------------------| 
| ↓ | ↓ ↓ ↓ | ↓ | ↓ | ↓ |
| 29,000 | + 4% · | 29,000 | = | $p$ |

**Solve**.

$$29,000 + 0.04 \cdot 29,000 = p$$
$$29,000 + 1160 = p$$
$$30,160 = p$$

**Check.** The new population will be 104% of the original population. Since 104% of $29,000 = 1.04 \cdot 29,000 = 30,160$, the answer checks.

**State.** After a year the population will be 30,160.

**78.** To find the average age we add the ages and divide by the number of addends.
$$\frac{18+21+26+31+32+18+50}{7} = \frac{196}{7} = 28$$
The average age is 28.

To find the median we first arrange the numbers from smallest to largest. The median is the middle number.

$$18, 18, 21, 26, 31, 32, 50$$
$$\uparrow$$
Middle number

The median is 26.

The number 18 occurs most frequently, so it is the mode.

**79.** $18^2 = 18 \cdot 18 = 324$

**80.** $7^3 = 7 \cdot 7 \cdot 7 = 343$

**81.** $\sqrt{9} = 3$

The square root of 9 is 3 because $3^2 = 9$.

**82.** $\sqrt{121} = 11$

The square root of 121 is 11 because $11^2 = 121$.

**83.** $\sqrt{20} \approx 4.472$    Using a calculator

**84.** $\frac{1}{3}$ yd $= \frac{1}{3} \times 1$ yd
$$= \frac{1}{3} \times 36 \text{ in.}$$
$$= \frac{36}{3} \text{ in.}$$
$$= 12 \text{ in.}$$

**85.** 4280 mm = _____ cm

Think: To go from mm to cm in the table is a move of 1 place to the left. Thus, we move the decimal point 1 place to the left.

4280    428.0.

4280 mm = 428 cm

**86.** 3 days $= 3 \times 1$ day
$$= 3 \times 24 \text{ hr}$$
$$= 72 \text{ hr}$$

**87.** 20,000 g = _____ kg

Think: To go from g to kg in the table is a move of 3 places to the left. Thus, we move the decimal point 3 places to the left.

20,000    20.000.

20,000 g = 20 kg

**88.** 5 lb $= 5 \times 1$ lb
$$= 5 \times 16 \text{ oz}$$
$$= 80 \text{ oz}$$

**89.** 0.008 cg = _____ mg

Think: To go from cg to mg in the table is a move of 1 place to the right. Thus, we move the decimal point 1 place to the right.

0.008    0.0.08

0.008 cg = 0.08 mg

**90.** 8190 mL $= 8190 \times 1$ mL
$$= 8190 \times 0.001 \text{ L}$$
$$= 8.19 \text{ L}$$

**91.** 20 qt $= 20$ qt $\times \dfrac{1 \text{ gal}}{4 \text{ qt}}$
$$= \frac{20}{4} \times 1 \text{ gal}$$
$$= 5 \text{ gal}$$

**92.** $a^2 + b^2 = c^2$    Pythagorean equation
$$5^2 + 5^2 = c^2$$
$$25 + 25 = c^2$$
$$50 = c^2$$
$$\sqrt{50} = c \quad \text{Exact answer}$$
$$7.071 \approx c \quad \text{Approximation}$$
The length of the third side is $\sqrt{50}$ ft, or approximately 7.071 ft.

**93.** $d = 2 \cdot r = 2 \cdot 10.4$ in. $= 20.8$ in.

$C = 2 \cdot \pi \cdot r$
$C \approx 2 \cdot 3.14 \cdot 10.4$ in. $= 65.312$ in.

$A = \pi \cdot r \cdot r$
$A \approx 3.14 \cdot 10.4$ in. $\cdot 10.4$ in. $= 339.6224$ in$^2$

**94.** $P = 2 \cdot (l + w)$
$P = 2 \cdot (10.3 \text{ m} + 2.5 \text{ m})$
$P = 2 \cdot (12.8 \text{ m})$
$P = 25.6 \text{ m}$

$A = l \cdot w$
$A = (10.3 \text{ m}) \cdot (2.5 \text{ m})$
$A = 10.3 \cdot 2.5 \cdot \text{m} \cdot \text{m}$
$A = 25.75 \text{ m}^2$

**95.** $A = \frac{1}{2} \cdot b \cdot h$
$A = \frac{1}{2} \cdot 10$ in. $\cdot 5$ in.
$A = 25$ in$^2$

**96.** $A = b \cdot h$

$A = 15.4 \text{ cm} \cdot 4 \text{ cm}$

$A = 61.6 \text{ cm}^2$

**97.** $A = \dfrac{1}{2} \cdot h \cdot (a + b)$

$A = \dfrac{1}{2} \cdot 8.3 \text{ yd} \cdot (10.8 \text{ yd} + 20.2 \text{ yd})$

$A = \dfrac{8.3 \cdot 31}{2} \text{ yd}^2$

$A = 128.65 \text{ yd}^2$

**98.** $V = l \cdot w \cdot h$

$V = 10 \text{ m} \cdot 2.3 \text{ m} \cdot 2.3 \text{ m}$

$V = 23 \cdot 2.3 \text{ m}^3$

$V = 52.9 \text{ m}^3$

**99.** $V = Bh = \pi \cdot r^2 \cdot h$

$V \approx 3.14 \cdot 4 \text{ ft} \cdot 4 \text{ ft} \cdot 16 \text{ ft}$

$V = 803.84 \text{ ft}^3$

**100.** $V = \dfrac{1}{3} \cdot \pi \cdot r^2 \cdot h$

$V \approx \dfrac{1}{3} \cdot 3.14 \cdot 4 \text{ cm} \cdot 4 \text{ cm} \cdot 16 \text{ cm}$

$= 267.94\overline{6} \text{ cm}^3$

**101.**
$$7 - x = 12$$
$$7 - x - 7 = 12 - 7$$
$$-x = 5$$
$$-1 \cdot x = 5$$
$$-1 \cdot (-1 \cdot x) = -1 \cdot 5$$
$$x = -5$$

The number $-5$ checks. It is the solution.

**102.** $-4.3x = -17.2$

$\dfrac{-4.3x}{-4.3} = \dfrac{-17.2}{-4.3}$

$x = 4$

The number 4 checks. It is the solution.

**103.**
$$5x + 7 = 3x - 9$$
$$5x + 7 - 3x = 3x - 9 - 3x$$
$$2x + 7 = -9$$
$$2x + 7 - 7 = -9 - 7$$
$$2x = -16$$
$$\dfrac{2x}{2} = \dfrac{-16}{2}$$
$$x = -8$$

The number $-8$ checks. It is the solution.

**104.**
$$5(x - 2) - 8(x - 4) = 20$$
$$5x - 10 - 8x + 32 = 20$$
$$-3x + 22 = 20$$
$$-3x + 22 - 22 = 20 - 22$$
$$-3x = -2$$
$$\dfrac{-3x}{-3} = \dfrac{-2}{-3}$$
$$x = \dfrac{2}{3}$$

The number $\dfrac{2}{3}$ checks. It is the solution.

**105.** $12 \times 20 - 10 \div 5 = 240 - 2 = 238$

**106.** $4^3 - 5^2 + (16 \cdot 4 + 23 \cdot 3) = 4^3 - 5^2 + (64 + 69)$

$= 4^3 - 5^2 + 133$

$= 64 - 25 + 133$

$= 39 + 133$

$= 172$

**107.** $|(-1) \cdot 3| = |-3| = 3$

**108.** $17 + (-3)$

The absolute values are 17 and 3. The difference is $17 - 3$, or 14. The positive number has the larger absolute value, so the answer is positive.

$17 + (-3) = 14$

**109.** $\left(-\dfrac{1}{3}\right) - \left(-\dfrac{2}{3}\right) = -\dfrac{1}{3} + \dfrac{2}{3} = \dfrac{1}{3}$

**110.** $(-6) \cdot (-5) = 30$

**111.** $-\dfrac{5}{7} \cdot \dfrac{14}{35} = -\dfrac{5 \cdot 14}{7 \cdot 35} = -\dfrac{5 \cdot 2 \cdot 7}{7 \cdot 5 \cdot 7} = -\dfrac{2}{7} \cdot \dfrac{5 \cdot 7}{5 \cdot 7} = -\dfrac{2}{7}$

**112.** $\dfrac{48}{-6} = -8$      Check: $-8 \cdot (-6) = 48$

**113.** Let $y =$ the number; $y + 17$, or $17 + y$

**114.** Let $x =$ the number; $38\%x$, or $0.38x$

**115.** *Familiarize.* Let $s =$ the amount Susan paid for her rollerblades. Then $s + 17 =$ the amount Sam paid for his.

*Translate.*

| Amount Susan paid | plus | Amount Sam paid | is | Total amount |
|:---:|:---:|:---:|:---:|:---:|
| ↓ | ↓ | ↓ | ↓ | ↓ |
| $s$ | $+$ | $(s + 17)$ | $=$ | $107$ |

*Solve.*
$$s + (s + 17) = 107$$
$$2s + 17 = 107$$
$$2s + 17 - 17 = 107 - 17$$
$$2s = 90$$
$$\dfrac{2s}{2} = \dfrac{90}{2}$$
$$s = 45$$

We were asked to find only $s$, but we also find $s + 17$ so that we can check the answer.

If $s = 45$, then $s + 17 = 45 + 17 = 62$.

**Check**. $62 is $17 more than $45 and $45 + $62 = $107. The answer checks.

**State**. Susan paid $45 for her rollerblades.

**116. Familiarize**. Let $P =$ the amount originally invested. Using the formula for simple interest, $I = P \cdot r \cdot t$, we know the interest is $P \cdot 8\% \cdot 1$, or $0.08P$, and the amount in the account after 1 year is $P + 0.08P$, or $1.08P$.

**Translate**.

$$\underbrace{\text{Amount in the account after 1 yr}}_{1.08P} \underset{=}{\text{is}} \underset{1134}{\$1134}$$

**Solve**.

$$1.08P = 1134$$
$$\frac{1.08P}{1.08} = \frac{1134}{1.08}$$
$$P = 1050$$

**Check**. $1050 \cdot 0.08 \cdot 1 = \$84$ and $\$1050 + \$84 = \$1134$, so the answer checks.

**State**. Originally, there was $1050 in the account.

**117. Familiarize**. Let $x =$ the length of the first piece, in meters. Then $x + 3 =$ the length of the second piece and $\frac{4}{5}x =$ the length of the third piece.

**Translate**.

| Length of 1st piece | plus | Length of 2nd piece | plus | Length of 3rd piece | is | Total length |
|---|---|---|---|---|---|---|
| $x$ | $+$ | $(x + 3)$ | $+$ | $\frac{4}{5}x$ | $=$ | $143$ |

**Solve**.

$$x + (x + 3) + \frac{4}{5}x = 143$$
$$\frac{14}{5}x + 3 = 143$$
$$\frac{14}{5}x + 3 - 3 = 143 - 3$$
$$\frac{14}{5}x = 140$$
$$\frac{5}{14} \cdot \frac{14}{5}x = \frac{5}{14} \cdot 140$$
$$x = \frac{5 \cdot 140}{14} = \frac{5 \cdot 14 \cdot 10}{14 \cdot 1}$$
$$x = \frac{14}{14} \cdot \frac{5 \cdot 10}{1}$$
$$x = 50$$

If $x = 50$, then $x + 3 = 50 + 3 = 53$ and $\frac{4}{5}x = \frac{4}{5} \cdot 50 = 40$.

**Check**. The second piece is 3 m longer than the first piece, and the third piece is four-fifths as long as the first piece. Also, 50 m + 53 m + 40 m = 143 m, so the answer checks.

**State**. The length of the first piece of wire is 50 m, the length of the second piece is 53 m, and the length of the third piece is 40 m.

**118.**
$$\frac{2}{3}x + \frac{1}{6} - \frac{1}{2}x = \frac{1}{6} - 3x$$
$$\frac{4}{6}x + \frac{1}{6} - \frac{3}{6}x = \frac{1}{6} - 3x$$
$$\frac{1}{6}x + \frac{1}{6} = \frac{1}{6} - 3x$$
$$\frac{1}{6}x + \frac{1}{6} + 3x = \frac{1}{6} - 3x + 3x$$
$$\frac{1}{6}x + \frac{1}{6} + \frac{18}{6}x = \frac{1}{6}$$
$$\frac{19}{6}x + \frac{1}{6} = \frac{1}{6}$$
$$\frac{19}{6}x + \frac{1}{6} - \frac{1}{6} = \frac{1}{6} - \frac{1}{6}$$
$$\frac{19}{6}x = 0$$
$$\frac{6}{19} \cdot \frac{19}{6}x = \frac{6}{19} \cdot 0$$
$$x = 0$$

The number 0 checks. It is the solution.

**119.**
$$29.966 - 8.673y = -8.18 + 10.4y$$
$$29.966 - 8.673y + 8.673y = -8.18 + 10.4y + 8.673y$$
$$29.966 = -8.18 + 19.073y$$
$$29.966 + 8.18 = -8.18 + 19.073y + 8.18$$
$$38.146 = 19.073y$$
$$\frac{38.146}{19.073} = \frac{19.073y}{19.073}$$
$$2 = y$$

The number 2 checks. It is the solution.

**120.**
$$\frac{1}{4}x - \frac{3}{4}y + \frac{1}{4}x - \frac{3}{4}y = \frac{1}{4}x + \frac{1}{4}x - \frac{3}{4}y - \frac{3}{4}y$$
$$= \left(\frac{1}{4} + \frac{1}{4}\right)x + \left(-\frac{3}{4} - \frac{3}{4}\right)y$$
$$= \frac{2}{4}x + \left(-\frac{6}{4}y\right)$$
$$= \frac{1}{2}x - \frac{3}{2}y$$

Answer C is correct.

**121.** $8x + 4y - 12z = 4 \cdot 2x + 4 \cdot y - 4 \cdot 3z$
$$= 4(2x + y - 3z)$$

Answer B is correct.

**122.** $-\frac{13}{25} \div \left(-\frac{13}{5}\right) = -\frac{13}{25} \cdot \left(-\frac{5}{13}\right) = \frac{13 \cdot 5}{25 \cdot 13} = \frac{13 \cdot 5 \cdot 1}{5 \cdot 5 \cdot 13} = \frac{13 \cdot 5}{13 \cdot 5} \cdot \frac{1}{5} = \frac{1}{5}$

Answer D is correct.

**123.** $-27 + (-11)$

We have two negative numbers. Add the absolute values, 27 and 11, getting 38. Make the answer negative.

$-27 + (-11) = -38$

Answer A is correct.

**124.** *Familiarize.* The difference of the numbers is 40, so one number is 40 more than the other. Let $x =$ the smaller number. Then $x + 40 =$ the larger number.

*Translate.* The sum of the numbers is 430, so we have

$$x + (x + 40) = 430.$$

*Solve.*

$$x + (x + 40) = 430$$
$$2x + 40 = 430$$
$$2x = 390$$
$$x = 195$$

If $x = 195$, then $x + 40 = 235$.

*Check.* The sum of the numbers is $195 + 235$, or 430, and their difference is $235 - 195$, or 40. The answer checks.

*State.* The numbers are 195 and 235.